THEORETICAL METHODS FOR DETERMINING THE INTERACTION
OF ELECTROMAGNETIC WAVES WITH STRUCTURES

NATO ADVANCED STUDY INSTITUTES SERIES

Proceedings of the Advanced Study Institute Programme, which aims at the dissemination of advanced knowledge and the formation of contacts among scientists from different countries.

The series is published by an international board of publishers in conjunction with NATO Scientific Affairs Division

A	Life Sciences	Plenum Publishing Corporation
B	Physics	London and New York
C	Mathematical and Physical Sciences	D. Reidel Publishing Company Dordrecht and Boston
D	Behavioural and Social Sciences	Sijthoff & Noordhoff International Publishers B.V.
E	Applied Sciences	Alphen aan den Rijn, The Netherlands and Rockville, Md., USA

Series E: Applied Sciences – No. 40

THEORETICAL METHODS FOR DETERMINING THE INTERACTION OF ELECTROMAGNETIC WAVES WITH STRUCTURES

edited by

J. K. SKWIRZYNSKI
GEC – Marconi Electronics, Ltd.
Great Baddow Research Laboratories
Chelmsford, Essex, UK

Sijthoff & Noordhoff 1981
Alphen aan den Rijn, The Netherlands
Rockville, Maryland, USA

Proceedings of the NATO Advanced Study Institute on
Theoretical Methods for Determining the Interaction of Electromagnetic Waves with Structures
Norwich, U.K.
July 23-August 4, 1979.

Also supported by:
- National Science Foundation – USA
- European Research Office of the US Army, UK
- Office of the Naval Research,
  Department of the Navy, USA
- Institution of Electrical Engineers, UK
- GEC – Marconi Electronics, Ltd. UK

ISBN 978-94-011-8131-0   ISBN 978-94-011-8129-7 (eBook)
DOI 10.1007/978-94-011-8129-7

Copyright © 1981 Sijthoff & Noordhoff International Publishers B.V., Alphen aan den Rijn, The Netherlands.
Softcover reprint of the hardcover 1st edition 1981

All rights reserved. No part of this publication may be reproduced, stored in a retrieval system, or transmitted, in any form or by any means, electronic, mechanical, photocopying, recording, or otherwise, without the prior permission of the copyright owner.

PREFACE

This volume contains almost complete proceedings of the NATO Advanced Study Institute (ASI) organised in 1979 to bring together principal innovators and numerous users of mathematical techniques for analysing the interaction of electromagnetic waves with engineering and biological structures. The mathematical disciplines which can be brought to bear on these problems necessitate examination of effectiveness, convergence and robustness of the derived analytic and numerical algorithms.

The aim of this ASI was to give a clear and up-to-date tutorial presentation of available techniques, and to bring together interested scientists, engineers and mathematicians, to discuss together their experience and to ensure wider familiarity with the subject.

Our programme consists of three distinct yet related parts.

The first two of these reflect two somewhat different methods applicable for different ranges of L/λ, where L represents a characteristic dimension of a structure and λ is a representative wavelength of radiation. The third part deals with the specific problem of biological interaction.

In the first part (Low and Intermediate Frequency Applications) we offer tutorial texts and user-oriented discussions on main techniques and problems concerning:
- radiation,
- scattering,
- aperture penetration,
- inverse scattering,

using moment methods and their developments.

The approach to the high frequency applications forms the subject of the second part of this volume, concentrating mainly on the geometrical theory of diffraction (GTD). There are three main variants of the GTD:
- uniform theory of diffraction (UTD),
- uniform asymptotic theory (UAT),
- spectral theory of diffraction (STD).

Thus the picture here is rather confusing to a newcomer and our aim has been to provide a balanced and hopefully unbiased presentation. While emphasis is laid on UTD, which enjoys most widespread applications, an opportunity is taken to examine the advantages of the other methods mentioned. In addition we offer reports on applications of GTD from organisations in Europe and the U.S.A.

The theoretical prediction methods used in electromagnetic-biological applications have much in common with techniques originally developed for engineering problems, which are the main concern of the first two parts of this volume. We aim at achieving a mutual benefit by bringing together experts in both these fields. The concluding discussion contains reports of numerous approaches and successes.

We hope that a particularly valuable feature of these proceedings will be the complete, almost verbatim reproduction of panel discussions on carefully selected problems. Unforseen difficulties in transcribing and editing these texts, and of validating them by actual speakers, have been the main reason for the delay in publication of this volume. The editor wishes to apologise here to contributors of these texts for initiating an untested and logistically difficult reproduction process, hoping nevertheless that the resulting reports of discussion have been worth waiting for.

The contents of the proceedings are now described in some detail.

LOW AND INTERMEDIATE FREQUENCY SCATTERING METHODS

This part of our programme was organised by Professor R.F. Harrington who undertook the task of choosing the lecturers and identified three important themes for the panel discussions.

The text begins with a tutorial survey by Professor C.M. Butler of the method of moments and its application to solution of integral equations. The method has a long history and is related to Rayleigh-Ritz variational and Galerkin methods. Here the author provides a balanced approach to solving integral equations of the EM theory which are singular in nature. The numerical techniques are shown to be efficient, rapidly convergent and easy to implement.

The wire grid methods are introduced by Dr. E.K. Miller. They present an outstanding example of applying the moment theory to practical problems. It is not only that many antenna problems of current interest involve wire-like geometries, but also wire grids can be employed to model continuous conducting surfaces.

In the latter case this approximation is attractive computationally since simplified results from the theory of electromagnetic interaction with a thin wire can be used. Applications illustrate radiation and scattering problems, near and far field effects, impedance characteristics and interaction with the ground.

Professor R.F. Harrington and Dr. J.R. Mautz point out the well known fact that the usual H-field and E-field surface integral equations for conducting bodies fail to give unique solutions when a surface forms a resonant cavity. The combined field and combined source formulations, which overcome this difficulty, are presented, together with representative computations for dielectric bodies.

Next we have the joint contribution by Professors J. Van Bladel and C.M. Butler. They begin with the study of fundamental electromagnetic boundary value problems involving apertures in conducting plane surfaces. Frequently problems involving scattering through apertures are difficult to handle numerically; here is presented a thorough discussion of available analytic techniques with carefully argued degrees of approximation; beginning with simple examples of plane screens with small apertures, the theory is extended to coupling between multiple apertures and apertures with near-by objects, and to apertures in cavities and waveguides.

Dr. E.K.Miller follows with an introduction to the natural mode method (complex resonance frequency) in both frequency and time domain analysis. The method for location of these modes and for determination of their residues is developed by application of Prony's method. Illustrated results are given for many scattering problems with arbitrary excitation.

The next two papers by Dr. E.K. Miller with Dr. J.A. Landt, and by Dr. C.L. Bennett present direct time domain techniques which offer computational advantages over the more familiar frequency domain techniques. The first paper concentrates on wire objects excited as antennas or scatterers, the second deals with surfaces and composite bodies. Results are illustrated by several scattering problems and required impulse responses.

The final tutorial introduction by Dr. C.L. Bennett gives a review of the techniques for the inverse problem of determining the characteristics of a scattering target. Time and frequency domain techniques are compared.

This part of the proceedings concludes with full reproduction of three panel discussions:

VIII

'Modelling of Three-Dimensional Bodies - Solid Patches and Wire Grids' was chaired by Professor R.F. Harrington who succeeded in collecting an interesting choice of approaches to this modelling problem. Discussion is opened by Dr. C.L. Bennett with a brief classification of relevant problems which deserve attention. Mr. T.W. Armour relates his experience with the CHAOS program which he found to provide powerful and versatile methods for predicting surface current densities on aircraft structures due to external harmonic or transient excitation. Dr. A.R. Djordjevic introduces a rapid method for analysis of wire structures, based on a vector-scalar potential equation and on a special choice of economic trial polynomials. Dr. T. Slivnik provides a simple method for solution of integral equations, based on the Nystrom method. The report of this panel discussion concludes with a short paper by Professor D.L. Jaggard on isoperimetric inequalities and quotients which provide simple quantitative estimates and bounds of aperture transmission coefficients, scattering cross-sections etc.

'Comparison and Evaluation of Different Formulations for Scattering Problems - Integral Equations, Differential Equations and Combined Methods' was chaired by Professor R. Mittra, the uniquely able adjudicator of controversial questions. The discussion is opened by Professor S. Strom with a presentation of the basic features of the 'Extended Boundary Condition' method which is not included explicitly in our tutorial sessions. In the ensuing exchange several controversial problems are raised, such as policies adopted in point matching, treatment of re - entrant surfaces and listing of problems for which this technique is particularly suitable. The discussion is continued by Professor P.W. Barber who considers the application of the extended boundary condition method to inhomogeneous scattering and absorption problems, which calls for the combination of this method with finite element techniques. This leads to a lively argument on some niceties of particular numerical approaches and to questions of computer memory capacity for practical situations. The presentation closes with a contribution by Professor L.W. Pearson - a short introduction of the mathematical formalism leading to the complex resonance description of electromagnetic scattering and hence to the singularity expansion method.

'Availability and Limitations of General Purpose Computer Programs' was chaired by Dr. E.K. Miller of the Lawrence Livermore Laboratory, where practical comparisons are made of many computer programs in this field. Professor L.W. Pearson opens the discussion by summarising various approaches for wire grid representation of plate structures, including degree of user-orientation and of documentation of available software. Dr. C.L. Bennett lists types of structures for which time domain integral equations have been programmed. Professor P.W. Barber concentrates on programs using the extended boundary condition or T - matrix approach. Professor R.F. Harrington discusses briefly the reason for choice of various testing functions for E - field and H - field integral equation formulations. Mr. T.W. Armour relates the experience of using the well - known AMP and CHAOS programs for various practical situations. Dr. P.A. Ramsdale discusses relative merits of using Pocklington or Hallen integral equations and proposes the latter as a suitable choice for practical antenna problems. The discussion concludes with a brief survey by Dr. T.K. Sarkar of function - analytic fundamentals of numerical procedures considered in this presentation.

HIGH FREQUENCY SCATTERING METHODS AND APPLICATIONS

This section of our programme was organised by Professor H. Bach whose Department at the Technical University of Denmark has the most extensive experience in Europe in using the GTD and other methods, particularly for the satellite scenario.

The presentation is opened by Dr. S. Cornbleet who provides basic foundations of the geometrical theory of diffraction (GTD) from Fermat to Keller, and is followed by Professor L.B. Felsen who derives fundamental relations for the ray-optical field.

Professor P.L. Christiansen derives solutions for some important canonical problems, together with extraction of GTD scattering coefficients.

The fundamental theory is continued by Professor G.A. Deschamps who presents a mathematically satisfying approach to uniform theories of diffraction. In particular, the uniform asymptotic theory (UAT) provides a valid field description in some awkward transition regions.

Professor R.G. Kouyoumjian and Dr. P.H. Pathak (with Dr. W.D. Burnside) present another uniform formulation which has proved to be an extremely flexible and versatile method for practical applications to important problems, particularly for predicting scattering from carefully modelled aircraft structures. This is exhibited by many illustrations, including some which compare the GTD with eigenfunction and moment methods for solution.

Professor H. Bach provides a neat summary of geometry required for ray tracing, including that part of the differential geometry which is essential for determining ray parameters. The practical development of GTD computer programs is also discussed.

His colleague, Dr. K. Pontoppidan, presents applications of GTD techniques for detailed analysis of polar diagrams of reflector antennas. These techniques are essential for wide - angle sidelobe analysis and for adequate analysis of sub - reflector radiation when physical optics methods are not adequate or are computationally cumbersome.

Dr. F.A. Molinet begins with hybrid methods which combine the moment methods with asymptotic techniques in circumstances when either of them alone has obvious limitations. He then considers some special cases of scattering from bodies of revolution for which GTD formulae can be developed directly and concludes with a description of computer programs at LCT, France, for analysing any kind of antenna mounted on an arbitrary structure consisting of plates, cylinders, spheres and cones.

The next two papers are concerned with the Spectral Theory of Diffraction (STD) in which the scattered field is represented by the Fourier transform of the induced currents on a scatterer. Professor R. Mittra and Dr. Y. Rahmat - Samii (with Dr.W.L. Ko) present the method formally, compare it with other uniform theories and provide iterative techniques for improving asymptotic solutions. Extensive discussion of scattering by a strip provides an illustration for bridging the gap between the low and high frequency regions by combining integral equation methods with asymptotic techniques.

Dr. Y. Rahmat - Samii presents elegant asymptotic formulations for the study of radiation from very large reflector antennas.

Professor L.B. Felsen (with Dr. T. Ishihara and Dr. A. Green) explores alternative representations of fields excited by a line source on a concave surface. Included are ray - optical, canonical - integral and 'whispering gallery' mode solutions. The physical interpretation of results is emphasised.

Professor G.A. Deschamps presents an introduction to the theory of gaussian beams, i.e. beams of electromagnetic radiation which remain substantially close to a central line and yet cannot be represented by a pencil of parallel rays because of field spreading. In fact gaussian beams can be envisaged as pencils of complex rays and thus can be processed by the laws of geometrical optics.

Professor L.B. Felsen reviews the developments of techniques for the study of evanescent fields which are similar to the local ray tracing methods used extensively for non-evanescent fields.

MECHANISMS OF BIOLOGICAL INTERACTION AND MICROWAVE BIOEFFECTS

The third part of this Institute was organized by Professor A.W. Guy who has extensive experience in bringing his subject to a wide audience. His contribution presents a survey of both the theoretical and measurement tools which have been developed for assessing the effect of electromagnetic waves on biological tissues.

Professor C. Yeh provides a critical discussion of several theoretical techniques which may be used to predict electromagnetic absorption in man and animal bodies. Two preferred methods (extended boundary condition and the global - local finite element) are discussed in greater detail.

The final contribution by Professor M.F. Iskander (and Professor C.H. Durney) concerns medical diagnosis and imaging using electromagnetic techniques.

The panel discussion on Experimental Methods for the Quantitation of Absorption Patterns in Biological Structures was chaired by Mr. F. Harlen. It was opened by Professor A.W. Guy who among other issues clarifies the importance of effects of sensors on absorption measurements in live tissues, and discusses methods for minimising these. Dr. E. Burdette reports on his work on measuring dielectric properties of lossy biological materials and deals with applications of volume - cell moment methods for solving integral equations for dielectric bodies to the problem of electromagnetic dosimetry. Dr. A.D. Yaghjian complements this contribution with detailed analytic treatment and Dr. E. Burdette illustrates the results with different models of animals. Professor M.F. Iskander summarises results at the University of Utah on near field irradiation of biological models. His approach provoked an extensive discussion. A short argument by Professor E.H. Grant concerns the differences between thermal and non - thermal effects. Dr. C.W. Smith reviews the recent work on measurement of dielectric biological materials. Finally Dr. G. Crosta proposes a theoretical approach for predicting and optimising tumor therapy using tools derived from control theory.

As director of the ASI and editor of these proceedings, I would like to acknowledge the help given to me during the two years it took to organise the Institute, and subsequently to prepare the contents of this volume for publication. On the technical side, I would like to thank my co - directors, Professor H. Bach, Professor R.F. Harrington and Professor A.W. Guy who gave their advice unsparingly on various aspects of the programme organisation. In particular I would like to pay tribute to the efforts of my fourth co - director and colleague, Mr. E.A. Pacello, who gave up a great deal of his time in advising me on scientific details and to editing the texts of panel discussions. The suggestion that electromagnetic - biological effects should be included in the programme came originally from him. In the task of preparing the panel discussion texts he was assisted by Dr. D.C. Brewster, Dr. P.K. Harbottle and Mr. S. Cant. Mr. R.W. Hines produced the figures for the texts, in some cases from the roughest of sketches. All are thanked here and complimented on the success of their work.

On the administration and organisation side of the Institute I would like to acknowledge gratefully, the work of my assistant Miss Marjorie Sadler, and of Miss Ann Williamson. I also commend my secretary, Mrs. S. Carnell for carrying out her clerical duties with patience and diligence. In addition I would like to thank Mr. Roger Lloyd, his assistant Mrs. J. Greenland and other staff at the University of East Anglia for making our stay in Norwich a pleasant and memorable experience.

I am grateful to my wife, Yvonne, for the tolerance and understanding she has shown during evenings and weekends when, over the two year period, I was occupied with technical details and correspondence.

Finally, I would like to thank Mr. G.D. Speake, the Technical Director of the GEC - Marconi Research Laboratories for giving me permission to undertake the organisation of this NATO Advanced Study Institute.

J.K. SKWIRZYNSKI

Director of the Institute

Great Baddow
17th October 1980

TABLE OF CONTENTS

PREFACE

Part I. LOW AND INTERMEDIATE FREQUENCY SCATTERING METHODS

1. Chalmers M. BUTLER, Introduction to moment methods with simple applications   3
2. E.K. MILLER, Wires and wire grid models for radiation and scattering including ground effects   57
3. R.F. HARRINGTON and J.R. MAUTZ, Surface integral equations for conducting and dielectric bodies   97
4. J. VAN BLADEL and C.M. BUTLER, Aperture problems   117
5. E.K. MILLER, Natural mode methods in frequency and time domain analysis   173
6. E.K. MILLER and J.A. LANDT, Direct time-domain techniques for transient radiation and scattering - wires   213
7. C. Leonard BENNETT, Time domain solutions via integral equations - surfaces and composite bodies   255
8. C. Leonard BENNETT, Inverse scattering   277

PANEL DISCUSSION, "Modelling of three dimensional bodies - solid patches and wire grids"   299

9. D.L. JAGGARD, On using isoperimetric inequalities and isoperimetric quotients in electromagnetic theory and applied physics   319

PANEL DISCUSSION, Comparison and evaluation of different formulations for scattering problems - integral equations, differential equation and combined methods.   343

PANEL DISCUSSION, Availability and limitations of general purpose computer programs   369

Part II. HIGH FREQUENCY SCATTERING METHODS AND APPLICATIONS

1. S. CORNBLEET, Introduction to geometrical optics   423
2. Leopold B. FELSEN, Foundations of the geometrical theory of diffraction   439
3. Peter L. CHRISTIANSEN, Canonical problems and diffraction coefficients   455
4. G.A. DESCHAMPS, Uniform theories of diffraction by edges   477
5. R.G. KOUYOUMJIAN, P.H. PATHAK and W.D. BURNSIDE A uniform GTD for the diffraction by edges, vertices and convex surfaces   497

6. H. BACH, Introduction to GTD applications	563
7. Knud PONTOPPIDAN, GTD applied to reflector antenna analysis	595
8. F.A. MOLINET, GTD applications at LCT	619
9. Raj MITTRA, Yahya RAHMAT-SAMII and Wai Lee KO, Transform approach to electromagnetic scattering	649
10. Yahya RAHMAT-SAMII, Some useful asymptotic and numerical techniques for solving high-frequency scattering problems	697
11. T. ISHIHARA, L.B. FELSEN and A. GREEN, High-frequency fields excited by a line source located on a perfectly conducting concave cylindrical surface	721
12. Georges A. DESCHAMPS, Gaussian beams: paraxial theory	745
13. Leopold B. FELSEN, Evanescent waves	767

Part III. MECHANISMS OF BIOLOGICAL INTERACTION AND MICROWAVE BIOEFFECTS

1. Arthur W. GUY, History and state of the art on the quantitation of the interaction of electromagnetic fields with biological structures	781
2. C. YEH, Theoretical and numerical methods of quantitation of absorption patterns in man and animal bodies	817
3. M.F. ISKANDER and C.H. DURNEY, Medical diagnosis and imaging using electromagnetic techniques	835
PANEL DISCUSSION, Experimental methods of the quantitation of absorption patterns in biological tissues	855
LIST OF LECTURERS AND DELEGATES	911

Part I

LOW AND INTERMEDIATE FREQUENCY SCATTERING METHODS

Organised by Professor R.F. Harrington
Syracuse University, USA

INTRODUCTION TO MOMENT METHODS WITH SIMPLE APPLICATIONS

Chalmers M. Butler

Department of Electrical Engineering
University of Mississippi
University, Mississippi 38677   USA

1. INTRODUCTION

The method of moments [1] is introduced from a tutorial point of view and its application to integral equations is outlined. Methods for numerically solving simple integral equations from electrostatics are discussed, and numerical results are compared with analytical solutions in selected cases. The moment method is applied to the problems of the TM and TE illuminated cylinders and strips of infinite length as well as to that of wire antennas and scatterers. The presentation is introductory in nature and emphasizes basic techniques. The reader is referred to the literature [1-5] for other treatments of problems discussed here as well as for analyses of more complex problems.

2. SOME INTEGRAL EQUATIONS FROM ELECTROSTATICS

In Fig. 1 is illustrated a surface S which is embedded in a homogeneous space of infinite extent, electrically characterized by the permittivity ε and on which exists a surface electric charge of density q_s (Coulombs/meter^2). From the basic principles of electrostatics and the definition of electric scalar potential Φ, one can show that the potential due to q_s is

$$\Phi(\bar{r}) = \frac{1}{4\pi\varepsilon} \iint_S q_s(\bar{s}') \frac{1}{R} dS' \tag{1}$$

where $\bar{s}'$ locates the source point (x', y', z') on S and where $\bar{r}$ locates the field point (x, y, z). In (1) the distance $R = |\bar{r} - \bar{s}'|$ is

$$R = \left[(x - x')^2 + (y - y')^2 + (z - z')^2\right]^{\frac{1}{2}}. \tag{2}$$

The expression for Φ in Eq. (1) incorporates the assignment of the value zero to the scalar potential at infinity.

Often, situations arise in which quantities of interest in electrostatics do not vary along a given fixed direction, or, at least, their variation is so slight that they can be treated as being constant insofar as a computation at hand is concerned. In this case, the problem under consideration is said to be two-dimensional and is viewed as one in which all quantities are of infinite extent and constant along a coordinate axis parallel to the direction of invariance. Fig. 2 shows a segment of a surface S which extends from $-\infty$ to $+\infty$ in the z direction and whose cross section is the same in all planes parallel to the xy plane. One can show readily that, due to a surface charge of density $q_s(x, y)$ on the cylindrical surface,

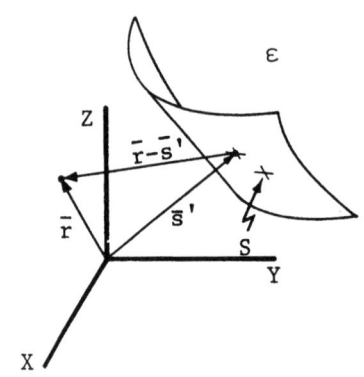

Fig. 1. Surface in homogeneous infinite space.

$$\Phi(\bar{\rho}) = -\frac{1}{2\pi\varepsilon} \int_C q_s(\bar{\rho}')\ln|\bar{\rho} - \bar{\rho}'| d\ell' \tag{3}$$

where the contour integral along the path C is over $\ell'\varepsilon(\ell_1, \ell_2)$ as suggested in Fig. 3. Notice that $\bar{\rho}' = \bar{\rho}'(\ell')$ locates points on C and is a function of ℓ', the displacement along the arc measured from an origin $\ell' = 0$. In two-dimensions, the potential is unbounded at infinity so one cannot assign to it the value zero there. Rather, it is usual to select a convenient reference point and compute Φ relative to its value at this reference. Since ultimate interest usually is in this potential difference or in $\bar{E}$ and q_s, the constant value of potential is not included in (3).

On the surface S of a perfect electric conductor (PEC), the tangential (to S) electric field must be zero everywhere, a constraint which requires the scalar potential Φ to be constant at all points on S. This property enables one to derive an integral equation for the unknown charge density induced on the surface of a perfectly conducting body when such a body is charged to a specified potential. If the surface illustrated in Fig. 1 is a PEC surface charged by means of a dc source to a potential V, the potential at a point $\bar{r}$ due to the surface charge q_s induced

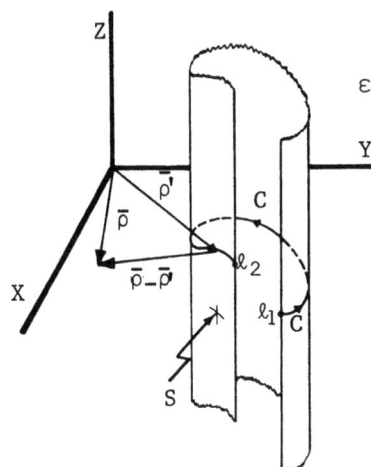

Fig. 2. Section of infinite two-dimensional cylindrical surface in homogeneous infinite space.

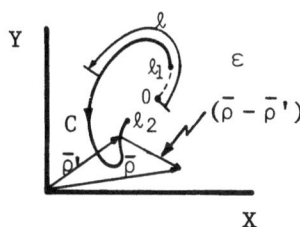

Fig. 3. Cross section of cylindrical surface.

on S is $\Phi(\bar{r})$ of Eq. (1). If now the field point $\bar{r}$ is constrained to fall on the surface S, then the potential Φ at all points on S must be equal to the constant value V impressed upon the surface. That is, for $\bar{r} = \bar{s}$, where $\bar{s}$ represents the location of a point on S,

$$\frac{1}{4\pi\varepsilon} \iint_S q_s(\bar{s}') \frac{1}{R} dS' = V, \quad \bar{s}\varepsilon S \quad (4)$$

where $R = |\bar{s} - \bar{s}'|$. This is an integral equation for the unknown surface charge density q_s. In (4) $1/R = 1/[|\bar{s} - \bar{s}'|]$ is called the kernel and V is called the forcing function of the integral equation. Eq. (4) is an integral equation because the unknown appears under the integral sign, integrated against the kernel. If one is interested in the problem in which S is a closed PEC surface or is the exterior surface of a solid conducting body, then (4) is still the appropriate integral equation.

Now we turn attention to the class of two-dimensional bodies and surfaces charged to specified potentials. If the surface of Fig. 2 is perfectly conducting and is charged to a potential V, the integral equation for q_s follows immediately from the requirement that Φ of (3) be equal to V on C:

$$-\frac{1}{2\pi\varepsilon} \int_C q_s(\bar{\rho}') \ln|\bar{\rho} - \bar{\rho}'| d\ell' = V, \quad \bar{\rho}\varepsilon C. \quad (5)$$

Eq. (5) must be enforced at all points on C from $\ell = \ell_1$ to $\ell = \ell_2$ but, because all quantities are independent of z, it is not necessary that the equation be enforced over the entire surface S.

3. A SIMPLE SOLUTION PROCEDURE

In this section we investigate, from the viewpoint of the pragmatist, a simple and direct procedure for solving integral equations which arise in electrostatic problems, with emphasis on the step-by-step procedures of obtaining a solution. As an initial example, we consider the integral equation

$$\int_{x'=-w}^{w} u(x')g(x, x')dx' = f(x), \quad x\epsilon(-w, w), \tag{6}$$

in which $u(x)$ is the unknown to be determined, $f(x)$ is the known forcing function, and $g(x, x')$ is the known kernel. If $g(x, x') = -\ln|x - x'|$, $f(x) = 1$, and $u(x) = q_s(x)/2\pi\epsilon V$, then (6) is Eq. (5) specialized to the case of an infinite PEC strip (flat) of width $2w$.

To illustrate how (6) may be solved, we outline a procedure known as collocation or point matching, which is direct, simple, and quite powerful when applied intelligently. The method comprises three major steps.

I. Approximate the unknown $u(x)$ by means of a linear combination of known functions u_n (called basis functions):

$$u(x) = \sum_{n=1}^{N} U_n u_n(x). \tag{7}$$

II. Establish a well-conditioned system of linear equations by enforcing equality of Eq. (6) (subject to the approximation of $u(x)$ above) at discrete points x_m (called match points).

III. Solve the system of linear equations resulting from II and thereby obtain $U_1, U_2, \ldots, U_N$ from which an approximation to $u(x)$ may be constructed (see I).

The set $\{u_n(x)\}$ of basis functions postulated in I is restricted only in that it must be linearly independent. Otherwise it is general and is subject only to constraints imposed by the particular nature of the equation being solved and, in some cases, by boundary conditions which the unknown must satisfy. As pointed out below, the selection of $\{u_n(x)\}$ should be guided by the complexity it introduces into the computations.

As an initial step, we approximate u over the interval $x\epsilon(-w, w)$ by means of a piecewise constant function (stair steps) as illustrated in Fig. 4. The interval is divided into N segments of length $\Delta = 2w/N$, with the center of the nth segment designated x_n:

$$x_n = -w + \Delta\left(n - \frac{1}{2}\right), \quad n = 1, 2, \ldots, N. \tag{8}$$

So that u can be incorporated into the numerical analysis, it must be represented mathematically. A convenient way to do this is by means of the following linear combination of pulse functions:

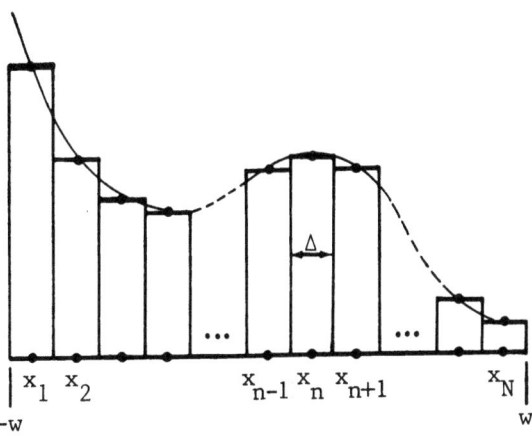

Fig. 4. Piecewise constant approximation.

$$u(x) = \sum_{n=1}^{N} U_n p_n(x) \tag{9a}$$

where, as stated above, the U_n are unknown constants and the so-called pulse functions $p_n(x)$ are defined

$$p_n(x) = \begin{cases} 1, & x \varepsilon (x_n - \Delta/2, x_n + \Delta/2) \\ 0, & \text{otherwise} \end{cases} \tag{9b}$$

From (9), as well as from Fig. 4, one observes that, within the nth segment $(x_n - \Delta/2, x_n + \Delta/2)$, $u(x)$ is approximated by the constant U_n; that is, $u(x) = U_n$, $x \varepsilon (x_n - \Delta/2, x_n + \Delta/2)$.

The next step is to substitute the approximation (9) of u into the integral equation (6) to obtain

$$\int_{x'=-w}^{w} \left\{ \sum_{n=1}^{N} U_n p_n(x') \right\} g(x, x') \, dx' \doteq f(x), \quad x \varepsilon (-w, w), \tag{10}$$

which, in view of the linearity of the integral operator, can be written

$$\sum_{n=1}^{N} U_n \int_{x'=-w}^{w} p_n(x') \, g(x, x') dx' \doteq f(x). \tag{11}$$

Since the nth pulse is unity for $x \varepsilon (x_n - \Delta/2, x_n + \Delta/2)$ and is zero outside this interval, (11) reduces to

$$\sum_{n=1}^{N} U_n \int_{x'=x_n-\Delta/2}^{x_n+\Delta/2} g(x, x')dx' \doteq f(x), \qquad (12)$$

which, for convenience, is expressed as

$$\sum_{n=1}^{N} U_n \delta_n(x) = U_1 \delta_1(x) + U_2 \delta_2(x) + \ldots + U_N \delta_N(x) \doteq f(x) \qquad (13)$$

where

$$\delta_n(x) = \int_{x'=x_n-\Delta/2}^{x_n+\Delta/2} g(x, x')dx' \, . \qquad (14)$$

Eq. (13) reflects the fact that the problem of determining u(x) at all points x in (-w, w) has been replaced by that of computing the N constants U_n. Clearly, (13) is a single equation containing N unknowns and it cannot be solved for the constants U_n in its present form.

The exact solution u(x) of the integral equation ensures that the right- and left-hand sides be equal in some sense at every point x in (-w, w), but any approximate solution does not ensure equality at all points. We seek a solution which is as nearly equal as possible to u(x) in the sense that, at specified match points in the interval (-w, w), Eq. (13), which is an approximation to (6), is forced to hold exactly. The center point x_n of each segment ($x_n - \Delta/2$, $x_n + \Delta/2$) is selected as a match point but, to distinguish it from the x_n employed in (8)-(14), the match point is designated x_m:

$$x_m = -w + \Delta\left(m - \frac{1}{2}\right), \quad m = 1, 2, \ldots, N. \qquad (15)$$

Now, at each match point x_m, (13) is enforced:

$$U_1\delta_1(x_1) + U_2\delta_2(x_1) + \ldots + U_n\delta_n(x_1) + \ldots + U_N\delta_N(x_1) = f(x_1)$$

$$U_1\delta_1(x_2) + U_2\delta_2(x_2) + \ldots + U_n\delta_n(x_2) + \ldots + U_N\delta_N(x_2) = f(x_2)$$

$$\vdots \qquad \vdots \qquad \vdots \qquad \vdots \qquad \vdots$$

$$U_1\delta_1(x_m) + U_2\delta_2(x_m) + \ldots + U_n\delta_n(x_m) + \ldots + U_N\delta_N(x_m) = f(x_m)$$

$$\vdots \qquad \vdots \qquad \vdots \qquad \vdots \qquad \vdots$$

$$U_1\delta_1(x_N) + U_2\delta_2(x_N) + \ldots + U_n\delta_n(x_N) + \ldots + U_N\delta_N(x_N) = f(x_N)$$

$$(16)$$

Note that δ_n is a function only of x and that, with a discrete value x_m substituted into this function, $\delta_n(x_m)$ is a known value which can in principle be computed from (14). Hence, (16) is a set of N linear equations with constant coefficients $\delta_n(x_m)$ and with unknowns U_n. Since the $f(x_m)$ are known, the U_n can be determined by standard techniques.

Eq. (16) can be written

$$\sum_{n=1}^{N} U_n S_{mn} = F_m, \quad m = 1, 2, \ldots, N, \tag{17}$$

where both

$$S_{mn} = \delta_n(x_m) = \int_{x' = x_n - \Delta/2}^{x_n + \Delta/2} g(x_m, x') dx' \tag{18}$$

and

$$F_m = f(x_m) \tag{19}$$

are known quantities. Alternatively, it can be written as a matrix equation

$$[S_{mn}][U_n] = [F_m] \tag{20}$$

where $[U_n]$ and $[F_m]$ are column vectors with N elements and $[S_{mn}]$ is an N × N square matrix. Since $[U_n]$ can be determined from the solution of (20),

$$[U_n] = [S_{mn}]^{-1} [F_m], \tag{21}$$

where $[S_{mn}]^{-1}$ is the inverse of $[S_{mn}]$, the approximate solution of (6) is available as

$$u(x) \doteq [U_n]^T [p_n(x)], \tag{22}$$

where $[U_n]^T$ is the transpose of $[U_n]$, or as

$$u(x) \doteq \sum_{n=1}^{N} U_n p_n(x). \tag{23}$$

The pulse functions p_n are introduced in (9) above in order that the solution technique could be illustrated in its simplest form. However, $u_n(x)$ of (7) could be, for example, a piecewise linear function in $(x_n - \Delta/2, x_n + \Delta/2)$ and zero outside this subinterval. In fact, in the procedure outlined above, $u_n(x)$ could be any subdomain expansion function, where, as the name implies, a subdomain function is one which is zero outside a subdomain of the entire domain of the unknown $u(x)$. Any $u_n(x)$ which is zero for $x \notin (x_n - \Delta/2, x_n + \Delta/2)$ is an appropriate candidate for use in the above procedure. The only modification needed to accommodate a u_n different from p_n is that (18) would be replaced by

$$S_{mn} = \int_{x' = x_n - \Delta/2}^{x_n + \Delta/2} b_n(x') \, g(x_m, x') dx' \quad (24)$$

where b_n is zero outside $(x_n - \Delta/2, x_n + \Delta/2)$.

There are a number of important points involved in the above discussions which deserve greater attention. For example, it is obvious that $[S_{mn}]$ must be well-conditioned in order that it may be inverted numerically. Salient properties of $[S_{mn}]$ depend both upon the nature of the kernel in Eq. (6) and upon the selected expansion functions $u_n(x)$. Moreover, one sees immediately that the set $\{u_n\}$ must be linearly independent, if $[S_{mn}]$ is to be non-singular. One should realize that the match point location can readily influence the resulting approximate solution which one obtains through these techniques.

3.1 Example I (charged strip)

To adapt the solution method above with pulse expansion functions to the integral equation

$$-\int_{x' = -w}^{w} u(x') \ln|x - x'| dx' = f(x), \quad x \varepsilon (-w, w), \quad (25)$$

one has merely to compute S_{mn} of (18) with g replaced by the logarithmic kernel and to follow the steps leading to (23). The matrix elements are computed from

$$S_{mn} = \delta_n(x_m) = -\int_{x' = x_n - \Delta/2}^{x_n + \Delta/2} \ln|x_m - x'| dx' \quad (26)$$

which, subject to the change of variable $x' = x_n + \zeta$, can be converted to

$$S_{mn} = -\int_{\zeta = -\Delta/2}^{\Delta/2} \ln|(x_m - x_n) - \zeta| d\zeta. \quad (27)$$

The integral of (27) can be evaluated analytically to obtain

$$S_{mn} = -\left\{(x_m - x_n)\ln\left(\frac{|x_m - x_n + \Delta/2|}{|x_m - x_n - \Delta/2|}\right)\right.$$
$$\left. + \frac{\Delta}{2}\ln\left|(x_m - x_n)^2 - \frac{\Delta^2}{4}\right| - \Delta\right\} \tag{28}$$

which, in view of (8) and (15), reduces to

$$S_{mn} = -\Delta\left\{(m - n)\ln\left(\frac{|m - n + 1/2|}{|m - n - 1/2|}\right) + \ln\Delta\right.$$
$$\left. + [1/2]\ln\left|(m - n)^2 - 1/4\right| - 1\right\}. \tag{29}$$

With F_m determined according to (19), one can apply the solution technique to the case of the integral equation of (25). With proper interpretation of terms, (25) is the equation for a flat strip of width 2w charged to a specified potential.

Numerical solutions of (25) with $f = 1$, x, and x^2 are presented graphically in Fig. 5, together with the exact solutions [6] for comparison. One should observe the numerical convergence of the results as the number N of terms in (9a) is increased. Notice also that, when f is an even (odd) function, u is even (odd) too, which is a property of the integral equation (25).

Inspection of (29) reveals that $[S_{mn}]$ is diagonally strong, i.e., the terms S_{mm} ($m = n$) along the main diagonal of the matrix are much larger in magnitude than others. This is a desirable feature which enhances inversion stability. Also, $[S_{mn}]$ is symmetric $S_{mn} = S_{nm}$ as can be seen from (29).

3.2 General cylinder

The integral equation (5) for a PEC cylindrical shell is written below in general form with a variable forcing function,

$$-\int_C u(\bar{\rho}')\ln|\bar{\rho} - \bar{\rho}'|d\ell' = f(\bar{\rho}), \quad \bar{\rho}\epsilon C, \tag{30}$$

where C is the shell contour illustrated in Fig. 6. The pulse/collocation method is well-suited for obtaining a numerical solution of this equation. The contour is partitioned into N segments (sub-contours) as suggested in Fig. 6 with the length of the nth segment designated ΔC_n. With the unknown represented by pulses

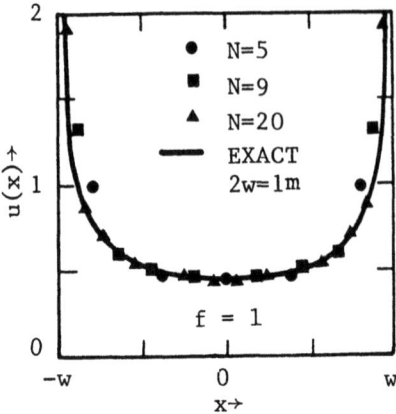

$$u(\bar{\rho}) = \sum_{n=1}^{N} U_n p_n(\bar{\rho}) \qquad (31a)$$

where now

$$p_n(\bar{\rho}) = \begin{cases} 1, & \bar{\rho} \varepsilon \, \Delta C_n, \\ 0, & \text{otherwise} \end{cases}, \qquad (31b)$$

and, with the equation enforced at the match points $\bar{\rho}_m$ located at the segment centers, the

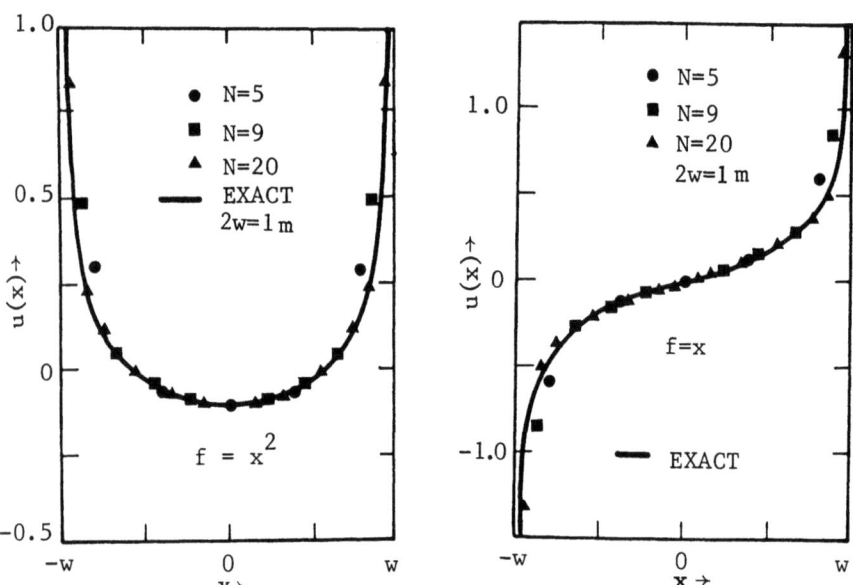

Fig. 5. Solution of eq. (25) for different forcing functions.

matrix elements S_{mn} are

$$S_{mn} = -\int_{\Delta C_n} \ln|\bar{\rho}_m - \bar{\rho}'| d\ell' . \qquad (32)$$

In solving problems with curved contours like that under discussion here, one usually finds it most convenient to take N sufficiently large to allow the segment ΔC_n to be well-represented by a straight

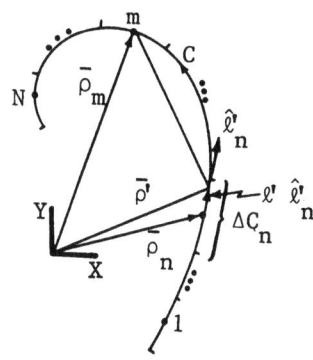

Fig. 6. Cross section of cylinder.

line segment. In (32), $\bar{\rho}'$ falls on ΔC_n and, with ΔC_n straight, it can be replaced by

$$\bar{\rho}' = \bar{\rho}_n + \ell'\hat{\ell}_n', \quad \ell' \epsilon (-\Delta C_n/2, \Delta C_n/2) \quad (33)$$

where $\hat{\ell}_n'$ is the unit vector, tangent to C at $\bar{\rho}_n$ and having the sense of C as illustrated in Fig. 6. Now, (32) becomes

$$S_{mn} = -\int_{\ell'=-\Delta C_n/2}^{\Delta C_n/2} \ln|\bar{\rho}_m - \bar{\rho}_n - \ell'\hat{\ell}_n'| d\ell' \quad (34)$$

which is readily evaluated. The straight-line condition imposed on ΔC_n causes $S_{mm}(m = n)$ of (34) to be the same as S_{mm} of (29) with Δ replaced by ΔC_n:

$$S_{mm} = S_{nn} = \Delta C_n [1 - \ln(\Delta C_n/2)]. \quad (35)$$

For $m \neq n$, S_{mn} can be computed easily by numerical integration, and, under the condition that $|\bar{\rho}_m - \bar{\rho}_n|/\Delta C_n \gg 1$, (34) can be approximated accurately by

$$S_{mn} = -\Delta C_n \ln|\bar{\rho}_m - \bar{\rho}_n|, \quad |\bar{\rho}_m - \bar{\rho}_n| \gg \Delta C_n. \quad (36)$$

Typically, for $m \neq n$, (34) should be evaluated by numerical integration when $|\bar{\rho}_m - \bar{\rho}_n|/\Delta C_n < 3$, with (36) used otherwise.

In many instances C is a sufficiently simple curve that (32) can be evaluated readily without the need to approximate ΔC_n by a straight segment. In such cases, effects of contour curvature can be accounted for directly and accurate solutions often can be obtained with N a relatively small number.

3.3 Example II (charged, finite-length tube)

The electrostatic problems considered to this point are two dimensional but now attention is turned to the three-dimensional problem of a finite-length PEC cylindrical tube charged to a potential $\Phi = V$. The tube, illustrated in Fig. 7, is in a homogeneous medium of infinite extent characterized by ϵ. Its walls are vanishingly thin and its length and radius are 2L and a, respectively. Applied to this problem, Eq. (4) becomes

$$\frac{1}{4\pi\epsilon} \int_{z'=-L}^{L} q_s(z') \int_{\phi'=-\pi}^{\pi} \frac{1}{R} a d\phi' dz' = V, \quad z \epsilon (-L, L) \quad (37)$$

where

$$R = \left[2a^2(1 - \cos[\phi - \phi']) + (z - z')^2\right]^{\frac{1}{2}}. \quad (38)$$

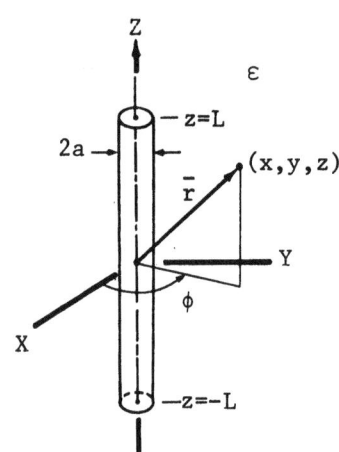

Fig. 7. Finite-length, PEC cylindrical tube in homogeneous infinite space.

As implied in (37), q_s is independent of ϕ due to the tube's circular symmetry and to its being isolated electrostatically in an infinite homogeneous medium. Consistent with ϕ-invariance of the physical problem, R is evaluated at $\phi = 0$; the value 0 is selected for ϕ because it is both convenient and representative of all other values. With the introduction of the total axial (linear) charge density (Coulombs/meter) defined by

$$q_\ell(z) = 2\pi a \, q_s(z) \quad (39)$$

and of the transformation $(1 - \cos \phi') = 2 \sin^2 \phi'/2$ in R, (37) can be written

$$\frac{1}{4\pi\varepsilon} \int_{z'=-L}^{L} q_\ell(z') \, k(z - z') \, dz' = V, \quad z\varepsilon(-L, L) \quad (40)$$

or, in generalized form,

$$\int_{z'=-L}^{L} u(z') \, k(z - z') dz' = f(z), \quad z\varepsilon(-L, L) \quad (41)$$

where the kernel k is

$$k(\xi) = \frac{1}{2\pi} \int_{\phi'=-\pi}^{\pi} \frac{1}{R} d\phi' \quad (42a)$$

with

$$R = \left[\xi^2 + 4a^2 \sin^2 \frac{\phi'}{2}\right]^{\frac{1}{2}}. \quad (42b)$$

Eq. (40) or (41) is an integral equation, having the form of (25), but its solution requires more labor than that for the flat strip because $k(z - z')$ is available only as the integral (42) which cannot be integrated analytically (with respect to z') as is possible with the logarithmic kernel. It should be evident that a procedure to solve (41) must incorporate a method to perform double numerical integration, which, unless done accurately and efficiently, can lead to major difficulties. Such difficulties can be particularly acute when the kernel is singular as is true of k of (42) and of others encountered in electromagnetics.

Here we outline an effective way to treat k which begins with simple variable transformations that enable one to express (42a) in the form

$$k(\xi) = \frac{2}{\pi} \int_{\theta=0}^{\pi/2} \frac{1}{[\xi^2 + 4a^2 - 4a^2 \sin^2\theta]^{\frac{1}{2}}} d\theta \qquad (43)$$

that is recognized to be

$$k(\xi) = \frac{1}{\pi a} \beta K(\beta) \qquad (44)$$

where $K(\beta)$ is the complete elliptic integral of the first kind [7],

$$K(\beta) = \int_{\theta=0}^{\pi/2} \frac{1}{[1 - \beta^2 \sin^2\theta]^{\frac{1}{2}}} d\theta, \qquad (45)$$

with modulus β given by

$$\beta = \frac{1}{\sqrt{1 + (\xi/2a)^2}} . \qquad (46)$$

The modulus has values $\beta \leq 1$, since $(\xi/2a)^2 > 0$. We next investigate $k(z - z')$ over the range of values of $[\overline{(z - z')}/2a]^2$. For $[(z - z')/2a]^2 \gg 1$, $R \doteq |z - z'|$ and

$$k(z - z') \doteq \frac{1}{|z - z'|} , \quad |z - z'| \gg 2a \qquad (47)$$

which can be integrated easily (in (41)). The other extreme is of more concern, since, as $\xi = (z - z')$ approaches zero, β approaches 1 and $K(\beta)$ exhibits a logarithmic singularity. In fact, one can show that

$$K(\beta) \xrightarrow[\beta \to 1]{} -\frac{1}{2}\ln\left(\frac{1-\beta}{8}\right) \tag{48}$$

and, subsequently, that

$$k(z - z') \xrightarrow[(z-z') \to 0]{} -\frac{1}{\pi a}\ln\left(\frac{|z-z'|}{8a}\right). \tag{49}$$

Due to its tenuous nature, direct numerical integration of a singular function like $k(z - z')$ is to be avoided so we proceed with a simple scheme that obviates such a task. The small-argument limit (49) is subtracted from, and added to, $k(z - z')$:

$$k(z-z') = \left\{k(z-z') + \frac{1}{\pi a}\ln\left(\frac{|z-z'|}{8a}\right)\right\} - \frac{1}{\pi a}\ln\left(\frac{|z-z'|}{8a}\right). \tag{50}$$

Obviously, the term within the braces, taken in concert, is very slowly varying as $(z - z')$ approaches zero, and the product of u_n and it can be integrated numerically with ease. The singular behavior of $k(z - z')$ manifests itself in the remaining term whose product with u_n can be integrated analytically for many commonly-employed expansion sets $\{u_n\}$.

Returning to the problem of solving (41), we apply the pulse-expansion/point-matching technique. The tube length is partitioned into N segments of length $\Delta = 2L/N$, and the pulse centers z_n and match points z_m are located at

$$z_n = -L + \Delta\left(n - \frac{1}{2}\right), \quad n = 1, 2, \ldots, N \tag{51}$$

and

$$z_m = -L + \Delta\left(m - \frac{1}{2}\right), \quad m = 1, 2, \ldots, N. \tag{52}$$

As before, we determine $[U_n]$ from (21) in which $F_m = f(z_m)$ and

$$S_{mn} = \int_{z'=z_n-\Delta/2}^{z_n+\Delta/2} k(z_m - z')dz'. \tag{53}$$

The kernel representation of (50) is substituted into (53) and, after integrating the term outside the braces, one arrives at

$$S_{mn} = \frac{1}{\pi a} \int_{\zeta = -\Delta/2}^{\Delta/2} \left\{ \beta_{mn} K(\beta_{mn}) + \ln(|\Delta[m-n] - \zeta|/8a) \right\} d\zeta$$

$$- \frac{\Delta}{\pi a} \left[\ln(\Delta/8a) + (m-n)\ln\left(\frac{\left|m - n + \frac{1}{2}\right|}{\left|m - n - \frac{1}{2}\right|}\right) \right.$$

$$\left. - 1 + \frac{1}{2} \ln \left| (m-n)^2 - \frac{1}{4} \right| \right] \tag{54}$$

in which use is made of (44) and in which, from the definition of β in (46), β_{mn} is seen to be

$$\beta_{mn} = \frac{1}{\sqrt{1 + \left(\frac{\Delta[m-n] - \zeta}{2a}\right)^2}} \quad . \tag{55}$$

The integral in (54) has a smoothly varying integrand and can be evaluated readily by an elementary numerical procedure. Of course, since for m = n both the elliptic integral and the logarithm function in the braces are unbounded as ζ → 0, the numerical integration method employed must be one which does not require evaluation of the integrand at ζ = 0 (identically). The elliptic

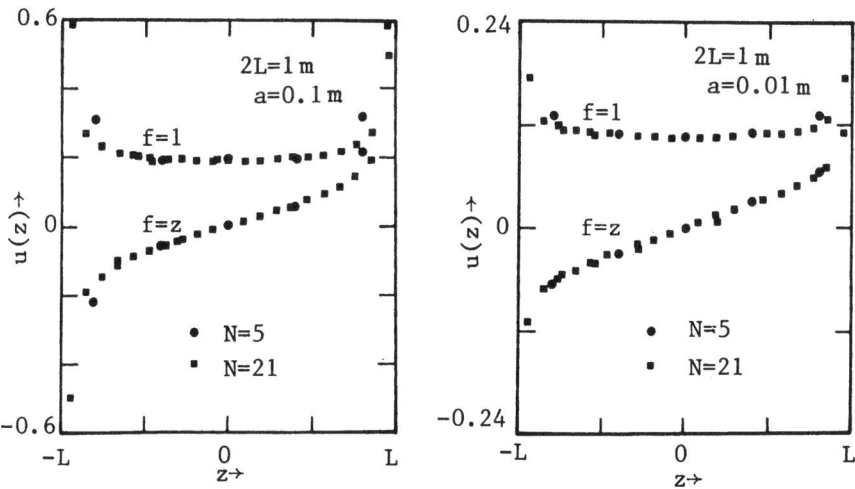

Fig. 8. Solution of Eq. (41) for different radii and forcing functions.

integral equivalent of the kernel from (44) is used in (54) because there is available [7] a simple and accurate approximation of K.

Results obtained by solving Eq. (41) with pulses and point matching are depicted in Fig. 8. Notice again the correlation between odd and even properties of f and u.

4. PROPERTIES OF $[S_{mn}]$

The matrix elements S_{mn}, encountered in a solution of an integral equation whose kernel g depends on x and x' according to the special relationship $|x - x'|$, exhibit interesting properties which can be used to advantage, especially when the elements of the basis set $\{u_n\}$ are of the subdomain type and when each u_n is an even function with respect to x_n, the center of its subdomain. The matrix elements S_{mn} resulting from the application of collocation to (6), having a kernel $g = g(|x - x'|)$, are computed from

$$S_{mn} = \int_{\zeta=-\Delta/2}^{\Delta/2} u_n(x_n + \zeta)\, g(|x_m - x_n - \zeta|)\, d\zeta \tag{56}$$

where Δ is the length of the subdomains and where $u_n(x_n + \zeta)$ is the nth basis function shifted to the interval $(-\Delta/2, \Delta/2)$ as indicated by its dependence on position expressed as displacement ζ from the nth subdomain center x_n. For basis functions which are the same, i.e., u_n is the same function with respect to x_n as u_p is with respect to x_p, $u_n(x_n + \zeta)$ can be replaced by $u(\zeta)$, representative of all u_n shifted to the interval $(-\Delta/2, \Delta/2)$. With this replacement and with the displacement $(x_m - x_n)$ from subdomain center x_n to match point x_m given by $\Delta(m - n)$ as per (8) and (15), (56) reduces to

$$S_{mn} = \int_{\zeta=-\Delta/2}^{\Delta/2} u(\zeta)\, g(|\Delta(m - n) - \zeta|)\, d\zeta. \tag{57}$$

We wish to investigate (57) subject to interchanging m and n:

$$S_{nm} = \int_{\zeta=-\Delta/2}^{\Delta} u(\zeta)\, g(|\Delta(n - m) - \zeta|)\, d\zeta. \tag{58}$$

Since $g(|\Delta(n - m) - \zeta|) = g(|\Delta(m - n) + \zeta|)$, the substitution $\zeta = -\zeta$ enables one to convert (58) to

$$S_{nm} = \int_{\zeta=-\Delta/2}^{\Delta/2} u(-\zeta)\, g(|\Delta(m - n) - \zeta|)d\zeta \tag{59}$$

from which it can be concluded that, if $u(\zeta) = u(-\zeta)$, $S_{nm} = S_{mn}$. In other words, for a set of basis functions that are even functions with respect to their subdomain centers, the matrix $[S_{mn}]$ is symmetric $S_{mn} = S_{nm}$.

Eq. (57) and $S_{mn} = S_{nm}$ imply that S_{mn} depends on $|m - n|$ ($= |n - m|$) rather than on $(m - n)(\neq (n - m))$. This symmetry and dependence upon $|m - n|$ are exemplified by S_{mn} of (29) and of (54). We emphasize these properties of S_{mn} by writing (57) as

$$S_{mn} = \int_{\zeta=-\Delta/2}^{\Delta/2} u(\zeta)\, g(|\Delta(|m - n|) - \zeta|)d\zeta, \text{ for } u(\zeta) = u(-\zeta). \tag{60}$$

Since $|m - n| = 0, 1, \ldots, N - 1$, experiences its full excursion in a single row or column of $[S_{mn}]$, the entire matrix can be filled from knowledge of the elements of a single row or column. This is a great savings in labor when S_{mn} must be computed by means of numerical integration. A matrix whose elements depend upon $|m - n|$ is said to be Toeplitz [1].

5. SYMMETRY

Frequently, as is pointed out in preceding sections, solutions of integral equations exhibit certain symmetry properties. The properties are quite useful as a basis for checking results obtained from numerical methods, and often one can utilize symmetry properties to lessen the amount of work to be done or to reduce computer storage requirements of the numerical solution procedure. For example, if it is known that u(x) is an even function of x, one should be able to determine u(x) in the interval (0, w) rather than in (-w, w), which suggests that the number of terms needed in Eq. (7) to produce a solution over (0, w) should be one-half the number needed over (-w, w). The reduction in computer storage space which can be achieved is obvious, if u is either an odd or an even function, but often space can be saved even when the unknown possesses no symmetry. Such is usually possible when the structure under investigation has geometric symmetry but, due to the nature of the excitation, the unknown quantity of interest possesses none. The space saving is accomplished by partitioning the problem into subproblems, whose superposition yields the original. Of course, we do not get something for nothing. The savings in space is traded for other work and more computer time needed to solve the problem.

We now establish properties of

$$L(u; x) = \int_{x'=-w}^{x} u(x') \, g(|x - x'|) dx', \quad x \in (-w, w) \tag{61}$$

which can be written

$$L(u; x) = \int_{x'=0}^{w} [u(x') \, g(|x - x'|) + u(-x') \, g(|x + x'|)] dx'. \tag{62}$$

For u an even function, $u(x) = u(-x)$ [denoted $u^e(x)$], and for u an odd function, $u(x) = -u(-x)$ [denoted $u^o(x)$], it follows from (62) that

$$L(u^{e}_{o}; x) = \int_{x'=0}^{w} u^{e}_{o}(x')[g(|x - x'|) \pm g(|x + x'|)] dx'. \tag{63}$$

Due to the dependence of g upon the absolute value of its argument and the relationship (63), the operator satisfies

$$L(u^{e}_{o}; x) = \pm L(u^{e}_{o}; -x). \tag{64}$$

Hence, in the integral equation,

$$L(u; x) = f(x), \tag{65}$$

an even (odd) forcing function f produces an even (odd) solution u, since both sides of the equation obviously must exhibit the same dependence on x. If (65) were an integro-differential equation with $[d/dx]L$ on the left side, an even (odd) f would produce an odd (even) u. The properties of $[d^2/dx^2]L = f$ would be the same as those of (65).

Summary

Even-function excitation: $f(x) = f(-x)$ [denoted $f^e(x)$]

$$\int_{x'=0}^{w} u^e(x') \, g^e(x, x') dx' = f^e(x), \quad x \in (0, w), \tag{66a}$$

where

$$g^e(x, x') = [g(|x - x'|) + g(|x + x'|)] \qquad (66b)$$

Odd function excitation: $f(x) = -f(-x)$ [denoted $f^o(x)$]

$$\int_{x'=0}^{w} u^o(x') \, g^o(x, x') dx' = f^o(x), \quad x \in (0, w), \qquad (67a)$$

where

$$g^o(x, x') = [g(|x - x'|) - g(|x + x'|)] \qquad (67b)$$

If f is either an even or an odd function, then (65) can be replaced by (66) or (67). The unknown in both (66) and (67) exists over $(0, w)$ as opposed to over $(-w, w)$ in (65), so u^e or u^o of (66) and (67) can be approximated by an expansion like (18) with one-half the number of terms needed to approximate u of (65). Thus, the matrices $[S_{mn}^e]$ and $[S_{mn}^o]$ associated with a numerical solution of (66) and (67) have one-fourth as many elements as that associated with the solution of (65). Computation of the elements S_{mn}^e or S_{mn}^o is no more demanding than is the determination of S_{mn} of (65), since the contribution to the value of an element due to the presence of $g(|x + x'|)$ in g^e and g^o is actually an appropriately identified element of S_{mn}.

Because a function can be expressed as the sum of an even function and an odd function, it follows that (66) and (67) can be used to calculate u in response to any general excitation f. One simply partitions the excitation f into odd and even constituents according to

$$f^e_o(x) = \frac{1}{2} [f(x) \pm f(-x)], \qquad (68)$$

solves (66) and (67) independently for u^e due to f^e and for u^o due to f^o, respectively, and synthesizes the solution as

$$u(x) = u^e(x) + u^o(x). \qquad (69)$$

Clearly, in such a two-step solution scheme, one trades analysis and computer time for computer storage space.

6. COUPLED INTEGRAL EQUATIONS

With relatively minor extensions, the solution methods presented in preceding sections can be applied to coupled integral equations. To illustrate methods for solving two integral equations in two unknowns, one considers the problem suggested in Fig. 9 where two infinite, PEC parallel strips are shown in cross section. Strip A is charged to $\Phi = V_a$ and Strip B to $\Phi = V_b$. In order to be able to utilize as much as possible of the analysis of the single strip, and thereby systematize the coupled strip analysis, we find it advantageous to base the geometric features of strip-to-strip coupling on the parameters defined below:

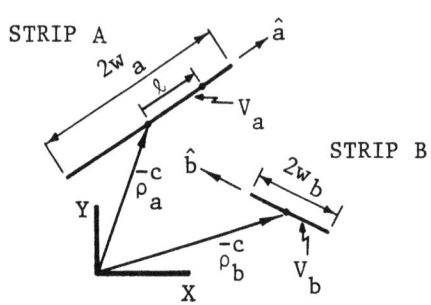

Fig. 9. Cross section of coupled strips.

$\bar{\rho}_a^c(\bar{\rho}_b^c)$ -- vector from origin to center of Strip A (Strip B)

$\hat{a}(\hat{b})$ -- unit vector specifying orientation and positive sense of transverse displacement for Strip A (Strip B)

$2w_a(2w_b)$ -- width of Strip A (Strip B)

The parameters above are computed from knowledge of the given strip edge coordinates. For example, if the "upper" and "lower" edge coordinates for the jth strip are (x_j^+, y_j^+) and (x_j^-, y_j^-), respectively, in a specified coordinate system, the above parameters are determined by means of

$$2w_j = [(x_j^+ - x_j^-)^2 + (y_j^+ - y_j^-)^2]^{\frac{1}{2}}, \quad (70a)$$

$$\hat{j} = \frac{(x_j^+ - x_j^-)\hat{x} + (y_j^+ - y_j^-)\hat{y}}{2w_j}, \quad (70b)$$

and

$$\bar{\rho}_j^c = \frac{1}{2}[(x_j^+ + x_j^-)\hat{x} + (y_j^+ + y_j^-)\hat{y}]. \quad (70c)$$

In terms of (70), the vector $\bar{\rho}^j$ from the origin to a point on the jth strip is

$$\bar{\rho}^j = \bar{\rho}_j^c + \ell\hat{j}, \quad \ell\varepsilon(-w_j, w_j) \quad (71)$$

where ℓ is a measure (in the direction $\hat{j}$) of displacement from the strip center to the point located by (71), as illustrated on Strip A of Fig. 9.

The surface charge induced on the strips must be of such distribution and density that the total potential Φ due to this induced charge be V_a on Strip A and V_b on Strip B. If Φ_j is the potential due to q_{sj}, the surface charge density on the jth strip, then the above statement is equivalent to

$$\Phi_a(\bar{\rho}^a) + \Phi_b(\bar{\rho}^a) = V_a, \quad \ell\epsilon(-w_a, w_a) \tag{72a}$$

$$\Phi_a(\bar{\rho}^b) + \Phi_b(\bar{\rho}^b) = V_b, \quad \ell\epsilon(-w_b, w_b) \tag{72b}$$

where $\bar{\rho}^a$ and $\bar{\rho}^b$ locate points on Strip A and Strip B according to (71). Φ_a and Φ_b at a point $\bar{\rho}$ are represented by

$$\Phi_j(\bar{\rho}) = -\frac{1}{2\pi\epsilon} \int_{\ell'=-w_j}^{w_j} q_{sj}(\ell') \ln|\bar{\rho} - \bar{\rho}^{j'}| d\ell'. \tag{73}$$

It is of interest to observe that $\Phi_j(\bar{\rho})$ evaluated on the jth strip ($\bar{\rho} = \bar{\rho}^j$) simplifies to

$$\Phi_j(\bar{\rho}^j) = -\frac{1}{2\pi\epsilon} \int_{\ell'=-w_j}^{w_j} q_{sj}(\ell') \ln|\ell - \ell'| d\ell', \tag{74}$$

since, in view of (71), $|\bar{\rho}^j - \bar{\rho}^{j'}| = |(\bar{\rho}_j^c + \ell\hat{j}) - (\bar{\rho}_j^c + \ell'\hat{j})| = |\ell - \ell'|$. The form of (74) is what one expects from previous discussions, since $\Phi_j(\bar{\rho}^j)$ is nothing more than the portion of the total potential on the jth strip that is due to the charge on the same j^{th} strip. In other words, (74) is the "self potential" of the jth strip. The expression for $\Phi_j(\bar{\rho})$ with $\bar{\rho}$ on the kth strip ($j \neq k$) obviously is

$$\Phi_j(\bar{\rho}^k) = -\frac{1}{2\pi\epsilon} \int_{\ell'=-w_j}^{w_j} q_{sj}(\ell') \ln|\bar{\rho}_k^c - \bar{\rho}_j^c + \ell\hat{k} - \ell'\hat{j}| d\ell',$$
$$\ell\epsilon(-w_k, w_k). \tag{75}$$

Substitution of (74) and (75), with appropriate specification of j and k in terms of a and b, into (72) leads explicitly to the pair of coupled integral equations for the two strips of Fig. 9:

$$-\frac{1}{2\pi\varepsilon} \int_{\ell'=-w_a}^{w_a} q_{sa}(\ell')\ln|\ell - \ell'|d\ell'$$

$$-\frac{1}{2\pi\varepsilon} \int_{\ell'=-w_b}^{w_b} q_{sb}(\ell')\ln|\bar{d}_{ab}^c + \ell\hat{a} - \ell'\hat{b}|d\ell' = V_a, \quad (76a)$$

$$\ell\varepsilon(-w_a, w_a)$$

$$-\frac{1}{2\pi\varepsilon} \int_{\ell'=-w_a}^{w_a} q_{sa}(\ell')\ln|\bar{d}_{ba}^c + \ell\hat{b} - \ell'\hat{a}|d\ell'$$

$$-\frac{1}{2\pi\varepsilon} \int_{\ell'=-w_b}^{w_b} q_{sb}(\ell')\ln|\ell - \ell'|d\ell' = V_b, \quad (76b)$$

$$\ell\varepsilon(-w_b, w_b)$$

where the vector $\bar{d}_{jk}^c$ between strip centers is defined by

$$\bar{d}_{jk}^c = \bar{\rho}_j^c - \bar{\rho}_k^c. \quad (77)$$

To solve the coupled equations of (76) by the pulse/collocation method described for the isolated strip problem, one partitions the jth strip into N_j subsegments of length $\Delta_j = 2w_j/N_j$ and represents the unknown charge density on each strip by a piecewise constant (pulse) scheme like that suggested in Fig. 10. On the jth strip the charge density is expressed as

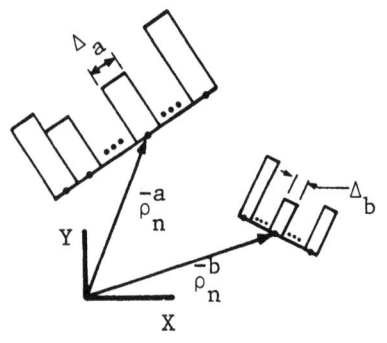

Fig. 10. Pulses on coupled strips.

$$q_{sj}(\ell)/2\pi\varepsilon = u_j(\ell) = \sum_{n=1}^{N_j} U_n^j p_n^j(\ell) \quad (78)$$

where U_n^j is the unknown magnitude of of p_n^j, the nth pulse on the jth strip of Fig. 10. The pulse centers on the jth strip are located by

$$\bar{\rho}_n^j = \bar{\rho}_j^c + [-w_j + \Delta_j(n - 1/2)]\hat{j}, \; n = 1, 2, \ldots, N_j. \quad (79)$$

Match point locations on the jth strip are computed by means of (79) with n replaced by m. With appropriate assignment of a and b to j, substitution of (78) into (76) and enforcement at match points -- (76a) at $\bar{\rho}_m^a$ on Strip A and (76b) at $\bar{\rho}_m^b$ on Strip B -- yield

$$\sum_{n=1}^{N_a} U_n^a S_{mn}^{aa} + \sum_{n=1}^{N_b} U_n^b S_{mn}^{ab} = F_m^a, \quad m = 1, 2, \ldots, N_a, \tag{80a}$$

$$\sum_{n=1}^{N_a} U_n^a S_{mn}^{ba} + \sum_{n=1}^{N_b} U_n^b S_{mn}^{bb} = F_m^b, \quad m = 1, 2, \ldots, N_b, \tag{80b}$$

where $F_m^j = f_j(\bar{\rho}_m^j)$ is used in (80) to imply that the solution method is valid for variable forcing functions. In the present case, of course, $f^j = V_j$. The elements S_{mn}^{kj} are

$$S_{mn}^{kj} = - \int_{\ell'=-\Delta_j/2}^{\Delta_j/2} \ln|\bar{\rho}_m^k - \bar{\rho}_n^j - \ell'\hat{j}| d\ell' \tag{81}$$

where, from (79),

$$\bar{\rho}_m^k - \bar{\rho}_n^j = \bar{d}_{kj}^c - [w_k - \Delta_k(m - 1/2)]\hat{k}$$

$$+ [w_j - \Delta_j(n - 1/2)]\hat{j} . \tag{82}$$

When $k = j$, $S_{mn}^{kj} = S_{mn}^{kk}$ becomes the same as S_{mn} of (29) with Δ replaced by Δ_k as one anticipates from (81) and the discussion of self potential:

$$S_{mn}^{kk} = \Delta_k \left\{ 1 - \ln\Delta_k - \frac{1}{2} \ln\left|(m - n)^2 - \frac{1}{4}\right| \right.$$

$$\left. - (m - n) \ln \left(\left|\frac{m - n + 1/2}{m - n - 1/2}\right|\right) \right\} . \tag{83}$$

In general, for $k \neq j$, (81) must be evaluated numerically but, for $|\bar{\rho}_m^k - \bar{\rho}_n^j| \gg \Delta_j$, it can be approximated by

$$S_{mn}^{kj} \doteq - \Delta_j \ln|\bar{\rho}_m^k - \bar{\rho}_n^j|, \quad |\bar{\rho}_m^k - \bar{\rho}_n^j| \gg \Delta_j . \tag{84}$$

If $\Delta_k = \Delta_j$ <u>and</u> $|\bar{\rho}_m^k - \bar{\rho}_n^j| \gg \Delta_j$, then $S_{mn}^{kj} = S_{nm}^{jk}$. However, in general, S_{mn}^{kj} does not exhibit this symmetry. For special orientations of one strip relative to the other, e.g., co-planar strips, the integration of (81) can be performed analytically and $S_{mn}^{kj} = S_{nm}^{jk}$.

The coupled strip solution procedure is presented here in a way that is amenable to extension to more than two strips. For example, in the case of three strips, designated A, B, and C, Eq. (80) would be, in matrix form,

$$\begin{bmatrix} [S_{mn}^{aa}] & [S_{mn}^{ab}] & [S_{mn}^{ac}] \\ [S_{mn}^{ba}] & [S_{mn}^{bb}] & [S_{mn}^{bc}] \\ [S_{mn}^{ca}] & [S_{mn}^{cb}] & [S_{mn}^{cc}] \end{bmatrix} \begin{bmatrix} [U_n^a] \\ [U_n^b] \\ [U_n^c] \end{bmatrix} = \begin{bmatrix} [F_m^a] \\ [F_m^b] \\ [F_m^c] \end{bmatrix}. \quad (85)$$

The elements S_{mn}^{kj} in each sub-matrix are computed from (83) for "self" sub-matrices and from (81), or (84) if applicable, for sub-matrices representing strip-to-strip coupling. It should be clear that a computer program developed to implement the solution procedure for any number of strips would have only two major subroutines: one to compute self sub-matrices $[S_{mn}^{kk}]$ from (83) and another to compute coupling sub-matrices $[S_{mn}^{kj}]$ from (81) and (84).

6.1 Example I (identical, co-planar strips)

For the two co-planar identical strips of width 2w shown in Fig. 11, the desired coupled equations result directly from substitution into (76) of $w_a = w_b = w$ and the terms indicated in the arguments of the kernels. If $V_b = V_a = V$, then from physical reasoning one concludes that $q_{sa}(\ell) = q_{sb}(-\ell)$, while, if $-V_b = V_a = V$, $q_{sa}(\ell) = -q_{sb}(-\ell)$, where the variable ℓ is measured in the $\hat{u}$ direction from the center of Strip A (Strip B) for $q_{sa}(q_{sb})$. In either case, only one equation is needed since the charge density on one strip is known in terms of that on the other:

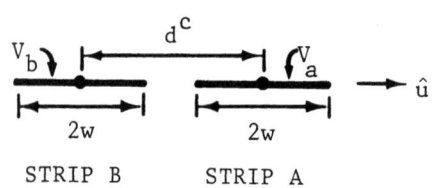

Fig. 11. Co-planar, coupled strips.

$$-\frac{1}{2\pi\varepsilon} \int_{\ell'=-w}^{w} q_{sa}(\ell')[\ln|\ell - \ell'| \pm \ln|\ell + d^c + \ell'|]d\ell' = V \quad (86)$$

The plus sign applies to the case of symmetric excitation ($V_b = V_a = V$) and the negative sign to the antisymmetric case ($-V_b = V_a = V$). Eq. (86) can be solved by the pulse/collocation (P/C) method with matrix elements

$$S_{mn}^{\pm} = S_{mn} \pm S'_{mn} \quad (87)$$

where S_{mn} are given in (29) and where

$$S'_{mn} = -\int_{\ell'=-\Delta/2}^{\Delta/2} \ln|[d^c - 2w + \Delta(m + n - 1)] - \ell'|d\ell' \quad (88)$$

which is seen to be (27) with $(x_m - x_n)$ replaced by $[d^c - 2w + \Delta(m + n - 1)]$. Hence, S'_{mn} can be determined directly from (28) by this replacement of terms.

One should realize that, due to the linearity of the equations for the coupled strips, superposition can be invoked to simplify analyses. In particular, the two solutions (symmetric and antisymmetric cases) for the coupled strips of Fig. 11 can be used to synthesize the resultant charge densities for any combination of values of V_a and V_b.

The charge density on two coupled, closely spaced, co-planar strips, both charged to 1 volt, is illustrated in Fig. 12. The properties of q_{sa} and q_{sb} discussed above are clearly exhibited in Fig. 12. Note, also, the evidence of repulsion of like charges.

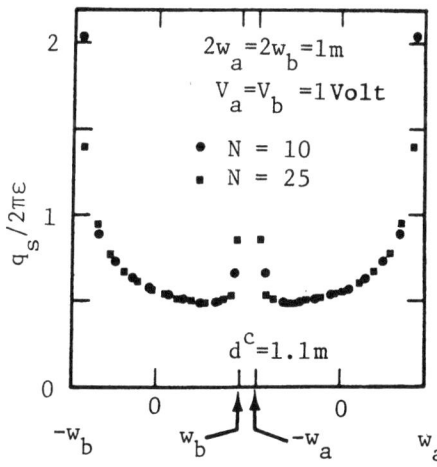

Fig. 12. Charge density on coupled co-planar strips.

6.2 Example II (bent strip)

In Fig. 13 are illustrated two equal-width strips joined at edges

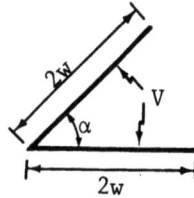

to form a bent strip having a "V" cross section. The bent strip is charged to a potential of $\Phi = V$ and resides in a medium characterized by ε. The charge density on the two halves must be the same and it satisfies (See (76))

Fig. 13. Bent strip.

$$-\frac{1}{2\pi\varepsilon} \int_{\ell'=-w}^{w} q_s(\ell') \cdot \left\{ \ln|\ell - \ell'| \right.$$

$$\left. + \ln\left[[(w+\ell)^2 + (w+\ell')^2 - 2(w+\ell)(w+\ell')\cos\alpha]^{1/2} \right] \right\} d\ell' = V,$$

$$\ell\varepsilon(-w, w), \quad (89)$$

which easily can be solved by the P/C method. The matrix elements are

$$S_{mn}^{v} = S_{mn} + S_{mn}^{ab} \quad (90)$$

where S_{mn} are the self elements of (29) and S_{mn}^{ab} are found from (81) and (82) to be

$$S_{mn}^{ab} = -\int_{\zeta=-\Delta/2}^{\Delta/2} \ln\left[[\Delta^2(m-1/2)^2 + \Delta^2(n-1/2)^2 + 2\Delta\zeta(n-1/2) + \zeta^2 \right.$$

$$\left. - 2\Delta(m-1/2)[\Delta(n-1/2) + \zeta]\cos\alpha]^{1/2} \right] d\zeta. \quad (91)$$

The integration indicated in (91) is not difficult to perform numerically, but, when $m = n = 1$ and $|\alpha| < \pi/2$, care must be exercised. For $m = n = 1$,

$$S_{11}^{ab} = -\Delta \int_{\zeta=0}^{1} \ln\left[\Delta \sqrt{1/4 + \xi(\xi - \cos\alpha)} \right] d\xi \quad (92)$$

whose integrand peaks sharply at $\xi = \frac{1}{2}\cos\alpha$ for $|\alpha| < \pi/2$ and Δ small. To efficiently integrate numerically such a peaked function, one should integrate from $\xi = 0$ to the peak ($\xi = \frac{1}{2} \cdot \cos\alpha$) and then from this peak to 1. For $|\alpha| > \pi/2$, the integrand

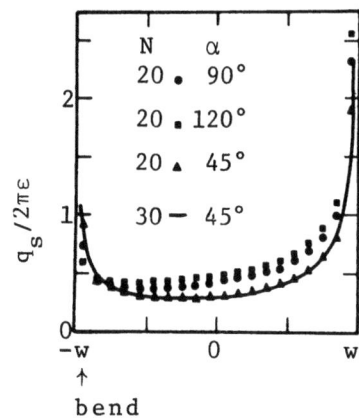

Fig. 14. Charge density on bent strip.

of (92) peaks at $\xi = 0$ and is simple to integrate.

In the bent strip problem of Fig. 13, one may choose the bend as the origin of coordinates and derive an integral equation that is slightly simpler than (89). This simpler equation is obtained, if in (89) one changes the integration limits to (0, 2w) and replaces w by 0 where it appears in the integrand. However, the matrix elements for this alternate equation are identical to S_{mn}^v.

The charge density on a bent strip is depicted in Fig. 14 for different bend angles α. If the number N of unknowns is increased sufficiently, the solutions exhibit the edge condition at the bend remarkably well.

7. SOME EQUATIONS FROM ELECTRODYNAMICS

In an infinite homogeneous medium characterized by the parameters (μ, ε), in which exist surface electric currents and charges having densities $\bar{J}_s$ and q_s, the electric and magnetic fields $(\bar{E}, \bar{H})$ due to these sources can be determined from

$$\bar{H} = \frac{1}{\mu} \nabla \times \bar{A} \tag{93}$$

and

$$\bar{E} = -j\omega\bar{A} - \nabla\Phi \tag{94}$$

where $\bar{A}$ is the magnetic vector potential

$$\bar{A}(\bar{r}) = \frac{\mu}{4\pi} \iint_S \bar{J}_s(\bar{s}') \, G(\bar{r} - \bar{s}') \, dS' \tag{95}$$

and Φ is the electric scalar potential

$$\Phi(\bar{r}) = \frac{1}{4\pi\varepsilon} \iint_S q_s(\bar{s}') \, G(\bar{r} - \bar{s}') \, dS' \, . \tag{96}$$

S is the surface on which $\bar{J}_s$ and q_s reside, and G is the open space Green's function

$$G(\bar{r} - \bar{s}') = \frac{e^{-jk|\bar{r} - \bar{s}'|}}{|\bar{r} - \bar{s}'|} \tag{97}$$

where $\bar{r}$ is the point at which the above quantities are observed and where $\bar{s}'$ locates a source point on S. In (97) $k = \omega\sqrt{\mu\varepsilon} = 2\pi/\lambda$ is the propagation constant, where λ is the wavelength in the medium* at the angular frequency ω of the assumed harmonic time variation $e^{j\omega t}$. $\bar{A}$ and Φ of (95) and (96) are integrals over the sources. They are those particular integral solutions of the equations governing $\bar{A}$ and Φ that cause $\bar{H}$ and $\bar{E}$, through (93) and (94), to satisfy the radiation condition. As an alternative to (94), one can explicitly† invoke the Lorentz condition, $\nabla \cdot \bar{A} + j(k^2/\omega)\Phi = 0$, and express $\bar{E}$ as a function of $\bar{A}$, hence, of $\bar{J}_s$, alone:

$$\bar{E} = -j\frac{\omega}{k^2}(k^2\bar{A} + \nabla\nabla \cdot \bar{A}) . \tag{98}$$

Also, conservation of charge implies that

$$\nabla_s \cdot \bar{J}_s + j\omega q_s = 0 \tag{99}$$

which can be used to eliminate q_s from (96), if desired.

In two-dimensional§ problems in which all quantities are invariant with respect to a linear coordinate - say, the z axis, $\bar{A}$ is independent of z and can be written

$$\bar{A}(\bar{\rho}) = \frac{\mu}{4\pi}\int_C \bar{J}_s(\bar{\rho}')\int_{z'=-\infty}^{\infty}\frac{e^{-jk[|\bar{\rho} - \bar{\rho}'|^2 + z'^2]^{1/2}}}{[|\bar{\rho} - \bar{\rho}'|^2 + z'^2]^{1/2}}dz'd\ell' \tag{100}$$

where $\bar{\rho}$ locates a field point in the xy plane and $\bar{\rho}'$ locates a source point on the contour C over the cross section of the

---

* If the medium is lossy, ε is replaced by $(\varepsilon - j\frac{\sigma}{\omega})$.

† The wave equations from which $\bar{A}$ and Φ of (95) and (96) are determined incorporate the Lorentz condition.

§ See Section 2.

infinitely long surface S of Fig. 2. Making use of the well-known integral representation of the zero-order Hankel function of the second kind $H_0^{(2)}$,

$$H_0^{(2)}(k\xi) = \frac{j}{\pi} \int_{z=-\infty}^{\infty} \frac{e^{-jk[\xi^2 + z^2]^{1/2}}}{[\xi^2 + z^2]^{1/2}} dz, \quad (101)$$

one converts $\bar{A}$ of (100) to

$$\bar{A}(\bar{\rho}) = -j\frac{\mu}{4} \int_C \bar{J}_s(\bar{\rho}') H_0^{(2)}(k|\bar{\rho} - \bar{\rho}'|) d\ell'. \quad (102)$$

Similarly, for the two-dimensional problem, Φ is

$$\Phi(\bar{\rho}) = -j\frac{1}{4\varepsilon} \int_C q_s(\bar{\rho}') H_0^{(2)}(k|\bar{\rho} - \bar{\rho}'|) d\ell'. \quad (103)$$

The fields due to two-dimensional, z-invariant source distributions can be computed from (93) and (94) (or (98)) applied to the potentials of (102) and (103). Of course, the z-invariance must be borne in mind when the differential operators are applied.

It is well known that the component of the <u>total</u> electric field tangential to the surface S of a PEC object <u>must</u> vanish everywhere on S. This is true for both a solid PEC object and an open or closed shell. To formulate an equation governing $\bar{J}_s$, the surface current induced on the surface of a PEC object (Fig. 15) excited by a known impressed or incident field ($\bar{E}^i$, $\bar{H}^i$), one simply enforces the condition that the tangential component of total electric field on S be zero

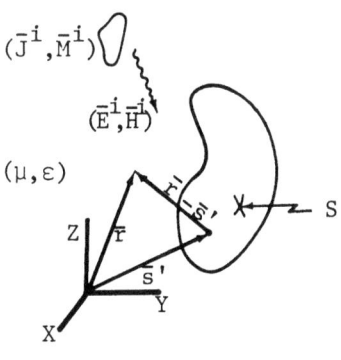

$$\bar{E}_t^s + \bar{E}_t^i = \bar{0} \text{ on S} \quad (104)$$

where the subscript "t" denotes the tangential (to S) component and where $\bar{E}^s$ is the so-called scattered field produced by the induced sources. The incident field ($\bar{E}^i$, $\bar{H}^i$) is that electromagnetic field which would be produced in the infinite medium by the known electric and magnetic source currents ($\bar{J}^i$, $\bar{M}^i$) with the object removed. By appealing to the equivalence principle

Fig. 15. PEC object in infinite space, illuminated by specified sources.

[8], one can show that the scattered field ($\bar{E}^s, \bar{H}^s$) can be computed from $\bar{J}_s$ by means of the potentials as if $\bar{J}_s$ existed in an infinite homogeneous space filled with (μ, ε). In this open space, $\bar{J}_s$ resides on S, the mathematical surface bounding the region occupied by the (removed) object. Hence, Eq. (104) can be written as either

$$(j\omega\bar{A} + \nabla\Phi)_t = \bar{E}_t^i \text{ on S} \tag{105}$$

or

$$j\frac{\omega}{k^2}(k^2\bar{A} + \nabla\nabla \cdot \bar{A})_t = \bar{E}_t^i \text{ on S,} \tag{106}$$

where $\bar{A}$ and Φ are given in (95) and (96) or (102) and (103).

7.1 Equation for TM-illuminated cylinder

If the infinite PEC cylinder of Fig. 2 is located in an infinite medium with parameters (μ, ε) and is excited by a z-invariant source which produces an excitation $\bar{E}^i = E_z^i\hat{z}$ and $\bar{H}^i = H_x^i\hat{x} + H_y^i\hat{y}$, the surface current $\bar{J}_s$ induced on the cylinder will have only a z component ($\bar{J}_s = J_{sz}\hat{z}$) as will the scattered electric field ($\bar{E}^s = E_z^s\hat{z}$). The cylinder and excitation are depicted in Fig. 16 which serves to define various quantities of interest. Since the components of $\bar{H}^i$ are entirely transverse to z ($\bar{H}_z^i = 0$), this excitation is classified as transverse magnetic (to z) and is referred to as TM_z illumination, or, briefly, TM illumination with the subscript "z" suppressed when the direction of the cylinder axis is understood.

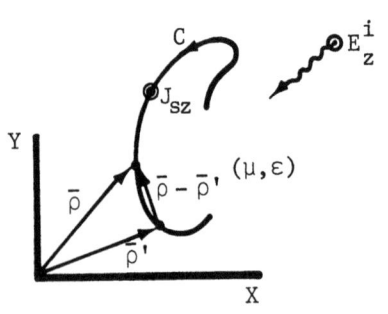

Fig. 16. TM-illuminated PEC cylinder.

Applied to the present two-dimensional problem, Eq. (105) reduces to

$$\frac{k\eta}{4}\int_C J_{sz}(\bar{\rho}') H_0^{(2)}(k|\bar{\rho} - \bar{\rho}'|)d\ell' = E_z^i(\bar{\rho}), \quad \bar{\rho}\varepsilon C \tag{107}$$

for the unknown surface current J_{sz} induced on the cylinder. Of course E_z^i is the known excitation or forcing function and it is evaluated on C. Subject to the identifications, $u = J_{sz}/(4/k\eta)$,

$f = E_z^i$, and $g(\bar{\rho}, \bar{\rho}') = H_0^{(2)}(k|\bar{\rho} - \bar{\rho}'|)$, Eq. (107) is of the form of the integral equations discussed in electrostatics. In addition to being of the same form, (107) and the earlier equations share the significant common feature of possessing logarithmically singular kernels.

Specialized to the case of the TM-illuminated PEC strip of width 2w, (107) becomes

$$\frac{k\eta}{4} \int_{x'=-w}^{w} J_{\hat{s}z}(x') H_0^{(2)}(k|x - x'|) dx' = E_z^i(x), \quad (108)$$
$$x \in (-w, w).$$

7.2 Equation for TE-illuminated cylinder

If the PEC cylinder is illuminated by an incident field in which $\bar{E}^i$ has no z-component, the excitation is classified as transverse electric (TE), and, in the z-invariant case considered here, $\bar{H}^i = H_z^i \hat{z}$. Applying (105) and (106) to the problem depicted in Fig. 17, we arrive at

$$j\omega \bar{A} \cdot \hat{\ell} + \frac{\partial}{\partial \ell} \Phi = \bar{E}^i \cdot \hat{\ell}, \text{ on } C \quad (109)$$

and

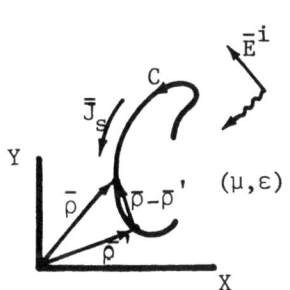

Fig. 17. TE-illuminated PEC cylinder.

$$j \frac{\omega}{k^2} [k^2 \bar{A} \cdot \hat{\ell} + \frac{\partial}{\partial \ell} (\nabla \cdot \bar{A})] = \bar{E}^i \cdot \hat{\ell} \text{ on } C \quad (110)$$

where $\hat{\ell} = \hat{\ell}(\bar{\rho})$ is the unit vector tangent to C at $\bar{\rho}$ with the sense of C as illustrated. The current is directed along C and has no z component. If the cylinder is solid or a closed shell, the current is continuous over the contour C, but, if it is an open shell of vanishing thickness like that suggested in Fig. 17, J_s approaches zero as $C_1\sqrt{\ell-\ell_1}$ for $\ell \to \ell_1$ and as $C_2\sqrt{\ell_2-\ell}$ for $\ell \to \ell_2$. In numerical solutions of (109) or (110), one usually enforces the conditions $J_s(\ell_1) = J_s(\ell_2) = 0$ to improve efficiency.

For the TE-excited flat strip of width 2w, (109) and (110) specialize to

$$\frac{\eta}{4k}\left\{k^2 \int_{x'=-w}^{w} J_{sx}(x')H_0^{(2)}(k|x-x'|)dx'\right.$$

$$\left. + \frac{d}{dx}\int_{x'=-w}^{w} \frac{d}{dx'} J_{sx}(x')H_0^{(2)}(k|x-x'|)dx'\right\} = E_x^i(x), \quad (111)$$

$$x\varepsilon(-w, w)$$

with $J_s(\pm w) = 0$ and

$$\frac{\eta}{4k}\left[\frac{d^2}{dx^2} + k^2\right]\int_{x'=-w}^{w} J_{sx}(x') H_0^{(2)}(k|x-x'|)dx' = E_x^i(x),$$

$$x\varepsilon(-w, w), \quad (112)$$

to which one appends the condition below in numerical solutions:

$$J_{sx}(\pm w) = 0. \quad (113)$$

Again each kernel is the Hankel function possessing the logarithmic singularity.

An alternate equation, the so-called magnetic field integral equation [9], is available for the TE-excited cylinder and often is easier to solve than are (109) and (110). However, the H-field equation holds only for closed contours which enclose non-zero area, whereas (109) and (110) apply to both open and closed contours.

7.3 Equation for thin cylindrical tube

Classical thin-wire antennas and scatterers have attracted the attention of both the practitioner and the theoretician for many years. An ideal thin wire is viewed as a PEC cylindrical tube of vanishing wall thickness whose radius a is much much less than the wavelength λ of operation (a << λ) and whose length 2L is far greater than its radius (L >> a).

Extrapolating from eigenfunction solutions of the TE- and TM-excited infinite-length PEC cylinder for which a/λ << 1, one assumes the following to hold for the thin, finite-length tube: (1) the axially-directed current so dominates the circumferentially-directed current that the latter can be ignored and (2) the axially-directed current is circumferentially invariant. Since, in the infinite-length cylinder analysis, the dominant current is due to the axially-directed component of incident electric field, one claims that current due to the other components of $\bar{E}^i$ can be ignored. Hence, in reference to Fig. 18, the only current of any

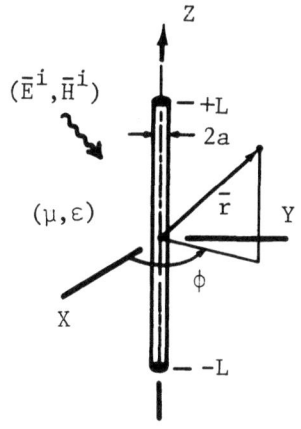

Fig. 18. PEC cylinder illuminated by incident field.

consequence is J_{sz} and it is independent of ϕ; also, J_{sz} is due entirely to E_z^i which is itself assumed constant on the cylindrical surface at a fixed z.

In usual thin-wire theory applied to a solid PEC cylinder, the caps at the ends are ignored on the grounds that their combined area is very small relative to λ^2 and is a very small fraction of the total surface area of the cylinder, which justifies the adoption of the tube model in this case too. If desired, however, the effects of the ends can be included in the analysis.

The simplifying assumptions introduced above are helpful in reducing the complexity of the analysis of the thin-wire scatterer, because they allow one to invoke ϕ independence of J_{sz} and to ignore E_ϕ^i. It is emphasized, however, that the analysis which follows is valid for any size PEC cylindrical tube, if, in the situation of interest, $E_\phi^i = 0$ and E_z^i is independent of ϕ.

At a point (a, ϕ, z) on the cylindrical surface, the vector and scalar potentials due to the induced sources are $(\bar{A} = A_z \hat{z})$

$$A_z(a, \phi, z) = \frac{\mu}{4\pi} \int_{z'=-L}^{L} J_{sz}(z') \int_{\phi'=-\pi}^{\pi} \frac{e^{-jkR}}{R} a d\phi' dz' \qquad (114)$$

and

$$\Phi(a, \phi, z) = \frac{j}{4\pi\omega\varepsilon} \int_{z'=-L}^{L} \frac{d}{dz'} J_{sz}(z') \int_{\phi'=-\pi}^{\pi} \frac{e^{-jkR}}{R} a d\phi' dz' \qquad (115)$$

where R is given in (38). The transformations employed to arrive at (42) in the electrostatic tube problem are applicable in the present case and allow one to write the potentials as

$$A_z(z) = \frac{\mu}{4\pi} \int_{z'=-L}^{L} I(z') K(z - z') dz' \qquad (116)$$

and

$$\Phi(z) = \frac{j}{4\pi\omega\varepsilon} \int_{z'=-L}^{L} \frac{d}{dz'} I(z') K(z-z') \, dz' \tag{117}$$

in which I is the total axial current (Amperes) on the cylinder,

$$I(z) = 2\pi a \, J_{sz}(z), \tag{118}$$

and K is the kernel

$$K(z-z') = \frac{1}{2\pi} \int_{\phi'=-\pi}^{\pi} \frac{e^{-jkR}}{R} \, d\phi' \, . \tag{119}$$

R in (119) given in (42b) with ξ replaced by $(z - z')$.

Enforcing (105) and (106) on the tube, one obtains, respectively, two integro-differential equations

$$\frac{j}{4\pi} \frac{\eta}{k} \left\{ k^2 \int_{z'=-L}^{L} I(z') K(z-z') dz' \right.$$

$$\left. + \frac{d}{dz} \int_{z'=-L}^{L} \frac{d}{dz'} I(z') K(z-z') dz' \right\} = E_z^i(z), \tag{120}$$

$$z \in (-L, L)$$

and

$$\frac{j}{4\pi} \frac{\eta}{k} \left(\frac{d^2}{dz^2} + k^2 \right) \int_{z'=-L}^{L} I(z') K(z-z') dz' = E_z^i(z), \tag{121}$$

$$z \in (-L, L)$$

satisfied by I. To solve either numerically, the boundary condition below is helpful:

$$I(\pm L) = 0 \, . \tag{122}$$

As in the case of the TE-excited strip, I on the tube actually approaches zero as $D_+\sqrt{L - z}$ does at $z = L$ and as $D_-\sqrt{L + z}$ does at $z = -L$. Also, the similarity in the TE-excited strip equation and the above wire equation should be noted.

For a perfectly conducting, straight-wire antenna driven by a voltage generator of V volts located at a point z_g on the wire, the impressed electric field caused by the voltage source must be determined and introduced into (120) and (121). If one adopts the standard slice generator source model [10], E_z^i is represented by

$$E_z^i(z) = V\delta(z - z_g), \tag{123}$$

where $\delta(z)$ is the familiar delta function.

8. SOLUTION PROCEDURE FOR TM-EXCITED STRIP EQUATION

Eq. (108) for the TM-excited strip is of the form of Eq. (25) and the kernels of these two integral equations exhibit the same singularity. Hence, the P/C solution method outlined for (25) applies to (108).

With the pulse expansion of (9)

$$J_{sz}(x) = \sum_{n=1}^{N} J_n P_n(x), \tag{124}$$

and match point location of (15), one determines the coefficients J_n by solving

$$\sum_{n=1}^{N} J_n Z_{mn} = E_m, \quad m = 1, 2, \ldots, N, \tag{125}$$

in which $E_m = E_z^i(x_m)$ and

$$Z_{mn} = \int_{\zeta = -\Delta/2}^{\Delta/2} H_0^{(2)}(k|\Delta(m - n) - \zeta|)d\zeta \tag{126}$$

where Z_{mn} is obtained by a sequence of steps similar to those leading to (27). For $|x_m - x_n| \gg \Delta$, or $|m - n| \gg 1$, (126) can be approximated by

$$Z_{mn} \doteq \Delta H_0^{(2)}(k\Delta|m - n|) \tag{127}$$

but, otherwise, it must be evaluated carefully. Below, we present a method for evaluation of (126) paralleling that discussed in the electrostatic tube problem.

The small argument form of $H_0^{(2)}(\xi)$ is [7]

$$H_0^{(2)}(\xi) \xrightarrow[\xi \to 0]{} 1 - j\frac{2}{\pi}\left(\gamma + \ln\frac{\xi}{2}\right) \tag{128}$$

where $\gamma \; (\doteq 0.5772)$ is Euler's constant. This small argument approximation is subtracted from and added to the integrand of (126), and the singular part is analytically integrated to obtain

$$Z_{mn} = \Delta\left[1 - j\frac{2}{\pi}\left(\gamma + \ln\frac{k}{2}\right)\right] + j\frac{2}{\pi}S_{mn}$$

$$+ \int_{\zeta=-\Delta/2}^{\Delta/2}\left\{H_0^{(2)}(k|\Delta(m-n)-\zeta|)\right.$$

$$\left. - \left[1 - j\frac{2}{\pi}\left(\gamma + \ln\left[\frac{k}{2}|\Delta(m-n)-\zeta|\right]\right)\right]\right\} d\zeta \qquad (129)$$

where S_{mn} is given in (29). As in the case of the elliptic integral, there is available [7] an accurate polynomial representation for the Hankel function $H_0^{(2)}$ in (129). One finds that the integrand above is very smooth and can be integrated numerically by means of any simple technique which does not place a point of evaluation at $\zeta = 0$, a precaution which must be observed only when $m = n$.

The TM-illuminated strip problem is now no more difficult to solve than is the electrostatic strip or cylindrical tube problem. In fact, with a subroutine available for evaluating (129), one can use the program developed for solving electrostatic problems. Results obtained in this manner are presented in Figs. 19 and 20. In Fig. 19, numerically computed current is compared with that

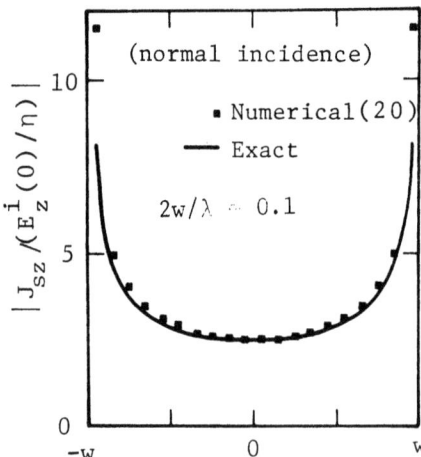

Fig. 19. Normalized surface current on narrow, TM-excited strip.

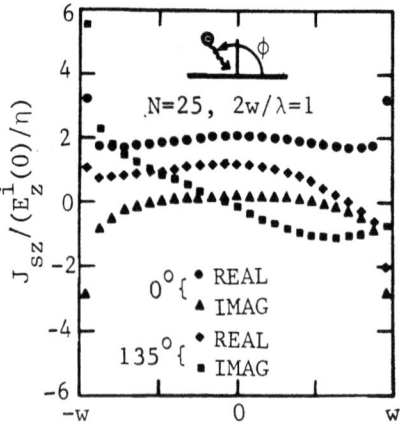

Fig. 20. Normalized surface current on wide, TM-excited strip.

determined from an exact solution [6] to the equation for a narrow strip. In Fig. 20, the current magnitude near the center of the wide strip is observed to be very close to $2|H_x^i(\bar{0})|$ as one would predict from physical optics principles.

9. SOLUTION PROCEDURE FOR INTEGRO-DIFFERENTIAL EQUATION

In this section is presented an efficient technique for solving the integro-differential equation of the wire and TE strip problems. The technique is relatively easy to apply—no more difficult to implement than are the collocation methods discussed in previous sections—and it yields rapidly convergent solutions. It is generally applicable for all types of excitation, including the slice generator of wire antennas, without special modifications needed to accommodate a given source. A special case of the method is the Galerkin procedure [1], and it includes, with still another specialization, the so-called "sinusoidal reaction matching" method [2].

Eqs. (112) and (121) are represented by

$$\left[\frac{d^2}{dz^2} + k^2\right] A(z) = f(z), \quad z\epsilon(-L, L) \tag{130a}$$

where

$$A(z) = \int_{z'=-L}^{L} I(z') K(z - z') dz' \tag{130b}$$

and where the unknown satisfies

$$I(-L) = I(+L) = 0. \tag{130c}$$

The identification of quantities above in terms of what they represent in the TE strip and wire problems is obvious.

The solution technique presented here is similar to that outlined for previous equations with the exception that the harmonic differential operator in (130a) must be accommodated. Also, a change in the sequence of the method's usual steps allows one to gain interesting insight into the nature of the numerical procedure. The alteration alluded to is that one performs the "numerical enforcement[*]" of the equation to be solved before the unknown is approximated as a linear combination of the elements of the basis set. The numerical enforcement procedure introduced

---

[*] See Equations (13) - (17).

below is designed to enable one to treat effectively and efficiently, the harmonic operator of the integro-differential equation. In the vernacular of the method of moments, we refer to enforcement in general as testing.

One easily can imagine situations in which a numerical method employing point-matching would lead to nonsense. A simple example is the case of an equation whose forcing function happens to pass through zero at several points selected for match points. One remedy for this difficulty is to replace enforcement of an equation at a match point by enforcement of the average at several nearby points. That is, the mth equation of (16) would be replaced by one which equates the average of the left member of (13) to that of the right member, with both averages determined from values computed at points very near x_m. Equating the weighted average over a subinterval of one side of an equation to that of the other side is an obvious but important generalization of the scheme described above. Used in conjunction with subdomain expansion functions, such a weighted average enforcement of equality leads to a matrix which is diagonally dominant, a desirable feature that enhances the likelihood of stability. Equating of weighted averages is an interpretation* of a special case of the testing procedure of the method of moments. Testing, as applied to Eq. (130), is outlined below in a fairly general setting. To ensure the stability of the anticipated system of linear equations to be formed in the solution procedure, only subdomain testing is considered here.

9.1 Testing

We postulate a set of M linearly independent weighting or testing functions $\{w_m\}$, chosen to be as general as possible but with particular attention given to one's desire to lessen the complexity of subsequent calculations involving w_m. The set $\{w_m\}$ is of the subdomain type with each $w_m(z)$ identically zero outside the interval (z_{m-1}, z_{m+1}). Testing of Eq. (130a) with $\{w_m\}$, which means that both sides are multiplied by w_m and integrated over $(-L, L)$, yields

$$\int_{z=z_{m-1}}^{z_{m+1}} \left[\left(\frac{d^2}{dz^2} + k^2 \right) A(z) \right] w_m(z) dz = \int_{z=z_{m-1}}^{z_{m+1}} f(z) w_m(z) dz \quad (131)$$

for each m. Clearly, (131) is a statement of equality of the

---

* Another interpretation from linear space theory is the equating of the projections upon a testing space.

weighted averages of the two sides of (130a). Parenthetically, we mention that, if $w_m = \delta(z - z_m)$, (131) reduces to what would be obtained by applying point-matching to (130a).

Two testing sets, which are particularly amenable to the numerical solution of (130), are the piecewise sinusoidal and piecewise linear sets whose elements are, respectively,

$$\Lambda_m^S(z) = \begin{cases} \dfrac{\sin k(\Delta - |z - z_m|)}{\sin k\Delta}, & z \varepsilon (z_{m-1}, z_{m+1}) \\ 0, & z \notin (z_{m-1}, z_{m+1}) \end{cases} \tag{132}$$

and

$$\Lambda_m^L(z) = \begin{cases} \dfrac{\Delta - |z - z_m|}{\Delta}, & z \varepsilon (z_{m-1}, z_{m+1}) \\ 0, & z \notin (z_{m-1}, z_{m+1}) \end{cases} \tag{133}$$

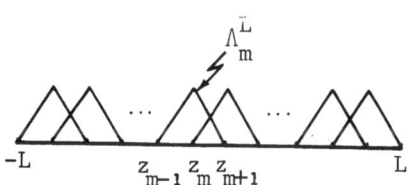

Fig. 21. Triangular pulses.

where $2\Delta = (z_{m+1} - z_{m-1})$. The latter set above is depicted in Fig. 21 where one observes overlapping triangles, uniformly spaced along the interval (-L, L), with each existing only over a subinterval. An illustration of the former set would be similar to what appears in Fig. 21 but with curved sides in place of the straight sides of the triangles. Often Λ_m^S and Λ_m^L are referred to as sinusoidal and triangular pulses.

Testing (130a) with Λ_m^S, i.e., replacing w_m by Λ_m^S in (131), leads to

$$\frac{k}{\sin k\Delta} \left[A(z_{m+1}) - 2\cos k\Delta \, A(z_m) + A(z_{m-1}) \right]$$

$$= \int_{z=z_{m-1}}^{z_{m+1}} f(z) \Lambda_m^S(z) \, dz \tag{134}$$

while testing (130a) with Λ_m^L leads to

$$\frac{1}{\Delta}\left[A(z_{m+1}) - 2A(z_m) + A(z_{m-1})\right] + k^2 \int_{z=z_{m-1}}^{z_{m+1}} A(z)\Lambda_m^L(z)dz$$

$$= \int_{z=z_{m-1}}^{z_{m+1}} f(z)\Lambda_m^L(z)dz, \qquad (135)$$

as one can demonstrate by integrating the left side of (131) twice by parts with w_m replaced by Λ_m^S and Λ_m^L, respectively. The simplicity of (134) and (135) should suggest to the reader why the functions of (132) and (133) are selected for testing (130a).

At this point, it is convenient to introduce approximations which simplify (134) and (135). For any function F, sufficiently smooth over the interval (z_{m-1}, z_{m+1}),

$$\int_{z=z_{m-1}}^{z_{m+1}} F(z)\Lambda_m^S(z)dz \doteq \frac{2}{k} \cdot \frac{(1-\cos k\Delta)}{\sin k\Delta} F(z_m) \qquad (136)$$

and

$$\int_{z=z_{m-1}}^{z_{m+1}} F(z)\Lambda_m^L(z) \doteq \Delta F(z_m), \qquad (137)$$

which can be incorporated into (134) and (135) to arrive, respectively, at

$$\frac{k}{\sin k\Delta}\left[A(z_{m+1}) - 2\cos k\Delta\, A(z_m) + A(z_{m-1})\right]$$

$$\doteq \frac{2}{k} \cdot \frac{1-\cos k\Delta}{\sin k\Delta} f(z_m) \qquad (138)$$

and

$$\frac{1}{\Delta}\left[A(z_{m+1}) - 2\left(1 - \frac{(k\Delta)^2}{2}\right)A(z_m) + A(z_{m-1})\right] \doteq \Delta f(z_m) \qquad (139)$$

To obtain (139) from (135), the approximation (137) is used twice, whereas (138) follows from application of (136) to (134) only once. In this sense of fewer approximated terms, Eq. (138) resulting from piecewise sinusoidal testing is more nearly exact than is (139) resulting from piecewise linear testing.

However, for Δ/λ sufficiently small, Λ_m^L and Λ_m^S obviously do not differ significantly, and, therefore, operations with one testing element produce results similar to those achieved via the other. Specifically, within the approximations, $\sin k\Delta \doteq k\Delta$ and $\cos k\Delta \doteq 1 - (\Delta k)^2/2$, which are quite accurate whenever $k\Delta \ll 1$, i.e., $2\pi(\Delta/\lambda) \ll 1$, (138) and (139) are identical.

Of additional significance and interest is the observation that, subject to division by Δ, (139) is the difference equation approximation to (130a), arrived at by replacing the second derivative operator by the second central difference operator. Thus, piecewise linear testing of Eq. (130a) is approximately equivalent to replacing it by the corresponding difference equation.

9.2 Approximation of I

The next step in the outline of the solution procedure is the introduction of an approximation of the unknown $I(z)$:

$$I(z) \doteq \sum_{n=1}^{N} I_n i_n(z). \qquad (140)$$

The I_n's are unknown complex constants to be determined ultimately, and the i_n's are the known basis functions. With I of (130b) approximated by (140), A can be represented by

$$A(z) \doteq \sum_{n=1}^{N} I_n A_n(z) \qquad (141a)$$

where

$$A_n(z) = \int_{z'=-L}^{L} i_n(z') K(z - z') dz' . \qquad (141b)$$

To arrive at a set of linear equations from which the coefficients I_n can be determined, one simply substitutes (141) into the tested equation ((138) or (139)).

In Fig. 22, a piecewise constant (pulse) approximation of $I(i_n = p_n)$ and a triangle testing scheme are illustrated. One counts nine (N = 9) pulses weighted with corresponding unknown coefficients I_n and nine testing triangles. The pulse and triangle centers are located, respectively, at

$$z_n = -L + n\Delta, \quad n = 1, 2, \ldots, N \tag{142}$$

and at

$$z_m = -L + m\Delta, \quad m = 1, 2, \ldots, N \tag{143}$$

where $\Delta = 2L/(N + 1)$. Observe the geometric relationship of the locations of the testing functions Λ_m^L and of the pulses. Observe, too, that at each end I is represented by half-pulses, each of which is made zero so that the boundary condition $I(\pm L) = 0$ is satisfied from the outset. The zero half-pulse is more in keeping with the condition that I be zero precisely at the ends than would be a zero full-pulse, since the latter would lengthen the region over which I is constrained to zero.

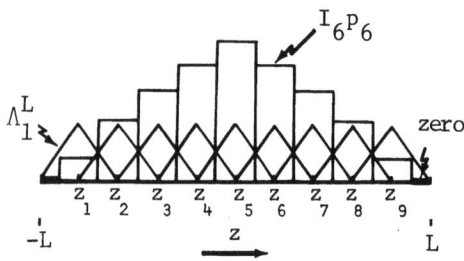

Fig. 22. Triangles (testing) and pulses (basis).

I may be represented by a piecewise linear approximation ($i_n = \Lambda_n^L$, where Λ_n^L is the triangular pulse (133) with m replaced by n. Triangles for testing and expansion are illustrated in Fig. 23, where the expansion triangles are seen to be arranged in such a way that the boundary conditions are satisfied conveniently by the end-most triangles. The peak of a given weighted triangle is equal to the value of $I(z)$ at the point where the peak occurs, i.e., the peak of of $I_n \Lambda_n^L$ is $I(z_n)$. Also an approximation of the unknown by a sum of weighted triangles is equivalent to piecewise linear interpolation as suggested by the dashed lines of Fig. 23.

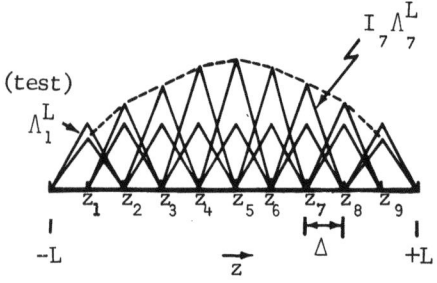

Fig. 23. Triangles for testing and expansion.

The use of the piecewise sinusoidal function for testing or for expansion of I should be clear from the two examples and figures above.

For a subdomain basis element $i_n(z)$ which exists over $z \varepsilon (z_n - \delta, z_n + \delta)$, where $\delta = \Delta/2$ for $i_n = p_n$ and $\delta = \Delta$ for $i_n = \Lambda_n^S$ or $i_n = \Lambda_n^L$, and is zero outside this interval, A_n of (141b) evaluated at z_m becomes

$$A_n(z_m) = \int_{z'=z_n-\delta}^{z_n+\delta} i_n(z') K(z_m - z') dz' . \tag{144}$$

Subject to the change of variable $z' = z_n + \zeta$ and to replacement of z_n and z_m by their values given in (142) and (143), A_n evaluated at z_m becomes

$$A_n(z_m) = \int_{\zeta=-\delta}^{\delta} i_n(z_n + \zeta) K(\Delta[m-n] - \zeta) d\zeta . \tag{145}$$

In (145) $i_n(z_n + \zeta)$ is the nth basis element shifted to the interval $(-\delta, \delta)$ which we represent simply as $i(\zeta)$ and interpret as the basis element with center at $z_n = 0$. Now $A_n(z_m)$ can be written as

$$A_n(z_m) = \int_{\zeta=-\delta}^{\delta} i(\zeta) K(\Delta[m-n] - \zeta) d\zeta . \tag{146}$$

9.3 Construction of $[Z_{mn}^{TB}]$ and formulation of linear equation

The unknown coefficients I_n can be determined from

$$\sum_{n=1}^{N} I_n Z_{mn}^{TB} = F_m^T, \quad m = 1, 2, \ldots, N \tag{147}$$

where Z_{mn}^{TB} represents the left side of (138) or (139) (with substitutions made from (140) - (146)) and where F_m^T stands for the corresponding right sides. The superscript "T" denotes the testing set employed to construct a term (T = S implies the use of Λ_m^S for testing, etc.) and the superscript "B" denotes what basis set is utilized. Hence, from (138) and (139) F_m^T is specialized as $F_m^S = 2[(1 - \cos k\Delta)/k \sin k\Delta] f(z_m)$ and $F_m^L = \Delta f(z_m)$.

Examples of matrix elements are

$$Z_{mn}^{LP} = \frac{1}{\Delta} \int_{\zeta=-\Delta/2}^{\Delta/2} \left[K([m-n+1]\Delta - \zeta) - 2\left(1 - \frac{(k\Delta)^2}{2}\right) K([m-n]\Delta - \zeta) \right.$$
$$\left. + K([m-n-1]\Delta - \zeta) \right] d\zeta \tag{148}$$

for the piecewise linear testing (T = L) and pulse expansion (B = P) scheme of Fig. 22 and

$$Z_{mn}^{LL} = \frac{1}{\Delta} \int_{\zeta=-\Delta}^{\Delta} \Lambda^L(\zeta) \left[K([m-n+1]\Delta - \zeta) - 2\left(1 - \frac{(k\Delta)^2}{2}\right) K([m-n]\Delta - \zeta) \right.$$

$$\left. + K([m-n-1]\Delta - \zeta) \right] d\zeta \tag{149}$$

for the piecewise linear testing and expansion (T = L and B = L) scheme of Fig. 23. Λ^L of (149) is Λ_m^L shifted to the interval $(-\Delta, \Delta)$. Observe that, if Λ^L is replaced by Λ^S in (149), then the resulting expression would be Z_{mn}^{LS}, the matrix element for piecewise linear testing and piecewise sinusoidal expansion.

The sequence of steps for constructing the matrix element Z_{mn}^{LP} is as follows. A_n of (146) (with i = p) is substituted into (141a) which is, in turn, put into the left side of (139); then Z_{mn}^{LP} is identified as the coefficient of I_n. Of course, Z_{mn}^{LL} is constructed in a similar way.

9.4 Summary of matrix elements

Below is presented a summary of Z_{mn}^{TB} formed by various combinations of testing and expansion sets. In the interest of compactness, the following definitions are introduced:

$$A_\nu^P = \int_{\zeta=-\Delta/2}^{\Delta/2} K(\nu\Delta - \zeta) \, d\zeta; \tag{150}$$

$$A_\nu^B = \int_{\zeta=-\Delta}^{\Delta} \Lambda^B(\zeta) \, K(\nu\Delta - \zeta) d\zeta; \tag{151}$$

$$D^S = k/\sin k\Delta, \quad C^S = 2 \cos k\Delta; \tag{152}$$

and

$$D^L = 1/\Delta, \quad C^L = 2[1 - (k\Delta)^2/2]. \tag{153}$$

From (143) and the definitions (150) - (153), we can express the matrix elements as

$$Z_{mn}^{TP} = D^T[A_{m-n+1}^P - C^T A_{m-n}^P + A_{m-n-1}^P] \tag{154}$$

and

$$Z_{mn}^{TB} = D^T[A_{m-n+1}^B - C^T A_{m-n}^B + A_{m-n-1}^B] . \tag{155}$$

D^T and C^T with T replaced by S or L are given by (152) and (153). Z_{mn}^{TP} of (154) represents the matrix elements with I approximated by pulses and with either piecewise sinusoidal or piecewise linear functions used for testing. Z_{mn}^{TB} of (155) is slightly more general in that it is the matrix resulting from any combination of sinusoidal and triangular pulses employed for testing and expansion.

9.5 Matrix properties

Appealing to a procedure closely related to that presented in Section 4, one can show that the matrix Z_{mn}^{TB}, with T = S or L and with B = S, L, or P, is symmetric: $Z_{mn}^{TB} = Z_{nm}^{TB}$. Also, Z_{mn}^{TB} is Toeplitz allowing one to fill the entire matrix from knowledge of either a column or a row. In addition, the reciprocity theorem enables one to demonstrate that $Z_{mn}^{SL} = Z_{mn}^{LS}$: the matrix resulting from piecewise sinusoidal testing and piecewise linear expansion is the same as that resulting from piecewise linear testing and piecewise sinusoidal expansion.

Reciprocity leads to even more remarkable properties. Namely, one can show that $Z_{mn}^{SP} = Z_{mn}^{PS}$. To this point, nothing has been said about pulse testing but now we see that the matrix resulting from piecewise sinusoidal testing and pulse expansion and that from pulse testing and piecewise sinusoidal expansion are identical. Similarly, $Z_{mn}^{LP} = Z_{mn}^{PL}$. Of major interest in these equivalences is the observation that, in one solution procedure, I can be approximated by a continuous function, i.e., sum of triangles, while in another it can be approximated by a discontinuous function, i.e., sum of pulses, yet the matrices of the two methods can be identical.

Furthermore, in a pulse testing scheme, the right side of (142) would be denoted F_m^P and, in the spirit of the approximations of (136) and (137), would have value $F_m^P \doteq \Delta f(z_m)$, which is the same as F_m^L and which is approached by F_m^S as Δ/λ becomes small. Hence, numerical solutions obtained by methods employing flat pulses for expansion and triangular or sinusoidal pulses for testing are essentially equal to those obtained by methods in which the roles of these basis and testing sets are interchanged.

9.6 Computation of matrix elements

In the TE-excited strip and thin-wire problems, the kernels of the integro-differential equations are logarithmically singular, and the matrix elements must be determined by means of numerical integration. The now-familiar approach of subtracting and adding an analytically-integrable singular term, employed in the electrostatic tube and TM-excited strip problems, is utilized again. The two kernels to be integrated are $H_0^{(2)}$ (strip equation) and K of (119) (wire equation). We demonstrate how to integrate these kernels against pulse and piecewise linear basis elements and, thereby, how to evaluate A_ν^P of (150) and A_ν^L of (151). With an algorithm for computing A_ν^P and A_ν^L available, any of the solution procedures above that incorporate pulses or triangles as basis functions can be implemented.

Due to the logarithmic singularity of the kernels, the integrals below are needed subsequently:

$$\delta_\nu^P = \int_{\zeta=-\Delta/2}^{\Delta/2} \ln|\nu\Delta - \zeta| d\zeta \tag{156a}$$

$$= \Delta\left[\left(\nu + \frac{1}{2}\right)\ln\left|\nu + \frac{1}{2}\right| - \left(\nu - \frac{1}{2}\right)\ln\left|\nu - \frac{1}{2}\right| + \ln\Delta - 1\right] \tag{156b}$$

and

$$\delta_\nu^L = \int_{\zeta=-\Delta}^{\Delta} \Lambda^L(\zeta) \ln|\nu\Delta - \zeta| d\zeta \tag{157a}$$

$$= \Delta\left[\ln\Delta - \nu^2\ln|\nu| - \frac{3}{2}\right.$$
$$+ (\nu + 1)\ln|\nu + 1| - (\nu - 1)\ln|\nu - 1|$$
$$\left.+ \frac{1}{2}[\nu^2 - 1]\ln|\nu^2 - 1|\right] \tag{157b}$$

For the TE strip problem with Hankel function kernel, one computes the terms in (154) and (155) by means of A_ν^P and A_ν^L of (150) and (151) with $K(\nu\Delta - \zeta) = H_0^{(2)}(k|\nu\Delta - \zeta|)$. By adding and subtracting the small argument approximation of (128) and integrating the product of the logarithm function and the basis function, one obtains

$$A_\nu^P = \Delta\left[1 - j\frac{2}{\pi}\left(\gamma + \ln\frac{k}{2}\right)\right] - j\frac{2}{\pi}\delta_\nu^P$$

$$+ \int_{\zeta=-\Delta/2}^{\Delta/2} \left\{ H_0^{(2)}(k|\nu\Delta - \zeta|) \right.$$

$$\left. - \left[1 - j\frac{2}{\pi}\left[\gamma + \ln\left(\frac{k}{2}|\nu\Delta - \zeta|\right)\right]\right] \right\} d\zeta \qquad (158)$$

and

$$A_\nu^L = \Delta\left[1 - j\frac{2}{\pi}\left(\gamma + \ln\frac{k}{2}\right)\right] - j\frac{2}{\pi}\delta_\nu^L$$

$$+ \int_{\zeta=-\Delta}^{\Delta} \Lambda^L(\zeta) \left\{ H_0^{(2)}(k|\nu\Delta - \zeta|) \right.$$

$$\left. - \left[1 - j\frac{2}{\pi}\left[\gamma + \ln\left(\frac{k}{2}|\nu\Delta - \zeta|\right)\right]\right] \right\} d\zeta \qquad (159)$$

In both (158) and (159) the integrands are slowly varying and can be integrated numerically quite rapidly. The Hankel functions in (158) and (159) should be computed by means of the polynomial approximation [7] mentioned in Section 8. When $|\nu| \gg 1$, A_ν^P and A_ν^L can be approximated by

$$A_\nu^P \doteq A_\nu^L \doteq \Delta H_0^{(2)}(k\Delta|\nu|), \quad |\nu| \gg 1 \quad . \qquad (160)$$

The thin-wire kernel of (119) approaches the kernel k of (42) encountered in the electrostatic tube problem, when kR becomes very small, and, hence, its singularity is that of k expressed in (49). By subtracting $1/R$ from and adding it back to the integrand of (119), one writes the thin-wire kernel as

$$K = \frac{1}{2\pi} \int_{\phi'=-\pi}^{\pi} \frac{e^{-jkR} - 1}{R} d\phi' + k \quad . \qquad (161)$$

The integrand above is smooth and is easily integrated numerically while the singularity of K is contained in k, which can be treated by the method presented in Section 3.3. With k of (161) represented as in (50) and (54), A_ν^P and A_ν^L for the thin-wire problem are found to be

$$A_\nu^P = \frac{\Delta}{\pi a} \ln(8a) - \frac{1}{\pi a} s_\nu^P$$

$$+ \frac{1}{\pi a} \int_{\zeta=-\Delta/2}^{\Delta/2} \left\{ \beta_\nu K(\beta_\nu) + \ln[|\nu\Delta - \zeta|/8a] \right\} d\zeta$$

$$+ \frac{1}{2\pi} \int_{\zeta=-\Delta/2}^{\Delta/2} \int_{\phi'=-\pi}^{\pi} [(e^{-jkR} - 1)/R] d\phi' d\zeta \qquad (162)$$

and

$$A_\nu^L = \frac{\Delta}{\pi a} \ln(8a) - \frac{1}{\pi a} s_\nu^L$$

$$+ \frac{1}{\pi a} \int_{\zeta=-\Delta}^{\Delta} \Lambda^L(\zeta) \left\{ \beta_\nu K(\beta_\nu) + \ln[|\nu\Delta - \zeta|/8a] \right\} d\zeta$$

$$+ \frac{1}{2\pi} \int_{\zeta=-\Delta}^{\Delta} \left\{ \Lambda^L(\zeta) \int_{\phi'=-\pi}^{\pi} [(e^{-jkR} - 1)/R] d\phi' \right\} d\zeta \qquad (163)$$

where

$$R = \left[(\nu\Delta - \zeta)^2 + 4a^2 \sin^2 \frac{\phi'}{2} \right]^{\frac{1}{2}} \qquad (164)$$

and

$$\beta_\nu = \left[1 + [(\nu\Delta - \zeta)/2a]^2 \right]^{-\frac{1}{2}}. \qquad (165)$$

K in (162) and (163) is the complete elliptic integral of the first kind and it should be computed by means of the polynomial given in [7]. All integrands of the integrals in (162) and (163) are very smooth. Of course, when $|ka| \ll 1$, $a/\Delta|\nu| < 1$, and $|\nu| \gg 1$, the following approximation is valid and saves computational labor:

$$A_\nu^P \doteq A_\nu^L \doteq \frac{1}{|\nu\Delta|} e^{-jk|\nu\Delta|} . \qquad (166)$$

Other techniques for integrating the product of the wire kernel and various basis functions can be found in the literature [11-14].

9.7 Sample results

Sample results for the current induced on a TE-excited flat strip and that induced on a thin-wire scatterer by an incident plane wave are presented in Figs. 24 - 26. In the wire scatterer problem, the incident electric field E_z^i on the wire is

$$E_z^i(z) = E_0^i \cos \psi \sin \theta \, e^{jkz \cos \theta} \qquad (167)$$

where θ is the polar angle measured from the positive z axis. $E_0^i \cos \psi$ is the projection of $\bar{E}^i$ upon the plane defined by the z axis and the radial line, through the wire center, along which the incident plane wave propagates.

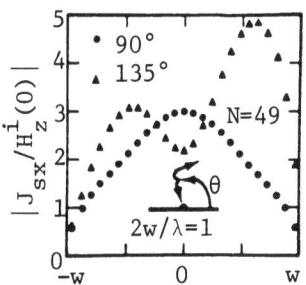

Fig. 24. Normalized Current on TE-excited strip.

An acceptable "rule of thumb" concerning convergence of current on a thin-wire scatterer is that N/λ, the number of terms in the basis set/wavelength, should be seven or greater to assure that the data are within five percent of that obtained for very large N/λ. Convergence rates depend upon wire radius and length, as well as upon excitation. Obviously, N must be sufficiently large to ensure adequate sampling of the excitation. Wire antenna data are depicted in Fig. 27.

9.8 Comments

A class of solution methods for solving the integro-differential equation (130) is presented above. These methods are efficient, rapidly convergent, and easy to implement. In fact, computer programs for such implementation differ from those for the simple point-matching techniques of Section 3 only in relatively minor details.

Numerous other methods for solving Eq. (130) have been developed in recent years and summaries can be found elsewhere [1-5]. Special-purpose techniques and computer programs have been developed to solve problems involving complex wire structures [5].

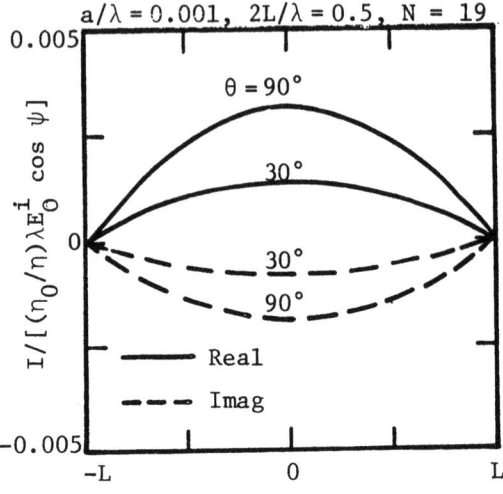

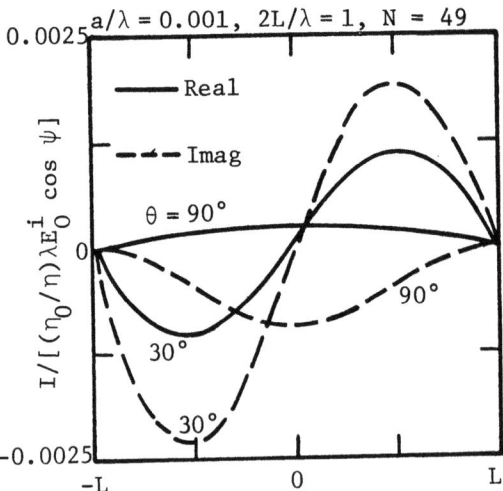

Fig. 25. Normalized current on thin-wire scatterer.

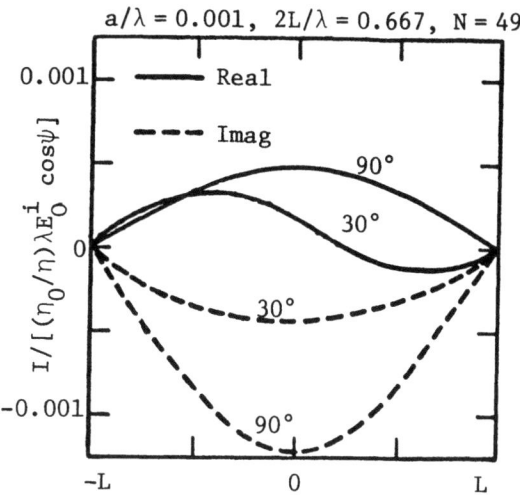

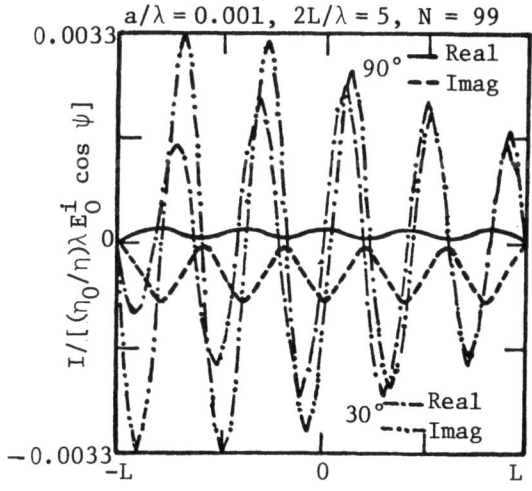

Fig. 26. Normalized current on thin-wire scatterer.

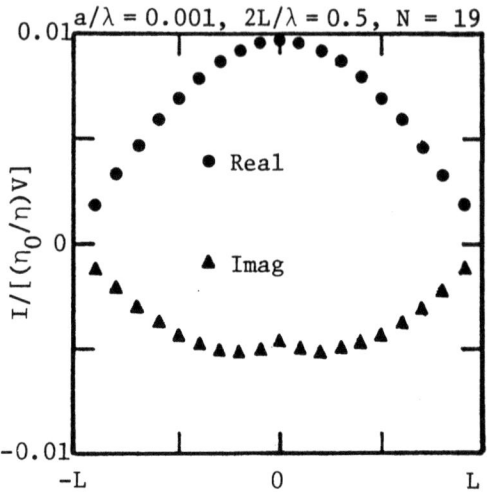

Fig. 2.7. Normalized current on center-driven, thin-wire antenna.

REFERENCES

1. R. F. Harrington, Field Computation by Moment Methods, Macmillan, New York, 1968.

2. R. Mittra, (ed.), Computer Techniques for Electromagnetics, Pergamon Press, New York, 1973.

3. R. Mittra, (ed.), Numerical and Asymptotic Techniques in Electromagnetics, Springer-Verlag, New York; 1975.

4. C. M. Butler and D. R. Wilton, IEEE Trans. AP, 23, 534.

5. D. R. Wilton and C. M. Butler, IEEE Trans. AP, 24, 83.

6. C. M. Butler and D. R. Wilton, IEEE Trans. AP, (accepted for publication: "General Analysis of Narrow Strips and Slots").

7. M. Abramowitz and I. A. Stegun, Handbook of Mathematical Functions, National Bureau of Standards, Washington, D.C., 1964.

8. R. F., Harrington, Time-Harmonic Electromagnetic Fields, McGraw-Hill Book Company, New York, 1961.

9. J. Van Bladel, Electromagnetic Fields, McGraw-Hill Book Company, New York, 1964.

10. R. W. P. King, Linear Antennas, Harvard University Press, Cambridge, Mass., 1956.

11. F. M. Tesche, IEEE Trans. EMC, 16, 209.

12. C. M. Butler, IEEE Trans. AP, 23, 293.

13. L. W. Pearson, IEEE Trans. AP, 23, 256.

14. W. A. Davis, Ph.D. Thesis, University of Illinois, Urbana, Ill., 1974.

WIRES AND WIRE GRID MODELS FOR RADIATION AND SCATTERING INCLUDING GROUND EFFECTS*

E. K. Miller

Electronics Engineering Department
Lawrence Livermore Laboratory
University of California
Livermore, California

ABSTRACT. An overview of wire-antenna computer modeling is given, with emphasis on the interface problem. The formulation and numerical-solution methods are summarized, and applications are demonstrated with numerous examples.

1. INTRODUCTION

Numerical methods based on integral-equation formulations are receiving increasing acceptance for application to real-life electromagnetic radiation and scattering problems. Computer codes have been developed and validated for both surface and wire geometries in both the frequency and time domains for modeling infinite, homogeneous medium problems. Some of these basic procedures have also been extended to the analysis of structures located near a planar interface. In this presentation we will discuss the general topic of computer models for wire antennas from a frequency-domain viewpoint with emphasis directed to antennas located near the ground-air interface. Some preliminary considerations are discussed in Section 2, followed by a brief summary of a specific formulation and numerical treatment in Section 3, with sample numerical results given in Section 4.

---

*This work was performed under the auspices of the U.S. Department of Energy by the Lawrene Livermore Laboratory under contract #W-7405-ENG-48.

2. PRELIMINARY CONSIDERATIONS

The derivation of an integral equation for a wire structure can be accomplished in many ways. What is basically involved is the writing of Maxwell's equations in integral form so that the scattered or secondary fields are given in terms of integrals over induced source distributions. By expressing the secondary field over loci of points where the behavior of the total field (incident or primary plus secondary) is known via boundary or continuity conditions, an integral equation for the induced source is obtained in terms of the primary field. Two broad general classes of integral equations are obtained, depending upon whether the forcing function (primary field) is electric or magnetic. An electric forcing function gives rise to a Fredholm integral equation of the first kind, in which the unknown appears only under the integral. A magnetic forcing field gives rise to a Fredholm integral equation of the second kind, in which the unknown also appears outside the integral. While derivatives of the unknown may occur as well, these equations are commonly called integral equations, rather than integro-differential equations as would be strictly correct.

Generally speaking, it has been found that the magnetic field type of integral equation is better suited for smooth, closed surfaces than it is for thin-plate or shell geometries and wires.[1] The converse is generally true of the electric field type of integral equation. It is the latter then that is more commonly employed for treating wire structures. Also involved in developing wire integral equations are the approximations that (1) the circumferential current is negligible; (2) the circumferential variation of the longitudinal current can be ignored; and (3) the thin-wire or reduced kernel can be used in place of the actual surface integration.

Many analytically equivalent integral equations for wires based upon the electric field can be derived. Three of the most commonly employed are the Hallen or vector potential type,[2] the mixed potential version,[3] and the Pocklington integral equation.[4] All are solved within the framework of the moment (or matrix) method but each exhibits distinctive characteristics which must be taken into account in its numerical treatment. The Hallen equation, for example, can produce results using a pulse current basis of accuracy comparable to those obtained from the Pocklington equation solved with a three-term (constant, sine, and cosine) basis for simple structures.[5] The Hallen equation is not, however, readily extendable to the complex geometries that the Pocklington equation can handle.[6]

Although pulse-current[4,7] and linear-current[8] bases have been quite widely used, and can under suitable circumstances be essentially equivalent, they are not as efficient for modeling traveling wave structures, regardless of the integral equation employed, as are sinusoidal bases which possess non-constant derivatives and which can closely resemble the actual current solution. Sinusoidal bases have appeared in subsectional or subdomain form in both the three-term expansion mentioned above and in the piecewise sinusoidal[9] or two-term form. Fourier series have also been studied as complete domain sinusoidal bases, but have not been widely adopted because they require more integration effort than subsectional bases and can lead to ill-conditioned matrices.[4]

The weight or test functions most often used have been delta functions, although Galerkin's method with both linear[8] (two-term) and sinusoidal[9] (two-term) functions has also been quite widely applied. The term "point matching" refers to the use of delta-function weights. A comparison of numerical convergence rates for several common methods applied to a straight-wire scatterer is shown in Fig. 1.[11]

In addition to the problem of choosing basis and weight functions, there are other special aspects of the numerical development which must be considered when selecting a code for computer modeling. Three of these aspects are discussed below.

2.1 Junction treatment

Any subsectional approach which employs finite-difference operators in the integral equation or other than a pulse current basis necessitates special consideration of both simple (two wires) and multiple (three or more wires) junctions. What is essentially required is a way to relate in some physically and mathematically reasonable way the current basis of each subsection (segment) to those of its neighbors. When pulse bases are used in the mixed potential integral equation, the finite-difference operator spans two segments and thus leads to a charge which involves the two- corresponding pulse-current samples.[3] For two or three-term bases, the condition of current amplitude and slope continuity at each simple junction leads to equations which permit all the constants in the current expansion to be given in terms of current samples at the segment junctions or centers.[5] A slightly different handling of the three-term basis was developed by Yeh and Mei,[12] in which the current is extrapolated from a given segment to the adjacent segment centers, but which is otherwise basically the same.

When a multiple junction is concerned, the treatment can get considerably more involved. The pulse-basis approach mentioned above was extended to the multiple junction[7] by dividing the total junction charge between the junction segments according to the ratio of their individual areas to the total area. It has since been demonstrated that a logarithmic area ratio is more appropriate.[13] The two-term expansions have been applied to multiple junctions by overlapping M - 1 of the bases a pair at a time at an M-segment junction.[8] Application of the three-term expansion to the multiple junction was accomplished by MBAssociates using the Yeh and Mei simple-junction procedure by incorporating a composite segment having the averaged length and total current of the M - 1 connected segments.[14] More recently, the Curtis-Wu-King approach was adapted to the three-term expansion for multiple junctions, while a spline treatment is used to handle simple junctions.[15]

A more elaborate multiple-junction approach has been developed for the three-term expansion by Andreason and Harris.[16] Their procedure apparently is the only one in which the junction geometry plays an explicit role in establishing the current relationships at the junction. Although all of these approaches evidently can produce satisfactory results, there is little or no direct evidence of their validity in terms of the junction current and charge distributions. It should be noted that the numerical results have been found in some cases to be quite sensitive to the juction treatment.[5] Further, the above list by no means exhausts all possible alternatives for the junction treatment.

2.2 Source Models

Determination of quantities such as absolute gain, efficiency, radiated power, input power, etc. requires not only the antenna current distribution but also the input characteristics, particularly the feedpoint impedance (or admittance). The feedpoint admittance can be found in various ways, but when using the integral equation approach one usually defines it in terms of source-region current per unit of terminal voltage. In order to calculate this quantity, a realistic source model is needed that not only provides an appropriate means for numerically exciting the antenna but also permits ready evaluation or specification of the effective terminal voltage. Thus if, as in a point-matching procedure, the excitation arises as a tangential field on the source segment of length Δ, the driving voltage might be assumed to be $-E^I\Delta$ if E^I is constant on the source segment and zero elsewhere. This assumption may be invalid, however, with the result that the actual voltage can only be obtained by integrating the tangential field in the vicinity of the source segment.[5] Somewhat less ambiguity should arise from Galerkin-type approaches where the boundary conditions are integrated, so that the classical delta function source might be numerically approximated. An alternative source model for point matching is provided as a current slope discontinuity, which also approximates a delta-function source field.[17] The current bases, junction treatment, and weight functions can all influence the usefulness of these alternative source models. In case of uncertainty, once the current distribution has been found, the impedance can be computed from the classical EMF method, although at the expense of the additional integration which this entails.

2.3 Integration

Integration is understandably an essential part of the moment method, being involved in applying the integral operator to the current bases and, in a Galerkin method, evaluating the inner product of this result with the weight functions. For most wire programs, these operations, which lead to the generalized impedance matrix, dominate the total solution time for numbers of unknowns less than ~200. It is thus important that the integration time be minimized consistent with the overall accuracy requirements.

One way to approach this goal is to choose appropriate bases and weight functions. The two-term sinusoidal current basis, for example, requires no numerical integration when the Pocklington integral equation is used together with point marching. This particular combination is not very accurate, however.[10] By adding the constant term, much better results are obtained, with the slight additional expense of the numerical integration required to find the longitudinal field of this current term; the radial component can be analytically expressed. Alternatively, use of a sinusoidal weight function[9] also gives much improved results and surprisingly requires numerical integration, at most of sine and cosine integrals. The piecewise linear basis used with the mixed potential equation cannot be analytically integrated, but good results are obtained with four-point rectangular integration of both the operator and inner-produce integrals. In addition, instead of applying a numerical integration to the self-term, a series expansion which gives a closed-form expression is used.[3] When numerical integration is resorted to, various adaptive routines and special techniques are available to improve efficiency.[18]

3. WIRE ANTENNA ANALYSIS

3.1 Infinite, homogeneous, isotropic media

It can be appreciated that there are many options available to the analyst concerning the integral equation to be selected and its numerical treatment in developing a computer model for application to wire antennas. In order to limit this discussion to a reasonable length, our attention will be primarily directed to an approach based on the Pocklington integral equation solved using a three-term subsectional basis (constant, sine, and cosine) and point matching. Unless otherwise indicated, antenna sources are introduced as tangential electric fields, with the Yeh and Mei[12] form of current extrapolation used for simple junctions and the MBAssociates extension of this extrapolation method to multiple junctions.[11,14] Both the source model and junction (simple and multiple) treatment used in this code may exhibit deficiencies, but when applied with care (e.g., equal segment lengths near sources and at multiple junctions) the code has proved to be valid and reliable. A brief overview of the relevant equations and numerical treatment used for free space and various interface theories and some special topics is given in this section. Numerical results follow in Section 4.

The Pocklington-type integral equation for a wire structure of contour $C(\bar{r})$ can be expressed in the form

$$\hat{s} \cdot \bar{E}^I(s) = \frac{i\omega\mu}{4\pi} \int_{C(\bar{r})} I(s') G_0(s,s') \, ds'; \tag{3.1}$$

$$s \in C(\bar{r}),$$

where

$$G_0(s,s') = \left[\hat{s} \cdot \hat{s}' + \frac{1}{k^2}(\hat{s} \cdot \nabla)(\hat{s}' \cdot \nabla)\right] \times g_0(\bar{r},\bar{r}'),$$

$$g_0(r,r') = \frac{e^{-ikR}}{R},$$

$$R = |\bar{r} - \bar{r}'| \geq a(\bar{r}),$$

$$\hat{s} = \frac{\nabla C(\bar{r}')}{|\nabla C(\bar{r})|},$$

and

$$\hat{s}' = \frac{\nabla C(r')}{|\nabla C(r')|},$$

where as usual k is the infinite-medium wave member, the

permeability and permittivity are denoted by μ and ϵ, $a(\bar{r})$ is the wire radius at $\bar{r}$, and $\bar{E}^I$ is the incident electric field.

Reduction of this equation to matrix form involves these seven steps:

(1) Approximating $C(\bar{r})$ as a piecewise linear sequence of N segments of length Δ_i, $i = 1, \ldots, N$, so that

$$C(\bar{r}) = \sum_{i=1}^{N} \Delta_i \hat{s}_i,$$

with $\hat{s}_i$ the unit tangent vector to $C(\bar{r})$ at $\bar{r} = \bar{r}_i$ (use of straight segments is not mandatory, but it is very convenient in simplifying the current integration);

(2) Introducing the subsectional bases

$$I_i(s') = A_i + B_i \sin[k(s' - s_i)]$$
$$+ C_i \cos[k(s' - s_i)]$$

to represent the unknown current (the final unknowns will be the N-sampled current values $I_i = A_i + C_i$, $i = 1, \ldots, N$, at the center of each of the N segments);

(3) A current interpolation procedure whereby the individual A_i, B_i, and C_i constants are expressed in terms of the sampled current values;

(4) Use of the N delta-function weights $\delta(s - s_j)$, $j = 1, \ldots, N$, to obtain an Nth-order impedance matrix of N independent field equations (note that the weight functions sample the field at the segment centers, and are thus "collocated" with the current sample locations);

(5) Specification of the N incident or primary field vector components $E_j = \bar{E}^I(s_j) \cdot \hat{s}_j$, $j = 1, \ldots, N$, which are the tangential fields at the N segment centers;

(6) Matrix manipulation to obtain an admittance equivalent of the impedance matrix; and

(7) Computation of the current distribution and whatever field components are desired.

The total computer solution time is well approximated by $AN^2 + BN^3$, where the "A" term corresponds to step (4) and the "B" term to step (6). For the code under the consideration here and for a CDC-7600 computer, $A \approx 4 \times 10^{-4}$ sec and $B \approx 2 \times 10^{-6}$ sec.

3.2 Perfectly conducting half-space

Equation (3.1) as written applies to wire structures excited as antennas or scatterers and located in infinite, isotropic, homogeneous media of arbitrary (possibly lossy) permittivity and permeability. It can easily be extended to permit the modeling of magnetic or electric image planes. For example, the perfectly conducting ground analog of Eq. (3.1) is, for an antenna elevated above a ground plane at $z = 0$,

$$s \cdot \overline{E}^I(s) = \frac{i\omega\mu}{4\pi}$$

$$\times \int_{C(r)} I(s')[G_0(s,s') + G_I(s,s'*)] \, ds', \qquad (3.2)$$

where

$$g_I = \frac{e^{-ikR*}}{R*},$$

$$R* = |\overline{r} - \overline{r}'*|,$$

$$\overline{r}'*(x,y,z) = \overline{r}'(x,y,-z),$$

$$s'* = \frac{\nabla C(\overline{r}'*)}{|\nabla C(\overline{r}'*)|}.$$

Similar forms can be written for a magnetic interface and for an interior right-angle corner. If the corner angle is otherwise arbitrary but related to π as an integer multiple, a discrete spectrum of angular images is obtained, but the essence of the integral equation form is preserved.[19] Precisely the same line of approach can also be used for interior problems where the wire structure is located between two parallel magnetic or electric planes.[20]

3.3 Imperfectly conducting half-space

A problem which is not so computationally simple to handle, but which is of perhaps greater practical interest, is that of an antenna located (buried or elevated) near the ground-air interface. This is a topic of considerable longevity in electromagnetics; a formal solution was worked out for this problem in 1909 by Sommerfeld.[21] The numerical complexity of evaluating the Sommerfeld integrals (which appear in the integral-equation kernel) for arbitrary source and observation-point locations and ground parameters, however, has prevented the Sommerfeld theory from being routinely used for such problems. Consequently, while some progress has been made in applying the Sommerfeld theory, alternative approaches to the antenna-ground problem have also been pursued. These various methods are briefly discussed below.

3.3.1 The Sommerfeld theory

Details of the steps in deriving the Sommerfeld integrals may be found elsewhere.[22] Here we will simply write one version of Eq. (3.1) which accounts for the interface reflected field via the Sommerfeld theory; alternative forms are also available, differing essentially in how the perfect-ground image terms are handled. It is

$$\hat{s} \cdot \overline{E}^I(s) = \frac{i\omega\mu}{4\pi} \int_{C(\overline{r})} I(s') \, ds'$$

$$\times \left\{ G_0(s,s') + G_I(s,s'*) \right.$$

$$+ \left(\cos\beta + \frac{1}{k^2} \frac{\partial^2}{\partial s \partial z} \right) \sin\beta' \, g_{Hz} - \cos\beta' \, g_{Vz}$$

$$+ \sin\beta' \left[\sin\beta \cos(\alpha - \alpha') \right.$$

$$\left. \left. + \frac{1}{k^2} \frac{\partial^2}{\partial s \partial t'} \right] g_{Ht} \right\}, \quad (3.3a)$$

where $\alpha = \alpha(\overline{r})$ and $\beta = \beta(\overline{r})$ are the direction angles of the wire at $\overline{r}$, $\hat{t}'$ is the horizontal projection of $\hat{s}'$, J_n is the Bessel function of order n, and

$$g_{Ht} = 2 \int_0^\infty \frac{\lambda}{\gamma + \gamma_E} J_0(\lambda\rho) e^{-\gamma(z+z')} d\lambda, \tag{3.3b}$$

$$g_{Hz} = \frac{-\cos(\phi - \alpha')}{k^2} \int_0^\infty \frac{\gamma - \gamma_E}{\epsilon_E \gamma + \gamma_E} \times J_1(\lambda\rho) e^{-\gamma(z+z')} \lambda^2 d\lambda, \tag{3.3c}$$

$$g_{Vz} = 2 \int_0^\infty \frac{\gamma_E}{\epsilon_E \gamma + \gamma_E} \times J_0(\lambda\rho) e^{-\gamma(z+z')} \frac{\lambda}{\gamma} d\lambda, \tag{3.3d}$$

$$\rho = \sqrt{(x - x')^2 + (y - y')^2 + a^2},$$

$$\phi = \tan^{-1}[(y - y')/(x - x')],$$

$$\gamma = \sqrt{\lambda^2 - k^2},$$

$$\gamma_E = \sqrt{\lambda^2 - \epsilon_E k^2},$$

with ϵ_E the lower half-space permittivity relative to the upper.

The presence of the double integral in Eq. (3.3), particularly the Sommerfeld portion, makes it quite time-consuming and sensitive to evaluate. In spite of that, the basic moment method can be used to solve it, but, in addition to the usual constraints imposed on current sampling, it is necessary to take into account the source distance from the interface.

3.3.2 Sommerfeld interpolation--Much effort has been devoted to reducing the computational expense of evaluating the Sommerfeld integrals which appear in Eq. (3.3). For the most part, this has involved analytical manipulation of the integrals and is intended to obtain forms more suitable for computation or approximation. Such approaches usually have had limited success because the constraints employed in deriving them restrict their scope of applicability.

A quite different approach which exploits the inherent smoothness of the Sommerfeld integral contributions to the field, however, has demonstrated that efficiency, accuracy and flexibility are not mutually incompatible. This approach is based on interpolation in the two dimensional space $(\rho, z + z')$

of the Sommerfeld field values which, as shown in Eqs. (3.3b - 3.3d), are dependent spatially only on ρ and $z + z'$. By computing and prestoring Sommerfeld field values on a grid of points which contains the antenna of interest, it is possible to perform the integration of Eq. (3.3a) without further resort to the Sommerfeld integrals themselves. Fig. 2 depicts one of the field components on a rectangular interpolation space.[23]

Although initial tests of the interpolation procedure employed the rectangular grid, additional study has shown a polar grid to work even better.[24] And, the computer times involved are indeed remarkable, considering that results obtained using this approach are essentially rigorous insofar as the Sommerfeld field contributions are concerned. For example, suppose the computer time needed to model a given antenna in free space is T, then the time needed to model the same antenna over ground, using this approach is $\sim(2-4)T$, once the interpolation grid has been computed, which itself uses a time $\sim T$. Furthermore, the grid can be stored for subsequent re-use for other heights or orientations of the same antenna, or for modeling other antennas. Finally, the interpolation procedure works well for wires within a thousandth of a wavelength or less of the ground plane.[24]

3.3.3 Modified image theory--In many cases, although they may not be always easy to identify a priori, the rigor represented by Eq. (3.3) is unnecessary; various approximations will be found adequate. The accuracy actually required of the computer model may be debatable, but it is probably reasonable to seek something on the order of experimental error. One approach which has been found, for simple antennas, to agree within 10-15% of the Sommerfeld results for input impedance, and so which appears useful in view of the above observation, is the reflection coefficient approximation.[25,26] It involves representing the interface-reflected fields in terms of their perfect-ground images multiplied by the Fresnel plane-wave reflection coefficients for the TE and TM field components evaluated at the specular reflection point. This approximation leads to the integral equation given below:

$$\hat{s} \cdot \overline{E}^I(s) = \frac{i\omega\mu}{4\pi} \int_{C(\overline{r})} I(s')$$

$$\times \left[G_0(s,s') + R_M G_I(s,s'*)\right.$$

$$+ \left(R_E - R_M\right) \sin \beta \sin \beta'$$

$$\left.\times \sin(\phi - \alpha) \sin(\phi - \alpha') g_I(\bar{r},\bar{r}'*)\right] ds', \tag{3.4}$$

where R_E and R_M are the TE and TM reflection coefficients given by

$$R_E = \frac{\sqrt{\epsilon_E - \sin^2 \theta} - \cos \theta}{\sqrt{\epsilon_E - \sin^2 \theta} + \cos \theta},$$

$$R_M = \frac{\epsilon_E \cos \theta - \sqrt{\epsilon_E - \sin^2 \theta}}{\epsilon_E \cos \theta + \sqrt{\epsilon_E - \sin^2 \theta}},$$

with Θ the angle of incidence with respect to vertical. (Although Eqs. (3.3) and (3.4) are written expressly for the reflected field, similar expressions also hold for the field transmitted across the interface.)

Since the reflection-coefficient integral equation (3.4) differs only trivially from that for the perfect-ground case given by (2), it may be appreciated that its numerical solution is obtained with almost equal efficiency, in marked contrast to the situation which holds for the Sommerfeld theory in its usual form. The reflection coefficient approximation is, in addition, applicable to a laterally inhomogeneous ground with little further complication. Layered grounds can also be handled using this approach.

3.3.4 Interface source distribution--The Sommerfeld theory is not the only rigorous formulation which can be derived for the antenna-ground problem. Some variations of that approach, which still involve integral-type elementary source solutions, are summarized by Baños.[27] A completely different method of treatment which results in integration in real space (the interface) rather than wave-space integration (the Sommerfeld or λ integral) can be postulated. One way is to treat the interface tangential fields as unknown in addition to the antenna current distribution itself. One can then solve for these surface sources together with the antenna current by applying the moment method to the coupled integral equations which result. An obvious disadvantage of this approach is that many more unknowns require consideration, an infinite number in principle, but finite in practice since only the region near the antenna needs be modeled. Advantages are that the computer-time penalty imposed by the interface-related calculation is relatively independent of antenna size, and the Sommerfeld integrals are entirely circumvented, with no nested numerical integrals being encountered.··

The integral equation which results from this treatment has the form

$$\hat{s} \cdot \overline{E}^I(s) = \frac{i\omega\mu}{4\pi} \int_{C(\overline{r})} I(s') G_0(s,s') ds'$$

$$- \frac{1}{4\pi} \hat{s} \cdot \int_A \left\{ i\omega\mu [\hat{z} \times \overline{H}(\overline{r}')] g_0(\overline{r},\overline{r}') \right.$$

$$- [\hat{z} \times \overline{E}(\overline{r}')] \times \nabla' g_0(\overline{r},\overline{r}')$$

$$\left. - [\hat{z} \cdot \overline{E}(\overline{r}')] \nabla' g_0(\overline{r},\overline{r}') \, da' \right\} ; \; s \in C(\overline{r}) \tag{3.5a}$$

$$\hat{z} \times \frac{i\omega\mu}{4\pi} \int_{C(\overline{r})} I(s') \overline{\overline{G}}_0(\overline{r},s') ds' = \frac{1}{4\pi} \hat{z} \times \int_A i\omega\mu [\hat{z} \times \overline{H}(\overline{r}')] g_1(\overline{r},\overline{r}')$$

$$- [\hat{z} \times \overline{E}(\overline{r}')] \times \nabla' g_1(\overline{r},\overline{r}')$$

$$\left. - [\hat{z} \cdot \overline{E}(\overline{r}')] \nabla' g_2(\overline{r},\overline{r}') \right\} da' ; \; r \in A \tag{3.5b}$$

$$\hat{z} \times \nabla \times \frac{1}{4\pi} \int_{C(\bar{r})} I(s') \bar{\bar{G}}_0(\bar{r},s') ds'$$

$$= \frac{1}{4\pi} \hat{z} \times \int_A \left\{ i\omega\epsilon [\hat{z} \times \bar{E}(\bar{r}')] \, g_2(\bar{r},\bar{r}') \right.$$

$$\left. + [\hat{z} \times \bar{H}(\bar{r}')] \times \nabla g_1(\bar{r},\bar{r}') + [\hat{z} \cdot \bar{H}(\bar{r}')] \nabla' g_1(\bar{r},\bar{r}') \right\} ds'; \quad \bar{r} \in A$$

where

$$g_1 = g_0 + g_E, \qquad g_2 = g_0 + \epsilon_E g_E,$$

$$g_E = \frac{e^{\left[-ik\sqrt{\epsilon_E} R\right]}}{R}$$

Also A is an area on the $z = 0$ ground plane under the antenna, $\bar{E}(\bar{r}')$ and $\bar{H}(\bar{r}')$ are the ground-plane source distributions, and

$$\bar{\bar{G}}_0(\bar{r},s') = \left[\hat{s}' + \frac{1}{k^2} \nabla(\hat{s}' \cdot \nabla) \right] g_0(\bar{r},\bar{r}').$$

Note that the surface integrals in the latter two of the above equations must be evaluated in a principal-value sense.

3.3.5 Surface source approximations--In the same way that the reflection-coefficient approximation follows in a straight-forward way from the rigorous Sommerfeld theory, approximations to the interface source distribution analysis discussed above naturally suggest themselves. Two we consider here are the surface-impedance and physical-optics approximations.

3.3.6 Surface-impedance approximation--Under the condition that the surface impedance concept is valid $[(\sin^2\theta)/\epsilon_E \ll 1]$,[28] the tangential components of the electric and magnetic fields at the surface are to a good approximation related as

$$E_{tan} = -H_{tan} Z_{surf},$$

where

$$Z_{surf} = \eta/\sqrt{\epsilon_E},$$

with η the upper medium wave impedance.

Upon employing this relationship in (3.5) the surface integral equations decouple and the number of surface unknowns is decreased by half. A corresponding reduction in the order of the overall linear system is thus achieved, with a potential significant saving in both the computer storage and solution time. Either of the two surface equations can be retained for use together with the structure-related integral equation. For convenience we might select the electric-field equation since then all required interaction coefficients involve the electric field only.

3.3.7 Physical-optics approximation--Even further simplification of (3.5) can be achieved by invoking a physical-optics type of approximation for the surface fields. But in contrast to the usual physical-optics magnetic field, which is given by $H_{tan} = 2H_{tan}^{inc}$, we instead use

$$\mathbf{H_{tan}} = (1 + R)H_{tan}^{inc},$$

with R the reflection coefficient together with the surface impedance approximation for E_{tan} to allow for a finite ground conductivity. This permits the total-surface field distribution to be expressed in terms of the currents on the wire structure and leads to the same number of unknowns as for the Sommerfeld theory, but with the advantage of a much simpler integral-equation kernel. A perfect-ground type of integral equation could be derived by decomposing the surface source contribution to the fields on the structure into a part due to the perfect ground (due to $2H_{tan}^{inc}$), which can be analytically integrated to give the usual perfect ground image, and a perturbation term $(R - 1)H_{tan}^{inc}$, which will require numerical integration. While in general we might use a Fresnel reflection coefficient for each incremental source and surface path, it would be simpler, and not inconsistent with approximations already employed, to everywhere approximate R by its normal incidence form. Note that by appropriate pairing of source and observation points on the wire structure and use of a single reflection coefficient evaluated at their specular point, we would obtain the reflection coefficient approximation already discussed.

3.3.8 The compensation theorem--Application of the compensation theorem to ground-plane problems has received considerable attention.[29-31] It has been used to determine the input impedance of vertical monopoles located over various ground configurations, including determining the effect of ground-screen size. However, more general antenna problems have evidently not been attempted with this theory. The reason for this lies, apparently, not in limitations inherent in the theory itself, but in its numerical implementation. A ground plane integral is involved, which, for all but the simplest situations, requires numerical evaluation.

The compensation theorem "is essentially an exact perturbation technique in which the fields in the unperturbed state are known."[31] If the unperturbed state is the case of a perfectly conducting ground plane and the perturbed state is the actual ground problem of interest, then we obtain for the antenna input impedance

$$Z' = Z + \frac{1}{I^2} \int_A \bar{H} \cdot \hat{z} \times \bar{E}' \, da,$$

where the primes denote perturbed quantities and I is the feedpoint current. Since the perfect-ground magnetic-field distribution can be accurately solved for, evaluation of Z' hinges on finding $\bar{E}'$. This is usually accomplished by using the surface impedance approximation, i.e., $E'_{tan} = -H'_{tan} \times Z_{surf}$, and then assuming $H'_{tan} \simeq H_{tan}$. These steps facilitate the calculation and permit use of the perfect-ground result as a sort of canonical solution to find the antenna impedance for the finitely conducting ground.

3.3.9 Ground-wave propagation--Another important aspect of the antenna-ground problem is that of determining the propagation characteristics of waves launched along the interface. Such surface waves are an important mechanism for short and intermediate-distance communication. A common problem which arises in assessing the path loss of a surface-wave mode link is that presented by an inhomogeneous path due to surface impedance changes or profile variations from the ideal smooth, curved earth.

A general integral-equation computer code for analyzing this problem has been developed by Ott.[32] The rigor of Ott's approach, while permitting the accurate modeling of quite complex paths, is not needed for many of the so-called mixed-path problems, where surface impedance changes only are of interest. For these situations, which typically involve propagation across a coastline, an approximate procedure derived by Eckersley[33] and Millington[34,35] is quite adequate. It involves use of the standard Norton attenuation functions and leads to the following generalized expression for the vertical surface-wave field E^V over an N-part mixed path as

$$E^V = \frac{i\omega\mu}{4\pi}\left[\left(\sum_{i=1}^{N_B} F^B_{Li} F_{zi} + \sum_{i=N_B+1}^{N_B+N_E} F^E_{Li}\right) I_i \Delta_i f_i\right] F_p f_R, \quad (3.6)$$

where superscripts B and E denote buried and elevated antenna segments, respectively, there being a total of $N_B + N_E$, and I_i is the current at the center of segment i of length Δ_i, coordinates x_i, y_i, z_i and direction angles α_i and β_i with respect to the x-y (ground) plane and the x axis, respectively. Further,

$$F^B_{Li} = \frac{\sin \alpha_i}{n^2}$$
$$+ \sqrt{n^2 - \sin^2 \theta_i} \cos(\phi - \beta_i) \cos \alpha_i$$

and

$$F^E_{Li} = \sin \alpha_i \sin \theta$$
$$+ \sqrt{n^2 - \sin^2 \theta_i} \cos(\phi - \beta_i) \cos \alpha_i$$

are the surface-wave launching efficiencies of the buried and elevated segments, where n is the refractive index of the lower half-space at the antenna location, and

$$F_{zi} = \exp\left[-ik_0 z_i\left(\sqrt{n^2 - 1} + 1\right)\right],$$

$$f_i = \exp\left\{-ik_0\left[\sin \theta_i(x_i \cos \phi + y_i \sin \phi) - z_i \cos \theta_i\right]\right\},$$

$$f_r = \frac{e^{-ik_0 R_T}}{R_T},$$

$$\theta_i = \tan^{-1}\left(\frac{R_T}{z+z_i}\right),$$

$$\phi = \cos^{-1}\left(\frac{R_T \cdot \hat{x}}{R_T}\right),$$

$$R_T = \sqrt{z^2 + r^2} \sim r,$$

with ϕ the observation angle with respect to the x axis, z the observation height, and r the radial distance to the observation point from the origin. Finally, F_{pi} is the mixed-path propagation factor, and has the form

$$F_{pi} = 2\left(\frac{\prod_{i=1}^{N} F_{i,R_i} \prod_{i=1}^{N} F_{i,R-R_{i-1}}}{\prod_{i=1}^{N-1} F_{i+1,R_i} \prod_{i=1}^{N-1} F_{i,R-R_i}}\right),$$

where $F_{i,R}$ is the TM-mode Norton attenuation function[36] calculated for a path of length R and having electrical parameters ϵ_i and σ_i, and with R_i given by

$$R_i = \sum_{j=1}^{i} r_j,$$

where r_j is the path length across part j of the N-part path. The above expression is written for a flat earth and assumes source and observation points close enough to the interface that the Fresnel plane-wave reflection coefficients are -1. Implicit in its derivation is the fact that the attenuation rate of a surface wave is primarily determined by the local wave-front curvature and local surface impedance.

3.4 Special topics

In addition to the above topics, there are other problem areas concerning wire-antenna computer modeling that deserve attention. Some of them are summarized here.

3.4.1 Wire grids--Hardly had the initial wire integralequation models been developed, than were they used for modeling wire-grid objects.[4,37] Two types of wire-grid applications naturally occur. One is where the object is actually a wire grid, as in a wire-grid reflector for an antenna, for example. The other, and the one more frequently employed, is where a solid conducting body or shell, for modeling purposes, is represented by a wire grid. This latter application involves two approximations, the physical one of replacing a continuous surface with a grid, and the numerical one involved in a computational solution. Of the two approximations, the physical one is probably the more limiting, although the errors of both can be reduced as the grid size is reduced or the number of unknowns is increased.[1]

One uncertainty of modeling continuous surfaces with wire grids is that of choosing the most appropriate wire size to use. The surface impedance of a wire grid is a function of both the wire radius and grid opening. Thus, changing either of these parameters can affect the electromagnetic behavior of the surface.[1] One guideline for the most appropriate ratio of segment length to wire radius is 2π, for which value the surface areas of the grid and the surface being modeled are equal. The ratio that works best seems to depend upon both the analytical formulation and numerical treatment employed.[38]

3.4.2 Ground screens--The compensation theorem has been employed in various ways to analyze ground-screen effects as mentioned above. The reflection-coefficient approximation has also been used for this purpose. It offers an easily implemented procedure for analyzing a broad class of ground-screen configurations with greater efficiency than available in general from the compensation theorem.

What is essentially required in order to include the ground-screen influence in the reflection-coefficient calculation is a modified reflection coefficient which takes into account the reflecting properties of the screen-ground combination. This is possible if the surface impedance of the combination is known. For ground screens whose wires are in good electrical contact with the soil, the effective surface impedance Z'_{surf} may be taken to be[39]

$$Z'_{surf} \approx \frac{Z_{surf} Z_{screen}}{Z_{surf} + Z_{screen}},$$

where Z_{screen} is the screen impedance. For a radial screen having N wires of radius a, the screen impedance at distance ρ from the center is given by [39]

$$Z_{screen} \approx \frac{i\mu\omega\rho}{N} \ln(\rho/Na).$$

A corresponding formula for a parallel grid of wires, whose center spacing is d, is

$$Z_{screen} \approx \frac{i\mu\omega d}{2\pi} \ln(d/2\pi a).$$

Meshes consisting of locally orthogonal wires having different spacings might be treated as anisotropically conducting planes whose principal-direction impedances are obtained from the parallel-wire formula using their corresponding spacings. From Z'_{surf} we infer an effective ground permittivity for use in computing the Fresnel reflection coefficients, and are thus able to include the screen in the integral-equation calculation. The anisotropic case requires decomposition of the TE and TM fields into components along the orthogonal screen wires.[40] Note that this method fails for vertical antennas located at the center of a radial screen.

An alternative possibility is offered by the work of Astrakhan[41] who derived reflection coefficients for infinite-plane wire grids. His results, given in terms of TE-TE, TM-TM, TE-TM, and TM-TE reflection coefficients can be modified to include the effect of the ground itself and used in the reflection-coefficient approximation. Other useful results are given by Wait[42] who studied the properties of a wire grid located near ground.

3.4.3 The layered ground--Reflection coefficients are of course available for a layered ground. For the special case of only two layers, and where the surface impedance approximation holds, the effective surface impedance is given by[43]

$$Z'_{surf} \approx Z_{surf} \frac{\sqrt{\epsilon_1} + i\sqrt{\epsilon_2} \tan kh\sqrt{\epsilon_1}}{\sqrt{\epsilon_2} + i\sqrt{\epsilon_1} \tan kh\sqrt{\epsilon_1}},$$

with ϵ_1 and ϵ_2 the relative permittivities of the two layers and h the thickness of layer 1.

3.4.4 Geometrical theory of diffraction--The geometrical theory of diffraction (GTD) does not have obvious application to antenna-ground problems. There are however, two areas where GTD may be beneficial: (1) ground-screen edge effects (diffraction) on input impedance and low-angle radiation; and (2) effects of largescale terrain variations, e.g., diffraction at a cliff. Application of GTD to both areas has been studied by Thiele.[44] His approach was to combine GTD with the moment method to find the effect of the edge-diffracted field on the current distribution of a monopole antenna located on a wedge. This leads to an integral equation modified from that for free space by inclusion of the diffracted fields, given in terms of the antenna current, in the total tangential electric field on the antenna. Thus, no additional unknowns are involved. The far field is treated in a similar manner. Results obtained to date are encouraging, although use of the technique to analyze a real ground screen awaits derivation of diffraction coefficients for a perfectly conducting half-plane lying on a lossy interface.

3.4.5 Backscreen evaluation--It is fairly common practice to employ backscreens to improve HF antenna performance. Backscreens are typically constructed of arrays of parallel, vertical wires whose spacing and diameter are selected to maximize the antenna's front-to-back ratio or some other measure of its performance. The backscreen parameters so chosen have necessarily been based on design criteria derived from experimental measurement and some simplified analysis for a limited number of cases,[45] and are thus unlikely to truly optimize the resulting antenna characteristics. Computer modeling offers some possibilities for improving this aspect of antenna design.

One approach that might be considered would be that of including the backscreen wires in the computer model. This could provide a very realistic representation of the overall antenna-backscreen system, but at a considerable increase in computer time, especially if extensive parametric studies were to be performed.

As an alternative, the parametric evaluation could be instead based on a two-dimensional integral equation, using infinite wires or strips for the backscreen and antenna members. The effects of spacing and size of the backscreen elements, backscreen width, antenna position, frequency, etc. could be much more efficiently studied, while many physical features important in the actual three-dimensional configuration could be retained. After identifying parameter values of greatest apparent interest, it might be then useful to perform limited calculations for the three-dimensional geometry to ensure the essential validity of the two-dimensional results. Another possible alternative which is worth mentioning is the application of image theory to the backscreen as well as to the interface. This is basically equivalent to two-dimensional screen model, while it retains the three-dimensional aspects of the actual antenna.

3.4.6 Impedance loading--In many cases of interest, the antenna may be connected to impedance loads of various kinds, or may even itself be lossy enough so that it cannot be accurately modeled as being perfectly conducting. These situations can be accommodated in the computer model by subtracting an appropriate voltage drop $Z_{ij}^{(L)} I_j$ from the source term E_i, where $Z_{ij}^{(L)}$ is the load impedance. When there are no mutual impedance effects, such as those due to transformer or transmission line interconnection, then $Z_{ij}^{(L)} = \delta_{ij} Z_{ij}^{(L)}$, i.e., the $\bar{\bar{Z}}$ matrix becomes diagonal Lumped loads are simply specified in terms of their resistive and reactive components. Their treatment is similar to that accorded sources, since the two can be viewed as mathematically equivalent. Distributed loads which might be used to model wire losses can be derived from the wire properties.[46]

3.4.7 Sheathed wires--Another problem of relevance, especially for antennas located in lossy media such as gorund or sea water, is that of a wire coated by a dielectric layer. One possible approach might be to model the sheath in the same way as a lossy wire, by a suitably derived impedance load.[46] An alternative, more rigorous approach has been taken by Richmond,[47] who models the sheath with a radially directed polarization current, reasoning that the tangential field, being much smaller, is by comparison of negligible import. Since the radial sheath fields which determine this current are known in terms of the charge density on the wire, no additional unknowns are introduced. One simply obtains a modified integral equation which can be solved in the usual way.

3.4.8 Time domain analysis--Previous discussion has dealt exclusively with frequency-domain formulations. It is worthwhile to point out that these problems can also be attacked from a time-dependent or time-domain viewpoint.[48-51] As one outcome of such an effort, there can be derived time-dependent integral equations which correspond closely to their frequency-domain counterparts. The solution procedure, while also developed from the moment method, is significantly different in that a solution is obtained as an initial-value problem via time stepping. This leads to results which are valid for only a single incident field or source configuration but over a band of frequencies, in contrast with the more familiar frequency-domain approach of which the converse is true. Solutions may consequently be obtained more efficiently in the time domain than in the frequency domain for certain types of problem, especially for wire structures analyzed as antennas. The modeling of nonlinear loads is also well suited to a time-domain approach.[52]

4. NUMERICAL RESULTS

In the context of practical applications, judgment on the relative merits of a particular computer model must ultimately depend upon the comparison of calculated results with independent data, preferably experimental, although other theoretically derived results may suffice. Unfortunately, reliable experimental data is not always available so that in many instances we may have to resort to various "computer experiments" and physical intuition when attempting to validate a numerical procedure. In the discussion which follows, we present numerical results for a variety of problems, accompanied where possible by measured data, to demonstrate the general applicability of the computer modeling approach for wire antennas. The order of presentation will follow that of Section 3 above.

4.1 Infinite, homogeneous, isotropic media

Problems which involve isolated antennas in infinite, homogeneous media are not frequently encountered, since, more often than not, environmental influences due to ground planes or the installation are important. Nevertheless, this kind of problem does provide a good, controlled check on the accuracy of computer calculations. Examples of such results are shown in Figs. 3 through 6.[53] The input impedance as a function of frequency near resonance of a capacitively loaded circular loop antenna is compared with measured data in Fig. 3. Results are shown in Fig. 4 for the input impedance resonance frequency of a zig-zag antenna as a function of the wire angle, also compared with experiment. A comparison of two computed near-field results for a circular loop antenna, one obtained from a moment-method solution of Eq. (3.1) and the second from an alternative analysis[54] are presented in Fig. 5. Finally, in Fig. 6 we show a comparison of computed and measured pattern results for a 19-element foreshortened log-periodic antenna.[10]

4.2 Perfectly conducting half-space

Some results for antennas located near perfectly conducting boundary planes are shown in Figs. 7 and 8. The computed input impedance for the top-loaded monopole, a LORAN-C antenna, is compared with measured results in Fig. 6.[10] Although the actual antenna is located over a finitely conducting ground, its ground screen (120 radial wires) simulates a perfect ground quite well, as demonstrated by the computed results, which are for a perfect ground. The input admittance for a half-wave antenna located midway between infinite, parallel, perfectly conducting plates is shown as a function of plate separation in Fig. 7, with the antenna angle relative to the plates a parameter[55] ($\Theta = 90°$ corresponds to perpendicular orientation).

4.3 Imperfectly conducting half-space

Representative results for antennas over lossy grounds are given in Figs. 9 through 15. Sommerfeld theory and reflection-coefficient approximation results for vertical and horizontal half-wave dipoles are compared in Fig. 9.[25,26] In Fig. 10 are presented computed results for a 3-m monopole antenna driven against a two-wire ground screen parallel to, and at a height of 0.01 λ above a lossy ($\sigma = 10^{-3}$ mhos, $\varepsilon_r = 4$, $f = 10$ MHz) ground.[56] These computations were performed using the interpolation procedure described in Section 3.3.2. Note that the ground wires point in the direction of pattern maxima; statement in reference 56 is incorrect. Results are included in Fig. 11 for the current distribution, impedance, and nearfield variation of a Beverage antenna.[57] Comparison of current distributions obtained from image theory with that computed using the Sommerfeld approach reveals the potential limitations of the former for this case. The graph of Fig. 13 illustrates use of the approximate surface-wave mixed-path model, which is compared with the rigorous calculation due to Ott.[32] Fig. 14 compares computed (using the reflection coefficient approximation) and measured impedances of the sectionalized LORAN transmitting (SLT) antenna, including the effect of a radial-wire ground screen in the computations.[58] The computed (reflection-coefficient approximation) and measured simulated EMP response of the fan-doublet antenna are shown in Fig. 15.[59] A compensation-theorem result for a vertical half-wave dipole is also included in Fig. 9.

4.4 Special topics

An example of how elaborate wire-grid models can become is illustrated in Fig. 16.[60] An 829-segment model of a ship is shown together with a comparison of computed and measured (scale-model) data for the radiation pattern of the twin-whip antenna at the ship's stern. Two measured patterns are used to demonstrate their frequency sensitivity. All major lobes at the higher frequency appear in the computed pattern, although with considerably different amplitudes (dB scale) for some of them.

One of the potential problems associated with wire-grid modeling is to introduce the possibility for transmission line-mode currents. This is shown in Fig. 17 for a "cage" model of a thick dipole antenna.[61] Although large circulating currents can flow on the individual wires of the model, their sum approximates the current on the equivalent antenna.

The effects of various wire ground screens and the influence of a subsurface sea-water layer on the impedance of the SLT antenna are shown in Fig. 18.[40] The effect of an octagonal ground plane on the input impedance of a monopole antenna computed using a combined moment-method and GTD approach due to Thiele[43] is illustrated in Fig. 19.

The dependence of the front-to-back ratio of wire backscreens (two-dimensional model) with various numbers of wires upon wire spacing and size is demonstrated in Figs. 20 and 21.[62] Fig. 22, which depicts the current distribution on a two-wire transmission line terminated in a matched load, is included to show an impedance load calculation.[40] The graph in Fig. 23 depicts results for a coated wire dipole due to Richmond.[47]

The concluding results of Fig. 24 demonstrate the transient feedpoint current response of a conical spiral antenna when it is excited by a Gaussian pulse and its input admittance derived from a Fourier transform of the current.[63]

5. ACKNOWLEDGMENTS

Besides the specific contributions made to this article by the individuals explicitly mentioned in the references and figures, various members of the LLL Electromagnetics and Systems Research Group have been instrumental in developing the computer codes and methods referred to or for which data was presented. They are R. W. Adams, Jr., R. M. Bevensee, G. J. Burke, F. J. Deadrick, J. A. Landt, D. L. Lager, R. J. Lytle, and A. J. Poggio. Appreciation is also due to Margie Hamilton who typed the manuscript and to Peggy Lorton for arranging the illustrations.

REFERENCES

1. A. J. Poggio and E. K. Miller, "Integral Equation Solutions of Three-Dimensional Scattering Problems," in *Computer Techniques for Electromagnetics*, R. Mittra, Ed., Pergamon Press, New York, 1973.
2. K. K. Mei, "On the Integral Equation of Thin Wire Antennas," *IEEE Trans. Antennas Propagat.* AP-13, 1965, p. 374.
3. R. F. Harrington, *Field Computation by Moment Methods*, Macmillan, New York, 1968.
4. J. H. Richmond, "Digital Computer Solutions of the Rigorous Equations for Scattering Problems," *Proc. IEEE* 53, 1965, p. 796.
5. E. K. Miller and F. J. Deadrick, "Some Computational Aspects of Thin-Wire Modeling", Lawrence Livermore Laboratory, Rept. UCRL-74818, 1973a.
6. C. M. Butler, "Currents Induced on a Pair of Skewed Crossed Wires," *IEEE Trans. Antennas Propagat.* AP-20, 1972, p. 731.
7. W. L. Curtis, Boeing Space Center, Seattle, Wash., private communication, 1972.
8. H. H. Chao and B. J. Strait, *Computer Programs for Radiation and Scattering by Arbitrary Configurations of Bent Wires*, Syracuse University, Scientific Rept. No. 7 on Contract No. F19628-68-C-0180; AFCRL-70-0374, September 1970.
9. J. H. Richmond, *Computer Analysis of Three Dimensional Wire Antennas*, Ohio State University, Department of Electronics Engineering, ElectroScience Laboratory, Tech. Rept. 2708-4, 1969.
10. MBAssociates, *Polar Class Icebreaker Antenna Analysis*, San Ramon, Calif., Rept. MB-P-70/77, November 12, 1970.
11. E. K. Miller, R. M. Bevensee, A. J. Poggio, R. Adams, and F. J. Deadrick, "Comparison of Wire Structure Computer Codes for Application to EMP Vulnerability Assessment of Aircraft," Lawrence Livermore Laboratory, UCRL-75556, 1973c.
12. Y. S. Yeh and K. K. Mei, "Theory of Conical Equiangular Spiral Antennas: Part I--Numerical Techniques," *IEEE Trans. Antennas Propagat.* AP-15, 1967, p. 634.
13. T. T. Wu and R. W. P. King, "The Tapered Antenna and Its Application to the Junction Problem for Thin Wires," *IEEE Trans. Antennas Propagat.* AP-24, 1976, p. 47.

14. S. Gee, E. K. Miller, A. J. Poggio, E. S. Selden, and G. J. Burke, "Computer Techniques for Electromagnetic Scattering and Radiation Analyses," invited paper presented at The Electromagnetic Compatibility Meeting, Philadelphia, Pa., 1971.
15. G. J. Burke, "NEC - Numerical Electromagnetics Code for Frequency Domain Analysis," Tech. Report UCRL-80942, Lawrence Livermore Laboratory, 1978.
16. M. G. Andreasen and F. B. Harris, Jr., Analysis of Wire Antennas of Arbitrary Configuration by Precise Theoretical Numerical Techniques, Granger Associates, Palo Alto, Calif., Contract DAAB07-67-C-0631, Tech. Rept. ECOM 0631-F, 1968.
17. A. J. Poggio, R. W. Adams, and E. K. Miller, "A Study of Antenna Source Models," Tech. Report UCRL 51693, Lawrence Livermore Laboratory, 1974.
18. E. K. Miller and G. J. Burke, "Numerical Integration Methods," IEEE Trans. Antennas Propagat. AP-17, 1969, p. 669.
19. K. K. Chan, L. B. Felsen, S. T. Ping, and J. Schmoys, "Diffraction of the Pulsed Field from an Arbitrarily Oriented Electric or Magnetic Dipole by a Wedge," AFWL Sensor and Simulation Note 202, 1973.
20. C. D. Taylor, "Thin Wire Receiving Antenna in a Parallel Plate Waveguide," IEEE Trans. Antennas Propagat. AP-15, 1967, p. 572.
21. A. Sommerfeld, "Uber der Ausbreitung der Willen in der Drahtlosen Telegraphic," Ann. Physik 28, 1909, p. 663.
22. A. Sommerfeld, Partial Differential Equations in Physics, Academic Press, New York, 1964.
23. J. N. Brittingham, E. K. Miller, and J. T. Okada, "A Bivariate Interpolation Approach for Efficiently and Accurately Modelling Antennas Near a Halfspace," Electron. Lett., 13, 1977, pp. 690-691.
24. G. J. Burke, "NEC - Numerical Electromagnetics Code - II," to be published.
25. E. K. Miller, A. J. Poggio, G. J. Burke, and E. S. Selden, "Analysis of Wire Antennas in the Presence of a Conducting Half Space: Part I. The Vertical Antenna in Free Space," Can. J. Phys. 50, 1972a, p. 879.
26. E. K. Miller, A. J. Poggio, G. J. Burke, and E. S. Selden, "Analysis of Wire Antennas in the Presence of a Conducting Half Space: Part II. The Horizontal Antenna in Free Space," Can. J. Phys. 50, 1972b, p. 2614.
27. A. Baños., Dipole Radiation in the Presence of a Conducting Half-Space, Pergamon Press, New York, 1966.
28. R. J. King, "Electromagnetic Wave Propagation over a Constant Impedance Plane," Radio Sci. 4, 1969a, p. 225.

29. G. D. Monteath, "Application of the Compensation Theorem to Certain Radiation and Propagation Problems," *Proc. IEE* **98** Part IV, 1951, p. 23.
30. R. Mittra, *Vector Form of Compensation Theorem and Its Application to Boundary Value Problems*, Department of Electronics Engineering, University of Colorado, Boulder, Scientific Rept. 2, AFCRL 575, 1961.
31. R. J. King, "On the Surface Impedance Concept," *Proc. Conf. Environmental Effects on Antenna Performance*, Boulder, Colo., J. R. Wait, ed., 1969b.
32. R. H. Ott, *A New Method for Predicting HF Ground Wave Propagation over Inhomogeneous Irregular Terrain*, U. S. Department of Commerce, OT/ITSRR 7, 1971.
33. P. David and J. Voge, *Propagation of Waves*, Pergamon Press, New York, 1969.
34. G. Millington, "Ground-Wave Propagation over an Inhomogeneous Smooth Earth," *Proc. IEE* **96** Part III, 1949, p. 53.
35. G. Millington, and G. A. Isted, "Ground Wave Propagation over an Inhomogeneous Smooth Earth: Part 2--Experimental Evidence and Practical Implications," *Proc. IEEE* **97** Part III, 1950, p. 209.
36. R. W. P. King, *The Theory of Linear Antennas*, Harvard University Press, Cambridge, Massachusetts, 1956.
37. R. L. Tanner and M. G. Andreasen, "Numerical Solution of Electromagnetic Problems," *IEEE Spectrum*, **4**, No. 9, 1967, p. 53.
38. K. S. H. Lee, L. Marin, and J. P. Castillo, "Limitations of Wire-Grid Modeling of a Closed Surface," *IEEE Trans. on EMC*, EMC-18, 3, 1976, p. 123.
39. J. R. Wait, "Characteristics of Antennas over Lossy Earth," in *Antenna Theory*, R. E. Collin and F. J. Zucker, eds., McGraw-Hill, New York, 1969, pp. 386-437.
40. E. K. Miller, and F. J. Deadrick, *Computer Evaluation of LORAN-C Antennas*, Lawrence Livermore Laboratory, Rept. UCRL-51464, 1973b.
41. M. I. Astrakhan, "Reflecting and Screening Properties of Plane Wire Grids," *Radio Engineering* **23** (1), (translated from the Russian), 1968, p. 76.
42. J. R. Wait, "Theories of Scattering from Wire Grid and Mesh Structures," *Electromagnetic Scattering*, Academic Press, New York, 1978, p. 253.
43. J. R. Wait, *Electromagnetic Waves in Stratified Media*, Pergamon Press, Macmillan Co., New York, 1962.
44. G. A. Thiele, "An Introduction to Low Frequency Numerical Methods," Ohio State Short Course on Application of GTD and Numerical Techniques to the Analysis of Electromagnetic and Acoustic Radiation and Scattering, Ohio State University, 1973.

45. E. B. Moullin, Radio Aerials, Oxford University Press, New York, ch. 5 and 11, 1949.
46. E. S. Cassidy and J. Fainberg, "Backscattering Cross Sections of Cylindrical Wires of Finite Conductivity," IRE Trans. Antennas Propagat. AP-8, 1960, p. 1.
47. J. H. Richmond, "Radiation and Scattering by Thin-Wire Structures in the Complex Frequency Domain," Ohio State University, Department of Electronics Engineering, ElectroScience Laboratory, Rept. 2902-10, 1973.
48. C. L. Bennett and W. L. Weeks, "Electromagnetic Pulse Response of Cylindrical Scatterers," G-AP Symposium, Boston, Massachusetts, 1968. (See also A Technique for Computing Approximate Electromagnetic Impulse Response of Conducting Bodies, Purdue University Rept. TR-EE68-11.)
49. E. P. Sayre and R. F. Harrington, "Time-Domain Radiation and Scattering by Thin Wires," Appl. Sci. Res. 26 (6), 1972, pp. 413-444.
50. E. K. Miller, A. J. Poggio, and G. J. Burke, "An Integro-Differential Equation Technique for the Time Domain Analysis of Thin Wire Structures, Part I. The Numerical Method," J. Computational Phys. 12, 1973a, p. 24.
51. A. J. Poggio, E. K. Miller, and G. J. Burke, "An Integro-Differential Equation Technique for the Time-Domain Analysis of Thin Wire Structures. II: Numerical Results," J. Computational Phys. 12, 1973, p. 210.
52. J. A. Landt, E. K. Miller, and F. J. Deadrick, "Time-Domain Modeling of Nonlinear Loads," Lawrence Livermore Laboratory, Rept. UCRL-52172, 1976.
53. S. Gee, E. K. Miller, and E. S. Selden, "Computer Analysis of Loaded Loop and Conical Spiral Antennas," MBAssociates, San Ramon, Calif., Tech. Rept. MB-R-70/38, 1970.
54. R. L. Fante, J. J. Otazo, and J. T. Mayhan, "The Near Field of the Loop Antenna," Radio Sci. 4, 1969, p. 697.
55. F. M. Tesche, On the Behavior of Thin-Wire Scatterers and Antennas Arbitrarily Located Within a Parallel Plate Region. I. The Formulation, EMP Sensor and Simulation Notes, Note 135, AFWL, Kirtland AFB, N.M., 1971
56. E. K. Miller, J. N. Brittingham, and J. T. Okada, "Explicit Modelling of Antennas with Sparse-Ground Screens," Electron. Lett., 14, 1978, pp. 627-629.
57. R. J. Lytle, D. L. Lager, E. K. Miller, and F. J. Deadrick, The Beverage Antenna--A Multiconductor Antenna that Significantly Interacts with the Ground, Lawrence Livermore Laboratory, Rept. UCRL-75449, 1974.
58. E. K. Miller, F. J. Deadrick, and W. O. Henry, "Computer Evaluation of Large, Low-Frequency Antennas," IEEE Trans. Antennas Propagat. AP-21, 1973b, p. 386.

59. J. A. Landt, F. J. Deadrick, E. K. Miller, and R. Kirchofer, <u>Computer Analysis of the Fan Doublet Antenna</u>, Lawrence Livermore Laboratory, Rept. UCRL-74846 Preprint, 1973.
60. E. S. Selden and G. J. Burke, "DLGN-38 Final Report," MBAssociates, San Ramon, Calif., Tech. Rept. MB-R-75/16, 1975.
61. G. J. Burke and E. S. Selden, private communication, 1972.
62. R. M. Bevensee, <u>Design Considerations for Parasitic Screen Antennas</u>, Lawrence Livermore Laboratory, 1974, to be published as UCRL report.
63. J. A. Landt and E. K. Miller, <u>Transient Characteristics of the Conical Spiral</u>, Lawrence Livermore Laboratory, Rept. UCID-16461, 1974.

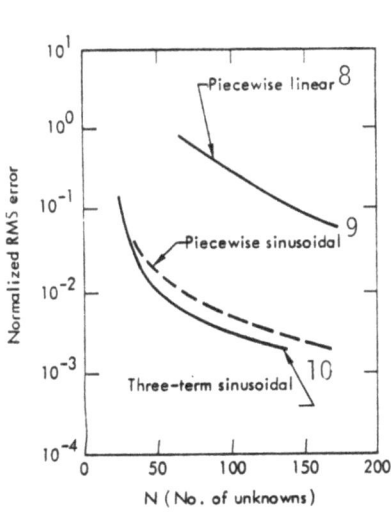

Fig. 1. Convergence rate for several solution methods.[11]

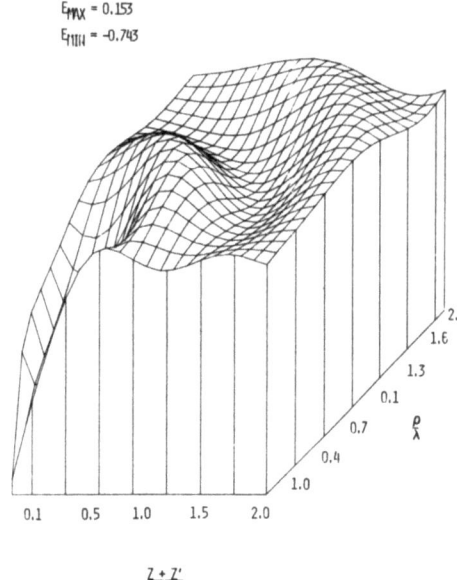

Fig. 2. A three dimensional aspect plot of the Sommerfeld-integral portion of the radial electric field for a vertical dipole is shown here. The real-part of the electric field is plotted on the vertical axis as a function of $z + z'$ and ρ. For this figure, the electrical parameters were $f = 10^7$ Hz, $\varepsilon_r = 9.0$, and $\sigma = 10^{-2}$ mhos per meter, while $0.1\,\lambda \leq \rho \leq 2.0\,\lambda$ and $.1\,\lambda \leq z + z' \leq 2.0\,\lambda$.

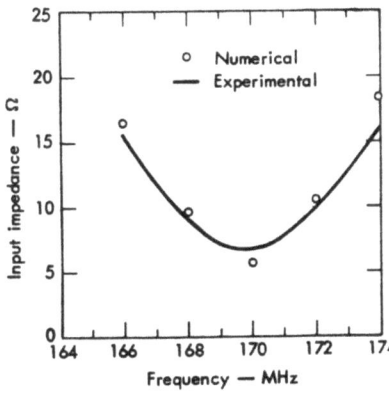

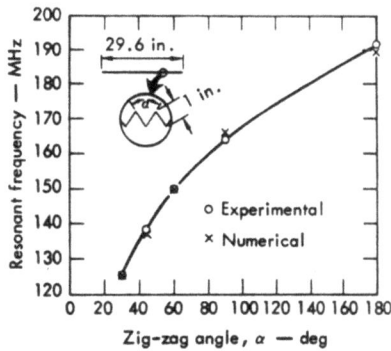

Fig. 3. Input impedance of a capacitively loaded circular loop antenna.[53]

Fig. 4. Resonance frequency of a zig-zag dipole.[53]

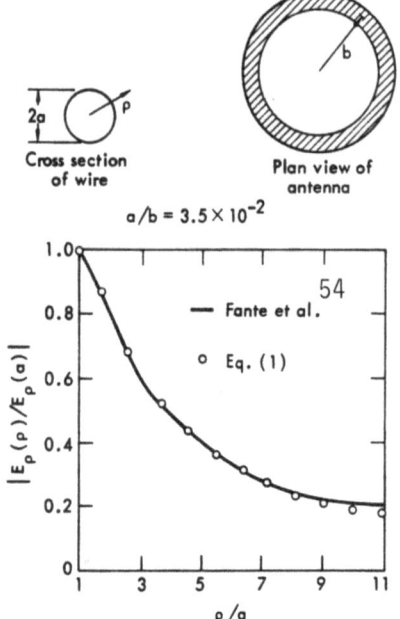

Fig. 5. The near field of a circular loop antenna.[53]

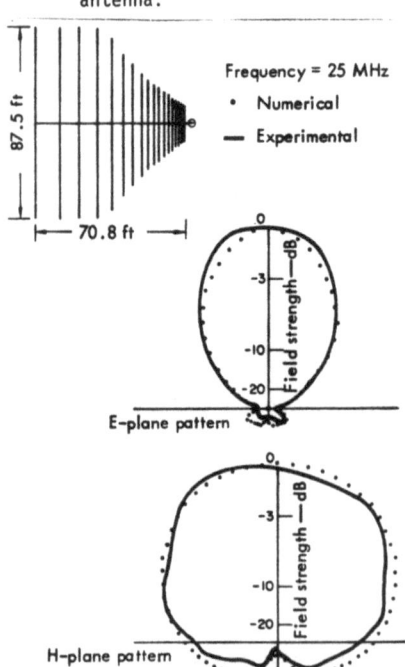

Fig. 6. Radiation patterns of a fore-shortened 19-element log-periodic antenna.[10]

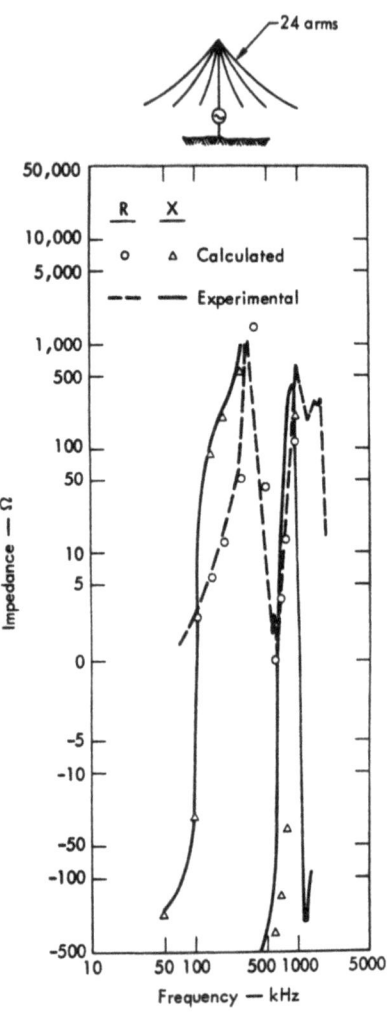

Fig. 7. Input impedance of a top-loaded monopole antenna.[10]

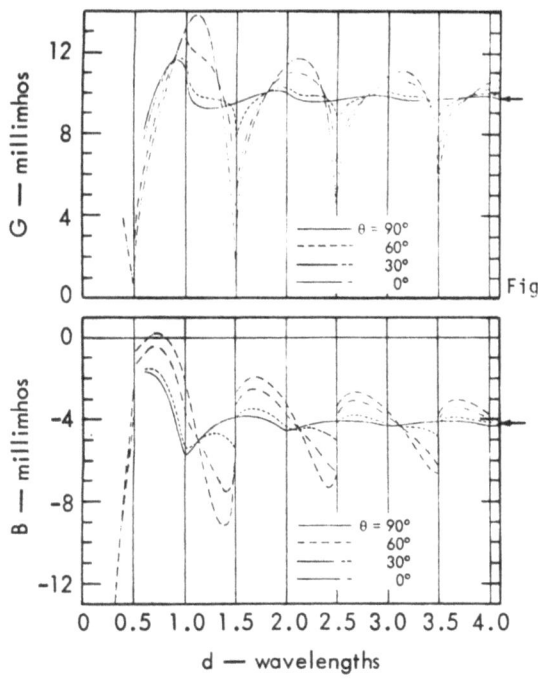

Fig. 8. Input impedance of a half-wave dipole located midway between two infinite parallel plates.[55] The arrows denote free-space values

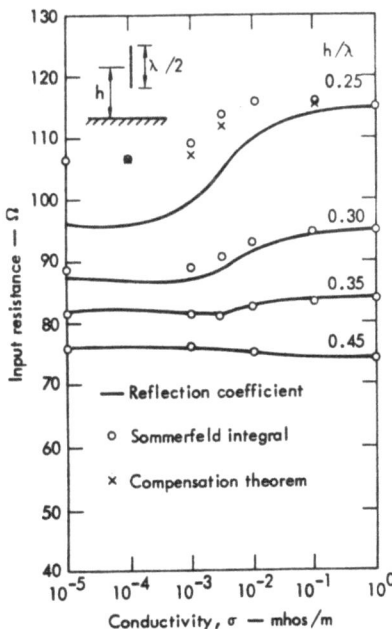

Fig. 9a. Results for a vertical half-wave dipole over a ground plane.[25]
$f = 3.0$ MHz; $\varepsilon_r = 10$; $a = 5 \times 10^{-4}\lambda$.

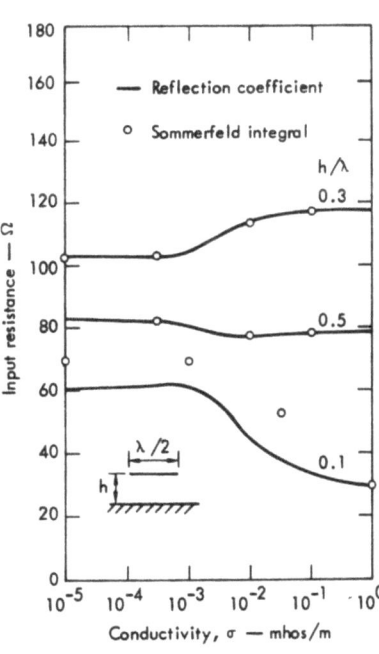

Fig. 9b. Results for a horizontal half-wave dipole over a ground plane.[26]
$f = 3.0$ MHz; $\varepsilon_r = 10$; $a/\lambda = 5 \times 10^{-3}$.

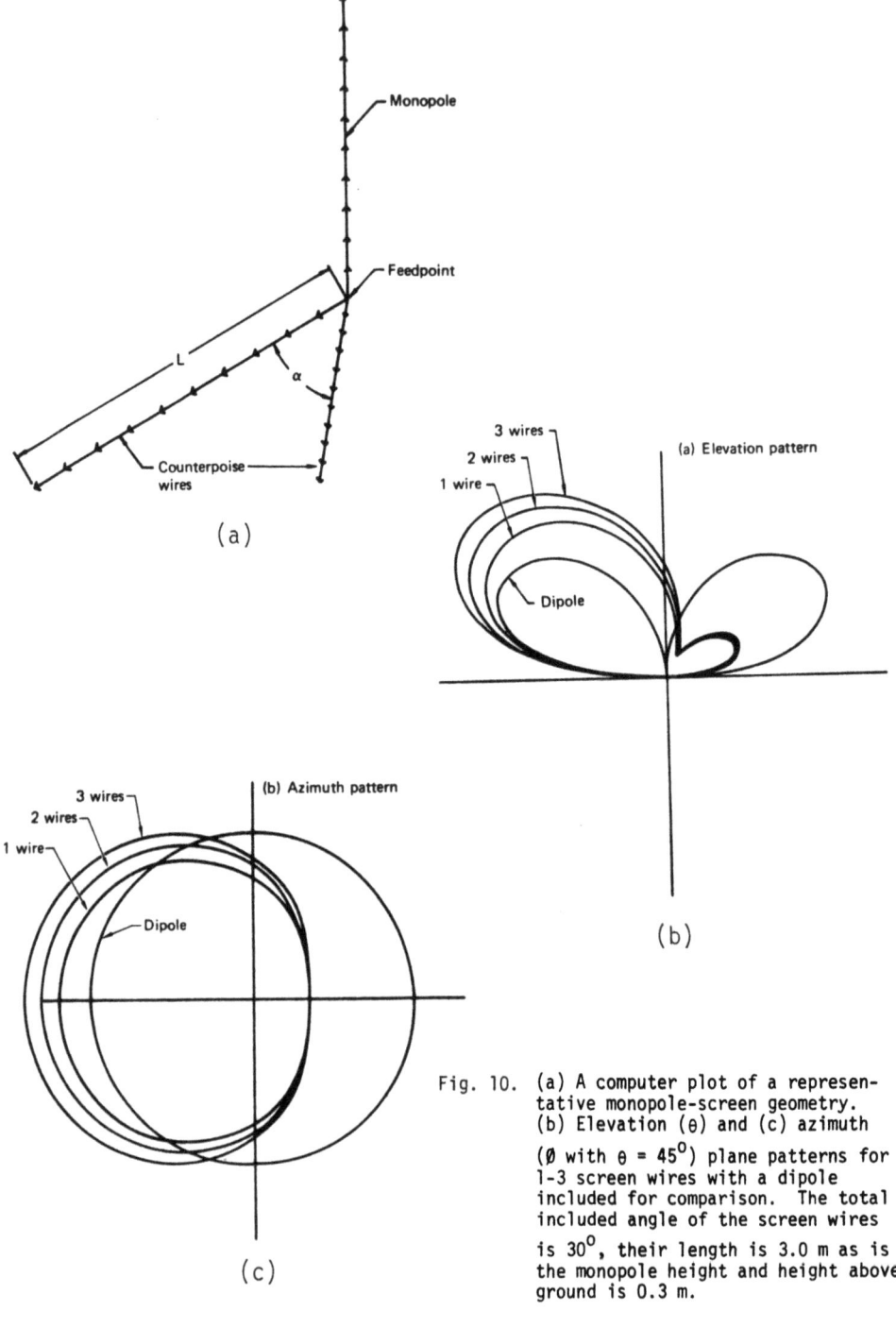

Fig. 10. (a) A computer plot of a representative monopole-screen geometry. (b) Elevation (θ) and (c) azimuth (∅ with θ = 45°) plane patterns for 1-3 screen wires with a dipole included for comparison. The total included angle of the screen wires is 30°, their length is 3.0 m as is the monopole height and height above ground is 0.3 m.

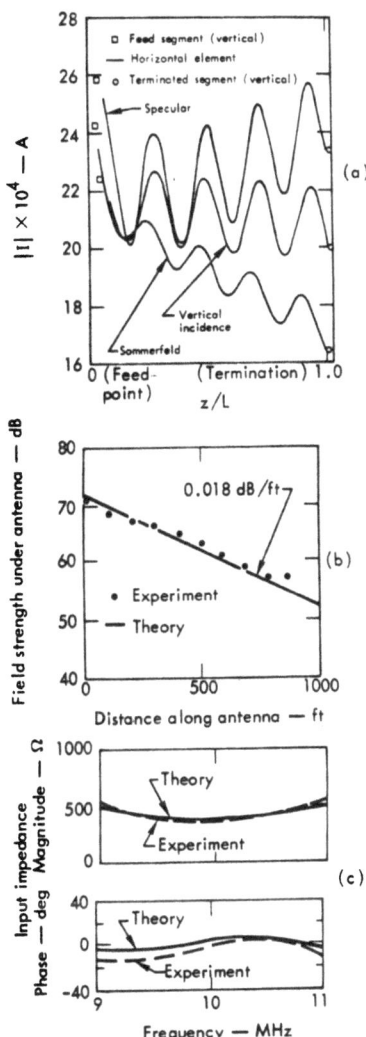
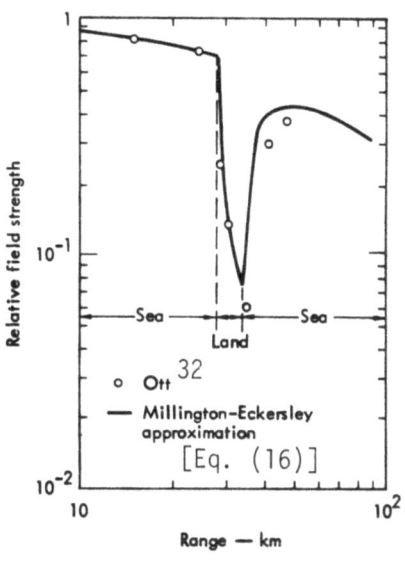

Fig. 13. Surface-wave field strength over a sea-land-sea mixed path. f = 10 MHz. For land: $\sigma = 0.002$; $\epsilon_r = 15$. For water: $\sigma = 2$; $\epsilon_r = 81$.

Fig. 11. Results for a Beverage antenna.[56]
(a) Computed current distributions. f = 10 MHz; L = 215 ft; h = 2 m; $Z_T = 452\ \Omega$; $\epsilon_r = 25$; $\sigma = 0.03$.
(b) Field strength. f = 10 MHz; L = 1000 ft; h = 2 m; $Z_T = 452\ \Omega$.
(c) Impedance.

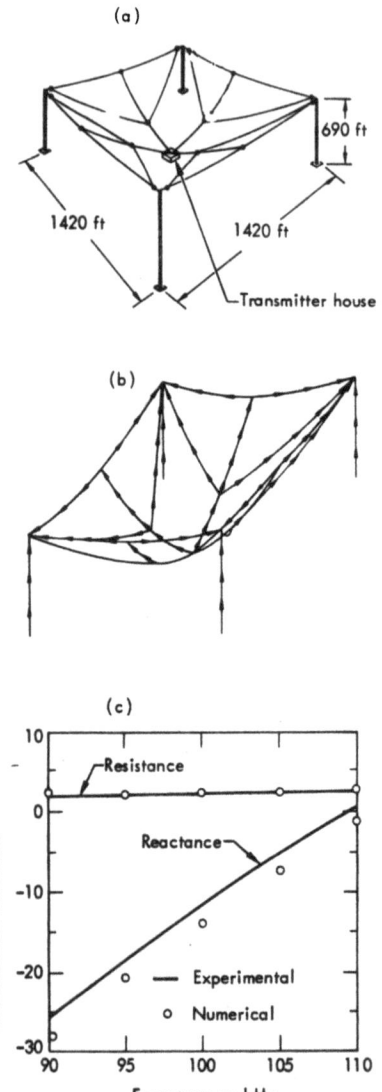
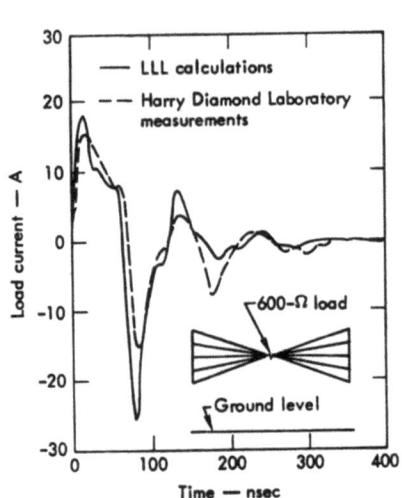

Fig. 15. EMP simulated response of the fan doublet antenna.[58]

Fig. 14. (a) Sectionalized LORAN Transmitter antenna geometry. (b) Computer model. (c) Experimental and numerical comparison of SLT antenna impedance.[57]

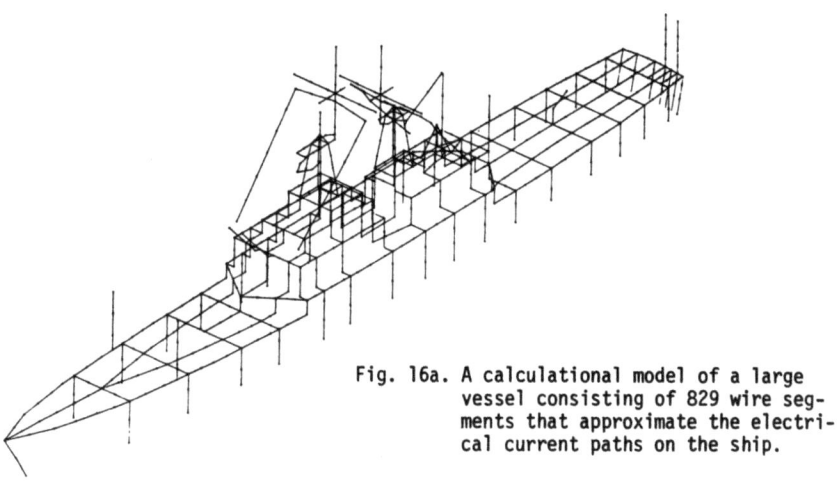

Fig. 16a. A calculational model of a large vessel consisting of 829 wire segments that approximate the electrical current paths on the ship.

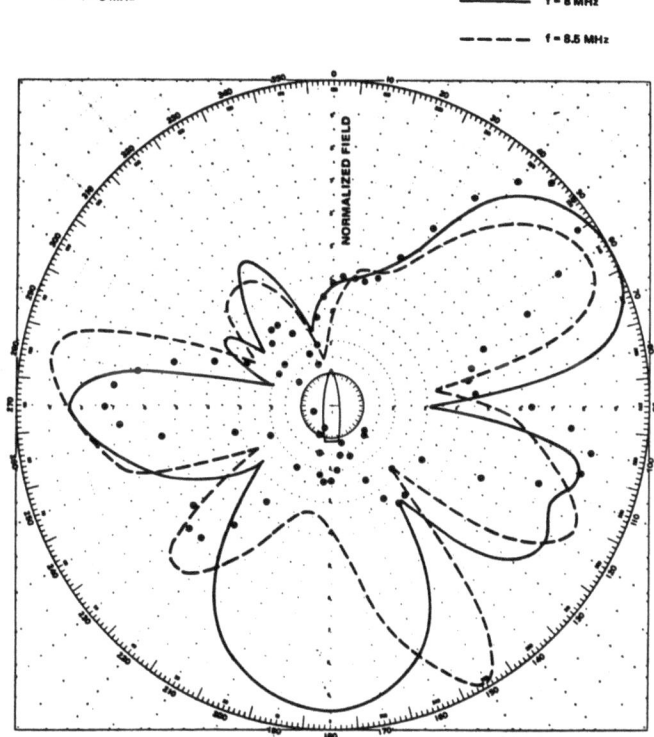

Fig. 16b. Azimuthal pattern for the twin whip antenna on the 829 segment grid model of Fig. 16a at 8 MHz compared with experiment.

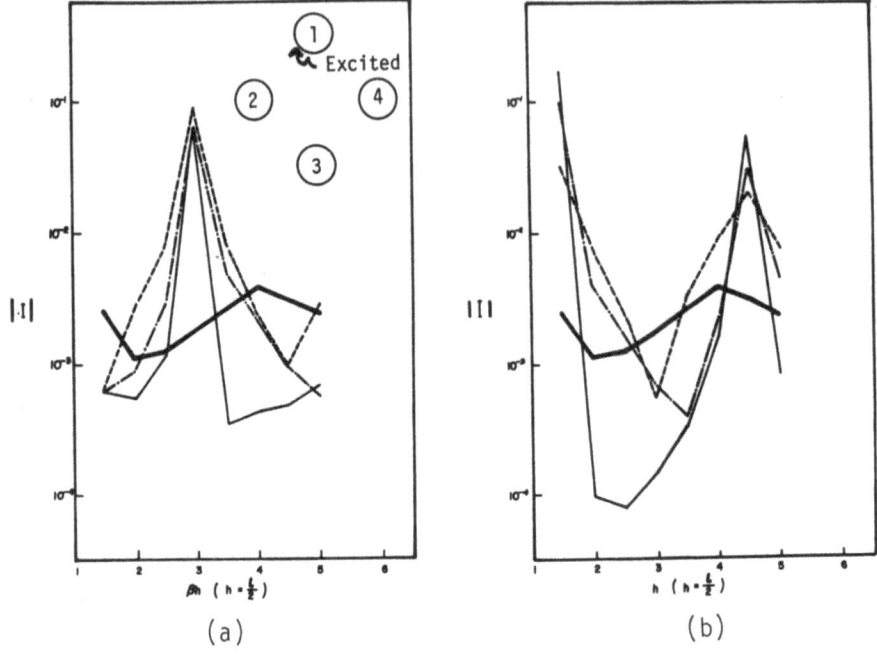

Fig. 17. Current on 4-wire dipole $\Omega = 6$:
(a) wire connected at ends, (b) wires open at ends.

―――― Sum of currents
------ Current on 1
―・― Current on 2
――― Current on 3

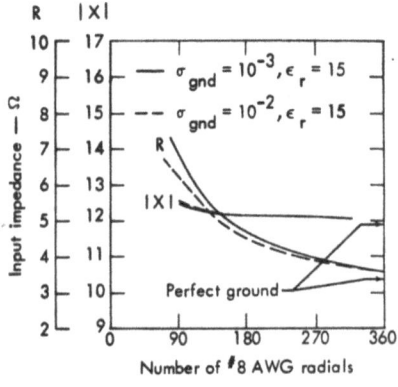

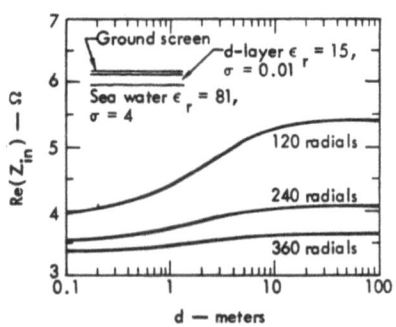

Fig. 18a. Influence of the ground screen on the sectionalized LORAN transmitting antenna.[40]

Fig. 18b. Effect of subsurface sea water on the sectionalized LORAN transmitting antenna.[40]

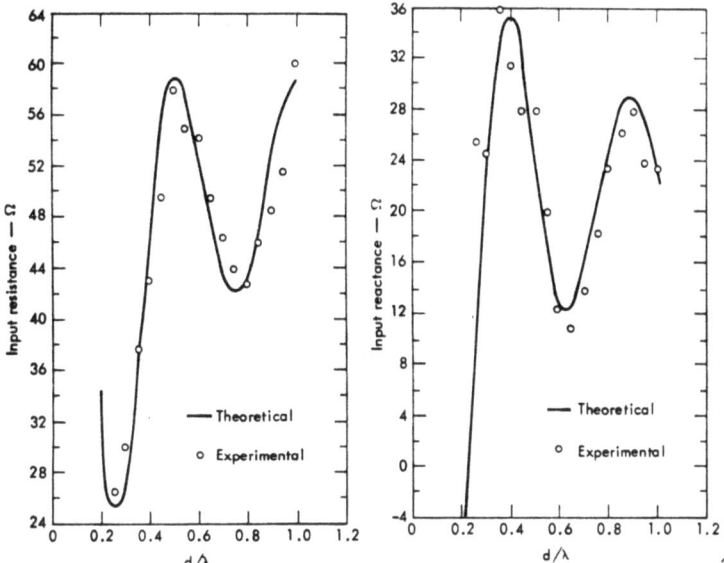

Fig. 19. Application of the combined moment-method and GTD technique.[43] These results are for an octagonal plate with a monopole at the center (fixed physical radius). h = 0.25λ; a = 0.1524 cm.

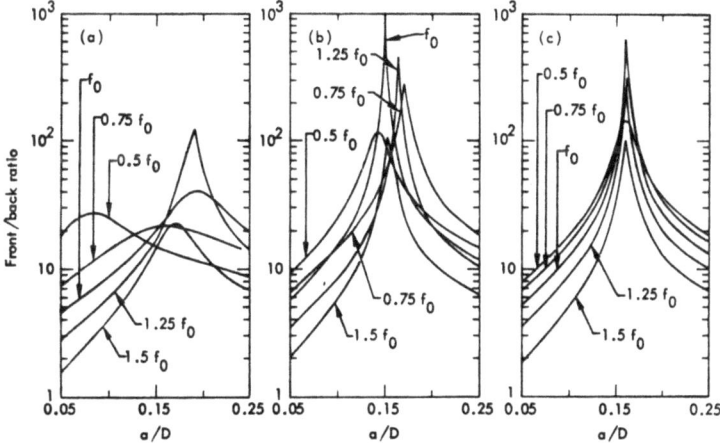

Fig. 20. Some results from a backscreen optimization study:[61] Front/back ratio of screen antenna vs a/D, at various frequencies about f_0. The parameters are defined in Fig. 21. (a) 10 wires in screen. (b) 30 wires in screen. (c) 50 wires in screen. Note the front-to-back ratio peaks at a/D ~ 1/2 π, where the screen wires have a surface area approximately equal to the frontal area of the screen.

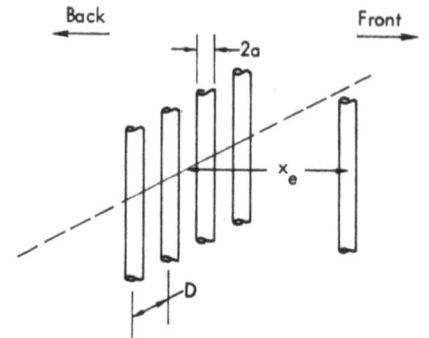

Fig. 21. Definition of screen-antenna parameters plotted in Fig. 20. At f_0, $D/\lambda_0 = 0.16$, $x_e/\lambda_0 = 0.25$.

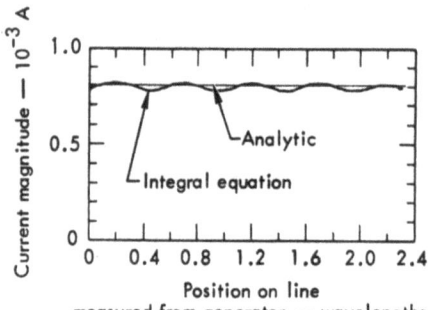

Fig. 22. Integral-equation results for a matched load on a two-wire transmission line.[40]

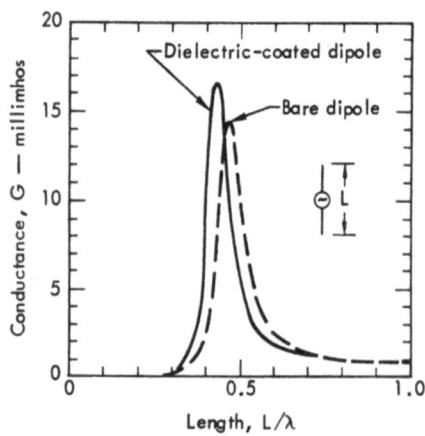

Fig. 23. Results for a sheathed dipole.[47] Wire diameter $d = L/100$; outer diameter of dielectric shell $D = 2d$; permittivity of shell $\epsilon = 4\epsilon_0$.

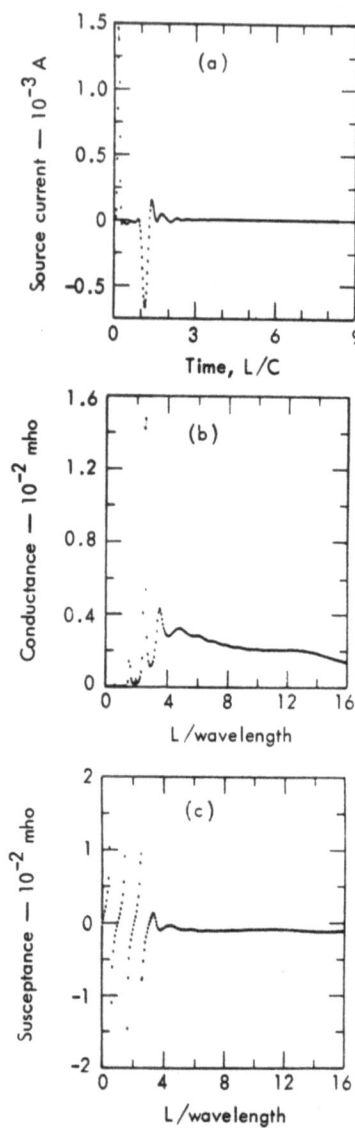

Fig. 24. Time-domain computer results for a conical spiral: (a) Transient feedpoint current excited by a Gaussian pulse from a 350-Ω generator. (b) Conductance. (c) Susceptance. (b) and (c) are obtained from a Fourier transform of (a) with the generator impedance removed.[62]

SURFACE INTEGRAL EQUATIONS FOR CONDUCTING AND DIELECTRIC BODIES

R. F. Harrington
J. R. Mautz

Department of Electrical and Computer Engineering,
Syracuse University, Syracuse, New York 13210, U.S.A.

ABSTRACT. The problems of electromagnetic radiation and scattering by conducting and dielectric bodies are discussed in terms of surface integral equations. It is shown that the usual H-field and E-field surface integral equations for conducting bodies fail to give unique solutions whenever the surface, when covered by an electric conductor, forms a resonant cavity. Two formulations which overcome this difficulty are the combined-field and combined-source formulations. It is shown that these two formulations always have a unique solution. Computations are given which serve to illustrate these theoretical predictions.

For dielectric bodies, it is shown that formulations in terms of surface integral equations can be obtained using the same E-field and H-field operators as used for conducting bodies. It is merely necessary to apply these operators twice, once for an infinite region with constitutive parameters equal to those external to the body in the original problem, and once for an infinite region with constitutive parameters equal to those internal to the body in the original problem. It is also shown that an infinite set of pairs of surface integral equations can be obtained for the dielectric body. One particular pair of equations is called the natural formulation because it arises from a direct application of the equivalence principle to the tangential components of E and H on the surface of the dielectric body. Another particular pair of equations is that obtained by Müller, which have the property that the singularity of the kernels of the integral equations is minimized. Computations are given for the case of a finite dielectric cylinder for both the natural formulation and the Müller formulation to illustrate the solution.

1. INTRODUCTION

Surface integral equations are particularly convenient for the computation of electromagnetic radiation and scattering from conducting and dielectric bodies. This is because, for a three dimensional body, the integral equations are over only the two-dimensional surface. Hence, the dimensionality of the problem is one less than it would be for a volume formulation.

A general discussion of the H-field, E-field, and combined-field integral equations for conducting bodies is given by Poggio and Miller in [1]. Solutions to the H-field equation for bodies of revolution have been given by Oshiro et al [2] and by Uslenghi [3]. Solutions to the E-field equation for bodies of revolution have been given by Mautz and Harrington [4]. Solutions to the combined-field equation for a sphere have been given by Mitzner et al [1,5], but no general solution for bodies of revolution was reported. The combined-field equation is a linear combination of the H-field and E-field equations, and it is solved using a linear combination of the matrices obtained for the H-field and E-field solutions. The combined-source formulation uses both electric and magnetic currents as equivalent sources in an E-field formulation. It can be shown [15] that the combined-source formulation is related by reciprocity to the combined-field formulation, and its solution again results in matrices related to those obtained for the H-field and E-field solutions.

Both the H-field and the E-field formulations fail to give unique solutions at eigenfrequencies corresponding to resonant frequencies of a conducting cavity having the same boundary surface as the scatterer. Neither the combined-field nor the combined-source formulations has this deficiency. This is the primary reason for using these latter two formulations.

The surface integral equations for homogeneous dielectric and magnetic bodies involve the same H-field and E-field operators as for conducting bodies. The problem is formulated in terms of equivalent electric and magnetic currents over the body surface. Application of boundary conditions leads to a set of four integral equations to be satisfied. Linear combinations of these four equations lead to a coupled pair of equations to be solved. One choice of combination constants gives the formulation described by Poggio and Miller [1]. This formulation has been applied to material cylinders by Chang and Harrington [6], and to material bodies of revolution by Wu [7]. We will call this choice the natural formulation. Another choice of combination constants gives the formulation obtained by Müller [8]. This formulation has been applied to dielectric cylinders by Solodukhov and Vasil'ev [9] and by Morita [10], and to bodies of revolution by Vasil'ev and Materikova

[11]. Conditions for the uniqueness of solutions can be established in terms of the combination constants [14,22]. It is found that solutions to both the natural formulation and to Müller's formulation are unique at all frequencies.

Computer programs for all of the above solutions are available for bodies of revolution in a series of research reports [12-16]. These programs are fully documented with sample input-output data included. Copies of these reports can be obtained from the authors while the supply lasts.

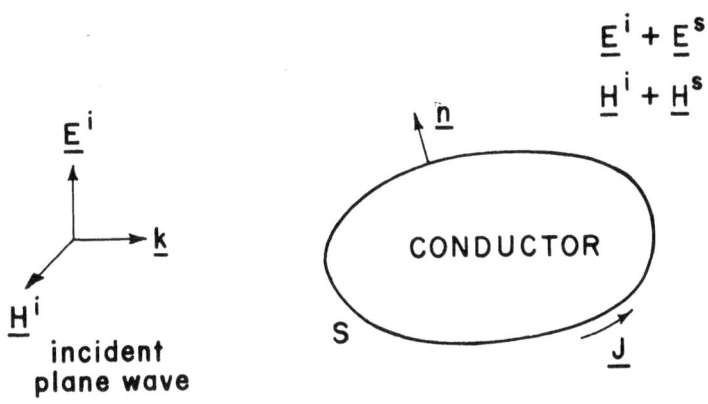

Fig. 1. A perfectly conducting body excited by an incident plane wave.

2. H-FIELD FORMULATION

We seek the electric surface current and the far scattered field of a perfectly conducting body excited by an incident plane wave, as shown in Fig. 1. The H-field integral equation is derived by setting the component tangential to S of the total magnetic field equal to zero just inside S, that is

$$-\underline{n} \times \underline{H}^s = \underline{n} \times \underline{H}^i \quad \text{just inside S} \tag{1}$$

where $\underline{n}$ is the unit outward normal vector to S, $\underline{H}^s$ is the magnetic field due to the electric surface current on S and $\underline{H}^i$ is the incident magnetic field. By expressing $\underline{H}^s$ in terms of the magnetic vector potential integral, and taking the limit as the field point approaches the surface from the interior of S, we obtain the H-field equation [1]

$$L_H(\underline{J}) = \underline{n} \times \underline{H}^i \qquad (2)$$

where

$$L_H(\underline{J}) = \frac{1}{2} \underline{J}(\underline{r}) - \iint_S \underline{n} \times \underline{\nabla} \times [\underline{J}(\underline{r}') \frac{e^{-jk|\underline{r}-\underline{r}'|}}{4\pi|\underline{r}-\underline{r}'|}]ds' \qquad (3)$$

Here $\underline{J}$ is the surface current on S, k is the propagation constant, $\underline{r}'$ is a source point on S, and $\underline{r}$ is the field point also on S. The improper integral in (3) is convergent.

Details of the application of this formulation to conducting bodies of revolution can be found in [12] and [17]. The solution was obtained by the method of moments [18]. Computer programs for the solution are given in [13].

3. E-FIELD FORMULATION

The problem being considered is the same one as for section 2, illustrated by Fig. 1. The E-field equation is derived by setting the component tangential to S of the total electric field equal to zero on S, that is,

$$-\underline{E}^s_{tan} = \underline{E}^i_{tan} \quad \text{on S} \qquad (4)$$

Here $\underline{E}^s$ is the electric field due to the electric surface current on S and $\underline{E}^i$ is the incident electric field. The subscript tan denotes tangential components on S. By expressing $\underline{E}^s$ in terms of the magnetic vector potential and electric scalar potential, we obtain a common form of the E-field equation [18]

$$L_E(\underline{J}) = \underline{E}^i_{tan} \qquad (5)$$

where

$$L_E = (j\omega \underline{A} + \underline{\nabla}\Phi)_{tan} \qquad (6)$$

$\underline{A}$ is the magnetic vector potential integral, Φ is the electric scalar potential integral, and ω is the angular frequency.

Again details of the application of this formulation to conducting bodies of revolution can be found in [12] and [17]. Computer programs for the solution obtained by the method of moments are given in [13].

4. COMBINED-FIELD FORMULATION

We shall see in section 6 that both the H-field and E-field formulations fail to yield unique solutions at the resonant frequencies of a cavity formed by a perfect conductor covering the surface S. To circumvent this difficulty, we consider the combined-field solution described in [1] and [5].

The true solution to the problem must satisfy both conditions (1) and (4), that is, both $\underline{E}_{tan}$ and $\underline{H}_{tan}$ must be zero just inside S. It must also satisfy a linear combination of (1) and (4), which can be written as

$$-\underline{n} \times \underline{H}^s(\underline{J}) - \frac{\alpha}{\eta} \underline{E}^s_{tan}(\underline{J}) = \underline{n} \times \underline{H}^i + \frac{\alpha}{\eta} \underline{E}^i_{tan} \quad \text{just inside S} \quad (7)$$

where α is a constant to be chosen and η is the intrinsic impedance of free space. It is shown in section 6 that the combined-field formulation has a unique solution at all frequencies if α is a positive real number. The representation of $\underline{n} \times \underline{H}^s$ as an integral operator is the same as in section 2 above, and the representation of $\underline{E}^s_{tan}$ as an integral operator is the same as in section 3 above.

Again details of the application of this formulation to conducting bodies of revolution can be found in [12] and [17]. Computer programs for the method of moments solution of this formulation are given in [13].

5. COMBINED-SOURCE FORMULATION

An alternative method for circumventing the failure of the H-field and E-field formulations at eigenfrequencies is to use an equivalent source consisting of both electric and magnetic currents on the surface S. This is an extension of the scalar combined-source solution suggested in [19], and implemented numerically in [20]. Application of this solution to two-dimensional electromagnetic problems, which can be formulated as scalar problems, is given in [21]. We here consider the general formulation of the method for three-dimensional electromagnetic fields.

In the combined-source formulation, we simulate the scattered field outside S in Fig. 1 by placing simultaneously an electric current $\underline{J}$ and a magnetic current $\underline{M}$ on S. Then the E-field equation becomes

$$-\underline{E}_{tan}(\underline{J},\underline{M}) = \underline{E}^i_{tan} \quad \text{just outside S} \quad (8)$$

This is one equation in two unknowns, $\underline{J}$ and $\underline{M}$. To complete the formulation we require a relationship between $\underline{J}$ and $\underline{M}$. In particular, we choose

$$\underline{M} = \alpha \, \underline{n} \times \underline{J} \qquad (9)$$

where $\underline{n}$ is the unit normal to S and α is a constant to be chosen. Equation (8) will have a unique solution at all frequencies if the real part of α is positive, as shown in section 6.

More details of this formulation can be found in [15]. Solution by the method of moments and a documented computer program for conducting bodies of revolution are given in [16].

6. UNIQUENESS OF SOLUTIONS

The solution to a linear operator equation is not unique if the corresponding homogeneous equation has a nontrivial solution. In particular, the solution to the E-field equation (5) will not be unique if $L_E(\underline{J}) = 0$ has a nontrivial solution. In terms of differential equations this means that the solution is not unique whenever there is a nontrivial solution to

$$\underline{\nabla} \times \underline{\nabla} \times \underline{E}^s(\underline{J}) = k^2 \underline{E}^s(\underline{J}) \qquad \text{within S} \qquad (10)$$

$$\underline{n} \times \underline{E}^s(\underline{J}) = 0 \qquad \text{on S} \qquad (11)$$

But these are the equations for the modes of a resonator formed by a perfect electric conductor covering S. The resonant frequencies of these modes are those frequencies at which S, when covered by a perfect electric conductor, forms a resonant cavity. Hence, the solution to the E-field equation is not unique at any of these resonant frequencies.

At a resonant frequency the solution for $\underline{J}$ is not unique in that it contains an undetermined amount of the resonant mode current. However, the resonant mode current produces no field external to S, and therefore the field $\underline{E}^s$ external to S determined from the E-field equation theoretically should be unique. Unfortunately, when numerically computing the solution, the matrix representing L_E becomes ill-conditioned and the solution for $\underline{E}^s$ external to S may degenerate at and in the vicinity of a resonant frequency.

For the H-field equation (2), the solution will not be unique if $L_H(\underline{J}) = 0$ has nontrivial solution. In terms of differential equations this means that the solution will not be unique whenever there is a nontrivial solution to

$$\underline{\nabla} \times \underline{\nabla} \times \underline{H}^s(\underline{J}) = k^2 \underline{H}^s(\underline{J}) \qquad \text{within S} \qquad (12)$$

$$\underline{n} \times \underline{H}^s(\underline{J}) = 0 \qquad \text{just inside S} \qquad (13)$$

These are the equations for the modes of a resonator formed by a perfect magnetic conductor covering S. However, (12) and (13) are mathematically the same form as (10) and (11), and have the same resonant frequencies. Therefore the frequencies for which the solution to the H-field equation is not unique are the frequencies at which the solution to the E-field equation is not unique.

The current associated with the $\underline{H}^s$ of (12) and (13) is obtained from

$$\underline{J} = \underline{n} \times (\underline{H}^s_{ext} - \underline{H}^s_{int}) \qquad (14)$$

where $\underline{H}^s_{ext}$ denotes $\underline{H}^s$ just outside S and $\underline{H}^s_{int}$ denotes $\underline{H}^s$ just inside S. Since $\underline{n} \times \underline{H}^s_{int}$ is zero, $\underline{n} \times \underline{H}^s_{ext}$ is not zero and hence there will be a field $\underline{H}^s$ external to S at the resonant frequencies. Thus, in contrast to the E-field equation, the H-field equation theoretically does not have a unique solution for the field external to S at resonant frequencies.

We now show that the solution to the combined-field formulation (7) is unique whenever α is a positive real number. The solution to (7) is unique if

$$-\underline{n} \times \underline{H}^s(\underline{J}) - \frac{\alpha}{\eta} \underline{E}^s_{tan}(\underline{J}) = 0 \qquad \text{just inside S} \qquad (15)$$

implies that $\underline{J} = 0$. The complex power flow P associated with $\underline{E}^s_{tan}(\underline{J})$, $\underline{H}^s_{tan}(\underline{J})$ into the region interior to S is given by

$$P = -\iint_S (\underline{E}^s(\underline{J}) \times \underline{H}^{s*}(\underline{J})) \cdot \underline{n} \, ds \qquad (16)$$

In view of (15)

$$P = -\frac{\alpha^*}{\eta} \iint_S |\underline{E}_{tan}(\underline{J})|^2 ds = -\frac{\eta}{\alpha} \iint_S |\underline{H}_{tan}(\underline{J})|^2 ds \qquad (17)$$

Because there are no sources inside S,

$$\text{Re}(P) \geq 0 \qquad (18)$$

If the real part of α is positive, then comparison of (17) with (18) gives

$$\underline{H}^s_{tan}(\underline{J}) = 0 \quad \text{just inside S} \tag{19}$$

$$\underline{E}^s_{tan}(\underline{J}) = 0 \quad \text{on S} \tag{20}$$

Since the region external to S has no resonances, (18) implies that $\underline{H}^s_{tan}(\underline{J}) = 0$ just outside S. Using this fact and (19) in (14) shows that $\underline{J} = 0$, so that the solution to the combined-field equation is unique at all frequencies.

That the combined-source formulation leads to a unique solution when the real part of α is positive can be shown as follows. The solution to (8) is unique if

$$\underline{E}_{tan}(\underline{J},\underline{M}) = 0 \quad \text{just outside S} \tag{21}$$

implies that $\underline{J} = \underline{M} = 0$. If (21) is true, then $(\underline{J},\underline{M})$ radiates an electromagnetic field $(\underline{E},\underline{H})$ which satisfies

$$\left. \begin{array}{l} \underline{n} \times \underline{H} = -\underline{J} \\ \\ \underline{E} \times \underline{n} = -\underline{M} \end{array} \right\} \text{just inside S} \tag{22}$$

The first of equations (22) follows from the fact that, since the region external to S has no resonances, (19) implies that $\underline{n} \times \underline{H}$ is zero just outside S. The complex power flow P associated with $(\underline{E},\underline{H})$ into the region interior to S is given by

$$P = -\iint_S \underline{E} \times \underline{H}^* \cdot \underline{n} \, dS \tag{23}$$

In view of (9), substitution of (22) into (23) gives

$$P = -\alpha \iint_S |\underline{J}|^2 ds = -\frac{1}{\alpha^*} \iint_S |\underline{M}|^2 ds \tag{24}$$

Because there are no sources inside S,

$$\text{Real}(P) \geq 0 \tag{25}$$

If the real part of α is positive, (24) and (25) require that $\underline{J} = \underline{M} = 0$ so that the solution to the combined-source formulation is unique at all frequencies.

7. REPRESENTATIVE COMPUTATIONS FOR CONDUCTING BODIES

Computations have been made for several different bodies of revolution using the general computer programs [12-16]. Graphs of

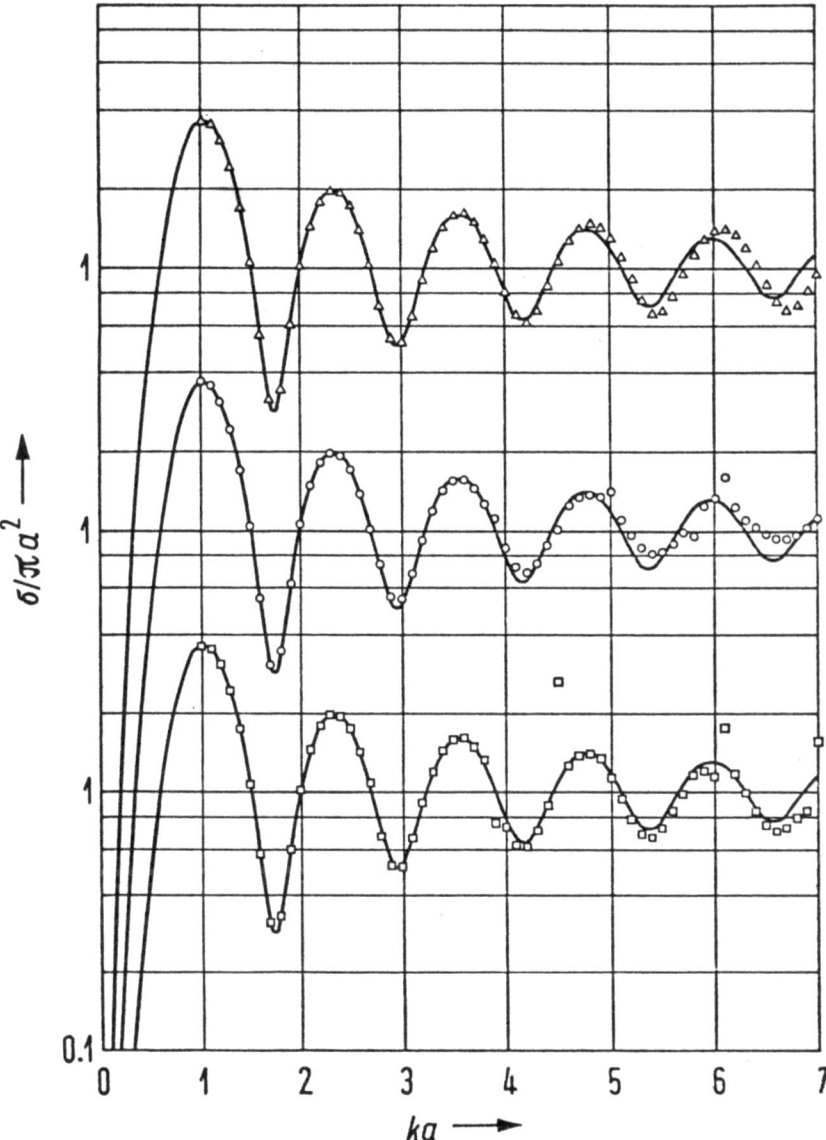

Fig. 2. Normalized radar cross section $\sigma/\pi a^2$ versus ka for a conducting sphere of radius a. Solid line denotes exact solution, squares denote H-field solution, circles denote E-field solution, and triangles denote combined field solution.

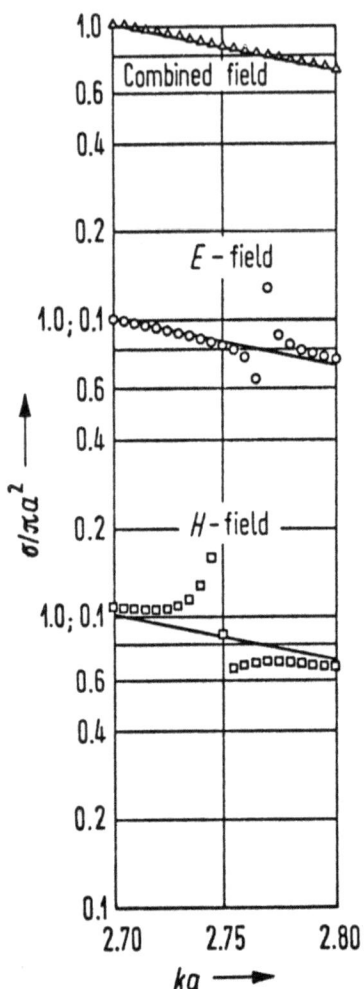

Fig. 3. Normalized radar cross section vs. ka for a conducting sphere in the vicinity of first resonance (ka ≈ 2.744).

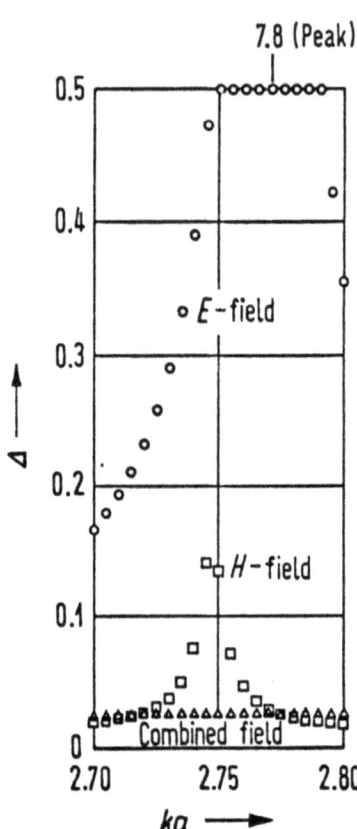

Fig. 4. Electric current error Δ vs. ka for a conducting sphere in the vicinity of first resonance (ka ≈ 2.744).

computed values of electric current error and normalized radar cross section for spheres, closed cylinders, and flat-back cones using the H-field, E-field, and combined-field formulations can be found in [17]. Graphs for the current on a sphere and on a cone-sphere excited by a plane wave, and for the current and radiation pattern of a slot in a sphere and a finite cylinder, can be found in [15]. We here give only a few of these results to illustrate the characteristics of such solutions.

Figure 2 shows the normalized radar cross section vs. ka for a conducting sphere of radius a as computed by the H-field, E-field, and combined-field formulations. The results using the combined-source formulation are the same as those using the combined-field formulation. The exact solution is shown by the solid curve. Except for a few points on the H-field and E-field curves, the computations compare well with the exact solution. However, if we expand the ka scale in the vicinity of the first spherical cavity resonant frequency (ka $\approx$ 2.744) we find that both the H-field and the E-field solutions are in error. This is shown in Fig. 3 for the normalized radar cross section, and in Fig. 4 for the electric current error. Note that the H-field and E-field solutions are in error over only a very small bandwidth. However, for an arbitrary body, the eigenfrequencies are unknown, and hence the places at which the H-field and E-field solutions are in error are unknown. Also, as the body becomes larger, the resonant frequencies become numerous and close together, and the H-field and E-field solutions become undependable. The combined-field and combined-source solutions continue to remain valid at all frequencies.

For aperture radiation problems, one usually specifies the tangential electric field in the aperture and takes it to be zero over the conducting part of the surface. Hence, to use this information directly, we must use either the E-field solution or the combined-source solution. The H-field solution and the combined-field solution could be used if the original problem was replaced by the equivalent external problem obtained by closing the aperture with a perfectly conducting surface and placing a sheet of magnetic current on this surface.

Figure 5 shows (by ×'s) the combined-source solution for the radiation gain pattern of a rotationally symmetric aperture at the equator of a conducting sphere at the first internal resonance (ka $\approx$ 2.75). The aperture is excited by a uniform electric field across it, which is roughly equivalent to a voltage applied across it. The exact solution is shown solid for comparison. Figure 6 shows (by ×'s) the radiation gain pattern for the same aperture as calculated by the E-field formulation. As predicted by theory, the solution is greatly in error, as compared to the exact solution shown solid.

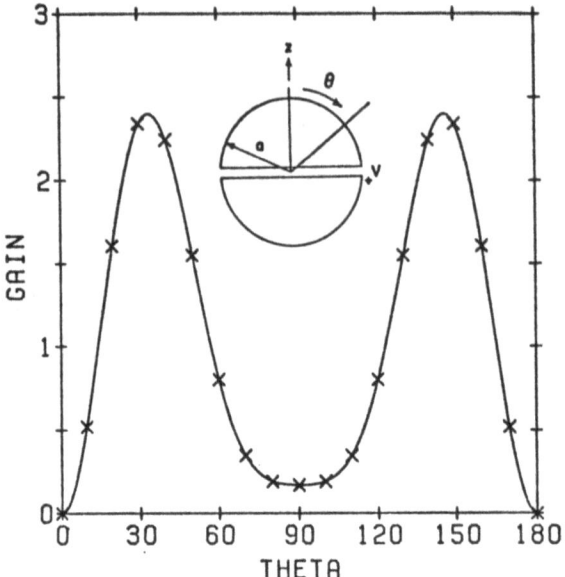

Fig. 5. Radiation gain pattern of an aperture at the equator of a conducting sphere, ka = 2.75. Combined-source solution shown by ×'s, exact solution shown solid.

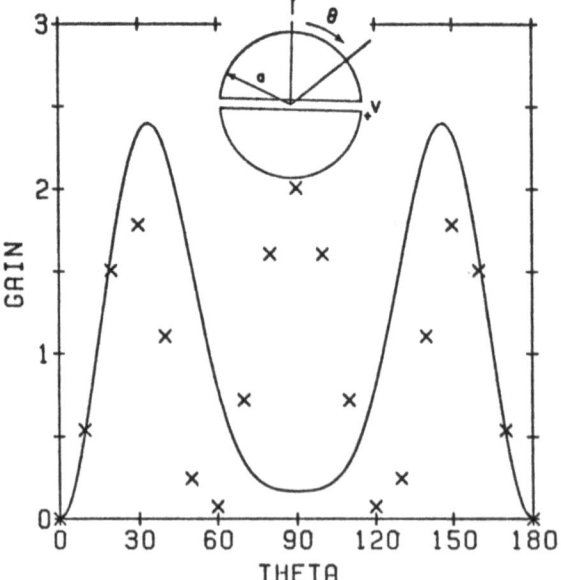

Fig. 6. Radiation gain pattern of an aperture at the equator of a conducting sphere, ka = 2.75. E-field solution shown by ×'s, exact solution shown solid.

8. FORMULATION FOR DIELECTRIC BODIES

We here give only a short development of the integral equations for scattering by dielectric and magnetic bodies. A more complete derivation can be found in [14] and [22].

The problem to be considered is that of an electromagnetic field propagating in a homogeneous medium of permeability μ_e and permittivity ε_e incident on the surface S of a homogeneous obstacle of permeability μ_d and permittivity ε_d. The subscript e denotes exterior medium and the subscript d denotes diffracting medium. We wish to calculate the scattered electromagnetic field $\underline{E}^s$, $\underline{H}^s$ outside S and the diffracted electromagnetic field $\underline{E}$, $\underline{H}$ inside S in terms of the electromagnetic field $\underline{E}^i$, $\underline{H}^i$ which would exist on S in the absence of the obstacle. This original problem is shown in Fig. 7 where $\underline{J}^i$, $\underline{M}^i$ are the electric and magnetic sources of $\underline{E}^i$, $\underline{H}^i$ and $\underline{n}$ is the unit normal vector which points outward from S.

The equivalence principle is used to piece together an outside situation consisting of medium μ_e, ε_e and field $\underline{E}^s$, $\underline{H}^s$ outside S and an inside situation consisting of medium μ_e, ε_e and field $-\underline{E}^i$, $-\underline{H}^i$ inside S. This composite situation is shown in Fig. 8. Since $\underline{E}^s$, $\underline{H}^s$ is source-free outside S and $\underline{E}^i$, $\underline{H}^i$ is source-free inside S, the only sources in Fig. 8 are the equivalent electric surface current $\underline{J}$ and the equivalent magnetic surface current $\underline{M}$ on S.

As a second application of the equivalence principle, we combine an outside situation consisting of medium μ_d, ε_d and zero field with an inside situation consisting of medium μ_d, ε_d and field $\underline{E}$, $\underline{H}$. This combination of situations is shown in Fig. 9. Since $\underline{E}$, $\underline{H}$ is source-free inside S, the only sources in Fig. 9 are the equivalent electric surface current $-\underline{J}$ and the equivalent magnetic surface current $-\underline{M}$ on S. By expressing the surface currents in Figs. 8 and 9 in terms of the discontinuities of tangential fields across them and by using

$$\underline{J} = \underline{n} \times (\underline{H}_e - \underline{H}_d) \tag{26}$$

$$\underline{M} = (\underline{E}_e - \underline{E}_d) \times \underline{n} \tag{27}$$

to express the surface currents in terms of the discontinuities of the tangential fields across S and by using

$$\underline{n} \times \underline{E} = \underline{n} \times (\underline{E}^s + \underline{E}^i) \tag{28}$$

$$\underline{n} \times \underline{H} = \underline{n} \times (\underline{H}^s + \underline{H}^i) \tag{29}$$

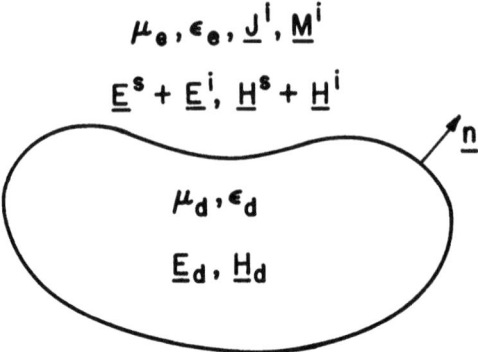

Fig. 7. Original problem.

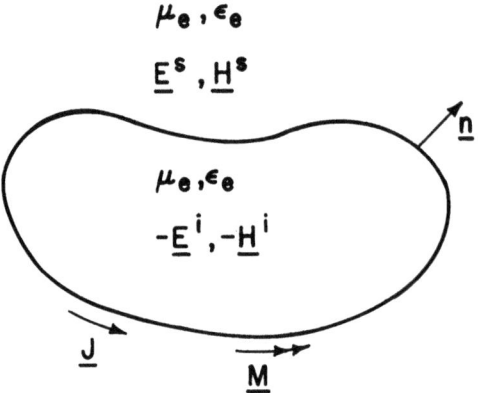

Fig. 8. Outside equivalence.

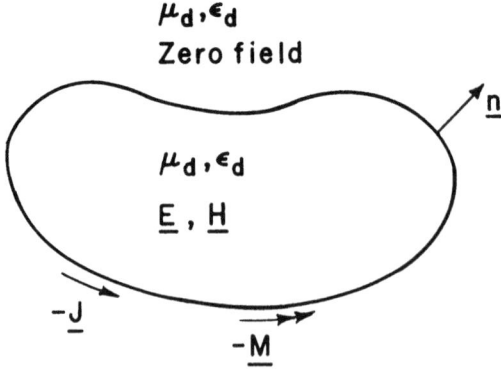

Fig. 9. Inside equivalence.

on S, the interested reader can verify that the surface currents in Fig. 9 are indeed the negatives of those in Fig. 8. Equations (28) and (29) are the boundary conditions that the tangential components of the fields in the original problem as shown in Fig. 7 are continuous across S.

The scattered field $\underline{E}^s$, $\underline{H}^s$ outside S and the diffracted field $\underline{E}$, $\underline{H}$ inside S could easily be calculated if $\underline{J}$ and $\underline{M}$ were known because the medium into which $\underline{J}$ and $\underline{M}$ radiate is homogeneous in Figs. 8 and 9. We have to determine $\underline{J}$ and $\underline{M}$. The equivalence principle states that there exist $\underline{J}$ and $\underline{M}$ which radiate the fields in Figs. 8 and 9, but the equivalence principle does not tell what $\underline{J}$ and $\underline{M}$ are. The equivalence principle does state that

$$\underline{J} = \underline{n} \times \underline{H} \tag{30}$$

$$\underline{M} = \underline{E} \times \underline{n} \tag{31}$$

but this is not very useful because $\underline{E}$ and $\underline{H}$ are unknown.

From Figs. 8 and 9,

$$-\underline{n} \times \underline{E}_e^- = \underline{n} \times \underline{E}^i \tag{32}$$

$$-\underline{n} \times \underline{H}_e^- = \underline{n} \times \underline{H}^i \tag{33}$$

$$-\underline{n} \times \underline{E}_d^+ = 0 \tag{34}$$

$$-\underline{n} \times \underline{H}_d^+ = 0 \tag{35}$$

where
- $\underline{E}_e^-$ is the electric field just inside S due to $\underline{J},\underline{M}$, radiating in μ_e, ε_e
- $\underline{H}_e^-$ is the magnetic field just inside S due to $\underline{J},\underline{M}$, radiating in μ_e, ε_e
- $\underline{E}_d^+$ is the electric field just outside S due to $\underline{J},\underline{M}$, radiating in μ_d, ε_d
- $\underline{H}_d^+$ is the magnetic field just outside S due to $\underline{J},\underline{M}$, radiating in μ_d, ε_d

The equivalent currents $\underline{J}$, $\underline{M}$ which appear in Figs. 8 and 9 satisfy (32) to (35) because (32) to (35) were obtained from Figs. 8 and 9. It is shown in [14] and [22] that the solution to (32) to (35) is unique. Therefore, these equations uniquely determine the equivalent currents $\underline{J}$, $\underline{M}$ of Figs. 8 and 9.

Equations (32) to (35) form a set of four equations in the two unknowns $\underline{J}$ and $\underline{M}$. The usual methods of equation solving apply only when the number of equations is equal to the number of unknowns. We want to reduce this set of four equations to two equations. One way to do this is to form the linear combination

$$-\underline{n} \times (\underline{E}_e^- + \alpha \underline{E}_d^+) = \underline{n} \times \underline{E}^i \qquad (36)$$

of (32) and (34) and the linear combination

$$-\underline{n} \times (\underline{H}_e^- + \beta \underline{H}_d^+) = \underline{n} \times \underline{H}^i \qquad (37)$$

of (33) and (35) where α and β are complex constants. It is shown in [14] and [22] that the solution to (36) and (37) is unique if $\alpha \beta^*$ is real and positive.

If $\alpha = \beta = 1$, then (36) and (37) become

$$-\underline{n} \times (\underline{E}_e^- + \underline{E}_d^+) = \underline{n} \times \underline{E}^i \qquad (38)$$

$$-\underline{n} \times (\underline{H}_e^- + \underline{H}_d^+) = \underline{n} \times \underline{H}^i \qquad (39)$$

The set of equations (38) and (39) is the coupled pair of surface integral equations described by Poggio and Miller [1]. We call these equations the natural equations for the problem because they can be obtained directly from the equivalence principle by requiring only the continuity of tangential $\underline{E}$ and $\underline{H}$ across S. The equations have been obtained independently by several different investigators [1,6].

If

$$\alpha = -\frac{\varepsilon_d}{\varepsilon_e} \qquad (40)$$

$$\beta = -\frac{\mu_d}{\mu_e} \qquad (41)$$

then (36) and (37) becomes

$$-\underline{n} \times (\underline{E}_e^- - \frac{\varepsilon_d}{\varepsilon_e} \underline{E}_d^+) = \underline{n} \times \underline{E}^i \qquad (42)$$

$$-\underline{n} \times (\underline{H}_e^- - \frac{\mu_d}{\mu_e} \underline{H}_d^+) = \underline{n} \times \underline{H}^i \qquad (43)$$

The set of equations (42) and (43) is the coupled pair of surface integral equations obtained by Müller [8]. The singularity that the kernels of the integral equations (42) and (43) exhibit as the source point passes through the field point is not as pronounced as the singularity of the kernels of (38) and (39). It is shown in [14] and [22] that the solution to (42) and (43) is unique even if the product $\alpha \beta^*$ implied by (40) and (41) is complex.

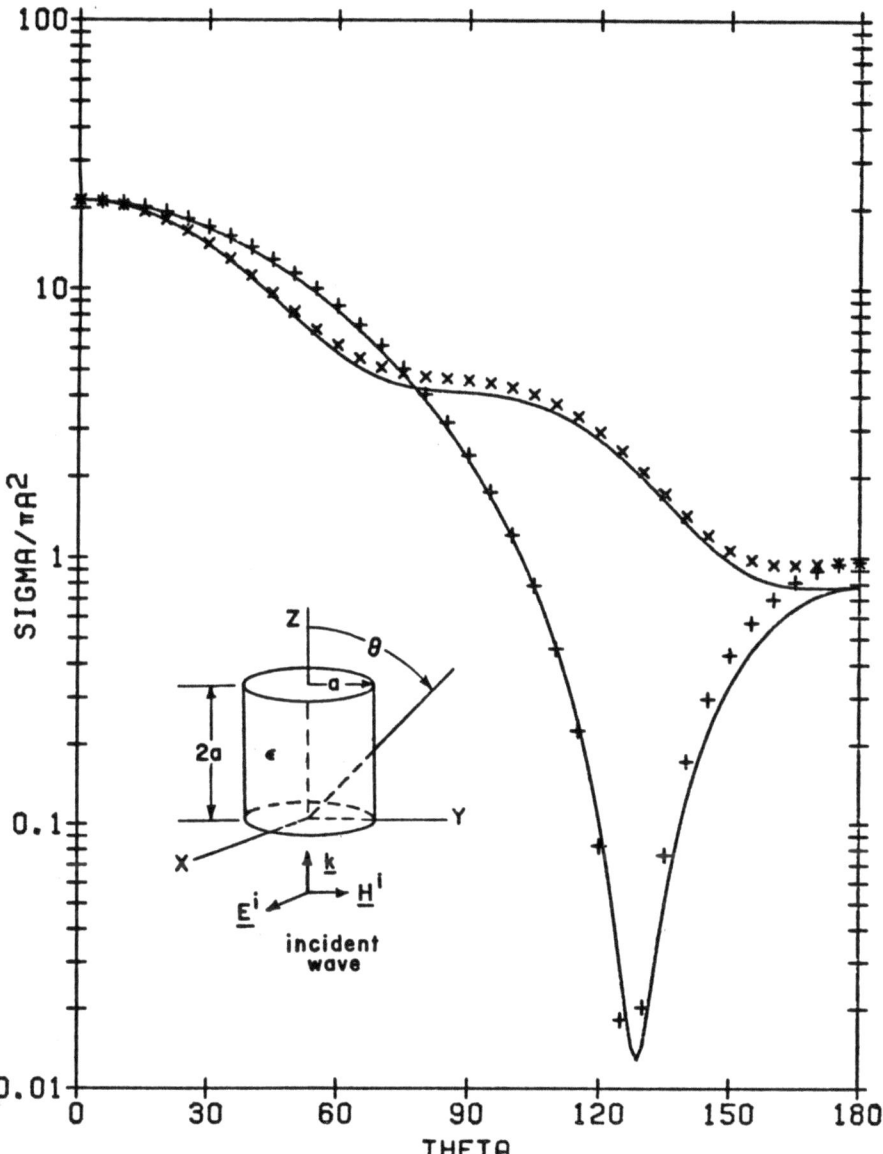

Fig. 10. Normalized bistatic radar cross section $\sigma/\pi a^2$ for a dielectric cylinder of radius a, height 2a, $\varepsilon_r = 4$, where $a = \lambda_0/4$. Cross section calculated from the natural formulation shown in the E-plane by ×'s, in the H-plane by +'s. Solid curves are calculated from the Müller formulation.

10. REPRESENTATIVE COMPUTATIONS FOR DIELECTRIC BODIES

A computer program for a method of moments solution to either (38) and (39) or to (42) and (43) for material bodies of revolution is given in [14]. Computations for the equivalent electric and magnetic currents of dielectric spheres are graphed and compared to the exact solutions in [14]. Computations for the plane wave scattering patterns (bistatic radar cross sections) are given for dielectric spheres and finite cylinders in [14] and [22]. We here give one such graph to illustrate typical computations.

Figure 10 shows the plane wave scattering pattern for a finite dielectric cylinder of radius 0.25 λ_o and height 0.5 λ_o, where λ_o is the free-space wavelength. The relative dielectric constant of the cylinder is 4. The excitation is an axially incident plane wave (incident on the flat end of the cylinder). The normalized bistatic radar cross section in the E-plane (shown by ×'s) and in the H-plane (shown by +'s) is shown versus θ, the angle measured from the cylinder axis. $\theta = 0$ in the forward scattering direction. The +'s and ×'s show the solution to the natural formulation; the solid curves show the solution to Müller's formulation.

REFERENCES

1. A. J. Poggio and E. K. Miller, "Integral Equation Solutions of Three-dimensional Scattering Problems," Chap. 4 of <u>Computer Techniques for Electromagnetics</u>, edited by R. Mittra, Pergamon Press, 1973.

2. F. K. Oshiro and K. M. Mitzner, "Digital Computer Solution of Three-dimensional Scattering Problems," <u>Symposium Digest</u>, IEEE International Symposium on Antennas and Propagation, Ann Arbor, Michigan, October 1967, pp. 257-263.

3. P.L.E. Uslenghi, "Computation of Surface Currents on Bodies of Revolution," <u>Alta Frequenza</u>, vol. 39, No. 8, 1970, pp. 1-12.

4. J. R. Mautz and R. F. Harrington, "Radiation and Scattering from Bodies of Revolution," <u>Appl. Sci. Res.</u>, vol. 20, June 1969, pp. 405-435.

5. F. K. Oshiro, K. M. Mitzner, and S. S. Locus et al., "Calculation of Radar Cross Section," Air Force Avionics Laboratory Tech. Rept. AFAL-TR-70-21, Part II, April 1970.

6. Yu Chang and R.F. Harrington, "A Surface Formulation for Characteristic Modes of Material Bodies," Report TR-74-7, Dept. of Electrical and Computer Engineering, Syracuse University, Syracuse, N.Y., October 1974.

7. T. K. Wu, "Electromagnetic Scattering from Arbitrarily-Shaped Lossy Dielectric Bodies," Ph.D. Dissertation, University of Mississippi, 1976.

8. C. Müller, Foundations of the Mathematical Theory of Electromagnetic Waves, Springer-Verlag, 1969, p. 301, Equations (40)-(41). (There are some sign errors in these equations.)

9. V. V. Solodukhov and E. N. Vasil'ev, "Diffraction of a Plane Electromagnetic Wave by a Dielectric Cylinder of Arbitrary Cross Section," Soviet Physics - Technical Physics, vol. 15, No. 1, July 1970, pp. 32-36.

10. N. Morita, "Analysis of Scattering by a Dielectric Rectangular Cylinder by Means of Integral Equation Formulation," Electronics and Communications in Japan, vol. 57-B, No. 10, October 1974, pp. 72-80.

11. E. N. Vasil'ev and L. B. Materikova, "Excitation of Dielectric Bodies of Revolution," Soviet Physics - Technical Physics, vol. 10, No. 10, April 1966, pp. 1401-1406.

12. J. R. Mautz and R. F. Harrington, "H-Field, E-Field, and Combined Field Solutions for Bodies of Revolution," Interim Technical Report RADC-TR-77-109, Rome Air Development Center, Griffiss Air Force Base, New York, March 1977.

13. J. R. Mautz and R. F. Harrington, "Computer Programs for H-Field, E-Field, and Combined Field Solutions for Bodies of Revolution," Interim Technical Report RADC-TR-77-215, Rome Air Development Center, Griffiss Air Force Base, New York, June 1977.

14. J. R. Mautz and R. F. Harrington, "Electromagnetic Scattering from a Homogeneous Body of Revolution," Technical Report TR-77-10, Dept. of Electrical and Computer Engineering, Syracuse University, Syracuse, N.Y. 13210, November 1977.

15. J. R. Mautz and R. F. Harrington, "A Combined-source Solution for Radiation and Scattering from a Perfectly Conducting Body," Technical Report TR-78-3, Dept. of Electrical and Computer Engineering, Syracuse University, Syracuse, N.Y. 13210, April 1978.

16. J. R. Mautz and R. F. Harrington, "Application of the Combined-source Solution to a Conducting Body of Revolution," Technical Report TR-78-6, Dept. of Electrical and Computer Engineering, Syracuse University, Syracuse, N. Y. 13210, June 1978.

17. J. R. Mautz and R. F. Harrington, "H-field, E-field, and Combined-field Solutions for Conducting Bodies of Revolution," AEÜ (Germany), vol. 32, pp. 157-164, April 1978.

18. R. F. Harrington, Field Computation by Moment Methods, Macmillan Co., New York, 1968.

19. H. Brakhage and P. Werner, "Über das Dirichletsche Aussenraumproblem für die Helmholtzsche Schwingungsgleichung," Archiv d. Math. vol. 16, pp. 325-329, 1965.

20. D. Greenspan and P. Werner, "A Numerical Method for the Exterior Dirichlet Problem for the Reduced Wave Equation," Arch. Rational Mech. Anal. vol. 23, pp. 228-316, 1966.

21. J. Bolomey and W. Tabbara, "Numerical Aspects on Coupling Between Complementary Boundary Value Problems," IEEE Trans. Antennas Propagat., vol. AP-21, pp. 356-363, May 1973.

22. J. R. Mautz and R. F. Harrington, "Electromagnetic Scattering from a Homogeneous Material Body of Refolution," AEÜ (Germany), vol. 33, in press, 1979.

APERTURE PROBLEMS

J. Van Bladel★ and C.M. Butler★★

★ Laboratorium voor Elektromagnetisme en Acustica
  Rijksuniversiteit-Gent, Ghent, B9000, Belgium
★★ Department of Electrical Engineering
  University of Mississippi, University, MS38677,USA.

1. APERTURES IN A PLANE METALLIC SCREEN

1.1 General integro-differential equations

(a) outline of the method

    This section provides a foundation for the study of the fundamental electromagnetic boundary value problems involving apertures or holes in conducting planar surfaces. The results are general in that they are also relevant for small apertures in curved surfaces discussed in Sec. 4.

    The fundamental problem to be considered here is that of the electromagnetic interaction of specified impressed sources ($\bar{J}^i, \bar{M}^i$) (or incident field ($\bar{E}^i, \bar{H}^i$)) and a planar conducting screen having a hole (aperture) cut in it. As shown in Fig. 1, the screen is in a homogeneous medium of infinite extent that is electrically characterized by (μ, ε). If desired, the medium may be lossy, which condition may be accounted for by replacing ε by $(\varepsilon - j\frac{\sigma}{\omega})$. The screen is located, for convenience, in the xy-plane of a Cartesian coordinate system.

    To facilitate the analysis which follows, the planar screen is assumed to be perfectly conducting, vanishing thin, and of infinite extent. As usual, the sources shown in Fig. 1 vary harmonically in time according to $e^{j\omega t}$, which factor is suppressed in all subsequent equations.

    The tangential component of the total electric field must be zero on the screen (on both sides), and both the electric and magnetic fields must be continuous along any path passing through

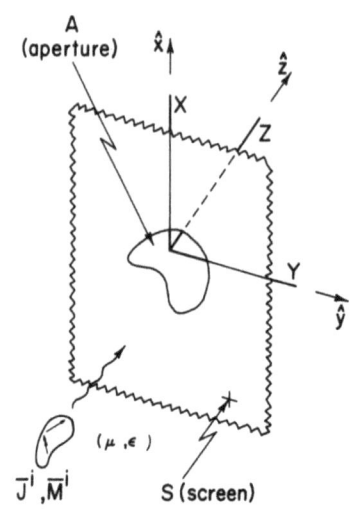

Fig. 1
Aperture in conducting screen

the aperture. We identify the transverse aperture electric field $\overline{E}_t^a$, the component of electric field in the aperture parallel to the screen, as the unknown to be determined. Once $\overline{E}_t^a$ is known, the fields are uniquely determined (taking the radiation condition into account), and they can be computed by suitable mathematical processes. A functional equation for $\overline{E}_t^a$ can be obtained by deriving expressions for the magnetic field based upon electromagnetic potentials, which ensures us that Maxwell's equations and the radiation condition are satisfied in both half spaces, and, also, upon image theory which ensures us that the boundary conditions on the screen itself (both sides) are satisfied.

Equating, in the aperture, the transverse component of magnetic fields, calculated from considerations of each half space individually, leads to the desired equations.

(b) the magnetic field

From the equivalence principle [1], we know that we can short the aperture, i.e., make the screen continuous, and place appropriate magnetic surface currents over the shorted aperture on both sides of the continuous screen without altering the fields in the two half spaces. The magnetic currents must have values, $\overline{E}_t^a \times (-\hat{z})$ on the illuminated side of the shorted screen in A and $\overline{E}_t^a \times \hat{z}$ on the shadow side. Thus, the two needed magnetic currents are equal in complex magnitude and are oppositely directed, as suggested schematically in Figs. 2 and 3.

Here we interject that the introduction of equivalent magnetic current is a convenience and is not necessary. The entire formulation could be based upon $\overline{E}_t^a$ alone, but the concept of equivalent magnetic current is helpful as an aid to one's thinking through the derivations and, thus, is deemed worthwhile.

To compute the magnetic field in the illuminated region (z<0), consider the equivalent problems shown in Fig. 2. In Fig. 2b the aperture is shorted and, over the region A on the side of the continuous conducting screen facing the left half space, we place an equivalent surface magnetic current of density $\overline{M}_s = \hat{z} \times \overline{E}_t^a$. The screen can be removed provided image sources

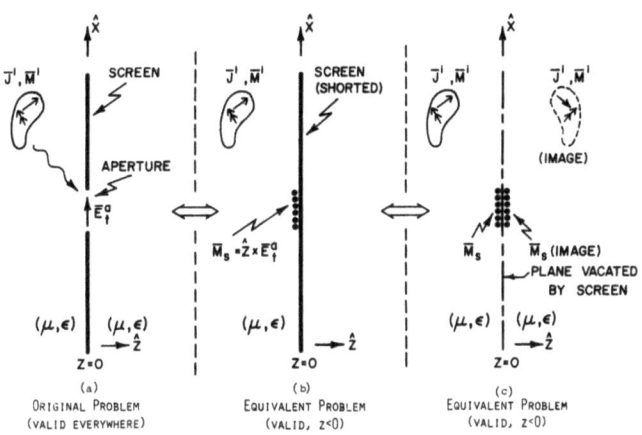

Fig. 2   Illuminated-side equivalences

are introduced to ensure that the tangential electric field remains zero in the xy plane. This leads to the equivalent problem of Fig. 2c. Now, since all sources radiate in an infinite homogeneous space - with no boundaries, one can determine fields from expressions involving only the simple open space Green's function. Using the electric vector potential $\bar{F}$, we can write the total magnetic field in the left half space $\bar{H}^-$ as

$$\bar{H}^-(\bar{r}) = \bar{H}^{sc}(\bar{r}) - j\frac{\omega}{k^2}\left[k^2\bar{F}(\bar{r}) + \nabla(\nabla\cdot\bar{F}(\bar{r}))\right], \quad z<0 \qquad (1)$$

where k is $2\pi/\lambda$. $\bar{H}^{sc}$ is called the short-circuit magnetic field and is that field which would exist in the illuminated region, if the aperture were not present in the screen. This field can be calculated by elementary means. The potential $\bar{F}$ is given by

$$\bar{F}(\bar{r}) = \frac{\varepsilon}{4\pi} \iint_A 2\bar{M}_s(\bar{r}')G(\bar{r},\bar{r}')ds' \qquad (2)$$

where

$$G(\bar{r},\bar{r}') = \frac{e^{-jk|\bar{r}-\bar{r}'|}}{|\bar{r}-\bar{r}'|} \qquad (3)$$

with

$$\bar{r} = x\hat{x} + y\hat{y} + z\hat{z} \qquad (4a)$$

and

$$\bar{r}' = x'\hat{x} + y'\hat{y} \quad , \quad (x',y') \epsilon A \qquad (4b)$$

In the shadow region, similar calculations lead to the equivalent problem depicted in Fig. 3. The magnetic field for z>0 is

$$\vec{H}^{+}(\bar{r}) = +j\frac{\omega}{k^2}(k^2\bar{F}(\bar{r}) + \nabla(\nabla \cdot \bar{F}(\bar{r}))) \quad , \quad z>0 \qquad (5)$$

where $\bar{F}$ is given by (2). The leading positive sign on the right side of (5) accounts for the negative sign on M_s of the shadow-side equivalence.

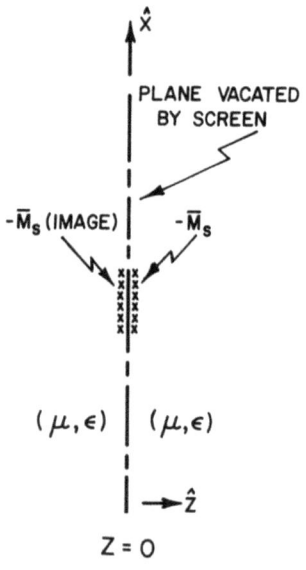

Fig. 3
Shadow-side equivalence
(valid, z>0)

(c) continuity conditions and final equation

As shown in Figs. 2 and 3, the electric field $\vec{E}_t$ created by the magnetic currents is automatically continuous. The requirement that the magnetic field be continuous through A is achieved by enforcing the following:

$$\lim_{z\uparrow 0}(\vec{H}^{-}(\bar{r}) \times \hat{z}) = \lim_{z\downarrow 0}(\vec{H}^{+}(\bar{r}) \times \hat{z}), \quad (x,y)\epsilon A \qquad (6)$$

In view of (1) and (5), (6) becomes

$$j2\frac{\omega}{k^2}(\nabla_t\nabla_t \cdot \bar{F} + k^2\bar{F}) \times \hat{z} = \bar{H}^{sc} \times \hat{z} \qquad (7)$$

where ∇_t is the transverse Laplacian operator. Since, on the screen, $\bar{H}^{sc} \times \hat{z} = 2\bar{H}^i \times \hat{z}$, one may choose to view the incident field as the forcing function of the equation and convert (7) to

$$j\frac{\omega}{k^2}(\nabla_t\nabla_t \cdot \bar{F} + k^2\bar{F}) \times \hat{z} = \bar{H}^i \times \hat{z}, \quad \text{in A} \qquad (8)$$

The simple relationship above between incident and short-circuit magnetic field holds only in the case of a planar screen of infinite extent, whereas Eq. (7) is valid for curved surfaces if an appropriate Green's function is used in place of that given in (3). In this sense, $\bar{H}^{sc}$ can be looked upon as a more fundamental forcing function than $\bar{H}^i$.

Since (7) is an integro-differential equation, additional information must be available before it can be solved for $\bar{M}_s$. This information is provided by the edge condition [2] which tells us that the component of $\bar{M}_s$ normal to the aperture/screen edge must approach zero as the square root of the distance from the edge to the point of observation, while the component tangential to the edge must approach infinity as the reciprocal of the square root of this distance. With this boundary condition available, (7) can, in principle, be solved for $\bar{M}_s$ (or, equivalently, for $\bar{E}_t^a$).

Since $\bar{M}_s$ is in the xy-plane, it has no z-component, from which it follows that $\bar{F}$ has no z-component. Thus, (7) embodies two scalar, coupled integro-differential equations with the two transverse (to z) components of $\bar{M}_s$, or, equivalently, of $\bar{E}_t^a$, as the unknown quantities.

In Cartesian coordinates, (7) takes on the convenient form

$$j2\frac{\omega}{k^2}\left[\left(\frac{\partial^2}{\partial x^2} + k^2\right)F_x + \frac{\partial^2}{\partial x \partial y}F_y\right] = H_x^{sc} \qquad (9)$$

$$(x,y) \ A$$

$$j2\frac{\omega}{k^2}\left[\left(\frac{\partial^2}{\partial y^2} + k^2\right)F_y + \frac{\partial^2}{\partial y \partial x}F_x\right] = H_y^{sc} \qquad (10)$$

where the inhomogeneous terms are the components of the short-circuit magnetic field evaluated in A and where F_x and F_y are the components of $\bar{F}$ at z=0 with $(x,y) \in A$:

$$F_{x \atop y}(x,y) = \frac{\varepsilon}{2\pi}\iint_A M_{s_{x \atop y}}(x',y')\frac{e^{-jkR}}{R}dx'dy' \qquad (11)$$

with

$$R = [(x-x')^2 + (y-y')^2]^{1/2} \qquad (12)$$

Once $\bar{M}_s$ is known, $\bar{H}$ can be determined from (1) and (5), and $\bar{E}$ from

$$\bar{E}^-(\bar{r}) = \bar{E}^{sc}(\bar{r}) - \frac{1}{\varepsilon}\nabla \times \bar{F}(\bar{r}) \qquad z<0 \qquad (13)$$

and

$$\bar{E}^+(\bar{r}) = \frac{1}{\varepsilon}\nabla \times \bar{F}(\bar{r}) \qquad z>0 \qquad (14)$$

where, of course, $\overline{E}^{sc}(\overline{r})$ is the short-circuit electric field in the illuminated region. It is to be noticed that continuity of the normal component of $\overline{E}$ requires that

$$(\overline{E}^{sc} - \frac{1}{\varepsilon} \nabla \times \overline{F}) \cdot \hat{z} \Big|_{z=0^-} = (\frac{1}{\varepsilon} \nabla \times \overline{F}) \cdot \hat{z} \Big|_{z=0^+} , \quad (x,y) \varepsilon A \quad (15)$$

or

$$\overline{E}^{sc} \cdot \hat{z} = (\frac{2}{\varepsilon} \nabla \times \overline{F}) \cdot \hat{z} = 2 \overline{E}^+ \cdot \hat{z} \quad \text{at} \quad z = 0 \text{ in } A \quad (16)$$

Since $\overline{E}^{sc} \cdot \hat{z} = 2 \overline{E}^i \cdot \hat{z}$ at $z = 0$, the above reduces to

$$\overline{E}^+ \cdot \hat{z} = \overline{E}^i \cdot \hat{z} \quad \text{at} \quad z = 0 \text{ in } A . \quad (17)$$

Since $\overline{E}^+ \cdot \hat{z}$ is the z component of the total electric field in A, the last equality says that the total normal component of electric field in A is equal to $\overline{E}^i_z$ there. Similarly, one can show that the transverse component of the total magnetic field in A is equal to the transverse component of $\overline{H}^i$ there.

$$\overline{H}^+ \times \hat{z} = \overline{H}^i \times \hat{z} \quad \text{at} \quad z = 0 \text{ in } A \quad (18)$$

By elementary means, one can show that the scattered field $(\overline{E}^s, \overline{H}^s)$ exhibits the following symmetry properties:

$$\hat{z} \times \overline{E}^s(x,y,z) = \hat{z} \times \overline{E}^s(x,y,-z) \quad (19)$$

$$\hat{z} \cdot \overline{E}^s(x,y,z) = -\hat{z} \cdot \overline{E}^s(x,y,-z) \quad (20)$$

and

$$\hat{z} \times \overline{H}^s(x,y,z) = -\hat{z} \times \overline{H}^s(x,y,-z) \quad (21)$$

$$\hat{z} \cdot \overline{H}^s(x,y,z) = \hat{z} \cdot \overline{H}^s(x,y,-z) \quad (22)$$

The transverse component of the scattered electric field is symmetric (an even function of z) and its normal component is antisymmetric (an odd function). On the other hand, the transverse component of the scattered magnetic field is antisymmetric (an odd function) while its normal component is symmetric (even). The reader is cautioned to recall that the scattered field at a point is the total field less the incident field there. In the present aperture/screen problem, the scattered field is that radiated by the electric current and charge induced on the screen. We note that $(\overline{E}^+, \overline{H}^+)$ is the shadow-side field resulting from penetration through the aperture in the screen and that $(\overline{E}^- - \overline{E}^{sc}, \overline{H}^- - \overline{H}^{sc})$ is the illuminated-side field which can be attributed to the presence of the hole in the screen; from (1), (5), (13) and (14), it can be demonstrated readily that these field components exhibit the properties (19) - (22). That is, the transverse component of the electric field and normal component of the magnetic, both due to the presence of the hole in the screen, are symmetric with respect to the screen, while the normal component of the electric field and the transverse component of the magnetic field are antisymmetric with respect to the screen.

A final property of the aperture field, one which is of practical use, is that, for a sufficiently large aperture relative to the wavelength, the presence of the screen should have minimal effect on the field in regions of A remote from the aperture/screen boundary. In the portions of the aperture remote from the screen, the incident field is overwhelmingly larger than the field scattered from the screen to such a location. Hence, in a large aperture, the field remote from the screen edges is approximately equal to the incident field there. In fact, often one can assume that the aperture field is equal to the incident field over all of A in calculations involving a large aperture.

(d) Duality. Babinet's principle

If one compares (7) with the corresponding integro-differential equation appropriate for the problem of the flat plate scatterer (perfectly conducting), he finds them to be of the same form. Moreover, subject to the correspondences below, the above-mentioned equations are identical.

Aperture	Plate
$\frac{2}{\eta} \bar{M}_s$ (or $\frac{2}{\eta} \hat{z} \times \bar{E}_t^a$)	$\eta \bar{J}_s$
$\frac{1}{2} \bar{H}^{sc} \times \hat{z}$	$\bar{E}^i \times \hat{z}$

$$(\eta = \sqrt{\mu/\varepsilon})$$

(23)

In view of the duality of the plate and aperture problems, one can bring to bear his experience with the plate problem upon that of the aperture in a screen. In fact a solution of one applies for the other with proper interpretation of physical quantities. This remark allows us to make use of the store of solved "plate" problems and immediately translate them into the dual aperture configuration.

(e) extensions [3,4]

If an impressed source is located in the right as well as in the left half space, one can readily modify (7) to render it applicable to this case of impressed sources on both sides of the screen. Eq. (1) remains as is but $\bar{H}^{sc+}$, the right half space short-circuit magnetic field must be added to (5). Denoting the left half space short-circuit magnetic field by $\overline{H}^{sc-}$, one expresses the magnetic field in the two half spaces as

$$\vec{H}^{\pm} = \overline{H}^{sc\pm} \pm j \frac{\omega}{k^2}\left[k^2\overline{F} + \nabla\nabla\cdot\overline{F}\right] \qquad (24)$$

and, by enforcing the continuity condition (6), one finds that (7) becomes

$$j2\frac{\omega}{k^2}[\nabla_t\nabla_t\cdot\overline{F} + k^2\overline{F}]\times\hat{z} = [\overline{H}^{sc-} - \overline{H}^{sc+}]\times\hat{z}, \quad \overline{r}\epsilon A \qquad (25)$$

for the two-source problem.

Another interesting extension [4] of the aperture/screen problem discussed above is the case in which the two half spaces are filled with different media. In this situation, we write the two magnetic fields in the form

$$\vec{H}^{\pm} = \overline{H}^{sc\pm} - j\frac{\omega}{k_\pm^2}\left[k_\pm^2\overline{F}^{\pm} + \nabla\nabla\cdot\overline{F}^{\pm}\right] \qquad (26)$$

where $k_\pm^2 = \omega^2\mu_\pm\epsilon_\pm$, with the subscript identifying a quantity with a half space, and where the electric vector potential on the two sides of the screen is

$$\overline{F}^{\pm}(\overline{r}) = \mp\frac{\epsilon_\pm}{4\pi}\iint_A 2\overline{M}_s(\overline{r}')\frac{e^{-jk_\pm|\overline{r}-\overline{r}'|}}{|\overline{r}-\overline{r}'|}dS'. \qquad (27)$$

The transverse component of the magnetic field is constrained to be continuous through A (Eq. (6)), which leads to the integro-differential equation for the two-media aperture/screen problem :

$$\left\{j\frac{\omega}{k_-^2}[\nabla_t\nabla_t\cdot\overline{F}^- + k_-^2\overline{F}^-] - j\frac{\omega}{k_+^2}[\nabla_t\nabla_t\cdot\overline{F}^+ + k_+^2\overline{F}^+]\right\}\times\hat{z}$$

$$= [\overline{H}^{sc-} - \overline{H}^{sc+}]\times\hat{z}, \quad \text{in A.} \qquad (28)$$

1.2 Low frequency appproximation. Zero order terms.

(a) Helmholtz theorem

Equations (2) and (8) show that the aperture problem leads to an integro-differential equation, the unknown of which is the two-dimensional vector function $\overline{M}_s$. The x and y components of

$\overline{M}_s$ satisfy (9) to (11) and, in addition, the boundary condition

$$\hat{n} \cdot \overline{M}_s = 0 \quad \text{on (c)} \tag{29}$$

which is a statement of the fact that the electric field in the aperture is perpendicular to the metallic edge (c). Instead of M_{sx} and M_{sy}, we can choose two other scalar functions as unknowns of our problem, namely the scalar potentials Γ and Θ which appear in the two-dimensional Helmholtz "splitting" of $\overline{M}_s$. Thus,

$$\overline{M}_s = \hat{z} \times \overline{E}_t^a = \nabla_t \Gamma + \hat{z} \times \nabla_t \Theta \tag{30}$$

The use of the potentials has the great advantage of leading to easily enforced boundary conditions. From (29), these are

$$\begin{cases} \Theta = 0 \\ \dfrac{\partial \Gamma}{\partial n} = 0 \end{cases} \quad \text{on (c)} \tag{31}$$

It is important, in order to solve (8) by the technique under discussion in this Section, to express $\nabla \cdot \overline{F}$ in terms of Γ and Θ. Thus, taking (29) into account,

$$\begin{aligned} \nabla_t \cdot \overline{F} &= \frac{\varepsilon}{2\pi} \iint_A \nabla_t \cdot \left[\overline{M}_s(\overline{r}') G(\overline{r}, \overline{r}') \right] dS' \\ &= -\frac{\varepsilon}{2\pi} \iint_A \overline{M}_s(\overline{r}') \cdot \nabla_t' G(\overline{r}|\overline{r}') dS' \\ &= \frac{\varepsilon}{2\pi} \iint_A G(\overline{r}|\overline{r}') \nabla_t' \cdot \overline{M}_s(\overline{r}') dS' \\ &= \frac{\varepsilon}{2\pi} \iint_A \nabla_t'^2 \Gamma \frac{e^{-jk|\overline{r}-\overline{r}'|}}{|\overline{r}-\overline{r}'|} dS' \quad \overline{r} \text{ in A} \end{aligned} \tag{32}$$

This expression for $\nabla_t \cdot \overline{F}$ can now be introduced in (8). The second term in the left hand member of (8) must also be expressed in terms of Θ and Γ. One easily finds that

$$\overline{F} \times \hat{z} = \frac{\varepsilon}{2\pi} \nabla_t \iint_A \Theta(\overline{r}') G(\overline{r}|\overline{r}') dS' + \frac{\varepsilon}{2\pi} \iint_A \nabla_t' \Gamma \times \hat{z} \, G(\overline{r}|\overline{r}') dS' \tag{33}$$

From (14) and (17), the divergence of the left-hand member is given by

$$\nabla_t \cdot (\overline{F} \times \hat{z}) = \hat{z} \cdot \nabla_t \times \overline{F} = \varepsilon \, \hat{z} \cdot \overline{E}^i \quad \text{in A} \tag{34}$$

Taking the divergence of both members of (33) and comparing the result to (34) yields

$$\nabla_t^2 \iint_A \Theta \frac{e^{-jk|\overline{r}-\overline{r}'|}}{|\overline{r}-\overline{r}'|} dS' = 2\pi \hat{z} \cdot \overline{E}^i + \int_c \frac{\partial \Gamma}{\partial c'} \frac{e^{-jk|\overline{r}-\overline{r}'|}}{|\overline{r}-\overline{r}'|} dc', \quad \overline{r} \text{ in A} \tag{35}$$

Expressions (32) and (33) should be inserted in (8), and the resulting equations for Θ and Γ solved numerically. The solution is particularly simple at low frequencies, i.e., at wavelengths which are large with respect to the aperture's dimensions. For such cases, an iterative procedure can be applied.

(b) Low-frequency approximation. Zero-order terms.

At low frequencies it pays to expand the fields in the small parameter jk. Thus,

$$\bar{F} = \bar{F}_o + jk\bar{F}_1 + \ldots$$
$$\bar{E} = \bar{E}_o + jk\bar{E}_1 + \ldots = \frac{1}{\varepsilon} \nabla \times \bar{F}_o + \frac{jk}{\varepsilon} \nabla \times \bar{F}_1 + \ldots \quad (36)$$
$$\bar{M}_s = \bar{M}_o + jk\bar{M}_1 + \ldots$$

In the corresponding expression for $\bar{H}$, we introduce a function

$$\psi = \frac{1}{j\omega\varepsilon\mu} \nabla \cdot \bar{F} \quad (37)$$

in terms of which the expansion for $\bar{H}$ can be written as

$$\bar{H}^+ = -\nabla\psi_o + jk(c\bar{F}_o - \nabla\psi_1) + \ldots \quad (38)$$

where $c = (\varepsilon\mu)^{1/2}$ is the speed of light in the medium. The magnetic field being solenoidal, ψ_o is a harmonic function, and its value for $z > 0$ can be found from the value of $\frac{\partial\psi_o}{\partial z}$ in the aperture ($\frac{\partial\psi_o}{\partial z}$ vanishes on the screen as $\bar{H}$ is tangent to the metal.). Setting the field point in the aperture yields the equation

$$\bar{H}_o = -\nabla_t \underbrace{\left[-\frac{1}{2\pi} \iint_A \frac{\partial\psi_o}{\partial z'} \frac{1}{|\bar{r}-\bar{r}'|} dS' \right]}_{\psi_o \text{ in A}} = (\bar{H}_o^i)_t, \quad \bar{r} \text{ in A} \quad (39)$$

from which the unknown $\frac{\partial\psi_o}{\partial z}$ can be determined. To make the solution unique, it is necessary to impose the additional condition

$$\iint_A \frac{\partial\psi_o}{\partial z} dS = 0 \quad (40)$$

which expresses that the magnetic flux vanish in the aperture. It is seen that the determination of $\bar{H}_o$ reduces to a magnetostatic problem. Similarly, the determination of $\bar{E}_o$ turns out to be an electrostatic problem. We first notice that (37) implies $\nabla \cdot \bar{F}_o = 0$ everywhere. As a result, from (32),

$$\iint_A \nabla_t^2 \Gamma_o \frac{1}{|\bar{r}-\bar{r}'|} dS' = 0, \quad \bar{r} \text{ in A} \quad (41)$$

This equation requires $\nabla_t^2 \Gamma_o$ to vanish. As $\frac{\partial \Gamma_o}{\partial n} = 0$ on (c), Γ_o must reduce to a constant, and the Helmholtz splitting consists of a single term

$$\overline{M}_o = \hat{z} \times \nabla_t \Theta_o$$
$$\overline{E}_{to} = \nabla_t \Theta_o \quad \text{in A} \quad (42)$$

The function Θ_o must satisfy the zero-order version of (35), which is

$$\begin{cases} \nabla_t^2 \dfrac{1}{2\pi} \iint_A \Theta_o(\overline{r}') \dfrac{1}{|\overline{r}-\overline{r}'|} dS' = \hat{z}.\overline{E}_o^i, \quad \overline{r} \text{ in A} \\ \Theta_o = 0 \quad \text{on(c)} \end{cases} \quad (43)$$

Once Θ_o is known in the aperture, a harmonic function Θ_o can be calculated in the right-hand space $z > 0$ through the equation

$$\Theta_o(\overline{r}) = \dfrac{1}{2\pi} \iint_A \Theta_o(\overline{r}') \dfrac{\partial}{\partial z'} \left(\dfrac{1}{|\overline{r}-\overline{r}'|}\right) dS' \quad (44)$$

The zero-order field is simply $\overline{E}_o = \nabla \Theta_o$. Eq. (44) shows [2] that $\overline{E}_o$ is produced by an electric dipole-layer of density $(-2\varepsilon_o \Theta_o)$ in the aperture, the screen being removed. At large distances, the layer can be replaced by an electric dipole of moment

$$\overline{P}_e = -2\varepsilon_o \iint \Theta_o dS \, \hat{z} \quad . \quad (45)$$

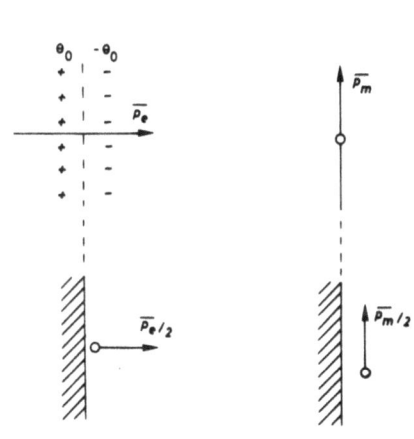

Fig. 4
Equivalent dipole moments of an aperture

located in empty space (Fig. 4). Equivalently, the aperture can be replaced by a dipole of moment $\overline{P}_e/2$ located next to the metallized (short-circuited) aperture plane. Similar arguments hold for the magnetic field. The value of ψ_o given in (39) is associated with a surface magnetic charge of density $(-2\mu_o \dfrac{\partial \psi_o}{\partial z})$. Because of (40), however, the total charge vanishes, hence the potential at large distances is produced by a magnetic dipole of moment (Fig. 4)

$$\overline{P}_m = 2\iint_A \left(-\dfrac{\partial \psi_o}{\partial z}\right) \overline{r} \, dS \quad (46)$$

This dipole lies in the aperture plane. Equivalently, from image theory, one can locate a magnetic dipole of moment $(\overline{P}_m/2)$ in front of the metallized aperture plane.

(c) Zero-order terms. Plane wave incidence.

Fig. 5a shows the incident plane wave and its two basic polarization waves 1 and 2. Let the complex amplitudes of these waves be A and B, respectively. The corresponding right-hand members in (39) and (43) are

$$\hat{z}.\overline{E}_o^i = -B \sin \theta_i \qquad (47)$$

$$\eta\, (\overline{H}_o^i)_t = -A \cos\theta_i \hat{t} + B\, \hat{z} \times \hat{t}$$

It is seen that the second members are constant. Looking first

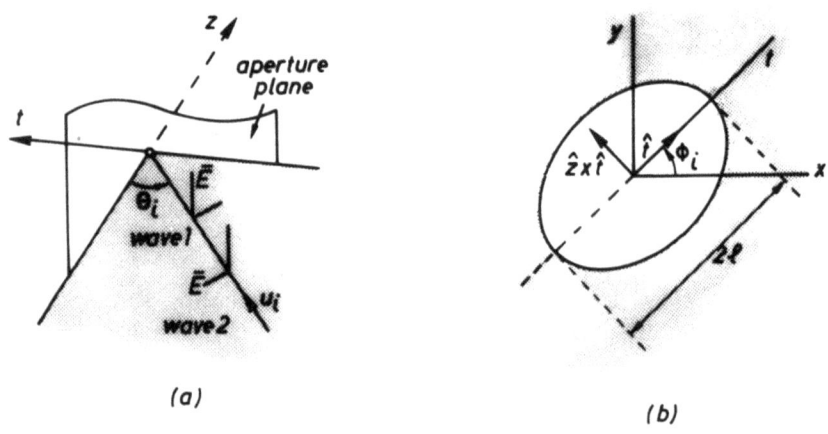

(a)   (b)

Fig. 5

(a) basic polarisations of the incident wave (b) coordinates in the aperture plane

at (43), we see that the basic problem reduces to the solution of the normalized equation

$$\begin{cases} \nabla_t^2 \dfrac{1}{2\pi} \displaystyle\iint_A \dfrac{\tau_o(\overline{r}')dS'}{|\overline{r}-\overline{r}'|} = -\dfrac{1}{\sqrt{S}}, & \overline{r} \text{ in } A \\ \tau_o = 0 & \text{on (c)} \end{cases} \qquad (48)$$

where S is the area of the aperture, and τ_o is dimensionless. The function τ_o for a circle of radius a is [5]

$$\tau_o = \dfrac{2}{\pi\sqrt{\pi}} \sqrt{1 - \dfrac{r^2}{a^2}} \qquad (49)$$

The variation of τ_o through the cross-section is shown in Fig. 6a. This variation is typical, in that for shapes which do not differ too much from the circle (such as the rectangle or the diamond),

τ_o peaks to a value of about 0.30 at the center and varies only slowly in the vicinity of this point [6]. Close to the rim (c), τ_o plunges to zero according to a law $\sqrt{d}$, where $\underline{d}$ is the distance

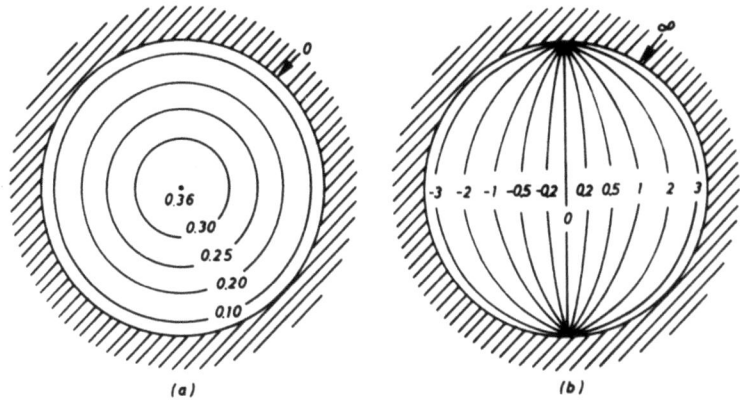

Fig. 6 Some characteristic functions for the circular aperture
(a) τ_o (b) ρ_x

to the rim. This behavior near (c) holds for contours of arbitrary shape.

Having solved (48), we can write the solution for Θ_o as

$$\Theta_o(\bar{r}) = -\sqrt{S}(\hat{z}\cdot\bar{E}_o^i)\tau_o(\bar{r}) = B\sin\Theta_i\sqrt{S}\,\tau_o(\bar{r}) \tag{50}$$

The electric field in the aperture and the electric dipole moment $\bar{P}_e$ are

$$(\bar{E}_o)_t = \nabla_t\Theta_o = B\sin\Theta_i\sqrt{S}\,\nabla_t\tau_o(\bar{r})$$

$$\bar{P}_e = 2B\sin\Theta_i\sqrt{S}\,\varepsilon_o\iint\tau_o dS\,\hat{z} = 2B\sin\Theta_i(S)^{3/2}\varepsilon_o(\tau_o)_{av}\hat{z} \tag{51}$$

where $(\tau_o)_{av}$ is the average value of τ_o over the aperture. For a circle, $(\tau_o)_{av} = 0.24$. It is convenient to write the moment $\bar{P}_e$ of the induced dipole in terms of the short-circuit field $\bar{E}^{sc}$ in the aperture. $\bar{E}^{sc}$ is directed along the normal to the screen, and is related to $\bar{E}_o^i$ by the relationship $\hat{z}\cdot\bar{E}_o^{sc} = 2\,\hat{z}\cdot\bar{E}_o^i$. Thus,

$$(n\,\bar{P}_e/2 = \varepsilon_o\pi_e\,\bar{E}_z^i) \tag{52}$$

where π_e, which has the dimension m^3, is the <u>electric polarizability</u> of the aperture. From (51),

$$\pi_e = \sqrt{S}\iint_A\tau_o\,dS = S^{3/2}(\tau_o)_{av} \tag{53}$$

For a circle, π_e is $\frac{4}{3} a^3$. Data for other shapes are given in Sec. 1.4.

If we now turn to the solution of (39), we notice that two basic problems must be solved, corresponding to resp. $\hat{x}$ and $\hat{y}$ of the right-hand member. We write

$$\frac{1}{2\pi} \nabla_t \iint_A \rho_x(\bar{r}') \frac{1}{|\bar{r}-\bar{r}'|} dS' = \hat{x}$$

$$\frac{1}{2\pi} \nabla_t \iint_A \rho_y(\bar{r}') \frac{1}{|\bar{r}-\bar{r}'|} dS' = \hat{y}$$
(54)

The basic functions, ρ_x and ρ_y, are dimensionless. They must both satisfy condition (40) and are clearly, to within a factor $2\varepsilon_o$, the charge densities which appear on the uncharged metallized aperture A, when A is immersed in an incident electrostatic field $\hat{x}$ or $\hat{y}$. The two functions can be combined into a single vector

$$\bar{\rho} = \rho_x \hat{x} + \rho_y \hat{y}$$
(55)

For a circular aperture, it can be shown [5] that

$$\bar{\rho} = \frac{4}{\pi} \frac{r}{\sqrt{a^2-r^2}} \hat{r}$$
(56)

In terms of $\bar{\rho}$ we can write

$$\frac{\partial \psi_o}{\partial z} = \bar{\rho} \cdot \bar{H}_o^i$$

$$\bar{P}_m = -2\iint_A (\bar{\rho} \cdot \bar{H}_o^i) \bar{r} \, dS$$
(57)

If we remember that $(\bar{H}^i)_t = \frac{1}{2} \bar{H}^{sc}$ (where $\bar{H}^{sc}$ lies in the aperture plane) and if we introduce the (symmetric) dyadic

$$\bar{\bar{\pi}}_m = \iint_A \bar{r}\, \bar{\rho}\, dS$$
(58)

we can express the dipole moment as

$$\bar{P}_m = -\bar{\bar{\pi}}_m \cdot \bar{H}_o^s$$
(59)

For a circle, for example,

$$\bar{\bar{\pi}}_m = \frac{8}{3} a^3 \bar{\bar{I}}$$
(60)

where $\bar{\bar{I}}$ is the unit dyadic. As $\bar{\bar{\pi}}_m$ is symmetric, it can be reduced to its principal axes $\underline{x}$ and $\underline{y}$, upon which it takes the form

$$\bar{\bar{\pi}}_m = \pi_{mx} \hat{x}\hat{x} + \pi_{my} \hat{y}\hat{y}$$
(61)

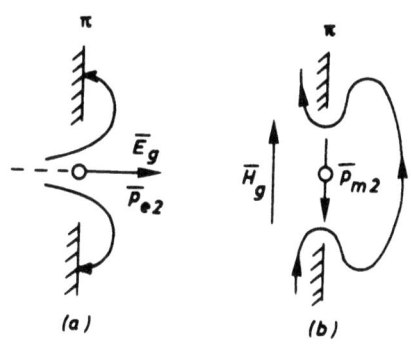

Fig. 7 Field penetration through an aperture

Some data on the magnetic polarizabilities π_{mx} and π_{my} are given in Sec. 1.4. The negative sign in formula (59) becomes intuitively clear from inspection of Fig. 7. The variation of ρ_x is given in Fig. 6b for a circle. One notices that the average value of ρ_x is zero, as required, and that ρ_x becomes infinite as $\frac{1}{\sqrt{d}}$ near the rim. This behavior, which is clearly evidenced in (56), is quite general and also holds for ρ_y.

1.3 Low frequency approximation. First-order terms.

(a) First-order terms

Determination of the next terms in the power expansion in jk allows one to extend the validity of the low-frequency approach toward the "resonance" region, where λ becomes of the order of the dimensions. This is important, for example, in the study of pulse-penetration through apertures, when the Fourier spectrum cuts off below the resonance region. We shall only consider the first-order terms and discuss them very briefly [7]. In plane incidence, the "forcing functions" are (Fig. 5)

$$\hat{z}\cdot\bar{E}_1^i = B\sin^2\Theta_i\cos\phi_i\,x + B\sin^2\Theta_i\sin\phi_i\,y$$

$$\eta(\bar{H}_1^i)_t = (A\sin\Theta_i\cos\Theta_i\cos\phi_i\,x + A\sin\Theta_i\cos\Theta_i\sin\phi_i\,y)\hat{t} \qquad (62)$$

$$- (B\sin\Theta_i\cos\phi_i\,x + B\sin\Theta_i\sin\phi_i\,y)\hat{z} \times \hat{t}$$

The integro-differential equation for $\frac{\partial\psi_1}{\partial z}$ is, from (8) and (38),

$$\nabla_t \frac{1}{2\pi} \iint_A \frac{\partial\psi_1}{\partial z'}\frac{1}{|\bar{r}-\bar{r}'|}\,dS' = (\bar{H}_1^i)_t - \frac{1}{\sqrt{\varepsilon\mu}}\bar{F}_o \qquad (63)$$

The value of $(\bar{H}_1^i)_t$ is given in (62), and $\bar{F}_o$ can be derived from (2) and (42). Thus,

$$c\bar{F}_o = - \frac{\sqrt{S}\,B\sin\Theta_i}{2\pi\eta}\,\hat{z} \times \nabla_t \iint_A \frac{\tau_o(\bar{r}')\,dS'}{|\bar{r}-\bar{r}'|} \qquad (64)$$

Having found $\frac{\partial\psi_1}{\partial z'}$ in A, one can determine ψ_1 in all space and,

hence, $(-\nabla \psi_1)$ which, according to (38), is one of two terms which $\overline{H}_1$ comprises.
Turning now to the scalar potentials Γ_1 and Θ_1, we notice, from (8) and (32), that

$$\nabla_t \frac{1}{2\pi} \iint_A {\nabla'}_t^2 \Gamma_1 \frac{1}{|\overline{r}-\overline{r}'|} dS' = -\eta \overline{H}_o^i, \qquad \overline{r} \text{ in } A \qquad (65)$$

This equation shows, in view of (47) and (54), that Γ_1 is the solution of the Neumann problem

$$\begin{cases} \nabla_t^2 \Gamma_1 = \overline{\rho} \cdot [A\cos\Theta_i \hat{t} - B \hat{z} \times \hat{t}] & \text{in } A \\ \dfrac{\partial \Gamma_1}{\partial n} = 0 & \text{on (c)} \end{cases} \qquad (66)$$

Once this problem is solved, we can focus attention upon the solution of Θ_1. From (35), this requires consideration of the integro-differential equation

$$\nabla_t^2 \iint_A \Theta_1(\overline{r}') \frac{1}{|\overline{r}-\overline{r}'|} dS' = 2\pi \hat{z} \cdot \overline{E}_1^i + \int_c \frac{\partial \Gamma_1}{\partial c'} \frac{1}{|\overline{r}-\overline{r}'|} dc' \qquad (67)$$

where the value of $\hat{z} \cdot \overline{E}_1^i$ is given in (62). We shall not dwell upon the numerical solution of this equation, which involves, for example, solution of equations of the type of (48) with right-hand members $(-\frac{x}{\sqrt{S}})$ and $(-\frac{y}{\sqrt{S}})$. For (63), it is necessary to solve equations of the type of (54), with right-hand members $\frac{x\hat{x}}{\sqrt{S}}$, $\frac{x\hat{y}}{\sqrt{S}}$, $\frac{y\hat{x}}{\sqrt{S}}$ and $\frac{y\hat{y}}{\sqrt{S}}$. It is to be noticed that the inclusion of the first order terms modifies the radiation pattern of the aperture in that the two dipole terms are no longer sufficient. Higher-order terms, involving quadrupole contributions, are needed [7]. It turns out that $\frac{\partial \psi_1}{\partial z}$ does not appear in the expression for these moments, hence, that $\frac{\partial \psi_1}{\partial z}$ is needed only for the calculation of the near field.

(b) Aperture in screen separating different half-spaces

The configuration of interest is shown in Fig. 8. Sources are present on both sides of the screen; they produce short-circuit fields $\overline{E}_1^{sc}$, $\overline{H}_1^{sc}$, $\overline{E}_2^{sc}$, $\overline{H}_2^{sc}$ on the metallized aperture. Assume that these fields are, in the zero order approximation, constant over A. This would be the case for an incident plane wave. A rather straightforward extension of the previous theory shows [8] that the fields in the aperture are now

$$D_z^a = \frac{\varepsilon_1 \varepsilon_2}{\varepsilon_1 + \varepsilon_2} (E_{1z}^{sc} + E_{2z}^{sc})$$

$$\overline{E}_t^a = (\frac{\varepsilon_2}{\varepsilon_1 + \varepsilon_2} E_{2z}^{sc} - \frac{\varepsilon_1}{\varepsilon_1 + \varepsilon_2} E_{1z}^{sc}) \sqrt{s} \, \nabla_t \tau_o$$

$$B_z^a = \frac{\mu_1 \mu_2}{\mu_1 + \mu_2} \overline{\rho} \cdot (\overline{H}_2^{sc} - \overline{H}_1^{sc})$$

$$\overline{H}_t^a = \frac{\mu_1}{\mu_1 + \mu_2} \overline{H}_1^{sc} + \frac{\mu_2}{\mu_1 + \mu_2} \overline{H}_2^{sc}$$

(68)

The additional fields produced by the aperture, in its vicinity and in the radiation region, are the result of equivalent dipoles. In medium 1, the moments of these dipoles are

$$\overline{P}_{e1} = \varepsilon_1 \, \overline{\overline{\pi}}_e \, (\frac{2\varepsilon_2}{\varepsilon_1 + \varepsilon_2} \overline{E}_2^{sc} - \frac{2\varepsilon_1}{\varepsilon_1 + \varepsilon_2} \overline{E}_1^{sc})$$

$$\overline{P}_{m1} = \frac{2\mu_2}{\mu_1 + \mu_2} \overline{\overline{\pi}}_m \cdot (\overline{H}_1^{sc} - \overline{H}_2^{sc})$$

(69)

The dipoles are assumed located in an infinite medium of (ε_1, μ_1) characteristics (the screen being removed). Similarly, in medium 2,

$$\overline{P}_{e2} = \varepsilon_2 \, \overline{\overline{\pi}}_e \, (\frac{2\varepsilon_1}{\varepsilon_1 + \varepsilon_2} \overline{E}_1^{sc} - \frac{2\varepsilon_2}{\varepsilon_1 + \varepsilon_2} \overline{E}_2^{sc})$$

$$\overline{P}_{m2} = \frac{2\mu_1}{\mu_1 + \mu_2} \overline{\overline{\pi}}_m \cdot (\overline{H}_2^{sc} - \overline{H}_1^{sc})$$

(70)

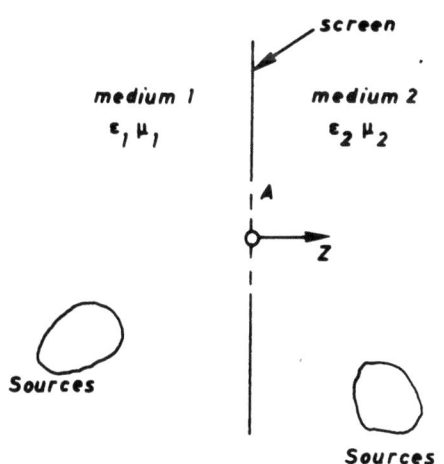

Fig. 8 Screen separating different media

Here, the dipoles are assumed located in an infinite medium of the (ε_2, μ_2) type.

Under the influence of sources in both-half spaces, currents $\overline{J}_s$ and charges ρ_s appear on the short-circuited aperture. It is convenient to rewrite some of the previous expressions in terms of these charges and currents. For example:

$$\overline{E}_t^a = \frac{\rho_s}{\varepsilon_1 + \varepsilon_2} \sqrt{s} \, \nabla_t \tau_o$$

$$B_z^a = \frac{\mu_1 \mu_2}{\mu_1 + \mu_2} \overline{J}_s \cdot (\hat{z} \times \overline{\rho})$$

(71)

Analogous modifications can be carried out in (69) and (70).

1.4 Some numerical data

(a) Polarizabilities

The electric polarizability π_e has the dimension m^3. In Fig. 9a, numerical results are given for the dimensionless ratio $(\pi_e/S^{3/2})$, equal to $(\tau_o)_{av}$ [6][9]. The curve for the ellipse is based on the analytical formula

$$(\tau_o)_{av} = \frac{2}{3\sqrt{\pi}\, E(e)} \sqrt{\frac{w}{\ell}} \tag{72}$$

where

$$e = \sqrt{1 - \frac{w^2}{\ell^2}}$$

$$E(e) = \int_0^{\pi/2} \sqrt{1 - e^2 \sin^2\Theta}\, d\Theta \tag{73}$$

For a very thin ellipse (an elliptical slot), for which $\frac{w}{\ell} \ll 1$:

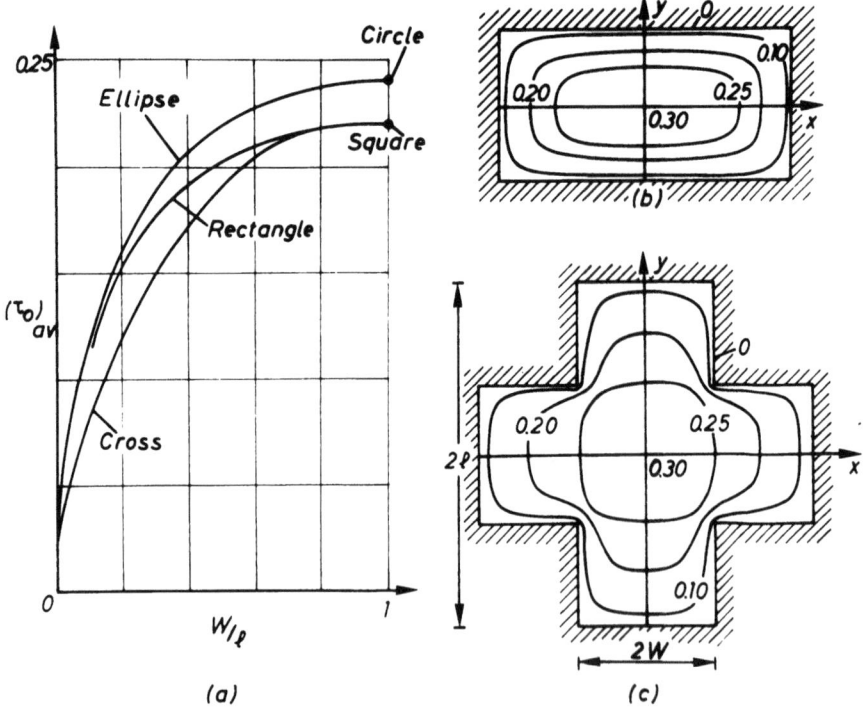

Fig. 9 (a) electric polarizability (b) (c) variation of τ_o in an aperture

$$(\tau_o)_{av} = \frac{2}{3}\sqrt{\frac{w}{\pi \ell}} \tag{74}$$

A few typical plots of τ_o are shown in Figs. 9b and 9c. The magnetic polarizabilities, π_{mx} and π_{my}, have the dimension m^3 too. Some data are plotted in Fig. 10 for the dimensionless

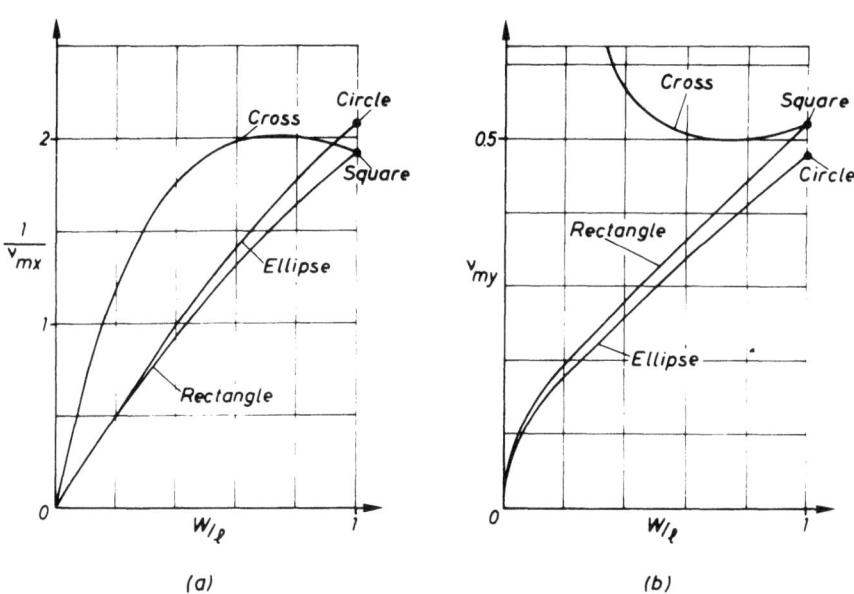

Fig. 10 Magnetic polarizabilities

ratios $\nu_{mx} = (\pi_{mx}/S^{1.5})$ and $\nu_{my} = (\pi_{my}/S^{1.5})$. For the ellipse, the analytical formulas are

$$\nu_{mx} = \frac{2e^2}{3\sqrt{\pi}} \left(\frac{\ell}{w}\right)^{1.5} \frac{1}{K(e)-E(e)}$$

$$\nu_{my} = \frac{2e^2}{3\sqrt{\pi}} \left(\frac{w}{\ell}\right)^{0.5} \frac{1}{E(e)- \frac{w^2}{\ell^2} K(e)} \tag{75}$$

where

$$K(e) = \int_0^{\pi/2} \frac{d\Theta}{\sqrt{1-e^2\sin^2\Theta}} \tag{76}$$

For a thin ellipse :

$$\nu_{mx} = \frac{2}{3}\sqrt{\frac{3}{\pi w^3}} \frac{1}{\ln(\frac{4\ell}{w})-1}$$

$$\nu_{my} = \frac{2}{3}\sqrt{\frac{w}{\pi \ell}}$$

(77)

Typical plots of ρ_x and ρ_y are given in Fig. 11

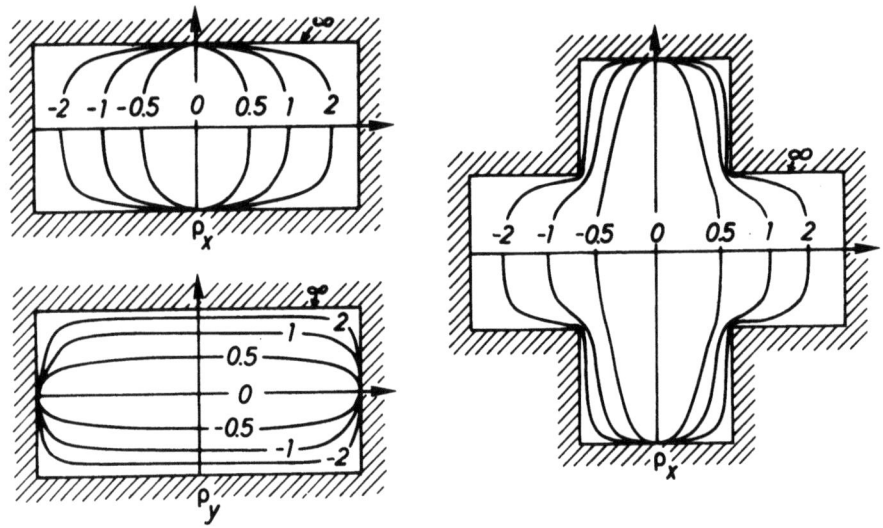

Fig. 11   Variation of ρ_x and ρ_y in a few apertures

(b) fields in the aperture and its vicinity

Sample results for $\overline{M}_s$ in square apertures subject to normally incident, plane wave illumination are presented in Fig. 12 [10][11]. The incident electric field is $\overline{E}^{i-} = -\hat{y}$ at $z = 0$. The value of the magnetic current at a point (x,y) in A is proportional to the height of the surface at this point, and, on each figure, reference heights are marked and a numerical value given so that one can judge the value of $\overline{M}_s$ at a desired point. These figures serve to convey to the reader an intuitive appreciation of the distribution of the equivalent magnetic current in the two apertures. Notice that the normal component of $\overline{M}_s$ at a aperture/screen edge is zero while the tangential component becomes large (infinite, if the numerical method were able to deal with such). When the results are compared with those for a $0.15\lambda \times 0.15\lambda$ square, it is found that $Re(\overline{M}_s)$ decreases considerably there, and that $Im(\overline{M}_s)$ increases to a value of the order of

0.6 at the center. In fact, a continuous evolution takes place from high to low frequencies. At very high frequencies $\bar{E}_t^a$ is equal to $\bar{E}_t^i$, as mentioned in Par. 1.1, and is therefore equal to the "real" vector $(-\hat{y})$. At very low frequencies, on the other hand, (51) shows that there is no zero-order term in $\bar{E}_t^a$ at <u>normal</u> incidence, and that the presence of an aperture field is <u>a purely</u> "imaginary" first-order effect, given by

$$\bar{E}_t^a = jk\nabla_t \Gamma_1 \times \hat{z} + jk\nabla_t \Theta_1 \tag{78}$$

The functions Θ_1 and Γ_1 can be found from (66). They are known in analytical form for a circular aperture [5][7], for which the relevant lines of force of $\bar{E}$ are shown in Fig. 13a. For a square,

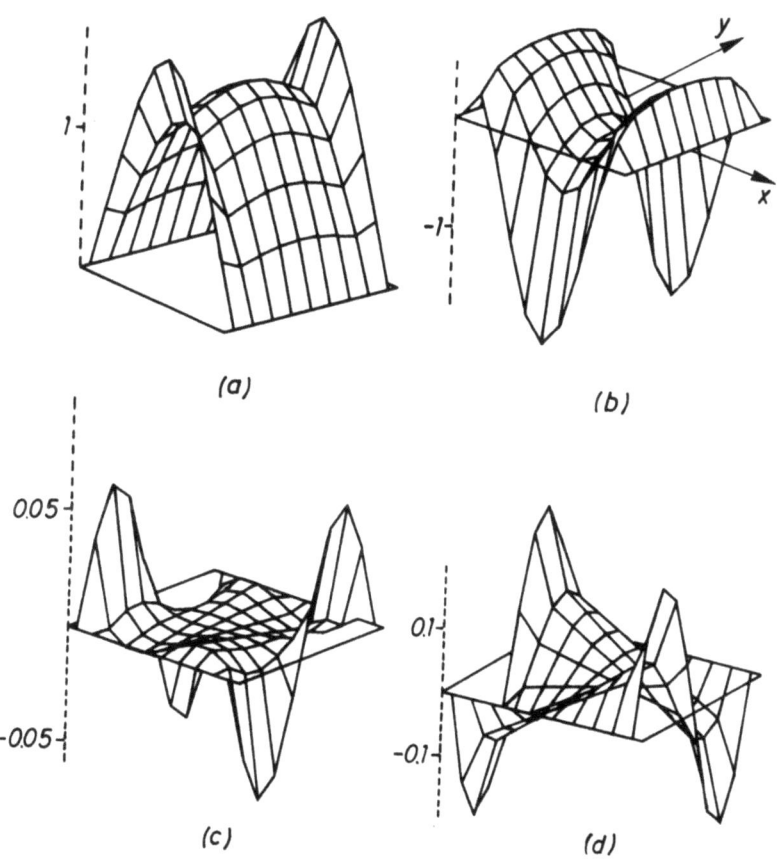

Fig. 12 Equivalent magnetic surface current $\bar{M}_s$ in a square aperture (1λ × 1λ), with $\bar{E}^{i-} = -\hat{y} \, e^{-jkz}$
(a) Re(M_{sx}) (b) Im(M_{sx}) (c) Re(M_{sy}) (d) Im(M_{sy})

the problem must be solved numerically [9]. Curves for the dimensionless functions $\Gamma_1' = (\Gamma_1/S)$ and $\Theta_1' = (\Theta_1/S)$ are shown in Fig. 13 b and c. From these curves it can be seen, for example, that $\overline{E}_t^a \approx j\frac{ka}{2}$ at the center.

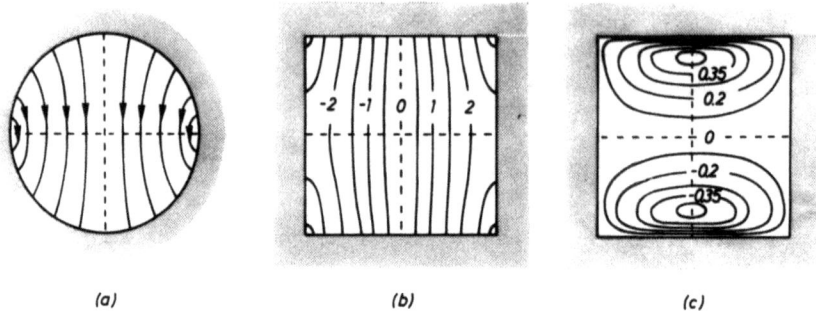

(a)　　　　　　　(b)　　　　　　　(c)

Fig. 13 (a) lines of force of $\overline{E}_t = jk\overline{E}_1$ in a circular aperture (b) the function Γ_1' in a square (c) the function Θ_1' in a square

A few data concerning the variation of $\overline{E}$ in the vicinity of the aperture are shown in Fig. 14.

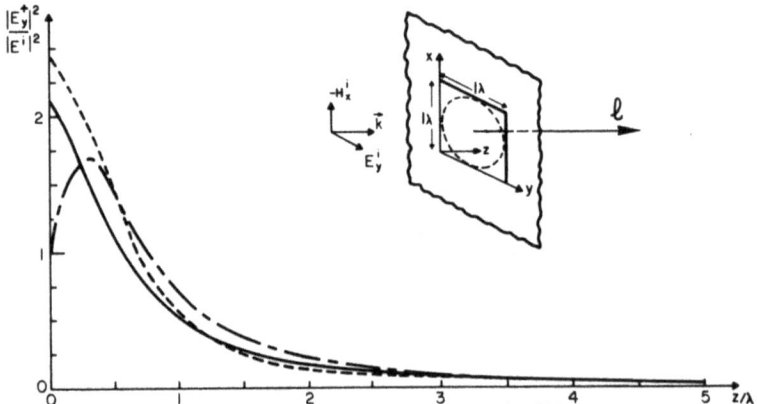

Fig. 14 Intensity distribution of E_y-field sampled along a line ℓ parallel to the z-axis and passing through the center of square and circular apertures. Integral equation (—) solution for square aperture. Kirchhoff approximation for square aperture (– –). Experimental result (---) for circular aperture from Andrews [12].

2. EXTENSIONS AND PARTICULAR CASES

2.1 Slots in planar screen.

(a) the two basic polarizations

If the aperture in the screen is an infinite slot of uniform width 2w as illustrated in Fig. 15, one may specialize Eq.(7), or Eq.(25) in the two-source situation, to obtain appropriate equations for the slot. We consider two types of excitation : (1) transverse electric (TE) in which the incident electric field

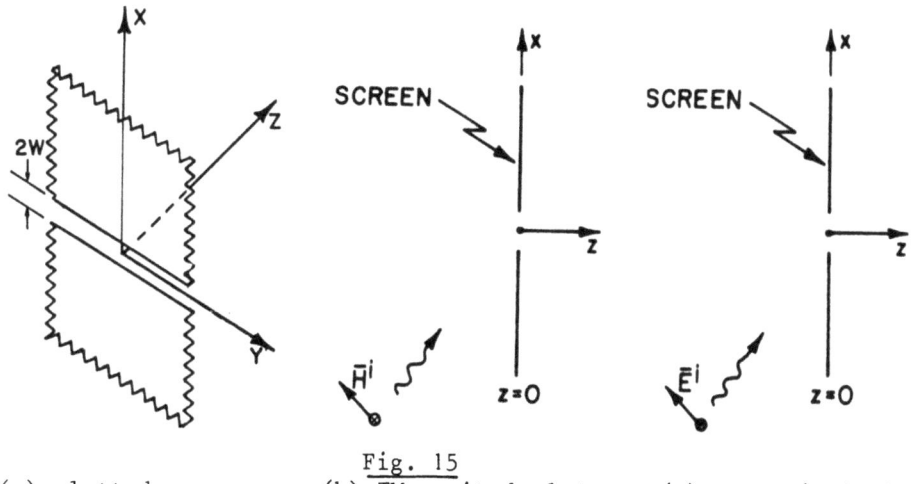

Fig. 15
(a) slotted screen   (b) TM-excited slot   (c) TE-excited slot

is entirely transverse to the slot axis (y axis) and (2) transverse magnetic (TM) in which the incident magnetic field is entirely transverse to the slot axis (Fig. 15). In both polarizations, the illumination is assumed here to be invariant along the slot axis (y axis), even though the present analysis can be extended readily to include excitation which does vary with y.

(b) TE Excitation

It should be clear to the reader that, with TE excitation of the uniform-width slot in which the illumination is independent of y, the total transverse electric field $\bar{E}^a_t$ in the aperture (slot) has only an x component and depends only upon $x (\bar{E}^a_t = E^a_x(x)\hat{x})$. Since $\bar{E}^a_t$ has only an x component, the equivalent magnetic current $\bar{M}_s$ has only a y component :

$$\bar{M}_s = \hat{z} \times \bar{E}^a_t = E^a_x(x)\hat{y} = M_{sy}(x)\hat{y} \tag{79}$$

Since $\bar{M}_s$ has only a y component, F_x of Eq. (10) is zero everywhere, and, since $\bar{M}_s = M_{sy}\hat{y}$ does not depend upon y, $\frac{\partial}{\partial y} F_y$ is zero everywhere too. With $F_x \equiv 0$ and $\frac{\partial}{\partial y} F_y \equiv 0$, Eq. (10) simplifies to

$$j\omega F_y = \frac{1}{2} H_y^{sc} \quad \text{(in slot)} \tag{80}$$

In addition, specialization of F_y of (11) to the slot geometry leads to

$$F_y(x) = \frac{\varepsilon}{2\pi} \int_{x'=-w}^{w} M_{sy}(x') \int_{y'=-\infty}^{\infty} \frac{e^{-jk[(x-x')^2+y'^2]^{1/2}}}{[(x-x')^2+y'^2]^{1/2}} dy'dx' \tag{81}$$

in which, since F_y is independent of y, R of (11) is evaluated at $y = 0$ in (81) with no sacrifice in generality. Using the well-known integral representation of the Hankel function of the second kind

$$H_0^{(2)}(k\xi) = \frac{j}{\pi} \int_{\zeta=-\infty}^{\infty} \frac{e^{-jk[\xi^2+\zeta^2]^{1/2}}}{[\xi^2+\zeta^2]^{1/2}} d\zeta \tag{82}$$

one can convert (81) to

$$F_y(x) = -j\frac{\varepsilon}{2} \int_{x'=-w}^{w} M_{sy}(x') H_0^{(2)}(k|x-x'|) dx' \tag{83}$$

With F_y of (83) in (80), one arrives at the integral equation below for the TE slot problem :

$$\frac{k}{\eta} \int_{x'=-w}^{w} M_{sy}(x') H_0^{(2)}(k|x-x'|) dx' = H_y^{sc}(x), \quad -w < x < +w \tag{84}$$

Equation (84) enables one to determine the unknown equivalent magnetic current or the slot electric field E_x^a due to the excitation H_y^{sc} in the slot. We note that Eq. (84) is of the same form as that for the corresponding TM strip problem (duality), and it can be solved by the techniques applied to the TM strip equation. Finally, we remark that (84) is an integral equation of the first kind and no boundary condition is needed to solve it.

(c) TM Excitation

With the incident electric field parallel to the slot axis and having no y variation in the xy plane, we see that $\bar{E}_t^a$ must be independent of y and parallel to the slot axis ($\bar{E}_t^a = E_y^a(\hat{x})\hat{y}$). Hence, $\bar{M}_s = M_{sx}(x)\hat{x}$, $F_y = 0$, and F_x is independent of y. Therefore, Eq.(9) reduces to the single integro-differential equation,

$$j \frac{\omega}{k^2}\left(\frac{d^2}{dx^2} + k^2\right) F_x = \frac{1}{2} H_x^{sc} \quad \text{(in slot)} \tag{85}$$

In view of (11) and (82),(85) becomes

$$\frac{1}{k\eta}\left(\frac{d^2}{dx^2} + k^2\right) \int_{x'=-w}^{w} M_{sx}(x') H_0^{(2)}(k|x-x'|) dx' = H_x^{sc}(x), \quad -w < x < +w \tag{86}$$

This is an integro-differential equation for the unknown magnetic current M_{sx} (or slot electric field). Notice that H_x^{sc} in the slot is the forcing function. Finally, we observe that (86) is of the form of the equation for the electric current induced on a conducting strip by TE illumination.

Since E_y^a in the slot is parallel to the perfectly conducting screen, it must be zero on the screen edges, i.e., along the lines $(w,y,0)$ and $(-w,y,0)$; thus, since $M_{sx} = -E_y^a$, M_{sx} must be zero on these edges too. Furthermore, we know that E_y^a, hence M_{sx}, must approach zero as $\sqrt{w-x}$ does at the upper edge and as $\sqrt{w+x}$ at the lower edge. In numerical solution procedures, one usually enforces the simplified boundary conditions $M_{sx}(\pm w) = 0$.

(d) screen separating half spaces of different media

For different media on the two sides of the conducting screen, Eq. (28) can be readily specialized to the slot problem :

TE-Excitation

$$\frac{1}{2} \int_{x'=-w}^{w} M_{sy}(x') \left[\frac{k_-}{\eta_-} H_0^{(2)}(k_-|x-x'|) + \frac{k_+}{\eta_+} H_0^{(2)}(k_+|x-x'|)\right] dx'$$

$$= \left[H_y^{sc-}(x) - H_y^{sc+}(x)\right], \quad x \in (-w,w) \tag{87}$$

and

TM-Excitation

$$\frac{1}{2}\left(\frac{d^2}{dx^2} + k_-^2\right) \int_{x'=-w}^{w} M_{sx}(x') \frac{1}{k_-\eta_-} H_0^{(2)}(k_-|x-x'|) dx' + \frac{1}{2}\left(\frac{d^2}{dx^2} + k_+^2\right)$$

$$\int_{x'=-w}^{w} M_{sx}(x') \frac{1}{k_+\eta_+} H_0^{(2)}(k_+|x-x'|) dx' = \left[H_x^{sc-}(x) - H_x^{sc+}(x)\right], \quad x \in (-w,w) \tag{88}$$

where the subscripts "+" and "-" refer to quantities in the half-spaces $z > 0$ and $z < 0$, respectively. Techniques for solving these equations are discussed in [4] where also is found an outline of their development.

(e) slot data

Data for slotted screens separating the same and different

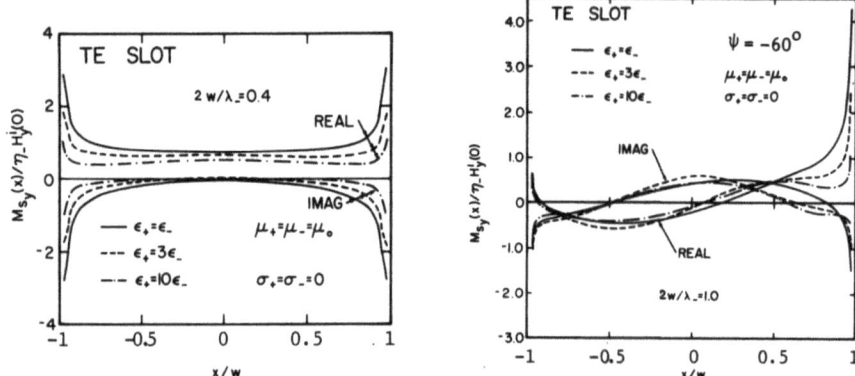

Fig. 16   TE equivalent magnetic current in 1.0 wavelength slot for different right half-space permittivities
(a) normal incidence   (b) -60 degree incidence; $\overline{H}^{i+} = \overline{0}$)

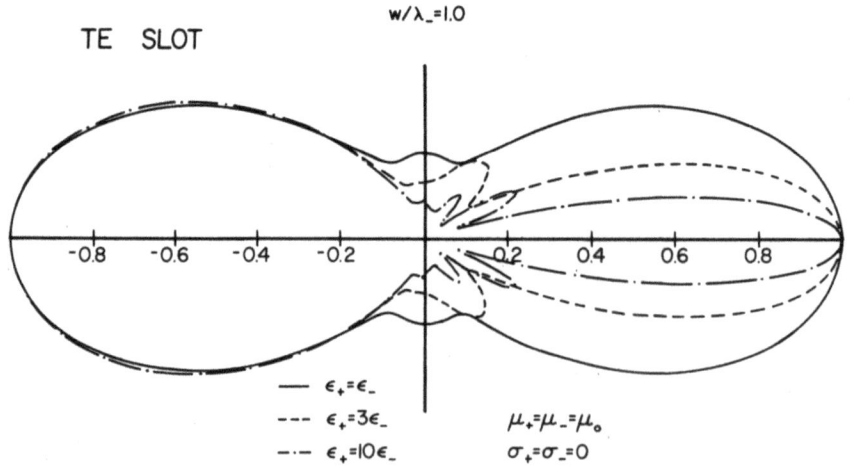

Fig. 17   Far magnetic field due to presence of TE-excited slot in screen for different right half-space permittivities (normal incidence; $\overline{H}^{1+} = \overline{0}$)

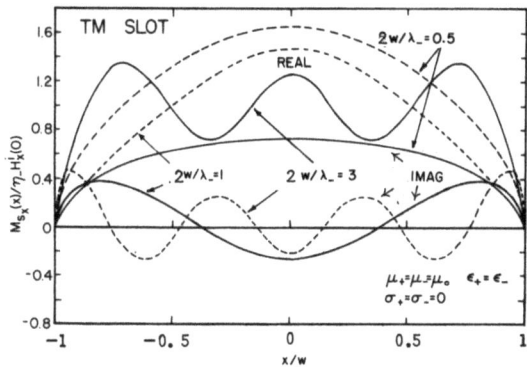

Fig. 18 TM equivalent magnetic current in 0.5-, 1.0, and 3.0- wavelength slots in screen separating half-spaces of the same media (normal incidence)

media are available in [4] and are reproduced here. Results are normalized by the intrinsic impedance η_- and the appropriate component of $\vec{H}^i(0)$, the incident magnetic field at the slot center ($\vec{H}^i = \vec{H}^{i-} - \vec{H}^{i+}$), and slot widths are measured relative to λ_-, the wavelength in the left half-space. Equivalent magnetic current is shown for TE illumination in Fig. 16. In Fig. 17 is illustrated the pattern of the far-zone magnetic field diffracted by the slot. In the region $z < 0$ of Fig. 17, $\vec{H}^{sc-}$ has been subtracted from the total magnetic field. Fig. 18 gives data for M_s due to TM excitation.

(f) narrow slot approximation [13]

For a slot that is narrow relative to the wavelength in both media ($|k_\pm w| << 1$), $|k_\pm(x-x')| << 1$, and one can replace the Hankel functions in Eq. (87) by their small argument approximations to arrive at

$$-j \frac{2}{\pi} \frac{k_e}{\eta_e} \int_{x'=-w}^{w} M_{sy}(x') \ln|x-x'| dx' - j \frac{2}{\pi} \frac{k_e}{\eta_e} C_e \int_{x'=-w}^{w} M_{sy}(x') dx'$$
$$= \left[H_y^{sc-}(x) - H_y^{sc+}(x) \right] \qquad |k_\pm w| << 1 \qquad (89)$$

for TE excitation. In this equation

$$C_e = \gamma + j\frac{\pi}{2} + \frac{1}{2} \frac{\varepsilon_-}{\varepsilon_e} \ln\left(\frac{k_-}{2}\right) + \frac{1}{2} \frac{\varepsilon_+}{\varepsilon_e} \ln\left(\frac{k_+}{2}\right) \qquad (90)$$

where $\gamma (= 0.5772)$ is Euler's constant and where effective quantities are defined viz. $\varepsilon_e = \frac{1}{2}(\varepsilon_- + \varepsilon_+)$, $\frac{1}{\mu_e} = \frac{1}{2}\left(\frac{1}{\mu_-} + \frac{1}{\mu_+}\right)$, $k_e = \omega\sqrt{\mu_e \varepsilon_e}$, and $\eta_e = \sqrt{\frac{\mu_e}{\varepsilon_e}}$. Since the slot is narrow, the excitation does not vary greatly over $x\varepsilon(-w,w)$ and one can approximate it by a two-term Taylor series to obtain

$$\left[H_y^{sc-}(x) - H_y^{sc+}(x)\right] \doteq \left[H_y^{sc-}(0) - H_y^{sc+}(0)\right] + j\frac{k_e}{\eta_e}\left[\frac{\varepsilon_-}{\varepsilon_e} E_z^{sc-}(0) - \frac{\varepsilon_+}{\varepsilon_e} E_z^{sc+}(0)\right] x \quad (91)$$

where $E_z^{sc\pm}$ is the short-circuit electric field normal to the screen. With the forcing function (91) in (89), the latter can be solved exactly [13] for the equivalent magnetic current M_{sy}:

$$M_{sy}(x) \doteq \frac{1}{2}\left\{ j\frac{\eta_e}{k_e} \cdot \frac{H_y^{sc-}(0) - H_y^{sc+}(0)}{\left(\gamma + j\frac{\pi}{2} + \frac{1}{2}\left[\frac{\varepsilon_-}{\varepsilon_e}\ell n\left(\frac{k_-w}{4}\right) + \frac{\varepsilon_+}{\varepsilon_e}\ell n\left(\frac{k_+w}{4}\right)\right]\right)} \right.$$

$$\left. + \left[\frac{\varepsilon_-}{\varepsilon_e} E_z^{sc-}(0) - \frac{\varepsilon_+}{\varepsilon_e} E_z^{sc+}(0)\right] x \right\} \frac{1}{\sqrt{w^2 - x^2}} \quad (92)$$

As demonstrated in [13], the corresponding TM-excited, narrow slot problem can be solved exactly, too:

$$M_{sx}(x) \doteq j\frac{k_e \eta_e}{2}\left\{\left[H_x^{sc-}(0) - H_x^{sc+}(0)\right]\right.$$

$$\left. + \frac{1}{2}\left[H_x^{sc-}(0) - H_x^{sc+}(0)\right]' x \right\} \sqrt{w^2 - x^2} \quad (93)$$

where $[H(0)]' = [\frac{d}{dx} H(x)]_{x=0}$. Knowledge of $\overline{M}_s$ from (92) and (93) enables one to determine the total field in both half spaces for the narrow slot problem.

2.2 Coupled slots and slots of finite length

(a) coupled slots

As a simple example of the analysis of multiple apertures in a screen, we consider the problem of two infinite, parallel slots excited by TE illumination as depicted in cross section in Fig. 19. Eq. (84) for the single TE-excited slot can be modified readily to apply to the present two-slot case. First, we point out that the equivalent magnetic current M_{sy} of Eq. (84) is zero outside the two slots, since the tangential component of electric field must be zero on the conducting screen. Thus, we introduce the definition

$$M_{sy}(x) = \begin{cases} M_{sy}^{(1)}(x) &, \quad x\varepsilon(x_{c_1}-w_1, x_{c_1}+w_1) \\ M_{sy}^{(2)}(x) &, \quad x\varepsilon(x_{c_2}-w_2, x_{c_2}+w_2) \\ 0 &, \quad \text{otherwise} \end{cases} \quad (94)$$

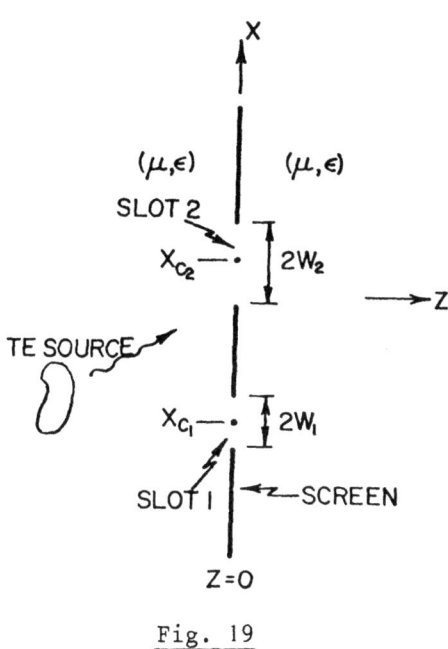

Fig. 19
Coupled slots

Also, we call attention to the fact that (84) is valid only in a slot -- not on the conducting surface. This is obvious, if one recalls that in the general derivation leading to (7), upon which (84) is founded, a key step is the enforcement of continuity, along any path through the aperture, of the tangential component of the total magnetic field (Eq. (6)). This continuity holds only in the two slots, not on the conductor. In fact, the tangential component of magnetic field is discontinuous along a path passing through the screen by the amount of the surface electric current density induced on the conducting surface.

A moment's reflection should suggest the following generalization of Eq. (84) to the two-slot case :

$$L_{11}(x) + L_{12}(x) = H_y^{sc}(x) \quad , \quad x \epsilon (x_{c_1}-w_1, x_{c_1}+w_1)$$
$$L_{21}(x) + L_{22}(x) = H_y^{sc}(x) \quad , \quad x \epsilon (x_{c_2}-w_2, x_{c_2}+w_2)$$
(95)

where the operators in (95) are defined by

$$L_{pq}(x) = \frac{k}{\eta} \int_{x'=x_{c_q}-w_q}^{x_{c_q}+w_q} M_{sy}^{(q)}(x') H_0^{(2)}(k|x-x'|) dx', \quad x \epsilon (x_{c_p}-w_p, x_{c_p}+w_p)$$
(96)

In (95), one has a pair of coupled integral equations for the unknowns $M_{sy}^{(1)}$ and $M_{sy}^{(2)}$. L_{11} is the magnetic field in slot 1 due to $M_{sy}^{(1)}$, while L_{12} is the magnetic field in slot 1 due to $M_{sy}^{(2)}$. Extension of the above to more than two slots is easily achieved.

(b) finite-length narrow slot
_ _ _ _ _ _ _ _ _ _ _ _ _ _ _ _ _

In Fig. 20 is depicted a finite-length slot of uniform width in a screen separating two half-spaces of the same medium (μ,ϵ). The slot length 2ℓ is much greater than its width $2w$ and the latter is much less than the wavelength ($w \ll \ell$, $w \ll \lambda$).

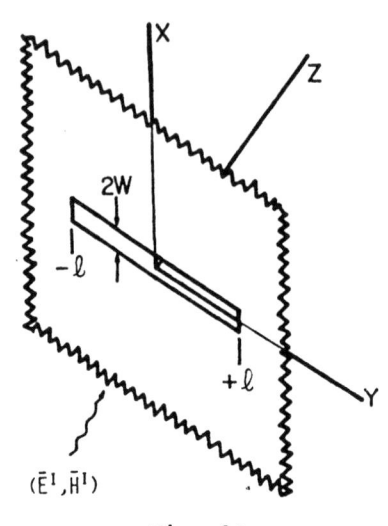

Fig. 20
Finite-length narrow slot

From a study of Eqs. (92) and (93), specialized to the single medium case, one finds that for equal excitation the magnetic current in the infinite slot is much greater in the TE case than in the TM case. Furthermore, the second term in Eq. (92) is less significant than the first unless $|E_z^{sc-}(0) - E_z^{sc+}(0)|$ is far greater than $|H_y^{sc-}(0) - H_y^{sc+}(0)|$, which for plane-wave excitation takes place only at incidence angles near grazing. Therefore, in almost all cases of practical interest, the first term of Eq. (92) is dominant.

Returning to the finite-length narrow slot, we assume that any TM component in the excitation can be ignored, that the significant component of equivalent magnetic current is M_{sy}, and that the transverse variation of M_{sy} is $1/\sqrt{w^2 - x^2}$, the same as the dominant term in (92). In view of these observations, we represent M_{sy} by

$$M_{sy}(x,y) = \frac{\frac{1}{\pi}}{\sqrt{w^2 - x^2}} K(y) \tag{97}$$

where $K(y)$ is the unknown axial variation of M_{sy} and where $1/\pi$ is a normalization factor introduced so that

$$\int_{x=-w}^{w} M_{sy}(x,y) dx = K(y) \tag{98}$$

Associated with this magnetic current is an electric vector potential F_y:

$$F_y(x,y,z) = \frac{\varepsilon}{2\pi} \int_{y'=-\ell}^{\ell} \int_{x'=-w}^{w} M_{sy}(x',y') \frac{e^{-jk[(x-x')^2+(y-y')^2+z^2]^{1/2}}}{[(x-x')^2+(y-y')^2+z^2]^{1/2}} dx' dy' \tag{99}$$

With M_{sx} assumed zero, $F_x = 0$ and it follows from Eq. (10) that F_y satisfies

$$j \frac{\omega}{k^2} \left[\frac{\partial^2}{\partial y^2} + k^2 \right] F_y = \frac{1}{2} \left[H_y^{sc-}(0,y,0) - H_y^{sc+}(0,y,0) \right] \quad (100)$$

in the slot.
Introducing (98) into (99) and (100) and making the simple transformation

$$x' = w \sin \frac{\alpha}{2} , \quad (101)$$

one enforces (100) along the slot axis to arrive at the following integro-differential equation for K(y)

$$\frac{j}{k\eta\pi} \left(\frac{d^2}{dy^2} + k^2 \right) \int_{y'=-\ell}^{\ell} K(y') \, G\left(y-y'; \frac{w}{2}\right) dy' = H_y^{sc}(y), \quad y \in (-\ell,\ell) \quad (102)$$

where
$$H_y^{sc}(y) = \left[H_y^{sc-}(0,y,0) - H_y^{sc+}(0,y,0) \right] \quad (103)$$

$$G(y-y';a) = \frac{1}{2\pi} \int_{\alpha=-\pi}^{\pi} \frac{e^{-jk\left[(y-y')^2 + 4a^2 \sin^2 \frac{\alpha}{2}\right]^{1/2}}}{\left[(y-y')^2 + 4a^2 \sin^2 \frac{\alpha}{2}\right]^{1/2}} d\alpha \quad (104)$$

We observe that G of (104) is the familiar exact kernel of cylindrical thin wire theory and, consequently, that (102) is identical in form to the well-known integro-differential equation of thin-wire theory. It is of importance to note that w/2 (one-fourth of slot width) appears in G of (102) where the radius of the cylindrical wire is found in the kernel of the thin-wire equation. Of course, since Eq. (102) differs from the above-mentioned equation of thin-wire theory only in the symbols used, methods for solving the latter apply directly for obtaining K(y) of (102). Data for the thin-wire are applicable for the finite-length, narrow slot under discussion here subject to proper symbol identification.

2.3 Pulse penetration through apertures

From knowledge of the equivalent dipoles one can compute the field which reaches the shadow side of the perforated surface. When the surface is an infinite plane, image theory applies and the shadow-side field is equivalent to that radiated by dipoles of moments $\overline{P}_e$ and $\overline{P}_m$ located at 0 in a uniform space of infinite extent characterized by (μ,ε). Radiation from such dipoles is easy to compute [2]. With $g = (e^{-jkr}/r)$, one finds

$$\vec{E}^+(\vec{r}) = - \frac{1}{4\pi\varepsilon} \nabla \times [\overline{P}_e \times \nabla g] + j\omega \frac{\mu}{4\pi} \overline{P}_m \times \nabla g \quad (105)$$

and

$$\vec{H}^+(\vec{r}) = -j\omega \frac{1}{4\pi} \overline{P}_e \times \nabla g - \frac{1}{4\pi} \nabla \times [\overline{P}_m \times \nabla g] \quad (106)$$

From the known short-circuit field and known aperture polarizabilities, one determines the moments $\bar{P}_e$ and $\bar{P}_m$, and then one employs the above equations to compute the shadow-side field. Eqs. (105) and (106) are exact for dipoles, but the representation of the aperture effect by equivalent dipoles leads to inaccuracies at points very close to A.

Knowing the excitation and polarizabilities of a small aperture, one can compute the time history of the field which passes through such a hole by making use of integral transform techniques. The polarizabilities of a small aperture are frequency independent, so in view of Eqs. (52) and (59), transform theory tells us that the temporal behavior of the dipole moments is the same as that of the short-circuit field. It follows directly from (105) and (106) that

$$\vec{e}^+(r,t) = \frac{1}{4\pi\varepsilon} \nabla \times \left\{ \frac{1}{r} \left[\frac{1}{r} \bar{P}_e(t - r/c) + \frac{1}{c} \dot{\bar{P}}_e(t - r/c) \right] \times \hat{r} \right\}$$
$$- \frac{\mu}{4\pi r} \left[\frac{1}{r} \dot{\bar{P}}_m(t - r/c) + \frac{1}{c} \ddot{\bar{P}}_m(t - r/c) \right] \times \hat{r} \quad (107)$$

and
$$\vec{h}^+(r,t) = \frac{1}{4\pi r} \left[\frac{1}{r} \dot{\bar{P}}_e(t - r/c) + \frac{1}{c} \ddot{\bar{P}}_e(t - r/c) \right] \times \hat{r}$$
$$+ \frac{1}{4\pi r} \nabla \times \left\{ \frac{1}{r} \left[\frac{1}{r} \bar{P}_m(t - r/c) + \frac{1}{c} \dot{\bar{P}}_m(t - r/c) \right] \times \hat{r} \right\} \quad (108)$$

where $(\vec{e}^+, \vec{h}^+)$ denotes the shadow-side, time-domain field, where $\bar{p}_e(t)$ and $\bar{p}_m(t)$ are the time histories of the dipole moments, where c is the speed of light in the medium characterized by (μ,ε), and where "·" signifies differentiation with respect to t. Eqs. (107) and (108) are valid only when A is small over the Fourier spectrum

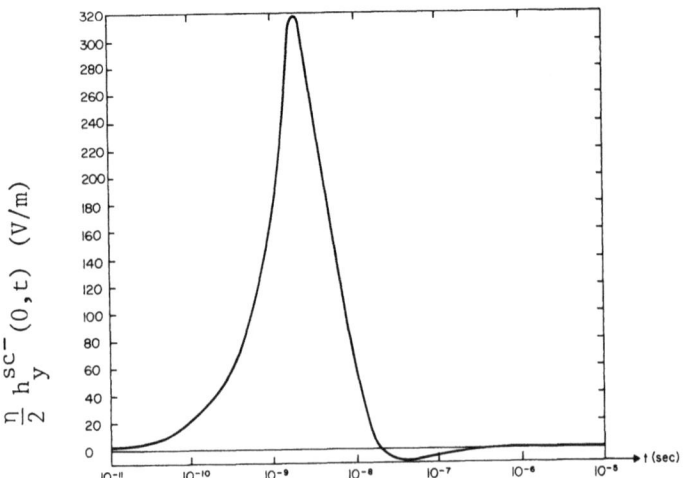

Fig. 21
Time history of $h_y^{sc-}(0,t)$

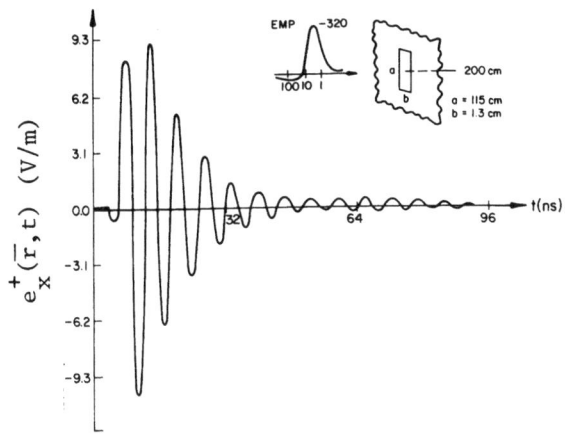

Fig. 22 Time history of the penetrated electric field $e_x^+(\bar{r},t)$ at a point on the Z axis two meters from the slot

the excitation. The reader is reminded that $\bar{p}_e(t)$ and $\bar{p}_m(t)$ are proportional (through the polarizabilities) to the known time histories of the short-circuit electric and magnetic fields, respectively. For an example illustrative of the EMP field which might pass through a finite-length slot, $h_y^{sc-}(0,t)$ is selected to be constant along the slot axis with the specified pulse time variation shown in Fig. 21. Fig. 22 displays the time history of the electric field $e_x^+(t)$ (at a point two meters behind the screen) which passed through a 115 cm x 1.3 cm slot [3]. These time-domain data were determined by computing the spectrum of $h_y^{sc-}(0,t)$, solving the slot problem with this spectrum as excitation, and then calculating the desired field components by Fourier inversion [3].

3. OBJECT NEAR AN APERTURE IN SCREEN

3.1 Introduction

In this section is presented an investigation of the coupling between an object and an aperture in a screen, which problem is depicted in Fig. 23. Integro-differential equations are formulated for the case of a perfectly conducting scatterer and an aperture, both of which are of arbitrary shape, with full account taken of the coupling between the aperture and the scatterer. The general equations are specialized to apertures and scatterers of interest, and selected data are presented.

3.2 General Equations

The derivation of equations for the aperture equivalent magnetic current $\bar{M}_s$ and the surface current $\bar{J}_s$ induced on the scatterer is based on the point of view that $\bar{M}_s$ and $\bar{J}_s$ can be treated as partially dependent sources. The development begins with Eq. (25) which is valid, of course, only when the object is not present and with a corresponding equation for the scatterer above a ground plane with no aperture. With no aperture in the screen, the usual requirement that the tangential component of electric field on the surface S_B of the scatterer be zero leads to

$$j \frac{\omega}{k^2}\left[k^2 \overline{A} + \nabla_s \nabla \cdot \overline{A}\right] \times \hat{n} = \overline{E}^{sc+} \times \hat{n}, \text{ on } S_B \quad (109)$$

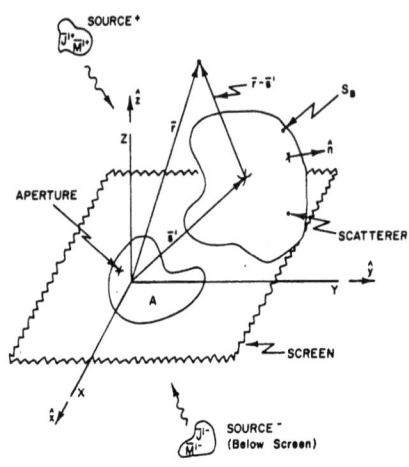

Fig. 23
Scatterer in the presence of an aperture-perforated, planar conducting screen

where $\overline{E}^{sc+}$ is the short-circuit electric field for $z > 0$ with the object removed and where $\hat{n}$ is the outward normal unit vector on S_B. The vector potential $\overline{A}$ due to $\overline{J}_s$ radiating above a ground plane is given by

$$\overline{A}(\overline{r}) = \frac{\mu}{4\pi} \iint_{S_B} \overline{J}_s(\overline{s}') \cdot \overline{\overline{G}}(\overline{r},\overline{s}') \, dS' \quad (110)$$

where

$$\overline{\overline{G}}(\overline{r},\overline{s}') = \overline{\overline{I}} \frac{e^{-jk|\overline{r}-\overline{s}'|}}{|\overline{r}-\overline{s}'|} + (2\hat{z}\hat{z}-\overline{\overline{I}}) \frac{e^{-jk|\overline{r}-\overline{s}' + 2(\overline{s}'\cdot\hat{z})\hat{z}|}}{|\overline{r}-\overline{s}' + 2(\overline{s}'\cdot\hat{z})\hat{z}|} \quad (111)$$

with the unit dyadic written as $\overline{\overline{I}} = \hat{x}\hat{x} + \hat{y}\hat{y} + \hat{z}\hat{z}$.
Now we generalize Eqs. (25) and (109) to the present problem of Fig. 23 by introducing terms to account for the coupling between the object and the aperture/screen. For this purpose, the field scattered back to the aperture due to $\overline{J}_s$ induced on the scatterer is looked upon as being a dependent part of the forcing function of Eq. (25). With $\overline{h}^{sc}$ defined as the upper-region short-circuit magnetic field due to $\overline{J}_s$,

$$\overline{h}^{sc}(\overline{r}) = \frac{1}{\mu} \nabla \times \overline{A}(\overline{r}), \quad (112)$$

we modify the right-hand side of Eq. (25) to arrive at

$$j \frac{\omega}{k^2}(k^2\overline{F} + \nabla_t \nabla_t \cdot \overline{F}) \times \hat{z} = \frac{1}{2}(\overline{H}^{sc-} - \overline{H}^{sc+} - \overline{h}^{sc}) \times \hat{z}, \text{ in } A \quad (113)$$

Since $\overline{J}_s$ is due in part to fields diffracted by the aperture and since $\overline{h}^{sc}$ is entirely due to $\overline{J}_s$, we view this portion of the forcing function of (113) as excitation resulting from a dependent "generator." Similarly, $\overline{e}^{ex}$ given by

$$\overline{e}^{ex}(\overline{r}) = \frac{1}{\varepsilon} \nabla \times \overline{F}(\overline{r}), \qquad (114)$$

is the electric field for $z > 0$ due to $(-\overline{M}_s)$; $\overline{e}^{ex}$ is included in Eq. (109) as part of the illumination of the object :

$$j \frac{\omega}{k^2} \left[k^2 \overline{A} + \nabla_s \nabla \cdot \overline{A} \right] \times \hat{n} = (\overline{E}^{sc+} + \overline{e}^{ex}) \times \hat{n}, \text{ on } S_B \qquad (115)$$

Eqs. (113) and (115), which can, in principle, be solved simultaneously for $\overline{M}_s$ and $\overline{J}_s$, completely account for coupling between the object and the aperture/screen. That $\overline{M}_s$ enters Eq. (115) through $\overline{e}^{ex}$ and that $\overline{J}_s$ enters Eq. (114) through $\overline{h}^{sc}$ are called to the attention of the reader.

3.3 Small Apertures

In the practically important situation where the aperture of interest is small, one can significantly simplify the analysis by availing himself of the body of knowledge of small aperture theory (Section 1.2). In view of the dependent source concept discussed above, the dipole moments of (70), specialized to the single-medium case, become

$$\begin{aligned} P_e \hat{z} &= \varepsilon \overline{\overline{\pi}}_e \left[E_z^{sc-}(\overline{0}) - E_z^{sc+}(\overline{0}) - e_z^{sc}(\overline{0}) \right] \hat{z} \\ \overline{P}_m &= -\overline{\overline{\pi}}_m \cdot \left[\overline{H}^{sc-}(\overline{0}) - \overline{H}^{sc+}(\overline{0}) - \overline{h}^{sc}(\overline{0}) \right] \end{aligned} \qquad (116)$$

where $(\overline{e}^{sc}, \overline{h}^{sc})$ is, as discussed above, the field above the shorted screen due to $\overline{J}_s$:

$$e_z^{sc}(\overline{0}) = -j \frac{\eta}{2\pi k} \hat{z} \cdot \left\{ (k^2 + \nabla\nabla \cdot) \iint_{S_B} \overline{J}_s(\overline{s}') g(\overline{r}-\overline{s}') dS' \right\}_{\overline{r}=\overline{0}} \qquad (117)$$

and

$$\overline{h}^{sc}(\overline{0}) = \frac{1}{2\pi} [\hat{x}\hat{x} + \hat{y}\hat{y}] \cdot \left\{ \nabla \times \iint_{S_B} \overline{J}_s(\overline{s}') g(\overline{r}-\overline{s}') dS' \right\}_{\overline{r}=\overline{0}} \qquad (118)$$

For a small aperture, $\overline{e}^{ex}$ of (114) is simply

$$\overline{e}^{ex} = \overline{e}_e + \overline{e}_m \qquad (119)$$

where $\overline{e}_e$ and $\overline{e}_m$ are the electric fields of the equivalent electric and magnetic dipoles, respectively (Eq. (105)) :

$$\begin{aligned} \overline{e}_e &= -\frac{1}{4\pi\varepsilon} P_e \nabla \times (\hat{z} \times \nabla g) \\ \overline{e}_m &= j \frac{k\eta}{4\pi} \overline{P}_m \times \nabla g \end{aligned} \qquad (120)$$

Of course, $\bar{e}^{ex}$ computed by means of (119) and (120) is an accurate representation of the electric field due to the small aperture only at large distances from A, measured relative to the maximum dimension across A.

With $\bar{e}^{ex}$ defined by (119) and (120), Eq. (115) reduces to

$$j \frac{\omega}{k^2} (k^2 \bar{A} + \nabla_s \nabla \cdot \bar{A}) \times \hat{n} = [\bar{E}^{sc+} + \bar{e}_e + \bar{e}_m] \times \hat{n}, \text{ on } S_B \quad (121)$$

for small apertures. The only unknown in (121) is $\bar{J}_s$ as one can readily see since $\bar{E}^{sc+}$ is an independent forcing function while $\bar{e}_e$ and $\bar{e}_m$ depend upon $\bar{J}_s$ through ($\bar{e}^{sc}$, $\bar{h}^{sc}$) of (116). Clearly, from the presence of ($\bar{E}^{sc\pm}$, $\bar{H}^{sc\pm}$) in (116), an independent term is contributed to (121) via $\bar{e}_e$ and $\bar{e}_m$. For small apertures, Eq. (113) is not needed, since the aperture size and shape are characterized by means of the polarizabilities, and one has only to solve (121).

3.4 Sample Results

As examples illustrative of the concepts discussed above, we present results depicting the current induced on a finite-length wire, parallel to and behind an aperture-perforated screen. Two apertures are considered; one is a finite-length, narrow slot and the other is a small elliptic hole. These data have been taken from [3,14,15].

In Fig. 24 is displayed the wire-slot geometry and the wire cur-

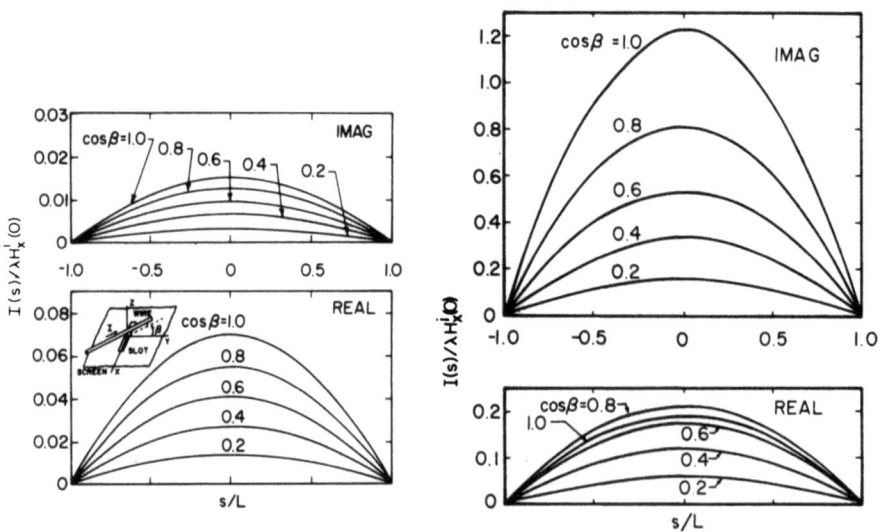

Fig. 24 Current on wire illuminated through slotted screen
($2w/\lambda = 0.05$; $a/\lambda = 0.001, 2L/\lambda = 0.5$; $x_c/\lambda = 0$, $y_c/\lambda = 0$, $z_c/\lambda = 0.25$; normal incidence)
(a) $2\ell/\lambda = 0.25$ (b) $2\ell/\lambda = 0.5$

rent for different relative orientations of wire and slot axes. The center of the slot is at (0,0,0) while that of the wire is given by (x_c, y_c, z_c); wire length (2L) and radius (a) together with slot length (2ℓ) and width (2w) are given in the figure caption. The excitation is a normally incident plane wave whose magnetic field at (0,0,0) is $H_x^1(0)$. As seen in Fig. 24, the current on the half-wavelength wire is of resonant shape and, as expected, is larger for $2\ell = \lambda/2$ than for $2\ell = \lambda/4$. Fig. 25 shows the wire current for the same excitation but for $2L = \lambda$. With the wire center over the slot center (Fig. 25a), the wire

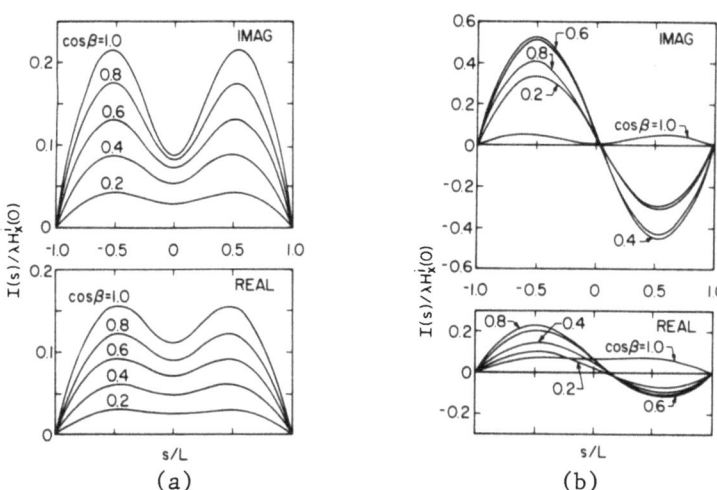

Fig. 25 Current on wire illuminated through slotted screen
($2w/\lambda = 0.05$, $2\ell/\lambda = 0.5$; $a/\lambda = 0.001$, $2L/\lambda = 1.0$; $y_c/\lambda = 0$, $z_c/\lambda = 0.125$; normal incidence)
(a) $x_c/\lambda = 0$ (b) $x_c/\lambda = 0.25$

excitation is an even function and no antiresonant current can be excited, whereas the large antiresonant current is excited in Fig. 25b with the wire center displaced for the z axis. Next we consider the problem described above for the half-wavelength wire but with the slot replaced by a small elliptic aperture oriented with semimajor axis along the x axis (see insert of Fig. 24a). The wire radius is 0.001λ, its center is $\lambda/4$ above that of the ellipse, and the excitation is normally incident upon the screen. The current at the wire center is tabulated below for various aperture sizes and for $\beta = 0°$. Subject to rotation of the wire, the current at a point on the wire varies proportionally to $\cos \beta$.

TABLE 1

Current at center of thin wire, excited by field which penetrates an elliptic aperture in a planar, conducting screen

Aperture Size	Wire Current (Center)
$\frac{w}{\lambda} = 0.02$ $\frac{\ell}{\lambda} = 0.05$	$\dfrac{I(0)}{\lambda H_x^i(0)} = -(2.31 + j0.53)10^{-3}$
$\frac{w}{\lambda} = 0.02$ $\frac{\ell}{\lambda} = 0.067$	$\dfrac{I(0)}{\lambda H_x^i(0)} = -(4.66 + j1.07)10^{-3}$
$\frac{w}{\lambda} = 0.02$ $\frac{\ell}{\lambda} = 0.10$	$\dfrac{I(0)}{\lambda H_x^i(0)} = -(12.78 + j2.99)10^{-3}$

ℓ = length of semimajor axis
w = length of semiminor axis

Additional data are available in [3,14,15].

4. APERTURES IN CAVITIES AND WAVEGUIDES

4.1 Admittance-matrix representation

Fig.26 shows two arbitrary regions, 1 and 2, coupled through an aperture A, with each region containing arbitrary volume sources. The field in 1 can be written as [16]

$$\bar{H}_1 = \bar{H}_1^{sc} + \bar{H}_1^a(\bar{M}_s) \qquad (122)$$

where $\bar{H}_1^{sc}$ is the field in the presence of the short-circuited aperture_ and $\bar{H}_1^a$ is the contribution in region 1 due to the tangential $\bar{E}$ in the aperture, or to the equivalent magnetic current,

$$\bar{M}_s = \hat{n} \times \bar{E} \qquad \text{in A} \qquad (123)$$

introduced previously, where $\hat{n}$ is the normal directed <u>outward</u> from region 1. The functional relationship $\bar{H}_1^a(\bar{M}_s)$ has been given previously, in (1) and (2), for a half-infinite space with plane boundary. It will be given in 4.2 for a cavity and in 4.3 for a waveguide. Similarly, in region 2,

$$\bar{H}_2 = \bar{H}_2^{sc} + \bar{H}_2^a(-\bar{M}_s) = \bar{H}_2^{sc} - \bar{H}_2^a(\bar{M}_s) \qquad (124)$$

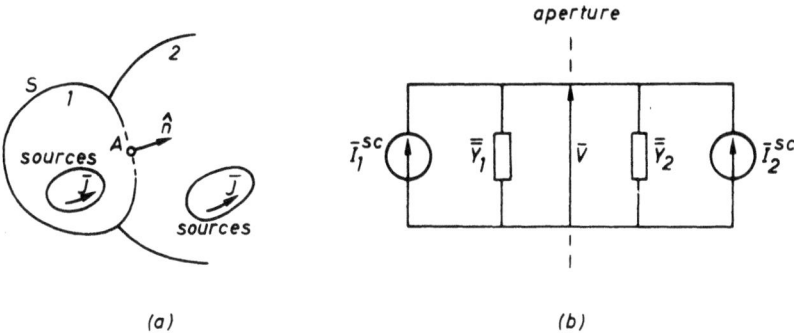

Fig. 26 (a) Regions coupled through an aperture
(b) Equivalent circuit

where the dependence of $\overline{H}_2^a$ is upon the magnetic current $(-\hat{n}) \times \overline{E}$, i.e. the negative of that in (123). Enforcing continuity of $\overline{H}_t$ in the aperture yields

$$\left[\overline{H}_1^a(\overline{M}_s) + \overline{H}_2^a(\overline{M}_s)\right]_t = \left[\overline{H}_2^{sc} - \overline{H}_1^{sc}\right]_t \quad \overline{r} \text{ in A} \quad (125)$$

This is the functional equation for $\overline{M}_s$, which is deceivingly simple in appearance but which leads to laborious calculations in practice. Computations typically are based upon the method of moments. One writes

$$\overline{M}_s = \sum_1^N V_n \overline{M}_n \quad (126)$$

where $\overline{M}_n$ is a set of vector expansion functions and the V_n's are unknown constants. From (125)

$$\left[\sum_1^N V_n \overline{H}_1^a(\overline{M}_n) + \sum_1^N V_n \overline{H}_2^a(\overline{M}_n)\right]_t = \left[\overline{H}_2^{sc} - \overline{H}_1^{sc}\right]_t \quad (127)$$

Introducing a set of N vector testing functions $\overline{W}_m$ and taking scalar products yields

$$\sum_1^N V_n \langle \overline{W}_m, \overline{H}_1^a(\overline{M}_n) \rangle + \sum_1^N V_n \langle \overline{W}_m, \overline{H}_2^a(\overline{M}_n) \rangle = \langle \overline{W}_m, \overline{H}_2^{sc} \rangle - \langle \overline{W}_m, \overline{H}_1^{sc} \rangle \quad (128)$$

for each $\overline{W}_m$ in the testing set (typically N). Notice the absence of the subscript "t" in (128). Such is not needed because the $\overline{W}_m$ are vectors tangential to the surface of the metallized aperture and the scalar product above, $\langle a, b \rangle$, embodies the dot product of $\overline{a}$ and $\overline{b}$. The N equations of the type of (128) can be written in matrix form as

$$\overline{\overline{Y}}_1 \cdot \overline{V} + \overline{\overline{Y}}_2 \cdot \overline{V} = \overline{I}_1^{sc} - \overline{I}_2^{sc} \quad (129)$$

where the mn element of $\overline{\overline{Y}}$ is $-\langle \overline{W}_m, \overline{H}^a(\overline{M}_n) \rangle$. The minus sign is chosen on the basis of power consideration [16]. The column vectors $\bar{V}$ and $\bar{I}^{sc}$ have respective components V_n and $\langle \bar{W}_m, \bar{H}^{sc} \rangle$. Matrix equation (129) can be given the intuitive network interpretation shown in Fig. 26b, which should greatly clarify matters. It shows, for instance, that the problem has been divided into two mutually exclusive parts. Once Y_1 has been determined for a given aperture and region 1 (half space, waveguide, cavity...), it can be used for problems involving any region 2. The second region is accounted for through its Y_2, without any influence on Y_1. It is also clear that the sources do not influence Y but are introduced only through the $\bar{I}$ vectors. The application of the equivalent circuit concept to a rectangular aperture in a screen separating two half-spaces can be found in [17].

4.2 Aperture in the wall of a cavity

(a) normal mode expansion

The expression $\bar{H}_1^a(\bar{M}_s)$ takes the following form in a volume bounded by perfectly conducting walls (region 1 in Fig. 26)

$$\bar{H}_1^a = -\frac{1}{j\omega\mu} \sum_m \bar{g}_m(\bar{r}) \iint_A \bar{M}_s \cdot \bar{g}_m \, dS + j\omega\varepsilon \sum_m \frac{\bar{h}_m(\bar{r})}{(k^2-k_m^2)} \iint_A \bar{M}_s \cdot \bar{h}_m \, dS \quad (130)$$

The symbols $\bar{g}_m$ and $\bar{h}_m$ denote, respectively, the irrotational and solenoidal magnetic eigenvectors of the cavity [2]. The vector $\bar{g}_m$ is $\nabla \psi_m$, where ψ_m is a Neumann eigenfunction, satisfying

$$\nabla^2 \psi_m + \nu_m^2 \psi_m = 0$$
$$\frac{\partial \psi_m}{\partial n} = 0 \quad \text{on } S \quad (131)$$

It is seen, from (130), that the terms in $\bar{g}_m$ do not produce any resonant effects. Resonances are associated with the $\bar{h}_m$, which satisfy

$$-\nabla \times \nabla \times \bar{h}_m + k_m^2 \bar{h}_m = \bar{0}$$
$$\hat{n} \times \nabla \times \bar{h}_m = \bar{0} \quad \text{on } S \quad (132)$$

Both $\bar{g}_m$ and $\bar{h}_m$ are assumed normalized, in the sense that $\iiint_V |\bar{g}_m|^2 dV = 1$ and $\iiint |\bar{h}_m|^2 dV = 1$. Eq. (130) can be put in the form

$$\bar{H}_1^a(\bar{r}) = \iint_A \bar{M}_s(\bar{r}') \cdot \mathcal{H}_1(\bar{r}|\bar{r}') dS' \quad (133)$$

where the "source" dyadic $\bar{\bar{\mathcal{H}}}_1$ has an evident representation in terms of the $\bar{g}_m$ and $\bar{h}_m$ derived for region 1. In the vicinity of a (sharp) resonance, as $k = \omega\sqrt{\varepsilon\mu}$ approaches k_m, the sum in (130) can be well-approximated by a single term (the dominant term):

$$\overline{H}_1 = F \, \overline{h}_m \tag{134}$$

where F has the general form [8]

$$F = -\frac{1}{2k_m} \frac{\iiint_V \overline{J}.\nabla \times \overline{h}_m \, dV}{\Delta k - \frac{jZ}{\eta} \iint_\Sigma |\overline{h}_m|^2 dS} + \frac{j}{2\eta} \frac{\iint_A \overline{M}_s . \overline{h}_m \, dS}{\Delta k - \frac{jZ}{\eta} \iint_\Sigma |\overline{h}_m|^2 dS} \tag{135}$$

In this formula :

- $\Delta k = k - k_m$ represents, to within a factor $\frac{2\pi}{c}$, the deviation of the operating frequency from the resonant value k_m.
- $\overline{J}$ represents the volume sources in the cavity. The first term in (135), therefore, is the coefficient of the short circuit field $\overline{H}_1^{sc}$.
- Z is the impedance of a patch of material covering an area Σ of the boundary surface S, while $\eta = \sqrt{\mu/\varepsilon}$ is the characteristic resistance of the medium filling the cavity. The imaginary part of Z introduces a shift in the resonant frequency; the real part is associated with losses and is best incorporated by means of the Q-factor concept. This is apparent when the denominator in (135) is rewritten as

$$\Delta k - \frac{jZ}{\eta} \iint_\Sigma |\overline{h}_m|^2 dS = \Delta k - \Delta_2 - \frac{jk_m}{2Q} \tag{136}$$

- Losses in the dielectric can be accounted for by writing

$$k^2 = \omega^2 \varepsilon \mu - j\omega\mu\sigma = \omega^2 \varepsilon \mu (1 - \frac{j}{Q_d}) \tag{137}$$

in (130). The symbol Q_d stands for the quality factor ($\omega\varepsilon/\sigma$) of the material, and now in (136) one writes

$$\frac{1}{Q} = \frac{1}{Q_d} + \frac{1}{Q_w} \tag{138}$$

where Q_w is associated with the losses in the wall.

The eigenvectors are known in closed analytical form for a few shapes only, such as the sphere, the coaxial cavity and the rectangular parallelepiped [18]. In the vicinity of a resonance it suffices to know the configuration of a <u>single</u> mode, the resonant one. In certain applications, however, the cavity is fairly large with respect to λ. It is known (and this easily can be checked in the simple example of the parallelepiped) that the resonant frequencies crowd together at higher frequencies. In other words, the number of resonances "per Hz" increases. It follows that the operating frequency, in the higher range of the spectrum, always lies in the vicinity of several resonant frequencies. As a result, multimode excitation occurs, and the strong field maxima and minima which occur in monomode excitation (the "standing wave pattern") tend to disappear, as the strong fields of a given mode tend to compensate for the low fields of another. One therefore obtains a fairly uniform distribution of energy density throughout the cavity, which is a desirable feature for a microwave oven for example. The uniformity can be further increased by inserting rotating blades in the cavity, the effect of which is to reshuffle the field distribution of the modes as the blades rotate.

The lower end of the spectrum, from D.C. to the lowest resonant frequency, is of importance for e.g. the (undesirable) penetration of MHz-range signals through an opening in a Faraday cage [19], or the penetration into an aircraft of pulses whose spectrum is rich at the lower end. The low frequency problem can be split into two parts. First, there is the electrostatic problem of the cavity immersed in the zero-order electric field $\bar{E}_o^1 = -\nabla\phi_o^1$. The field in the vicinity of the cavity is of the form $\bar{E} = -\nabla\phi$, where ϕ is constant along the wall S_W. The penetration of the electric field into the cavity is suggested in Fig. 27a. Second, there is

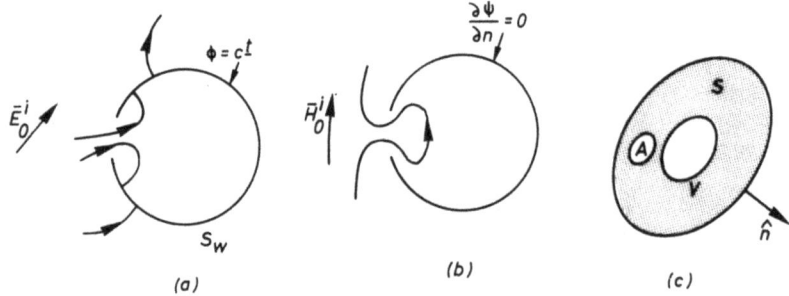

Fig. 27 Aperture-coupled cavities

the magnestotatic problem of determining $\bar{H} = -\nabla\psi$, subject to $\hat{n}.\bar{H} = 0$ on S_W. The problem can be solved by expanding ψ in the Neumann eigenfunctions ψ_m defined in (131). Alternatively, the low frequency $\bar{H}$ can be found from the first term of (130). This term is clearly dominant as ω approaches zero, and it is easy to show that it does not become infinite at $\omega = 0$, although it contains a coefficient $\frac{1}{\omega\mu}$. Indeed, from classical vector analysis [2],

$$\iint_A \bar{M}_s . \bar{g}_m dS = \iint_{A+S_W} \nabla\psi_m . (\hat{n} \times \bar{E}) dS =$$

$$= \iint_{A+S_W} \nabla.(\psi_m(\hat{n} \times \bar{E})) dS - \iint_{A+S_W} \psi_m \nabla.(\hat{n} \times \bar{E}) dS$$

$$= \iint_A \psi_m \hat{n}.\nabla \times \bar{E} \, dS = -j\omega\mu \iint_A \psi_m \hat{n}.\bar{H} \, dS \qquad (139)$$

It is seen that the $\omega\mu$ factors cancel. From (139), it is a simple matter to show that the terms in $\bar{g}_m$ in (130) represent $(-\nabla\psi)$, as expected.

A few additional remarks are in order. First, the $\bar{g}_m$ family in a doubly connected cavity (a coaxial resonator, or the accelerator "doughnut" shown in Fig. 27c) must include a harmonic vector $\bar{g}_o$ satisfying [2,20]

$$\nabla \cdot \bar{g}_o = 0 \quad \text{in } V$$
$$\nabla \times \bar{g}_o = \bar{0} \quad \text{in } V \quad (140)$$
$$\hat{n} \cdot \bar{g}_o = 0 \quad \text{on } S$$

Second, it is well-known that the fields scattered by the outer cavity wall can be attributed to induced dipole moments $\bar{P}_e$ and $\bar{P}_m$. The presence of an aperture modifies these moments in a manner which can be determined theoretically [21].

(b) the coupling problem

Eq. (130) gives the $\bar{H}^a(\bar{M}_s)$ relationship for the interior of a cavity. To obtain an equation for $\bar{M}_s$, it is necessary to derive a similar expression for the region outside the cavity. When the second region is another resonator, (130) is again the desired relationship. If the exterior region, however, is an open region of arbitrary boundary, the solution must be obtained by a (exceedingly difficult) numerical procedure. A case in point is the "rectangular" cavity, for which separation of variables yields the interior normal modes, but not the exterior fields. The problem becomes tractable, however, when the cavity aperture radiates into a half-space, for which the $\bar{H}^a(\bar{M}_s)$ relationship is

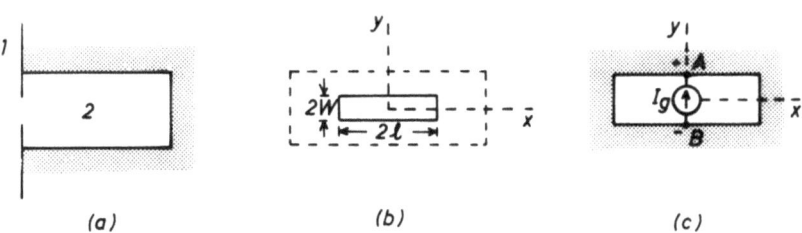

Fig. 28 Rectangular cavity with rectangular aperture

available (Fig. 28a). The cavity/half-space configuration is more than a mathematical nicety. It represents an array element suitable for aircraft and space vehicle antennas, as it is readily flush-mounted and exhibits low mutual coupling between elements. The cavity itself can be loop-or probe-fed from strip or coaxial lines and the coupling aperture is often rectangular (Fig. 28b). For a slot of arbitrary size and shape, one must solve an integral equation for $\overline{M}_s$, e.g. by the method of moments [22]. To obtain good convergence it is necessary to include not only a sufficient number of basis function in the slot but also, away from resonance, a sufficient number of normal modes in the expansion for $\overline{H}$ (Eq.(130)). This latter number must increase as the size of the aperture decreases.

Consider now a slot which is excited by a coaxial line, the conductors of which are connected at A and B (Fig. 28c). The aperture excitation can be represented by a current generator of surface current density $\overline{J}_s = I_g \delta(x)\hat{y}$. This current creates a discontinuity of $\overline{H}_{tang}$ across the aperture, viz. [23]

$$H_{1x} - H_{2x} = I_g \delta(x) \tag{141}$$

If the slot is very narrow, the tangential electric field can be written as (See (97))

$$\overline{E} = -V(x) \frac{1}{\sqrt{w^2 - y^2}} \hat{y} \tag{142}$$

The polarity of V is as indicated in Fig. 28c. The unknown of the problem is now the scalar function $V(x)$, for which an integro-differential equation can be obtained by expressing $\overline{H}_1^a$ and $\overline{H}_2^a$ in terms of $V(x)$ and applying condition (141). The resulting equation is of the type encountered in the theory of linear antennas [23]. It follows that $V(x)$ varies as the current along the antenna. For a center-fed slot, for example, the sinusoidal law

$$V(x) = V(0) \sin k(\ell - |x|) \tag{143}$$

is a good approximation for slot lengths 2ℓ smaller than λ. For a (passive) illuminated slot, $V(x)$ will be given by $I(x)$ on a passive antenna. If $\overline{H}$ along the slot is uniform, a law $V(x) = V(0) \cos \frac{\pi x}{2\ell}$ will be satisfactory, in agreement with Babinet's principle (see Par. 1.1). The problem now reduces to the determination of $V(0)$ through enforcement of (141) at a single point [24]. The relationship between $V(0)$ and I_g at that point is of the form

$$I_g = Y V(0) \tag{144}$$

where Y is the (scalar) admittance of the region under consideration. The admittance can be obtained from

$$\frac{1}{2} \iint_A \hat{n} \cdot (\overline{E} \times \overrightarrow{H}^*) dS = \frac{1}{2} VI^* = \frac{1}{2} Y^* |V|^2 \tag{145}$$

where $\hat{n}$ is the unit vector pointing into the region. The left-hand member i.e., the flux of the complex Poynting vector, can be found from an application of the divergence theorem. For a lossless cavity, for example,

$$\frac{1}{2} \iint_A \hat{n} \cdot (\overline{E} \times \overline{H}^*) dS = j\omega \iiint_v \left[\frac{\mu}{2}|\overline{H}|^2 - \frac{\varepsilon}{2}|\overline{E}|^2 \right] dV \qquad (146)$$

The fields $\overline{E}$ and $\overline{H}$ are given by formulas such as (130), where $\overline{M}_s$ is expressed in terms of (142) and (143). Numerical and experimental results can be found in [24]. The equivalent circuit of Fig. 26 now becomes a classical (scalar) circuit.

A structure where symmetry of revolution allows a degree of simplification is shown in Fig. 29. Here, the various fields, together with the magnetic current $\overline{M}_s$, can be expanded in a Fourier series with respect to the azimuth angle ϕ. Fig. 29 illustrates

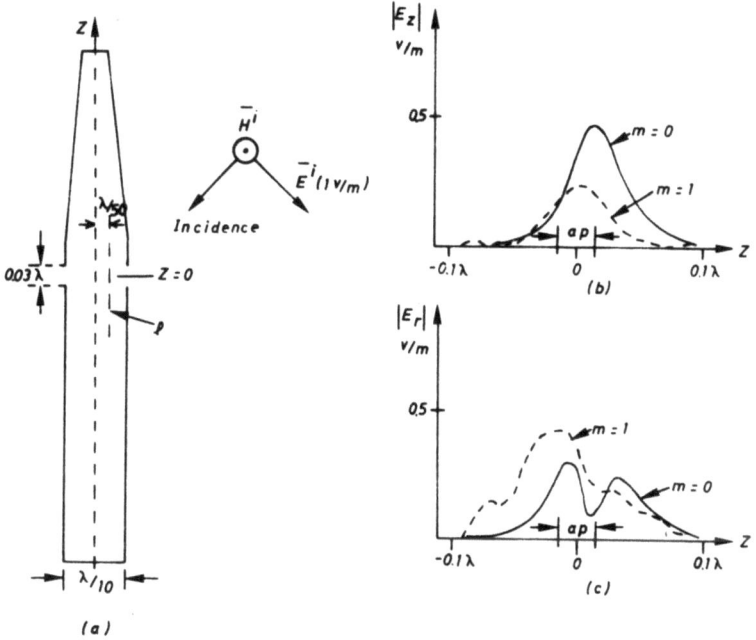

Fig. 29 (a) Body of revolution and incident wave; (b) Axial electric field along the straight line ℓ; (c) Radial electric field along ℓ. (from Ref. 25).

data indicative of the penetration of $\overline{E}$ into the cavity through a ring shaped aperture [25]. The Fourier components of interest are m = 0 (ϕ independent) and m = 1. Higher order terms in $e^{\pm jm\phi}$ are found to be more than 20dB below those for m = 0 and m = 1.

In certain cases (ellipsoids, spheroids, spheres,...) separation of variables can be applied to interior and exterior field analyses, and the internal and external expansion coefficients can be obtained by field matching in the aperture. This method has been applied to a sphere with a circular aperture (Fig. 30a), in which case expansions in spherical harmonics can be utilized [26]. To

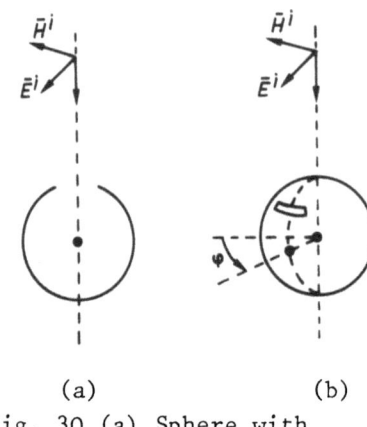

Fig. 30 (a) Sphere with circular aperture
(b) Sphere with azimuthal aperture

accelerate convergence, it is important to take into account the known behavior of the fields near the edge of the aperture. In the configuration of Fig. 30b, the aperture is a thin slot which might be used, for example, as a telemetry slot antenna on the surface of a satellite. The tangential $\bar{E}$ in the thin slot is approximated as in (142) and (143), the unit vector being now $\hat{\theta}$ and the sinusoidal variation being with respect to ϕ. Here again, radiation and cavity admittances, relating E_θ to H_ϕ, can be determined [27].

4.3 Apertures in waveguides

(a) fields in the waveguide

The magnetic field in the waveguide of Fig. 31a is given by [2]

$$\bar{H}_{tr} = \underbrace{\Sigma I_m(z) \hat{z} \times \nabla\phi_m}_{\text{TM modes}} + \underbrace{\Sigma I_n(z) \nabla\psi_n + \Sigma \mathcal{J}_n(z) \psi_n \hat{z}}_{\text{TE modes}} \quad (147)$$

where ϕ_m denotes a Dirichlet eigenfunction of the cross-section and ψ_n a Neumann eigenfunction. The eigenvectors are normalized, i.e., $\iint_S |\nabla\phi_m|^2 dS = 1$ and $\iint_S |\nabla\psi_n|^2 dS = 1$. In a doubly-connected cross-section, e.g., a coaxial line, the eigenvector $\hat{z} \times \nabla\phi_o$, which represents the magnestotatic field in the classical two-wire line, must be included. For arbitrary time-dependence, I_m, I_n and $\mathcal{J}_n$ depend on z and t and must satisfy partial differential equations [2]. For a sinusoidal time dependence there are the characteristic

$$\frac{d^2 I_m}{dz^2} + (k^2 - \mu_m^2) I_m = \mu_m^2 \iint_S (\bar{J} \cdot \hat{z}) \phi_m dS - \frac{d}{dz} \iint_S \bar{J} \cdot \nabla\phi_m dS$$
$$+ j\omega\epsilon \int_c (\hat{n} \times \bar{E}) \cdot (\hat{z} \times \nabla\phi_m) dc \quad (148)$$

$$\frac{d^2 I_n}{dz^2} + (k^2 - \nu_n^2) I_n = -\frac{d}{dz} \iint_S \bar{J} \cdot (\nabla\psi_n \times \hat{z}) dS - \frac{k^2 - \nu_n^2}{j\omega\mu} \int_c (\hat{n} \times \bar{E}) \cdot \nabla\psi_n dc$$
$$- \frac{\nu_n^2}{j\omega\mu} \frac{d}{dz} \int_c (\hat{n} \times \bar{E}) \cdot \psi_n \hat{z} \, dc \quad (149)$$

numbers μ_m^2 are defined by (131) while ν_n^2 are defined by the same differential equation but subject to the boundary condition $\phi_m = 0$ for the Dirichlet eigenfunctions. The solution of (148) and (149) requires knowledge of the boundary conditions at the end planes $z = z_1$ and $z = z_2$. With short circuits there, for example, one must require $\overline{E}_{tr} = \overline{0}$, the enforcement of which involves the use of an expansion like (147) for $\overline{E}_{tr}$ [2]. Once $\underline{I}_n$ is found, $\mathcal{J}_n$ follows from

$$\mathcal{J}_n = \frac{\nu_n^2}{k^2-\nu_n^2} \left[j\omega\varepsilon \int_c (\hat{n} \times \overline{E}) \cdot \psi_n \hat{z} \, dc - \iint_S \overline{J} \cdot (\nabla\psi_n \times \hat{z}) dS - \frac{dI_n}{dz} \right] \quad (150)$$

The terms in $\overline{J}$ in the riht-hand members of (148) and (149) generate the short circuit fields $\overline{H}^{sc}$, while the terms in $\hat{n} \times \overline{E}$ produce the field $\overline{H}^a$ due to the aperture. It is through solution of (148) to (150) that the $\overline{H}^a(\hat{n} \times \overline{E}) = \overline{H}^a(\overline{M}_s)$ relationship can be enforced.

Consider first an aperture in the side-wall of a waveguide (Fig. 31a) and assume that a single mode propagates. For a coaxial line, this would be the TEM mode in $\hat{z} \times \nabla\phi_o$; for a simply connected cross-section, the TE mode in $\nabla\psi_1$. For the latter, the mode-current in an infinite guide, to the left of the aperture i.e., $z < z_a$, is given by

$$I_1(z) = -\frac{1}{2\omega\mu} e^{-j\gamma_1|z|} \int_{z_a}^{z_b} (\gamma_1 A_1 + j\nu_1^2 B_1) e^{-j\gamma_1 z'} dz' \quad (151)$$

where $\gamma_1 = (k^2-\nu_1^2)^{1/2}$ and where the "forcing functions" are

$$A_1(z) = \int_c (n \times E) \cdot \nabla\psi_1 dc \qquad B_1(z) = \int_c (\hat{n} \times \overline{E}) \cdot \psi_1 \hat{z} dc$$

These vanish outside the aperture region. To the right of the aperture, i.e., for $z > z_b$,

$$I_z(z) = -\frac{1}{2\omega\mu} e^{-j\gamma_1 z} \int_{z_a}^{z_b} (\gamma_1 A_1 - j\nu_1^2 B_1) e^{j\gamma_1 z'} dz' \quad (152)$$

Expressions (151) and (152) allow one to compute the radiated fields, to the left and right of the aperture, in terms of

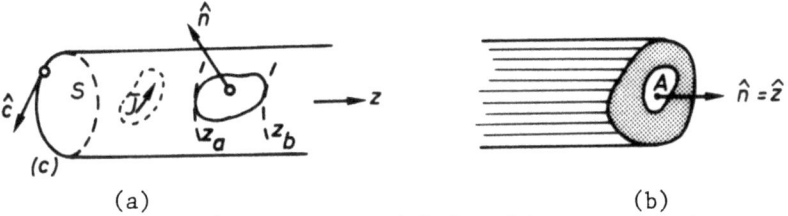

(a)   (b)

Fig. 31 Waveguide with aperture (a) in side wall;(b) in end face.

$\hat{n} \times \overline{E}$ in A. For field matching purposes, however, one must have an expression for I_1 <u>in the aperture region</u> itself, i.e., between z_a and z_b:

$$I_1(z) = -\frac{1}{2\omega\mu} e^{-j\gamma_1 z} \int_{z_a}^{z} (\gamma_1 A_1 - j\nu_1^2 B_1) e^{j\gamma_1 z'} dz'$$
$$-\frac{1}{2\omega\mu} e^{j\gamma_1 z} \int_{z}^{z_b} (\gamma_1 A_1 + j\nu_1^2 B_1) e^{-j\gamma_1 z'} dz' \quad (153)$$

The corresponding expression for a damped TE mode is

$$I_n(z) = -\frac{1}{2j\omega\mu} e^{-\delta_n z} \int_{z_a}^{z} (\delta_n A_n + \nu_n^2 B_n) e^{\delta_n z'} dz'$$
$$-\frac{1}{2j\omega\mu} e^{\delta_n z} \int_{z}^{z_b} (\delta_n A_n - \nu_n^2 B_n) e^{-\delta_n z'} dz' \quad (154)$$

where $\delta_n^2 = \nu_n^2 - k^2$. In a TM mode, the forcing function is

$$C_m(z) = \int_c (\hat{n} \times \overline{E}) \cdot (\hat{z} \times \nabla\phi_m) dc = -\int_c \frac{\partial \phi_m}{\partial n} (\hat{z} \cdot \overline{E}) dc \quad (155)$$

in terms of which

$$I_m(z) = -\frac{j\omega\varepsilon}{2\delta_m} e^{-\delta_m z} \int_{z_a}^{z} C_m e^{\delta_m z'} dz' - \frac{j\omega\varepsilon}{2\delta_m} e^{\delta_m z} \int_{z}^{\infty} C_m e^{-\delta_m z'} dz' \quad (156)$$

where $\delta_m^2 = \mu_m^2 - k^2$. In a coaxial line, where the lowest mode is the TEM mode in $\hat{z} \times \nabla\phi_0$, the mode coefficient is

$$I_o(z) = -\frac{1}{2\eta} e^{-jkz} \int_{z_a}^{z} C_o e^{jkz'} dz' - \frac{1}{2\eta} e^{jkz} \int_{z}^{z_b} C_o e^{-jkz'} dz' \quad (157)$$

Here k and η are for the medium filling the coaxial structure. When the aperture is located in the end-plane of a semi-infinite waveguide (Fig. 31b), the transverse magnetic field in the guide is of the form

$$\overline{H}_{tr} = -\sum_m \frac{j\omega\varepsilon}{\delta_m} \left[\iint_A (\hat{z} \times \overline{E}) \cdot (\hat{z} \times \nabla\phi_m) dS \right] e^{\delta_m z} (\hat{z} \times \nabla\phi_m)$$
$$- \sum_{n \neq 1} \frac{\delta_n}{j\omega\mu} \left[\iint_A (\hat{z} \times \overline{E}) \cdot \nabla\psi_n dS \right] e^{\delta_n z} \nabla\psi_n$$
$$- \frac{\gamma_1}{\omega\mu} \left[\iint_A (\hat{z} \times \overline{E}) \cdot \nabla\psi_1 dS \right] e^{j\gamma_1 z} \nabla\psi_1 \quad (158)$$

The last term represents the propagating wave, which is responsible for power dissipation and which contributes the real part in the Y dyadic of the waveguide. Eq. (158) easily can be written in the form of (133), with $\mathcal{H}$ having an obvious representation in terms of the eigenvectors $\nabla\psi_1$, $\nabla\psi_n$, and $\hat{z} \times \nabla\phi_m$.

(b) coupling problems

Using the $\overline{H}^a(\overline{M}_s)$ relationships for half-spaces, resonators, and waveguides, one finds it possible to solve a large number of coupling problems. The actual formulation is invariably tedious, so we omit details and merely discuss solutions available in the abundant literature. Fig. 32 shows examples of waveguide-to-half-space couplings [28,29].

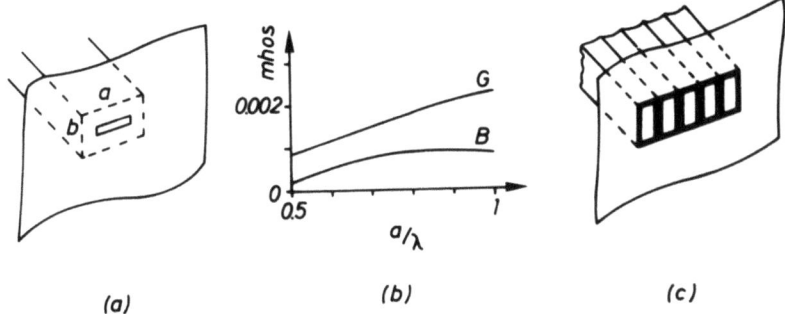

(a)   (b)   (c)

Fig. 32 (a) Rectangular waveguide, end-coupled to a half space
(b) Load admittance   (c) End-face array

Fig. 32b gives the equivalent aperture admittance seen by the dominant mode in the guide of Fig. 32a, assumed open-ended and of dimensions $a/b = 2.25$. In the wall-slot radiator shown in Fig. 33a, the thickness $\underline{t}$ of the wall can be taken into account, but this leads to the solution of a pair of coupled integral equations [30].
Two waveguides coupled through a hole in a common wall or end face also can be analyzed by the techniques of this section. If the aperture is a narrow slot [31,32,33], the tangential slot field

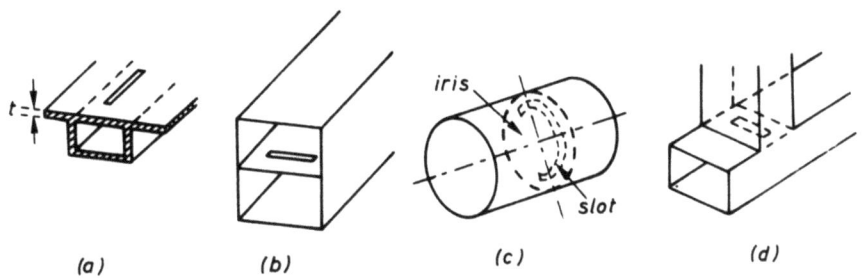

(a)   (b)   (c)   (d)

Fig. 33 Slot coupling of waveguides
       (a) to half space   (b) side-wall coupling
       (c) end-wall coupling   (d) mixed coupling

can be written in a form similar to (142) and (143), which is introduced into (153) for the side-wall coupling or into (158) for the end-plane coupling.

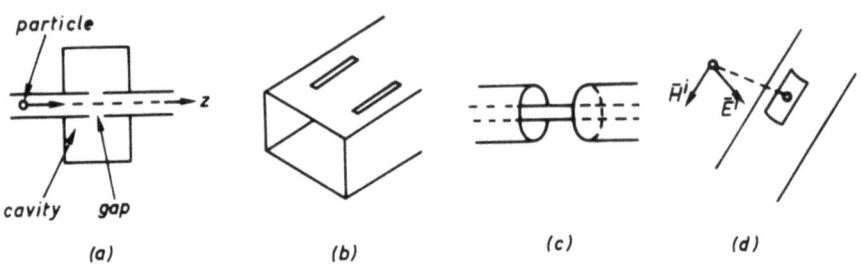

Fig. 34 Side wall coupling to waveguide

A cavity coupled to a waveguide is another configuration for which $\overline{H}^a(M_s)$ is available in both regions. An early publication [34] examines an infinite circular guide, e.g., an accelerator "doughnut," surrounded by a coaxial RF-cavity (Fig. 34a). The main problem is to investigate the penetration of the fields into the waveguide, which is below cut-off at the first resonance of the cavity, with the main unknown being E_z in the aperture. A particle experiences a momentum kick when it flies through the field region, concentrated in the vicinity of the gap.

When the waveguide is coupled to a general unbounded region, the numerical problem becomes exceedingly difficult. The difficulty can be circumvented when the outer wall, in which the hole is cut, is a plane section as in Fig. 34b. When the aperture is sufficiently far from the edge and the plane section is not small relative to λ, one can assume, with reasonable accuracy, that the plane section is infinite in extent. The problem is now of the kind shown in Fig. 33a, and it becomes possible to investigate configurations with thin slots, as in Fig. 34b, and to determine the self and mutual admittances of the slots [35], e.g., Y_{ii} and Y_{ik}. For a few waveguide cross-sections, the exterior problem can be solved by separation of variables. An obvious example is the coaxial line with a circumferential slot in the outer shell (Fig. 34c). Assume that the line carries its lowest (TEM) mode. When the gap is narrow with respect to λ, E_z in the gap can be assumed constant [36], or, better, of the general form of (142). With V the voltage across the gap, the scalar admittance formulation again can be utilized [37]. The gap may exist inadvertently, destroying the shielding effect of the line, or it can be created intentionally to launch a guided wave along the external shield [36]. The exterior problem is actually that of the gap-excited, infinite antenna, which has been studied extensively. The problem becomes more complicated when the aperture in the outer shell is of arbitrary shape [2]. A solution has been given for the structure shown in Fig. 34d, for the interior region a coaxial cavity [38].

4.4 Small apertures in cavities and waveguides

The influence of the smallness of the aperture can best be understood by starting from relationships such as (130) and (133). The integrals in these equations contain terms of the form $\iint_A \overline{M}_s \cdot \overline{a}\, dS$ where $\overline{a}$ varies slowly in the aperture when the latter's dimensions are small with respect to both λ and the radii of curvature of the wall. For such a case, the multipole expansion can be utilized. It yields [39]

$$\iint_A \overline{M}_s \cdot \overline{a}\, dS = \overline{a} \cdot \iint_A \overline{M}_s\, dS + \frac{1}{2} \nabla \times \overline{a} \cdot \iint_A \overline{r} \times \overline{M}_s\, dS \qquad (159)$$

Here, $\overline{a}$ and $\nabla \times \overline{a}$ are the values at a reference point in the aperture. From (42) and (45),

$$\iint_A \overline{r} \times \overline{M}_s\, dS = \iint_A \overline{r} \times (\hat{n} \times \overline{E})\, dS = -\frac{\overline{P}_{e2}}{\varepsilon_2} \qquad (160)$$

Similarly, from (46) and Maxwell's equations,

$$\iint_A \overline{M}_s\, dS = j\omega \cdot \frac{1}{j\omega} \iint_A (\hat{n} \times \overline{E})\, dS = jk\eta \frac{\overline{P}_{m2}}{2} \qquad (161)$$

Combining (159), (160) and (161) yields

$$\iint_A \overline{M}_s \cdot \overline{a}\, dS = jk\eta\, \overline{a} \cdot \frac{\overline{P}_{m2}}{2} - \frac{1}{\varepsilon_2} \nabla \times \overline{a} \cdot \frac{\overline{P}_{e2}}{2} \qquad (162)$$

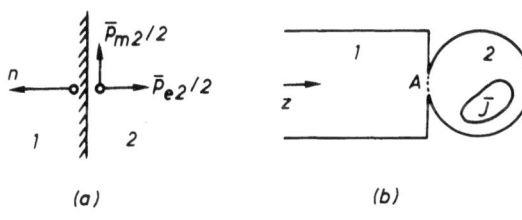

Fig. 35 (a) Equivalent dipole moments
(b) Cavity coupled to end plane of a waveguide

If (162) is introduced into (130) and if the resulting fields are compared with those produced by dipoles $\overline{P}_e$ and $\overline{P}_m$ [2], it will be seen that the aperture may be replaced, with respect to region 2, by moments $(\overline{P}_{e2}/2)$ and $(\overline{P}_{m2}/2)$, located as shown in Fig. 35a. The argument can be repeated for waveguides and other regions.

It remains to determine $\overline{P}_e$ and $\overline{P}_m$. This must, by necessity, be based on Eqs. (125) and (133) and must rely heavily on the singular behavior of the dyadic at $\overline{r} = \overline{r}'$. The various terms of the dyadic become infinite according to some power of $\dfrac{1}{|\overline{r}-\overline{r}'|}$, in a manner which is known from the particular case of the dyadic for a half space [8][40]. In the latter case, the rest of the dyadic is bounded and regular and contributes only higher order terms to the fields in the aperture. The singular terms, however, lead directly to Eqs. (39) and (43) and are therefore responsible for the zero-order fields in the aperture.

Consider now the waveguide-cavity system of Fig. 35b. Away from resonance, the dyadic evidences the usual $|r-r'|$ singularity and the remaining terms, which contain the effects of curvature for example, are bounded. As a result, equations of the type (39) and (43) are still valid, as are

$$\overline{P}_{e2} = -\varepsilon_o \overline{\overline{\pi}}_e \cdot \overline{E}_2^{sc} = -\overline{P}_{e1}$$
$$\overline{P}_{m2} = \overline{\overline{\pi}}_m \cdot \overline{H}_2^{sc} = -\overline{P}_{m1} \tag{163}$$

The short-circuit fields are produced, in this case, by the $\overline{J}$ sources in the cavity. The situation changes in the vicinity of a resonance. The difficulties there can be investigated in a number of ways [41]. It is possible, for example, to start with (130), (134) and (135). We remark that the source dyadic evidences, for perfectly conducting walls, a singularity in $\frac{1}{\Delta k}$ in addition to that in $\frac{1}{|r-r'|}$ and we proceed by writing the cavity fields as [2]

$$\overline{E}_2 = -j\eta F \frac{1}{k_m} \nabla \times \overline{h}_m = -j\eta F \overline{e}_m$$
$$\overline{H}_2 = F \overline{h}_m \tag{164}$$

Here, F is the level of excitation of the mode. To evaluate this quantity, we replace (43) by

$$\nabla_t^2 \frac{1}{2\pi} \iint_A \Theta_o(\overline{r}') \frac{1}{|r-r'|} dS' = \frac{1}{2} \hat{z} \cdot \overline{E}_2 = -\frac{j\eta F}{2} \hat{z} \cdot \overline{e}_m \tag{165}$$

Taking (36) and (42) into account yields

$$(\overline{E}_o)_t = \frac{j\eta F \sqrt{S}}{2} (\hat{z} \cdot \overline{e}_m) \nabla_t \tau_o \tag{166}$$

This expression gives $\hat{n} \times \overline{E}_o = \overline{M}_s$, which can then be inserted into (135), where the term in $\overline{M}_s$ represents the influence of the $\frac{1}{\Delta k}$ singularity. As $\overline{M}_s$ is proportional to F, (135) can be solved for F, the level of excitation of the resonant mode. Once F is known, we write

$$\overline{P}_{e2} = -\overline{P}_{e1} = \frac{j}{c} F \overline{\overline{\pi}}_e \overline{e}_m$$
$$\overline{P}_{m2} = -\overline{P}_{m1} = F \overline{\overline{\pi}}_m \cdot \overline{h}_m \tag{167}$$

The value of F obtained above does not take the transfer of energy into the waveguide into account. This represents losses to the cavity which can be treated in a subsequent approximation by an additional dissipation factor in (138), viz. [8]

$$\frac{1}{Q_{rad}} = \frac{\gamma_1}{4} (\nabla\psi_1 \cdot \overline{\overline{\rho}} \cdot \overline{h}_m)^2 \qquad (168)$$

The steps outlined above allow one to solve a series of problems involving coupled resonant or non-resonant structures. In the case of Fig. 36, for example, they yield the equivalent circuit

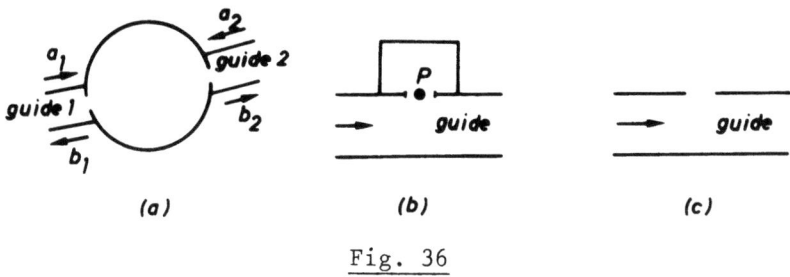

Fig. 36

(a) Cavity filter (b) Sidewall resonator (c) Coupling to free-space

of the cavity loading the waveguide. They also make it possible to find the power transmission coefficient of the cavity filter of Fig. 36a, or the equivalent T and π circuits of the sidewall--coupled cavity shown in Fig. 36b [8]. In the T-circuit, the series impedance turns out to be proportional to $(\nabla\psi_1 \cdot \overline{\overline{\pi}} \cdot \overline{h}_m)$ where $\nabla\psi_1$ and $\overline{h}_m$ represent the values of the eigenvectors at point P. It follows, for example, that, for locations, shapes, and orientations of the aperture which make $(\nabla\psi_1 \cdot \overline{\overline{\pi}}_m \cdot \overline{h}_m)$ zero, the cavity can be represented by a <u>shunt</u> element. Similar methods can be applied to find the equivalent circuit of a small hole in a waveguide wall (Fig. 36c), together with the induced aperture moments $\overline{P}_e$ and $\overline{P}_m$. Often it is necessary to determine the associated radiated fields. If the outer surface is an infinite plane (or a planar surface of dimensions $>>\lambda$), the methods described in Par. 4.3 can be applied. If the surface is curved, a reciprocity theorem can be of assistance [17]. Technological applications are numerous : the design of aperture arrays [42], the study of the (unwanted) penetration of strong electromagnetic pulse through the braided shield of a transmission line [43], and the design of microwave couplers [44,45].

REFERENCES

The references below are by no means exhaustive. They are restricted to those books and articles which are cited in the text. A more extensive bibliography can be found in [3].

1. R.F. Harrington, Time-Harmonic Electromagnetic Fields, Mc Graw-Hill Book Company, New York, 1961, pp. 106-110.
2. J. Van Bladel, Electromagnetic Fields, Mc Graw-Hill Book Company, New York, 1964, pp. 382-387, 56, 297, 501, 619, 217.
3. C.M. Butler, Y. Rahmat-Samii and R. Mittra, "Electromagnetic penetration through apertures in conducting screens", IEEE Trans. Ant. and Prop., Vol. AP-26, pp. 82-93, 1978.
4. C.M. Butler and K.R. Umashankar, "Electromagnetic penetration through an aperture in an infinite, planar screen separating two half spaces of different electromagnetic properties", Radio Science, Vol. 11, pp. 611-619, 1976.
5. C.J. Bouwkamp, "Diffraction Theory", Rep. Progr. in Physics, vol. 17, pp. 35-100, 1954.
6. F. De Meulenaere and J. Van Bladel, "Polarizability of some small apertures", IEEE Trans. Ant. and Prop., Vol. AP-25, pp. 198-205, 1977.
7. J. Van Bladel, "Field penetration through small apertures : the first-order correction", Radio Science, vol. 14, pp. 319--331, 1979.
8. J. Van Bladel, "Small hole coupling of resonant cavities and waveguides", Proc. I.E.E., vol. 117, pp. 1098-1104, 1970.
9. R. De Smedt, "Laagfrekwent penetratie door aperturen : 1e orde korrektie-termen", E.E. thesis, Rijksuniversiteit Gent, 1978.
10. D.R. Wilton, A.W. Glisson and C.M. Butler, Numerical solutions for scattering by rectangular bent plates, Contract Report (No. N00123-75-C-1372) for Naval Electronics Laboratory Center, Department of Electrical Engineering, University of Mississippi, University, MS, Oct., 1976.
11. A.W. Glisson, On the development of numerical techniques for treating arbitrarily-shaped surfaces, Ph.D. Thesis, Department of Electrical Engineering, University of Mississippi, University, MS, June, 1978.
12. C.L. Andrews, "Diffraction pattern in a circular aperture measured in the microwave region", J. Appl. Phys., vol. 22, pp. 761-767, 1950.
13. C.M. Butler and D.R. Wilton, "General analysis of narrow strips and slots", accepted to IEEE Trans. Ant. and Prop.
14. C.M. Butler and K.R. Umashankar, "Electromagnetic excitation of a wire through an aperture-perforated, conducting screen", IEEE Trans. Ant. and Prop., vol. AP-25, pp. 456-462, July, 1976.
15. C.M. Butler, "Electromagnetic excitation of an object through an aperture in a conducting screen", IEE International Conference on Antennas and Propagation, Conference Proceedings (Part I), pp. 165-169, London, November, 1978.
16. R.F. Harrington and J.R. Mautz, "A generalized network formulation for aperture problems", IEEE Trans. Ant. and Prop., vol. AP-24, pp. 870-873, 1976.

17. R.F. Harrington and J.R. Mautz, "Electromagnetic transmission through an aperture in a conducting plane", AEU, vol. 31, pp. 81-87, 1977.
18. C.G. Montgomery, Technique of microwave measurements, Mc Graw-Hill Book Co., 1947, pp. 294-308.
19. H.A. Mendez, "On the theory of low-frequency excitation of cavity resonators", IEEE Trans. Microwave Theory and Techniques, vol. MTT-18, pp. 444-448, 1970.
20. K.S.H. Lee, F.C. Yang and K.C. Chen, Cavity excitation via apertures, AIP Memo 12, 1977.
21. J. Van Bladel, "Small apertures in cavities at low frequencies", AEU, vol. 26, pp. 481-486, 1972.
22. K. Suneli, "Numerical solution of the aperture field integral for a wide slot backed by a rectangular cavity", Proc. 7th European Microwave Conf., pp. 680-684, Copenhagen 1977.
23. J. Galejs, "Hallén's method in the problem of a cavity-backed rectangular slot antenna", J. Res. Nat. B Stand., vol. 67D, pp. 237-244, 1963.
24. C.R. Cockrell, "The input admittance of the rectangular cavity-backed slot antenna", IEEE Trans. Ant. and Prop., vol. AP-24, pp. 288-294, 1976.
25. H.K. Schuman and D.E. Warren, "Aperture coupling in bodies of revolution", IEEE Trans. Ant. and Prop., vol. AP-26, pp. 778-783, 1978.
26. T.B.A. Senior and G.A. Desjardins, "Electromagnetic field penetration into a spherical cavity", IEEE Trans. on Electromagn. Compat., vol. EMC-16, pp. 205-208, 1974.
27. M.A. Plonus, "Electromagnetic scattering by slots on a sphere", Proc. I.E.E., vol. 115, pp. 622-626, 1968.
28. J.R. Mautz and R.F. Harrington, Transmission from a rectangular waveguide into half-space through a rectangular aperture, Technical Report TR-76-5, Syracuse University, 1976.
29. J. Luzwick and R.F. Harrington, "A reactively loaded aperture antenna array", IEEE Trans. Ant. and Prop., vol. AP-26, pp. 543-547, 1978.
30. T.V. Khac and C.T. Carson, "Impedance properties of a longitudinal slot antenna in the broad face of a rectangular waveguide", IEEE Trans. Ant. and Prop., vol. AP-21, pp. 708-710, 1973.
31. V.M. Pandharipande and B.N. Das, "Coupling of waveguides through large apertures", IEEE Trans. Microwave Theory and Techniques, vol. MTT-26, pp. 209-212, 1978.
32. Kh.L. Garb and P.Sh. Fridberg, "Theory of weakly-radiating slots", Radio Engineering and Electronic Physics (USSR), vol. 15, pp. 599-606, 1970.
33. V.M. Pandharipande and B.N. Das, "Slot-coupled tee junction in rectangular guide E plane", IEEE Trans. Microwave Theory and Techniques, vol. MTT-27, pp. 65-69, 1979.

34. E.G. Cristal and J. Van Bladel, "Fields in cavity-excited accelerators", J. Appl. Phys., vol. 32, pp. 1715-1724, 1961.
35. B.N. Das and G.S. Sanyal, "Mutual impedance between two resonant-slot radiators", Proc. I.E.E., vol. 118, pp. 1535-1538, 1971.
36. P. Delogne and R. Liégeois, "Le rayonnement d'une interruption du conducteur extérieur d'un câble coaxial", Ann. Telecom., vol. 26, pp. 85-100, 1971.
37. D.C. Chang, "Equivalent-circuit representation and characteristics of a radiating cylinder driven through a circumferential slot", IEEE Trans. Ant. and Prop., vol. AP-21, pp. 792-796, 1973.
38. S. Safavi-Naini, S.W. Lee and R. Mittra, "Transmission of an EM wave through the aperture of a cylindrical cavity", IEEE Trans. on Electromagn. Compat., vol. EMC-19, pp. 74-81, 1977.
39. J. Van Bladel, "The multipole expansion revisited", AEÜ, vol. 31, pp. 407-411, 1977.
40. J. Van Bladel, "Small holes in a waveguide wall", Proc. I.E.E. vol. 118, pp. 43-50, 1971.
41. J.R. Mautz and R.F. Harrington, "Boundary formulations for aperture-coupling problems", Technical Report TR-79-1, Syracuse University, 1979.
42. A.J. Sangster, "Circularly polarized linear waveguide array", IEEE Trans. Ant. and Prop., vol. AP-21, pp. 704-705, 1973.
43. K.S.H. Lee and C.E. Baum, "Application of modal analysis to braided-shield cables", IEEE Trans. on Electromagn. Compat., vol. EMC-17, pp. 159-169, 1975.
44. M. Kumar and B.N. Das, "Multi-aperture directional coupler using dissimilar transmission lines", Proc. I.E.E., vol. 123, pp. 1299-1301, 1976.
45. M. Kumar and B.N. Das, "Coupled transmission lines", IEEE Trans. Microwave Theory and Techniques, vol. MTT-25, pp. 7-10, 1977.

NATURAL MODE METHODS IN FREQUENCY AND TIME DOMAIN ANALYSIS*

E. K. Miller

Lawrence Livermore Laboratory, Livermore,
California 94550, U.S.A.

1. INTRODUCTION

Electromagnetic fields, as is true of many physical phenomena, are described by second-order differential equations whose source-free solutions are exponential in the time domain. Their corresponding frequency-domain solutions are given by a series of poles in the complex frequency plane. This basic fact has led to the Singularity Expansion Method (SEM)[1,2] as a way to characterize both the transient and spectral EM fields of an object in terms of a set of common poles and residues. In principal, the SEM provides a means to do this for arbitrary excitation. But a condition necessary to implement the SEM is that a method must be available for finding the poles of the object, the realization of which is the primary topic of this discussion.

A number of techniques have been tested for obtaining the EM poles of antennas and scatterers. Some involve exploration of the complex frequency plane,[3] and as such, are not applicable to experimental data. Prony's Method however, is useful not only for computed or measured transient response data, but also, with slight modification, for spectral data as well. Its analytical development is outlined in Section II below. Some introductory numerical results are given in Section III and its application is demonstrated for various types of data and objects in Section IV.

---

*This work was performed under the auspices of the U.S. Department of Energy by the Lawrence Livermore Laboratory under contract #W-7405-ENG-48.

Use of poles for target identification from both transient and spectral data is considered in Section V. The discussion concludes in Section VI with a demonstration of other ways in which poles can arise and be useful in electromagnetics, such as pattern synthesis and the one-dimensional inverse problem.

2. ANALYTICAL BACKGROUND

The "late-time" EM transient response of a conducting object can be written[1,2]

$$f(t) = \sum_{\alpha=1}^{W} R_\alpha e^{s_\alpha t} , \qquad (2.1)$$

where s_α represents the "pole" frequencies and R_α the relative amplitudes with which each of the terms has been excited, with W the number of waveform poles. The frequency-domain equivalent of the exponential series in Eq. (2.1), as obtained from a LaPlace transformation, is

$$F(s) = \sum_{\alpha=1}^{W} R_\alpha / (s - s_\alpha) . \qquad (2.2)$$

Note that, because Eq. (2.1) represents a "real" waveform, the sum must consist of pairs of complex-conjugate terms. These could be arranged in any order. For convenience, we will assume that the first W/2 terms are the poles having positive-frequency components and the second W/2 terms (in the same order) are the complex-conjugate terms. Thus, $R_\alpha = R^*_{W/2+\alpha}$ and $s_\alpha = s^*_{W/2+\alpha}$. Note that, for W poles, there are 2W real unknowns - W for the (W/2) complex R_α's and the other W unknowns for the (W/2) complex s_α's.

2.1 Time-domain Prony

Derivation of the poles from Eq. (2.1) using the Prony technique involves sampling $f(t)$ at D points $t_i = i\delta$, $i = 0, 1, \ldots, D-1$ over a window of width $T = \delta(D - 1)$.[1] We thus rewrite Eq. (2.1) as[4,5]

$$f_i \equiv f(t_i) = \sum_{\alpha=1}^{P} \hat{R}_\alpha \hat{X}_\alpha^i \; ; \; i = 0, 1, \ldots, D-1 \qquad (2.3)$$

where

$$\hat{X}_\alpha = e^{\hat{s}_\alpha \delta} ,$$

P is the number of poles being sought and the carot denotes a parameter-value estimate.

Equation (2.3) can be written

$$\hat{R}_1 + \hat{R}_2 + \ldots + \hat{R}_P = f_0,$$
$$\hat{R}_1\hat{X}_1 + \hat{R}_2\hat{X}_2 + \ldots + \hat{R}_P\hat{X}_P = f_1,$$
$$\hat{R}_1\hat{X}_1^2 + \hat{R}_2\hat{X}_2^2 + \ldots + \hat{R}_P\hat{X}_P^2 = f_2,$$
$$\vdots$$
$$\hat{R}_1\hat{X}_1^{D-2} + \hat{R}_2\hat{X}_2^{D-2} + \ldots + \hat{R}_P\hat{X}_P^{D-2} = f_{D-2},$$
$$\hat{R}_1\hat{X}_1^{D-1} + \hat{R}_2\hat{X}_2^{D-1} + \ldots + \hat{R}_P\hat{X}_P^{D-1} = f_{D-1}. \quad (2.4)$$

Now, suppose X_α represents the roots of a P-th order polynomial, i.e.,

$$\sum_{\alpha=0}^{P} \hat{a}_\alpha X^\alpha = (X - \hat{X}_1)(X - \hat{X}_2) \ldots (X - \hat{X}_P)$$

$$= \prod_{\alpha=1}^{P} (X - \hat{X}_\alpha); \quad a_P = 1. \quad (2.5)$$

Upon multiplying the first equation in Eq. series (2.4) by $\hat{a}_0$, the second by $\hat{a}_1$, etc., and adding the first P of these equations together, a new equation can be found, which upon using Eq. (2.5), can be reduced to

$$\hat{a}_0 f_0 + \hat{a}_1 f_1 + \ldots + f_P = 0.$$

If this operation is repeated D - P - 1 times by successively beginning with the second equation of Eq. series (2.4), ..., up to the D-P-1 equation, the following D-P linear equations will result:

$$\bar{\bar{M}}_t \bar{\hat{a}} = -\bar{f}, \quad (2.6a)$$

where

$$\bar{\bar{M}}_t = \begin{bmatrix} f_0 & f_1 & f_2 & \cdots & f_{P-1} \\ f_1 & f_2 & f_3 & \cdots & f_P \\ \vdots & & & & \\ f_{D-P-1} & f_{D-P} & \cdots & & f_{D-2} \end{bmatrix}, \quad (2.6b)$$

$$\bar{f} = \begin{bmatrix} f_P \\ f_{P+1} \\ \vdots \\ f_{D-1} \end{bmatrix}, \quad (2.6c)$$

and the vector $\bar{\hat{a}}$ represents the coefficients of the polynomial Eq. (2.5) (sometimes called the characteristic equation).

The linear system in Eq. (2.6) provides the basis for obtaining the P coefficients in the polynomial in Eq. (2.5), from whose roots the P pole values are found as

$$\hat{s}_\alpha = \frac{1}{\delta} \ln \hat{x}_\alpha . \quad (2.7)$$

If $D = 2P$, the system is square of order P and can be solved directly for the P real a's. For $D > 2P$, the method of least squares can be used. Finally, the $\hat{R}_\alpha$'s can be found by various means, such as fitting the first P equations of Eq. series (2.4) or by applying the least-squares technique to the entire set.

2.2 Frequency-domain Prony

The treatment of Eq. (2.2) to derive the pole values proceeds in an analogous fashion. First, rewrite Eq. (2.2) as

$$F(s) = \sum_{\alpha=1}^{P} \hat{R}_\alpha \prod_{\beta=1}^{P}{}^\alpha (s - \hat{s}_\beta) / \prod_{\gamma=1}^{P} (s - \hat{s}_\gamma) , \quad (2.8)$$

where the superscript α on the first product signifies omission of the $\beta = \alpha$ term and again the carot denotes an estimated parameter. Upon recognizing that

$$\prod_{\gamma=1}^{P} (s - \hat{s}_\gamma) = \sum_{\gamma=0}^{P} \hat{a}_\gamma s^\gamma ; \ a_P = 1 \quad (2.9a)$$

and

$$\prod_{\beta=1}^{P}{}^\alpha (s - \hat{s}_\beta) = \sum_{\beta=0}^{P-1} \hat{b}_\beta^{(\alpha)} s^\beta ; \ b_{P-1}^{(\alpha)} = 1 , \quad (2.9b)$$

Eq. (2.8) can be expressed as

$$\sum_{\alpha=0}^{P-1}[\hat{a}_\alpha s^\alpha F(s) - \tilde{b}_\alpha s^\alpha] = -F(s)s^P, \qquad (2.10a)$$

where

$$\tilde{b} = \sum_{\alpha=1}^{N} \hat{R}_\alpha \hat{b}_\beta^{(\alpha)}.$$

and it should be noted that coefficients $\hat{a}_\alpha$ and $\tilde{b}_\alpha$ are real for spectra where $F(s) = F^*(s^*)$. This fact can be used to reduce the number of real unknowns in Eq. (2.10a) from 4P to 2P, and is assumed to be the case here.

If $D \geq 2P$ samples for $F_i = F(s = s_i)$ where $i = 0, 1, \ldots, D-1$ are available, Eq. (2.10a) provides the basis for obtaining the P real $\hat{a}_\alpha$ and $\tilde{b}_\beta$ from

$$\overline{\overline{M}}_s \overline{\hat{c}} = -\overline{F} \qquad (2.10b)$$

where

$$\overline{\overline{M}}_s = \begin{bmatrix} F_0 & s_0 F_0 & s_0^2 F_0 & \cdots & s_0^{P-1} F_0 & -1 & -s_0 & \cdots & -s_0^{P-1} \\ F_1 & s_1 F_1 & s_1^2 F_1 & \cdots & s_1^{P-1} F_1 & -1 & -s_1 & \cdots & -s_1^{P-1} \\ \vdots & & & & & & & & \\ F_{D-1} & s_{D-1} F_{D-1} & s_{D-1}^2 F_{D-1} & \cdots & s_{D-1}^{P-1} F_{D-1} & -1 & -s_{D-1} & \cdots & -s_{D-1}^{P-1} \end{bmatrix}$$

$$(2.10c)$$

$$\hat{\overline{c}} = \begin{bmatrix} \hat{a}_0 \\ \hat{a}_1 \\ \vdots \\ \hat{a}_{P-1} \\ \tilde{b}_0 \\ \tilde{b}_1 \\ \vdots \\ \tilde{b}_{P-1} \end{bmatrix}, \quad (2.10d) \qquad \overline{F} = \begin{bmatrix} s_0^P F_0 \\ s_1^P F_1 \\ \vdots \\ s_{D-1}^P F_{D-1} \end{bmatrix} \qquad (2.10e)$$

The P pole frequencies are then obtained as the roots of the Pth-order homogeneous polynomial Eq. (2.9a) from the $\hat{a}_\alpha$. Similarly, $\hat{R}_\alpha$ can be found using Eq. (2.9b) or, more directly, by returning to Eq. (2.2) and matching P points or using all D points to obtain a least-squares solution.

2.3 Discussion

Note that Prony's Method linearizes the problem of estimating the $\hat{R}_\alpha$ and $\hat{s}_\alpha$ in Eq. (2.1) or (2.2) by separating their solutions. The above steps (in Time Domain Prony) comprise Prony's original method of 1795[1] and were described in 1956 by Hildebrand.[5] It is worth noting that Eq. (2.6) is a difference equation whose form in the Z-domain is that of an all-pole model.[6]

In its simplest form, Prony's method requires that the user specify only two parameters for the time domain case, say δ and P, from which D (=2P) and T ($\delta(D-1)$) follow automatically. The result is a square system of linear equations with P unknowns, the coefficients of which are the sampled data, and with the general coefficient A_{ij} given by f_{i+j-2}, where $i,j=1,\ldots,P$.

A unique solution can be found only when $P \geq W$, and when $\delta < 1/2f_{max}$ where f_{max} is the highest frequency component of f(t). In general, because W may not be known a priori, experimentation using different values for P may be required to establish an acceptable choice for it. Similarly, in order to establish the best value of δ, its variation to the extent permitted by the data may be necessary. Whereas δ must be small enough to avoid aliasing (i.e., δ must satisfy the Nyquist criterion), it cannot be made too small without the risk of ill-conditioning the data matrix by decreasing the linear independency of neighboring rows and columns. Clearly, application of even the basic version of Prony's Method requires care if acceptable results are to be obtained.

If more data points (D') were available than the minimum 2W values required to compute the W poles of f(t), this would clearly result in an overdetermined system because the number of equations exceeds the number of unknowns. Two approaches to handling this situation are apparent. In the first, we could continue to use 2P' = D', i.e., compute as many poles as permitted by the number of data samples available. This number of poles would, of course, be greater than that actually present in f(t), and the data matrix in Eq. (2.6) should then be singular unless, as is invariably the case, the data are noisy or of otherwise limited accuracy. This approach has been followed in much of the early work in electromagnetics that employed Prony's Method.

Alternatively, we could use the original value of P and reduce the number of equations from P' to P by multiplying Eq. (2.6) by the transpose of the data matrix. This results in

$$\overline{\overline{M}}_t' \overline{\hat{a}} = -\overline{f}' \ , \qquad (2.11a)$$

where

$$\overline{\overline{M}}_t' = \begin{bmatrix} c_{00} & c_{01} & c_{02} & \cdots & c_{0,P-1} \\ c_{10} & c_{11} & c_{12} & \cdots & \cdot \\ c_{20} & c_{21} & c_{22} & \cdots & \cdot \\ \vdots & & & & \\ c_{P-1,0} & & \cdots & & c_{P-1,P-1} \end{bmatrix} \ , \qquad (2.11b)$$

$$\overline{f}' = \begin{bmatrix} B_0 \\ B_1 \\ \vdots \\ B_{P-1} \end{bmatrix} \ , \qquad (2.11c)$$

$$c_{ij} = \sum_{k=0}^{P'-1} f_{i+k} f_{j+k} \ , \qquad (2.11d)$$

and

$$B_i = \sum_{k=0}^{P'-1} f_{i+k} f_{P+k} \ . \qquad (2.11e)$$

The two approaches could be combined by choosing any value of P' between P and (D')/2.

Unlike $\overline{\overline{M}}_t$, $\overline{\overline{M}}_t'$ is not a Hankel matrix but has the form of an autocovariance matrix, the usual starting point for digital signal processing of transient waveforms. This form results if the minimization of the square of the error between the model and the data is achieved by varying the model parameters. Prony's Method does not consider this error explicitly, however, and, as originally formulated, it is essentially a point-matching procedure. The process of developing the new data matrix in Eq. (2.11b) through multiplying the original matrix by its transpose also minimizes the squared error; this is sometimes called regularization but is referred to more often as generating the pseudo inverse.

An important point to be noted is that Prony's Method contains features identical to a technique employed in digital signal processing called all-pole modeling.[6] Furthermore, a spectral estimation procedure called the maximum-entropy method is also based on an all-pole model[7] and is, therefore, also equivalent to Prony's method. It is thus a little puzzling to hear frequent claims that Prony's Method is more ill-conditioned, more sensitive to noise, etc., than more "modern" methods. These remarks evidently overlook the links of these other techniques to Prony's Method because, as originally formulated, Prony's Method did not include the pseudo-inverse approach. Clearly, however, Prony's Method is the forerunner of such techniques and deserves to be recognized as such.

There are additional ways to handle an overdetermined system. One we have found to be most useful, called the moving-window technique, processes the data using several sequential time windows that "move" along the waveform by $\Delta = n^{\pm 1}\delta$ (n an integer ≥ 1). In this way, the data are not used all at one time, and several sets of estimates are obtained for the poles. Authentic poles should cluster around an average value while spurious poles due to noise and curve fitting (if $P > W$) will appear randomly. Superimposing the plots of all the pole sets on one graph helps identify the valid poles. An interesting alternate procedure used by Van Blaricum[8] is to process the data in both forward and reverse directions with respect to time. The result is that genuine poles are reflected about the $j\omega$ axis, and artifacts remain predominantly in the negative half-plane.

3. INTRODUCTORY NUMERICAL RESULTS

Some test cases were run to evaluate the performance of Prony's Method when applied to analytically specified data. A few results are summarized here to highlight key points of its use.

3.1 Noise-free data

A 10-pole system having unit-amplitude, pure-real residues and pole values of $s_\alpha = \sigma \pm j\omega = -0.1 \pm j1.0, -0.25 \pm j2.5, -0.5 \pm j5$, $-0.75 \pm j7.5$, and $-1.0 \pm j10$, was used.[9] In Fig. 1 the average accuracy in digits of the poles and residues (denoted by $\overline{D}(\hat{S})$ and $\overline{D}(\hat{R})$ respectively) as estimated by Prony's Method is shown as a function of time step δ for $D = 20 = 2P = 2W$. Two sets of curves are displayed, one where the input data accuracy $D(f_i)$, is 10 digits, and the other, 6 digits. Use of too small a time step in sampling the data when $D = 2W$, as used here, can significantly reduce the accuracy of the estimated parameters. But when δ approaches $\delta_N = 1/2f_{max}$, or the Nyquist sampling rate, the parameter accuracy approaches that of the input data.

Fig. 2 demonstrates a different facet of this behavior. There, the eigen values of the data matrix, $\overline{M}_t$, are plotted for three different sampling intervals. As δ approaches δ_N, the eigenvalue range decreases, showing an improving numerical conditioning of the data matrix, or equivalently, the increasing linear independency of the waveform samples.

Results similar to those of Figs. 1 and 2 are presented in Figs. 3 and 4 for a pole set having the same real components as above, but for imaginary components given by $\pm\omega = 6, 7, 8, 9, 10$. The time step δ was fixed at 0.3 s and D was systematically increased from $2P = 20, 22, 24, \ldots$, with $D(f_i)$ supplied to computer accuracy (~12 digits). As can be seen in Fig. 3, the data window must approach $1/\Delta f$, with Δf the frequency separation of the two most closely spaced poles, before the parameter accuracy (average of both poles and residues shown here) approaches $D(f_i)$. The behavior of the eigen values in Fig. 4 is consistent with this result.

Another aspect of the influence on the performance of Prony's Method of the processing parameters used is illustrated in Fig. 5 for a different pole set and waveforms.[10] There the original waveform and its reconstruction from the parameter estimates are shown in part (a) (they are graphically indistinguishable) for $P = W = 4$. When however, $P = 3 < W$ as in part (b), a large error occurs in the reconstruction, due to invalid parameter estimates. In both cases, $D = 2P$ and the data is supplied to computer accuracy.

Aliasing is another problem which can arise due to improper sampling. This occurs when $\delta > 1/2f_{max}$. An example of aliasing is illustrated in Fig. 6, where the pole and residue estimates are displayed on an isometric plot.[10] Pole locations are indicated by the dots in the complex frequency plane, with the residue amplitudes proportional to the lengths of the vertical lines on a three-decade logarithmic scale. In part (a), the sampling frequency exceeds the frequency of the maximum pole by ~11 percent, and in part (b), the sampling rate is decreased by one half. A foldover of the higher frequency poles is seen in the latter plot.

3.2 Noisey data

The performance of a parameter-estimation procedure such as Prony's Method can be significantly affected by noise. Some examples of treating noise-contaminated waveforms are included here.

In Fig. 7, isometric pole-residue plots are presented for two waveforms.[11] One, in part (a), is noise free (a computed transient response for a Gaussian-pulse excited straight wire). The other, in part (b), is identical to the first but has additive, zero-mean, uniformly distributed random (white) noise whose peak amplitude is one percent of the waveform peak amplitude. It can be seen that the pole locations in the latter case are changed significantly from their noise-free values only for poles whose associated residue values are, relative to the maximum residue amplitudes, comparable to, or less than, the noise level.

A procedure that has been found to be useful to reduce noise effects is what has been called the moving window technique, as described above in Section II. One example of its application to a ten-pole waveform whose residue values are all equal $(0 + i1)$, and to which is added white noise is shown in Fig. 8.[11] The pole locations in the complex frequency plane are plotted for each of 10 successive windows and peak noise of ±0.014 in Part (a), and 100 windows and peak noise of ±.28 in part (b). Their imaginary components exhibit much less variation than the real or damping values in both cases. Averaging the clusters of part (a) provides improved pole-estimate accuracies on the order of a factor of 6 and 2, respectively, for the imaginary and real components.

The improvement in pole accuracy due to the moving window procedure is demonstrated in another way in Fig. 9. In this figure, the average cumulative pole accuracy of a 10-pole system (all equal residues) is shown as a function of window number for 10 different runs,[12] and a peak-signal-to-peak-noise ratio, S_p, of 10^3. The run-to-run variation of pole accuracy improvement ranges from 0.3 to 3.3 bits with an average improvement of ~1.5 bits. This value approximately equals the square root of 10, and is consistent with the noise model employed.

An important characteristic of Prony's Method is demonstrated in Fig. 10. Shown there are scatter plots of the relative accuracies of the input data, characteristic-equation coefficients, roots and poles for a number of independent one-window runs of the same 10-pole system as used in Fig. 9.[12] Note that the coefficient accuracy is consistently less than that of the roots and poles in spite of the fact the latter are derived from the former.

The average pole accuracy can be comparable to the peak-signal-

to-peak-noise level as seen in Fig. 11.[12] These results were
obtained from one-window calculations for a six-pole system
modeled with eight poles (i.e., P = 8, W = 6), and represent
average pole accuracies (not accuracies of averaged pole values)
from 29 independent waveforms. Some other attempts to handle
noise are reported elsewhere.[13]

4. RESULTS FROM REPRESENTATIVE APPLICATIONS

Much of the impetus for pursuing techniques to obtain poles from
temporal or spectral data was provided by initial work on the
Singularity Expansion Method (SEM).[1,2] SEM was intended to
permit easy development of an object's transient EM response via
its poles, which are source independent, and a set of coupling
coefficients, which do depend on the source. While SEM has found
somewhat limited application to date for that purpose, it has
promoted considerable interest in poles themselves, their determination, and their application. Various aspects of these topics
are discussed below.

One of the most useful properties of poles in EM is their independence of the excitation. This is demonstrated for transient
data in Fig. 12 where the backscattered field and resulting poles
are shown for a Gaussian pulse incident from two different angles
on a straight wire.[14] The slight variations in pole locations
are due to numerical inaccuracy which result from small residue
values. The residue amplitudes differ because of their dependence on the incident field. Note that only one set (or layer)
of poles are shown here. Pole extraction based on searching the
complex frequency plane reveals that additional layers of poles
having larger loss components exist,[15] but their contribution to
the objects' transient response generally are negligible. Therefore, they are not obtained in applying Prony's Method to the
transient waveform.

Pole extraction from spectral data is illustrated in Fig. 13.[16]
In this case the straight wire is excited at its center as an
antenna, and the feedpoint current and broadside far field are
used to obtain the poles. Again, the pole sets are very similar,
as are, in this case, the residue amplitudes as well. Note that
the pole sets of Figs. 12 and 13 are not plotted on the same
scales, accounting for their different appearance. Also observe
that the poles can be obtained from any observable of the object,
e.g. current and charge, near fields and far fields. More examples
of frequency-domain Prony are given below in Section V.

Fig. 14 shows results of either resistively loading a straight wire
at its center, or bending it at a right angle at its center.[16]
In the former case, the $\alpha = 1, 3, 5, \ldots$ poles are more lossy
than the $\alpha = 2, 4, 6, \ldots$ poles, while the converse is true in

the latter case. This occurs because the odd-numbered modes are even about the wire's center, so that the resulting current maxima at the resistive loads produce a large dissipative loss in the former case. But the even-numbered modes are odd and have charge maxima at the wire's center, thus producing a large radiative loss at the bend in the latter case.

Besides being useful to find the pole locations for SEM, Prony's method serves as well to store the transient waveform in a shorthand form. One interesting application of data handling is the ability to extrapolate both forward and backward in time.[17] Fig. 15a shows a portion of the computed backscattered field observed from the wire object shown. Fig. 15b plots the pole locations obtained using only that portion of the response indicated in Fig. 15a. The poles from Fig. 15b (and residues which are not shown) were then used to fill in the entire transient response. This extrapolated response is compared to the original computed response in Fig. 15c, with excellent agreement. Note that now every fourth pole is more lossy, rather than every second pole as in Fig. 14, because the bends are one-quarter of the wire's length from its ends.

Additional results are presented in Fig. 16 for three other wire objects.[16] Notice that the straight wire and loop of part (c) do not interact with each other; the waveform is simply a superposition of their individual waveforms. This result is included to show that the pole sets of independent objects can be separated via Prony's Method.

Transient data other than the EM variety can also be processed using Prony's Method. Two examples are presented in Figs. 17 and 18. That of Fig. 17 is the measured response of a building to an earthquake.[18] The pole clusters produced by the moving-window technique are quite compact. They have relatively small negative real parts, or even positive values, showing the effect of processing the data while energy is still being supplied to the structure. In the case of the EEG signal (Fig. 18),[19] which is essentially a colored random process that results from passing white noise through a filter, the windows are non-overlapping. The pole locations vary in time, and are thought to indicate the different anathesia levels of a patient.

Many other applications have been found for pole modeling, including speech analysis and synthesis,[6] spectral absorption characterization,[20] and vibration analysis.[21]

5. USE OF POLES FOR OBJECT IDENTIFICATION

Because of their independence of how an object is excited, the poles are especially attractive for use in radar target identification. This is because, rather than having to store transient waveforms or frequency spectra for many angles of incidence, only the poles need be stored, thus greatly condensing the data base. We consider here two approaches for object identification based on poles, one using frequency-domain and the other using time-domain, data. Although these two data types form a transform pair, and should contain equivalent information, it is useful to examine both. Furthermore, which approach is best in a given situation seems likely to depend on how the data is obtained and how efficiently the necessary computations can be done.

5.1 Frequency-domain application

The basic problem is to establish to which set of a library of pole sets the spectrum of an unknown target is most likely to belong. For test purposes, the three wire targets shown in Fig. 19 were modeled using a moment-method computer model.[22] Backscatter-field spectra for a fixed angle of incidence were computed over a frequency range of 0.125 to 10 Mhz. The frequency-domain equivalent of the moving-window method was used with Frequency-domain Prony to obtain the cluster plot of Fig. 20a for target 3. An alternate procedure to obtain a cluster or consensus, pole set which uses a fixed frequency range and various angles of incidence leads to the cluster plot of Fig. 20b, again for target 3. Either procedure helps to separate curve-fitting from actual target poles, although that was not done here. Instead, the pole set obtained from one angle of incidence and one frequency window was used for the target pole library.

Using pole sets obtained in this way, the identification of an object from a set of stored (previously obtained) library poles can be accomplished totally in the frequency domain. This process begins by separating the spectrum of the unknown object into two sets of data $F_M(\omega_q)$, $F_M(\omega_p)$ which interleave along the $j\omega$-axis. The stored pole sets are used to calculate a set of residue values $\hat{R}_n^L$ from the data by employing the equation

$$F_M(\omega_q) = \sum_{\alpha=1}^{P} \frac{\hat{R}_\alpha^L}{j\omega_q - s_\alpha^L} \quad ; \quad q = 1, \ldots, D/2 \qquad (5.1)$$

where s_α^L denotes the Lth set of library poles. The residues found from the solution of Eq. (5.1), along with the corresponding library pole set used to find them, are then used to calculate a predicted response on the ω-axis at the points ω_p by the expression

$$F_C^L(\omega_p) = \sum_{\alpha=1}^{P} \frac{\hat{R}_\alpha^L}{j\omega_p - s_\alpha^L} \quad p = 1, \ldots, D/2$$

The normalized r.m.s. error between the actual response F_M and the calculated response $\hat{F}_C^L$ is then calculated and a correlation number between the stored poles set and the unknown object is obtained by computing

$$R_L = \frac{1}{1 + \epsilon_L}$$

where ϵ_L is the normalized r.m.s. error.

To demonstrate this procedure, the poles of the three models shown in Fig. 19 were found and stored for an incidence angle of 60°. Then the same three backscattered responses were used in the above identification process. The results of this numerical identification test are shown in Table Ia. Note that each set of poles does correlate best with backscattering data from the model for which it was derived.

The result of using the same pole sets, but an incidence angle of 30° for the backscattered response are shown in Table Ib. Because the pole sets exhibit some numerical variation as the incidence angle changes, the correlation values of Table Ib are less than unity, but are still largest along the diagonal. This simple example shows the possibility, for a small set of simple targets, of using poles and backscattered field spectra for identification. Other correlation strategies than that used here could be employed, and noise effects also require consideration, before the practical utility of a pole-based, frequency-domain approach can be fairly assessed.

5.2 Time domain application

The basic problem in the time domain is to determine which set of a library of pole sets to which the transient response of an unknown target most likely belongs. For this problem, analytically specified sets of 10 poles (5 conjugate pairs; Table II) were used both as the library, and to generate transient waveforms to represent unknown target returns, for which all residue values were set at $1 + j0$.[23] Four different techniques (Table III) were evaluated for their ability to provide a correct identification as a function of additive noise level (zero mean, uniformly distributed).

Some results from this exercise are summarized in Table IV, with the peak-signal-to-peak-noise ratio a parameter. Shown there are the misidentifications or false alarm rates based on the individual correlation coefficients from each of 10 Monte Carlo runs (part A), and the average values of these correlation coefficients (part B). While waveform correlation works best here, note that

TABLE I. TARGET CORRELATION MATRICES R

	Case A $\theta_{Poles} = 60°$; $\theta_{RCS} = 60°$		
ℓ/w	1	2	3
1	1.000	0.329	0.637
2	0.316	1.000	0.701
3	0.523	0.536	1.000
	Case B $\theta_{Poles} = 60°$; $\theta_{RCS} = 30°$		
ℓ/w	1	2	3
1	0.760	0.330	0.389
2	0.235	0.561	0.340
3	0.461	0.531	0.736

TABLE II. POLE SETS USED FOR TIME-DOMAIN IDENTIFICATION CALCULATIONS.

Pole Set Number	Poles
1	$-.5 \pm j2,4,6,8,10$
2	$-.5 \pm j3,5,7,9,11$
3	$-.5 \pm j4,6,8,10,12$
4	$-.5 \pm j2,3,6,9,10$
5	$-.5 \pm j2,5,6,7,10$

TABLE III. METHODS USED FOR TIME-DOMAIN IDENTIFICATION

Method	Computation	Correlation
1. Linear predictor	$f_P^{(\ell,w)}(t_\beta) = \sum_{\alpha=1}^{P} a_{p,\alpha}^{(\ell)} f_M^{(w)}(t_\beta - \alpha\delta t)$	$r_{\ell w}^2 = \frac{1}{D} \sum_{\beta=1}^{D} [f_P^{(\ell,w)}(t_\beta) - f_M^{(w)}(t_\beta)]^2$
2. Residue Calculation	$f_M^{(w)}(t_\beta) = \sum_{\alpha=1}^{P} \hat{R}_\alpha^{(\ell,w)} e^{s_\alpha^{(\ell)} t_\beta}$	$r_{\ell w}^2 = \frac{1}{D} \sum_{\beta=1}^{D} \left\{ \sum_{\alpha=1}^{P} [\hat{R}_\alpha^{(\ell,w)} e^{s_\alpha^{(\ell)} t_\beta} - f_M^{(w)}(t_\beta)]^2 \right\}$
3. Pole calculation	$f_M^{(w)}(t_\beta) = \sum_{\alpha=1}^{P} \hat{R}_\alpha^{(w)} e^{\hat{s}_\alpha^{(w)} t_\beta}$	$r_{\ell w}^2 = \frac{1}{P} \sum_{\alpha=1}^{P} (\hat{s}_\alpha^{(w)} - s_\alpha^{(\ell)})_{min}^2$
4. Waveform correlation	—	$R_{\ell w} = \sum_{\beta=1}^{D} f_M^{(w)}(t_\beta) f_L^{(\ell)}(t_\beta) / \sum_{\beta=1}^{D} [f_M^{(w)}(t_\beta)]^2$

$R_{\ell w} = \frac{1}{1+r_{\ell w}}$ for methods 1, 2, 3.

$f_M^{(w)}(t)$ is w'th measured waveform

$f_L^{(\ell)}(t)$ is ℓ'th library waveform

$s_\alpha^{(\ell)}$ is ℓ'th library pole set

TABLE IV. FALSE ALARM RATES FOR TIME-DOMAIN METHODS OF TABLE III. PART A IS BASED ON INDIVIDUAL MONTE-CARLO RUN (TOTAL OF 10) RESULTS AND PART B IS BASED ON $r_{\ell w}$ AVERAGED OVER THE 10 RUNS. WHERE TIES OCCUR (TO TWO DECIMAL PLACES), THE ERROR IS TAKEN AS ONE HALF.

Part A

Method \ S_P	10^2	$10^{3/2}$	10^1	$10^{1/2}$	10^0	$10^{-1/2}$
1	0	0.14	0.56	0.70	--	--
2	0.02	0.06	0.28	0.58	--	--
3	0.06	0.14	0.50	0.74	--	--
4	--	--	--	--	--	--

Part B

Method \ S_P	10^2	$10^{3/2}$	10^1	$10^{1/2}$	10^0	$10^{-1/2}$
1	0	0.2	0.6	0.7	--	--
2	0	0.2	0.2	0.6	--	--
3	0	0	0.1	0.5	--	--
4	0	0	0	0	0	0.2

it would require storing much more library data than the other techniques since the waveforms are angle dependent. For purposes of this study, only the "correct angle" waveforms (those for residues of 1 + j0) were used in the library of the correlation approach.

6. OTHER REPRESENTATIONS INVOLVING POLES

Our discussion has thus far concentrated on "signals" which are exponential in the time domain and pole like in the frequency domain. But because so many physical phenomena are described via second-order differential equation, such behavior is encountered in many other situations. A few examples are considered here.

6.1 Radiation patterns

Consider the far field of a linear array of S isotropic radiators, which is given by

$$f(\theta) = \sum_{\alpha=1}^{S} R_\alpha e^{jkd_\alpha \cos\theta} \qquad (6.1)$$

where θ is measured from the array axis, and d_α is the position of the αth element. Eqs. (2.1) and (6.1) are clearly equivalent if the substitution $x = \cos\theta$ is made; therefore, the pattern $f(\theta)$ can be analysed by Prony's method. Note: (a) the requirement that the sampling be in uniform exponential changes implies that angular sampling varies in uniform steps in $\cos\theta$ rather than θ, and (b) whereas in Eq. (2.1) the poles in general can be complex, those of Eq. (6.1) evidently must be purely imaginary if the d_α

are to represent real-space locations. Alternatively, a real pole component in Eq. (6.1) may be viewed as representing source directivity.

Application of Prony's method to a radiation pattern derived from Eq. (6.1) can have two outcomes. If the number of sources S is less than the aperture rank A (roughly the aperture width in half wavelengths), the derived source distribution can be the actual one. The exercise in this case is essentially one of imaging. If, however, S > A, then the derived source distribution is only an equivalent one insofar as the pattern is concerned. This can be the situation when the source distribution approaches a continuous one in the limit $S \to \infty$, for a fixed A.

The pattern, on the other hand, may be a specified one whose description is not exponential. In this case, there is no assurance that its poles will represent a realizable source distribution (see note (b) above). Although not considered here, this possibility might be eliminated by constraining the derived poles to be purely imaginary. Our interest is in using Prony's method for synthesizing the sources of patterns given by sums or integrals of exponential terms.

Prony's method was applied to patterns given by Eq. (6.1) for several source distributions. Both S < A and S > A cases were examined as a function of P. Results showed that P must equal or exceed the pattern rank A to obtain the actual sources and that stable results are obtained for increasing values of P beyond A. Examples of both cases are shown in Fig. 21. When the aperture width was 3λ only eight equivalent sources of the 11 actual sources were found, but all sources were derived at a width of 5λ or greater. The latter case is an example of source imaging, whereas the former implies that an aperture pattern can be described with ~ $2A/\lambda$ equivalent sources. These findings were not dependent on the amplitude distribution or density of the discrete sources within the aperature. The pattern rank implied by the eigenvalues of the data matrix supports these results (Fig. 22). Cases (d) and (e) gave the most successful pole extractions. These curves had the smallest slopes and lowest ratio of the first to the eleventh eigenvalue.

An application of Prony's method to synthesis of a discrete source distribution that produces the pattern of a continuous 7λ aperture is shown in Fig. 23. The original pattern (a) gave the derived source distribution (b), which in turn produced the reconstructed pattern (c). The latter produced the constrained pattern (d), which is obtained by setting the real pole components of the derived sources to zero but using the same residue values.

Note that some of the poles have nonzero real components σ_α, which implies a source directivity of the form $\exp(\sigma_\alpha \cos\theta)$. For small σ_α, this is approximately realizable by placing a passive director-reflector near (within a few hundredths of a wavelength) a driven element. Thus, the constrained pattern represents a worst-case result, and the realizable pattern most likely lies somewhere between the reconstructed and constrained patterns. Note that the sources near the edges of the aperture are closer together.

Elliot[25] reported an iterative approach that contains some ingredients in common with the above, but which requires choosing a pirori the number and coordinates of the sources. These results compare well with his but are obtained more directly without prespecifying the array geometry.

Another linear-geometry source which is amenable to modeling via Eq. (2.1), and therefore suitable for Prony processing is a straight wire antenna or scatterer. Some representative results for this particular problem are presented in Figs. 24-26. Note that the largest residues (and hence strongest sources) are, for the antenna case, located at the wires ends and at the driving point. The outcome for the scattering mode can be more complicated. In either case, we might view the derived sources as providing an EM image of the wire, showing from where its far-field radiation originates. Since charge acceleration is the mechanism by which radiation occurs, which for the wire antenna would be its ends and exciting source region, these results seem physically plausible. Note that because the pattern of a sinusoidal current varies as $f(\theta)/\sin\theta$, the pattern data was multiplied by $\sin(\theta)$ prior to using Prony's method.

It is tempting to try extending these results to two and three dimensional source distributions. This does not seem possible, unfortunately, at least using Prony's Method, since uniform steps in all the exponential terms of the pattern function cannot be realized, as examination of Eq. (6.2) will verify.

$$f(\theta) = \sum_{\alpha=1}^{S} R_\alpha e^{ik\rho_\alpha \cos(\theta-\theta_\alpha)}$$

$$= \sum_{\alpha=1}^{S} R_\alpha e^{ik\rho_\alpha [\cos\theta\cos\theta_\alpha + \sin\theta\sin\theta_\alpha]} \quad (6.2)$$

Here, ρ_α and θ_α are the polar coordinates of source R_α. Were the source strengths to be independent of frequency however, then Eq. (6.2) would be suited for Prony processing as a function of frequency, using a fixed observation angle. It would then be necessary

generally, to use two different viewing angles, orthogonal for
example, to deduce both parts of the pole, i.e., ρ_α and θ_α.

The above process can be essentially reversed, by uniformly sampling
the fields spatially along a line or over a plane, of one or more
incident plane waves. This procedure also leads to a set of data
samples which can be Prony processed, to do direction finding for
example, or to separate the incident wave components.[26] It is
closely akin to imaging techniques being pursued in radio astronomy
and elsewhere. It provides an alternative to beam-forming and
steering methods which use phase shifters and adders to combine
the outputs of the individual elements in the array, but at the
expense of measuring separately each element's output.

6.2 One-dimensional scattering

Another type of problem to which a pole description applies is
that of TEM one dimensional scattering. This geometry could
involve, for example, a two-wire transmission line, a coaxial
cable, or layered media. The common ingredient in any of these
problems which can give rise to poles is the reflection due to
medium inhomogenety. Consider the specific example, shown in Fig.
27, where a planar layer of finite thickness is sandwiched between
two half spaces. An impulsive plane wave incident normally on the
layer from the free-space side produces a reflected field given
by[27]

$$r(t) = R_{12}\delta(t) + T_{12}T_{21}R_{23}(R_{21}R_{23})^{-2} \sum_{\alpha=-\infty}^{\infty} e^{j\omega_\alpha t} \quad (6.3a)$$

The exponent terms ω_α are the roots of

$$1 - R_{21}R_{23} \exp(-2j\omega\sqrt{\mu_2\epsilon_2}\, D_2) = 0 \quad (6.3b)$$

and we assume the medium is non-dispersive.

The corresponding reflected field in the spectral domain can also
be written in exponential from as

$$R(\omega) = R_{12} + R_{23}T_{12}T_{21} \sum_{\alpha=1}^{\infty} (R_{21}R_{23})^{\alpha-1}$$

$$\times e^{-j2\omega_\alpha \sqrt{\mu_2\epsilon_2} D_2} \quad (6.4)$$

It is intriguing to see that descriptions of the same problem
expressed in two different domains which form a transform pair
can, as in this case, both be exponential. Of interest to us in a

pole cotext is the fact that two different sets of poles result. From Eqs. (2.1), (6.3) and (6.4) they are seen to be

$$s_\alpha^{(t)} \equiv \omega_\alpha = \frac{-j \ln|R_{21}R_{23}| + \lfloor R_{21}R_{23} \pm \alpha 2\pi}{2D_2 \sqrt{\mu_2 \epsilon_2}} \quad (6.5a)$$

for the time-domain poles, and the frequency-domain poles are

$$s_\alpha^{(\omega)} = -j2\alpha\sqrt{\mu_2\epsilon_2} \, D_2 \quad (6.6a)$$

where the superscript denotes the domain in which the behavior is exponential.

The corresponding residues are given by

$$R_\alpha^{(t)} = T_{12}T_{21}R_{23}(R_{21}R_{23})^{-2} \quad (6.5b)$$

$$R_\alpha^{(\omega)} = T_{12}T_{21}R_{12}(R_{21}R_{23})^{\alpha-1} \quad (6.6b)$$

Note that from Eqs. (2.1-2.2) and (6.5-6.6), the reflected fields can be put in the form summarized in Table V.

TABLE V. REFLECTED FIELDS IN VARIOUS DOMAINS

Time	Frequency	Space
$\sum R_\alpha^{(t)} e^{s_\alpha^{(t)} t}$	$\sum \dfrac{R_\alpha^{(t)}}{j\omega - s_\alpha^{(t)}}$	---
---	$\sum R_\alpha^{(\omega)} e^{s_\alpha^{(\omega)} \omega}$	$\sum \dfrac{R_\alpha^{(\omega)}}{D - s_\alpha^{(\omega)}}$

Comparing Eqs. (6.5) and (6.6) we see that the poles and residues of both contain information about the layer's electrical and physical parameters. This information however, appears in different ways in the poles and residues. For example, $R_\alpha^{(t)}$ is a constant independent of α, whereas $R_\alpha^{(\omega)}$ exhibits a geometric series α dependence. And in contrast to $s_\alpha^{(t)}$, which is a function of the interface reflection coefficients electrical thickness and angle of incidence, $s_\alpha^{(\omega)}$ is a function of the latter two quantities only. Further, for a lossless (real μ and ϵ) layer, $s_\alpha^{(\omega)}$ is pure imaginary, while $s_\alpha^{(t)}$ can be complex in general. Perhaps most intersting is the differing α-dependence of both pole types, since $s_\alpha^{(t)}$ has a real part (for the lossless case) which is inversely proportional to the layer electrical thickness, while $s_\alpha^{(\omega)}$ has an imaginary part which is directly proportional to the electrical thickness.

Both sets of poles and residues must therefore lend themselves to solving the inverse problem, i.e., given r(t) or R(ω), the layer's parameters can be determined, in principle at least. However, the step of going from $s_\alpha^{(t)}$, $R_\alpha^{(t)}$ or $s_\alpha^{(\omega)}$, $R_\alpha^{(\omega)}$ to μ_2, ϵ_2 and D_2 may not be a trivial one and is not the primary subject of this discussion, although it will be touched on briefly below. We are more interested instead in the extraction of poles from reflected field data, for the viability of this approach to the inverse problem requires that such poles be available, for which some results are presented below.

Using both r(t) and R(ω) data, the pole sets which arise from a dielectric layer of relative permittivity ϵ_r = 25 and thickness D_2 = 0.125 m sandwiched between half-spaces of ϵ_r = 1 (free space) and ϵ_r = 10^4 are shown in Figs. 28 and 29. Also shown are the reflected fields r(t) and |R(ω)|. The incident field arrives from the ϵ_r = 1 half space normal to the planar surfaces.

In the time-domain case, the poles associated with the incident-field (a double exponential) are so large as to mask the residue magnitudes of the layer-caused poles. For the frequency-domain case however, the residue magnitudes are seen to fall off linearly on the vertical log scale, demonstrating their agreement with Eq. (6.6b). Also observable is the absence of loss in the $s_\alpha^{(\omega)}$ poles, whereas the $s_\alpha^{(t)}$ poles do exhibit loss through having non-zero real components. The remaining result will be for the frequency-domain poles only.

The reflection coefficient magnitude and pole set which results for normal incidence on two layers having relative permittivities of 9 and 16 and thicknesses of 0.208 m and 1 m respectively, which are sandwiched between the same two half spaces as for Figs. 28 and 29, are shown in Fig. 30. It may be seen that the pole set is composed of several groups. The first group is due to multiply reflected waves in the upper (ϵ_r = 9) layer (the pole at the origin comes from the field reflected at the first interface).

The second group arises from waves which make one round trip through the second (and electrically thicker) layer and are multiply reflected in the first layer. Subsequent groups are caused by higher order reflections in the lower layer. Note that the residues amplitudes for these groups do not monotonically decrease as in Fig. 29, because of the increasing number of reflection combinations which yield the same total electrical phase change as the number of "bounces" increases.

7. CONCLUDING REMARKS

This presentation has concentrated on poles, or resonance frequencies, and the way they are characteristic of many physical phenemona, with special emphasis on their role in electromagnetics. Among the topics discussed were ways by which poles can be obtained, problems in which they arise, and possible physical interpretations and applications for which they may be suited.

There are several methods that can be used to find poles in electromagnetics, with the choice dependent on the kind of data available. The two methods described here, time-domain Prony, and frequency-domain Prony, can be used for computed or measured data which is termwise exponential or termwise pole like, respectively. They are similar in implementation, proceeding from a matrix of data samples to a characteristic (polynomial) equation whose roots yield either the poles directly (frequency domain) or antilogs of the poles (time domain).

We have explored several types of problems in which poles occur. These included transient and spectral radar scattering, radiation patterns, and one-dimensional scattering. The latter is of special interest because both the transient and spectral response are exponential, but possessing different poles and residues.

One application discussed was object identification exploiting the source-independence of the poles. Another was radiation pattern synthesis and source imaging. Poles also can provide a way to interpret radiation phenomena through their loss dependence. And they may be useful to invert one-dimensional scattering data.

Overall, poles play a fundamental role in wave phenomena which is only now becoming appreciated.

REFERENCES

1. C. E. Baum, "Electromagnetic Transient Interaction with Objects with Emphasis on Finite Sized Objects and Some Aspects of Transient Pulse Production," in Proc. Spring URSI Meeting, Washington, D.C., 1972.
2. C. E. Baum, "The Singularity Expansion Method," in Transient

Electromagnetic Fields, L. B. Felsen, Ed., New York: Springer, 1976, Ch. 3.
3. B. K. Singaraju, D. V. Giri, and C. E. Baum, "Contour Integration Method of Evaluating the Zeros of Analytic Functions and Its Application in Finding Natural Frequencies of a Scatterer," in Proc. National Conf. on Electromagnetic Scattering, University of Illinois at Chicago Circle, June 15-18, 1976.
4. R. Prony, "Essai Experimental et Analytique sur les Lois de la Dilatabilite des Fluides Elastiques et sur Celles de la Force Expansive de la Vapeur de l'Alkool a Differentes Temperatures," J. l'Ecole Polytech., Paris, Vol. 1, 1795, pp. 24-76.
5. A. S. Householder, "On Prony's Method of Fitting Exponential Decay Curves and Multiple-Hit Survival Curves," Oak Ridge National Laboratory, Oak Ridge, Tennessee, Rept. ORNL-455, 1950.
F. B. Hildebrand, Introduction to Numerical Analysis, New York, McGraw-Hill, 1956, pp. 378-382.
M. C. Van Blaricam and R. Mittra, IEEE Trans. Antennas and Propagation, Vol. AP-23, No. 6, 1975, 777.
6. J. Makhoul, "Linear Prediction: A Tutorial Review," Proc. of the IEEE, Vol. 63, No. 4, 1975.
7. R. T. Lacoss, "Data Adaptive Spectral Aanlysis Methods," Geophysics, Vol. 35, No. 4, August 1971, pp. 661-675.
T. J. Ulrych and T. N. Bishop, "Maximum Entropy Spectral Analysis and Autoregressive Decomposition," Rev. Geophys. and Space Phys., 13, 1975, pp. 183-200.
8. M. Van Blaricum, Private Communication, 1978.
9. E. K. Miller and D. L. Lager, "Information Extraction Using Prony's Method, Lawrence Livermore Laboratory Rept. No. UCRL-52329, August 23, 1977.
10. A. J. Poggio, M. L. Van Blaricum, E. K. Miller, and R. Mittra, "Evaluation of a Processing Technique for Transient Data," IEEE Trans. Antennas and Propagation, Vol. AP-26, No. 1, January 1978.
11. H. G. Hudson and D. L. Lager, "Observations on the Operation of the Sempex Code," Lawrence Livermore Laboratory Rept. No. UCID-17440, September 22, 1976.
12. E. K. Miller, "Prony's Method Revisited," Lawrence Livermore Laboratory Rept. No. UCRL-52590, October 18, 1978.
13. D. G. Dudley, "Fitting Noisy Data with a Complex Exponential Series," Lawrence Livermore Laboratory Rept. No. UCRL-52242, March 7, 1977.
T. K. Sarkar, J. Nebat, and D. Weiner, "Suboptimal System Approximation/Identification with Known Error," Mathematics Note 49, September 1977.
J. T. Cordaro, "Comparison of Three Techniques for Calculating Poles and Residues from Experimental Data," Mathematics Note 61, August 1978.

14. E. K. Miller, F. J. Deadrick, H. G. Hudson, A. J. Poggio, and J. A. Landt, "Radar Target Classification Using Temporal Mode Analysis," Lawrence Livermore Laboratory Rept. No. UCRL-51825, May 27, 1975.
15. F. M. Tesche, "On the Analysis of Scattering and Antenna Problems Using the Singularity Expansion Technique," IEEE Trans. Antennas and Propagation, Vol. AP-21, January 1973, p. 53.
16. J. N. Brittingham, E. K. Miller, and J. L. Willows, "The Derivation of Simple Poles in a Transfer Function from Real-Frequency Information, Part 2: Results from Real EM Data," Lawrence Livermore Laboratory Rept. No. UCRL-52118, August 23, 1976.
17. E. K. Miller and J. A. Landt, "Direct Time-Domain Techniques for Transient Radiation and Scattering," Lawrence Livermore Laboratory Rept. No. UCRL-52315, July 1, 1976.
18. E. K. Miller, "Data Characterization and Compression," Lawrence Livermore Laboratory Rept. No. UCID-17511, July 6, 1977.
19. W. D. Smith and D. L. Lager, "Parametric Characterization of Random Processes Using Prony's Method," Lawrence Livermore Laboratory Rept., to be published.
20. W. J. Wiscombe and J. W. Evans, "Exponential Sum Fitting of Radiative Transmission Functions," J. Comp. Physics, Vol. 24, August 1977.
21. N. K. Gupta and J. F. Bohn, "A Technique for Measuring Rotorcraft Dynamic Stability in the 40 x 80 Foot Wind Tunnel, NASA Contractor Rept. NASACR-151955, 1977.
22. E. K. Miller, J. N. Brittingham, and J. L. Willows, "Identification of E.M. Spectrum by Known Pole Sets," Electronics Letters, Vol. 13, No. 25, December 1977, pp. 774-775.
23. E. K. Miller, "A Study of Target Identification Using Poles," Lawrence Livermore Laboratory Rept. No. UCRL-52685, March 15, 1979.
24. E. K. Miller and D. L. Lager, "Prony's Method for the Angle Domain," Lawrence Livermore Laboratory Rept. No. UCID-17502-Rev. 1, October 6, 1977.
25. R. S. Elliot, "On Discretizing Continuous Aperture Distributions," IEEE Trans. Antennas and Propagation, Vol. AP-25, 1977, pp. 617-621.
26. J. M. Kelso, "Measuring the Vertical Angles of Arrival of HF Skywave Signals with Multiple Modes," Radio Science, Vol. 7, No. 2, 1972, pp. 245-250.
27. R. J. Lytle and D. L. Lager, "Using the Natural-Frequency Concept in Remote Probing of the Earth," Radio Science, Vol. 11, No. 3, March 1976, pp. 199-209.
28. E. K. Miller and D. L. Lager, "Inversion of One Dimensional Scattering Data Using Prony's Method," Lawrence Livermore Laboratory Rept. No. UCRL-52667, February 12, 1979.

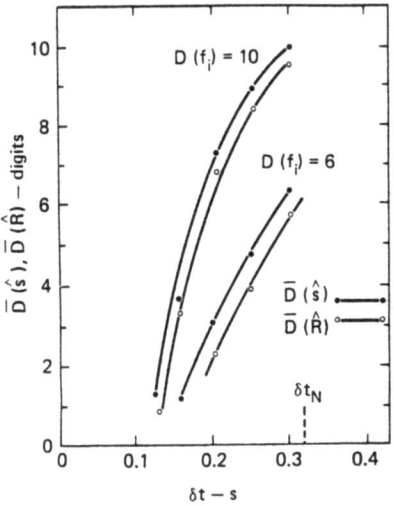

Fig. 1. The accuracy of Prony's method is very sensitive to the time interval δt between data samples, as illustrated here for noise-free data. We show the pole and residue accuracy as a function of δt for two values of $D(f_i)$. In this and the next three cases, the solution is for the square system, where $D = 2P = 2W$. (Ref. 9)

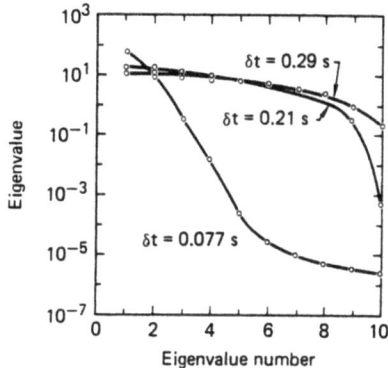

Fig. 2. Eigenvalues of the data matrix for three values of δt demonstrate the ill-conditioning of the data that can occur when the samples are too closely spaced. (Ref. 9)

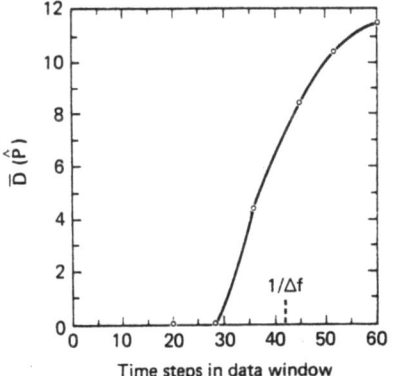

Fig. 3. When applied to exponential data, the accuracy of Prony's method is very sensitive not only to the interval between data samples, but also to the total data window available. As shown here, the data window must be comparable to the inverse beat frequency to obtain maximum accuracy. (Ref. 9)

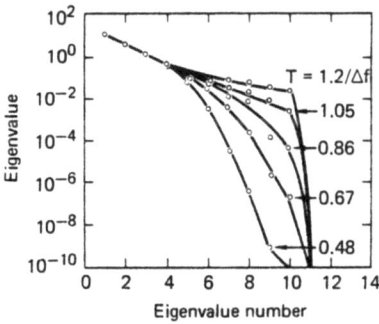

Fig. 4. The data window influences the range of eigenvalues of the data matrix. (Ref. 9)

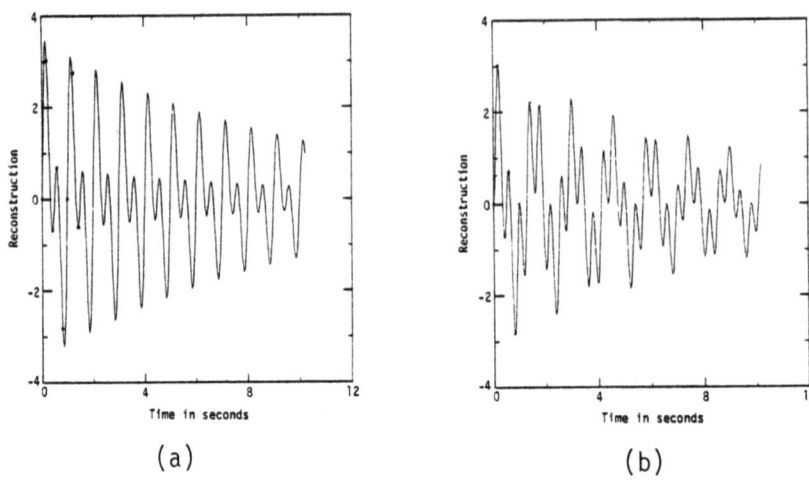

Fig. 5. Interpolatory and extrapolatory evaluation of the processing scheme (a) Original waveform and for P = W. (b) P < W. (Ref. 10)

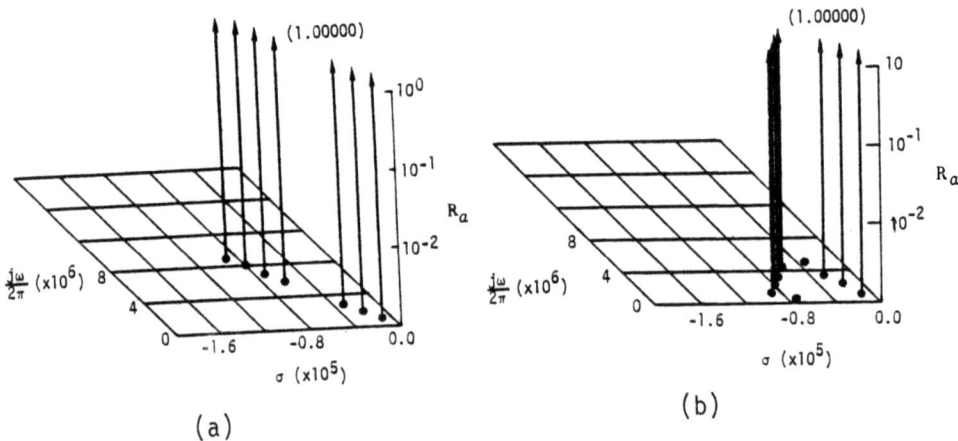

Fig. 6. Aliasing effects in the complex s - plane. (a) Sampling frequency = 20 MHz (correct results). (b) Sampling frequency = 10 MHz (aliased results). (Ref. 10)

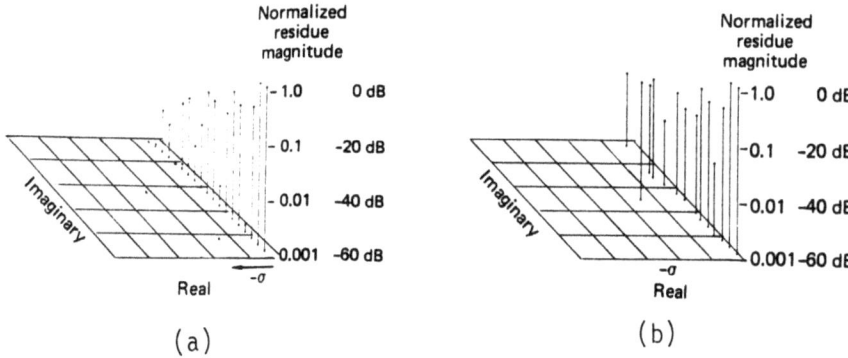

Fig. 7. (a) Three-dimensional plot of noise-free pole locus for 60-m dipole. (b) Three-dimensional plot of noisy pole locus for 60-m dipole. (Ref. 11)

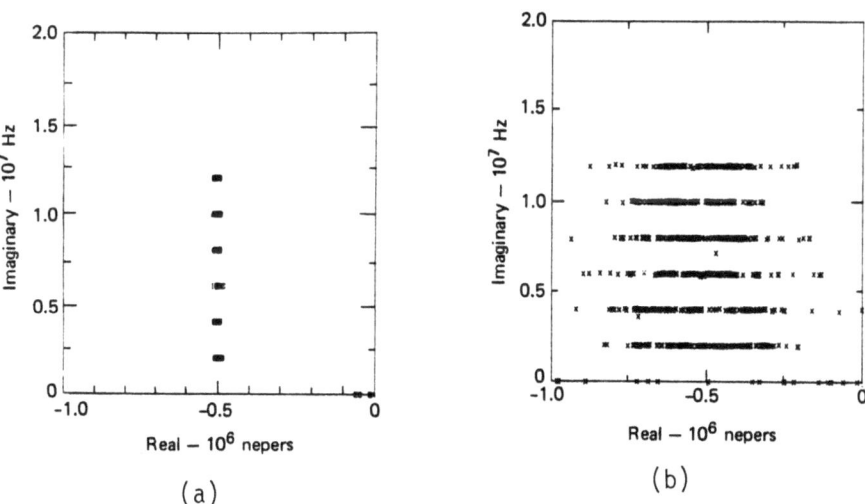

Fig. 8. (a) Pole clusters from ten iterations with ±0.14 per cent added noise. (b) Example of pole-cluster spreading with increasing noise; ±2.8 per cent noise added. (Ref. 11)

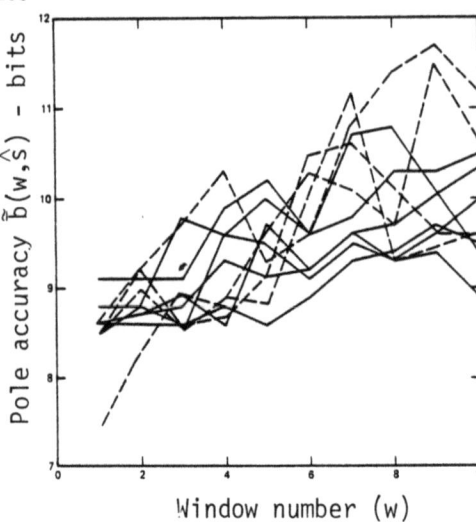

Fig. 9. Cumulative average pole accuracies, $\tilde{b}(w,\hat{s})$, obtained from moving-window procedure shown as function of window number. Ten different runs are shown for the pole set of Table 1; the parameters used are $W=P=10$, $D=20$, $\delta=10\Delta=0.0486$ s, and $S_p=10^3$. Accuracy improvement is not monotonic from window to window but exhibits an average value of 1.45 bits over 10 windows. (Ref. 12)

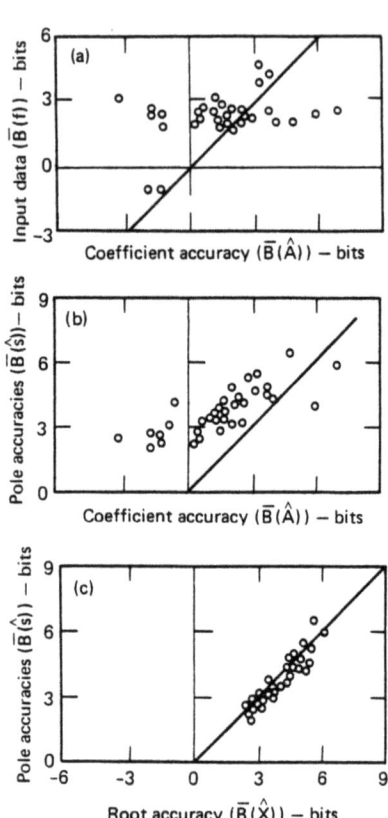

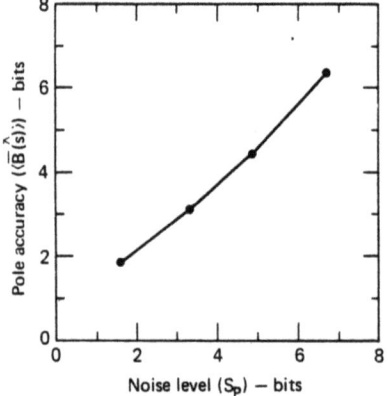

Fig. 10. Scatter plots which illustrate the relationships existing among input data, coefficients, roots, and poles. (a) Input data and coefficient accuracies appear essentailly uncorrelated. (b) Pole accuracies prove significantly better than coefficient accuracies. (c) Pole and root accuracies are closely correlated. Parameter values are $P=W=10$, $S_p=100$, $\delta=0.0467$ s, and $D=2P$. (Ref. 12)

Fig. 11. Extracted pole accuracy as function of peak-signal-to-peak-noise ratio. This correlation serves as a useful measure of the overall performance of Prony's Method in estimating the parameters of noisy exponential data. The agreement between input information, defined by the input data accuracy, and output information, represented by the bit accuracies of the poles, suggests that Prony's Method preserves essentially all of the information during the transformation from data samples to parameter estimates. Other parameters are: $P=D/2=8$, and $\delta=0.0467$ s. (Ref. 12)

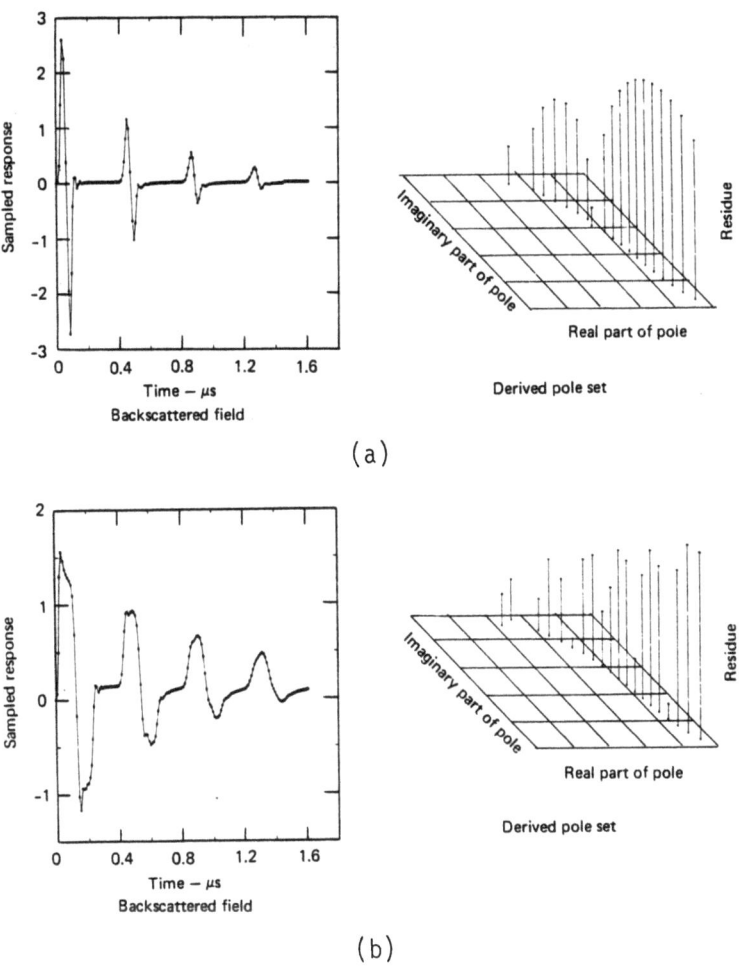

Fig. 12. (a) SEM poles can be directly extracted from transient waveforms using Prony's method. Results for the transient field scattered from a dipole illuminated by a Gaussian pulse at 30° incidence yield the pole set s_a, shown. (b) Results for the transient field scattered from a dipole illuminated by a Gaussian pulse at 60°. Note that, in spite of the extremely different waveforms, the pole locations are the same as in (a). All the waveforms difference is due to variations in the residues, R_a, presented here in magnitude as vertical lines on a 3-decade log scale. (Ref. 14)

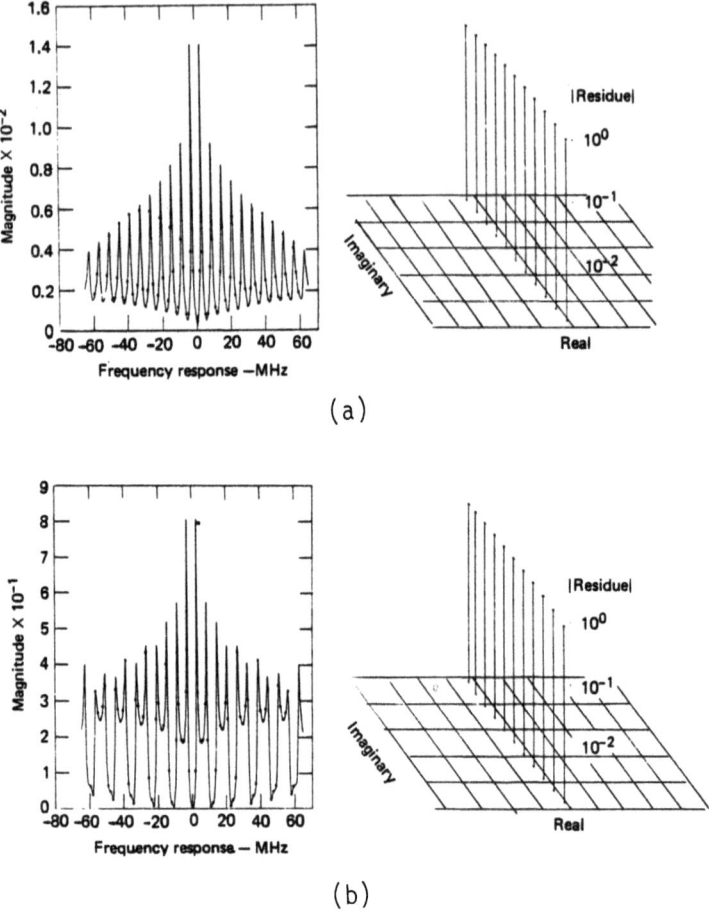

Fig. 13. Application of frequency-domain Prony to find the poles and residues for a 50-m wire. The figures show the magnitude of the original pole-residue spectrum with the data points denoted by asterisks and an isometric plot of the calculated poles and residues. Results from the feed-point current are shown in (a), while (b) shows results from the broadside far field where, in both cases, the antenna is center excited. (Ref. 16)

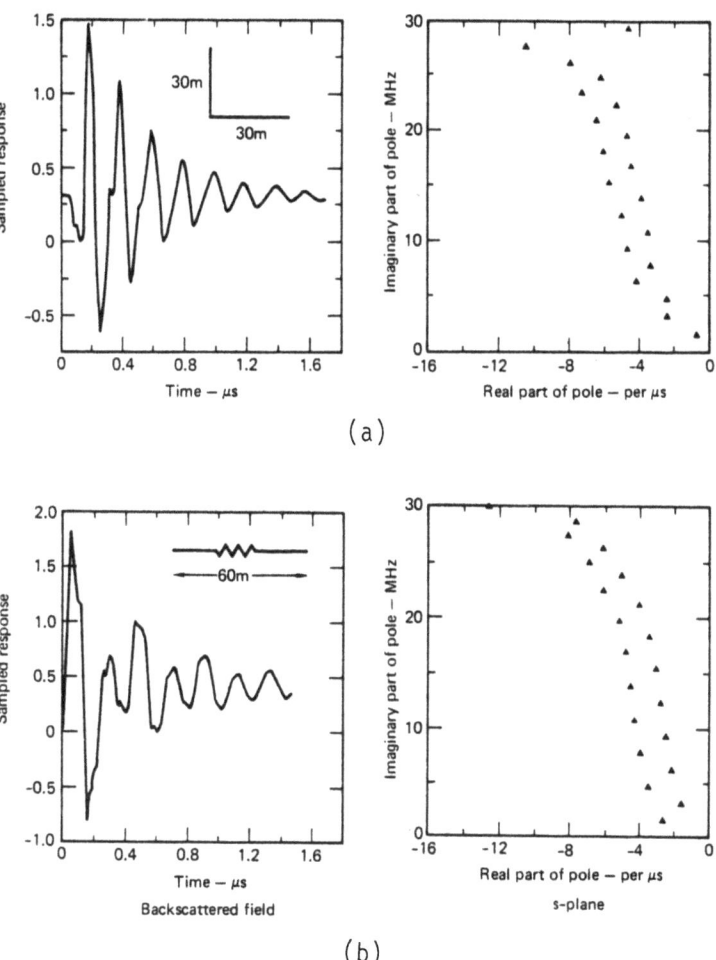

Fig. 14. (a) Energy loss mechanisms can affect the pole locations in the complex frequency plane. Here we present the pole sets for a 60-m wire, showing the result when the wire is straight and loaded with a 100-ohm resistance at its center. (b) here, the wire is unloaded but has a 90° bend at its center. The $\alpha = 1,3,...$ poles are more lossy for the loaded wire, due to dissipative loss of those modes which have current maxima at its center, and the $\alpha = 2,4,...$ modes are more lossy for the bent wire, because the charge maxima at the bend produce a greater radiation loss. (Ref. 17)

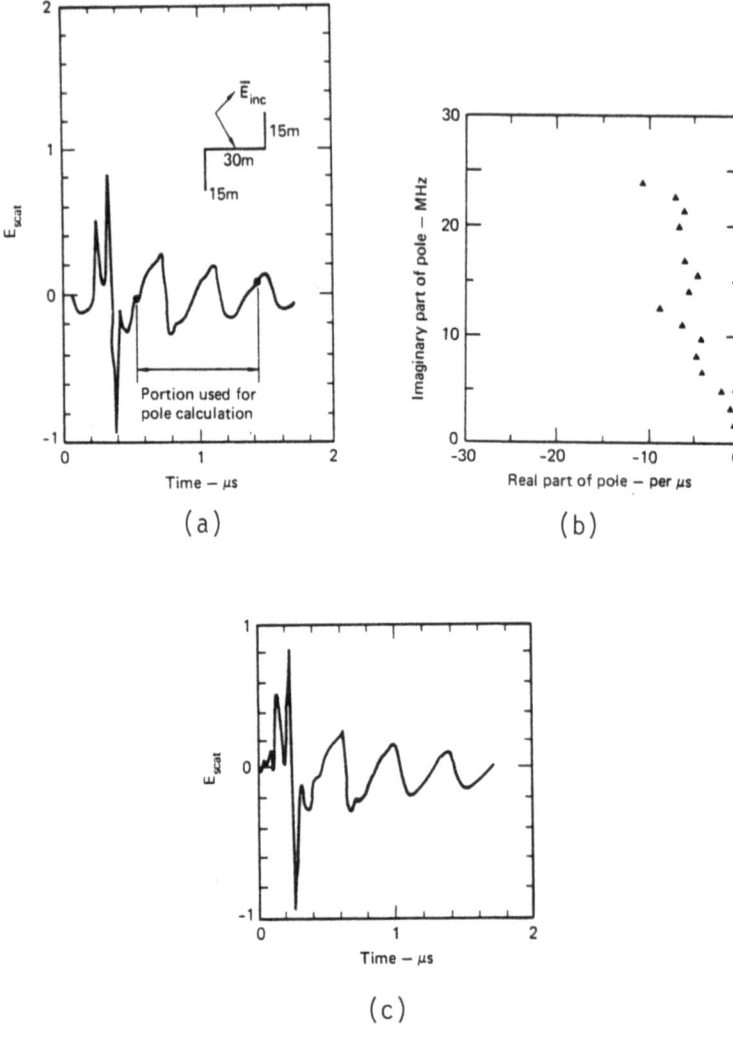

Fig. 15. (a) Extrapolation of a transient waveform can also be accomplished using the poles extracted from it. Here, part of a computed field scattered from the thin-wire object is illustrated. (b) Poles extracted from a portion of the transient waveform shown in (a). (c) The waveform reconstruction, which compares well with the original waveform both before and after the sampled portion. (Ref. 17)

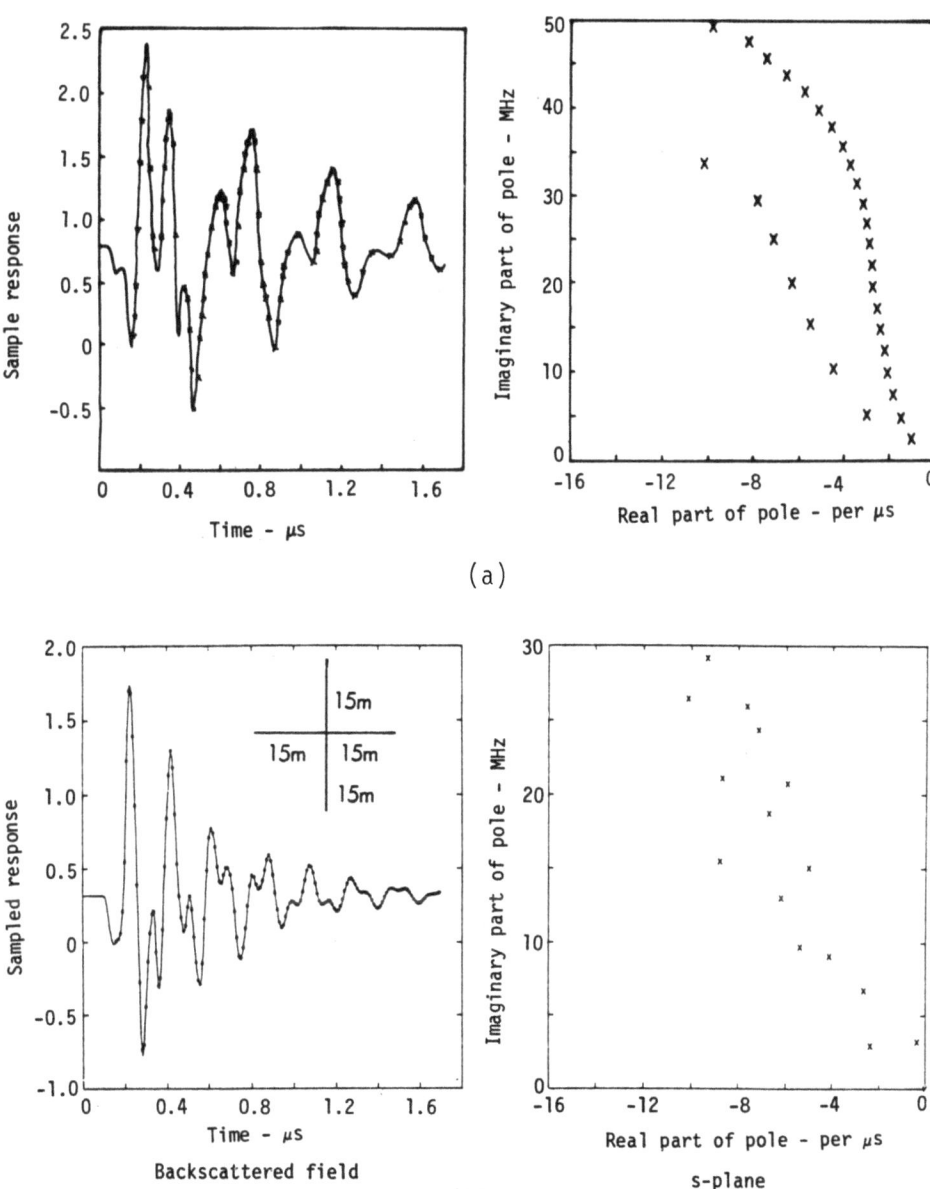

Fig. 16. Poles of loop and straight wire are separately extractable from their superimposed waveforms. Pole structure (a) of a 60 m loop differs markedly from that of straight and bent wires of same length in Figs. 14 and 15. Note number of poles is reduced to about half and their real components are larger, due to the presence of only odd modes on the loop, and the radiation loss caused by the loop curvature. A cross (b) having a total wire length of 60 m produces a more complex pole structure than those objects shown previously. Radiation due to the bend at the center of the cross is evidently the cause of the increased real-pole component.

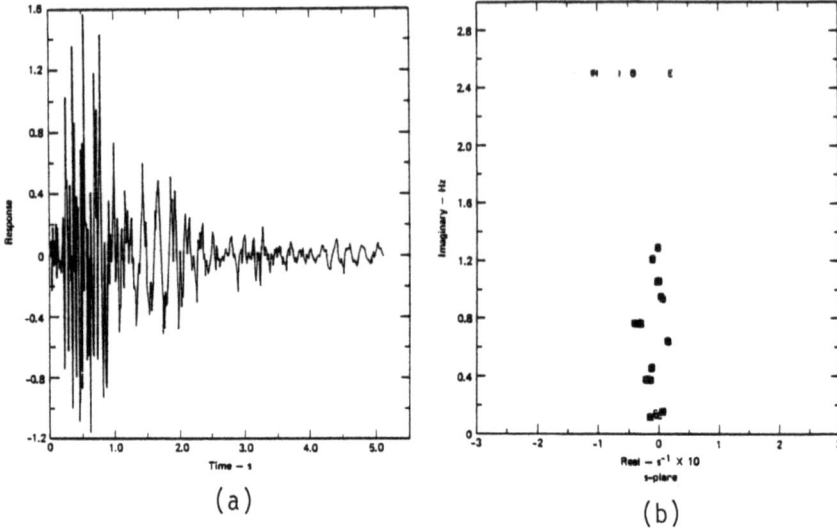

(a)

(b)

Fig. 17. Application of Prony's method to the parameterization of waveforms arising in structural dynamics. The accelerogram response (a) to a finite-length driving force (earthquake) was processed using the moving-window technique to obtain the complex natural frequencies of an existing building (b). One can use this information to determine the frequency of oscillation and the damping in each mode. By considering mode shapes, one can also interpret oscillations in terms of rigid body (soil-structure interaction) and elastic modes. (Ref. 18)

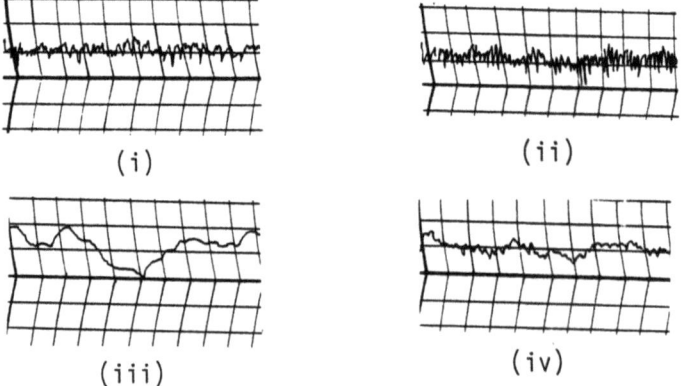

(i)

(ii)

(iii)

(iv)

Fig. 18. (a)

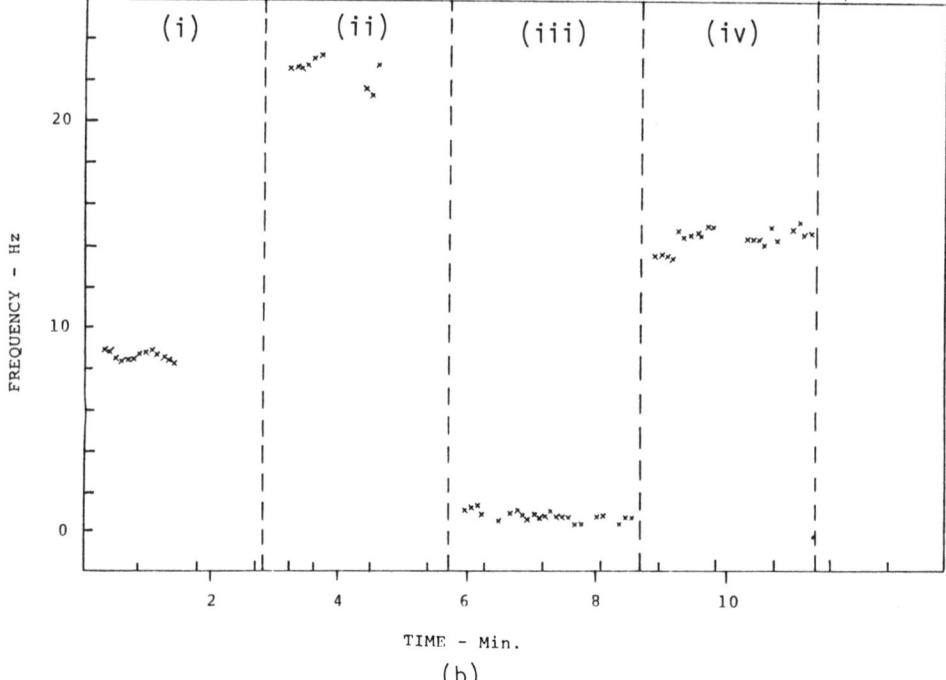

Fig. 18. (a) Representative EEG signals. (b) Plot of the dominant poles as a function of time, as obtained from D > 2P, for the EEG of a patient being anesthetized for surgery. Several distinct patient states can be observed, corresponding to an initial awake state (i), anagitated state (ii), over-anesthetized state (iii), and a steady state where surgery continued uneventfully (iv). Waveforms in (a) correspond in order to the four states whose poles are shown in (b). (Ref. 19)

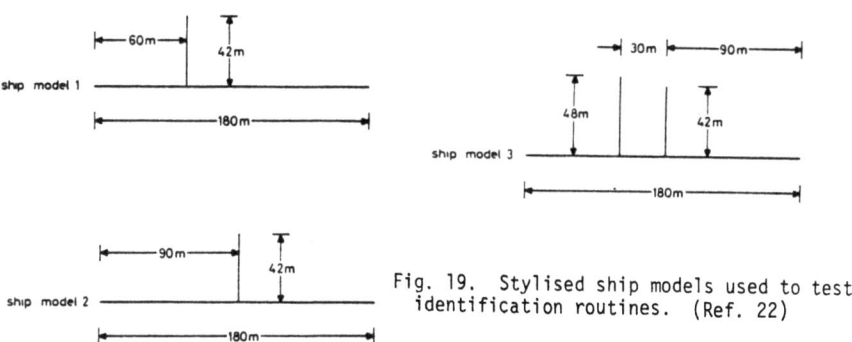

Fig. 19. Stylised ship models used to test identification routines. (Ref. 22)

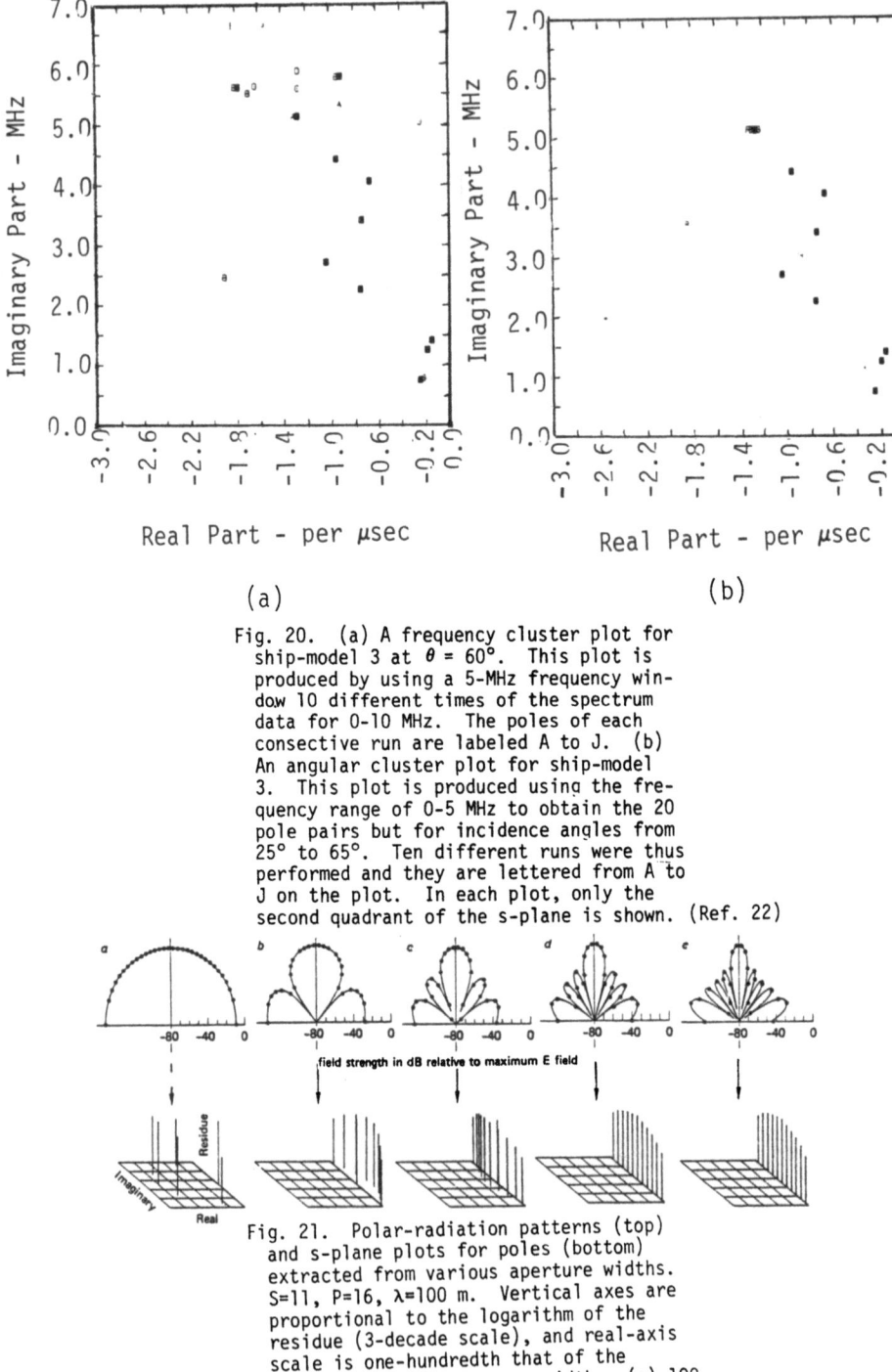

Fig. 20. (a) A frequency cluster plot for ship-model 3 at θ = 60°. This plot is produced by using a 5-MHz frequency window 10 different times of the spectrum data for 0-10 MHz. The poles of each consecutive run are labeled A to J. (b) An angular cluster plot for ship-model 3. This plot is produced using the frequency range of 0-5 MHz to obtain the 20 pole pairs but for incidence angles from 25° to 65°. Ten different runs were thus performed and they are lettered from A to J on the plot. In each plot, only the second quadrant of the s-plane is shown. (Ref. 22)

Fig. 21. Polar-radiation patterns (top) and s-plane plots for poles (bottom) extracted from various aperture widths. S=11, P=16, λ=100 m. Vertical axes are proportional to the logarithm of the residue (3-decade scale), and real-axis scale is one-hundredth that of the imaginary axis. Aperture width m (a) 100 (b) 300, (c) 500, (d) 700, (e) 900. (Ref. 24)

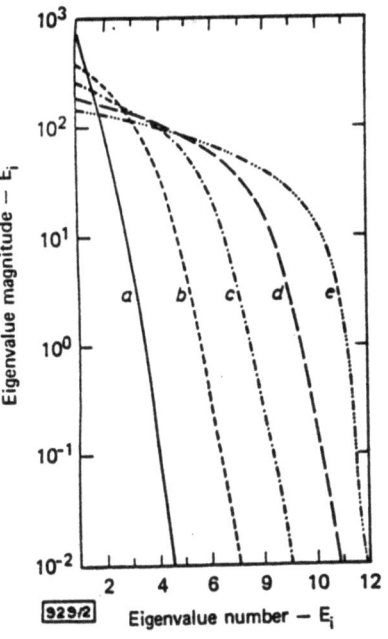

Fig. 22. Eigenvalues of the data matrix (Eq. (6b)) for sections (a) through (e) of Fig. 21. The most successful pole extractions occurred for (d) and (e). These curves had the smallest slope and the smallest ratio of the first to the 11th (since the system had 11 poles) eigenvalue. (Ref. 24)

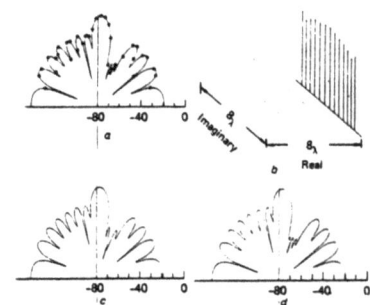

Fig. 23. Synthesis of a polar pattern (dB scale) of a continuous 7λ (700 m)-aperture. (a) original pattern, (b) its derived distribution, using this technique, (c) constrained pattern, (d) reconstructed pattern. Patterns given on dB scale. (Ref. 24)

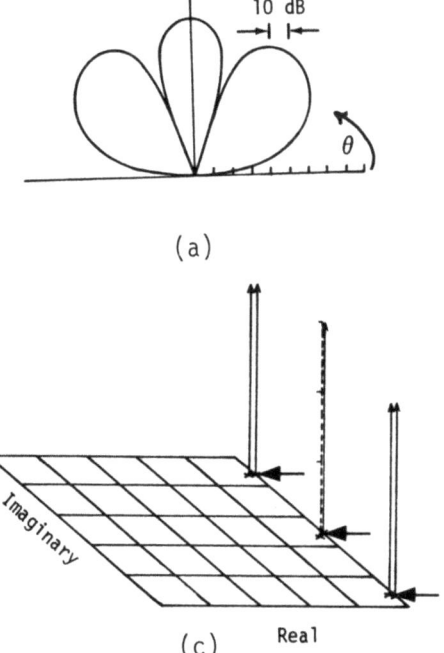

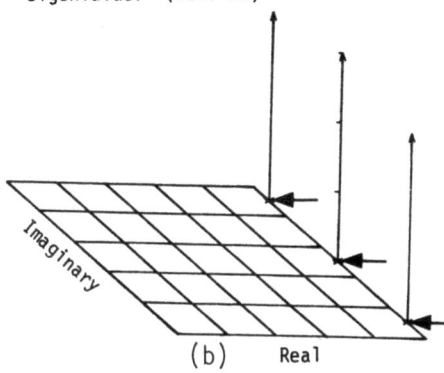

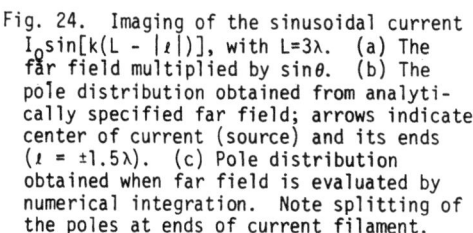

Fig. 24. Imaging of the sinusoidal current $I_0 \sin[k(L - |\ell|)]$, with $L=3\lambda$. (a) The far field multiplied by $\sin\theta$. (b) The pole distribution obtained from analytically specified far field; arrows indicate center of current (source) and its ends ($\ell = \pm 1.5\lambda$). (c) Pole distribution obtained when far field is evaluated by numerical integration. Note splitting of the poles at ends of current filament.

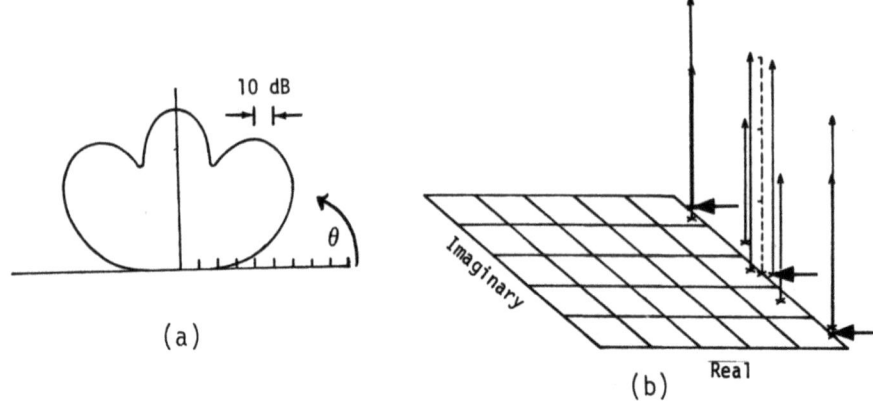

Fig. 25. Imaging of the current (computed from moment method) on a center-excited 3-λ wire antenna. (a) The far field multiplied by sinθ. (b) The pole distribution.

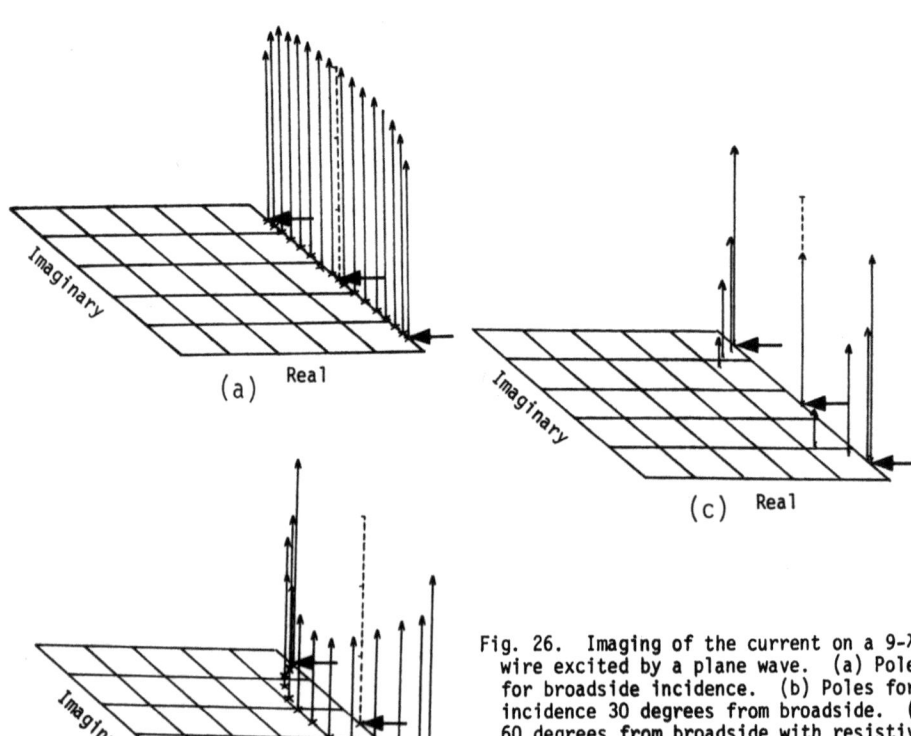

Fig. 26. Imaging of the current on a 9-λ wire excited by a plane wave. (a) Poles for broadside incidence. (b) Poles for incidence 30 degrees from broadside. (c) 60 degrees from broadside with resistive load added at wire's center. Note additional pole which arises due to the load.

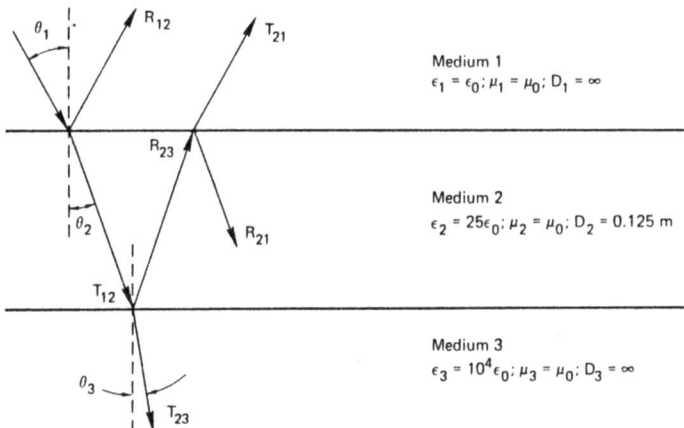

Fig. 27. Geometry and notation used for the one-layer problem. Plane wave is incident from medium 1 at angle θ_1. Letters R and T denote reflected and transmitted waves respectively, with the first subscript indicating the medium from which a wave originates the second the medium with which it has interacted.

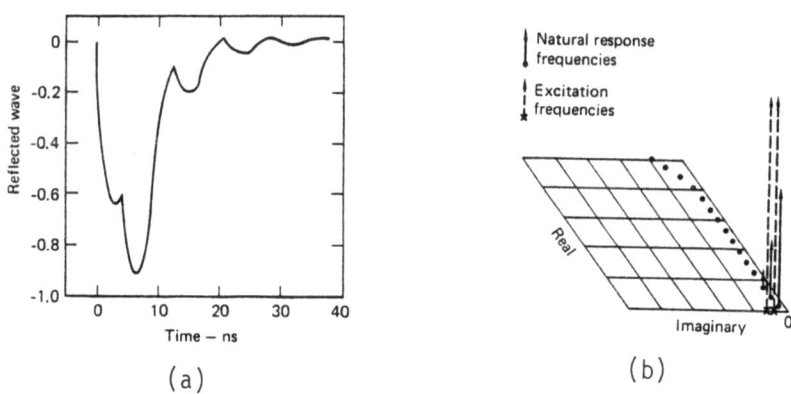

Fig. 28. Time domain reflected wave (a) and poles extracted from it using time-domain Prony (b), for layer of Fig. 27. Note the appearance of the incident-field poles, and that all poles are lossy. (Ref. 27)

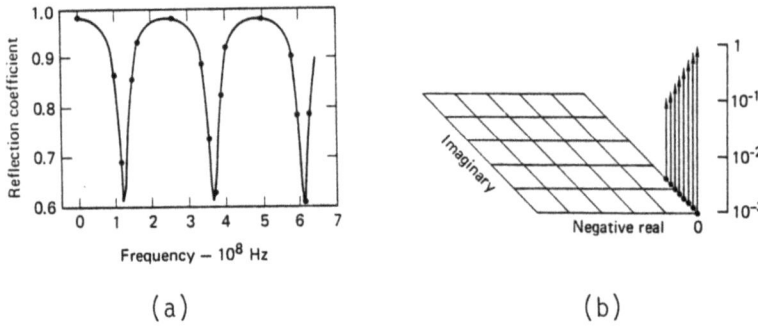

Fig. 29. Frequency-domain spectrum (a) and poles extracted from it using frequency-domain Prony (b) for the layer of Fig. 27. Here, in contrast to Fig. 28, all the poles are pure imaginary, or lossless. (Ref. 28)

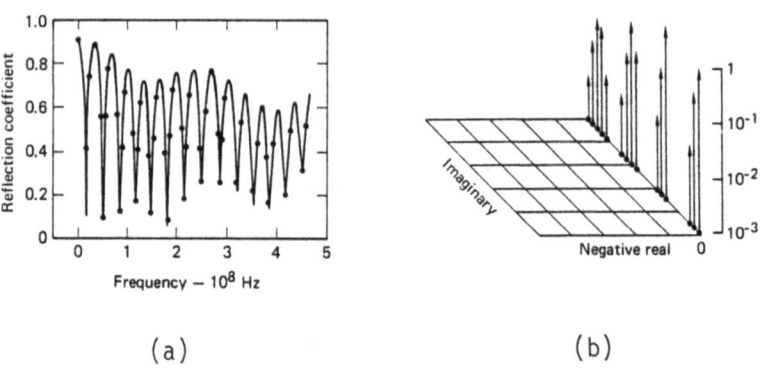

Fig. 30. Frequency-domain spectrum (a) and poles extracted from it using frequency-domain Prony (b), for a two-layer geometry. The poles remain lossless as in Fig. 29, but now occur in groups associated with a fixed number of round trips in the lower layer. (Ref. 28)

DIRECT TIME-DOMAIN TECHNIQUES FOR TRANSIENT RADIATION AND SCATTERING - WIRES*

E. K. Miller and J. A. Landt

Lawrence Livermore Laboratory, Livermore,
California 94550, U.S.A.
Los Alamos Scientific Laboratory, Los Alamos,
New Mexico, 87545, U.S.A.

ABSTRACT. This is a brief tutorial introduction to transient electromagnetics, focusing on direct time-domain techniques. We examine physical, mathematical, and numerical aspects of time-domain methods, with emphasis on wire objects excited as antennas or scatterers. Numerous computed examples illustrate the characteristics of direct time-domain procedures, especially where they may offer advantages over the more familiar frequency-domain techniques. These advantages include greater solution efficiency for many types of problems, the ability to handle nonlinearities, improved physical insight and interpretability, availability of wide-band information from a single calculation, and the possibility of isolating interactions among various parts of an object using time-range gating.

1. INTRODUCTION

Transient electromagnetics has interested scientists since Maxwell's equations were formulated, but our ability to obtain analytical or experimental results in this field is relatively recent. The two most important reasons for this new ability are first, the digital computer, which has made possible advanced computational and analytical work, and second, technological

---

*This work was performed under auspices of the U.S. Department of Energy by the Lawrence Livermore Laboratory under contract #W-7405-ENG-48.

developments in short-pulse hardware. These two factors have significantly increased the availability of transient results for an expanding variety of electromagnetics (EM) problems.

While appreciating the academic value of understanding transient electromagnetics, one might question the practical need for, and utility of, transient solutions. But developing technology in short-pulse hardware has motivated much analytical and computational work, which in turn has demonstrated the unique contributions that transient techniques can make to a more fundamental understanding of electromagnetics in general. Thus, transient techniques are emerging as a separate branch of electromagnetics.

Current applications for transient EM include space-object identification via short-pulse radar, nuclear electromagnetic pulse (EMP) effects, and nonlinear phenomena. These and other applications represent a significant departure from the monochromatic problems with which we may be more familiar. While some techniques used for transient or wide-band problems may not differ from those used for monochromatic problems, others may differ greatly.

This introductory, and necessarily sketchy, discussion of transient EM emphasizes engineering aspects of transients. It is hoped that the material will not only inform the reader about the current status of this developing technology, but also motivate him to exploit it for his own applications.

The presentation is organized as follows. First, we define our terms and provide an overview of the transient EM areas to be covered. Next, we discuss the techniques available for realizing transient behavior in EM. Then we consider the physical, mathematical, and numerical aspects of transient analysis. Finally, we demonstrate features and applications of transient EM. We demonstrate the variety of ways to characterize transient behavior for a given problem, as well as the variety of problem types that can be handled, and emphasize the physical interpretability of phenomena through transient analysis.

2. DEFINITION OF TERMS AND OVERVIEW

Transient electromagnetics may be broadly and qualitatively defined as all non-monochromatic EM phenomena. This discussion is limited to scatterers and radiators in linear, time-invariant media. Transient information can be obtained from transformed frequency-domain data or from a direct time-domain solution. We focus on the latter area, but refer to frequency-domain work to clarify the contrast between the two approaches.

More specifically, we concentrate on direct time-domain solutions for wire objects obtained from an integral-equation treatment. This narrow scope allows us to treat a few topics in depth, rather than many subjects superfically. Although this presentation is fairly long, it covers only a small part of the developing area of transient electromagnetics. For more detail, the interested reader may consult various references (in particular, 1-4) that provide an overview of the topic.

The terms "frequency domain" and "time domain" characterize the analytical or experimental procedure used to obtain the EM response desired. A frequency-domain procedure factors out the time dependence via an $e^{i\omega t}$ multiplier. A time-domain procedure, on the other hand, treats time as an explicit independent variable. The distinction can become blurred in pulsed-continuous-wave or swept-frequency systems, but in general the frequency domain uses monochromatic or continuous-wave excitation, and the time domain uses impulsive excitation.

When seeking transient information for a linear system, then, one must decide whether to employ a frequency-domain or time-domain approach. Most earlier work in transient EM analysis was based on a frequency-domain formulation,[5,6] because closed-form, time-domain solutions are almost impossible to obtain. Even frequency-domain problems were difficult to solve because of the extensive computational effort and many frequencies required to evaluate a problem and transform it to the time domain. Most results from early transient analysis involved acoustic scattering from infinite cylindrical structures, because two-dimensional problems are more easily computed than three-dimensional problems[7]. More recently, Rheinstain[8] has solved three-dimensional transient EM problems (the conducting and dielectric spheres) in the frequency domain, and numerous examples resulting from EMP studies have lately been developed[9,10].

The first time-domain approach to EM transient problems was based on physical optics to obtain the approximate backscatter-impulse response of a flat plate and spheroid.[11,12] This work was later extended to other geometries, such as the cone sphere. Subsequent direct time-domain work has concentrated on integral-equation techniques, the primary subject of attention below.[13-19]

Some advantages of these direct time-domain solutions over frequency-domain treatments of transient problems are:

1) Greater solution efficiency for many types of problems.

2) More convenient handling of non-linearities.

3) Improved physical insight and interpretability.

4) Availability of wide-band information from a single calculation.

5) Opportunity to isolate interactions, using time-range gating (e.g., pulse reflection from wire ends, bends, etc.).

6) Possibility for more directly and efficiently obtaining SEM poles.

These benefits require some trade-offs however. Foremost is the greater complexity of a time-domain code than of an equivalent frequency-domain version, with a resulting increase of difficulty in its development and use. The computing capability required can also be significantly larger.

3. TECHNIQUES FOR OBTAINING TIME-DOMAIN RESULTS

3.1 Analytical Techniques

Unfortunately, few EM transient solutions can be expressed in closed form in terms of standard functions. In spite of that limitation, some success has been achieved for a variety of problems. Wu[20] has worked out a time-domain solution for a step-exicited infinite cylindrical antenna. Chan et al.[21] have developed a closed-form solution for the diffraction of a pulsed field from an arbitrarily oriented dipole by a wedge. The latter problem belongs to a class of scattering problems for which pulse solutions take on a simpler form than do time-harmonic or frequency-domain solutions.[21] Franceschetti and Papas[3] developed a heuristic description of the general properties of transient radiation. But even those time-domain solutions that can be given in closed form may require extensive computation to obtain numerical results. Consequently, most time-domain solutions inevitably involve substantial computer processing.

3.2 Numerical Techniques

EM time-domain numerical analysis naturally proceeds from time-dependent Maxwell's equations. When we express those equations in differential form, we can solve them in terms of finite-difference* approximation, sampling the unknown fields in both space and time. Imposed spatial boundary conditions and temporal initial values, which are also sampled in a weighted sense, lead to a system of equations which can be solved for the sampled values over the space considered at a sequence of time steps.

---
*We include any method which approximates differentials by discrete samples in the finite-difference category, e.g., the finite-element technique, uni-moment method, etc.

Alternatively, we can integrate Maxwell's equations, using an appropriate (usually infinite-medium) Green's function. Imposed spatial boundary conditions and temporal initial values now lead to an integral equation in which the unknowns are the sources induced on the surfaces over which the boundary conditions are applied. A sampling of these unknowns in space and time and a weighted sampling of the boundary values again lead to a system of equations for the space-time sampled values of the unknown, which is solvable as an initial-value problem.

There are some important differences between these two approaches. First, in the differential-equation formulation, the unknowns are sampled at all points within and on the boundary(ies) of the solution space. In the integral-equation formulation, on the other hand, the sampling is done only over the boundary on which the boundary conditions are applied. Thus, the integral-equation method can result in substantially fewer unknowns.

Second, the integral-equation treatment requires a time- and space-dependent Green's function. This function allows surface sampling to replace the volume sampling of the differential-equation formulation. For lossless, non-dispersive, homogeneous, linear, isotropic media, such a Green's function is readily obtained, but otherwise substantial complications occur that can require a volume, rather than a surface, integration. Consequently, for these more general media, an integral-equation approach may not be suitable. Nevertheless, most EM time-domain analysis to data uses the integral-equation approach.

Frequency-domain analysis is also based Maxwell's equations in differential or integral form, but for an assumed $e^{i\omega t}$ time variation. Imposed conditions on the field behavior over spatial boundaries complete the analytical formulation of the problem. A subsequent spatial sampling of the unknown, and weighted sampling of known boundary values, lead to a system of equations for the unknown sampled values.

In both the frequency domain and time domain, then, a system of equations is developed for sampled values of the unknown. The numerical solutions of these equations are substantially different, however, due to fundamental differences in their respective formulations. In the integral-equation treatment, for example, the interactions between the unknown sampled values are global in the frequency domain; i.e., the total field at a given observation point is due to the unknown sources distributed over the entire boundary. The spatial separation between the source and its field is manifested by a geometric attenuation and phase shift which are distance dependent. Consequently, all the unknown samples are mutually dependent, and must be solved simultaneously. The solutions are usually accomplished via matrix factorization or inversion.

In a time-domain integral-equation treatment, on the other hand, the interactions between unknown samples are displaced in time by an amount equal to that required for a field to propagate between them at the speed of light. This displacement (time retardation) means that a particular unknown sample value at a given point in space and time is essentially determined by the exciting field at that same space-time point and by the scattered fields there from earlier, more distant locations. Consequently, the unknown samples can be solved at any time step, provided all sample values at earlier times are already known. The time-domain problem is thus solved via time-stepping and without matrix inversion, given the initial values of the unknown sample values.

The equations from the differential-equation formulations are treated much like those from the integral-equation approach. The frequency-domain version results in a spatial set of unknown samples, all of which can mutually interact, and which therefore require a simultaneous solution, although a given sample depends explicity on its nearest neighbors only. The time-domain approach, on the other hand, results in unknown samples whose separation produces a time retardation in their interaction, and so permits a time-stepping solution.

3.3 Singularity-Expansion Method (SEM)

We normally associate transformed frequency-domain data with real frequencies; i.e., the ω in $e^{i\omega t}$ is a real number. However, we can express the frequency-domain transfer function of a given problem in terms of a complex frequency.

The SEM exploits a special feature of scattering and radiation from three-dimensional objects: the simple (i.e., first-order) poles their transfer functions may possess in the complex-frequency plane. If we know the locations and amplitudes (or residues) of these poles, we can easily construct a transient response, which is simply a series of exponentially damped sinusoidal terms, one for each pole. Much early work in SEM used frequency-domain analytical techniques to find the poles.[23,24] More recent work shows that the poles are extractable from time-domain data.[25,26] Because SEM provides a simple relationship between the frequency domain and time domain, we regard it as a hybrid technique.

3.4 Asymptotic Techniques

Asymptotic techniques may involve either low- or high-frequency characteristics of the frequency-domain approach, or the corresponding late and early responses of the time-domain approach.[12] In either case, asymptotic techniques attempt to exploit what is analytically deducible about time or frequency behavior for the limits indicated. For example, we can show that the radiated far fields produced by a pulse-excited finite-sized radiator or scatterer must vanish as ω approaches 0. Therefore, we know that the time integral of the far-field waveform must also vanish.

Low-frequency asymptotic results may also be based on the Rayleigh law of scattering, wherein the fields go to zero with decreasing frequency as ω^2. In the high-frequency limit, we might employ physical optics, geometrical optics, or the geometrical theory of diffraction to obtain the asymptotic behavior.

3.5 Measurement

Transient-response data are measured primarily through direct time-domain procedures. Developments in short-pulse technology enable us to generate and measure high-amplitude ($>10^3$ v), fast-rise-time ($\leq$300 picoseconds) pulses.[27,28] By using these pulses to excite a test object such as a scatterer or antenna, and using a sampling oscilloscope to measure induced currents and scattered fields, instantaneous-measurement bandwidths of 10:1 and more are possible. Such data can validate time-domain calculations directly; transformed to the frequency domain, they can meet a variety of needs.

4. PHYSICAL ASPECTS OF TRANSIENT ANALYSIS

Perhaps the single most useful aspect of transient analysis is the opportunity it provides for more clearly depicting and interpreting the physical behavior of electromagnetic fields. For example, a short pulse propagating on an open-ended wire clearly demonstrates the effects of radiation damping, dispersion, and reflection from an impedance discontinuity. Because such effects are harder to interpret as a function of frequency, they must usually be indirectly inferred in the frequency domain. Of course, by transforming frequency-domain results, we can derive the transient behavior which illustrates such phenomena, but direct development in the time domain is generally more tractable, and the concepts are easier to visualize.

In this section we discuss fundamental physical aspects of
transient radiation and scattering from a heuristic viewpoint.
We first present typical results of time-domain computations.
Then we examine in some detail the phenomena thus illustrated.
Next we consider radiation fields as a manifestation of charge
acceleration.

4.1 Some Numerical Results

The calculations presented here precede a detailed account of
their mathematical and numerical aspects in the next two
sections. They are included now to provide a mental image of
some physical aspects involved.

Figure 1 and 3 show current and charge distributions on a
straight wire excited by a voltage source at its center as an
antenna, and excited by a tangential electric field of a normally
incident plane wave as a scatterer. In both cases, the time
variation of the exciting source is Gaussian, i.e., $\exp -g^2 t^2$,
and space distributions of current and charge are shown for
several instants of time. Figures 2 and 4 show the time
variation of the current at the wire's center and the resulting
far fields in the broadside direction.

In Figure 1 the current for the antenna divides into two pulses,
approximately Gaussian in form and with a small oscillatory
undershoot, which propagate outward from the source, accompanied
by oppositely signed charge pulses. The pulses (current and
charge) decrease in amplitude and spread out as they propagate
from the source. Nearing the end of the wire, the charge pulses
increase in amplitude, while the current pulses decrease, falling
to zero amplitude at the end. After the reflection, the
amplitude of both pulses diminishes, and the current reverses
sign because the charge flow reverses direction. Initially the
radiated field closely resembles the Gaussian shape of the
exciting pulse with which it coincides in retarded time (allowing
for propagation time), but then exhibits a slight negative
undershoot. A large negative pulse in the radiation field
coincides with the end reflection of the current-charge pulse.

The current for the scatterer distinctly differs from that for
the antenna. It is uniform over the entire wire except near the
ends, where it falls to zero, and where the charge is first
concentrated. The current amplitude in the central region of the
wire slowly decreases with time, while near the ends it reverses
sign as the boundaries of the pulse collapse inward due to the
charge reflection. In the radiation (or scattered) fields, an
initial return is similar in shape to the Gaussian exciting pulse
and coincides in retarded time with the current buildup. This

part of the scattered field is sometimes referred to as a specular flash. A sign reversal of the scattered field closely follows, with a peak value less than half that of the first maximum, beyond which the field decays with time. This part of the scattered field has a time variation similar to that of the decaying current.

4.2 Interpretation Of Numerical Results

The relationship observed between the antenna current and radiation field clearly shows that the onset of current flow is responsible for the first portion of the radiated waveform. The subsequent sequence of oppositely signed pulses in the field are as clearly due to the reflection of the current and charge from the wire ends. These results suggest that a radiated field is produced both when charge is accelerated (during turn-on of the current) and when charge is decelerated (during reflection from the wire ends). Note that the sense of the radiated field evidently depends upon the direction of the charge acceleration; the deceleration here corresponds to negative acceleration with respect to the original charge motion, and produces a field of opposite sign at the first end reflection. The intermediate part of the radiated field between these first two pulses is not so easily accounted for. It appears to correlate with the spreading out of the current pulse and with a corresponding decrease in its amplitude as it propagates down the wire. Thus, it could be ascribed to a shedding of energy during propagation.

The initial pulse in the scattered field is also clearly due to the initial charge acceleration associated with setting up the current. The nearly immediate sign change of the scattered field is evidently due to charge deceleration at the wire ends as the current begins to be reflected there. In this case the resulting opposite-sign radiation persists, although its value decreases with time, because current excited along the entire wire must eventually propagate to either end (recall that the antenna current divided into two oppositely propagating current pulses), where it is reflected. The decreasing amplitude of the scattered field in this portion of its waveform must be due primarily to the slow decay of the current reaching the ends, as demonstrated by the current behavior in the central area of the wire. This decrease in current amplitude provides further, although indirect, evidence that the current radiates as it propagates along the wire.

In both the antenna and scattering cases, then, we find evidence that accelerated charge causes radiation. This radiation occurs whether the charge is driven by an accelerating, exciting field which provides an external force, or by a decelerating, induced field which provides an "internal" or self force. The external force sets the charge in motion and produces the associated radiation fields. The internal force does no work on the charge but instead receives energy from it as the charge slows to zero velocity. As the charge accelerates in the opposite direction, not all the energy returns to it. The difference is lost in the form of radiation. An additional loss seems related to a spreading charge pulse propagating on a wire, a result directly observable in the antenna case. The loss mechanism in this case might be the charge deceleration which the pulse dispersion implies. We might conclude that this behavior arises because the trailing edge of the pulse propagates more slowly than the leading edge, whose speed is near that of a light wave in the medium. Propagation of the charge pulse in this situation resembles the behavior of electron bunches in a klystron.

4.3 Some Mathematical Relationships

It is indeed a fact that charge acceleration causes electromagnetic radiation. This fundamental principle provides a basis for a heuristic understanding of all radiation processes, both transient and steady state.

We can study accelerated-charge fields in two ways: the microscopic approach, which concerns fields of individual charges, and the macroscopic approach which concerns average fields over the charge distributions.[29] We primarily consider the latter approach here. We can distinguish between the two on the basis of whether the observation times and distances are smaller (microscopic) or larger (macroscopic) than the characteristic times and distances associated with the sources (e.g., the relaxation time in a metal or its skin depth). The microscopic approach involves equations of motion of individual electrons, and can include mass, relativistic, and quantum-mechanical effects. In the macroscopic approach, we are more interested in solving Maxwell's equations in a field description than in the physical details of their sources.

Using Maxwell's equations, the far-radiated fields due to an electric current distribution K over a surface S can be written

$$\overline{H}(\overline{r},t) = \frac{1}{4\pi rc} \int_S \frac{\partial}{\partial \tau} \overline{K}(\overline{r}',\tau) \times \frac{\overline{r}}{|\overline{r}|} \, ds' \qquad (4.1a)$$

$$\overline{E}(\overline{r},t) = \eta_0 \overline{H}(\overline{r},t) \times \frac{\overline{r}}{|\overline{r}|} \quad , \qquad (4.1b)$$

where $\bar{r}$ and $\bar{r}'$ are observation and source-point coordinate vectors, respectively, t is the observation time, τ (the retarded time) is $t - |\bar{r} - \bar{r}'|/c$, c is the speed of light, and η_0 is the wave impedance of free space. Equations (4.1a-4.1b) represent a macroscopic viewpoint of radiation.

Since $\bar{K}$ is by definition the flow of charge and charge is conserved, any change in $\bar{K}$ with time, can be due only to a change in velocity of the charge which carries the current. Equation (4.1a) therefore states that radiation is due to accelerated (or decelerated) charge, in agreement with the preceding discussion. The fields of a point charge q moving with (non-relativistic) velocity $\bar{v}$ and acceleration $\dot{\bar{v}}$ are[30]

$$\bar{H}(\bar{r},t) = \frac{q\mu_0}{4\pi}\left\{\frac{\bar{v}(\tau) \times \bar{R}}{R^3} + \frac{\bar{R} \times [\bar{R} \times (\bar{R} \times \dot{\bar{v}}(\tau))]}{cR^4}\right\} \quad (4.1c)$$

$$\bar{E}(r,t) = \frac{q}{4\pi\epsilon_0}\left\{\frac{\bar{R}}{R^3} + \frac{\bar{R} \times \bar{R} \times \dot{\bar{v}}(\tau)}{R^3 c^2}\right\}, \quad (4.1d)$$

where $\bar{R} = \bar{r} - \bar{r}'(\tau)$ and $R = |\bar{R}|$, while ϵ_0 and μ_0 are the permittivity and permeability of free space.

When $\dot{\bar{v}}$ is zero, $\bar{H}$ and $\bar{E}$ have $1/R^2$ static components, an elementary result, and $\bar{E}_{rad}$ changes sign when $\dot{\bar{v}}$ reverses direction. The rate of energy radiation is also given by

$$\frac{dW}{dt} = -\frac{q^2|\dot{\bar{v}}|^2}{6\pi\epsilon_0 c^3} \quad . \quad (4.1e)$$

The total amount of energy radiation during a given velocity change is proportional to the time integral of $|\dot{\bar{v}}|^2$, so the higher the acceleration, the greater the radiated energy. Equations (4.1c - 4.1e) represent a microscopic view of the radiation process.

In our time-domain solution, we can develop expressions which provide a macroscopic description of the radiation process in terms of the current and charge distributions on the object. For the particular case of a wire object having contour C, the total energy collected by the object from the incident field up to time t is given by

$$W_C = W(t)\Big|_{Collected} = \int_{-\infty}^{t}\int_C \bar{E}_{inc}(w',t') \cdot \bar{I}(w',t')dw'dt. \quad (4.1f)$$

Similarly the total energy dissipated due to resistive loss can be expressed up to time t as

$$W_D = W(t)\Big|_{Dissipated} = \int_{-\infty}^{t}\int_C I^2(w',t')R(w')dw'dt'. \quad (4.1g)$$

Finally, the total energy stored in the fields near the wire due to its current and charge density is, at time t, proportional to

$$W_W = W(t)\Big|_{Wire} \alpha \frac{1}{4\pi}\int_C \left[\mu_0 I^2(w',t) + Q^2(w',t)/\epsilon_0\right]dw' = W_I + W_Q, \quad (4.1h)$$

where W_I and W_Q are the current and charge contributions, respectively.

The energy in the fields outside the wire can then be represented as

$$W_R = W_C - W_D \alpha W_W, \quad (4.1i)$$

which represents the energy in both the radiation and the near fields. The quantity W_W usefully indicates the time-changing stored energy and, by implication, when and where radiation occurs.

5. MATHEMATICAL ASPECTS OF TRANSIENT ANALYSIS

From Maxwell's equations for a perfectly conducting closed surface we can derive two time-dependent integral equations based upon the electric and magnetic field, respectively, as the forcing functions:[2]

$$n(\bar{r}) \times \bar{E}_{inc}(\bar{r},t) = \frac{\hat{n}(\bar{r})}{4\pi\epsilon_0} \times \oint_S \left\{-\frac{\bar{R}}{R^2}\sigma(\bar{r}',\tau) - \frac{\bar{R}}{cR^2}\frac{\partial}{\partial\tau}(\bar{r}',\tau)\right.$$

$$\left. + \frac{1}{c^2} \frac{\partial}{\partial \tau} \overline{K}(\overline{r}',\tau) \right\} \frac{ds'}{R} \quad ; \overline{r} \in S \quad (5.1)$$

$$\overline{K}(\overline{r},t) = 2\hat{n}(\overline{r}) \times \overline{H}_{inc}(\overline{r},t) + \frac{\hat{n}(\overline{r})}{2\pi} \times \oint_S \left\{ \frac{1}{R} \overline{K}(\overline{r}',\tau) \right.$$

$$\left. + \frac{1}{c} \frac{\partial}{\partial \tau} \overline{K}(\overline{r}',\tau) \right\} \times \frac{\overline{R}}{R^2} ds' \quad ; \quad \overline{r} \in S \quad ,(5.2)$$

where $\oint_S$ denotes the integral $\lim_{\Delta S \to 0} \int_{S-\Delta S}$; $\overline{E}$ and $\overline{H}$, the electric and magnetic fields, respectively; subscript inc, the incident field; $\overline{K}$ and σ, the surface current and charge densities; S, the surface of the object; and $\hat{n}$, the outward-pointing surface normal.

Equations (5.1) and (5.2) are the time-domain versions of what are often called the electric-field integral equation (EFIE) and the magnetic-field integral equation (MFIE), respectively, after the incident-field terms which appear in them. They are mathematically classified as Fredholm integral equations of the first and second kind, respectively, with the unknown appearing only under the integral in the former, but outside it as well in the latter. Besides this important difference, which significantly affects their numerical treatment, they also differ in the order of the spatial singularities which occur when $R \to 0$. In the EFIE, the highest-order singularity is the $1/R^3$ of the $\sigma(\overline{r}',\tau)$ term, coming from the $\overline{R}/r^3$ factor, and the spatial derivative of $K(\overline{r}',\overline{\tau})$ which results from replacing $\sigma(\overline{r}',\tau)$ via the continuity equation. The highest-order term in the MFIE, by contrast, is $1/R^2$. Finally, the MFIE is not as well suited to wires long compared with their diameter, because $\hat{n}$, being nearly parallel to $\overline{K} \times \overline{R}$, can produce numerical ill-conditioning. The far fields corresponding to Eqs. (5.1) and (5.2) have been given in Eq. (4.1).

Most frequency-domain antenna analysis is based upon the thin-wire approximation, which involves replacing a two-dimensional surface integration with a one-dimensional line integration, and approximating the surface current an an axially directed filament. This same approximation is also useful in the time domain. When applied to Eq. (5.1), it leads to

$$\hat{w} \cdot \bar{E}_{inc}(\bar{r},t) = \frac{\mu_0}{4\pi} \int_C \left[\frac{\hat{w} \cdot \hat{w}'}{R} \frac{\partial}{\partial \tau} I(w',\tau) + \frac{c}{R^2} \hat{w} \cdot \bar{R} \frac{\partial}{\partial w'} I(w',\tau) \right.$$

$$\left. - \frac{c^2}{R^3} \hat{w} \cdot \bar{R} q(w',\tau) \right] dw' \; ; \; \bar{r} \epsilon C + a, \quad (5.3a)$$

where

$$q(w',\tau) = -\int_{-\infty}^{T} \frac{\partial}{\partial w'} I(w',t') dt' ,$$

$\hat{w}$ and $\hat{w}'$ are tangent vectors to the wire at $\bar{r}$ and $\bar{r}'$, respectively, C is the wire contour, a is the wire radius, and $\bar{r} \epsilon C + a$ denotes that the field is to be evaluated on the wire surface. Other time domain integral equations specialized to wire geometries can also be developed. Equation (5.3a) provides the basis for most of the discussion which follows. The far-field expression which corresponds to Eq. (5.3a) can be written

$$\bar{E}(\bar{r},t) = -\frac{\mu_0}{4\pi R} \int_C \left\{ \hat{w}' \frac{\partial}{\partial \tau} I(w',\tau) + c \frac{\bar{R}}{R} \frac{\partial}{\partial w'} I(w',\tau) \right\} dw' . \quad (5.3b)$$

The frequency-domain counterparts of the time-domain integral equations are given by [2].

$$n(\bar{r}) \times \bar{E}_{inc}(\bar{r}) = \frac{1}{4\pi i \omega \epsilon_0} \hat{n} \times \oint_S \left\{ \nabla' \phi \cdot \bar{K}(\bar{r}') \nabla' \phi \right.$$

$$\left. - \omega^2 \mu_0 \epsilon_0 \bar{K}(\bar{r}') \phi \right\} ds' \quad (5.4)$$

$$\bar{K}(\bar{r}) = \frac{1}{2\pi} \hat{n} \times \oint_S \bar{K}(\bar{r}') \times \nabla' \phi \, ds' + 2\hat{n} \times \bar{H}_{inc}(\bar{r}) , \quad (5.5)$$

where

$$\phi = \exp[-ikR]/R \text{ and } k = \omega\sqrt{\mu_0 \epsilon_0}.$$

6. NUMERICAL ASPECTS OF TRANSIENT ANALYSIS

For clarity, we illustrate the direct, time-domain solution procedure for a simpler, or prototype, time-domain integral equation in place of the actual Eq. (5.3a). We thus consider

$$g(x,t) = \int_{-h}^{h} f(x',t')K(x,x')dx' \quad ; \quad -h \leq x \leq h, \quad (6.1)$$

where

$$t'(x,x',t) = t - |x - x'|/c,$$

as the equation to be solved, where g is specified and f is to be found. Proceeding on an intuitive basis, we might decide to approximate Eq. (6.1) with a discrete sequence of samples for f as a function of both x and t. If we further choose some reasonable variation between these discretely sampled points (i.e., select an interpolation function) and also specify how the right and left sides of this sampled equation are related, then we can reduce Eq. (6.1) to a linear system in which the samples for f are the unknowns.

For practical purposes this procedure constitutes the moment method. The interpolation function mentioned, in the moment-method context, is called a basis function, and the relationship between the two sides of the reduced equation depends on a weighting or testing function. Let us now use a space-time pulse approximation for f, i.e.,

$$f(x',t') = \sum_{i'=1}^{N} \sum_{j'=1}^{j} U_{i'j'} A_{i'j'},$$

where $U_{i'j'} = 1$ if
$$\begin{cases} x_{i'} - \Delta/2 \leq x' < x_{i'} + \Delta/2 \\ t_{j'} - \delta/2 \leq t' < t_{j'} + \delta/2 \end{cases}$$

and is zero otherwise, and $x_{i'} = \Delta i'$, $t_{j'} = \delta j'$. Let us also point-match the integral equation at the space-time sample locations x_i, t_j. Then we formally obtain

$$g_{ij} \equiv g(x_i,t_j) = \int_{-h}^{h} \sum_{i'=1}^{N} \sum_{j'=1}^{j} U_{i'j'} A_{i'j'} K(x_i,x')dx'$$

$$= \sum_{i'=1}^{N} \int_{(i'-1/2)\Delta}^{(i'+1/2)\Delta} \sum_{j'=1}^{j} U_{i'j'} A_{i'j'} K(i\Delta, x') dx'$$

where
$$\begin{aligned} i &= 1,\ldots,N \\ j &= 1,\ldots,N_T \end{aligned} \qquad (6.2)$$

But since $t' = t - |x - x'|/c$, and upon using $\Delta = c\delta$, Eq. (6.2) simplifies to

$$g_{ij} = \sum_{i'=1}^{N} A_{i',j-|i-i'|} \int_{(i'-1/2)\Delta}^{(i'+1/2)\Delta} K(i\Delta, x') dx' \quad , \quad (6.3)$$

which associates with each $A_{i'j'}$, a spatial integral (or interaction coefficient) over segment i' at retarded time $j - |i - i'|$.

It is helpful to rewrite Eq. (6.3). Let us first denote the interaction coefficients by

$$Z_{ii'} = \int_{(i'-1/2)\Delta}^{(i'+1/2)\Delta} K(i\Delta, x') dx' \quad ,$$

so that

$$g_{ij} = \sum_{i'=1}^{N} Z_{ii'} A_{i',j'} = Z_{ii} A_{ij} + Z_{i,i-1} A_{i-1,j-1}$$

$$+ Z_{i,i-2} A_{i-2,j-2} + \cdots + Z_{i,1} A_{1,j-i+1}$$

$$+ Z_{i,i+1} A_{i+1,j-1} + Z_{i,i+2} A_{i+2,j-2}$$

$$+ \cdots + Z_{i,N} A_{N,j-N+i} \quad . \qquad (6.4)$$

We observe that the first term involves a sample at the time step j, while all other samples are for $j-1$, $j-2$, ... and thus are from earlier times. Upon solving Eq. (6.4) for A_{ij}, we then find

$$A_{ij} = \frac{g_{ij} - \sum_{i'=1}^{N} {}^{i} Z_{i,i'} A_{i',j-|i-i'|}}{Z_{ii}} \quad ; \quad \begin{array}{l} i = 1,\ldots,N \\ j = 1,\ldots,N_T \end{array}, \quad (6.5)$$

where the summation excludes the term $i' = i$.

Equation (6.5) should forcefully demonstrate that at time step j the integral Eq. 6.1) can be solved as an initial-value problem by time-stepping, for if all values of A_{ik} are known for $k \leq j - 1$, then A_{ij} is completely specified by them and the present value of the forcing function, g_{ij}. This equation shows the explicit effect of causality and the finite velocity at which EM fields propagate, as an increasing time delay between a source at i' and its influence being observed at i (through the time index $j - |i - i'|$). In principle this factor permits solving Eq. (6.1) without matrix inversion, as further emphasized by the Z_{ii}, coefficient matrix in Eq. (6.5).

We have of course simplified this discussion of the numerical solution by choosing an apparently simple integral equation and using pulse-basis and delta-weight functions. The actual computer code[31] used to generate the results presented here uses a nine-term polynomial basis, up to and including quadratic space-time variation, and delta-function weights. Nevertheless, because the integral equation used here represents that which would apply to a straight wire, it is realistic. It makes retarded time depend very simply on source-observation distance, i.e.,

$$t' = t - |x - x'|/c.$$

Therefore, as we integrate the source space (x'), the space-time path described by x' and t' is a straight line, as illustrated in Figure 5. This integration path furthermore passes diagonally through the center of each space-time sample "patch" (or A_{ij}), because the observation points are also located at the patch centers and $\Delta = c\delta$ was used everywhere. This path means that a single A_{ij} is associated with each $Z_{ii'}$. In general, the integration path can pass through two A_{ij}'s for integration over one space segment when using a pulse basis, so that a given $Z_{ii'}$ decomposes into two parts, one multiplying each respective A_{ij} value.

For example, if the space observation points are not co-linear (i.e., on the same line defined by the current sample), then t' and x' are not linearly, but instead

$$t' = t - \sqrt{(x - x')^2 + (y - y')^2 + (z - z')^2}/c$$
$$= t - \sqrt{\rho^2 + (s - s')^2}/c,$$

where s and ρ are cylindrical coordinates of the observation point relative to the line along which the source current flows. Figure 5 also shows other representative space-time integration paths.

The g term in Eq. (6.1) (or the g_{ij} terms in Eq. (6.5)) represents the actual excitation (or its sampled values) which excites the response f (or the sample values A_{ij}). Since we are dealing with an integral equation based on the electric field and specialized to a wire, g represents the tangential electric field distribution along the contour C of the wire. For a scattering calculation, and assuming a point-matching solution is employed (which is the case for all the results subsequently presented), g_{ij} then is simply the specified value of the incident electric field tangent to C at a sequence of space points $\bar{r}_i$ and time steps t_j. For an antenna calculation, the exciting field is limited to one or a few segments of length Δ_i centered at observation point $\bar{r}_i$ and is equivalent to an applied voltage $V_{ij} = -\Delta_i g_{ij}$. The computation difference between a scattering and radiation calculation is minor.

The effect of lumped or distributed loads is mathematically indistinguishable from the treatment of the exciting source. The load causes a voltage drop in opposition to (passive) or in the direction of (active) the exciting field. The actual load voltage depends on its V-I characteristics. For linear, passive loads we have (resistance R, inductance L, capacitance C)

$$V_{LOAD}(\bar{r},t) = I(\bar{r},t)R(r) + L(\bar{r})\frac{\partial}{\partial t}I(\bar{r},t) + C(\bar{r})\int_{-\infty}^{t} q(\bar{r},\tau)d\tau,$$

the effects of which can be readily included in the numerical treatment previously outlined. For example, this would result, in the case of our prototype integral equation, in

$$g(x,t) \to g(x,t) - V_{LOAD}(x,t),$$

and considering a resistive load only at x_L, Eq. (6.5) becomes

$$A_{Lj} = \frac{g_{Lj} - \sum_{i'=1}^{N} Z_{i,i'}^{L} A_{i',j - |i-i'|}}{Z_{LL} + R_L} \qquad (6.5)'$$

while all other A_{ij}'s remain as given by (6.5).

When the load is non-linear, if for example the value of R_L depends on I_L, then formally Eq. (6.5)' still applies. But since R_L then depends on A_{Lj}, which in turn depends on R_L, we generally must solve the equation by iteration at each time step. For the special case of an ideal diode, however, which is specified by only a forward and reverse resistance, we can readily solve Eq. (6.5)' by finding A_{Lj} for $R_L = 0$, then using the direction of current thus determined to establish R_L, since the load cannot in this case reverse the current flow. We might also consider time-varying loads and other more general non-linearities.[32]

7. REPRESENTATIVE RESULTS

In the preceding discussion, we have briefly summarized various physical mathematical and numerical aspects of direct, time-domain modeling. We continue our presentation by including below representative numerical examples to demonstrate features essentially unique to time-domain treatments, and to show various types of applications.

7.1. Features Unique To Direct Time-Domain Analysis

7.1.1 Nonlinear Capability - Solution formulation directly in the time domain permits analysis of nonlinear problems by a time-stepping procedure. This approach has been used with finite-difference techniques to handle the problem of a metallic object immersed in a nonlinear medium.[33] Schuman[34] and others [35,36] have considered a thin, straight wire loaded at its center by a diode. This case, as well as nonlinear loading of more general wire geometries, is possible using the approach developed by Miller, Poggio, and Burke.[16]

Here, we concentrate on the Miller, Poggio, and Burke approach. Figures 6-8 shows the response of a linear dipole antenna loaded with a diode. The diode was placed in series with a Gaussian voltage source. Figure 6 shows early-time history of the current and charge distribution along the structure. The dipole is initially biased in the forward-conducting direction, allowing a charge separation to build up across it. When the exciting voltage decreases to zero, the accumulated charge then reverse-biases the diode, allowing only a small leadkage current in the reverse direction. The dipole then responds as two shorter dipoles placed end-to-end but insulated from each other. Comparison of the center current for this case (Figure 7) and the unloaded case (Figure 2) illustrates that the initial current is essentially the same for both (there is a slight amplitude difference caused by different wire radii) until the current in the linear case changes sign. From then on, the two currents are distinctly different. The late-time far field radiated broadside from the dipole is predominantly a dampened sinusoid at twice the frequency of the unloaded dipole, as shown in Figure 8.

7.1.2 Time Gating - Time gating can be used in transient-electromagnetic measurements to eliminate the effects of reflections from scatterers physically separated from the target. This powerful technique makes possible wideband measurements without expensive anechoic chambers, etc. But time gating can also be used numerically, as demonstrated in Figures 9 and 10.[37] Here, a dipole of finite length is illuminated by the time derivative of a Gaussian plane wave at broadside incidence. Figure 9 shows the current at the center of the dipole. The calculation ended before end reflections arrived. Thus, time gating permits us to obtain an infinite-wire response from a finite-wire response. Transformation to the frequency domain establishes the validity of this approach for the transfer admittance. Figure 10 shows the analytic results and the FFT of Figure 9. We can find the frequency range over which this technique produces valid data in the same way as in transient measurements. The clear time determines the lowest frequency of valid data, and the sampling requirements determine the highest frequency.

Time gating can also be used to study the propagation of a current pulse along a wire, reflections from ends of wires, reflections from junctions of wires, and similar phenomena.

7.1.3 Self-Diagnosis - One may encounter many pitfalls in applying any numerical technique. A peculiarity of direct time analysis that can be used to advantage is the way the results display ill-conditioning. For example, experience shows that using the thin wire approximation (see Miller, Poggio and Burke[17]) that wire segments should be at least as long as their diameter. In Figures 11-12 we show the current on a dipole excited by a broadside-incident, Gaussian-pulse plane wave for two wire radii. In each case, the dipole length is 1 m, and the segment length (Δ) is 1.666 cm. The two curves in Figures 11-12 show the result of increasing the wire radius, by only ~5% with $\Delta/2a \sim$ unity. For the larger $\Delta/2a$, the current is well behaved and not a function of the number of segments. (Of course, if too few segments are used, the results will be inaccurate.) But an exponentially growing current is obtained below some critical value for $\Delta < 2a$, signifying that something has become ill-conditioned. The particular source of this phenomenon in the numerical procedure has not been found, but is apparently a result of the thin-wire approximation. While the impedance matrix or interaction coefficients display nothing obviously wrong, the growing currents clearly signal that something in the numerical model is in error, however. Similar instabilities have been observed if only a single segment is too short, or if the match point of one segment lies inside the volume of another segment, and thus serve as a diagnostic of an invalid numerical model.

7.1.4 Inherently Broadband Calculations - The frequency-domain analysis of a structure is limited to discrete-frequency samples. The user of the code must choose a priori those frequencies where he thinks the structure will have interesting or useful behavior. If he chooses incorrectly, he may need to continue the calculations until he obtains the needed information, and can thus be reasonably confident that essential aspects of the object's behavior have not been overlooked. This is especially the case when the narrow-band resonances of a high-Q structure are being studied.

The time-domain analysis of a structure, on the other hand, is inherently broadband. As long as the excitation contains sufficient energy in the frequency range of interest, the entire response over that range is automatically obtained. For example, consider a frequency-independent antenna, the conical spiral. However, this frequency independence occurs over some limited bandwidth where the antenna was designed to operate. If one needs characteristics over a wider band (for pulse applications or EMP coupling analysis), transient analysis conveniently provides them. In Figures 13-14 we show some characteristics of a typical conical spiral. Figure 13 includes the feed-point current resulting from a Gaussian excitation with a 350-ohm source impedance,[38] included to damp out the low frequency resonances that would otherwise cause the response to persist for many transit times. Transforming this current to the frequency domain yields the input impedance (corrected to remove the generator resistance) presented in Figure 14. The calculation produces low-frequency resonances of very high Q. The frequency-independent region of this particular conical spiral is fairly narrow, from 1.2 GHz to about 3 GHz. At low frequencies, the spiral acts very much like an open-circuit transmission line because it is basically a twisted pair of wires, with the wire separation increasing slowly. Consequently, the observed resonances are expected and qualitatively explainable from transmission-line theory.

7.1.5 Object-Pole Finding - A convenient procedure for finding the EM poles of an object is to apply Prony's method to some measure (current, charge, near field, far field) of its transient response. One especially useful characteristic of these poles is their independence of the excitation, a subject demonstrated elsewhere.[26] Here, we demonstrate some other interesting aspects of EM poles.

Figures 15 and 16 show results of resistively loading a dipole at its center, and bending it at a right angle at its center. The transient waveforms are the backscattered fields from the dipoles when illuminated by a Gaussian-pulse field. In the former case, the α = 1, 3, 5, ... poles are more lossy than the α = 2, 4, 6, ... poles, while the converse is true in the latter case. This occurs because the odd-numbered modes are even about the dipole's center, so that the resulting current maxima at the resistive loads produce a large dissipative loss, while the even-numbered modes are odd and have charge maxima at the center, thus producing a large radiative loss at the bend.

Besides being useful to find the pole locations for SEM, Prony's method serves as well to store the transient waveform in a shorthand form. One interesting application of data handling is the ability to extrapolate both forward and backward in time. Figure 17 shows a portion of the computed backscattered field observed from the wire object shown. Figure 18 plots the pole locations obtained using only that portion of the response indicated in Figure 17. The poles from Figure 18 (and residues which are not shown) were then used to fill in the entire transient response. This extrapolated response is compared to the original computed response in Figure 19, with excellent agreement.

7.1.6 Physical Insight - Transient methods can be advantageously used to help understand the characteristics of the electromagnetic response of structures. The physical insight thus made possible may be greatly enhanced by novel data-presentation techniques. For example, a motion picture of the currents along the structure and the radiated fields has been found useful in showing differences between radiation associated with end reflection compared with that from a curved wire.

A motion picture format is not always suitable, of course, but the presentation continuity it provides in both space and time can be obtained in other ways. One example is presented in Figures 20 and 21. Equal-current contours are shown on these graphs as a function of position along the wire and as a function of time. Distance (horizontal axis) and time (vertical axis) describes a straight line with a slope of unity.

The dipole response (Figure 20) illustrates that the current pulse undergoes little decay as it travels along the wire, as shown by the closing of all the contours near the wire's end, but that the current pulse reflected from the ends is slightly less than that incident; energy has been lost to radiation. The current along the conical spiral (Figure 21), however, is continually decaying as shown by closing of the contours along the entire length of the antenna. The explanation of both these responses is that accelerated charge provides the radiation. This also explains why the dipole produces linear polarization and why the conical spiral produces circular polarization. Also observe that the current contours on the straight wire are parallel, showing that leading and trailing edges of the pulse propagate with little dispersion. By contrast, the pulse on the spiral broadens due to dispersion, as shown by the nonparallel current contours.

7.2 Applications

7.2.1 Transient-Pulse Shaping - For some applications, it may be desired to radiate a specific transient-pulse shape. Physical constraints may prohibit the antenna from being frequency independent throughout the band of the desired pulse, and the antenna then functions as part of the wave-shaping network. For example, consider a zero-impedance voltage source connected to a conical spiral antenna (the same antenna used for the results in Figures 13-14).[38] Assume that we want the radiated field in one polarization plane to be the third derivative of the Gaussian, i.e.,

$$E_{rad}(t) \alpha \frac{d^3}{dt^3} \left[\exp(-a^2 t^2) \right] .$$

The desired quantity is the voltage required to be applied to the antenna to produce that far-field behavior. To find this quantity, we first calculate the transfer function between the radiated field and the applied voltage, which is shown in Figure 22 for the conical antenna previously considered. Next, we obtain the spectrum of the required voltage by dividing the spectrum of the desired radiated field by this transfer function. The spectrum of the desired field must go to zero with decreasing frequency faster than the transfer function, because the antenna cannot radiate a static field. We then find the required transient voltage (Figure 23) by transforming this spectrum to the time domain. To verify the result, we use this waveform in the TWTD code to obtain the radiated field shown in Figure 24 from the direct time-domain calculation. This waveform is the desired third-time derivative of the Gaussian as specified. A similar approach could also be used with measured antenna characteristics.

7.2.2 Wire-Grid Models - A reasonably complicated grid problem is depicted in Figure 25 of a wire-grid model of a light truck with a rear-mounted 108-inch whip antenna. The effect of the ground was included as a perfect image plane. Figure 26 shows the results for the input admittance obtained from a time-domain calculation. An input resistance of ~70 ohms occurs at 27 MHz, the operating frequency of the transmitter, which may be recognized as the CB frequency. This result has been verified by actual operation, demonstrating the utility of the wire-grid model.

A variety of other wire-grid calculations have also been performed using the time-domain approach. These include mesh models for plates, a fan-type antenna, and conical shell. Because wire grids are only physical approximations to closed surfaces, care must be exercised in their use, however.

7.2.3 Transient Response - Another obvious application of time-domain analysis is the calculation of coupling such as that due to EMP which of course is a transient phenomenon. For example, in Figures 27-29 we show the response of a 10-m straight wire resistively loaded at its center illuminated from broadside by an EMP pulse.

The ease of obtaining these responses conveniently permits efficient parameter studies. Figures 30 and 31 show the results of one such study. Here, wires of different lengths were exposed to an EMP pulse. The wires were loaded resistively at their centers, and the value of resistance was also considered a parameter. For each combination of length and resistance, the transient calculation was performed and the peak load current and total energy dissipated in the load were evaluated. The results of these calculations are shown as contour plots, where the loci describe parameter pairs that produce the same peak current on total energy. Such plots are invaluable for condensing and evaluating large amounts of data. These plots show, for example, that peak current is approximately proportional to wire length in the wire-length range from several meters to several hundred meters. Resistive loading of several ohms to several hundred ohms does not drastically affect peak currents or total energy. Calculations for reactive loads show that a single resistance in series with a resistance has little effect on total energy collected.

. CONCLUSION

In the above discussion, we have presented the rudiments of EM transient analysis via the use of direct time-domain integral-equation solutions, and demonstrated their utility for a variety of applications. We have emphasized the physical aspects of transient behavior and examined the numerical treatment in some depth, while introducing a minimum of mathematical detail. The single most important point we would like to leave with the reader is an appreciation for the practical utility of direct time-domain techniques in providing greater insight and understanding of electromagnetic phenomena. Direct techniques also offer greater solution efficiency than transform techniques for many types of transient problems, the ability to handle nonlinearities, the convenience of wide-bandwidth information from a single calculation, the opportunity to use time-range gating to isolate interactions, and the possibility for directly obtaining the complex resonances of objects excited by EM sources. In closing, we hope the reader will be encouraged to employ transient computation and measurement where appropriate as an additional tool for solving EM problems..

. ACKNOWLEDGMENTS

The authors appreciate the assistance of discussions they had with several colleagues while preparing this work. These colleagues include R. M. Bevensee, J. N. Brittingham, G. J. Burke, F. J. Deadrick, T. Lehman, and A. J. Poggio. They are also grateful to M. Schmidt and P. Lorton for their skill and diligence in typing the manuscript, and to Robert Waite for his patience and care in editing it.

REFERENCES

1. E. K. Miller, Some Computational Aspects of Transient Electromagnetics, Lawrence Livermore Laboratory, Livermore, Calif. UCRL-51276 (1972).

2. A. J. Poggio and E. K. Miller, "Integral Equation Solutions of Three-Dimensional Scattering Problems," in Computer Techniques for Electromagnetics, R. Mittra, Ed. (Pergamon Press, New York, 1973), Ch. IV.

3. G. Franceschetti and C. H. Pappas, "Pulsed Antennas," IEEE Trans. AP-S, 22 (5), 651 (1974).

4. L. Felsen, Ed., Transient Electromagnetics (Springer-Verlag, New York, 1976).

5. P. O. Brundell, "Transient Electromagnetic Waves Around a Cylindrical Transmitting Antenna," Erricsson Tech. 16 (1), 137-162 (1960).

6. O. Einarsson, "The Step-Voltage Current Response of an Infinite Conducting Cylinder," Trans. Roy. Inst. Technol. Stockholm 191 (1962).

7. F. J. Friedlaender, Sound Pulses (Cambridge University Press, London, 1958).

8. J. Rheinstein, "Backscatter from Spheres: A Short Pulse View," *IEEE Trans. Ant. and Prop.* 16, 89 (1968).

9. C. Flammer and H. E. Singhaus, "The Interaction of Electromagnetic Pulses with an Infinitely Long, Conducting Cylinder above a Perfectly Conducting Ground," *AFWL Interaction Note* 144 (1973).

10. P. R. Barnes and D. B. Nelson, "Transient Response of Low Frequency Vertical Antennas to High Altitude Nuclear Electromagnetic Pulse (EMP)," *AFWL Interaction Note* 160 (1974).

11. E. M. Kennaugh and D. L. Moffatt, "On the Axial Echo Area of the Cone Sphere Shape," *Proc. IRE (Correspondence)* 50, 199 (1962).

12. E. M. Kennaugh and D. L. Moffatt, "Transient and Impulse Response Approximations," *Proc. IEEE* 53, 983 (1965).

13. C. L. Bennett and W. L. Weeks, *Electromagnetic Pulse Response of Cylindrical Scatterers*, G-AP Symposium, Boston, Mass., 1968. See also *A Technique for Computing Approximate Electromagnetic Impulse Response of Conducting Bodies*, Purdue University, Lafayette, Ind., Report TR-EE68-11.

14. E. P. Sayre, *Transient Response of Wire Antennas and Scatterers*, Electrical Engineering Department, Syracuse University, Syracuse, N.Y., Technical Report TR-69-4 (1969).

15. C. L. Bennett and W. L. Weeks, "Transient Scattering from Conducting Cylinders," *IEEE Trans. Ant. and Prop.* 18 (5), 627-633 (1970).

16. E. P. Sayre and R. F. Harrington, "Time-Domain Radiation and Scattering by Thin Wires," *Appl. Sci. Res.* 26 (6) 413-444 (1972).

17. E. K. Miller, A. J. Poggio, and G. J. Burke, "An Integro-Differential Equation Technique for the Time-Domain Analysis of Thin-Wire Structure. Part I, the Numerical Method," *J. Comput. Phys.* 12 (1), 24-48 (1972), "Part II, Numerical Results," *J. Comput. Phys.* 12 (2), 210-233 (1972).

18. A. J. Poggio, *Space-Time and Space-Frequency Domain Integral Equations*, MBA Technical Memo MB-TM-69-63 (1969).

19. T. K. Lui and K. K. Mei, "A Time-Domain Integral Equation Solution for Linear Antennas and Scatterers," *Radio Sci.* 8 (8-9), 797-804 (1973).

20. T. T. Wu, "Transient Response of a Dipole Antenna," J. Math. Phys. 2 (6), 892-894 (1961).

21. K. K. Chan, L. B. Felsen, S. T. Ping, and J. Schmoys, "Diffraction of the Pulsed Field from an Arbitrarily Oriented Electric or Magnetic Dipole by a Wedge," AFWL Sensor and Simulation Note 202 (1973).

22. L. B. Felsen and N. Marcuvitz, Radiation and Scattering of Waves (Prentice-Hall, Inc., Englewood Cliffs, N.J., 1973).

23. L. Marin, "Natural Mode Representation of Transient Scattering from Rotationally Symmetric Bodies," IEEE AP-S Trans. 22 (2), 266 (1974).

24. K. R. Umashankar, T. H. Shumpert, and D. R. Wilton, "Scattering by a Thin-Wire Parallel to a Ground Plane Using the Singularity Expansion Method," IEEE AP-S Trans. 23 (2), 178 (1975).

25. M. L. Van Blaricum and R. Mittra, "A Technique for Extracting the Poles and Residues of a System Directly from Its Transient Response," IEEE AP-S Trans. 23 (6), 777 (1975).

26. E. K. Miller, F. J. Deadrick, H. G. Hudson, A. J. Poggio, and J. A. Landt, Radar Target Classification Using Temporal Mode Analysis, Lawrence Livermore Laboratory, Livermore, Calif., UCRL-51825 (1975).

27. A. M. Nicolson, C. L. Bennett, D. Lamensdorf, and L. Susman, "Applications of Time Domain Metrology to the Automation of Broad Band Measurements," IEEE Trans. Micro. Theory and Techniques 20 (1), 3-9 (1972).

28. R. M. Bevensee, F. J. Deadrick, E. K. Miller, and J. T. Okada, Validation and Calibration of the LLL Transient-Electromagnetic Measurement Facility, Lawrence Livermore Laboratory, Livermore, Calif., UCRL-52225 (1977).

29. F. Rohrlick, Classical Charged Particles (Addison-Wesley Publishing Co., Inc., Reading, Mass., 1965).

30. W. K. H. Panofsky and M. Phillips, Classical Electricity and Magnetism (Addison-Wesley Publishing Co., Inc., Reading Mass., 1956).

31. J. A. Landt, E. K. Miller, and M. L. Van Blaricum, WT-MBA.LL1B: A Computer Program for the Time-Domain Response of Thin-Wire Structures, Lawrence Livermore Laboratory, Livermore, Calif., UCRL-51585 (1974).

32. E. K. Miller, F. J. Deadrick, and J. A. Landt, <u>Time-Domain Analysis of Non-linear Loads</u>, Lawrence Livermore Laboratory, Livermore, Calif., to be published.

33. D. E. Merewether, "Transient Currents Induced on a Metallic Body of Revolution by an Electromagnetic Pulse," <u>IEEE Trans. Electromagnetic Compatability</u> <u>13</u> (2), 41-44 (1971).

34. H. K. Schuman, "Time-Domain Scattering from a Non-linearly Loaded Wire," <u>IEEE AP-S Trans.</u> <u>22</u> (4), 611 (1974).

35. T. K. Liu and F. M. Tesche, "Analysis of Antennas and Scatterers with Non-linear Loads," <u>IEEE AP-S Trans.</u> <u>24</u> (2), 131 (1976).

36. T. K. Sarkar and D. D. Weiner, "Scattering Analysis of Non-linearly Loaded Antennas," <u>IEEE AP-S Trans.</u> <u>24</u> (2), 125 (1976).

37. J. A. Landt and E. K. Miller, "Transient Response of the Infinite Cylindrical Antenna and Scatterer," <u>IEEE AP-S Trans.</u> <u>24</u> (2), 246 (1976).

38. E. K. Miller and J. A. Landt, "Short Pulse Characteristics of the Conical Spiral Antenna," <u>IEEE AP-S Trans.</u> (1977).

ACKNOWLEDGEMENT

Appreciation is due to Alice Blair for typing the manuscript and to Peggy Lorton for arranging the illustrations.

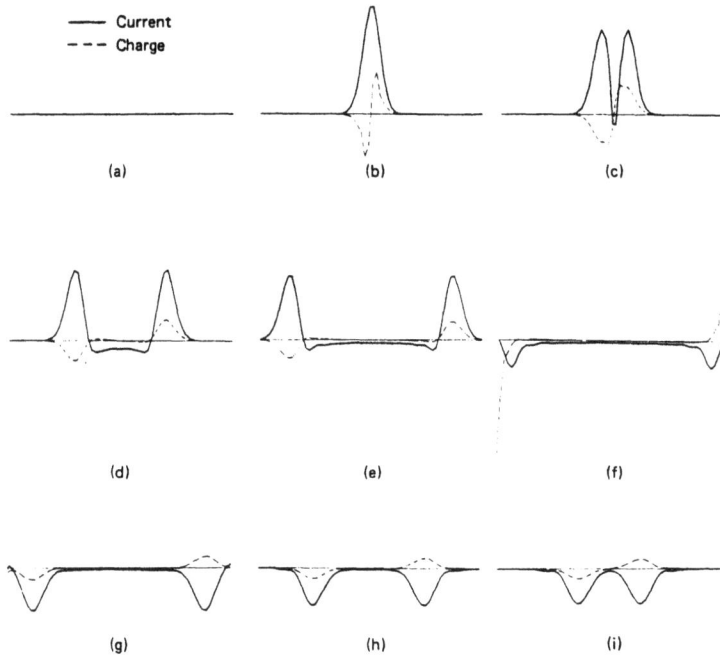

Fig. 1. Current and charge distributions at several instants of time for a straight-wire antenna excited by a Gaussian pulse. The antenna and numerical parameters are: length (L) = 1 m; radius (a) = 6.738 × 10^{-3} m (which gives $\Omega \equiv 2 \ln L/a = 10$); Gaussian-pulse parameter g = 5.556 × 10^{-10} where the pulse time variation is $\exp[-g^2 t^2]$; space-segment length (Δ) = L/60; and time interval (δ) = Δ/c. After its initial excitation, the charge-current pulse splits into two oppositely propagating pulses of oppositely signed charge, resulting in current pulses having the same sense. A slight decrease in the amplitude of the pulse can be observed as they progress down the wire, and a more noticeable decrease in amplitude occurs upon end reflection.

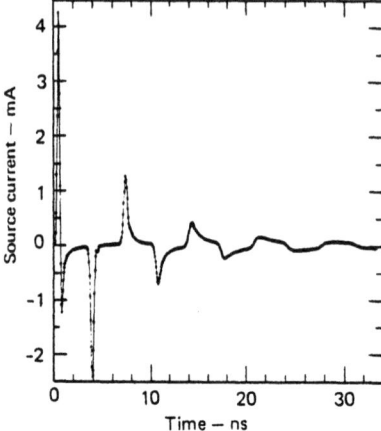

Fig. 2a. Feedpoint current as a function of time.

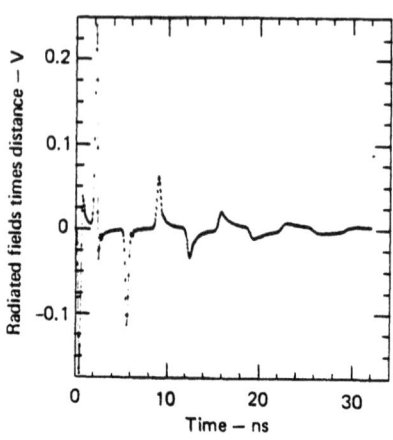

Fig. 2b. Broadside-radiated electric field as a function of time for a straight-wire antenna excited by a Gaussian pulse. Six hundred time steps are shown. The peaks in the radiated field appear to coincide with the initial excitation and end reflection (in retarded time).

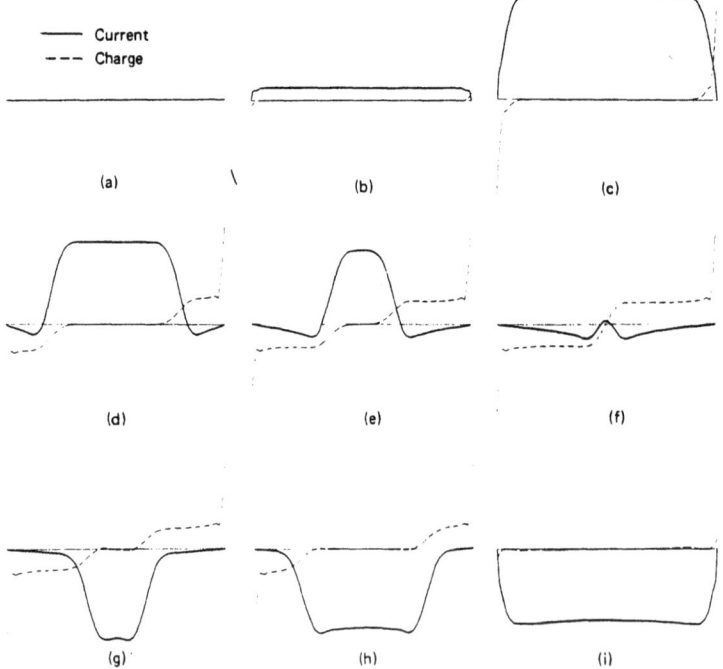

Fig. 3. Analog of Fig. 1 for the scattering case. The excitation is now a Gaussian-pulse plane wave incident from broadside. A uniform current is initially excited along the wire and collapses inward due to end reflection. No explanation has been developed for the small-amplitude oscillation in the current and charge of both Figs. 1 and 3.

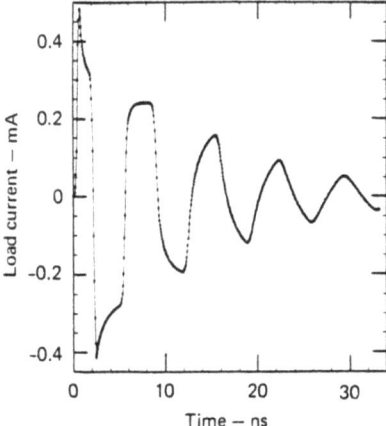

Fig. 4a. Transient current at the center of the long-wire scatterer.

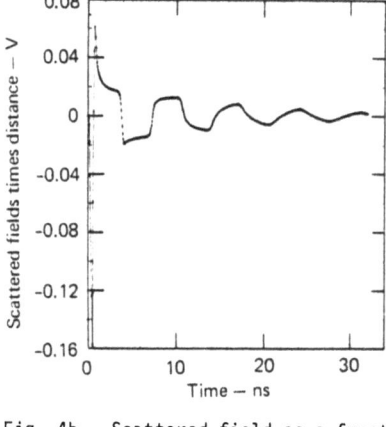

Fig. 4b. Scattered field as a function of time for a straight-wire scatterer excited by a Gaussian-pulse plane wave incident from broadside. The scattered field exhibits a specular-flash return as the current is first excited, then a longer lasting but slowly decaying portion of opposite sign.

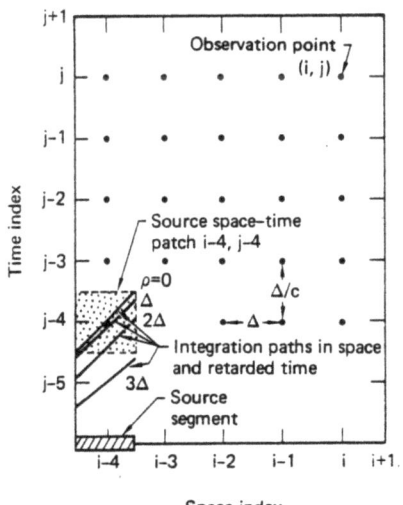

Fig. 5. Space-time integration paths in the source-coordinate (s',t') space for an observation point located at time t and cylindrical coordinates ρ and s relative to the center axis of the current filament (segment) being integrated. The retarded or source time t' is given by $t' = t - \sqrt{\rho^2 + (s - s')^2}/c$. Note: $s' = i'\Delta$; $s = i\Delta$; $t' = j'\delta$; $t = j\delta$. The uniformly spaced dots represent the centers of the space and time sample intervals for a straight wire, and the curves show the integration paths for a given source segment on that wire as ρ is changed. Note that unless $\rho = 0$ and s coincides with a space sample point (i.e., the observation point is collocated with the sample point on the wire), the integration over a single space segment can involve more than one time sample of the current on that segment. This shows one reason for using basis functions that smoothly vary over the entire space-time domain.

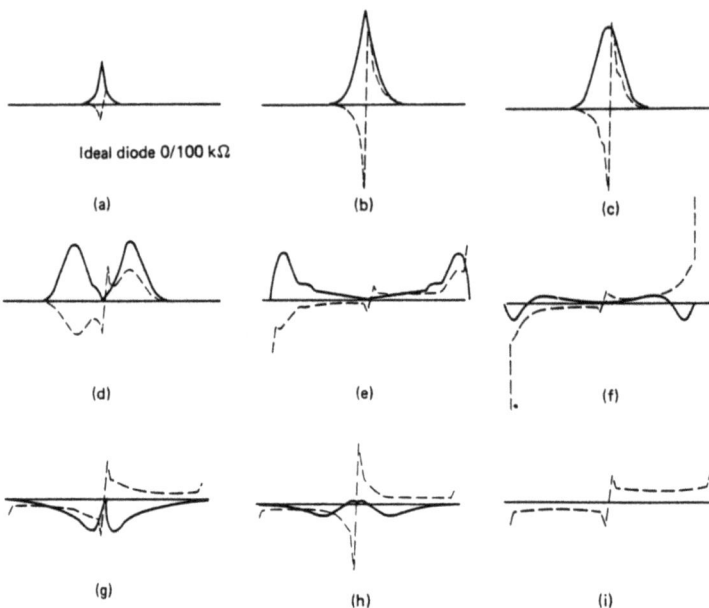

Fig. 6. Nonlinear loads can be easily handled in the time domain. Here the current (—) and charge (---) distributions on a 1-m center-fed dipole having a diode load in series with the generator are shown at several instants of time. Observe that the diode presents zero impedance only to current flowing in the initial direction, and that it effectively acts as an open circuit in the opposite direction (impedance = 100 kΩ). The two halves of the dipole thus retain a static charge distribution at late times.

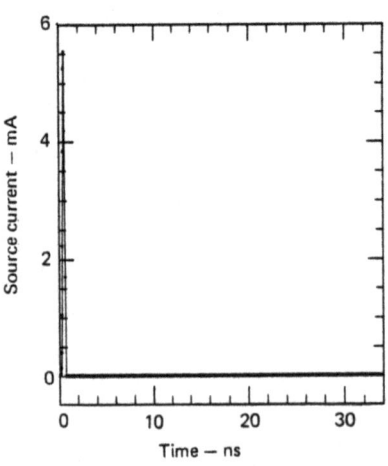

Fig. 7. The transient feedpoint current on the non-linearly loaded antenna of Fig. 6.

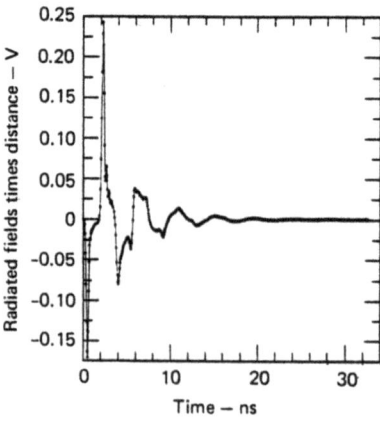

Fig. 8. The radiated field possesses a fundamental frequency twice that of an unloaded dipole.

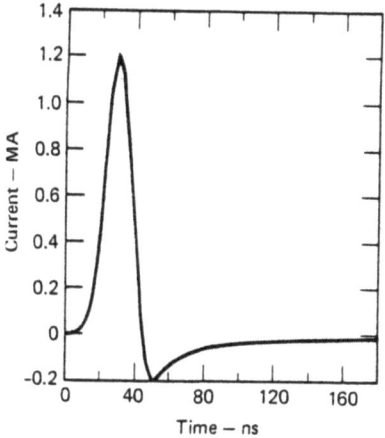

Fig. 9. Time-range gating can be used to separate reflection effects from other phenomena. In this example, the computed current prior to end reflection at the center of a long (120 m) center-fed dipole modeled with 120 segments is extrapolated in time to approximate the current that would be seen on an infinite antenna.

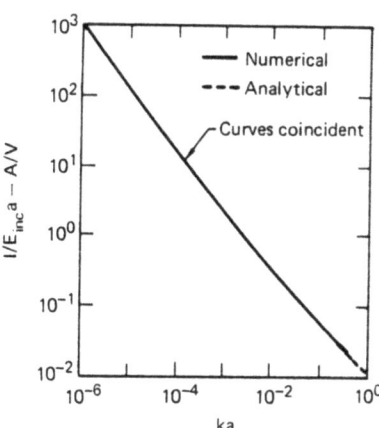

Fig. 10. The transfer function can be obtained by transforming the current from Fig. 9 to the frequency domain. The transfer function compares well with analytical results. Several calculations were required to span the ka range shown.[37]

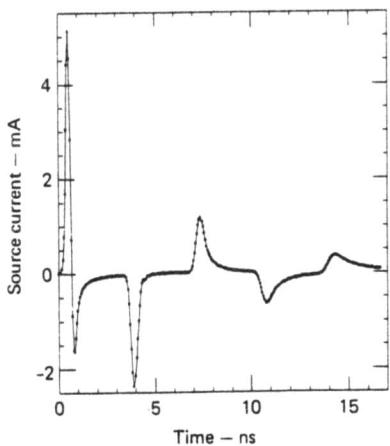

Fig. 11. Solution instability caused by use of too-short segments. The transient current on a 1 m straight wire modeled using 60 segments (Δ = 1.66 cm) is displayed here for a wire 9 mm in radius, so that $\Delta/2a = 0.926$.

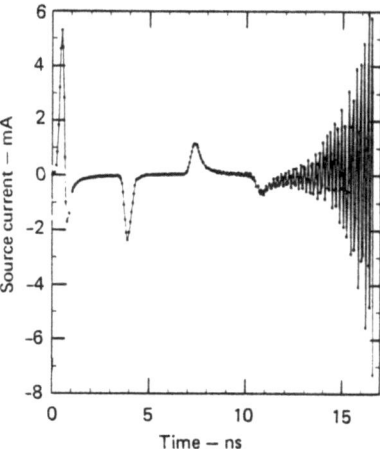

Fig. 12. Transient current for a wire 9.5 mm in radius where $\Delta/2a = 0.877$. The divergent solution that occurs is evidently due to numerical inaccuracy of the thin-wire approximation.

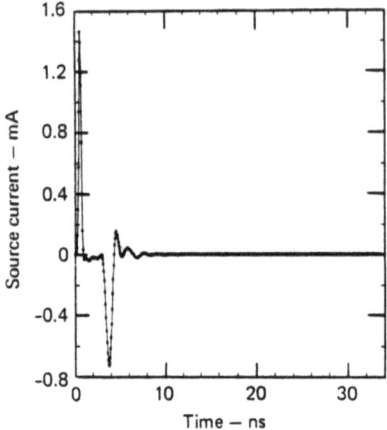

Fig. 13. Feedpoint current for a conical-spiral antenna excited by a Gaussian pulse. The effect of a 350-ohm series resistor is included. The numerical model consists of a piece-wise linear approximation (60 segments) to the actual antenna, whose total length is 1 m and radius is 1 mm, and which has two arms of 2 1/2 turns wound on a cone of half-angle 10° with small-end and large-end diameters of 3.36 cm and 9.16 cm, respectively.

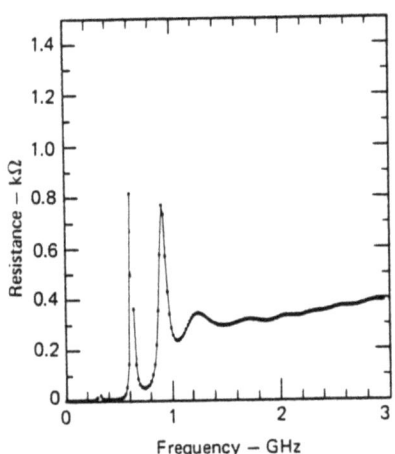
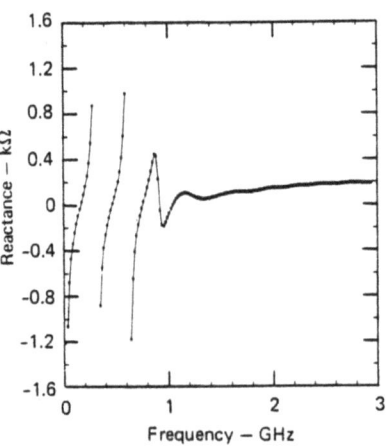

Fig. 14. Impedance for a conical-spiral antenna excited by a Gaussian pulse, obtained by transformation to the frequency domain. The late-time ringing of the spiral that would otherwise occur is damped out by the resistor and has been subtracted from the resistance curve.

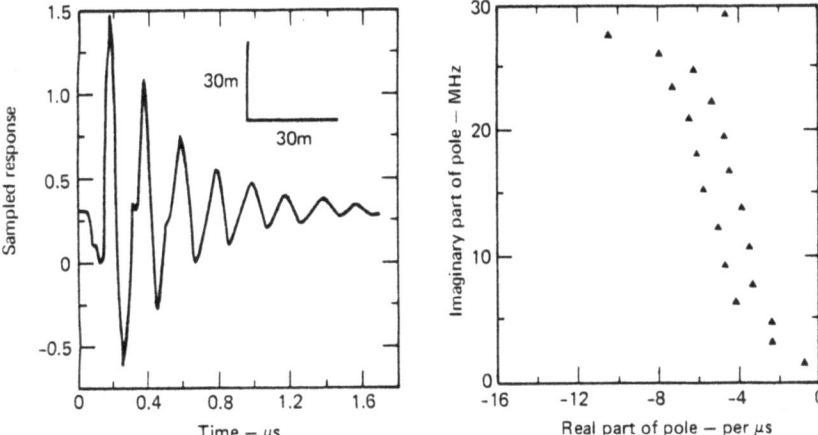

Fig. 15. Energy loss mechanisms can affect the pole locations in the complex frequency plane.[26] Here we present the pole sets for a 60-m wire, showing the result when the wire is straight and loaded with a 100-ohm resistance at its center.

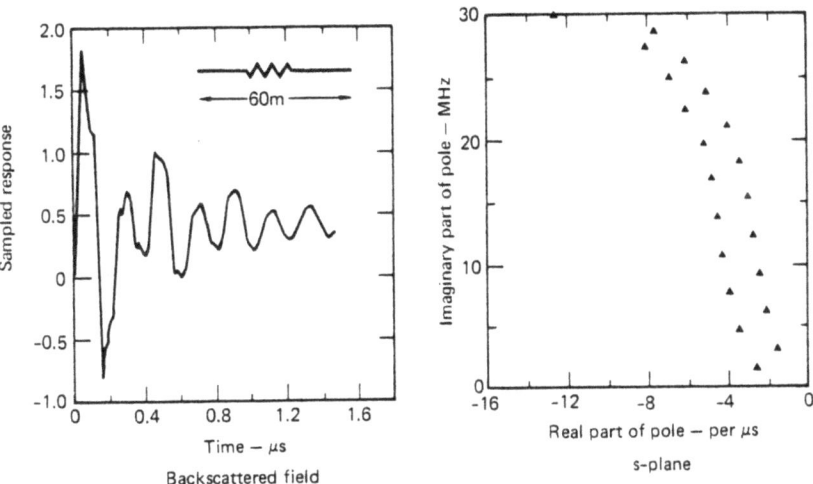

Fig. 16. Here, the wire is unloaded but has a 90° bend at its center. The $\alpha = 1, 3, \ldots$ poles are more lossy for the loaded wire, due to dissipative loss of those modes which have current maxima at its center, and the $\alpha = 2, 4, \ldots$ modes are more lossy for the bent wire, because the charge maxima at the bend produce a greater radiation loss.

248

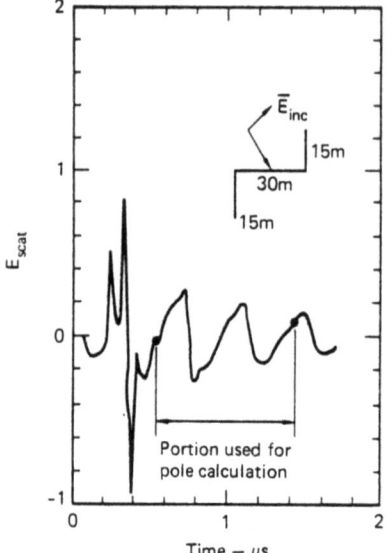

Fig. 17. Extrapolation of a transient waveform can also be accomplished using the poles extracted from it. Here, part of a computed field scattered from the thin-wire object is illustrated.

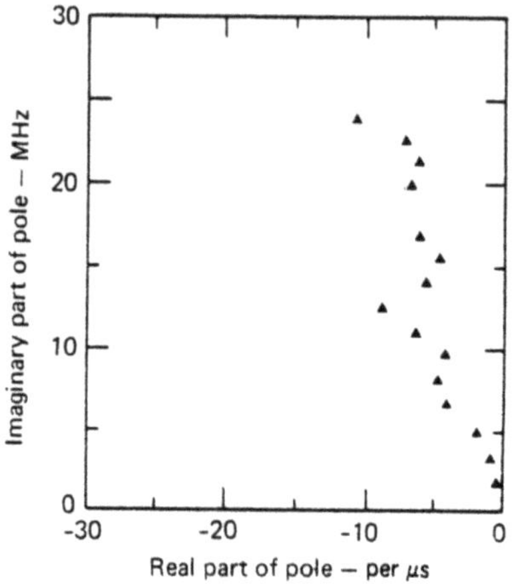

Fig. 18. Poles extracted from a portion of the transient waveform shown in Fig. 17.

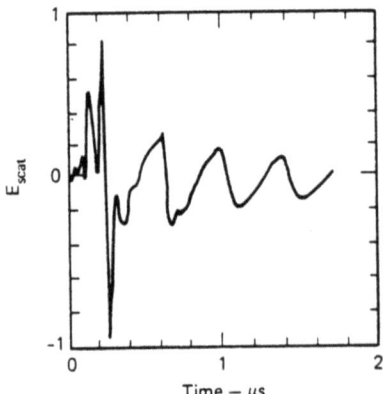

Fig. 19. The waveform reconstruction, which compares well with the original waveform both before and after the sampled portion.

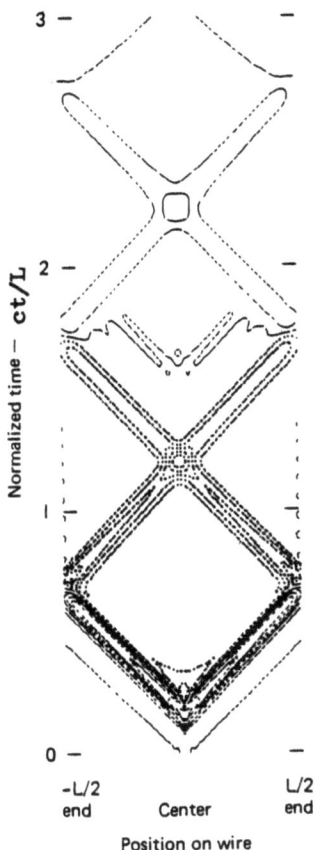

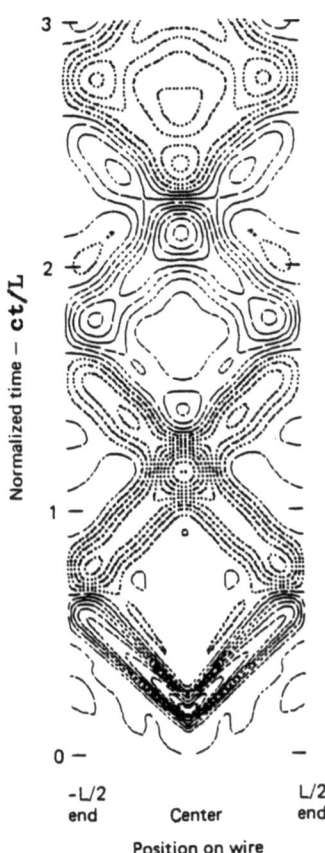

Fig. 20. Space-time contour plots of the current can convey a great deal of information concerning transient behavior. Results are presented here for a 1-m linear dipole antenna excited by a Gaussian voltage pulse. The time variation is shown on the vertical axis and the space variation is shown on the horizontal axis, where the scales are in the ratio of c, so a 45° straight line represents motion at the speed of light.

Fig. 21. Space-time contour plot of the current for a conical-spiral antenna of 1 m overall length, excited by a Gaussian voltage pulse. The difference in radiation mechanisms for the two antennas, and other features as well, can be deduced from these results, as discussed in the text.

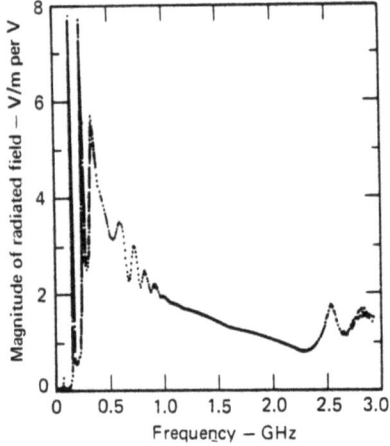

Fig. 22. Antenna pulse synthesis is straightforward using time-domain analysis.[38] Figures 22-24 illustrate one possible procedure by its application to a conical-spiral antenna (the same one considered in Fig. 13). Here we present the boresight radiation-field transfer function.

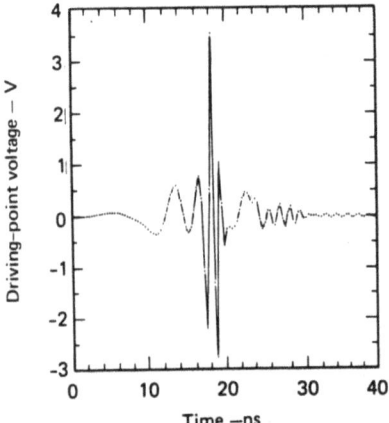

Fig. 23. The transient voltage waveform required for the antenna to radiate the third derivative of a Gaussian pulse.

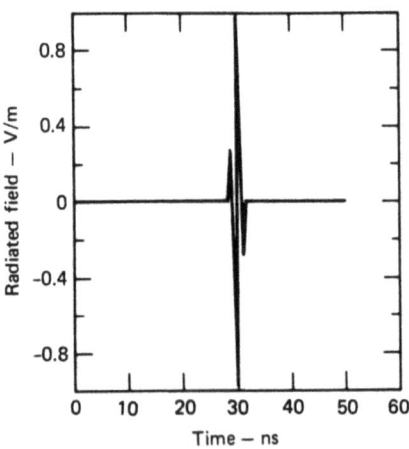

Fig. 24. The actual radiated field produced by application of the voltage pulse in Fig. 23.

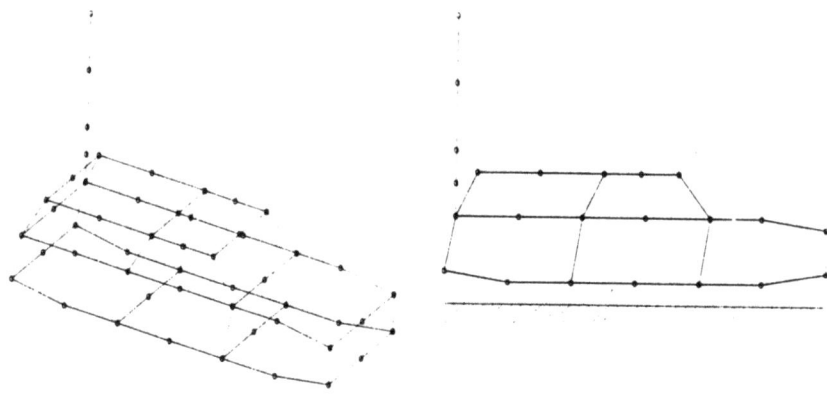

Fig. 25. Wire-grid model of a light truck located over a perfect ground plane, used to compute the impedance characteristics of a 108-inch whip antenna mounted as shown.

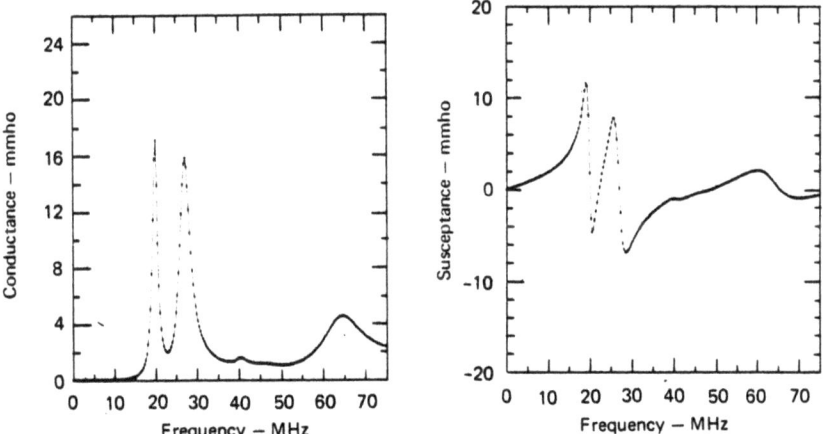

Fig. 26. The antenna admittance demonstrates an input impedance of ~70 ohms at 27 MHz, the frequency of operation for this emergency communication system (CB radio).

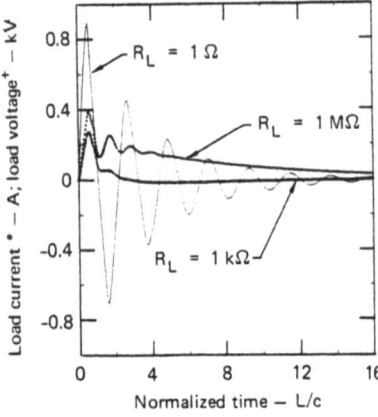

Fig. 27. EMP applications of time-domain solutions. A 10-m-long wire of radius 6.738×10^{-2} m and modeled with 30 equal-length segments is illuminated from broadside by a nominal EMP pulse given by $5.25 \times 10^4 [\exp(-4 \times 10^6 t) - \exp(-4.76 \times 10^8 t)]$v/m. The midpoint current and voltage that result for several midpoint load values are shown here.

*Applies to $R_L = 1\,\Omega$ and $R_L = 1\,k\Omega$ only.
+Applies to $R_L = 1\,k\Omega$ and $R_L = 1\,M\Omega$ only.

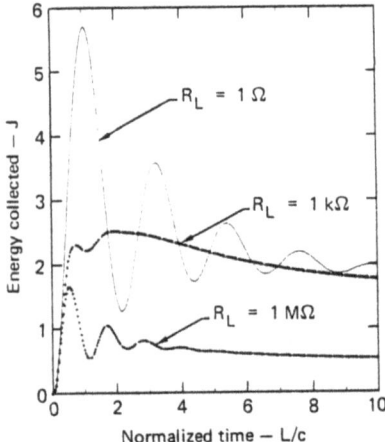

Fig. 28. Cumulative energy collected from the pulse by the wire, shown as a function of time (see Eq. 4.1). Observe that the collected energy is oscillatory in time, signifying the re-radiation of energy.

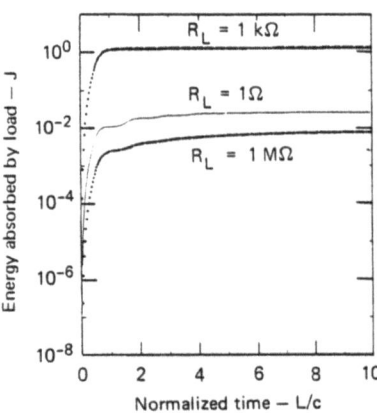

Fig. 29. Energy dissipated in the load as a function of time.

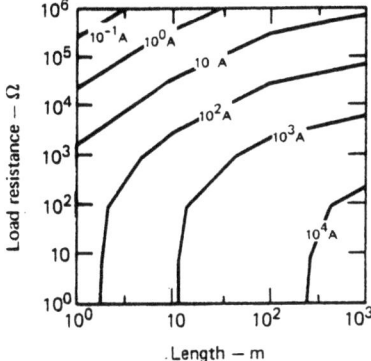

Fig. 30. The value of a time-domain computation in finding an efficient early-time solution. Here, the peak current excited on a dipole by a broadside-incident EMP pulse (see Fig. 27) is plotted as a constant contour value as a function of wire length (m) and center load-resistance value (ohms). The time-domain computation permits the peak current, which occurs early in time, to be found without requiring a complete transient waveform.

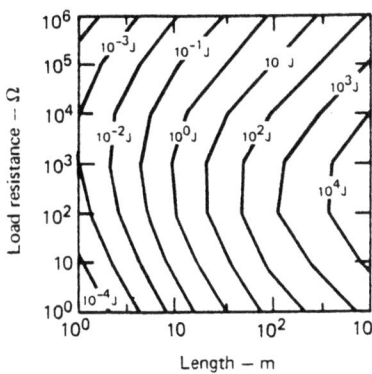

Fig. 31. Total load energy plotted as a constant contour value as a function of wire length (m) and center load-resistance value (ohms). The energy calculation requires that I^2R decay to a small value before terminating the calculation.

TIME DOMAIN SOLUTIONS VIA INTEGRAL EQUATIONS
- SURFACES AND COMPOSITE BODIES

C. Leonard Bennett

Sperry Research Center
Sudbury, Massachusetts USA

ABSTRACT. This paper discusses the time domain solution of scattering by conducting closed surfaces, open thin surfaces, surfaces with wires, and surfaces with fins. A brief description of the space-time integral equations that describe these problems will be followed by a discussion of the marching on in time numerical solution procedure. Smoothed impulse responses for selected targets in these classes will be presented, discussed, and compared with companion time domain measurements. Finally a description of the impulse response augmentation technique which yields an estimate of the impulse response will be presented.

1. INTRODUCTION

Before the advent of high-speed digital computers most of the electromagnetic scattering problems were solved in the frequency domain. The classical boundary value solution technique was applied to the problems where the target surface coincided with the coordinate surfaces of a separable coordinate system. This technique was limited to only a few target classes. Approximate techniques such as geometric optics, physical optics and the Rayleigh approximation were applied to other problems to obtain estimates of the scattered response. However, with the development of high-speed digital computers frequency domain integral equation techniques were developed [1,2] which allowed solution of the scattering problem for simple bodies of arbitrary shape from near zero frequency well into the resonance region. These integral equation solutions have provided many interesting and useful results. Shortly thereafter it was observed that an integral equation and its solution could also be obtained directly in the time domain

[3-6]. The direct time domain solution had advantages where the solution of transient electromagnetic scattering problems were desired. The calculations are carried out by illuminating the target with a smoothed impulse waveform. The resulting smoothed impulse response contains all the information about the electromagnetic scattering properties of the target over the frequency band that is defined by the spectrum of the incident smoothed impulse. The utility of using the impulse response as a characteristic signature was noted in 1965 [7]. This representation is useful for a number of reasons. First, all the electromagnetic information about the scattering properties of a target is contained in the impulse response. Second, the radar cross section or equivalently the frequency response can be obtained from the impulse response by a Fourier Transform. Thirdly, the response of the target due to any radar waveform can be obtained from the impulse response by a simple convolution procedure. For the purpose of target classification the impulse response is a particularly useful characterization since it is closely related to the actual target geometry. Finally, the impulse response can provide a better understanding of electromagnetic transient phenomena.

The choice of the frequency domain integral equation or the time domain integral equation approach must depend on the specifics of the problem that needs to be solved [8-10]. It also depends upon the cost of computer time and the size of computer memory. For the surface scattering problem, if the total time response is desired and only a moderate number of incident angles are required the time domain yields less running time. On the other hand if only the response at a single frequency is needed then the frequency domain approach is faster. The computer memory requirements for a given surface geometry are less for the time domain case. Since these time and memory results are algorithm and machine dependent the development of better algorithms in the future could change the balance.

The remainder of this paper presents a review of the space-time integral equation solution of scattering problems involving conducting surfaces or composite bodies. Scattering from closed surfaces, open-thin surfaces, bodies with wires, and bodies with fins is discussed. The impulse response augmentation technique, which extends the integral equation solution to shorter pulse widths or, equivalently, higher frequencies, is also discussed.

2. SPACE TIME INTEGRAL EQUATION

The derivation of the integral equation for the current density on the surface of a perfectly conducting scatterer is described in this section. In Fig. 1 the geometry of the problem is described. There is an electromagnetic wave incident on a target which sets

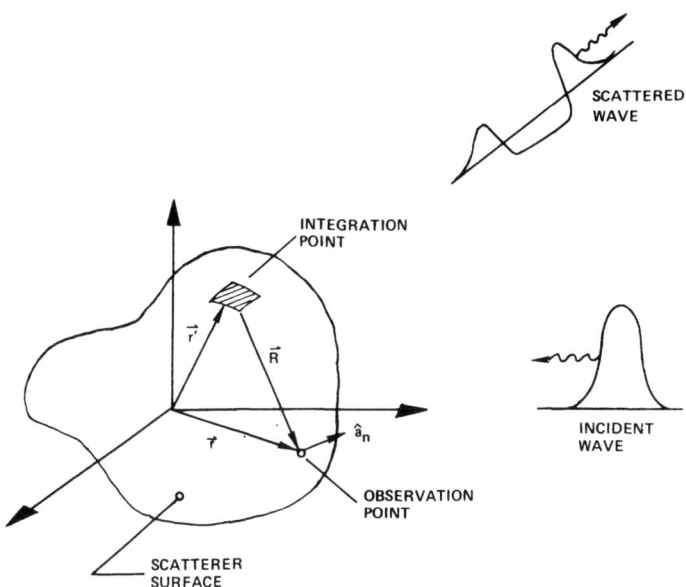

Fig. 1. Geometry of problem.

up currents on the surface of the target in such a way that the boundary condition on the surface is satisfied. These currents in turn produce a scattered field. Once the expression for the currents has been obtained, the problem is solved, since the field at any point in space can be computed from them.

The technique for obtaining an expression for the surface currents [3] is to start with the expression for the total field at an arbitrary point in space due to the incident field and the surface currents. This arbitrary point in space is then moved to a point on the surface of the scatterer using a limiting procedure. Next the boundary conditions on the tangential components of the fields at the surface are applied and a space-time integral equation results which can be solved numerically by a stepping on in time procedure. Once the numerical representation of the surface current has been found the field at any point in space, in particular the far field, can be calculated.

For the case of perfectly conducting scatterers being discussed here either the H-field boundary condition or the E-field boundary condition is sufficient to derive the integral equation. If the H-field boundary condition is used the H-field or Maue (1948) integral equation [11] results. This is a second kind equation and no space derivatives on the surface are required. From a numerical solution standpoint this equation is well suited for closed conducting bodies but gives problems on open thin surfaces or wires where the thickness goes to zero.

If the E-field boundary condition is used the E-field integral equation results. This is a first kind equation and space derivatives on the surface are required. For a numerical standpoint it is well suited for targets with zero body thickness and can handle thin wires and open thin surfaces.

3. SOLID CONDUCTING BODIES

The solution of scattering from closed conducting surfaces is accomplished by numerical solution of the H-field integral equation given as

$$\vec{J}(\vec{r},t) = 2\hat{a}_n \times \vec{H}^i(\vec{r},t)$$

$$+ \frac{1}{2\pi} \int_S \hat{a}_n \times \left\{ \left[\frac{1}{R^2} + \frac{1}{Rc} \frac{\partial}{\partial \tau} \right] \vec{J}(\vec{r}',\tau) \times \hat{a}_R \right\} dS' \quad (1)$$

$$\tau = t - R/c$$

where

$\vec{r}$ = position vector to the observation point

$\vec{r}'$ = position vector to the integration point

$R = |\vec{r} - \vec{r}'|$

$\hat{a}_R = \dfrac{\vec{r} - \vec{r}'}{R}$

t = time in seconds

c = the speed of light

$\hat{a}_n$ = unit normal at $\vec{r}$

S = surface of the target

$\vec{H}^i$ = incident magnetic field

$\vec{J}$ = surface current density

The first term in the right-hand side of eq. (1) may be considered the source term and represents the direct influence of the incident field on the current at the observation point $(\vec{r},t)$. Moreover, this term, when applied to the illuminated side of the scatterer, yields the familiar physical-optics approximation for the surface current. The integral term on the right-hand side of eq. (1) represents the influence of currents at other surface points on the current at $(\vec{r},t)$. The crucial observation here is that the influence of other currents on the current at $(\vec{r},t)$ is delayed by R/c, which makes the "marching on in time" numerical solution of (1) feasible.

Once the current density has been obtained, the field at any point in space can be computed. In most applications of interest, however, the quantity of interest is the far scattered field given by

$$\vec{H}^s(\vec{r},t) = \frac{1}{4\pi rc} \int_S \left\{ \frac{\partial \vec{J}(\vec{r}',\tau)}{\partial \tau} \right\} \times \hat{a}_r \, dS' \qquad (2)$$

$$\tau = t - R/c$$

where

$\vec{H}^s$ = far scattered H field

r = distance of observer from target

$\hat{a}_r$ = unit vector pointing from target to observer.

The technique for solution of the integral is to first write (1) in terms of its tangential components

$$J_u(\vec{r},t) = J_u^i(\vec{r},t) + I_u(\vec{r},t)$$

$$J_v(\vec{r},t) = J_v^i(\vec{r},t) + I_v(\vec{r},t)$$

where

$$J_u^i = \left(2\hat{a}_n \times \vec{H}^i \right)_u$$

$$J_v^i = \left(2\hat{a}_n \times \vec{H}^i \right)_v$$

I_u = u component of the second term on the right-hand side of (1)

I_v = v component of the second term on the right-hand side of (1).

Next the target surface is divided into patches and the current is computed at the center, $\vec{r}_i$, of each patch

$$J_u(\vec{r}_i,t) = J_u^i(\vec{r}_i,t) + \rho_{ui} J_u(\vec{r}_i,t) + \sum_{k \neq i} f_{uik} \Delta S_k \qquad (3)$$

where

$\rho_{ui} = \frac{1}{2}\sqrt{\frac{\Delta S_i}{4\pi}} \left(K_{ui} - K_{vi}\right)$ = contribution due to patch at $\vec{r}_i$

K_{ui}, K_{vi} = the principal curvatures at $\vec{r}_i$

ΔS_k = the area of the patch at $\vec{r}_k$

f_{uik} = the contribution to $J_u(\vec{r}_i,t)$ due to currents on other patches at earlier points in time.

Rearrangement of terms in (3) yields the recurrence relation in time for the surface current

$$J_u(\vec{r}_i,t) = \frac{J_u^i(\vec{r}_i,t) + \sum_{k \neq i} f_{uik} \Delta S_k}{1 - \rho_{ui}} \qquad (4)$$

The ideal excitation would be an electromagnetic impulse, however, it is not amenable to numerical solution. What has been used is a smoothed impulse or a regularization of an impulse. The Gaussian form

$$e(t) = \frac{a_n}{\sqrt{\pi}} \exp\left(-a_n^2 t^2\right) \qquad (5)$$

has been chosen because it has rapidly decaying frequency domain tails and because it is "time limited" and well suited for numerical solution. The resulting response computed using this excitation is the smoothed impulse response of the target.

For solution, the target geometry must be represented numerically. This is carried out by dividing the surface into patches which give a good geometry representation, are approximately curvilinear square, and have approximately equal areas. The minimum spacing between patch centers should be less than 1/8 the width of the incident smoothed impulse.

The time variable must also be divided into increments for numerical solution. In the work to date time has been divided into equal increments. The basic time increment should be less than 1/8 the width of the incident smoothed impulse. Moreover, the basic time increment should be chosen to be less than the minimum spacing between patch centers for stability reasons.

Equation (4) is solved on a digital computer for the current density by simply marching on in time. Once the current density has been obtained, the far scattered field is computed by performing the integration in (2) numerically. Procedures have been developed, implemented, and tested for computing the smoothed impulse response of multiple cylindrical scatterers, smooth convex bodies, and bodies with edges.

The case of twin cylinders with radius a and center-to-center spacing of 2a is considered [12] in Fig. 2. The TE smoothed impulse response of twin cylinders with end-on incidence is shown in Fig. 2(a). In this space snapshot and those that follow, all the space dimensions are to scale except the distance to the scatterer. The basic space unit a (the radius of the cylinder in this case), and both the space variation of the scattered field and the size of the scattering body are drawn to the same scale. The large semicircle is the distance reference of the scattered field and represents the locus of points in space that the peak of the incident pulse would have reached if it had been reflected from the origin. The specular return in the backscatter direction exhibits itself as an initial positive pulse. The initial part of the return is identical with that obtained from a single circular cylinder. This

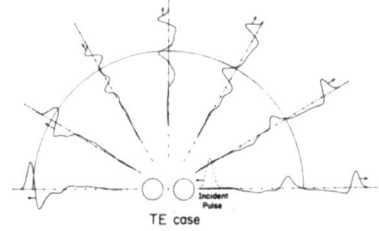

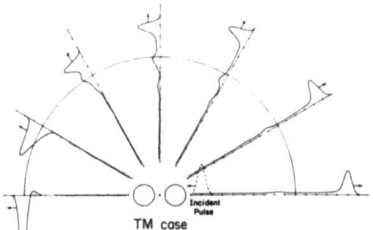

Fig. 2. Smoothed impulse response of twin cylinders with radius a (end-on incidence). (a) TE case. (b) TM case.

is expected since the incident wave does not reach the second cylinder until later in time. Although the second cylinder is "hidden" from the view of the observer in the backscatter direction, the second positive pulse that this observer sees can be attributed to a "reflection" of the incident wave from the front side of the back cylinder. The third positive pulse can be attributed to a wave traveling around the rear of the second cylinder.

The TM smoothed impulse response of twin cylinders with end-on incidence is shown in Fig. 2(b). The initial part of the return in the backscatter direction is exactly the same as that obtained for a single cylinder. As expected, there is very little return in the backscatter direction that can be attributed to the shadowed cylinder.

In Fig. 3(a) the smoothed impulse response of a blunt nose cone sphere with backside incidence is shown. The initial portion of the response in the near backscatter directions is the return from the large sphere cap and is the same as was obtained for the sphere. The response then becomes small until the return from the wave traveling around the rear appears. The smoothed impulse response of the blunt-nose-cone sphere with nose on incidence is displayed in Fig. 3(b). The small return from the spherical cap on the nose of the cone can be easily distinguished. Next the sides of the cone appear to produce a small, nearly constant return. Following this is a negative pulse that can be attributed to the join between the cone and the sphere cap. Finally, the wave traveling around the rear of the sphere exhibits itself as a positive pulse.

4. OPEN THIN SURFACES

The previous section described the integral equation, the solution technique used, and some results that have been obtained for transient scattering from closed surfaces. This section will treat the case of open thin surfaces. As mentioned earlier, the E field

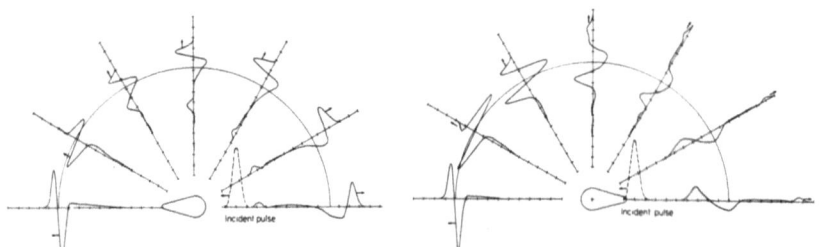

Fig. 3. E-plane smoothed impulse response of blunt-nose cone sphere.

integral equation is more appropriate from a numerical standpoint for solution of this type of problem. The general space-time integral equation for a conducting body is given by

$$0 = \varepsilon \frac{\partial \vec{E}^i_{tan}(\vec{r},t)}{\partial t} + \left\{ \left[\nabla(\nabla \cdot) - \frac{1}{c^2} \frac{\partial^2}{\partial t^2} \right] \left[\frac{1}{4\pi} \int_S \left\{ \frac{\vec{J}(\vec{r}',\tau)}{R} \right\} dS' \right] \right\}_{tan}$$

$$\tau = t - R/c$$

(6)

where

ε = permittivity of space

∇ = del operator

tan = subscript denoting the component of the vector which is tangent to the surface at $\vec{r}$.

This is an integral equation of the first kind since the unknown, $\vec{J}$, appears only inside the integral on the right-hand side.

As an example of the solution procedure for open thin surfaces, the case of a flat plate will be considered. For this case, eq. (6) can be simplified by noting that the surface current flows in only one plane. The geometry of a rectangular flat plate is displayed in Fig. 4 in which the plate is located in the y = 0 plane. For this geometry, the components of the surface current are obtained from (6) as

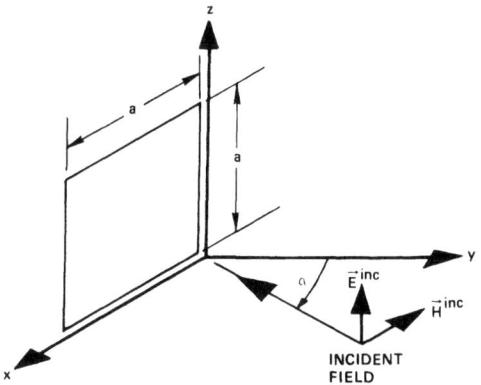

Fig. 4. Geometry of scattering from a rectangular flat plate.

$$\Box^2 J_x + \frac{\partial^2}{\partial_x \partial_z} J_z = \frac{1}{\gamma} \left[-\varepsilon \frac{\partial E_x^i}{\partial t} - \left(\Box^2 + \boxtimes\right) \frac{1}{4\pi} \int_{\substack{\text{Non} \\ \text{Self}}} \left\{ \frac{J_x}{R} \right\} dS' \right]_{\tau = t - R}$$

$$\Box^2 J_z + \frac{\partial^2}{\partial_z \partial_x} J_x = \frac{1}{\gamma} \left[-\varepsilon \frac{\partial E_z^i}{\partial t} - \left(\Box^2 + \boxtimes\right) \frac{1}{4\pi} \int_{\substack{\text{Non} \\ \text{Self}}} \left\{ \frac{J_z}{R} \right\} dS' \right]_{\tau = t - R}$$
(7)

where

$\boxtimes$ = the cross derivative terms in the $\nabla(\nabla \cdot)$ operator

$$\gamma = 2 \bigg/ \left(\sqrt{\frac{\Delta S}{\pi} - \left(\frac{\delta}{2}\right)^2} - \frac{\delta}{2} \right)$$

ΔS = patch area

δ = plate thickness.

The left-hand side of eq. (7) contains a wave operator operating on the major component and a mixed space derivative operator operating on the other current component. It is this mixed derivative operator that provides the coupling between the two components of surface current in (7). On the right-hand side are terms due to the incident field and due to currents at other points on the surface but at earlier points in time. For numerical solution, the plate is divided into curvilinear square patches, and the current is computed at the center of all interior patches by marching on in time using the numerical representation of (7). The current on the edge and corner patches is obtained by applying the boundary conditions. For the cases studied to date, the component of the current perpendicular to the edge is set equal to zero on the edge patches. The component of the current parallel to the edge is obtained by noting that it must vary as one over the square root of the distance from the edge. Finally, both components of the current at the corner patches are set equal to zero.

As an example of the results which have been obtained using this technique, the smoothed impulse response for a square flat plate is shown in Fig. 5. The calculated results for normal incidence are compared with results measured on the time domain scattering range and shown to be in good agreement in Fig. 5(a). Note that the initial response approximates a smoothed doublet as

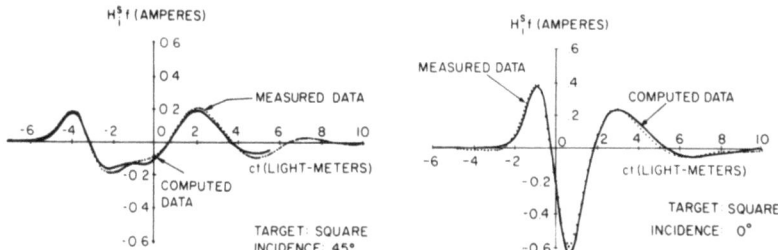

Fig. 5. Smoothed impulse response of a square flat plate.

expected. This is followed by a second positive pulse which is due to a wave traveling across the plate face and returning to the observer. The calculated and measured results for a 45° angle of incidence also agree well, as can be seen in Fig. 5(b). This technique has also been extended to the case of non-planar surfaces [13-15]. A sample result is displayed in Fig. 6 for a rectangular parabolic cylinder section. Again, the good agreement between calculations and measurements is apparent.

5. BODIES WITH WIRES

The problem of determining the scattering properties of targets with wire antennas or struts is important in aircraft and satellite surveillance. This problem has been addressed both in the frequency domain [16,17] and in the time domain [18,19]. This section will describe the time domain solution which yields the smoothed impulse response of the target.

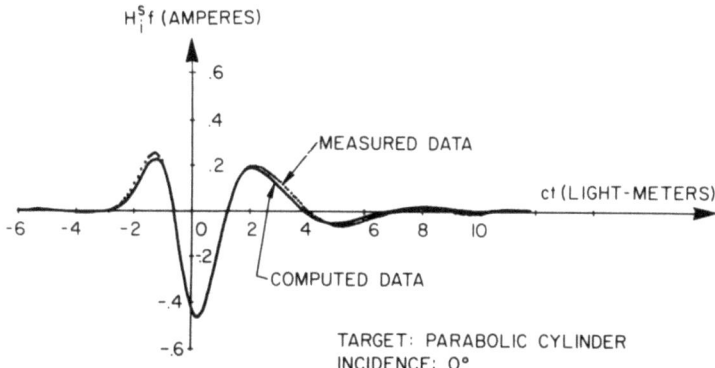

Fig. 6. Smoothed impulse response of rectangular parabolic cylinder section.

Since this problem contains both closed surfaces and thin wires, a hybrid integral equation approach is applied which uses the E field integral equation over the wire portions and the H field integral equation over the surface portions. The integral equations are for the wire current

$$\Box^2 I(\ell,t) = -\frac{1}{\alpha}\left[\sqrt{\frac{\epsilon}{\mu}}\frac{\partial E^i_\ell(\ell,t)}{\partial t} + \sqrt{\frac{\epsilon}{\mu}}\frac{\partial E_{s\ell}(\ell,t)}{\partial t}\right.$$

$$\left. + \frac{1}{4\pi}\Box^2 \int_{\substack{\text{Non}\\\text{Self}\\\text{Wire}}}\left\{\frac{I(\ell',\tau)}{R}\right\}d\ell'\right] \quad \tau = t-R \qquad (8)$$

where

ℓ = dimension along wire length

$E^i_\ell = \ell^{th}$ component of the incident E field

$E_{s\ell} = \ell^{th}$ component of the $\vec{E}$ field due to the surface currents

t = time in light meters

and for the surface current

$$\vec{J}(\vec{r},t) = 2\hat{a}_n \times \vec{H}^i(\vec{r},t) + 2\hat{a}_n \times \vec{H}_w(\vec{r},t)$$

$$+ \frac{1}{2\pi}\int_S \hat{a}_n \times \left\{\left[\frac{1}{R^2} + \frac{1}{R}\frac{\partial}{\partial\tau}\right]\vec{J}(\vec{r}',\tau) \times \hat{a}_R\right\}dS' \qquad (9)$$

$$\tau = t-R$$

where

$\vec{H}_w = \vec{H}$ field due to the wire currents

t = time in light meters.

The numerical solution of the coupled space time integral equations (8) and (9) is carried out by first dividing the surface into patches and the wires into segments. Next, the current at the center of each patch on the surface and at the end of each segment on the wire is computed by marching on simultaneously in time the numerical representation of the coupled equations (8) and (9) in a manner similar to that used for wires alone and surfaces alone. The current at both the wire free end and at the wire surface end are obtained by applying the boundary conditions for those two points.

As an example of an application of this technique, a model of the small scientific satellite (SSS) shown in Fig. 7 is considered. This target consists of a sphere located at the origin with four antennas attached, each a sphere diameter long, coincident with the x and y axis. Both the E plane and H plane smoothed impulse response of this target is displayed in Fig. 8 for the case of an

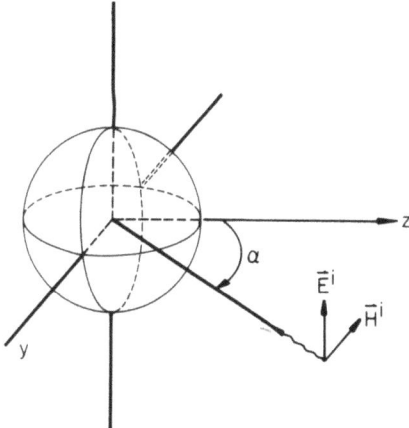

Fig. 7. SSS satellite model (length-to-diameter ratio of each antenna is 32).

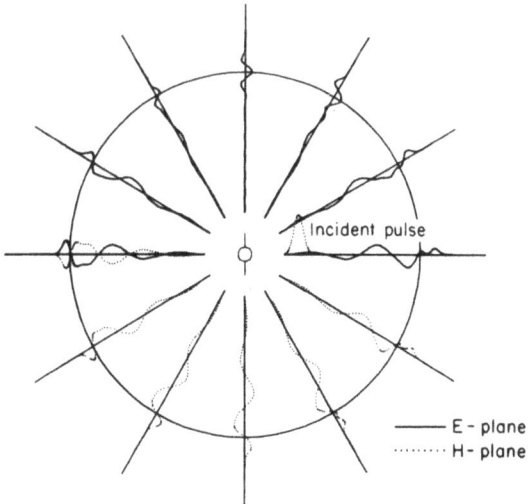

Fig. 8. Smoothed impulse response of SSS satellite model.

incident wave traveling in the negative z direction. It is interesting to relate the response in the backscatter direction to the target geometry. The specular return from the nose of the sphere appears first and is of the same value as was obtained for the case of scattering by a sphere alone. Next, there is a second peak that can be attributed mainly to the specular return from the wire antennas. This is then followed by a damped oscillation with a spacing between zero crossings that is approximately 2.5 sphere diameters. This is less than the distance between the opposite free tips of the wire (3 sphere diameters) but more than the length of an image representation of each antenna (2 sphere diameters). Moreover, the decay of these oscillations is much more rapid than would occur for the wires alone in free space with the same length-to-diameter ratio. This demonstrates a significant interaction between the antennas and the sphere surface.

6. BODIES WITH FINS

The problem of determining the scattering properties of targets with fins attached is important in missile detection and classification. This problem has been addressed directly in the time domain [20] by using a hybrid integral equation approach. This formulation used an E field integral equation over the fin portions and an H field integral equation over the surface portions. The integral equations are for the surface current

$$\vec{J}_S(\vec{r},t) = 2\hat{a}_n \times \vec{H}^i(\vec{r},t) + \frac{1}{2\pi} \int_S \hat{a}_n \times$$

$$\times \left\{ \left[\frac{1}{R^2} + \frac{1}{R} \frac{\partial}{\partial \tau} \right] \vec{J}_S(\vec{r}',\tau) \times \hat{a}_R \right\} dS' \bigg|_{\tau = t - R} \quad (10)$$

$$+ \frac{1}{2\pi} \int_F \hat{a}_n \times \left\{ \left[\frac{1}{R^2} + \frac{1}{R} \frac{\partial}{\partial \tau} \right] \vec{J}_F(\vec{r}',\tau) \times \hat{a}_R \right\} dS' \bigg|_{\tau = t - R}$$

where

$\vec{J}_S$ = surface current

$\vec{J}_F$ = fin current

and for the fin current

$$\square^2 \vec{J}_F(\vec{r},t) + \boxtimes \vec{J}_F(\vec{r},t) = \frac{1}{\gamma}\left\{-\varepsilon\frac{\partial \vec{E}^i}{\partial t} - \square^2\left[\vec{A}_{F_{NS}}(\vec{r},t) + \vec{A}_S(\vec{r},t)\right]\right.$$

$$\left. - \boxtimes\left[\vec{A}_{F_{NS}}(\vec{r},t) + \vec{A}_S(\vec{r},t)\right]\right\} \quad (11)$$

$$\vec{A}_S = \frac{1}{4\pi}\int_S \left\{\frac{\vec{J}_S(\vec{r}',\tau)}{R}\right\}_{\tau = t-R} dS'$$

$$\vec{A}_{F_{NS}} = \frac{1}{4\pi}\int_{\substack{\text{Non-}\\ \text{Self}\\ \text{Fin}}} \left\{\frac{\vec{J}_F(\vec{r}',\tau)}{R}\right\}_{\tau = t-R} dS'$$

$$\gamma = \frac{2}{\left[\sqrt{\frac{\Delta S}{\pi} - \left(\frac{\delta}{2}\right)^2} - \frac{\delta}{2}\right]}$$

ΔS = fin patch area

δ = fin thickness

The numerical solution of the coupled space-time integral equation given in (10) and (11) is carried out by marching on in time using a digital computer after first dividing the surface and fin into patches. The surface current is computed at the center of each surface patch using a numerical representation of (10) in a manner similar to that used for the solution by surfaces alone. The fin current is computed at the center of each interior fin patch using a numerical representation of (11) in a manner similar to that used for a flat plate. The currents on the free space edge patches and on the surface join patches of the fin are established by application of the appropriate boundary conditions.

As an example of an application of this technique the finned cylinder target shown in Fig. 9 is considered. This target consists of a right cylinder of radius a with length to diameter ratio of two-to-one with its axis coinciding with the z-axis and centered at the origin. Four fins (one cylinder diameter in a side) are arranged symmetrically about the cylinder body coinciding with the x = 0 and y = 0 planes. A vertically polarized smoothed impulse whose base width is equal to the cylinder length is used to illuminate this target. The smoothed impulse response that has been calculated for this target is compared in Fig. 10 for an axial angle

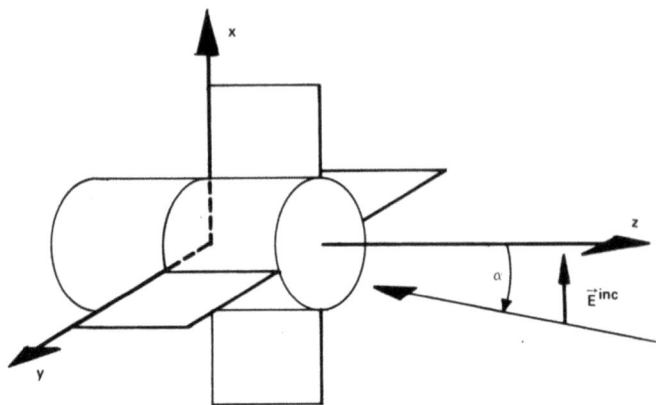

Fig. 9. Geometry of cylinder with fin attached.

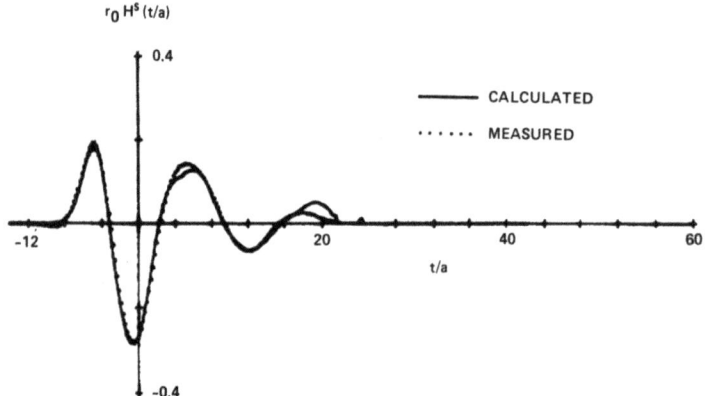

Fig. 10. Smoothed impulse response with cylinder with fin attached for axial incidence.

of incidence with the measured result. The agreement between the calculation and the measurement is good. The initial portion of the response is due to the return from the face of the cylinder and the leading vertical fin edges. This is followed by the return from the far edge of the vertical fins, a creeping wave return, and oscillation due to the interaction between the fins and the cylinder body.

7. IMPULSE RESPONSE AUGMENTATION TECHNIQUE

What has been described up to now in this paper is the space-time integral equation, its numerical solution, and results obtained using a smoothed impulse excitation. This numerical solution of the integral equation yields the smoothed impulse response of target sizes up to several pulse widths. If transformed into frequency domain terms, this technique yields results from zero frequency up well into the resonance region of the target. The impulse response augmentation technique, first suggested in 1968 [3] and later developed in 1973 [21,22], couples the calculated smoothed impulse response with knowledge about the singular portion of the impulse response, such as location, type, and size of singularity, and knowledge about the asymptotic variation in frequency to obtain an estimate of the impulse response or equivalently an estimate of the frequency response over the entire spectrum. The philosophy motivating this approach is to make use of all information available about the scattering problem in order to obtain an estimate of the "total" response. This may be viewed as a hybrid approach to the EM scattering problem which combines results obtained by numerical techniques with results obtained by asymptotic techniques. Using this technique, results have been obtained for smooth

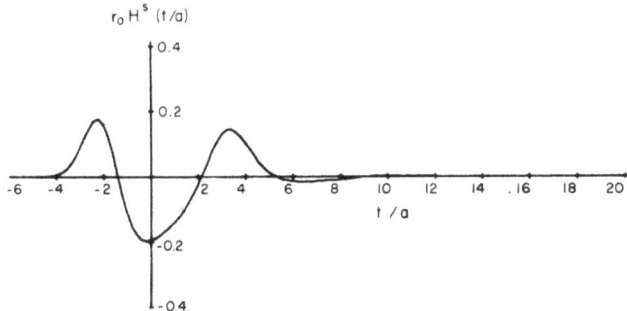

(a) Smoothed Impulse Response

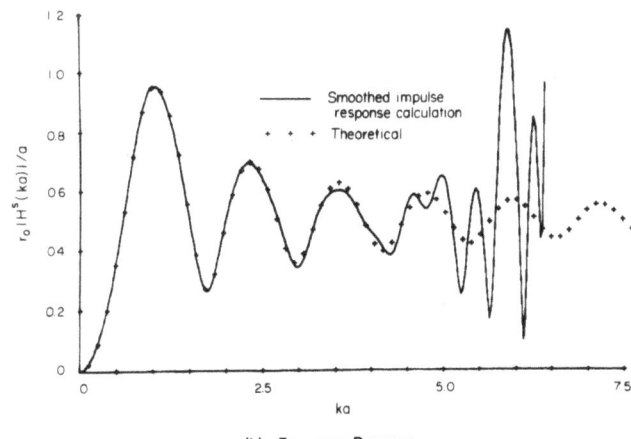

(b) Frequency Response

Fig. 11. Response of sphere with radius a.

convex bodies (sphere, sphere-capped cylinders, and prolate spheroid), for bodies with edges (right-circular cylinders) and for open thin surfaces (square flat plates) [13, 20-24]. To date, the technique has dealt only with the far field although it could also be applied to the near field or the surface currents.

As an example of the results obtained with this technique, the case of a sphere is considered. Figure 11(a) displays the far field smoothed impulse response of a sphere with radius a in the backscatter direction. The width of the incident pulse is twice the sphere diameter. Figure 11(b) displays a comparison between the frequency response obtained by transforming the calculated smoothed impulse response with the familiar Mie series solution. The two results agree well up to a ka of 4.5. Figure 12 compares the impulse response and frequency response obtained by use of the impulse response augmentation technique with the theoretical results and shows very good agreement. It is interesting to relate the features of the impulse response in Fig. 12(a) to the geometry of the target. The initial return is an impulse of amplitude 0.5

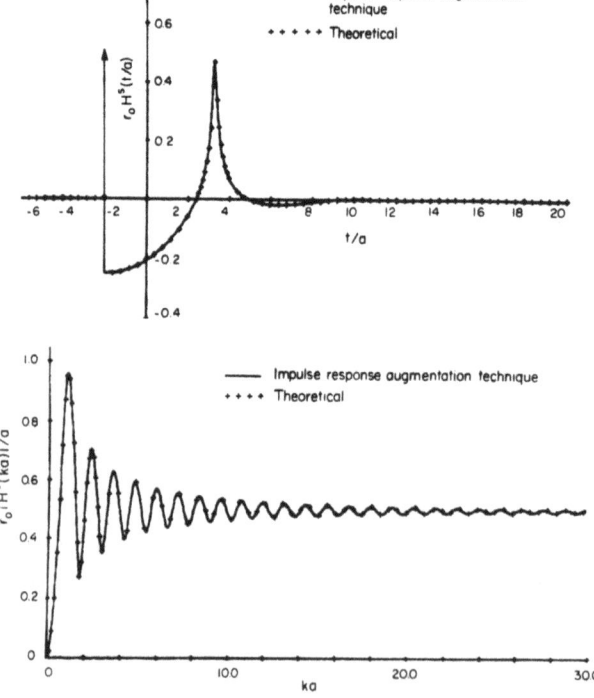

Fig. 12. Response of sphere with radius a.

that is due to the specular reflection from the nose of the sphere. This is followed by a negative step of amplitude 0.25 that is due to the return from the vicinity of the specular point and would be predicted by a physical optics approximation. It is interesting to note that the slope of the waveform following the negative step is a function of the difference in the principal curvatures at the specular point [20,23]. Finally, the second positive pulse at $t/a \simeq 3.1$ is due to the creeping wave that travels around the rear of the sphere and returns to the observer.

8. SUMMARY

The advances in the state-of-the-art software technology have brought to the engineer much more powerful tools for the solution of electromagnetic scattering problems. This paper presented a review of the time-domain solution for the direct scattering problem. The numerous examples presented demonstrate the power and versatility of a technique that has been made possible by the advent of the large, high-speed digital computer. A more detailed set of references are contained in [25-27].

REFERENCES

1. M.G. Andreasen, "Scattering from Parallel Metallic Cylinders with Arbitrary Cross-Section," IEEE Trans. Ant. and Prop., Vol. AP-12, pp. 746-754, November 1964.
2. M.G. Andreasen, "Scattering from Bodies of Revolution," IEEE Trans. Ant. and Prop., Vol. AP-13, pp. 303, March 1965.
3. C.L. Bennett, "A Technique for Computing Approximate Electromagnetic Impulse Response of Conducting Bodies," Ph.D. Thesis, School of Electrical Engineering, Purdue University, August 1968 (Univ. of Microfilms Order No. 69-7420).
4. C.L. Bennett and W.L. Weeks, "Electromagnetic Pulse Response of Cylindrical Scatterers," 1968 International Antenna and Propagation Symposium, Boston, MA, September 1968.
5. E.P. Sayre and R.F. Harrington, "Transient Response of Straight Wire Scatterers and Antennas," 1968 International Antenna and Propagation Symposium, Boston, MA, September 1968.
6. E.P. Sayre, "Transient Response of Wire Antennas and Scatterers," Ph.D. Thesis, Syracuse University, Syracuse, N.Y., June 1969.
7. E.M. Kennaugh and D.L. Moffatt, "Transient and Impulse Response Approximations," Proc. IEEE, Vol. 53, pp. 893-901, August 1965.
8. C.L. Bennett and E.K. Miller, "Some Computational Aspects of Transient Electromagnetics," 1972 Spring URSI Meeting, Washington, D.C., 13-15 April 1972.
9. E.K. Miller, "Some Computational Aspects of Transient Electromagnetics," Lawrence Livermore Lab. Rep. UCRL-51276, Sept. 1972.

10. C.L. Bennett and R.M. Hieronymus, "Time Domain vs. Frequency Domain Solution of EM Scattering Problems," 1975 APS/URSI Meeting, Urbana, Illinois.
11. A.W. Maue, "Zur Formulierung Lines Allgemeinen Beugungs: Problems Durch Eine Integralgleichung," Zeitschrift fur Physik, Bd 126, S. 601-618, 1949.
12. C.L. Bennett and W.L. Weeks, "Transient Scattering from Conducting Cylinders," IEEE Trans. Antennas and Propagation, Vol. AP-18, No. 5, September 1970.
13. C.L. Bennett, K.S. Menger, and R. Hieronymus, "Space-Time Integral Equation Approach for Targets with Edges," Final Report on Contract No. F30602-73-C-0124, SCRC-CR-74-3, March 1974.
14. K.S. Menger, C.L. Bennett, D. Peterson, and C. Maloy, "The Space-Time Integral Differential Equation Solution of Scattering by Open, Thin Surfaces," 1974 URSI Meeting, Atlanta, GA, 11-13 June 1974.
15. C.L. Bennett and R. Hieronymus, "Numerical Solution of Space-Time Integral Equation for Scattering by Open, Thin Surfaces," 1975 Annual URSI Meeting, Boulder, CO.
16. N.C. Albertsen, J.E. Hansen, and N. Jensen, "Numerical and Experimental Investigations of the Influence of Spacecraft Structures on Antenna Radiation Patterns," Report on Contract No. ESTEC 1340/71/AA, The Technical University of Denmark, Lyngby, February 1972.
17. N.C. Albertsen, J.E. Hansen, and N.E. Jensen, "Numerical Analysis of Radiation from Wire Antennas on Conducting Bodies," 1972 Spring URSI Meeting, Washington, D.C. 13-15 April 1972.
18. C.L. Bennett et al., "Integral Equation Approach to Wideband Inverse Scattering," Final Report on Contract No. F30602-69-C-0332; TR SCRC-CR-70-16, June 1970, Vol. 1, AD 879 849, Vol. 2 AD 876 627.
19. C.L. Bennett, A.M. Auckenthaler, and J.D. DeLorenzo, "Transient Scattering by Three-Dimensional Conducting Surfaces with Wires," 1971 International Antenna and Propagation Symposium, Los Angeles, CA, 22-24 September 1971.
20. C.L. Bennett and R.M. Hieronymus, "Integral Equation Solution," Final Report on Contract No. F30602-75-C-0040, Sperry Research Center, Sudbury, MA, December 1975.
21. C.L. Bennett, A.M. Auckenthaler, R.S. Smith, and J.D. DeLorenzc "Space-Time Integral Equation Approach to the Large Body Scattering Problem," Final Report on Contract No. F30602-71-C-0162, RADC-CR-73-70, AD 763 794, May 1973.
22. C.L. Bennett, "The Impulse Response Augmentation Technique," 1973 International IEEE/G-AP Symposium, Boulder, CO, August 1973.
23. C.L. Bennett, "Effects of Polarization on Electromagnetic Impulse Response," 1974 Annual URSI Meeting, Boulder, CO, 14-17 October 1974.

24. C.L. Bennett, "Application of the Impulse Response Augmentation Technique to Scattering by a Sphere-Capped Cylinder," 1975 IEEE/URSI Meeting, Urbana, IL, June 3-5, 1975.
25. C.L. Bennett, "The Numerical Solution of Transient Electromagnetic Scattering Problems," (Invited Paper), Proceedings National Conference on Electromagnetic Scattering, University of Illinois, Chicago, June 15-18, 1976, pp. 70-73.
26. C.L. Bennett, "The Numerical Solution of Transient Electromagnetic Scattering Problems," _Electromagnetic Scattering_, (ed. P. Uslenghi), Academic Press, 1978.
27. C.L. Bennett and G.F. Ross, "Time Domain Electromagnetics and Its Applications," Proc. IEEE, March 1978, pp. 299-318.

INVERSE SCATTERING

C. Leonard Bennett

Sperry Research Center
Sudbury, Massachusetts USA

ABSTRACT: The inverse scattering problem, briefly stated, is to find the shape and constitutive characteristics of a scattering target when given the knowledge of the incident field and the resulting scattered field data. Over the past two decades, interest has developed in the solution of the electromagnetic inverse scattering problem and more recently, this interest has been heightened by the quest to develop target recognition techniques using radar. In spite of the difficulty of this problem, progress has been made over the past two decades in solving some simpler cases. An elaborate survey of inversion techniques as applied in a variety of disciplines is compiled in a NASA Technical Memorandum [1]. More recently, an excellent State of the Art Review of Electromagnetic Inverse Scattering has been completed [2] and contains an extensive list of references.

It is the purpose of this paper to review the progress in inverse scattering over the past two decades by describing a representative number of techniques that have been developed. The use of the physical optics or Kirchhoff approximation as a basis for the solution of the problem is reviewed and discussed in both the time domain and in the frequency domain. "Exact" techniques in the frequency domain that require knowledge of the far scattered field over a substantial portion of the enclosing sphere are considered. Next, "exact" time domain techniques that are based on the time sequential solution of a space-time integral equation are described. Finally, a description of techniques that are based on currently available radar observables such as radar cross section at multiple frequencies are presented.

1. PHYSICAL OPTICS INVERSE SCATTERING - TIME DOMAIN

In 1958 Kennaugh and Cosgriff [3] presented the remarkable result that the impulse response of a scattering conducting body is simply the second derivative of the projected area function of the body if physical optics currents are postulated on the surface. This section will derive this result and discuss some of its implications.

The far scattered field of a conducting body is given in the time domain by:

$$\vec{H}^s(\vec{r},t) = \frac{1}{4\pi c r_o} \int_S \left\{ \frac{\partial \vec{J}(\vec{r}',\tau)}{\partial \tau} \times \hat{a}_r \right\}_{\tau = t - R/c} dS' \qquad (1)$$

where

$\vec{r}$ = position vector to the observation point

$\vec{r}'$ = position vector to the integration point

$R = |\vec{r} - \vec{r}'|$

$\hat{a}_r$ = unit vector in backscatter direction

t = time in seconds

c = speed of light

S = surface of target

$\vec{H}^s$ = far scattered magnetic field

$\vec{J}$ = surface current density

r_o = distance of far field observer from target.

In the physical optics or Kirchhoff approximation, the currents on the surface of the target are taken to be:

$$\vec{J}(\vec{r},t) = \begin{cases} 2 \hat{a}_n \times \vec{H}^i(\vec{r},t) & ; \quad \text{illuminated side} \\ 0 & ; \quad \text{shadow side} \end{cases} \qquad (2)$$

where

$\vec{H}^i(\vec{r},t)$ = incident magnetic field

$\hat{a}_n$ = unit vector normal to the surface

Furthermore, let the incident field be a unit ramp plane wave with the form

$$\vec{H}^i(\vec{r},t) = \hat{a}_H \, r_p\!\left(t - \frac{\hat{a}_r \cdot \vec{r}}{c}\right) \tag{3}$$

where

$$\hat{a}_H = \text{a constant unit vector}$$

$$r_p(\tau) = \begin{cases} \tau & ; \quad 0 < \tau \\ 0 & ; \quad \tau < 0 \end{cases}$$

Substitution of (2) and (3) into (1) and noting that $\hat{a}_r \cdot \hat{a}_H = 0$ yields:

$$r_o \vec{H}_R^s(\vec{r},t) = \frac{1}{2\pi c} \hat{a}_H \int_S \left\{ u\!\left(\tau - \frac{\hat{a}_r \cdot \vec{r}'}{c}\right) \right\} \hat{a}_r \cdot \vec{dS}' \Bigg|_{\tau = t - \frac{\hat{a}_r \cdot \vec{r}'}{c}} \tag{4}$$

where

$$\vec{dS}' = \hat{a}_n' \, dS'$$

$$u(\tau) = \begin{cases} 1 & ; \quad 0 < \tau \\ 0 & ; \quad \tau < 0 \end{cases}$$

$\vec{H}_R^s$ = far scattered magnetic field due to incident ramp wave,

which gives

$$r_o \vec{H}_R^s(\vec{r},t) = \frac{1}{2\pi c} \hat{a}_H \int_S u\!\left(t - \frac{\hat{a}_r \cdot \vec{r}'}{c/2}\right) dS_{proj} \tag{5}$$

This is evaluated by inspection to give:

$$r_o \vec{H}_R^s = \frac{1}{2\pi c} \hat{a}_H \, S(t_s) \tag{6}$$

where

$S(t_s)$ = The projected silhouette area of the scatterer as delineated by the incident plane wavefront assumed to be moving over the scatterer at one-half the free space velocity.

Thus, with (6) the relationship between the far scattered field and the projected area function of the target is given. It says that if the target is illuminated with an incident ramp plane wave, then the projected area function of the target is directly proportional to the far scattered ramp response. This relationship, however, is approximate since it is based on the assumption of physical optics currents. The more familiar expression [3] for the impulse response is obtained by differentiating (6) twice with respect to time to obtain:

$$r_o \vec{H}_I^S = \frac{1}{2\pi c} \hat{a}_H \frac{d^2 S(t_s)}{d t_s^2} \qquad (7)$$

where

$\vec{H}_I^S$ = far scattered magnetic field due to an incident impulse wave.

For the case where a target is rotationally symmetric and the direction of incidence is axial, the target geometry is determined uniquely, if only approximately, from (6) since

$$S = \pi \rho^2 \qquad (8)$$

where

ρ = the target contour function,

so that

$$\rho = \left[2 c r_o \left(\vec{H}_R^S \cdot \hat{a}_H \right) \right]^{1/2} \qquad (9)$$

This ramp response technique was first suggested in 1965 by Kennaugh and Moffatt [4] for radar target identification. In fact they predicted that satisfactory approximate ramp response signatures could be obtained with narrower bandwidth interrogating signals than for the impulse response. For example, Young [5] has shown that a ten-harmonic approximation provides a good representation of a sphere ramp response. Young [3] indicates that his data show that the "target profile function" which is obtained from ramp response measurements seems to hold in the illuminated region of the target and beyond the shadow boundary as well with slight loss in accuracy.

It has also been predicted from lower order moments of the ramp response waveform [5] that the integral of the ramp response is proportional to the Rayleigh coefficient. Thus, an approximate volume estimate can be obtained from the ramp response waveform. Although the volume estimate is polarization sensitive, an "almost constant" relationship of volume versus ramp waveform integral values has been reported [5] when the incident $\vec{E}$ field is perpendicular to the silhouette major axis.

Young [5] further extended the ramp response technique to arbitrarily shaped targets. The remainder of this section describes Young's technique for creating a target image using geometrical information inferred from the ramp-response waveforms. The geometric parameters inferred from the target ramp response obtained at three look angles are

(1) cross sectional area versus distance along the line of sight

(2) object length along the line of sight

(3) total object volume.

Orthogonal look angles greatly simplify the application of this technique but are not an inherent requirement. Because of the limited nature of the input parameters, a three dimensional image cannot be uniquely specified. However, it has been demonstrated [5] that the use of "approximate limiting surfaces" with suitable constraints do produce images which closely approximate targets from which the measured data were obtained. The approximate limiting surface which has produced good results is hyperbolic for each of the orthogonal look angles. For example, looking in the x-direction the limiting contour would be given by:

$$|yz| = \frac{1}{K_x} S\left(\frac{2x}{c}\right) \tag{10}$$

where S is the projected area function formed from the measured ramp response. If the constant K_x is taken to be 2π, then the limiting hyperbolic contour bounds all horizontally and vertically oriented ellipses with area S. The constant K can be varied to obtain optimum images for different object shapes. Young also found it necessary to provide adjustable parameters for rotating the hyperbolic limiting contours. The limiting contours are rotated by an angle α_i from the original position, and the major axis of each limiting surface is positioned at angles β_i and γ_i with respect to the line of sight. For example, in the x look direction the limiting contour in (10) would be given by

$$|y'' z''| = \frac{1}{K_x} S\left(\frac{2x}{c}\right) \tag{11}$$

where

$$y'' = y' \cos \alpha_x - z' \sin \alpha_x$$

$$z'' = y' \sin \alpha_x + z' \cos \alpha_x$$

$$y' = y + x \tan \beta_x$$

$$z' = z + x \tan \gamma_x$$

thus, it is seen that a total of twelve parameters (K_i, α_i, β_i, γ_i; i = x, y, z) are utilized in the actual image reconstruction. Young has had success in using a criterion he calls "profile function consistency" for selecting the values of these twelve parameters. In this process an iterative procedure was used with man-in-the-loop to obtain the final images. Images have been obtained for multiple-frustrum cylinder targets [5], which are in close agreement with the actual geometry.

2. PHYSICAL OPTICS INVERSE SCATTERING - FREQUENCY DOMAIN

The physical optics inverse scattering theory in the frequency domain is based on the identity obtained by Bojarski in 1967 [6]. The excellent introduction of this identity that was given by Lewis [7] in 1969 is adopted here.

Consider an incident-plane time harmonic electromagnetic wave with harmonic magnetic field

$$\vec{H}^i(\vec{r},t) = \hat{a}_H H_o e^{jk(\hat{p} \cdot \vec{r}) - j\omega t} \qquad (12)$$

where

ω = radian frequency

$k = \omega/c$

r_o = range

$\hat{p}$ = unit vector in direction of incidence ($\hat{p} = -\hat{a}_r$)

H_o = amplitude of incident magnetic field

$\hat{a}_H$ = polarization direction of $\vec{H}^i$

$\vec{H}^i$ = incident magnetic field

The far field backscattered from a conducting body is given by

$$\vec{H}^s(\vec{r}) = \frac{1}{4\pi r_o} \int_S jk(\vec{J} \times \hat{a}_r) e^{+jkR} \, dS' \tag{13}$$

where

$\vec{J}$ = the surface current

$R = |\vec{r} - \vec{r}\,'|$

$\vec{r}\,'$ = the integration point.

The physical optics approximation for the surface current is given by:

$$\vec{J}(\vec{r}) = \begin{cases} 2\hat{a}_n \times \vec{H}^i & ; \quad \text{illuminated side of surface} \\ 0 & ; \quad \text{shadow side of surface} \end{cases} \tag{14}$$

Using the same process as was employed in Section 1 after substituting (14) in (13) the following expression for the far scattered field is obtained:

$$\vec{H}^s(\vec{r}) = \frac{jk \, \hat{a}_H \, H_o \, e^{+jkr_o - j\omega t}}{2\pi r_o} \int_L \hat{p} \cdot \hat{a}_n \, e^{2jk \hat{p} \cdot \vec{r}\,'} \, dS' \tag{15}$$

where

L = the illuminated portion of the scatterer

$R = r_o + \hat{p} \cdot \vec{r}\,'$

Next let

$$\rho = -\frac{jk}{\sqrt{\pi}} \int_L \hat{p} \cdot \hat{a}_n \, e^{2jk \hat{p} \cdot \vec{r}\,'} \, dS' \tag{16}$$

Then (15) becomes

$$\vec{H}^s(\vec{r}) = \left(-\frac{\hat{a}_H \, H_o \, e^{jkr_o - j\omega t}}{2\sqrt{\pi} \, r_o} \right) \rho \tag{17}$$

The $\vec{k}$ is now introduced and defined as

$$\vec{k} = -2k\hat{p} = 2k\,\hat{a}_r \tag{18}$$

which upon substitution into (16) yields

$$\rho(\vec{k}) = \frac{j}{2\sqrt{\pi}} \int_{\vec{k}\cdot\hat{a}_n > 0} \vec{k}\cdot\hat{a}_n\, e^{-j\vec{k}\cdot\vec{r}'}\, dS' \tag{19}$$

From (19) it follows that

$$\rho(\vec{k}) + \rho^*(-\vec{k}) = \frac{j}{2\sqrt{\pi}} \int_S \vec{k}\cdot\hat{a}_n\, e^{-j\vec{k}\cdot\vec{r}'}\, dS' \tag{20}$$

where * denotes complex conjugate and the integral is over the entire surface S that bounds the volume V of the target. It follows from the divergence theorem that

$$\begin{aligned}\rho(\vec{k}) + \rho^*(-\vec{k}) &= \frac{j}{2\sqrt{\pi}} \int_V \nabla\cdot\left(\vec{k}\, e^{-j\vec{k}\cdot\vec{r}'}\right) dV \\ &= \frac{\vec{k}\cdot\vec{k}}{2\sqrt{\pi}} \int_V e^{-j\vec{k}\cdot\vec{r}'}\, dV\end{aligned} \tag{21}$$

Finally, the characteristic function γ of V is introduced

$$\gamma(\vec{r}) = \begin{cases} 1 & ;\ \vec{r}\ \text{in}\ V \\ 0 & ;\ \vec{r}\ \text{not in}\ V \end{cases}$$

and $\Gamma(\vec{k})$ is defined as

$$\Gamma(\vec{k}) = 2\sqrt{\pi}\,\frac{\rho(\vec{k}) + \rho^*(-\vec{k})}{k^2}$$

so that (21) becomes the famous Bojarski identity:

$$\Gamma(\vec{k}) = \int \gamma(\vec{r})\, e^{-j\vec{k}\cdot\vec{r}}\, dV \tag{22}$$

It shows that $\Gamma(\vec{k})$, which is defined in terms of $\rho(\vec{k})$ or equivalently the far scattered field, is the Fourier transform of the target's characteristic function $\gamma(\vec{r})$. It thus follows from (22) that

$$\gamma(\vec{r}) = \frac{1}{(2\pi)^3} \int \Gamma(\vec{k}) \, e^{j\vec{k}\cdot\vec{r}} \, d\vec{k} \tag{23}$$

If the backscattered field could be measured at all frequencies, k, and all aspects, $\hat{p}$, then $\Gamma(\vec{k})$ would be known for all $\vec{k}$ and (23) would yield the solution to the inverse scattering problem, i.e., the size and shape of V. It should be noted again that this identity is based on the assumption that the currents on the target are given by the physical optics approximation.

In general, $\rho(\vec{k})$ is only measurable for restricted frequencies and aspect angles, i.e., in a restricted domain D of $\vec{k}$-space. Lewis [7] showed how the restricted information about $\rho(\vec{k})$ can be used to yield complete or partial information about the target. Suppose that $\Gamma(\vec{k})$ can be obtained over some portion D of $\vec{k}$-space. Then defining the characteristic function

$$G(\vec{k}) = \begin{cases} 1 & ; \quad \vec{k} \text{ is in D} \\ 0 & ; \quad \vec{k} \text{ is not in D} \end{cases}$$

and its Fourier Transform

$$g(\vec{x}) = \frac{1}{(2\pi)^3} \int G(\vec{k}) \, e^{j\vec{k}\cdot\vec{r}} \, d\vec{k} \tag{24}$$

The function

$$F(\vec{k}) = G(\vec{k}) \, \Gamma(\vec{k})$$

can be formed and the function

$$f(\vec{r}) = \frac{1}{(2\pi)^3} \int F(\vec{k}) \, e^{j\vec{k}\cdot\vec{r}} \, d\vec{k} \tag{25}$$

can be constructed. It follows from the convolution theorem of Fourier transform that

$$f(\vec{r}) = \int \gamma(\vec{r}') \, g(\vec{r} - \vec{r}') \, d\vec{r}' \tag{26}$$

Since $f(\vec{r})$ can be constructed from measurements via (25) and $g(\vec{r})$ is known from (24), then (26) represents an integral equation for $\Gamma(\vec{r})$. Lewis [7] shows that (26) can be solved even when $G(\vec{k})$

vanishes in much of $\vec{k}$-space because the form of the characteristic function $\gamma(\vec{r})$ is known in advance, and in fact, he showed that (26) can be solved even when $\Gamma(\vec{k})$ is measured only on a surface in $\vec{k}$ space. A special case is obtained when

$$G(\vec{k}) = G^*(-\vec{k}) .$$

Then $g(\vec{r})$ is real and

$$f(r) = \frac{1}{2(\pi)^{5/2}} \text{Re} \int \frac{G(\vec{k}) \rho(\vec{k})}{k^2} e^{j\vec{k} \cdot \vec{r}} d\vec{k} \qquad (27)$$

and the determination of $g(\vec{r})$ and $f(\vec{r})$ is simplified. Further extensions can be found in the references [2,8,9].

3. RELATIONSHIP BETWEEN TIME AND FREQUENCY DOMAIN INVERSE SCATTERING

The application of Radon transform theory was first applied to the inverse scattering problem by Lewis in 1967 [10]. Independently following a different Ansatz, similar results were later obtained by Das and Boerner [11,12]. Boerner showed [2] that the size and the shape of a closed body can be obtained from its cross-sectional area. Since the projected area functions are related to the ramp response as shown in (6) and can be recovered for one aspect angle up to the shadow boundary, then curve fitting the projected area functions across the common shadow boundary for the two complementary view angles ($\vec{k}$ and $-\vec{k}$) will provide the input data. It should be noted that in forming Bojarski's identity (22), a similar Ansatz was used to define $\Gamma(\vec{k})$ in terms of $\rho(\vec{k})$ and $\rho^*(-\vec{k})$. In fact, Boerner [2] shows that both physical optics approaches, the Bojarski-Lewis frequency domain method and the Kennaugh-Cosgriff time-domain method, as extended by Das and Boerner [12], are identical in that they are related as a Fourier-Radon transform pair and for the purpose of reconstructing shape and size of a target they will require the same amount of data and a priori information.

Both of these techniques suffer from the same deficiencies

- physical optics approximation
- need for low frequency data
- loss of polarization information
- band limited and aspect limited data can lead to nonunique solutions.

4. "EXACT" INVERSE SCATTERING - FREQUENCY DOMAIN

In the previous sections the solution to the inverse scattering problem was derived using a physical optics approximation. In this section a technique will be discussed for solving the inverse scattering using an analytic continuation technique for the case of conducting targets. Weston and Boerner [13] show that the "inverse boundary condition" given by

$$|\vec{E}^i| - |\vec{E}^s| = 0$$

where

$\vec{E}^i$ = incident electric field

$\vec{E}^s$ = scattered electric field

is a necessary and sufficient condition to allocate the surface of a perfectly conducting scatterer. Imbriale and Mittra [14] demonstrated that the knowledge of the incident field and the scattered far field at one frequency and all bistatic angles may be employed to determine the size, shape, and location of a perfectly conducting two dimensional scatterer. A description of this work follows. Considering the TM case where the electric field is z-directed and target axis is parallel with the z-axis, the far scattered field may be represented as:

$$E_z^s(r,\phi) = \sum_{n=-\infty}^{+\infty} \left[\left(\frac{2j}{\pi k r} \right)^{\frac{1}{2}} e^{jkr} \right] a_n e^{jn\phi} \qquad (28)$$

where

E_z^s = the far scattered electric field

r = the distance of the far field observer

ϕ = the cylindrical coordinate system angle

a_n = the expansion coefficients.

Since E^s is given at a particular distance r, a particular frequency k, and all aspects ϕ it may be used to determine the expansion coefficients, a_n. Furthermore, it is known that outside the scatterer the scattered field E_z^s must satisfy the scalar Helmholtz equation, and therefore, it may be represented as

$$E_z^s(r,\phi) = \sum_{n=-\infty}^{+\infty} \left[-j^{-n} H_n^{(2)}(kr) \right] a_n e^{jn\phi} \qquad (29)$$

where

$H_n^{(2)}$ = an nth order Hankel function of the second kind.

The next step is to form

$$E_z^T(r,\phi) = E_z^i(r,\phi) + E_z^S(r,\phi) \tag{30}$$

and to search for the maximum value of r for which total electric field is zero. This locates a point on the scatterer surface and also is the minimum radius enclosing the body. For the next step a new origin for the coordinate system is chosen and the process is repeated. This iteration is continued until the desired target contour is obtained. This process represents the analytic continuation of the fields up to the target surface and good results have been obtained for convex targets using this technique [14]. For arbitrarily shaped contours, including concave sections, the process is augmented by use of the additional expansion about the origin

$$E_z^S(r',\phi') = \sum_{n=-\infty}^{+\infty} C_n J_n(kr') e^{jn\phi'} \tag{31}$$

to obtain the target contour in the concave regions. This has been applied with limited success in the reconstruction of two cylinders [14]. Although the extension to the three dimensional case appears possible [14], no results for this have appeared.

Ahluwalia and Boerner [15] in 1973 used a set of more general inverse boundary conditions to recover the size, shape, and averaged local surface impedance directly from the total near-field expressions. They verified the validity of these conditions for the two dimensional circular cylinder one and two body cases and provided the numerical approach for treating more sophisticated two dimensional conducting shapes. It should be noted that although these results are excellent, Boerner [2] reports that the inverse boundary conditions they used are not sufficient to uniquely determine simultaneously the size, shape, and material surface properties unless some a priori information about one of these three characteristic parameters is given. In searching for an additional condition Boerner's initial analysis resulted in the "vectorial impedance boundary condition" [16] that may hold promise in solving this problem.

The inverse source problem for the scalar wave equation was treated by Bleistein and Bojarski [17] where they showed that the source function is a solution of the Fredholm integral equation of the first kind. Their solution was further analyzed by Bleistein and Cohen [18] where they showed that the solutions to the resulting integral equations are not unique if "non-radiating sources" are present. This difficulty in the formulation of the direct problem leads to non-uniqueness in the inverse problem [18].

5. "EXACT" INVERSE SCATTERING - TIME DOMAIN

In this section, the inverse scattering problem is formulated as an inversion of the space-time integral equation. Consideration is limited to perfectly conducting scatterers. This approach to determining the response of the scatterer incorporates both the effects of the incident field directly and those due to the correction currents flowing on the body and was first suggested by Bennett et al. in 1974 [20] and later refined in 1977 [21]. By using a complete description of the surface current interactions, this formulation should yield a very close approximation to the target geometry. The presentation here follows that in [19,20] and is summarized in [21].

As presented in Section 1 the far scattered field is given in terms of the surface current $\vec{J}$ in the time domain by:

$$r_o \vec{H}^s(\vec{r},t) = \frac{1}{4\pi c} \int \left\{ \frac{\partial \vec{J}(\vec{r}',\tau)}{\partial \tau} \times \hat{a}_r \right\} dS' \qquad (32)$$
$$\tau = t - R/c$$

The surface current at a point $\vec{r}$ on the scatterer surface and time t is given [22] by

$$\vec{J}(\vec{r},t) = 2\hat{a}_n \times \vec{H}^i(\vec{r},t) + \vec{J}_C(\vec{r},t) \qquad (33)$$

where

$$\vec{J}_C(\vec{r},t) = \frac{1}{2\pi} \int \hat{a}_n \times \left\{ \left[\frac{1}{R^2} + \frac{1}{Rc} \frac{\partial}{\partial \tau} \right] \vec{J}(\vec{r}',\tau) \times \hat{a}_R \right\} dS' \qquad (34)$$
$$\tau = t - R/c$$

$\vec{H}^i(\vec{r},t)$ = incident magnetic field

The first term on the right-hand side of (33) may be considered the source term and represents the direct influence of the incident field on the current at the observation point (r,t). Moreover, this term, when applied to the illuminated side of the scatterer, yields the familiar physical optics approximation for the surface current. The second term on the right-hand side of (33) represents the influence of currents at other surface points on the current at (r,t). It is important to note that the influence of the current at other points on the surface on the current at $(\vec{r},t)$ is delayed in time by R, the distance between the two points. This allows

(33) to be solved by a "marching on in time" procedure rather than necessitating matrix inversion. Substituting the surface current expression (33) in (32) yields

$$r_o \vec{H}^s(\vec{r},t) = \frac{1}{4\pi c} \frac{\partial}{\partial \tau} \int \left\{ 2\hat{a}'_n \times \vec{H}^i(\vec{r}',\tau) \times \hat{a}_r \right\} dS' \bigg|_{\tau = t - R/c}$$

$$+ \frac{1}{4\pi c} \frac{\partial}{\partial \tau} \int \left\{ \vec{J}_c(\vec{r}',\tau) \times \hat{a}_r \right\} dS' \bigg|_{\tau = t - R/c}$$

(35)

If the incident field is a ramp, then the first term in (35) is simply the term that has been previously recognized in Section 1 to be proportional to the target area function. For the case the incident wave is an electromagnetic ramp, then (35) becomes

$$r_o \vec{H}^s_R(\vec{r},t) = \frac{1}{2\pi} S(t_s) \hat{a}_H + \frac{1}{4\pi c} \frac{\partial}{\partial \tau} \int_S \left\{ \vec{J}_{CR}(\vec{r}',\tau) \times \hat{a}_r \right\} dS' \bigg|_{\tau = t - R/c}$$

(36)

where

$r_o \vec{H}^s_R$ = the ramp response of the target

$\vec{J}_{CR} = \vec{J}_C$ that results from an incident ramp waveform.

$S(t_s)$ = the silhouette area of the scatterer as delineated by the incident ramp waveform assumed moving over the scatterer at one half the free-space velocity

r_o = distance of far-field observer from the origin

$t = t_s + r_o$

$\hat{a}_H = \dfrac{\vec{H}^i}{|\vec{H}^i|}$

Thus, by direct consideration of the space-time integral equation the exact relationship between the target response and the target geometry has been obtained. In particular, (36) gives the target ramp response in terms of both the target area function and the contribution due to the "correction currents", $\vec{J}_c$. Moreover, it is important to note that the correction currents as given in (36) are time-retarded functions of currents at other space points, and thus will be zero at the leading edge of the incident wavefront as it travels across the target. It is this feature, exclusive to the time domain formulation, that allows the determination of the target geometry from its ramp (or equivalently, impulse) response.

The specific problem class that has been solved [19-21] is that of a rotationally symmetric scattering problem which consists of a rotationally symmetric target illuminated by a wave from the axial direction as shown in Fig. 1. This target geometry can be represented simply by its contour function ρ as defined in Fig. 1. The space time integral equation for the solution of this inverse problem is given by

$$\rho(z,t) = \left[2r_o \, \hat{a}_H \, \vec{H}_R^s(\vec{r},t) - \frac{1}{2\pi} \hat{a}_H \int_S \left\{ \vec{J}_{CR}(\vec{r}',\tau) \times \hat{a}_r \right\} dS' \right]^{\frac{1}{2}}$$

$$\tau = t - R/c$$

(37)

where

ρ = contour function of the target

$\vec{H}_R^i$ = incident ramp H field

$\vec{J}_R$ = total current

$\vec{J}_{CR} = \vec{J}_R - 2\hat{a}_n \times \vec{H}_R^i$

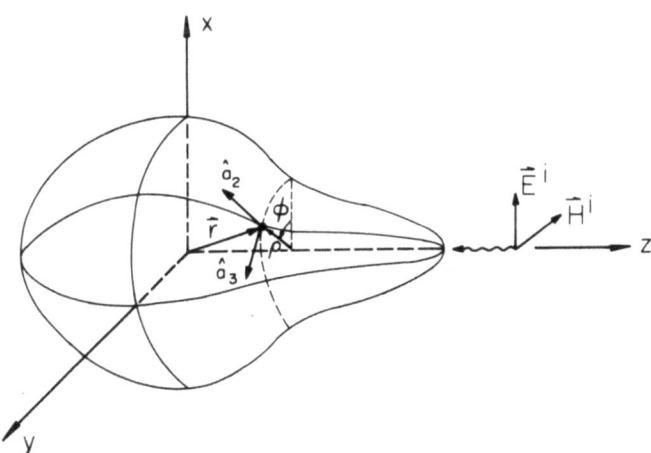

Fig. 1. Geometry of rotationally symmetric scattering problem.

This expression represents the exact relationship between the scatterer response, $\vec{H}_R^S$, and the target geometry, ρ. In particular the target contour function, ρ, is given in terms of the scattered far field $\vec{H}_R^S$ and in terms of "correction currents", $\vec{J}_{CR}$ that appear at earlier computed surface points at earlier points in time. This space-time integral equation has been solved by iteration of the marching on in time approach [20-22].

An example of the application of this technique is shown in Fig. 2 for the case of a sphere cap flatend cylinder when viewed from the sphere cap end. Note that the first iteration gives a

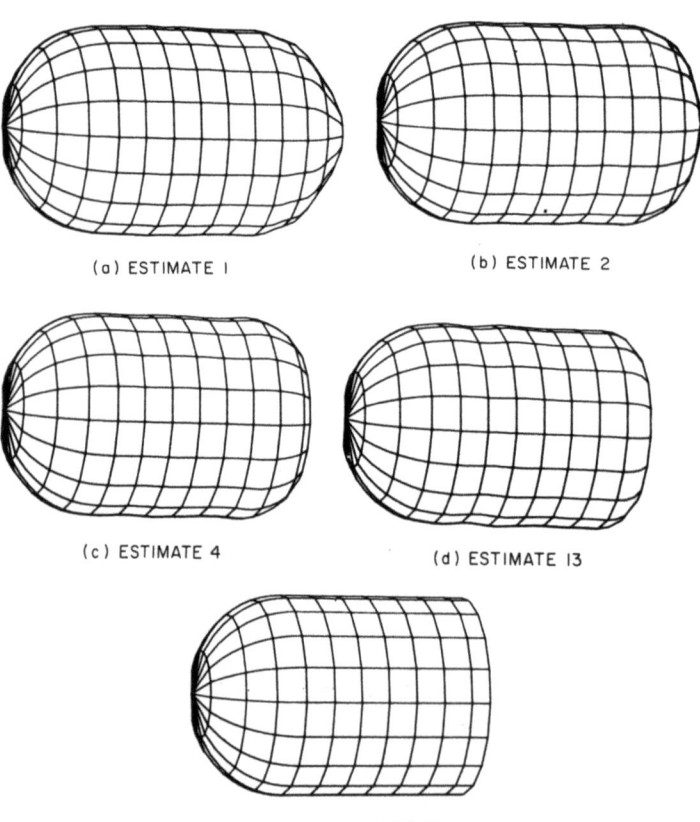

(a) ESTIMATE 1

(b) ESTIMATE 2

(c) ESTIMATE 4

(d) ESTIMATE 13

(e) ACTUAL CONTOUR

Fig. 2. Contours of sphere capped flatend cylinder obtained by time domain solution of inverse scattering problems.

good estimate of the spherical end cap and sides of the cylinder, but it took 13 iterations to give a reasonable estimate of the far end of the target. Although there is some rounding of the far edges, this technique can provide a powerful tool in obtaining information concerning the geometry of a target in the shadow region and in this case, shows good agreement with the actual contour in Fig. 2(e).

Recently, Bolomey, Durix, and Lesselier [23] have reported the solution of a time domain integral equation to determine the properties of an inhomogeneous dispersive slab by illuminating it with an incident TEM plane wave of arbitrary time dependence. The problem is one dimensional in space and the numerical solution process consists of a space discretization that is stepped on in time in a manner similar to that described above. Results were obtained for a number of inhomogeneous-lossy-nondispersive dielectric slabs for the case of normal incidence.

6. RADAR TARGET IDENTIFICATION

The mathematical difficulties associated with inverse scattering led to the consideration of a simpler, less inclusive, problem. This simpler approach allows only a finite number, M, of possible solutions (target shapes) whereas inverse scattering allows an essentially uncountably infinite number of solutions. The target identification problem may be viewed as a very crude form of inverse scattering which determines which of M prespecified body shapes is most consistent with the scattered field. Since radar is the most common device capable of sensing scattered electromagnetic fields, the discussion in this section will focus on radar systems.

Virtually all radar target identification methods are based on matching stored models of expected radar returns (signature libraries) with actual radar data. The stored model which is closest (in some well-defined sense) to the radar data is taken to be the target shape. This correlation operation is made more difficult by the fact that the target aspect angle(s) and the polarization vector orientation, which strongly influence the backscattered field, are usually not known in advance.

The fundamental considerations discussed in the previous sections indicate that wideband illumination at many aspect angles would be the most desirable incident field. Since simultaneous illumination at several aspect angles is beyond the capability of most radar systems, only the monostatic case is considered where the radar transmitter(s) and receiver(s) are collocated. With this restriction on the spatial properties of incident field, it remains to specify the polarization and temporal (or frequency domain) characteristics of the illumination.

The variation of the scattered field with the incident polarization is clearly an important feature for discriminating between body shapes. A simple example is the difference between the polarization scattering matrices for a sphere and a long thin rod. Object identification based on polarization measurements is discussed by Lowenschuss [25] and Kuhland Covelli [26].

More progress has been reported in the area of wideband illumination. With this approach, the frequency content of the incident field spans one or more octaves. The practical constraints of radar technology force this illumination to be narrowband, that is, concentrated at two or more carrier frequencies which span the overall frequency region. This illumination is better termed multi-frequency rather than wideband.

Goggins [26], Ksienski et al. [27], Lin and Ksienski [28] and Toomey and Bennett [29] have explored this approach in some detail. The results of this research indicate that the lower resonance region frequencies are most effective for target identification. It is believed that the relative enhancement of creeping waves at these wavelengths leads to better shape characterization.

The approach of Goggins [26] is somewhat unique in that both amplitude and phase of the backscattered radar returns are used for target identification. This is made possible by using a special property of Rayleigh region scattering: namely, that the target itself introduces no phase shift. For smooth, perfectly conducting bodies in the Rayleigh region, the phase shift experienced by the radar signal is solely due to the round-trip distance between the target and the radar. Thus, Rayleigh region illumination provides a phase reference from which the target-induced phase changes at higher frequencies can be measured. Goggins also describes a method for eliminating the effects of unknown polarization vector orientation by varying the transmitting or receiving antenna polarizations. Using Harrington's method of moments [30] to characterize the scattered fields from fairly simple targets, Goggins' simulations show good identification performance even with noisy radar data.

The work of Ksienski and his associates [27,28] considers the use of amplitude, phase, and polarization. They also compare the performance of various decision rules by Monte Carlo evaluation of identification error rate. In addition, the efficacy of the various radar observables (such as phase) is evaluated by this technique. Perhaps the most significant aspect of this work is the large assumed target set which consists of not only simple geometrical shapes, but also several different types of aircraft.

The Ksienski group examined the effect of using other radar observables such as trajectory and Doppler to obtain a rough estimate of aircraft aspect angle. The use of this aspect angle knowledge enabled low error rate identification with only a few frequencies. These results strongly indicate that aircraft identification by multifrequency illumination is feasible if not practical in an engineering sense.

The work of Toomey and Bennett [29] considers multifrequency target identification with an existing radar facility capable of simultaneously illuminating targets in the VHF, UHF, L, and S radar bands with circular polarization. The use of circular polarization eliminates any explicit polarization angle dependence. The identification processor structure was synthesized by applying the principles of communication receiver design to a mathematical model of the radar system. The result is believed to be the most general, powerful, and efficient classification algorithm yet developed. For example, the approach enabled the automatic inclusion of all radar system parameters (such as antenna gains, noise levels, range) in the identification logic. It also includes sequential multipulse processing for moving targets and automatic aspect angle estimation.

Although a general and practical solution to the inverse scattering problem is beyond the current state of the art, a review of recent work on multiple-frequency target identification indicates that a substantial amount of information concerning target shape can be obtained from appropriately designed radar systems.

REFERENCES

1. L. Colin, Editor, <u>Mathematics of Profile Inversion</u>, NASA Workshop, Ames Research Center, Moffatt Field, Calif., June 12-16, 1971, NASA TM X-62, 150, August 1972.
2. W-M. Boerner, "Polarization Utilization in Electromagnetic Inverse Scattering," University of Illinois at Chicago Circle, Communications Laboratory Report 78-3, October 1978.
3. E.M. Kennaugh and R.L. Cosgrift, "The Use of Impulse Response in Electromagnetic Scattering Problems, Proc. 1958 IRE National Convention Record, Part I, pp. 72-77.
4. E.M. Kennaugh and D.L. Moffatt, "Transient and Impulse Response Approximations," Proc. IEEE, Vol. 58, pp. 893-901, Aug. 1965.
5. J.D. Young, "Radar Imaging from Ramp Response Signatures," IEEE Trans. Ant. and Prop., Vol. AP-24, pp. 276-282, May 1976.
6. N.N. Bojarski, "Three-Dimensional Electromagnetic Short Pulse Inverse Scattering," Syracuse University Res. Corp., Syracuse, New York, February 1967.
7. R.M. Lewis, "Physical Optics Inverse Diffraction," IEEE Trans. Ant. and Prop., Vol. AP-17, pp. 308-314, May 1969.
8. N.N. Bojarski, "Electromagnetic Inverse Scattering," Final Report on Contract No. N00019-72-C-0462, April 1973.
9. N.N. Bojarski, "Inverse Scattering," Final Report on Contract No. N00019-73-C-0312, February 1974.

10. R.M. Lewis, "Theoretical Determination of Target Geometry for Short Pulse Data," Proc. of the GISAT II Symposium, 2-4 Oct. 1967, Bedford, Mass., Vol. II, Part I, pp. 195-207, AD 839 700.
11. Y. Das, "Application of Concepts of Image Reconstruction from Projections and Radon Transform Theory to Radar Target Identification," Ph.D. Dissertation, U. of Manitoba, Winnepeg, Manitoba, November 1977.
12. Y. Das and W-M. Boerner, "On Radar Target Shape Estimation using Algorithms for Reconstruction from Projections," IEEE Trans. Ant. and Prop., Vol. Ap-26, pp. 274-279, March 1978.
13. V.H. Weston and W-M. Boerner, "An Inverse Scattering Technique for Electromagnetic Bistatic Scattering," Can. J. Phys., Vol. 47, pp. 1177-1184, 1969.
14. W.A. Imbriale and R. Mittra, "The Two-Dimensional Inverse Scattering Problem," IEEE Trans. Ant. and Prop., Vol. AP-18, pp. 633-642, Sept. 1970.
15. H.P. Ahluwalia and W-M. Boerner, "Application of a Set of Inverse Boundary Conditions to the Profile Characteristics Inversion of Conducting Circular Cylindrical Shapes," IEEE Trans. Ant. and Prop., Vol. AP-21, pp. 663-672, Sept. 1973.
16. O.A. Aboul-Atta and W-M. Boerner, "Vectorial Impedance Identity for the Natural Dependence of Harmonic Fields on Closed Boundaries," Can. J. Phys., Vol. 53, pp. 1404-1407, 1975.
17. N. Bleistein and N.N. Bojarski, "Recently Developed Formulations of the Inverse Problem in Acoustics and Electromagnetics," University of Denver, Colorado, Report MS-R-7501, 1975.
18. N. Bleistein and J.K. Cohen, "Non-Uniqueness in the Inverse Source Problem in Acoustics and Electromagnetics," University of Denver, Colorado, Report MS-R-76-09, 1976.
19. C.L. Bennett, K.S. Menger, R. Hieronymus et al., "Space-Time Integral Equation Approach for Targets with Edges," Final Report, Contract F30602-73-C-0214, AD 725 120, July 1974.
20. C.L. Bennett, R.M. Hieronymus, and H. Mieras, "Impulse Response Target Study," Sperry Research Center, Sudbury, MA, Final Report, Contract No. F30602-76-C-0209, AD-A044-801, June 1977.
21. C.L. Bennett and G.F. Ross, "Time Domain Electromagnetics and Its Applications," Proc. IEEE, March 1978, pp. 299-318.
22. C.L. Bennett, "A Technique for Computing Approximate Electromagnetic Impulse Response of Conducting Bodies," Ph.D. Thesis, School of Electrical Engineering, Purdue University, August 1968 (Univ. Microfilms No. 69-7420).
23. J. Ch. Bolomey, Ch. Durix, and D. Lesselier, "Time Domain Integral Equation Approach for Inhomogeneous and Dispersive Slab Problems," IEEE Trans. Ant. and Prop., Vol. Ap-26, pp. 658-667, Sept. 1978.
24. O. Lowenschuss, "Scattering Matrix Application," Proc. IEEE, Vol. 53, No. 8, pp. 988-992, August.
25. F. Kuhl and R. Covelli, "Object Identification by Multiple Observations of the Scattering Matrix," Proc. IEEE, Vol. 53, pp. 1110-1115, August 1965.

26. W.B. Goggins, Jr., "Identification of Radar Targets by Pattern Recognition," Ph.D. Dissertation, Air Force Institute of Technology, June 1973.
27. A.A. Ksienski, et al., "Low Frequency Approach to Target Identification," Proc. IEEE, Vol. 63, pp. 1651-1659, December 1975.
28. Y.T. Lin and A.A. Ksienski, "Identification of Complex Geometrical Shapes by Means of Low Frequency Radar Returns," Rad. Elec. Eng., Vol. 46, No. 10, pp. 472-486, Oct. 1976.
29. J.P. Toomey, R.M. Hieronymus and C.L. Bennett, "Multiple Frequency Classification Technique," Sperry Research Center, Sudbury, Mass., Final Report, Contract No. F30602-76-C-0039, AD-A039-497, December 1976.
30. R.F. Harrington, <u>Field Computation by Moment Methods</u>, MacMillan Co., New York, 1968.

Panel Discussion

on

MODELLING OF THREE DIMENSIONAL BODIES – SOLID PATCHES AND WIRE GRIDS

Chairman: Professor R.F. Harrington — Syracuse University, New York, USA.

Panel:
- Dr C.L. Bennett — Sperry Rand Research Centre, Sudbury, Massachusetts, USA.
- Dr E.K. Miller — Lawrence Livermore Laboratory, CA 94550, USA.
- Mr T.W. Armour — A.W.R.E. Aldermaston, England.
- Professor D.L. Jaggard — University of Utah, USA.*
- Dr A.R. Djordjević — University of Belgrade, Yugoslavia.
- Dr T. Slivnik — Fakulteta za Elektrotehniko, Ljubljana, Yugoslavia.

Contributors:
- Professor S.J. Kubina — Concordia University, Montreal, Canada.
- Professor C. Butler — University of Mississippi, USA.
- Mr J.I.R. Owen — Royal Aircraft Establishment, Farnborough, England.
- Dr P.A. Ramsdale — Royal Military College of Science, Shrivenham, England.
- Dr S. Sensiper — P.O. Box 3102, Culver City, CA 90230, USA.

---

* This contribution, together with discussion is on pp.315-337.

Professor R.F. Harrington - The topic of our panel discussion is modelling three-dimensional bodies. I'm chairman and we've chosen panel members, Dr C.L. Bennett, Dr E.K. Miller, Professor D.L. Jaggard, Mr T.W. Armour, Dr A.R. Djordjevic and Dr T Slivnik. These people may or may not give short presentations and they will respond to questions as appropriate. There are also other people in the audience who have things to say and if we have enough time they will be able to participate. We may run out of time however. Now the proposed topic, is: How do you model a solid surface?

One thing you can do is divide it up into grids, rectangular or otherwise. In fact, I suggest the use of triangular grids because they are easier to fit to the surface. Once you do that you can use a wire mesh approximation to the surface. I don't think anybody really wants to defend that as being the best way. Logically, a better way to do it would be to model each one of the patches by currents over the surface. Once you do that you have many different approaches to take. For wires one uses the E field equations (the H field equation doesn't apply). This involves interior resonance problems of the wires. You will have problems in a wire grid model when the body is near an interior resonance. You will have trouble with the patch model even in that case. The E field equation and the H field equation can be put together in different ways to get the combined field and combined source equations. Apparently there are no good, user oriented programs for the E field equations for patches (at least I don't know of them), or for combined field or combined source equations. There may be some for the H field equation around. Anybody who has anything to contribute to that topic, is invited to talk first. I realise that people can't stick to the topic, so don't worry if you talk about other things. I know that people want to talk about aperture problems as well. But let's first talk about the modelling of three dimensional bodies. I think Dr Bennett has something to say on this.

Dr C.L. Bennett - Suppose you want to find the scattering from an aircraft, say a 747. You have to put together some simplified physical model for it, because it is much too complicated to model this aircraft exactly. You might consider putting together a stick model like Dr Miller showed us or you might consider modeling the fuselage as a prolate spheroid and putting some thin flat fins on it. Next this physical model (which is something you can build to measure) is modeled numerically with patches or wire-grids. Then you select a numerical solution technique and compute the result.

After a model has been developed, solution techniques selected, and results computed, it is extremely important that the total procedure be verified. One thing that could be done is to check your solution technique using your numerical wiregrid or patch model for some canonical problem, such as a sphere where the solution is known. Or you can apply your procedure to other problems that people have solved and reported in the literature. Comparing results will then give you a feel for the goodness of your solution procedure. But the real test is to compare with experimental measurements. Take your physical model, say your stick model or your prolate spheroid with fins on it's model, build it, measure its response and compare the measurements of your solution technique. Another thing that could be done is to build a plastic scale model of a 747 using an aeroplane kit, coat it with conducting paint, measure its response, and compare the result with the numerical solution techniques. It is important to have experimental measurements to verify your solution procedure in addition to verification with canonical problems. Finally, one should take experimental measurements for the canonical problems to determine the accuracy of the measurements.

Professor Harrington - The first topic is wire grid modelling. Is there anyone that has a contribution to the surface modelling problem? Apparently there has been some of use the H field equation for surface modelling, although I'm not familiar with it. The only thing I can say about that is it looks like a field for further investigation. I do know some work has been done at the University of Mississippi using triangular patches, but we don't have anybody here ready to talk about it. They primarily use the E field equation. They have also been working on the H field equation, and when they get those two done they will put them together in a combined field or a combined source solution. However, there is no assurance that this work will continue. There is a graduate student at the University of Mississippi by the name of Mr Rao who has used triangular patches in the E field equations to solve a few problems. He gets good answers when compared with other solutions for the sphere, and he's done a few other shapes. But he doesn't have a general program yet.

Professor S.J. Kubina - My recollection is that Goldirsh and Knepp, about eight years ago reported a surface patch model of the helicopter (a Bell helicopter) and reported fairly successful results. But one of their conclusions was because of the difficulty in treating edges they were going to try a wire mesh model of the same machine.

Professor Harrington - Yes, I remember seeing a paper on that and I think it was the H field equation, which has difficulty in treating edges. I also should call to attention the work done at the Technical University of Denmark. They did a patch model where they used the H field equation for part of the body. It was for modelling a satellite. Then they used the E field equation to model wire antennas on the body. This has been published in a paper and also in a report. If anybody is interested in that I would suggest they write to the Electromagnetics Laboratory of the Technical University of Denmark.

Professor C. Butler - The probable paper was published in the Journal Applied Physics by Leonard Tsai about three years ago. Leonard Tsai published a cube done by the H field equation. But Leonard deceased and the task was not finished. .

Dr C.L. Bennett - I'm going to talk about time domain solutions for scattering from surfaces. We've been developing solution techniques for these problems since about 1968. The solution techniques consist of solving a time domain integral equation. The target model which we use is a patch model where the surface is divided into patches and the current is computed in the center of each patch. The patch is given by its coordinate location, its normal vector, its principal curvatures, and the direction of these principal curvatures. We solved the closed conducting surface problem using H field integral equations applied to this patch model where the patch shapes were curvilinear square. The open conducting thin surface problem has been solved using an E field integral equation applied to this patch model. We have used both curvilinear square and triangular patch shapes for the open thin surfaces. We have also combined E field and H field integral equations for conducting surfaces with wires attached. We did it about the same time Professor Albertson did it in Denmark with his frequency domain integral equations. The surface currents were computed with an H field integral equation and the wire currents were computed with an E field integral equation. Curvilinear square patches were used to model the surface. Also fins have been attached to surfaces, again using patch models. Quite a bit of work has been done on patch models. It seems if you want to model a surface type target then the natural way to do it is with a patch model. There may be other features of wire grid models that make them more appealing. Perhaps they run in less computer time, I don't know, since I have no experience with it.

Mr J.I.R. Owen - Why do you say it is natural to model a structure such as that with patches?

Dr Bennett - Because the physical model is a closed structure and because the integral equation represents that closed structure. Then the approximations you are making are only in the solution of that integral equation, whereas in the wire grid model you are modelling a closed structure physically with holes in it. Now the result may be just as good, it may be even faster, I don't know. But the patch model is a more natural way of approaching the problem.

Professor Harrington - I think you have mentioned something important. You said that the wire grid model has holes in it. I think the wire grid model would be a poor method to use to calculate the internal field in a closed body. It doesn't seem to be a good model for that case. Mr Armour has used the wire grid model and could comment on some of his experiences.

Mr T.W. Armour - I shall be reporting on the use of wire-grid modelling to determine the response of aircraft to EMP. A number of wire-grid codes have been used to predict current and current density on simple structures and comparisons made with the results obtained by alternative methods such as those using patches, with classical analysis and also with experiment. Calculations made with the 'CHAOS' program seemed to be particularly satisfactory.

CHAOS uses the method of moments, overlapping triangular basis functions and a Galerkin type solution of the electric field integral equation. It can handle large dense matrices so that some 750 wire segments may be included in the model. One of its advantages is that it operates initially in the frequency domain so that any structure resonances may be easily identified. Constant excitation over a chosen spectrum is assumed. The frequency-domain amplitudes are then scaled to take account of the EMP driving function and finally transformed to the time-domain. Comparison of the predictions with experiment and alternative theoretical methods may thus be made in either frequency or time-domain.

The CHAOS program was developed initially by CHAO and STRAIT at Syracuse University. Further developments have been made at AWRE and a number of (wire) junction routines included in the program.

Several aircraft have been modelled and predictions made of the induced surface currents. These include the US F111 and the Anglo-French Jaguar. The F111 was among those chosen since measurements had been taken on this aircraft in the ARES Simulator at the Air Force Weapons laboratory, Albuquerque NM and published in a paper by Holland [1]. The length of the fuselage is

---

[1] R. HOLLAND - "THREDE - A Free-Field EMP Coupling and Scattering Code". Mission Research Corporation, Albuquerque, New Mexico, USA. Reference: AMRC-R-85.

22.4 metres and the wing span 19.2 metres. The total number of wire segments is 756.

Following the usual rule of wire-grid modelling the wire-radius is so chosen that the total surface area of all the wires is equal to that of the modelled surface. Current density is derived from the predicted wire currents on the assumption that the current is uniformly distributed over a rectangular patch of width equal to the wire separation.

When the E-field is parallel to the fuselage and the k-vector at right angles to both fuselage and wings a typical axial wire (on the top of the fuselage) shows a damped sinusoid with a resonant frequency for the F111 of about 8 MHz. There is, in addition a 'field dependent' term in the wave-form. This decay term is not present in the current wave-forms at the sides of the fuselage and is of reversed polarity for axial wires at the back of the fuselage. Peak (axial) currents predicted by the wire-grid code tend to be high. For example in one case the predicted current density is 400 A/m, that measured in ARES, 285. Predicted and measured wave-forms are very similar in shape.

When the E-field is parallel to the wing (the k-vector again at right angles to both wing and fuselage) the fuselage circumferential wire currents are now predominantly 'field-dependent' and their shapes are very similar to the EMP driving function. Complex interactions may occur, however, and some of the field dependent wave-forms show modulation by a 8 MHz wing resonance. Further examples are shown in [2].

It is interesting to note that Holland also uses a prediction code for comparison with the F111 experimental data. The THREDE code uses finite differences and operates in time-domain. Although radically different in their approach the two programs CHAOS and THREDE produce very similar results and this is certainly encouraging..

The Jaguar is a smaller aircraft than the F111 and the number of wires used is less - some 468 instead of 756 for the F111. A quarter scale Jaguar was also constructed and measurements taken. The agreement between prediction and experiment was very good. The model, some 4 metres long, was the largest which could be erected vertically in our travelling-wave simulator.

---

[2] T.W. ARMOUR - "Calculation of Surface Current Distributions on Aircraft Structures". 1978 Nuclear EMP Meeting. University of New Mexico, USA. 6-8 June 1978.

The very satisfactory results give confidence in our predictions for the full scale aircraft.

The next step in our aircraft studies is penetration through large apertures and the prediction of cavity fields behind the aperture. Initially apertures in spheres and cylinders will be considered and the results compared with classical analysis and also with experiment. The first predictions for a 1 metre diameter circular aperture in a 2 metre diameter sphere show good agreement with analysis for both electric and magnetic fields. The theoretical studies of Senior and Desjardins (University of Michigan Radiation Laboratory), Glisson and Wilton (University of Mississippi) are proving particularly helpful.

In summary we believe that the CHAOS program represents a powerful and versatile method for predicting surface current densities on aircraft structures. Galerkin's method is superior to 'point matching' or 'collocation' particularly at low frequencies where 'point-matching' has led to large errors. The system of overlapping wires at a junction originally proposed by CHAO and STRAIT seems to be the most consistent of those tested. The general accuracy of the program for large structures such as aircraft appears to be about 30% when compared with other theoretical methods or with experiment. A limitation is its inability to distinguish between the upper and lower surfaces of a wing and it is here that a patch program would be helpful. We believe, however, that a patch program should not be regarded as replacing wire-grid modelling but as complementary. Wing edges and undercarriage struts require wires for example.

Fundamental difficulties in representing a surface by means of a wire mesh are discussed by Lee, Marin and Castillo [3]. We believe the most practical approach to be the requirement that the total surface area of all the wires and that of the modelled surface should be equal. This ensures that, to a first approximation, the inductance of mesh and surface are equal. It is not clear, however, how best to incorporate the self-capacity term discussed in [3].

The prediction of cavity resonances is often stated to be a

---

[3] K.S.H. LEE   "Limitation of Wire-Grid Modelling of
L. MARIN   Closed Surfaces"
J.P. CASTILLO   IEEE Trans. EMC. EMC 18 No. 3 pp 123-129 August 1976.

limitation. This is not necessarily so. Most cavity resonances are often above the regions of interest in EMP studies.

Professor S.J. Kubina - May I ask what experience you have had with the choice of wire radius with regard to Dr Miller's comment on the surface equivalence considerations?

Mr T.W. Armour - In our program we match the total surface areas, exactly as Dr Miller has recommended. This is in some degree satisfying the recommendations of [3], but not completely, since we do not introduce the self-capacity term discussed by Lee, Marin and Castillo. By making the total surface areas equal we have achieved good agreement between experiment and theory for a wide variety of structures. (Our experience with plates and strips is, however, very limited).

Dr E.K. Miller - Yes, when you are comparing with Holland's results, how do you translate the wiregrid currents to equivalent surface currents?

Mr T.W. Armour - We predict wire currents and in order to transfer to a surface current density we open up the current over a patch extending half way to the next wire. Wherever possible we try to use an 'orthogonal' set of wires to cover the surface. We have actually built wire-grid structures so that there is a practical model which is an exact replica of the computer mathematical model. We then built the same model in copper sheet. This allows us to check physically that wire current can be transferred to surface current density.

Dr P.A. Ramsdale - I have a couple of comments. If you are modelling a strip by means of a wire you tend to use a radius which is one quarter the width of the strip. If you are wire-grid modelling a surface you always unravel the strips to fill the surface. The two methods are not consistent. I see nothing in the literature to say that you should unravel the wires in this way. Does anyone know the justifications for this procedure?

Mr T.W. Armour - The only reference we have is that of Lee, Marin and Castillo.

We have only limited experience with plates and strips. It is an area which we hope to explore in more detail. You will notice the difficulties we have had with aircraft wings and our inability to distinguish between the upper and lower surfaces. It is here we believe that a patch program would be helpful.

Dr E.K. Miller - That is a good question. I think part of the

reason the rationale has developed is that first of all you are
talking about an isolated strip versus a number of interacting
elements that are intended to simulate a solid surface. Based
on scattering measurements which certainly are not very conclusive in that they are a far-field measure of performance, I
think it has been pretty well established that a strip performs
as a thin wire of half the strip width. There is some experimental evidence going back to the 1930's regarding back screens and
design of back-screens for low frequency antennas in which it was
established and I believe this is reported in Jessic's book on
antennas, that the maximum front to back ratio, i.e. the maximum
directivity was obtained when the back screen wires did indeed
have this relationship that unrolled they would equal the projected area of the back screen. That has no content of numerical
analysis, analytical work or anything else. It is a physical
measurement which I think is reasonably conclusive in demonstrating that property. It is also mentioned in Busheys (Radio
Aerials) published just after the war.

Dr P.A. Ramsdale - This equivalence of wires and surface has
always worried me and it has never seemed very clear how best it
should be done. I have a further query concerning wire-grid
models. They introduce a lot of capacitance which tends to shift
the resonant frequency. When the figure of 30% was mentioned, had
a factor been added to allow for this extra capacitance or had
prediction and measurement been taken at the same frequency?

Mr T.W. Armour - The figure of some 30% holds generally for
cylinders and aircraft. It applies to comparisons between
prediction and measurement in both frequency and time-domain
(although most of our aircraft comparisons have been in time-
domain). Resonant frequencies are usually shifted downwards by
about 10%.

You may be interested in some calculations we are doing to
determine the cavity fields in a sphere with a large aperture at
the front. When we allow for the shift in resonant frequency we
get quite good agreement with the detailed calculations of Senior
and Desjardins. This applies to both the electric and magnetic
fields. Electric fields have been calculated by means of a
(frequency calibrated) dipole, magnetic fields using, again
frequency calibrated, magnetic loops. Our predictions tend to
be high deep in the cavity, rather low across the aperture.
Waveforms are very similar. Senior and Desjardins assume an
infinitely thin surface. Again patch programs would be expected
to give closer agreement.

Professor Chalmers Butler - Two comments on the concern over the equivalent area of a wire and a strip, that is width over four should equal the radius of the wire. This is based on two facts. First, one must take the total current, that is the current on the front and also the back of the strip. Now in the case of a wire one is talking purely about the current round the surface of the strip, so the area you may want to consider includes the back of the strip as well. The equivalence is based on the fact that the current on the strip has a singular behaviour and the current on the wire is assumed to be circumferentially uniform. One cannot therefore simply compare areas because it depends on these two factors, distribution of current and the fact that the strip has a front and a back current. I don't know how best to make the equivalence myself but this does not directly concern me.

Professor Harrington - Any other comments?

Dr S Sensiper - Yes, I have a question, Sensiper, U.S.A. Do I understand that when you have modelled for example the sphere and you calculate the resonant frequency that the frequency is different than it would be for a solid surface (i.e. a solid shell).

Mr T.W. Armour - We find that the change in resonant frequency occurs in all our wire-grid models. Smaller changes have also been noted when using patch programs, but this depends very much on the application.

Professor Harrington - Now we have Dr Djordjevic.

Dr A.R. Djordjević - I would like to present briefly a method for analysis of complex thin-wire structures which we developed at the Department of Electrical Engineering, University of Belgrade. Details about the method can be found in a recently published paper [1].

This method is capable of analysing structures assembled from arbitrarily located and interconnected straight-wire segments. The method is based on the so-called vector-scalar-potential equation (VSPE), which can be derived from the boundary condition for the total electric field at perfectly conducting surfaces,

$$(\vec{E} + \vec{E}_i)_{tang} = 0, \qquad (1)$$

where $\vec{E}$ is the electric field produced by currents and charges on wires, and $\vec{E}_i$ is the (known) excitation field. The first component can be expressed in terms of magnetic vector-potential $\vec{A}$

and electric scalar potential v as

$$\vec{E} = -j\omega\vec{A} - \text{grad } V, \qquad (2)$$

where ω is the angular frequency, and these potentials can be related to wire currents and charges. Applying the charge-conservation law one can finally obtain the VSPE:

$$\sum_{i=1}^{N} \int_{s_{ai}}^{s_{bi}} \left[\vec{i}_p \cdot \vec{i}_{s_i} I_i(s_i) + \frac{1}{k^2} \frac{dI_i}{ds_i} \frac{\partial}{\partial p} \left. \frac{\exp(-jkr)}{r} \right|_{p=0} \right] ds_i$$
$$= j \frac{4\pi}{\omega\mu_o} \vec{i}_p \cdot \vec{E}_i. \qquad (3)$$

In this equation the reduced kernel is used, resulting in single integrals only, I_i is the unknown current distribution along the i-th wire segment, N the overall number of segments, $\vec{i}_{s_i}$ the unit vector of the i-th segment axis, s_i the coordinate along this axis, s_{ai} and s_{bi} segment boundaries along this axis, r the distance between the element ds_i and the observation point, $\vec{i}_p$ the unit vector being tangential to the perfectly conducting wire surface at the observation point, the coordinate along this vector, and $k = \omega\sqrt{\varepsilon_o\mu_o}$ the free-space propagation coefficient.

Eqn. (3) is an integro-differential equation which involves both the unknown current distribution and its first derivative. We solve this equation numerically, using the point-matching technique |2| with the polynomial approximation of current, i.e. we take

$$I_i(s_i) \cong \sum_{m=0}^{n_i} a_{im} s_i^m, \quad i = 1, \ldots, N, \qquad (4)$$

---

[1] Djordjević, A.R. Popović, B.D. Dragović, M.B., "A method for rapid analysis of wire-antenna structures", Archiv für Elektrotechnik, Vol. 61, 1969, pp. 17-23.

[2] Harrington, R.F.; "Field Computation by Moment Methods". Macmillan, New York, 1968.

where a_{im} are unknown coefficients, and n_i is the adopted degree of the polynomial. From eqn. (4) one can readily obtain the approximation for the derivative dI_i/ds_i without any numerical differentiation.

In order to obtain a numerically stable solution it was found necessary to postulate that the first Kirchhoff law be satisfied at all wire functions and free wire ends, i.e.

$$\sum_i \pm I_i = 0 . \qquad (5)$$

In addition to it, at a junction of m wires (m > 2), (m - 1) integral conditions are imposed, having the form

$$\int_{L_i} (\vec{E} + \vec{E}_i) \cdot \vec{dl}_i = 0, \ i = 1, \ \ldots, \ (m-1), \qquad (6)$$

where L_i is a path along the wire axes in the vicinity of the junction, as shown in Fig. 1. This integral is evaluated numerically, involving up to four points, and thus represents up to

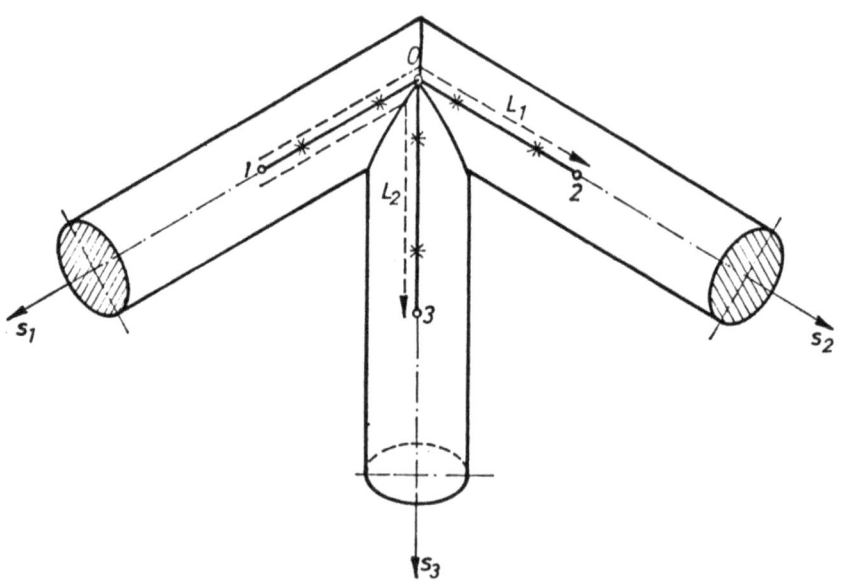

Fig. 1   Path of integration in eqn. (6). Asterisks denote points involved in numerical integration.

four point-matching equations added together, i.e. averaged. Eqns. (5) and (6) give exactly m conditions for each junction or free wire end. Therefore it is most suitable to choose (n_i -1) matching points at the i-th wire segment. These points are taken equidistantly, the distance between them being twice the distance between the segment end and the closest matching point. With all these precautions a very stable solution is obtained.

The method was tested on various examples, including some highly resonant structures. Theoretical results for the antenna input impedance, current distribution and radiation pattern were compared with experimental data. Very good agreement between these results was always obtained.

The memory requirements for the computer program are small because it suffices to take only about ten unknown coefficients in eqn. (4) per wavelength, compared to more than 30 with subdomain approximations of the current distribution. The running time of the program is given by the approximate formula

$$T = \alpha (N_e^2 + 0.017 N_e^3), \qquad (7)$$

where α is a constant, and N_e the overall number of the unknown coefficients. The first term is eqn. (7) corresponds to matrix element computation and the second to the solution of the linear equation system. It can be easily seen that for $N_e \sim 60$ these two terms become equal. This high speed was achieved by carefully designing the procedure for evaluation of integrals encountered in eqn. (3).

This method was designed primarily for the purpose of antenna synthesis |3|, but it is also very suitable for analysis of large and complex structures, such as wire-grid models of surfaces and solid bodies. As an example, wire-grid models of two aircraft

---

[3] Djordjević, A.R. Popović, B.D. "Synthesis of thin-wire antennas assembled from arbitrarily interconnected straight segments", <u>Proc. of Int. Conf. on Antennas and Propagation</u>, London, November 1978, pp. 403-407.

antennas are shown in Fig. 2. The first one, Fig. 2a, is a
triangular plate with a wire attached to it, and the second, 2b,
conical antenna. In both cases fairly good agreement between
theoretical and experimental results was observed.

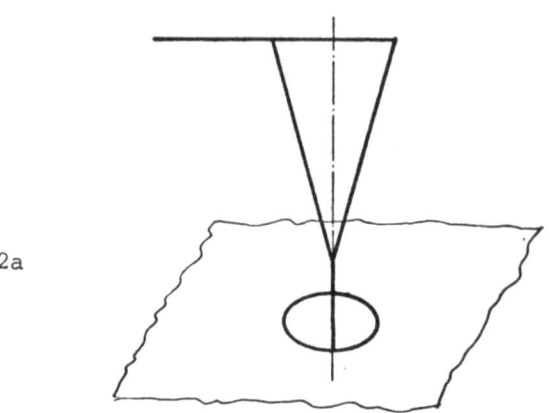

2a

Fig. 2. Examples of wire-grid modelling: (a) flat
triangular antenna with attached wire, (b)
eliptical-cone antenna.

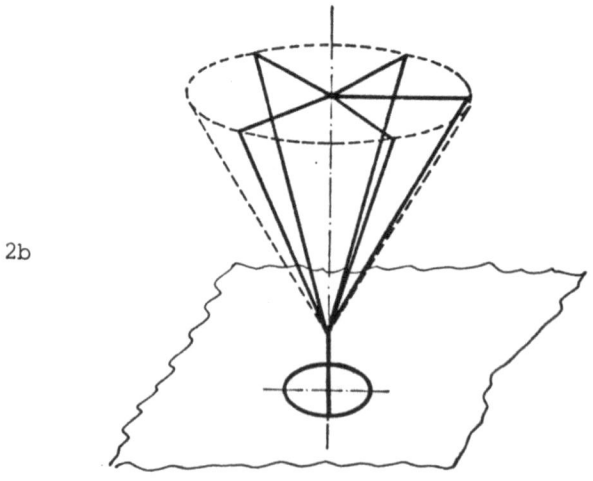

2b

Dr E.K. Miller - Do you use an entire-domain basis?

Dr A.R. Djordjević _ We use almost-entire-domain basis. On each segment of antennas shown in Fig. 2, for example, we use just one polynomial expansion.

Dr E.K. Miller - How many terms do you use in the polynomial expansion?

Dr A.R. Djordjević - We found it satisfactory to use up to 5 or 6 terms per half-wavelength segment. If the segment is longer, we cut it into two or more parts.

Dr E.K. Miller - And you then apply a condition of current continuity at each of the junctions of these polynomials. Is there any particular reason why you chose a polynomial basis?

Dr A.R. Djordjević - We have only a small-size computer (with only 16 K 16-bit words). Polynomials are an almost-entire-domain basis and we found it very suitable because only a small number of current-distribution coefficients is required. Thus we can analyse wire-structures up to 10 wavelengths long with our computer.

Dr E.K. Miller - It is intriguing here to consider the work done by Professor Richmond of Ohio State University 15 years or so ago. I guess he examined various kinds of polynomials (Legendre polynomials, Hermite polynomials and so on) and I believe he concluded that entire-domain bases were not too well suited to the problems he looked at, perhaps in terms of conditioning of the matrix and so on. Could you comment whether you ran into conditioning problems?

Dr A.R. Djordjević - There is a problem because when you have an entire-domain basis and you put in some non-physical condition then an unstable system is obtained. But if you try to avoid such conditions or do some sort of smoothing, you get extremely good results because the entire-domain bases are, in a sense, self-supporting. You have only to match the solution at a small number of points, and between these points it is fairly accurate. But you must not impose any nonphysical condition. With a pulse basis, if you put some strange condition, you destroy the pulses too. But, as the pulses are loosely coupled, you do not affect a whole wire, you affect only one of your subsegments.

Professor R.F. Harrington - I might say if you get a smart enough person almost any method seems to work. Of course I should not say smart enough because a lot of smart people try methods and they fail but somebody just finds the right way and they seem to work. I have seen that happen in many things that I thought were not good and people have got them to work by using ingenuity.

Professor C.M. Butler - Do you apply this condition (i.e. the integration of the electric field for a straight wire) at the ends, just for a straight wire not cut into segments?

Dr A.R. Djordjević - We apply it only for junctions, and not for wire ends.

Professor C.M. Butler - We have also tried numerous polynomials and after 5 or 6 terms typically we had terrible instability.

Dr A.R. Djordjević - For the case of a simple cylindrical antenna if we change the polynomial degree on a subsegment from 2 to 6, we get less than 0.1% difference in the input admittance. We had, however, to take precautions about the end effect, because we used the reduced kernel which is in fact the exact kernel for the extended boundary condition. If you put the extended boundary condition on the wire axis not taking into account the end effect, I mean the charge on the antenna tips, your system breaks down. When we took the end effect into consideration exactly, we had no more problems. I think this is the answer. If you impose some nonphysical conditions at the end of a segment, you do have troubles, but if you are careful, it works.

Professor R.F. Harrington - I might say, we at one time tried static problems of a cylinder and matching the field on the axis of the cylinder and if you did not close off the ends you got almost meaningless results.

Dr A.R. Djordjević - You are right.

Professor C.M. Butler - We have done that too but still with polynomials we have trouble.

Dr A.R. Djordjević - All I can say is that we are computing in extended precision on IBM 1130 (it means 9.2 significant digits) in all our computations. We went up to more than 30 complex unknown coefficients. In electrostatics we went up to about 50. We had no problem with instabilities due to polynomial approximations.

Professor R.F. Harrington - The last speaker is Dr Slivnik.

Dr T. Slivnik - In electromagnetics we often solve problems by means of integral equations. Analytical solutions are rarely possible, so we use numerical methods. If we consider two of the most commonly used methods: the moment method and the point matching method, we see that we need integrations for computing every matrix element. These integrations are normally done numerically and are very time consuming. Moreover matrix fill times for large problems are much greater than the inversion of the matrix required for the solution. Perhaps a simpler method for solution of integral equations is the Nyström method which doesn't require any integration for computing matrix elements. How does it work? Suppose that we have an integration formula (quadrature formula) for numerical computation of integral:

$$\int_a^b f(x)dx = \sum_{i=1}^n w_i f(x_i) ; \quad x_i \in [a,b]$$

Here w_i are weights of the quadrature for mula (known quantities). Because of numerical stability we prefer formulae with positive weights $w_i > 0$ ($i = 1,2,\ldots,n$). For instance Simpson formula has this property.

We want to solve integral equation:

$$\int_a^b K(x,y) f(y) dy = g(x) ; \quad x \in [a,b]$$

We approximate the left hand of the equation by quadrature formulae:

$$\int_a^b K(x,y) f(y) dy = \sum_{i=1}^n K(x,x_i) f(x_i) w_i$$

and to the resulting equations we apply point matching at points x_i, $i = 1,2,\ldots,n$. We get the system of equations:

$$\sum_{i=1}^n K(x_j,x_i) f(x_i) w_i = g(k_j) ; \quad j = 1,2,\ldots n$$

where $f(x_i)$ are unknown. As we can see, we need no integration for computing the matrix elements $w_i K(x_j,x_i)$. If the kernel $k(x,y)$ is symmetric, we can easily obtain the system of equations with symetric matrices, which is computationally preferable.

This method gives us results, which are comparable with other methods, but it has some advantages. We need no integration for computing matrix elements. We don't use this method, because kernels of our equations are usually singular - they have logarithmic singularity. If we want to extend availability of Nyström method to these cases, we must use some kind of integration formula for analytical integration of logarithmical singularity!

Consider the equation:

$$(A) \quad \int_{-a}^{a} f(x) \left[\ln(x-y) + G(x,y) \right] dx = g(y) ;$$

$$y \in [-a, a]$$

which we have met many times. Here $G(x,y)$ is a continuous function of both variables. For solving this equation we can use two types of approximations:

a) We approximate $f(x)$ by piecewise polynomial function and analytically compute the integrals:

$$\int_{-a}^{a} f(x) \ln[x-y] dx \approx \sum_{i=1}^{n} f(x_i) \gamma_i(y)$$

Here $\gamma_i(y)$ can be computed analytically.

b) We approximate the remaining integral by quadrature formula, which has the same order of accuracy as the approximation for $f(x)$ in a). We obtain:

$$\int_{-a}^{a} f(x) G(x,y) dx \approx \sum_{i=1}^{n} f(x_i) G(x_i,y) w_i$$

Clearly the singular part of the integral is approximated better than the continuous one, so we can expect that we shall get a method with the same properties as Nyström method for continuous levels.

If we match at points x_i, $o = 1,\ldots n$, we get the system of equations:

$$\sum_{i=1}^{n} f(x_i) \left[\gamma_i(x_j) + G(x_i, x_j)\, w_i\right] = g(x_j) \quad ; \quad j = 1,\ldots n$$

Here coefficients $\gamma_i(x_j)$ for all i can be computed once and used for the same equation with different $G(x,y)$!

This approach can be easily used for getting approximations of very high order of accuracy. If we use this method for computing singular solutions, we must use open quadrature rules! The described method has the same properties, as Nyström method for continuous levels. Its advantage is that we can very easily use high order approximations.

At this point we can ask one question more. Why not use approximations of higher order, with moment solutions of integral equations which are quite common with differential equations? Generally speaking, higher order methods give us more accurate results if the solutions are smooth enough. If solutions are discontinuous or have discontinuous derivatives, we shall obtain the best results with zero order methods - pulse approximations. (We can expect the same order of accuracy with step approximation as with piecewise linear approximation). Therefore - singularities in solutions are responsible for slow convergence and for using low order approximations. Sometimes we can remove singularities from the solutions in a very simple manner.

Consider equation (A). Sometimes it happens that the solution $f(x)$ is singular. In important cases we have already seen, we can suppose that the solution is of the form

$$f(x) = \frac{h(x)}{\sqrt{a^2 - x^2}}$$

If we want to solve this equation without considering this singularity by either method, we cannot achieve better results with higher order methods - the simplest approximations give us the best result. But we can avoid singularity by simple device. We put:

$$x = a \cos \alpha$$
$$y = a \cos \beta$$

and get

$$\int_0^\pi af(a\cos\alpha)\sin\alpha\left[\ln|a(\cos\alpha-\cos\beta)| + G(a\cos\alpha, a\cos\beta)\right]dx$$

If we observe that

$$p(\alpha) = af(a\cos\alpha)\sin\alpha = h(a\cos\alpha)$$

is a continuous function, we see that we have an integral equation with the continuous solution $p(\alpha)$. If we extract singularities, we get the equation:

$$\int_0^\pi p(\alpha)\left[\ln|\alpha-\beta| + \ln|\alpha+\beta| + \ln|2\pi-\alpha-\beta| + H(\alpha,\beta)\right]dx = g(a\cos\beta)$$

where $H(\alpha,\beta)$ is a continuous function. So we have pushed singularities from the solution to the kernels. This equation can be solved with extended Nyström method. In any case - removing of singularities from solutions is an essential step towards better convergence. When using this method with the Fredholm integral equations of the first kind one must be aware of numerical instabilities which can occur with higher order approximations, but this is an essential property of the integral equation of the first kind and not a property of the numerical method.

ON USING ISOPERIMETRIC INEQUALITIES AND ISOPERIMETRIC
QUOTIENTS IN ELECTROMAGNETIC THEORY AND APPLIED PHYSICS

D.L. Jaggard

Department of Electrical Engineering,
University of Utah,
Salt Lake City,
Utah 84112   USA

Abstract

   Here we present an alternative method of solving
boundary-value problems. The method, based on certain isoperimetric inequalities, produces two-sided bounds on the value
of physical quantities such as capacitance, aperture transmission coefficients, scattering cross sections, and the
equivalent radii of wires. The results, which are valid in
the low frequency regime, have the virtue of simplicity. In
addition, these bounds show how the boundary must be altered to
optimize the physical quantity under consideration.

I.   Introduction

       Traditionally one can solve boundary-value problems in
the low frequency regime through the technique of the separation
of variables, by variational schemes, or through numerical
methods. As an alternative, one can reformulate the problem so
that upper and lower bounds on the true solution and not the
true solution itself are sought. Although such reformulations
are sometimes available through pairs of variational expressions
[1-4], the calculations are often laborious. In addition,
the calculations have to be repeated each time the boundary
configuration is altered. Accordingly, it is of some interest
to try a simpler method of sandwiching the true solution between
upper and lower bounds.

       Here we consider the solution of boundary-value prob-
lems through the use of isoperimetric inequalities [5] which
may be easily used to establish two-sided bounds on many

physical quantities of interest. The bounds typically depend only on geometric descriptions of the boundaries such as area, perimeter, or volume. These calculations have the advantage of expressing the desired physical quantities in terms of simple parameters and of directly relating these quantities to the geometric properties of the structures under consideration.

II. Capacitance

As an illustration, we consider the electrostatic capacitance C of a conducting plate of infinitesimal thickness, area A, and perimeter P. Through considerations based on fundamental potential theory [6], it is clear that the capacitance of this plate is bounded by the capacitance of two circular plates, namely, that plate which inscribes and that plate which circumscribes the plate under consideration.

The notion of relating plate shape to capacitance is not new. J.C. Maxwell postulated that square plates possessed a smaller capacity than rectangular plates of the same area [7] and so realized that the ratio of perimeter to area played a role in the determination of capacitance. Lord Rayleigh observed the previously mentioned upper bound on capacitance based on the circumscribed plate and improved the lower bound based on the inscribed plate by noting that a circular plate of the same area provided a larger lower bound [8].

The above ideas were formalized by Pólya and Szegö [5, 9]. They proved that for all conducting plates of a given area A, the circular plate has a minimum capacitance C. Similarly, they conjectured that for all plates with a given perimeter P, the circular plate has the maximum capacitance C. Using these two concepts and the fact that the capacitance of a circular plate of radius a is given by $C = 8\varepsilon_o a$, we are led, respectively, to the following inequalities:

$$\frac{C}{8\varepsilon_o} \geq r_{in} \tag{1}$$

$$\frac{C}{8\varepsilon_o} \leq r_{out} \tag{2}$$

where the inner radius is defined by:

$$r_{in} = \sqrt{\frac{A}{\pi}}, \tag{3}$$

the outer radius by:

$$r_{out} = \frac{P}{2\pi}, \tag{4}$$

and $\varepsilon_o = (36\pi)^{-1} \times 10^{-9}$ farads per meter. From relations (1) – (4), we calculate the two-sided bounds on the capacitance of a convex conducting plate of area A and perimeter P as

$$r_{in} = \sqrt{\frac{A}{\pi}} \leq \frac{C}{8\varepsilon_o} \leq \frac{P}{2\pi} = r_{out}. \tag{5}$$

The plausibility of this result will be investigated in the following section.

III. Inner and Outer Radii and the Isoperimetric Quotient

The inner and outer radii defined above have additional uses other than those associated with the calculation of capacitance. For this reason it is useful to observe inner and outer radii for some typical shapes.

In Fig. 1, the inner and outer radii for several regular polygons are shown. It is immediately apparent that for convex figures, the inner radius is larger than the radius of the inscribed circle while the outer radius is smaller than the radius of the circumscribed circle.

It is also instructive to examine these radii for elongated structures. Consider the ellipse of Fig. 2 with semiaxes a and $b < a$ and eccentricity $e = \sqrt{1 - b^2/a^2}$. From definitions (3) and (4), we obtain the values

$$r_{in}^{ellipse} = \sqrt{\frac{A}{\pi}} \tag{6}$$

$$r_{out}^{ellipse} = \frac{2}{\pi} \sqrt{\frac{A}{\pi}} E(e^2) / (1 - e^2)^{1/4} \tag{7}$$

for an ellipse of area $A = \pi ab$. Here $E(e^2)$ is the complete elliptic integral of the second kind. In the small eccentricity regime (almost circular ellipses), these quantities take on the values

$$r_{in}^{ellipse} \xrightarrow[e \to 0]{} \sqrt{\frac{A}{\pi}} \tag{8}$$

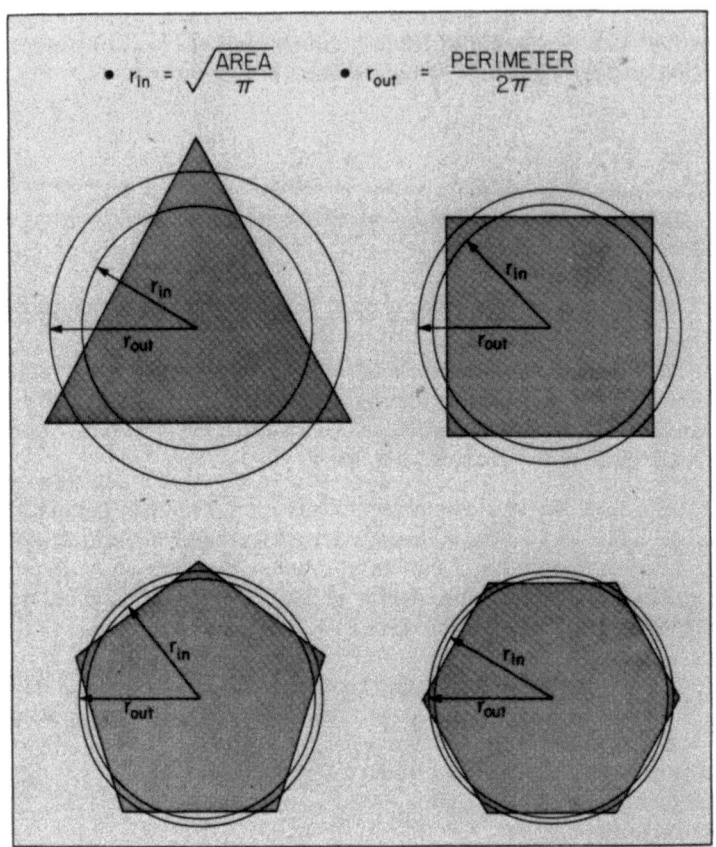

Fig. 1 Demonstration of inner and outer radii for regular polygons of three to six sides, each having the same area. Note that the outer radius approaches the inner radius as the number of sides increases.

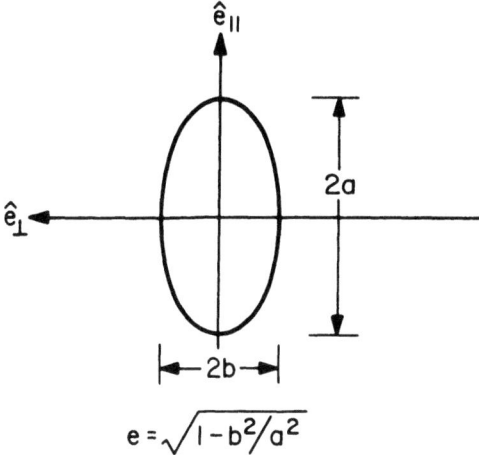

Fig. 2. Ellipse with major axis $2a$ minor axis $2b$ eccentricity $e = \sqrt{1 - b^2/a^2}$, and area $A = \pi ab$. Here the axes are oriented along the unit vectors $\hat{e}_{||}$ and $\hat{e}_{\perp}$.

$$r_{out}^{ellipse} \xrightarrow[e \to 0]{} \sqrt{\frac{A}{\pi}} \left[1 + \frac{3e^2}{64}\right] \tag{9}$$

In the regime where the eccentricity approaches unity (elongated ellipses), these radii become

$$r_{in}^{ellipse} \xrightarrow[e \to 1]{} \sqrt{\frac{A}{\pi}} \tag{10}$$

$$r_{out}^{ellipse} \xrightarrow[e \to 1]{} \frac{2}{\pi}\sqrt{\frac{A}{\pi}} \bigg/ (1 - e^2)^{1/4} \tag{11}$$

As a check on conjecture (5) concerning the capacitance of plates, we note that the capacitance of an elliptical plate of area A and eccentricity e is given by:

$$\frac{C^{ellipse}}{8\varepsilon_o} = \sqrt{\pi A} \bigg/ 2K(e^2)(1 - e^2)^{1/4} \tag{12}$$

where $K(e^2)$ denotes the complete elliptic integral of the first kind. For small eccentricities relation (12) takes on the value

$$\frac{C^{\text{ellipse}}}{8\varepsilon_o} \xrightarrow[e \to 0]{} \sqrt{\frac{A}{\pi}} \left(1 + \frac{e^2}{64}\right) \tag{13}$$

and is bounded by expressions (8) and (9) as predicted. Similarly, when the eccentricity approaches unity,

$$\frac{C^{\text{ellipse}}}{8\varepsilon_o} \xrightarrow[e \to 1]{} \sqrt{\pi A} \Big/ 2(1 - e^2)^{1/4} \ln\frac{4}{\sqrt{1-e^2}} \tag{14}$$

which is bounded by the quantities in (10) and (11). In fact, it can be shown numerically that the capacitance of (12) is bounded by the inner and outer radii of (6) and (7) for all values of eccentricity. This confirms the conjecture of inequality (5) for one class of shapes.

We turn now to a quantity, denoted the isoperimetric quotient or IQ, which is closely related to the inner and outer radii. The isoperimetric quotient is defined by the ratio

$$IQ = \frac{r_{in}^2}{r_{out}^2} = \frac{4\pi A}{P^2} \tag{15}$$

for closed planar figures of area A and perimeter P [10]. This quantity has its origin in the legend concerning Dido's problem. In this problem the goddess Dido desired to know the maximum area A that could be enclosed by a planar figure of perimeter P. As was known since 300 BC, the figure which solves Dido's problem is the figure with maximum IQ, namely, the circle. One is then led immediately to the inequality for the isoperimetric quotient.

$$IQ \leq 1 \tag{16}$$

From definition (15) one immediately finds that the IQ of a circle is unity, while the IQ for regular polygons with three and five sides is 0.60 and 0.86, respectively (see Fig. 1).

Combining equations (5) and (15), it becomes clear that a conducting plate with unity IQ possesses the smallest capacitance of any plate with a given area. Furthermore, from the calculations expressed by equations (8) and (9) for elliptical figures, it is apparent that the inner and outer radii, and consequently the isoperimetric quotient, are relatively insensitive to changes in shape for mild eccentricities. We can generalize this fact to other shapes and state that the IQ of a figure is close to unity whenever the figure is almost circular.

IV. <u>Aperture Transmission and Scattering of Acoustic Waves</u>

We consider now an acoustic plane wave which impinges on a thin planar screen containing a small,* arbitrarily shaped, convex aperture of area A and perimeter P as shown in Fig. 3.

For flexible screens (Dirichlet problem), the transmitted field is proportional to the electrostatic polarizability of an identically shaped aperture in a perfectly conducting thin screen, whereas for rigid screens (Neumann problem) the transmitted field is proportional to the electrostatic capacitance of a perfectly conducting plate which is in the shape of the aperture. Therefore, in the low frequency regime, the aperture transmission problem for acoustics is reduced to the calculation of two electrostatic quantities [11]. In keeping with our general approach to solving boundary-value problems, we will obtain two-sided bounds for these quantities and will therefore bound the transmission coefficient. The bounds for the capacitance have been given by the conjecture expressed by (5). Their plausibility has been demonstrated by the calculations involving the capacity of elliptical plates.

We now propose bounds on the electrostatic polarizability by invoking two conjectures. We first conjecture that circular apertures of radius a possess the largest electrostatic polarizability $\alpha = 8a^3/3$ of any aperture of a given area A. Therefore, we can write the inequality

$$\frac{\alpha}{A} \leq \frac{8}{3\pi} \sqrt{\frac{A}{\pi}} \tag{17}$$

The plausibility of this conjecture is supported by noting that the polarizability of an elliptical aperture of eccentricity e is given by

$$\frac{\alpha^{ellipse}}{A} = \frac{4}{3} \sqrt{\frac{A}{\pi}} \frac{(1-e^2)^{1/4}}{E(e^2)} \xrightarrow[e \to 0]{} \frac{8}{3\pi} \sqrt{\frac{A}{\pi}} \left(1 - \frac{3e^4}{64}\right) \tag{18}$$

---

* Here, as in the following section, "small" means that $kL \ll 1$, where k is the wave number of the incident wave and L is the maximum linear dimension of the aperture.

SCALAR (ACOUSTIC) DIFFRACTION

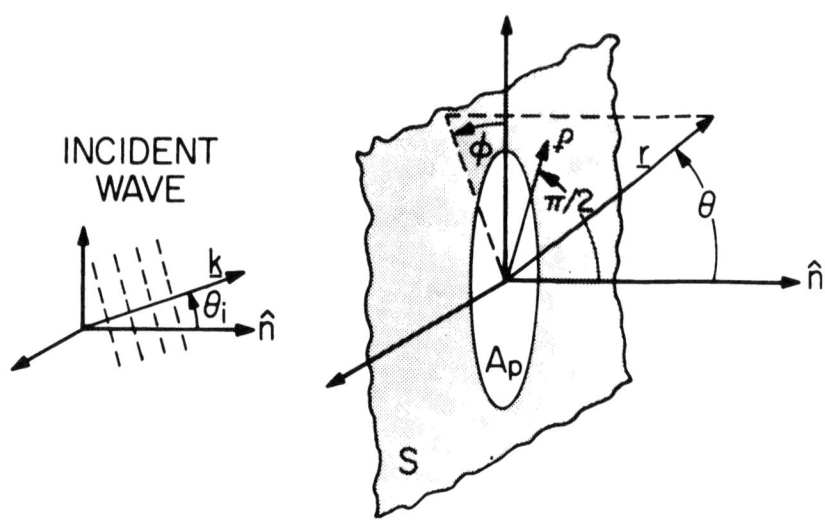

Fig. 3. Configuration for the transmission of acoustic waves through an aperture A_p in an infinitesimally thick screen S of infinite extent. The wave vector of the incident wave $\underline{k}$ forms an angle θ_i with the screen normal $\hat{n}$.

and indeed attains a maximum when the ellipse degenerates into a circle. Since it has been previously noted [4] that the functional dependence of the polarizability on shape is given by

$$\frac{\alpha}{A} \sim \frac{A}{P}, \qquad (19)$$

we seek a lower bound of this form. By noting that elongated rectangular apertures with sides a and $b > a$, area $A = ab$, and eccentricity $e = \sqrt{1 - b^2/a^2}$ have an electrostatic polarizability which is given by

$$\frac{\alpha^{\text{rectangle}}}{A} \xrightarrow[e \to 1]{} \frac{\pi}{4} \sqrt{A} (1 - e^2)^{1/4} \xrightarrow[e \to 1]{} \frac{\pi}{2} \frac{A}{P}, \qquad (20)$$

we are led to the second conjecture

$$\frac{\alpha}{A} \geq \frac{\pi}{2} \frac{A}{P}. \qquad (21)$$

Combining both (17) and (21), we propose the two-sided bounds

$$\frac{\pi}{4} \frac{r_{in}^2}{r_{out}} = \frac{\pi}{2} \frac{A}{P} \leq \frac{\alpha}{A} \leq \frac{8}{3\pi} \sqrt{\frac{A}{\pi}} = \frac{8}{3\pi} r_{in} \qquad (22)$$

for the electrostatic polarizability of convex apertures.

Returning to the acoustic aperture problem, we note the following transmission coefficients τ_ϕ and τ_ψ found by Lord Rayleigh for apertures in rigid and flexible screens [11]:

$$\tau_\phi = \frac{\alpha^2 k^4 \cos^2 \theta_i}{24\pi A} \quad \begin{pmatrix} \text{Rigid Screen or} \\ \text{Dirichlet Problem} \end{pmatrix} \qquad (23)$$

$$\tau_\psi = \frac{c^2}{8\pi \epsilon_o^2 A} \quad \begin{pmatrix} \text{Flexible Screen or} \\ \text{Neumann Problem} \end{pmatrix} \qquad (24)$$

Here θ_i is the angle of incidence (see Fig. 3). Applying the inequalities of (5) and (22) to the transmission coefficients above, we find

$$\frac{\pi A^3 k^4 \cos^2 \theta_i}{96 P^2} \leq \tau_\phi \leq \frac{8 A^2 k^4 \cos^2 \theta_i}{27 \pi^4} \qquad (25)$$

$$\frac{8}{\pi^2} \leq \tau_\psi \leq \frac{2P^2}{\pi^3 A} \qquad (26)$$

where k is the wave number of the impinging acoustic wave.

In Fig. 4 we picture the bounds expressed by (25) and (26) in addition to the exact values for the acoustic transmission coefficient of elliptical apertures with varying eccentricity. As observed from these plots, the transmission coefficient is relatively insensitive to eccentricity as long as the axial ratio of the aperture is 2:1 or less.

From the above inequalities and the definition (15) of isoperimetric quotient, it is apparent that acoustic apertures in flexible (rigid) screens with unity IQ possess the largest (smallest) transmission coefficient of any aperture of a given area. The bounds and exact values for the transmission coefficient of apertures of different shape can be found elsewhere [12].

The scattering cross sections σ_ψ and σ_ϕ for acoustic scattering from flexible (Neumann) or rigid (Dirichlet) plates are found from an application of Babinet's principle and inequalities (25)-(26). The results are:

$$\frac{16}{\pi^2} \leq \sigma_\phi \leq \frac{4P^2}{\pi^3 A} \quad \begin{pmatrix} \text{Rigid Plate or} \\ \text{Dirichlet Problem} \end{pmatrix} \tag{27}$$

$$\frac{\pi A^3 k^4 \cos^2 \theta_i}{48P^2} \leq \sigma_\psi \leq \frac{16 A^2 k^4 \cos^2 \theta_i}{27\pi^4} \quad \begin{pmatrix} \text{Flexible Plate or} \\ \text{Neumann Problem} \end{pmatrix} \tag{28}$$

V. **Aperture Transmission and Scattering of Electromagnetic Waves**

Apparently the first electromagnetics problems that have been recently investigated using isoperimetric inequalities are the scattering of waves from electrically small scatterers [13-15] and the transmission of waves through electrically small convex apertures [16-20]. Here we briefly discuss the latter problem.

The geometry under consideration is shown in Fig. 5 for incident plane electromagnetic waves of parallel ($||$) or perpendicular ($\perp$) polarization on a perfectly conducting screen of infinitesimal thickness. Since it is well known that the radiated fields from a small aperture are proportional to the (scalar) electrostatic polarizability α and the (dyadic) magnetostatic polarizability $\underline{\beta}$ [19], one can write the electromagnetic transmission coefficient τ_{EM} for the case where the principle axes of $\underline{\beta}$ correspond to $\hat{e}_{||}$ and $\hat{e}_\perp$ (see Figs. 2 and 5) as

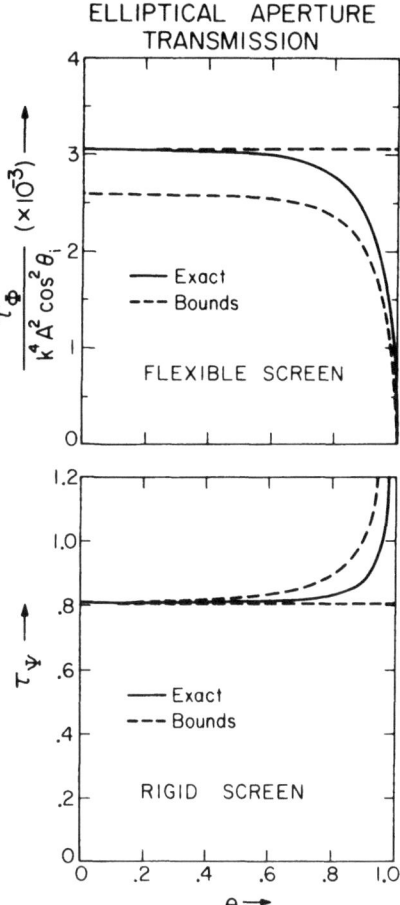

Fig. 4. Transmission coefficients τ_Φ and τ_Ψ for elliptical apertures of eccentricity e in flexible and rigid screens. The aperture area is A, the incident wave number is k, and θ_i is the angle of incidence.

$$\tau_{EM} = \frac{k^4}{12\pi A} \left[\alpha^2 \sin^2 \gamma \begin{pmatrix} 0 \\ 1 \end{pmatrix} \right.$$
$$\left. + (\beta_{11}^2 \sin^2 \chi + \beta_{22}^2 \cos^2 \chi) \begin{pmatrix} \cos^2 \gamma \\ 1 \end{pmatrix} \right] \quad (29)$$

for $\begin{pmatrix} \perp \\ \| \end{pmatrix}$ polarization. Here γ is the angle of incidence and χ is the angle between the incident magnetic field $\underline{H}^{inc}$ and $\hat{e}_\perp$ for $\|$

polarization and is the angle between the incident electric field E^{inc} and $\hat{e}_{||}$ for $\perp$ polarization (see Fig. 5). As before, α is the electrostatic polarizability, while β_{11} and β_{22} denote the diagonal components of the magnetostatic polarizability dyadic $\underline{\beta}$.

Since the quantity α appearing in relation (29) is simply the electrostatic polarizability of the previous section,

VECTOR (ELECTROMAGNETIC) DIFFRACTION

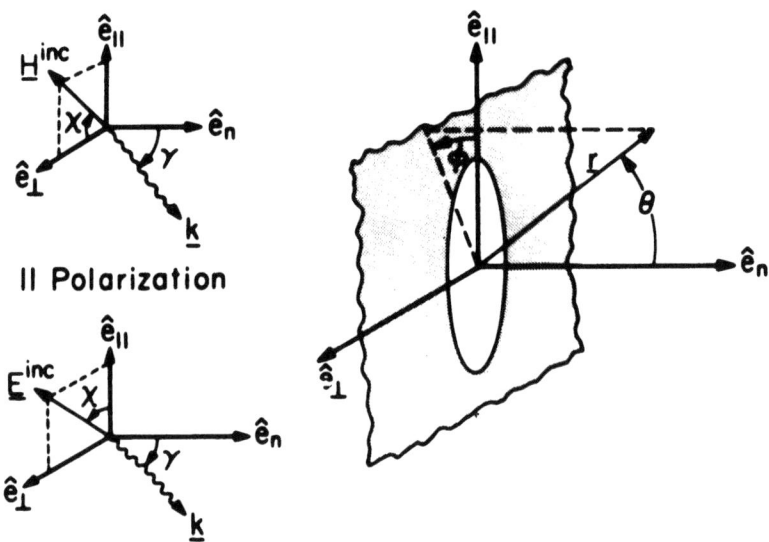

Fig. 5. Configuration for the transmission of electromagnetic waves through an aperture in a perfectly conducting planar screen of infinitesimal thickness. The unit vectors $\hat{e}_{||}$ and $\hat{e}_{\perp}$ lie in the aperture plane while $\hat{e}_n = \hat{e}_{||} \times \hat{e}_{\perp}$. For $||$ polarization $\underline{H}^{inc}$ is parallel to the aperture plane and makes an angle χ with respect to $\hat{e}_{\perp}$. For $\perp$ polarization $\underline{E}^{inc}$ is parallel to the aperture plane and makes an angle χ with respect to $\hat{e}_{||}$. Here γ is the angle of incidence.

we use the inequalities of (22) to bound this quantity. Next we consider bounds on the components of the magnetostatic polarizability dyadic.

By considering the (scalar) expression for the magnetostatic polarizability of a circular aperture of radius a, namely, $\beta = 16a^3/3$, and replacing the radius by the inner and outer radii (3) and (4), we are led to the conjecture

$$\frac{16}{3}\left(\frac{A}{\pi}\right)^{3/2} \leq \frac{\beta_{11} + \beta_{22}}{2} \leq \frac{16}{3}\left(\frac{P}{2\pi}\right)^3 \qquad (30)$$

for the mean magnetostatic polarizability of a convex aperture. Note that the quantity $(\beta_{11} + \beta_{22})$ is the invariant trace of the dyadic $\underline{\beta}$, and so inequality (30) is independent of aperture orientation.

Of particular interest in EMP calculations is the maximum possible transmission through small apertures regardless of polarization. Using (22), (29), (30), and the properties of trigonometric functions, we find that this quantity, denoted by τ_{EM}^{max}, satisfies the inequality

$$\tau_{EM}^{max} \leq \frac{68 \, (P/\lambda)^6}{27\pi^3 \, (A/\lambda^2)} \qquad (31)$$

Here, $\lambda = 2\pi/k$ is the wavelength of the incident radiation. It is immediately apparent that circular apertures with unity IQ possess the smallest electromagnetic **transmission coefficient** of any aperture of a given area. We note that there are significant differences between the acoustic and electromagnetic cases. In particular the wavelength and shape dependence of the transmission coefficients are functions of the type of incident wave (e.g. acoustic or electromagnetic), the boundary condition on the screen, and the type of polarization [12, 19].

As in the previous section, the use of Babinet's principle allows the results of this section to be used for the scattering problem involving electromagnetic waves and thin conducting convex plates of arbitrary shape.

VI. Equivalent Radii of Wires

The concept of the equivalent radius, first introduced by Hallén [22], has been useful in the solution of antenna and scattering problems which involve electrically thin wires of arbitrary cross section. Generally, the solution is first calculated for circular wires of arbitrary radius, and this radius is then replaced by the equivalent radius of a non-circular wire. In this way the original problem is reduced to that of finding the equivalent radius.

From electrostatic considerations Hallén noted that the equivalent radius was exactly equal to the radius of a circular wire whose capacitance was identical to that of the noncircular wire in question. In other words, the original problem, shown in Fig. 6(a), could be replaced by the equivalent problem of Fig. 6(b).

A calculation using the isoperimetric inequalities of Pólya and Szegö shows that the equivalent radius r_{eq} of electrically small wires is bounded by the inner and outer radii r_{in} and r_{out} [23, 24]. Therefore, we are led to the rigorous two-sided bound,

$$r_{in} = \sqrt{\frac{A}{\pi}} \leq r_{eq} \leq \frac{P}{2\pi} = r_{out}, \tag{32}$$

for electrically small wires with convex cross sections.

As an illustrative example, we note that the exact equivalent radius for an ellipse with eccentricity e and area A is (see Fig. 2):

$$r_{eq}^{ellipse} = \frac{1}{2}\sqrt{\frac{A}{\pi}} \left[(1 - e^2)^{1/4} + (1 - e^2)^{-1/4} \right] \tag{33}$$

This quantity takes on the values

$$r_{eq}^{ellipse} \xrightarrow[e \to 0]{} \sqrt{\frac{A}{\pi}} \left[1 + \frac{e^4}{32} \right] \tag{34}$$

and

$$r_{eq}^{ellipse} \xrightarrow[e \to 1]{} \sqrt{\frac{A}{\pi}} \Big/ 2(1 - e^2)^{1/4} \tag{35}$$

for the limits of small and large eccentricities. A comparison of expression (34) with (8) and (9) and (35) with (10) and (11) shows that inequality (32) is satisfied in these limiting cases.

To display the equivalent radius of an ellipse in relation to its bounds for all values of eccentricity, we show Fig. 7. We note that the arithmetic mean of the upper and lower bounds appears to closely approximate the exact value. In fact, this simple method of approximation is, in some cases, more accurate than the approximate but laborious expression given by Uda and Mushiake [25] for the equivalent radius. Similar results occur for wires of rectangular cross section.

It is apparent from the results of this section that

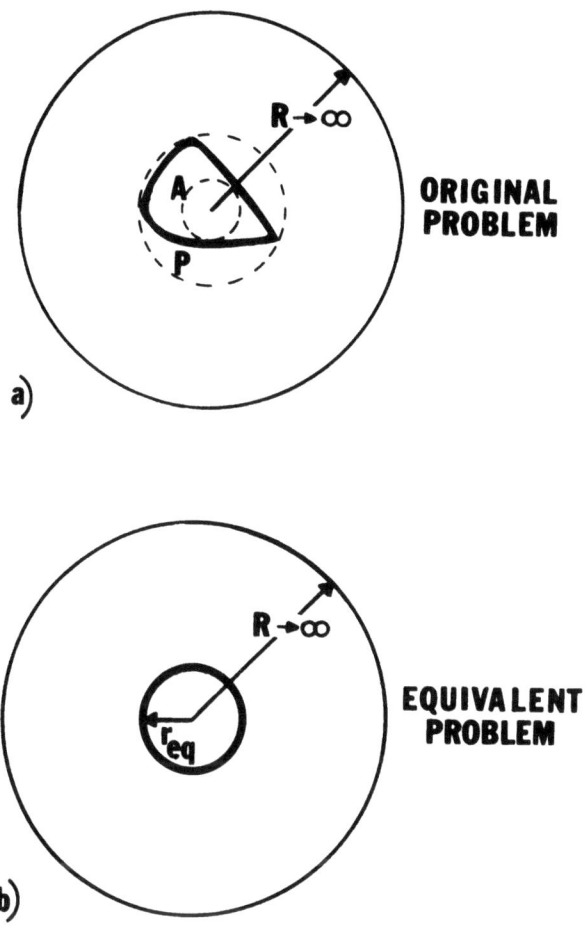

Fig. 6. (a) Cross section of a coaxial line with a center cylindrical conductor of arbitrary shape, area A, and perimeter P, that is used in a capacitance calcultation. (b) Equivalent problem in which the capacitance per unit length is identical to that of the original problem but the inner conductor is circular and of radius r_{eq}. In both cases the outer conductor radius R approaches infinity.

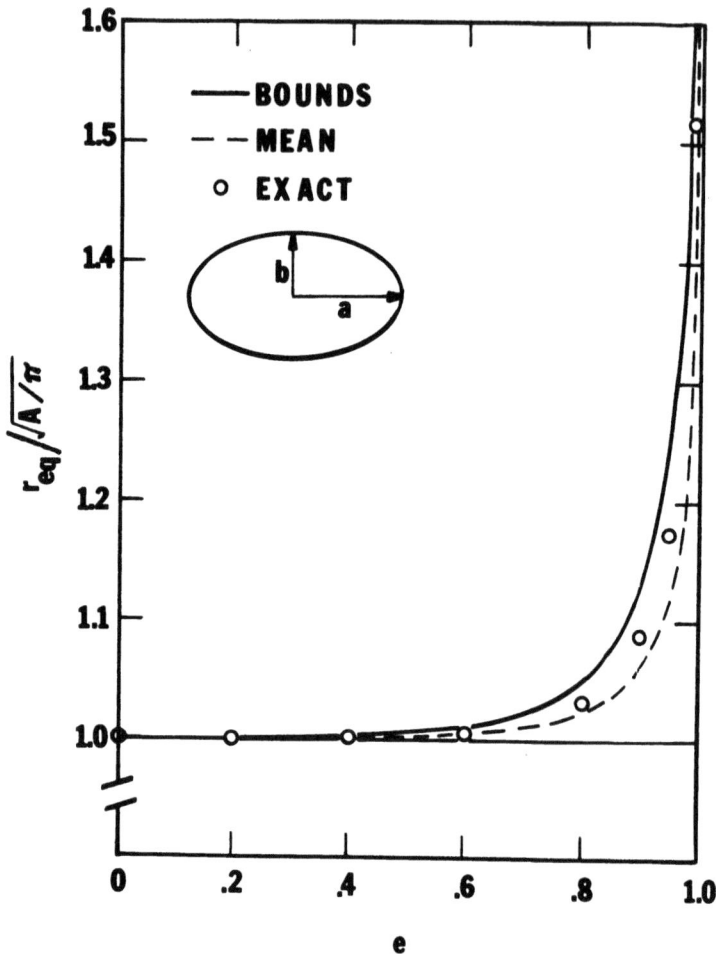

Fig. 7. Equivalent radius for an elliptical figure of area A and varying eccentricity e. Note the excellent agreement between the arithmetic mean of the bounds, denoted by the dashed line, and the exact values, denoted by the circles. The solid lines indicate two-sided bounds.

the equivalent radius of a wire is minimized by maximizing the isoperimetric quotient. The minimum is obtained whenever the wire is circular. These notions can be applied to the scattering of waves from electrically small cylinders or arbitrary shape and in shielding problems involving wire meshes [24].

VII.   Extensions and Conclusions

The bounds for the two-dimensional problems considered in this report were stated for convex figures only. However, in almost all cases, this restriction can be relaxed to include star-shaped figures and might, in fact, be further relaxed to include general nonconvex figures that are simply connected. As a general rule, as one progresses from convex to star-shaped and then to other nonconvex figures, the bounds become more widely spread and therefore become of less use. However, widely spaced bounds may still convey important information if they indicate the connection between changes in the shape of the structure and changes in the physical quantity under consideration.

The extension to three dimensions has occurred for several problems of interest including capacitance [5, 9, 26] and heat conduction [5]. As an example of the latter problem, we note that a three-dimensional isoperimetric quotient can be defined for closed structures by the relation

$$IQ_{3-d} = (36\pi)^{1/3} \frac{V^{2/3}}{A} \leq 1 \tag{36}$$

where A is the surface area and V is the enclosed volume. The equality in (36) holds for a sphere. It is well known that for a given material and volume, the heat loss increases with increasing surface area. Therefore, the most energy efficient design of a structure is one that maximizes the three-dimensional isoperimetric inequality of (36). For example, animals which want to cool themselves will stretch out (low IQ_{3-d}) while animals desiring warmth will curl up into a ball (high IQ_{3-d}). This is pictured in Fig. 8 for cats. In an analogous manner, energy efficient buildings are constructed so that their shape is cubical rather than that of an elongated rectangular parallelopiped. Similar considerations appear in the design of thermal contacts [28].

Next, we note some unsolved problems. First, several of the two-sided bounds presented here are conjectures and therefore need rigorous proofs. Second, the concepts explained in this report are restricted to the low frequency regime but need to be extended. For example, in the high frequency limit (i.e., the physical optics regime), the transmission coefficient of an aperture is simply proportional to the area of the aperture, whereas we have demonstrated here that the shape becomes important in the low frequency limit. The unsolved problem is that of obtaining general solutions, in the resonance region, which demonstrate clearly the connection between the aperture shape and the amount of energy which passes through the aperture.

3-D ISOPERIMETRIC INEQUALITIES

COOL CAT

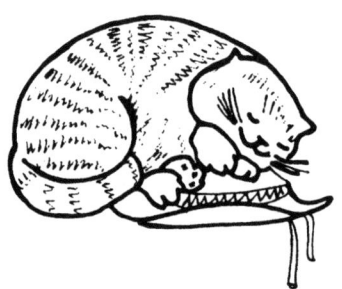

WARM CAT

Fig. 8 Sketch of cool cat (low IQ_{3-d}) and warm cat (high IQ_{3-d}).
(Figure on the top is after [27]).

Finally, we emphasize that the techniques discussed in this report possess two virtues in particular. The bounds we present are in terms of simple quantities which can be understood and measured by those who are not familiar with the techniques involved in the solution of boundary-value problems. In addition, these two-sided bounds often point the way towards optimizing certain parameters of interest by altering the shape of the associated structure.

A variety of additional problems have been attacked using similar concepts to those outlined here. These include problems involving transmission lines [20], waveguides [29], propagation in tunnels [30], and the shielding parameters of enclosures [31].

VIII. References

[1] P. Morse and H. Feshbach, *Methods of Theoretical Physics*, Vols. 1 and II, McGraw-Hill, New York (1953).

[2] F.E. Borgnis and C.H. Papas, *Randwertprobleme der Mikrowellenphysik*, Springer, Berlin (1955).

[3] R.E. Collin, *Field Theory of Guided Waves*, McGraw-Hill, New York (1960).

[4] R. Fikhmanas and P. Fridberg, *Sov. Phys. Doklady*, 14, 1155-1157 (1970); also *Radio Eng. Electron. Phys.*, 18, 824-829 (1973).

[5] G. Pólya and G. Szegö, *Isoperimetric Inequalities in Mathematical Physics*, Princeton University Press, Princeton (1951).

[6] O. Kellogg, *Foundations of Potential Theory* (1929), reprinted by Dover Publishers, New York (1953).

[7] J.C. Maxwell, ed., *The Scientific Papers of the Honorable Henry Cavendish*, Vol. I (1897), reprinted by Cambridge University Press, Cambridge (1921).

[8] Lord Rayleigh, *The Theory of Sound*, Vol. II (1896), reprinted by Dover Publishers, New York (1945).

[9] G. Pólya and G. Szegö, *Amer. J. Math.*, 67, 1-32 (1945).

[10] See, e.g., N. Kazarinoff, *Geometric Inequalities*, New Mathematics Library, The Mathematical Association of America, New York (1961).

[11] Lord Rayleigh, *Phil. Mag.*, XLIII, 259-272 (1879).

[12] For details see D.L. Jaggard, *Appl. Phys.*, 18 (Springer-Verlag), 149-154 (1979).

[13] R. Kleinman and T.B.A. Senior, *Radio Science*, 7, 937-942 (1972).

[14] J.B. Keller, R.E. Kleinman, and T.B.A. Senior, *J. Inst. Maths. Applics.*, 9, 14-22 (1972).

[15] T.B.A. Senior, *Radio Science*, 11, 477-482 (1976).

[16] C.H. Papas, "An Application of Symmetrization to an EMP Shielding Problem", Calif. Inst. Tech. Aut. Lab. Report No. 80, California Institute of Technology (December 1976).

[17] D.L. Jaggard, "Transmission through One or More Small Apertures of Arbitrary Shape", Calif. Inst. Tech. Aut. Lab. Report No. 83, California Institute of Technology (May 1977) and EMP Interaction Note 323.

[18] D.L. Jaggard, "Coupling through Small Apertures", presented at 1977 International IEEE AP-S, USNC/URSI Meeting and URSI International Electromagnetics Symposium, Palo Alto, California (June 20-24, 1977).

[19] For details see D.L. Jaggard and C.H. Papas, *Appl. Phys.*, 15 (Springer-Verlag), 21-25 (1978).

[20] D.L. Jaggard, "Isoperimetric Inequalities in Electromagnetic Theory", presented at the National Radio Science Meeting, Boulder, Colorado (January 9-13, 1978).

[21] H.A. Bethe, *Phys. Rev.*, 60, 163-182 (1942).

[22] E. Hallén, *Nova Acta Regiae Soc. Sci. Upsaliensis*, Ser. IV, Vol. II, 1-44 (1938).

[23] D.L. Jaggard, "An Application of Isoperimetric Inequalities to the Calculation of Equivalent Radii", presented at the National Radio Science Meeting, Boulder, Colorado (November 5-8, 1979).

[24] D.L. Jaggard, to appear in *IEEE Trans. Ant. Prop.*, May 1980.

[25] See S. Uda and Y. Mushiake, *Yagi-Uda Antenna*, Tohoku University, Sendi, Japan (1954), and C. Su and J. German, *Microwave Journal*, 9, 64-67 (1966).

[26] T.H. Shumpert and D.J. Galloway, *IEEE Trans. Ant. Prop.*, AP-25, 284-286 (1977).

[27] Dr. Seuss, *The Cat in the Hat*, Random House, New York (1957).

[28] See, e.g., M.M. Yovanovich, S.S. Burde, and J.C. Thompson, *AIAA Progress in Astronautics and Aeronautics: Thermophysics of Spacecraft and Outer Planet Entry Probes*, 56, ed. by A.M. Smith, New York, 127-140 (1977); also M.M. Yovanovich and S.S. Burde, *AIAA J.*, 15, 1523-1525 (1977).

[29] See, e.g., E. Kornhauser and I. Stakgold, *J. Math. Phys.*, 31, 45-54 (1952); also D.L. Jaggard, "On Using the IQ of Waveguides Intelligently", presented at 1979 International IEEE/AP Symposium and National Radio Science Meeting, Seattle, Washington (June 18-23, 1979).

[30] D.L. Jaggard and M.F. Iskander, "Bounding the Propagation Characteristics of TEM Modes in Tunnels of Arbitrary Shape", presented at 1979 International IEEE/AP Symposium and National Radio Science Meeting, Seattle, Washington (June 18-23, 1979).

[31] K.S.H. Lee and G. Bedrosian, *IEEE Trans. Ant. Prop.*, AP-27 194-198 (1979).

IX.   Acknowledgement.

The author was introduced to isoperimetric inequalities by Professor C.H. Papas (California Institute of Technology, USA). For this introduction and subsequent discussions the author is greatly indebted.

Discussion

<u>Professor R.F. Harrington</u>. Perhaps I should ask if this is written up somewhere?

<u>Professor D.L. Jaggard</u>. Yes, parts have appeared in <u>Applied Physics</u> (Springer-Verlag), Vol. 15, pp. 21-25 (1978), and Vol. 18, pp. 149-154 (1979).

<u>Professor J. van Bladel</u>. Not a question but a little comment which I made yesterday is that we tried all that to compare with our computer results for fun and we found, as we expected, that for a cross, for instance, it's not too good because the upper bound and lower bound are in the ratio of 5 to 1, something like that, as the perimeter is so enormous compared with the area.

<u>Professor D.L. Jaggard</u>. Precisely. The problem there is that as you go from convex to nonconvex planar figures, the perimeter becomes very large compared to the square root of the area and consequently the bounds become more widely spaced. However, these bounds may still show how the quantity of interest changes as the figure becomes elongated. For example, if a plane capacitor has a large perimeter, there are many places for the charges to sit. This means the capacitance will be large. This increase is predicted by the appropriate isoperimetric inequalities.

<u>Professor J. van Bladel</u>. I think the figures that were shown yesterday make us understand why these bounds are rather stationary when you go slightly away from the circle. As you might remember, the circle function called τ_o has a similar behaviour in the circle, and in the square, very much the same.

<u>Professor D.L. Jaggard</u>. Exactly.

<u>Professor C. Butler</u>. In the determination of the equivalent radius of a noncircular cross section, is there any account given for the variation in distribution of current on the surface? Because, for example, if one excites the strip, a narrow strip, with a uniform excitation, a TM strip, then we get the classic distribution which leads to the equivalent radius of 1/4 of the width. On the other hand, if one excites the strip in such a way that it has an odd function excitation, is there no way to get an equivalent radius of 1/4 of the width?

<u>Professor D.L. Jaggard</u>. My answer is no. I say this because Hallén originally defined the equivalent radius of a wire of arbitrary cross section to be the radius of a wire with

circular cross section and equal capacitance. Essentially, this calculation is a quasi-electrostatic or low-frequency one.

Panel Discussion

on

COMPARISON AND EVALUATION OF DIFFERENT FORMULATIONS
FOR SCATTERING PROBLEMS - INTEGRAL EQUATIONS,
DIFFERENTIAL EQUATION AND COMBINED METHODS.

Chairman: Professor R. Mittra   University of Illinois, USA.

Panel: Professor S. Ström   Institute of Theoretical Physics
S-41296 Goteborg, Sweden

Professor P.W. Barber   University of Utah, Salt Lake City, USA.

Professor L. Wilson Pearson   University of Kentucky, Lexington, USA.

Contributors:

Dr E.K. Miller   Lawrence Livermore Laboratory, Livermore CA 94550, USA.

Professor R.H. Harrington   Syracuse University, New York, USA.

Professor P.L. Christiansen   Technical University of Denmark, Lyngby, Denmark.

Professor J. van Bladel   Ghent, Belgium.

Professor C. Butler   University of Mississippi, USA.

Mrs M.L. Calvo   University of Complutense, Madrid, Spain.

Dr T.K. Sarkar   Rochester Institute of Technology, New York, USA.

Dr S. Sensiper   Culver City, USA.

Prof. Mittra - Before I go on, I would like to remind you, the audience, once more that it is really a discussion session and even though the name panel discussion might imply that only the panel members are going to talk, it really should be your show, i.e., that you, the audience, should be actively participating in the discussion and bringing up any questions that you might have. I would also like to caution you about one fact which holds true for many of the techniques discussed here. Of course, you have seen many applications of these techniques and how they apply to various types of geometries. However, I would like to warn you that you cannot always extrapolate these techniques to the problem that you yourself are interested in solving, unless, of course, the character of your own problem is very similar to the ones discussed by the speakers. I will now begin by asking Professor Ström to make a few comments on the extended boundary condition method.

Prof. Ström - I have been asked to present to you in just a couple of minutes the basic features of what is called the extended boundary condition method. It is perhaps not the best possible nomenclature for that method but the name has stuck so we will use it. In order to put it in relation to what has been discussed here so far, I think one should emphasise that this is not an application of the surface integral equation, which has been the main topic here. Most speakers have started from some surface integral formulation of the basic equations, but you can also use a volume integral formulation and use standard Fredholm methods, although this approach has not been discussed here very much. This method is based on a slightly different starting point. You start (I have illustrated here in the scalar case but the structure is essentially the same in the vector case) from the Green's theorem. We have here a scattering problem, and we have a scatterer bounded by the surface as in Fig.1. Then we form the integral relation and we have:

$$\left.\begin{array}{c}\psi(\vec{r})\\ 0\end{array}\right\} = \psi^i(\vec{r}) + \int_S \left[\psi^+ \nabla' G - \nabla' \psi^+ G\right] \hat{n} \, ds' \quad \text{for} \begin{cases}\vec{r} \text{ outside } S\\ \vec{r} \text{ inside } S\end{cases} \quad (1)$$

It gives the total field and it gives zero respectively, depending on whether you are considering a point outside the surface or a point inside. Starting from this integral representation you will have to use it in all of space and make use of both these two lines. The procedure is then that you introduce a specific expansion of the incoming field and of the scattered field:

$$\left.\begin{array}{l}\psi^i(\vec{r}) = \sum_m a_m \, \text{Re}\left(\psi_m(k_o\vec{r})\right) \\ \psi^{sc}(\vec{r}) = \sum_m f_m \, \psi_m(k_o\vec{r})\end{array}\right\} \Rightarrow \qquad (2)$$

$$a_m = -\int_S \left[\psi^+(\vec{r}') \nabla' \psi_m(k_o\vec{r}') - \nabla' \psi^+(\vec{r}') \psi_m(k_o\vec{r}')\right] \hat{n} \, ds' \qquad (3)$$

$$f_m = +\int_S \left[\psi^+(\vec{r}') \nabla' \text{Re} \, \psi_m(k_o\vec{r}') - \nabla' \psi^+(\vec{r}') \text{Re} \, \psi_m(k_o\vec{r}')\right] \hat{n} \, ds' \qquad (4)$$

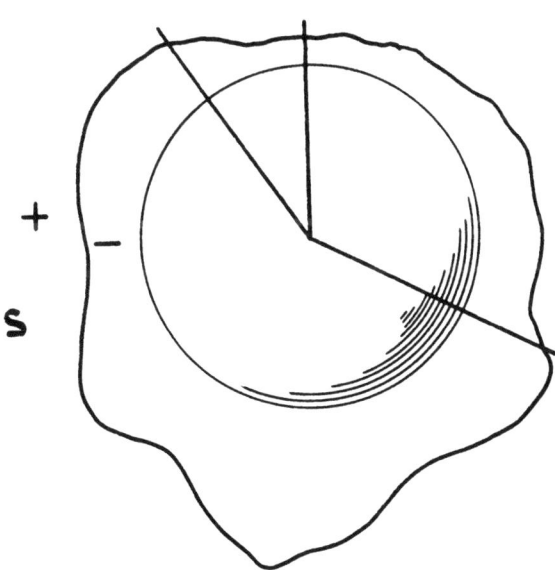

Fig.1.

The scattered field is given by the integral over the surface S, and for the scattered field you usually work with the outgoing spherical waves representation and that is given in this expansion. The expansion coefficient is f_m and ψ_m denotes outgoing spherical waves. Then, for the incoming field, you have some suitable expansion and a reasonable choice for the functions are the regular spherical waves. Now the source of that incoming field could be anywhere outside S, it need not be far away or anything like that. You introduce these expansions in this integral representation and you use the two cases of $\vec{r}$ inside S and outside S. Specifically, when you consider $\vec{r}$ inside S you go inside the inscribed circle and there you have a definite expansion of the Green's function which appears in the integral representation so that you could use the product expansion there, and out comes the first of these relations: the coefficient a_m equal to the surface integral in Eq. (3). By considering a point outside the circumscribed sphere you get the second relation. Now you want to use the boundary conditions to relate the values on the outside (that is ψ^+ and its normal derivative) to the inside values. Next, you want to eliminate these surface fields between equations (3) and (4) in order to solve the scattering problem. Next, we want to introduce some suitable expansion for the field on the inside and its normal derivative and this is usually done (well, almost always done) in terms of global functions. When global functions are chosen, they are usually chosen as the spherical functions and one makes use of their completeness properties. Thus, we have both the field and its normal derivatives and we need one more relation between these two in order to be able to eliminate them. Then one uses exactly the same integral representations for the inner region. If we have an idealised case, i.e. an acoustically soft or hard body, then this step is simplified. Then we can just have one of these quantities and we can expand them right away. Let us now come back to the question of how one chooses these surface functions, the surface field expansions, but let us look at the canonical case of a homogeneous scatterer. Then one can obtain a quantity which is here denoted $T_{nn'}$, which is the transition matrix referring to spherical waves.

$$T_{nn'} = \sum_m Q^1_{nm} [Q^2]^{-1}_{mn'} \qquad (5)$$

where

$$Q^{[1]}_{nm} \sim \int_S \left\{ \begin{bmatrix} \operatorname{Re} \psi_n(k_0 \vec{r}') \\ \psi_n(k_0 \vec{r}') \end{bmatrix} \nabla' \phi_m - \nabla' \begin{bmatrix} \operatorname{Re} \psi_n(k_0 \vec{r}') \\ \psi_n(k_0 \vec{r}') \end{bmatrix} \phi_m \right\} \hat{n} \cdot ds' \qquad (6)$$

It is given as a product between two matrices one of which is inverted. Both these Q matrices are given as integrals over the surface S and these integrals contain the spherical waves, the regular or the outgoing ones, and they contain the expansion system you have chosen for the surface fields on the inside. Everything in here is now explicit and can be computed in a straightforward way. If you want the surface field you stop with one of the intermediate steps in which you just relate the a_m's to the surface fields here and once you have a relation between the field on the inside and its normal derivative you can solve for the surface field also from this relation. This is essentially the scheme that is called the extended boundary condition method and it can be developed in a very similar way for acoustic electromagnetic and also elastic wave scattering, when we have a full three-dimensional vector problem.

Dr. Miller - Why go through these expansions? Why not take instead the usual integral expressions and simply put match points inside the boundary rather than worrying about evaluating sources on a surface? That is in terms of the usual integral equation with principal value part that Leonard Bennett talked about, match boundary conditions inside and compute surface currents. Has anyone compared how that might work for this technique which I believe is quite different?

Prof. Ström - In the scattering problem you never solve for the currents, but just go straight for the transition matrix.

Dr Miller - You understand what I mean?

Prof. Mittra - Yes indeed. You see in the conventional integral equation you have the unknown currents right here on the surface, and you apply the boundary condition also on the surface. Then you have to worry about the singularity and the principal value and so on. Now, an alternative approach will be to argue that anywhere inside of this surface, you could choose another surface, an enclosed surface, and on that surface you could still apply the same boundary condition, namely the scattered field as produced by these currents be negative of the incident field. Now what Dr. Miller is asking is why not leave the unknown current here on the surface and use the conventional expansion; namely, whatever basis function you choose whether it be pulse, triangular or whatever and calculate the fields generated by these currents on this surface and apply the boundary condition on this surface here, as opposed to doing what is done in the method that Prof. Ström described. The scattered field need not be expanded in terms of a set of functions (which for instance in the three dimensional case would be say spherical wave functions). Likewise the incident field expanded in terms

of the same functions such that the boundary condition is again
applied in effect on a surface which is internal to the scatterer.
When you apply the boundary condition and the total field is zero
you are really on an enclosed surface in a sense. So, the question
is why not solve for the currents in a conventional way. I think
the response that you gave was that by doing it in this manner
you will have the scattered field directly when you get it all
done - you do not have to integrate the current again in order to
find the scattered field - which really works out to be useful
when you are doing a multiple scattering type of problem which is
where the T-matrix comes in. In other words, an important aspect
of formulating a problem in this manner is that when you finish
you have actually solved for these coefficients f_m rather than
something that is related to the expansion of the current, and
you have the scattered field. Therefore, if you were to put
another scatterer in here you can let it interact and you can
generate the T-matrix that will allow you to implement the effect
of the interaction with the original scatterer.

Prof. Harrington - I do not think that this is what Dr Miller
asked. I will tell you what I think he said. What Professor
Ström says is fine and what Professor Mittra says is also true,
that the conventional method can be used. So what I understood
Dr. Miller to say is why not apply a bunch of points in here. We
have our formula, we can just point match throughout the volume
and solve it that way. Now how would that work?

Prof. Mittra - Actually you do not need to point match throughout
the volume. You only need to do it on a surface.

Prof. Harrington - I know, but what would happen if you did it in
the volume? Is this a legitimate way to do it? That is the
question, I do not have the answer.

Dr Miller - Well the reason I asked it is because the extended
boundary condition seems to depend on doing the field matching
inside the volume and this always seems to be implemented using
the Waterman approach of spherical wave expansions. I wonder if
anyone has looked at what would happen if we used a more conventional integral equation technique with interior point matching.
I have always been told it is not as well conditioned.

Prof. Mittra - That is true.

Dr Miller - I think the same argument would apply to this with the
same geometry. As I understand this technique it does not work
very well, say if we have a long narrow body.

Prof. Ström - Well, as perhaps you can gather from this short
presentation, it will work in the smoothest and easiest possible

way for something that is nearly a sphere. If we think in terms
of spherical waves for rather long wavelengths, you then have to
investigate how far you can push it and I think from experience it
works very well in the long wavelength region and well up in the
resonance region, but how far you can go beyond the resonance
region will depend on the geometry. Of course the trend is that
if you have a very elongated object you tend to get ill-conditioned
matrices, but there are two ways that have been discussed of
alleviating this; one is a type of regularisation procedure; the
second one has to do with another aspect of the method (which I did
not go into here very much) namely the choice of these functions.
The standard procedure is to choose spherical functions here also,
but there is no necessity of doing that and you could also choose
spheroidal functions; you would choose them in such a way that you
would fill up very much more of the interior of your scatterer and
that tends to improve your numerical analysis considerably. This
point has been discussed in considerable detail elsewhere.

<u>Prof. Mittra</u> - Last month there was a conference in Ohio State University at which both of these points were discussed. First, the choice of expansion function need not be spherical; in fact, there were several other examples given where spheroidal types of functions were used for elongated type of body. Second, to alleviate the problem of ill-conditioning, regularisation procedures have been used successfully and some very good results have been obtained.

<u>Dr Miller</u> - Could a multipole expansion be used as the expansion series here?

<u>Prof. Mittra</u> - That's what this is, but not multipole in the physical sense.

<u>Dr Miller</u> - The usual kind of thing in setting up a dipole, quadrupole, so on and so forth is equivalent to this? The reason I asked is I thought you might use the elongated multipoles rather than spherical multipoles.

<u>Prof. Mittra</u> - Then you have got this spheroidal type of function.

<u>Dr Miller</u> - But those wave functions are complicated to compute.

<u>Prof. Ström</u> - Do you mean multipoles in different positions? That has just been suggested but I do not think it has been tried yet because it sort of goes a little bit against the general scheme and one tries to keep the general scheme fixed but of course in going to elongated objects I think you will have to do some variations of that nature. It was discussed during the conference in Columbus.

Maybe I could make some more general remarks on this choice of expansion functions. In general, it turns out that if you have an ideal conductor you have a much greater freedom, whereas if you have a permeable scatterer vector problem it tends to sort of make these spherical ones more useful and more preferable. Still there would be this conditioning problem but then one could stick to the spherical ones and use other types of regularisation.

Prof. Mittra - Let me make the comment that spheroidal wave functions can in turn be expressed in terms of the spherical wave function and therefore by re-organising the conventional spherical wave expansion you can achieve convergence, so rather than enforcing the boundary condition at isolated points, what you are doing is minimising something using an othogonal basis set of functions. This gives better conditioning than taking an arbitrary set of functions. That is not to say that in general you would not still have the problem of ill-conditioning. You would, if the body shape is largely incompatible with the co-ordinate system associated with the expansion functions.

Prof. Christiansen - In order to ask my question I have to make a change in your drawing (see Fig.2). Could you do any problems

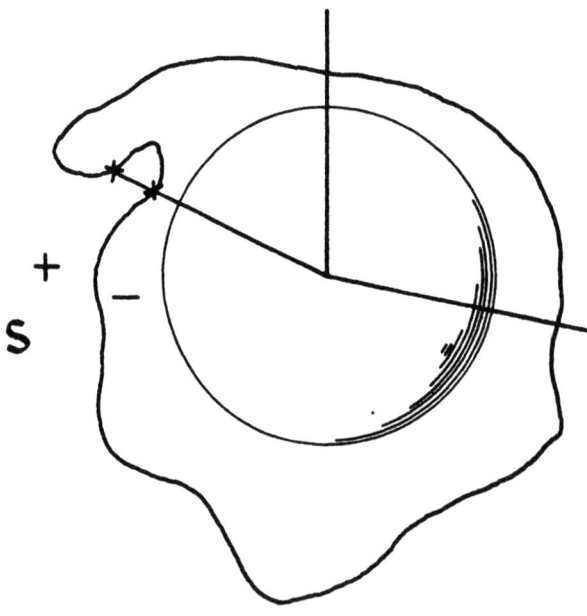

Fig.2.

where - you see here you have a local concavity. I think then you cannot always draw radii vectors like that where the contour only intersects each vector at one point. Have you done any problems where that is as bad as this one for example so that there will be two intersection points by one radius vector, because that would require as far as I can see a modification of the scattered field. You need those incoming and outgoing rays in that region.

Prof. Ström - The difference comes in here exactly on this point. It is a very good question, I was very brief here. The completeness properties of the spherical waves depend on your possibility of choosing an inner point from where you see all of the surface. So if you could find a set of functions, which have the corresponding completeness properties on this and what you would look for of course is the various function systems associated with simple separable coordinate systems or so and none of those I guess can be bent around that way. Thus, it it necessary to impose certain geometrical restrictions in this method.

Prof. Mittra - That is a very good point and I think the reason you probably raised it is because if I were to use a conventional expansion and keep expanding outward until I hit such points on the surface, then obviously I would run into problems because I could never get into this concave region. The point is that as long as it satisfies the condition on a surface such as this, then that means that the field everywhere here is also zero. Now from here making this as base, you can analytically continue the field into anywhere you like so as long as you have not crossed any sources. That is the only criterion and therefore in principle, analytically speaking, as long as this function set is complete as Professor Strom says they are, then you are OK. You may again have difficulties in the numerical sense but that is a separate issue.

Prof. Ström - In this context maybe I could mention in relation to Professor Butler's remark that if you introduce global wave functions you tend to get a dependence on the geometry, that truly we do that here but still for a given set of expansion functions like the spherical ones or spheroidal ones we have a fairly general class for which it works well. People have tried to put in local functions at various stages but they have found that for these particular relations one has here, it does not work very well. David Wall discussed that during the Colombus Conference and he gave explicit conditioning numbers. We do not get too bad a dependence in this respect although we use spherical wave functions.

Dr Miller - If you want to spend more time on this, I have another question. Why does this have any greater benefit for multiple targets than any other integral equation approach where interactions between the targets are via the diagonal parts of the matrix? The off-diagonal blocks, and individual parts of the multiple target can be modelled via their own matrices. Does this somehow have a distinct advantage over conventional integral equation techniques?

Prof. Mittra - What you are saying is if we have two objects, each one will have their sub-matrices and then you have the coupling matrix. So why is this formalism more convenient for dealing with multiple scatterers?

Prof. Ström - Well, I do not know whether one can say that in general that it is more convenient to begin with, but let me just tell you what my experience is. Since in this formulation you have a way of arriving at the transition matrix and that is what really describes the scattering and in a multiple scattering situation that is the only thing you really want to know about the individual scatterers. That means that in this formulation you could, for instance, express the total transition matrix for several scatterers in a multiple scattering problem in terms of the individual ones and that would be a sort of minimised solution in the sense that it only contains the physical quantities, namely the various individual scatterers, distributed in some way. It is of course true that in most other schemes that you use, you can als extract these same quantities. I think the question is how clearly do you see this? I think that perhaps in this context, it would be appropriate to make the general remark that if you want to apply this method to these more complicated situations where there are several scattering surfaces, there is a certain amount of analytic work to be done and you can do that in such a fashion so as to stay with the physically interpretable quantities.

Prof. Mittra - Basically because the scattered fields are directly represented, so it is just more convenient.

Prof. Ström - For the cases I have been involved in, there the results have been fairly transparent in a physical sense. Let me just give a brief list of problems to which this method has been applied (for scattering problems and surface field calculations):

1) Single scatterers

2) Sinusoidal interface

3) Piecewise homogeneous scatterers (several layers, several inclusions)

4) Several scatterers

    a) $N = 2$ (3)

    b)   Statistical distribution of scatterers

5) Buried scatterers
6) Obstacle in a waveguide
7) Antenna problems
8) Determination of resonance frequencies
9) Circular (spherical) physical optics approximation
10) Nonhomogeneous scatterers

I am not saying that in each case this method (EBC) is the best one to use, it is just that this has been done and a lot of development is taking place at this time. It has been said if you really want to do difficult problems and get the best out of a method, you ought to get into all the details and you have to be, in the words of Dr. Bennett, a 'skilled craftsman' in these fields.

<u>Prof. Mittra</u> - Thanks very much Prof. Ström. I would like to ask Peter Barber who has been using both this (EBC) and the finite element techniques to make some remarks.

<u>Prof. Barber</u> - We have been using the extended boundary condition method to solve problems in bio-physics and chemistry, to primarily look at the light scattering characteristics of small particles. So far it has been used primarily for homogeneous and layered models of particles. As Steffan Ström indicated, in the extended boundary condition method, you expand the incident and scattered fields in spherical harmonics, and then relate the fields using the extended boundary condition. The scattered field coefficients are given in terms of the incident field coefficients. The matrix which relates the two sets of coefficients, which is called the T-matrix, contains all the information about the scattering object.

Another type of problem which we would like to solve would be that involving inhomogeneous objects. Probably the most powerful method presently available for solving inhomogeneous scattering and absorption problems is the finite element method, which was developed at the University of California, Berkeley, by Michael Morgan, Steve Chang and Ken Mei. In the finite element method, the fields at node points inside the object, which is inhomogeneous, are related by the finite element differential formulation. However, in the finite element approach, the boundary condition at infinity is not naturally enforced in the formulation. The uni-moment solution to this problem is to encompass the object in a sphere, mesh the entire region inside the sphere, and then couple the interior solution with a spherical harmonic expansion on the surface.

When we consider the types of problems that the different methods can solve, the extended boundary condition method (and the surface integral equation method-of-moments) is restricted to homogeneous objects or at most layered objects, whereas the finite element method can be used for inhomogeneous objects. What we have tried to do is develop a method which will eliminate the disadvantages of each of these methods while retaining their advantages. An obvious disadvantage of the finite element method is the need for all the extra cells between the surface of the object and the enclosing sphere. This generally results in a huge matrix, and run-times on the order of 15 to 20 minutes on a Control Data 7600 are not unusual for problems of 2 to 3 wavelengths in size. On the other hand, looking at the surface approach of the extended boundary condition method, run-times are on the order of seconds for the same size of objects.

What we have done is combine the extended boundary condition method with the finite element method. At some point in the extended boundary condition development we obtain, as Professor Mittra has pointed out, the fundamental relation that the scattered field, which is generated by the surface currents, must cancel the incident field throughout an interior spherical volume. The scattered field is in terms of the surface currents $\bar{n} \times \bar{H}_+$ and $\bar{E}_+ \times \bar{n}$. We use the finite element method inside to generate tangential surface currents with unknown coefficients. We then use these expressions directly in the extended boundary condition and obtain a matrix equation relating the scattered field coefficients to the incident field coefficients. The extended boundary condition method is used only in the outside region where it is optimum. In other words, the extended boundary condition works best for conducting objects and the efficiency goes down as the dielectric constant goes up when working in dielectric regions. By using the extended boundary condition outside and the finite element method inside, the major disadvantages of each of the methods have been eliminated, while retaining their advantages. We have a method which should efficiently solve non-spherical inhomogeneous scattering and absorption problems.

I should make one other comment. On the question of expansion functions, Professor R.H.T. Bates has shown that for objects with a high aspect ratio, either very long objects or very flat objects, using spheroidal expansions rather than spherical expansions will reduce convergence problems that may occur when using the spherical functions. As far as generating the spheroidal expansion functions, Shaji Asano in Japan has spent a lot of time, probably the last five years, developing methods for obtaining these functions accurately on the computer and it does not seem to be a problem. We plan to use the spheroidal expansions for the extended

| FINITE ELEMENT (FEM) | EXTENDED BOUNDARY CONDITION (EBCM) |
| DIFFERENTIAL APPROACH | INTEGRAL APPROACH |

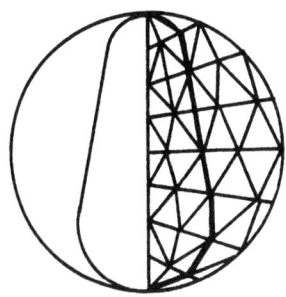

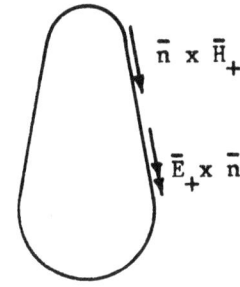

Relate the fields at the nodes and couple to a spherical harmonic expansion at the surface.

$$\bar{E}_i = \sum_{j=1}^{N} [a\bar{M}_j^1 + b\bar{N}_j^1]$$

$$\bar{E}_s = \sum_{j=1}^{N} [f\bar{M}_j^3 + g\bar{N}_j^3]$$

$$\begin{bmatrix} f \\ g \end{bmatrix} = - \begin{bmatrix} T \end{bmatrix} \begin{bmatrix} a \\ b \end{bmatrix}$$

EBCFEM

EBCM outside.

FEM inside.

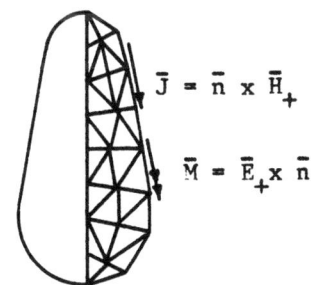

Eliminate the major disadvantages.

Retain the best features.

EBCFEM

An efficient method for nonspherical inhomogeneous objects.

boundary condition method in the outside region, which will permit a solution for very long objects and very flat objects that are inhomogeneous.

Prof. Mittra - I want to ask a question. Let me go back to the figure. Basically in the conventional approach, that you classified as the unimoment method, what we do is fill up this whole space in terms of cells, so the unknowns are all associated with each one of these nodes. Then we come to this outermost surface and we can relate all of the fields here, the unknowns in terms of the tangential fields. Next, we can apply continuity conditions from the inner surface to outer surface and we can go on from there. Now you said that as we use this approach and the body is somewhere here then we have all these cells where we have all these unknowns - so why not become clever and introduce the unknowns only inside and relate these unknowns to what might be the surface currents in general for dielectric scatterers. You then have both the electric and magnetic surface current and the particular advantage is retained that the scatterer could be inhomogeneous. This would not bother you because, for each cell, as long as it is locally homogeneous, you can then just write an equation which is valid for that cell. Now comes the question of applying the next step, namely the extended boundary condition. As you have seen in the case of a conducting scatterer, the extended boundary condition is applied somewhere inside, because the boundary conditions are such that the total field is to be zero inside. Now you say let us forget that because we do not have that kind of boundary condition here, rather we have some kind of a continuity conditions and we are expressing all these things in terms of surface currents. Another type of condition would be on some kind of external sphere, where we could then apply the continuity condition throughout, as was done in the unimoment method on that external sphere. You use the extended boundary condition this time in a different sense than it would be used for the perfectly conducting scatterer.

Prof. Barber - We still use the extended boundary condition method right down to the point where we need an expression for the tangential surface currents. We continue using the extended boundary condition formulation including the inside equivalence problem and generate the spherical expansions which are complete on the surface and then substitute those in with unknown coefficients. What we do now is, rather than continuing with the external spherical expansion, we go to the finite element method and generate the tangential fields from the finite element formulation so that we get away from the requirements on the internal spherical expansion that are inherent in the normal use of the spherical expansion method.

Prof. Mittra - Internally you do not use spherical expansion! Let me ask the question this way: for your extended boundary condition application are you still not using expansion functions? What kind are they?

Prof. Barber - We are using spherical expansion functions.

Prof. Mittra - Outside - the outside type.

Prof. Barber - Yes.

Prof. Mittra - This is where the difference is from the conducting body. In the conducting body case you are coming to a surface which is inscribed by the scatterer and you obviously use the inside type of functions. Anyway your unknowns are here and you relate those to the surface currents and then you use the external boundary condition type of approach. Right, before you go I should like to ask one more question namely, what in your experience is the size of the scatterer which you can handle today with the 7600?

Prof. Barber - Well actually, we have used a bigger machine than the 7600, but there does not seem to be any size limitation. Staffan Ström has pointed out, the real problem seems to be the aspect ratio. If the aspect ratio is too large or the internal characteristics are extreme, such as a very high complex dielectric constant, you may get matrix conditioning problems. However, if you are trying to solve, say a 2:1 Prolate spheroid with a dielectric constant of 2, no difficulties will be encountered.

Prof. Mittra - And so at the moment then, would you say that the size limitation would come primarily from the number of cells that you have to use and that number would of course be dependent upon the dielectric constant because you want to map the fields inside. Also, if the shape was nearly spherical then it would be better because there is no need to use that many cells? In fact, ideally say, if the shape was exactly a spherical one then it does not really matter how large it is and you can still handle it.

Prof. Barber - The matrix becomes diagonal at that point and you end up with conventional Lorenz-Mie theory, but the idea of using the spheroidal functions is that when the object is approaching the sphere these will collapse to the spherical functions and then as you go to either long finite cylinders or flat discs then you will choose the appropriate functions and fill up the interior as much as possible.

Dr Miller - Is there any application for this approach in general to shell-like objects? The problem of a sphere with a hole in it. Does this theory have any place in that problem?

Prof. Barber - I do not see that there would be any problem, it is just a multi-layered configuration. Steffan has worked out a theory for that and we have made numerical calculations for it.

Dr Miller - But the field inside is no longer zero.

Prof. Barber - No, that field inside being zero, that is a big source of confusion. This is a special problem which results because of the application of the equivalence principle. It is only zero inside if you have a perfect conductor and for the dielectric case you create a special problem just as Professor Harrington does in the method of moments - it is the same type of formulation.

Prof. Ström - It is incredible the kind of confusion that is created by this. That zero is independent of anything, any of the properties of the scatterer, completely independent of what type of scatterer you have, as long as you have a conducting scatterer. It is very dangerous to give too much physical significance in the sense that it is just the analytical continuation over a quantity you have on the right-hand side.

Prof. Mittra - What Ed is saying for instance is I have an open structure such as this, a shell.

Dr Miller - A body with a hole, so the field is no longer zero inside?

Prof. Mittra - The field now is zero only on the surface, where the metal is. Then this equation here, which is a statement of the total field being zero, will no longer apply.

Prof. Ström - No, that applies to the inside of a closed surface. I am sorry but my remark is perhaps relevant because I have seen so many confusing statements about some sort of physical significance.

Prof. Mittra - Now that we have got the question clarified, what is the answer? Can somebody solve this kind of shell problem?

Prof. Barber - A dielectric shell?

Prof. Mittra - No, a conducting shell. For the dielectric shell it would work because there is no reason in Peter's example why it would not. After all, his scatterer is inhomogeneous anyhow and he could just as well take a shell such as this and it will work for him, wouldn't it? Next, I have a question. Where would you draw the spheres which will continue the spheres in that example?

Prof. Barber - Well, for this example here, when you apply the extended boundary condition formulation, you'd probably pick the one inside the inscribed sphere.

Prof. Pearson - You are really talking about the analytical continuation of the field.

Prof. Barber - Yes.

Prof. Pearson - I mean this is a method that has been used before.

Prof. Mittra - Not exactly the way he is describing. I do not think that anybody has - has anybody done this?

Prof. Pearson - Not the finite element combination but I mean with a standard case.

Prof. van Bladel - Just a short question. We are speaking about very high dielectric constants of some of your applications. Have you run into trouble with the rather sharp resonances that occur with dielectric bodies that have very high epsilon and where those resonances become very close to each other?

Prof. Barber - Professor van Bladel, I know what you are referring to. We have not looked at those particular problems. We have encountered trouble at lower values than the ones which you are referring to.

Prof. Mittra - You have losses in your modelling?

Prof. Barber - Yes, typically we do have loss so that it might smooth out the resonances.

Prof. Butler - Do you have any difficulty applying the original technique, not necessarily this combination, if you have sources in contact with the surface?

Prof. Barber - Some work has been done for near fields, but as far as I know no work has been done for the source right at the surface.

Prof. Mittra - In other words what you are saying is that suppose one were to put a source right on the surface, like an aperture on some kind of a conducting body.

Prof. Butler - What I am getting to is that if you can do that you can solve an interior problem and an exterior problem in exactly the way you do it and handle the problem with the shell with the hole in it - aperture theory.

Prof. Ström - I do not think that has been done, I am not quite sure - anyway we will be interested in problems of that kind in the elastic case.

Prof. Mittra - Now let us take a case of a perfectly conducting scatterer with a hole, then as we were drawing this picture after Ed raised the question. The boundary conditions are now satisfied only on the surface, the true surface, and you cannot find another surface inside where the total field is zero.

Prof. Butler - No, you solve the interior problem and the exterior problem.

Prof. Mittra - I realise that, but you see suppose this thing was closed and you expand the incident field just like it says there in terms of spherical wave functions or whatever. You do the same thing with the scattered field and then you apply this business here, equate the scattered field to the negative of the incident field which you are doing effectively when you are matching this spherical wave term by term on a spherical surface. Now that is no longer true in this case.

Prof. Butler - If you close that hole entirely and put in an equivalent magnetic current, then in so far as the exterior problem is concerned, the fields inside are zero.

Prof. Mittra - So what you are saying is close the thing but put an magnetic current in there - but you do not know this magnetic current.

Prof. Butler - No, you have to solve for it in the process.

Prof. Mittra - All right, any others questions on this?

Mrs Calvo - The Green's functions are important but you have not talked about what kind of Green's functions you have used in different problems.

Prof. Barber - The reason that we use an inscribed sphere and circumscribed sphere is because of limitations with the Green's function. A spherical harmonic expansion can only be used in certain regions. Analytic continuation allows us to use it in the non-spherical region. The Green's function question is a good point which we have not talked about here.

Prof. Mittra - Let me ask a question to clarify what you were asking. Are you suggesting the use of a different Green's function or a different representation for the Green's function, since they are different things? When he writes this equation he will have the G something like exp(jkr)/r type of thing, i.e., just the free space type of Green's function. Now what you do with that Green's function, meaning how you represent it, whether you do it in terms of spherical waves, spheroidal wave or some other thing that is essentially dependent indirectly on the choice of your basis functions and the role they play in generating the elements of the matrix. But insofar as this Green's function is concerned it is just conventional exp(jkr)/r type of thing. How you expand it is a second step.

Prof. Butler - One more question of a similar nature. If one is interested in an interior boundary value problem, a source inside a dielectric body, can one project inside of a body that has a perfectly conducting enclosure? Is it possible to project onto an exterior sphere and solve an interior problem?

Prof. Barber - I do not see why it would be a problem. I do not know anyone who has looked at it.

Prof. Butler - Unless you can, you cannot solve a cavity problem.

Prof. Barber - There was a paper in the MTT within the past five years that solves for waveguide modes in guides with arbitrary cross sections; by doing a moment solution and enforcing the boundary condition on a circumscribed circle.

Dr. Sarkar - I would like to direct these questions to all members of the panel - yesterday Mr Armour presented some data and he said that with his wiregrid modelling you could obtain accuracy within 3o% of the experimental results. I would like to know from other members of the panel with their surface formulation; - first, have they tried to duplicate any experimental data; second, what type of accuracy they would expect with their methods - will it be better than 30 percent, if so by how much?

Prof. Mittra - That is a very general type of question but is anybody going to comment?

Prof. Harrington - Yes, we check out with experimental data. For example, we took Tom Senior's experimental data for penetration into a sphere with a circular hole in it; we used our standard body of revolution programme for a conductor and we got much closer results than Tom Senior did and I sent him this piece of paper, I think it was five days after I received his report. Of course we checked them with the sphere and that is much closer.

We had good results, there were as many figures as we want - from the Mie solution. We have Dick Mack's excellent measurements for wires and arrays, (small arrays of wires), and we have compared them there and we get excellent results, within except possibly a shift due to input impedance, we get results within one percent maybe and these things have been checked against many many different experimental results and Ken Mei checks his for example. They are highly accurate if you accurately model what you measure. Now Tom Armour said 30 percent, he is not too accurately modelling what he is measuring because he is modelling an aeroplane with a wire grid, I think 30 percent is very good. We got much better results when we accurately model.

Prof. Mittra - Tom could also get the figure much lower than 30 percent if he were modelling a simple structure, but modelling the aeroplane that is another story. Professor Butler, do you have any comments on this?

Prof. Butler - Well in any case that we are able to do so we make simple measurements on configurations that are readily machinable to check our numerical results and sometimes we get good results. I do not remember getting any results as good as 1 percent, but 5 percent is not uncommon.

Prof. Mittra - That is good enough for Government work.

Prof. Butler - 1 percent I am sure is good enough for Government work. We have made measurements on current distributions on bent wires and crossed cylinders, we have made measurements on interior surfaces of cavities (measuring from the outside in), and surfaces on crossed fat cylinders.

Prof. Mittra - I will ask a general question to all of you people. Tom Armour mentioned that he is now handling matrices, 700 x 700 dense matrices. How far can we go? What does the future look like in the immediate near future five years from now in terms of computer power?

Prof. Harrington - I will comment on that, I think we are getting way too big but I know someone who did 2000 x 2000 complex.

Prof. Mittra - Of course you pay.

Prof. Harrington - I would like to just ask a question here. We have got the standard boundary condition of a T-matrix approach and people say it does not work in long thin bodies or short fat bodies, but I do not know why people are pushing it for that because the integral equation method works excellently for those things, for inhomogeneous dielectrics. In fact people are using them, I think someone in Michigan is using that and it works extremely well for shells as Jack Richmond pointed out 15 years ago and would work very well for narrow bodies so I think you ought to really use the method which is natural and not try to push a method which is not natural for this type of problem.

<u>Prof. Barber</u> - Is it not one of the restrictions on the volume approach that you have to use cubical cells?

<u>Prof. Harrington</u> - No, any size or shape cells. In fact I think that a very efficient program could be written for the volume formulation and it is not that bad even for a spherical shape body, but we tend to say that the surface has less elements if you use the surface formulation than a volume formulation, but it is not all that different for the numbers of matrices you would want to handle. I would strongly recommend that that be looked at for thin bodies.

<u>Prof. Mittra</u> - Next, Prof. Pearson. Wilson, you are on.

<u>Prof. Pearson</u> - My purpose is to briefly sketch the mathematical formalism whereby one arrives at the complex resonance description of electromagnetic scattering. These are the resonances of which Dr. Miller has spoken in two earlier presentations.

We begin with a formal statement of a generic electromagnetic integral equation as

$$< \widetilde{\overline{\overline{\Gamma}}}(\bar{r},\bar{r}',s) \; ; \; \widetilde{\overline{J}}(\bar{r}',s) > = \widetilde{\overline{E}}(\bar{r},s)$$

The symmetric product notation for surface integration is used. Viz.

$$< \widetilde{\overline{\overline{\Gamma}}} \; ; \; \widetilde{\overline{J}} > = \int_S \widetilde{\overline{\overline{\Gamma}}}(\bar{r},\bar{r}',s) . \widetilde{\overline{J}}(\bar{r}',s) ds'$$

Here, $\widetilde{\overline{\overline{\Gamma}}}$ denotes the dyadic kernel for either the electric or magnetic integral equation, $\widetilde{\overline{J}}$ the surface current on the scatterer, and $\widetilde{\overline{E}}$ the tangential part of the excitation field appropriate to the integral equation formulation used. The frequency variable is s, the Laplace transform variable. It is determined by the substitution $s \to j\omega$. The tilde over symbols indicates a Laplace transformed quantity. The solution to the integral equation is given in terms of the spectral expansion of the resolvent kernels namely, the kernel $\widetilde{\overline{\overline{\Gamma}}}^{-1}$ such that

$$\widetilde{\overline{J}}(\bar{r},s) = < \widetilde{\overline{\overline{\Gamma}}}^{-1}(\bar{r},\bar{r}',s) \; ; \; \widetilde{\overline{E}}(\bar{r},s) >$$

The resolvent kernel can be expanded as

$$\widetilde{\overline{\overline{\Gamma}}}^{-1}(\bar{r},\bar{r}',s) = \sum_n 1/\lambda_n(s)\, \widetilde{\bar{L}}_n(\bar{r}',s)\, \widetilde{\bar{R}}_n(\bar{r},s)$$

where the eigenvalues and the "left-hand" and "right-hand" eigenvectors are given by

$$<\widetilde{\bar{L}}_n(\bar{r},s)\,;\,\widetilde{\overline{\overline{\Gamma}}}(\bar{r},\bar{r}',s)> = \lambda_n(s)\, \widetilde{\bar{L}}_n(\bar{r},s)$$

and

$$<\widetilde{\overline{\overline{\Gamma}}}(\bar{r},\bar{r}',s)\,;\,\widetilde{\bar{R}}_n(\bar{r},s)> = \lambda_n(s)\, \widetilde{\bar{R}}(\bar{r},s)$$

It can be shown that the $1/\lambda_n(s)$ factors are analytic in the right half of the complex plane.

The formal representation of the integral equation solution follows as

$$\widetilde{\bar{J}}(\bar{r},s) = \sum_n 1/\lambda_n(s)\, <\widetilde{\bar{L}}_n(\bar{r},s)\,;\,\widetilde{\bar{E}}(\bar{r},s)>\, \widetilde{\bar{R}}_n(\bar{r},s)$$

It can be shown that $\widetilde{\bar{J}}(r,s)$ is 1) analytic in the right half of the s-plane and (2) on a finite extend body situation in a lossless medium, $\widetilde{\bar{J}}$ possesses only pole singularities.

The pole singularities $\{s_{ni}\}$ are the values of s for which some eigenvalue vanishes. Viz

$$\lambda_n(s_{ni}) = 0$$

A residue expansion of the reciprocals of the eigenvalues yields

$$1/\lambda_n(s) = \sum_i \frac{[d\lambda_n(s_{ni})/ds]^{-1}}{s-s_{ni}} + \widetilde{b}_n(s)$$

where $\widetilde{b}_n(s)$ represents a branch integral contribution to the expansion. The branch integral terms are elusive and are associated with branch points at values of s where eigenvalues degenerate.

The singularity expansion representation of the current follows from inserting the residue expansion for the reciprocals of the eigenvalues into the eigenfunction expansion for $\tilde{\bar{J}}(\bar{r},s)$:

$$\tilde{\bar{J}}(\bar{r},s) = \sum_n \sum_i \beta_{ni} < \tilde{\bar{L}}_n(\bar{r},s_{ni}); \tilde{\bar{E}}(\bar{r},s_{ni}) > \frac{\tilde{\bar{R}}_n(\bar{r},s_{ni})}{s - s_{ni}}$$

We deduce that the $\tilde{b}_n(s)$ terms sum to zero since J has only pole singularities. The singularity expansion method terminology calls the $\{s_{ni}\}$ poles

$$\bar{C}_{ni}(\bar{r}) = \tilde{\bar{L}}_n(\bar{r},s_{ni}) \qquad \text{, coupling vectors}$$

$$\bar{J}_{ni}(\bar{r}) = \tilde{\bar{R}}_n(\bar{r},s_{ni}) \qquad \text{, natural modes, and}$$

$$\beta_{ni} = \left[\frac{d\lambda(s_{ni})}{ds}\right]^{-1} \qquad \text{, normalisation constants}$$

The factor $< \tilde{\bar{L}}_n(\bar{r},\bar{r}',s_{ni}); \tilde{\bar{E}}(\bar{r},s_{ni}) >$ may be viewed as one of the non-unique forms termed "coupling coefficients".

The time domain representation for $\bar{J}(\bar{r},t)$ results from inverse transforming the singularity expansion.

$$\bar{J}(\bar{r},t) = \sum_n \sum_i \beta_{ni} \eta_{ni}(t) \tilde{f}(s_{ni}) \bar{J}_{ni}(\bar{r}) e^{s_{ni}t}$$

Here, we have factored the time history of the incident wave as $\tilde{f}(s_{ni})$. The $\eta_{ni}(t)$ represents a generic coupling coefficient and is, in general, time varying during the forced portion of the object's response.

The pole identification from transient data of which Dr Miller has spoken is evidently the s_{ni} in the representation above.

The eigenmode associations of the poles are indicated qualitatively for a straight wire scatterer on the accompanying figure. All of the poles associated with a given eigenvalue lie along an arc on the figure. Qualitative sketches of some of the modal currents are given, as well. The SEM description may be determined by at least three means:

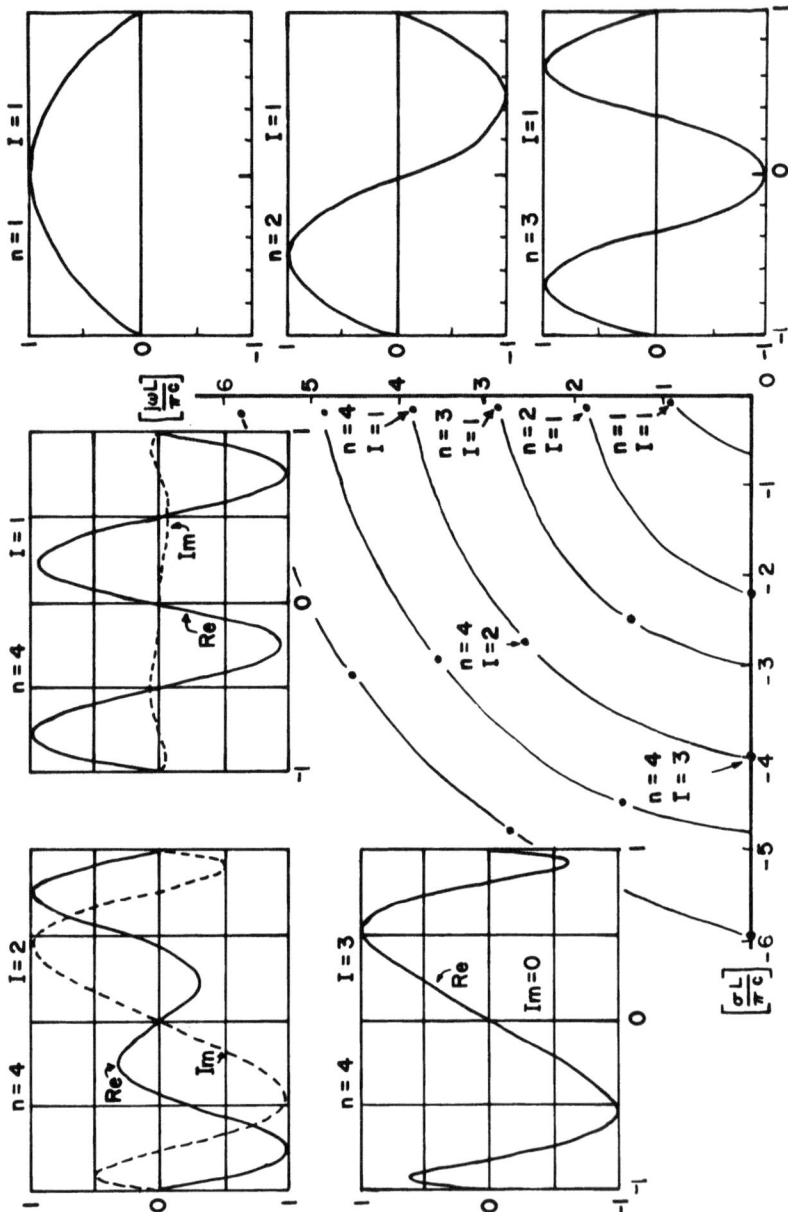

Eigenset associations of poles of a thin-wire scatterer with representative natural modes. All of the modes associated with the fourth eigenset are shown.

1. By boundary value analysis - strict or approximate - e,g, sphere, wire loop;
2. Searching numerically for the zeros of the determinant of the moment matrix, e.g. straight wires, bent wire, crossed wires, rectangular plate, prolate spheroid; or
3. Measurement of the transient response of the object at many spatial points and using pole extraction processing to determine the SEM quantities.

We are presently obtaining initial results for the third approach. It should prove useful for describing objects whose shape is too complex for computer modelling.

<u>Prof. Harrington</u> - A question. Why don't you assign anything with that pole on the negative real axis?

<u>Prof. Pearson</u> - I simply didn't choose to display the mode here. I picked one example of an imbedded pole here; it would have essential features very similar to this, it would be an anti-symmetric mode.

<u>Prof. Harrington</u> - That cannot be so; it is only one pole right in the middle.

<u>Prof. Pearson</u> - It would contain no imaginary part. It would have to be pure real because the modes occur in conjugate pairs.

<u>Prof. Harrington</u> - Isn't that the static case when ω is zero?

<u>Prof. Pearson</u> - It would be akin to static case but it is damped.

<u>Prof. Mittra</u> - It is critically damped and not the static case.

<u>Prof. Pearson</u> - The ways that these modes are obtained for separable problems, or problems which can be approximated in a separable fashion, analytic results have been obtained in boundary value solutions. This has been carried out already for a wire loop. In the moment method formulation we can use the techniques shown here earlier this week; extending the complex frequency by simply walking about the complex plane and looking for those values of s where the determinant of the moment matrix goes to zero and then solve with the corresponding homogeneous solutions to the matrix equation in order to get the natural modes. As you might guess this approach is computationally expensive. It has been carried out for several fairly simple structures. In a method that we are working on developing now involving measurements of waveforms at many points on the object followed by pole extraction, where we put probes over the object at many locations to probe the surface currents, we will get a collection of waveforms

containing special information about the object. From these waveforms through a pole identifier we obtain poles and residues and then by appropriate sorting of the residues we can arrange them into the mode description. This will give fairly crude results but can be particularly useful in things like EMP testing in that if we know simply the features of the modes on the structure we can narrow the data window over which full-scale test of an aircraft has to be carried out so that we can predict a range of angles in which we would anticipate worst case situations.

Prof. Mittra - Are there any questions for Wilson?

Prof. Harrington - Before you take the transform there are more than just simple poles. There is a coefficient that has some terms depending on s; that will give you some time dependent coefficients.

Mr Pearson - Yes. Actually, the time dependent coefficient has to do with the time which was permitted to close on the left the Bromwich contour to actually obtain residue terms.

Prof. Mittra - That is what this allows you to express, everything in terms of damped exponentials in that time window, beyond that this expansion is good.

Prof. Pearson - And it turns out that there are a range of values over which the left hand closure, well close right and we always get zero, close left and your attempts at which we can close left to get zero and so there is...

Prof. Harrington - Do you always have a zero contribution to the branch?

Prof. Pearson - Yes, the branch is not there in this series. The branch arises from a degeneracy in eigenvalues. So that arbitrariness in coupling coefficient comes in choosing a time in this ambiguous region.

Mr Sensiper - Is this written up here?

Prof. Pearson - The coupling coefficient representation is being written up right now; a paper with Professor Wilton, University of Mississippi.

Prof. Mittra - I think time is getting late so it is time to wrap up here. You have been a very good audience. Thank you all - the speakers and the audience.

Panel Discussion

on

AVAILABILITY AND LIMITATIONS OF
GENERAL PURPOSE COMPUTER PROGRAMS

Chairman: Dr E.K. Miller   Lawrence Livermore Laboratory CA 94550, USA.

Panel:
- Professor L.W. Pearson — University of Kentucky, Lexington, USA.
- Dr C.L. Bennett — Sperry Rand Research Centre, Sudbury, Massachusetts, USA.
- Dr P.W. Barber — University of Utah, Salt Lake City, USA.
- Professor R.F. Harrington — Syracuse University, New York USA.
- Mr T.W. Armour — A.W.R.E., Aldermaston, England.
- Professor C. Butler — University of Mississippi, USA.
- Mr J.I.R. Owen — Royal Aircraft Establishment Farnborough, England.
- Dr P.A. Ramsdale — R.M.C.S., Shrivenham, England.

Contributors:
- Dr Y. Rahmat-Samii — Jet Propulsion Laboratory, Pasadena, USA.
- Professor R. Mittra — University of Illinois, Urbana, USA.
- Dr N.A. Adatia — ERA Ltd., Leatherhead, England.
- Dr R.J. Chignell — ERA Ltd., Leatherhead, England.
- Mr J.E. Summers — Signals & Radar Establishment, Malvern, England.
- Dr T.K. Sarkar — Rochester Institute of Technology, New York, USA.
- Mr D.J. Brain — ESTEC, Noordwijk, The Netherlands.
- Mr W.L. Jones — R.M.C.S. Shrivenham, England.

Contribution on the solution of operator equations by:

Dr T.K. Sarkar   Rochester Institute of Technology, New York, USA.

<u>Dr E.K. Miller</u> - We will be discussing a topic which I think is very germane to the presentations given this far at this Institute: It is how we can obtain computer codes that we have heard can do all those wonderful things and perhaps realistically, what are the limitations. We hear often about the things that they do well and people do not tell us very readily or voluntarily of the things that they can't do. This morning we hope to uncover a few of those aspects of calculations and how these codes can be expected to perform best. Here are the people who have agreed to make a few remarks at this panel discussion. I have divided this into two parts:-

Professor L.W. Pearson	Mr T.W. Armour
Dr C.L. Bennett	Mr J.I.R. Owen
Professor C. Butler	Dr P.A. Ramsdale
Dr P.W. Barber	
Professor R.F. Harrington	

Those on the left side are the individuals or representatives of organisations which I would consider more to be developers of computer code technology. Lamentably perhaps, we don't have that many people who are actively engaged in applying these codes to real life problems. We do have however, three representatives from the United Kingdom here to tell us about some of the nasty kinds of problems that can be encountered in attempting to handle real-life situations. That does not mean of course that those on the left don't deal with these problems. They can actually tell us more in a very concise and summary fashion where the state of the art is and what might be expected over the next few years by way of additional development. Then the people on the right will discuss some of their current applications and concerns especially as relates to the limits of computer code modelling.

We will begin with Professor Wilson Pearson who will discuss primarily the difficulties of the attachment wires to plates.

<u>Professor L.W. Pearson</u> - The first thing I would like to bring up is basically the techniques and codes available for surface solution methods which we can apply to either plate-type or planar-aperture-type problems which Professor Butler will discuss later in a little more detail. To my knowledge, none of these codes are available in a well documented fashion intended for distribution, but I believe that by approaching the appropriate authors they would be available in some useable form. The first techniques which were applied to a plate structure used the integrated electric field integral equation. This is essentially the counterpart of the Hallen equation for a thin wire re-done, for a plate

or aperture. The work was done by Dr Rahmat Samii and Professor Mittra (Ref. [1] and [2]). Their approach to integrating the electric field integral equation introduced a Bessel function expansion for the homogeneous solution to the polar coordinate form of the harmonic equation. Slightly later Wilton and Dunaway (Ref. [3]) also approached the plate problem using an integrated electric field integral equation. Their expansion for the homogeneous part of the solution was in terms of a line source over the contour boundary. The codings of these two approaches are available respectively in a rectangular plate form and in an arbitrarily shaped plate form within the limits of a polygon approximation of a plate in the latter case. I would point out that some question has been raised about the completeness of the Bessel function expansion in the formal solution. The reason being that it results from truncating a spectral integral to eliminate the invisible part of the spectrum, whereas within the Wilton and Dunaway approach the line source expansion is complete; the line source is just discretised and solved self-consistently as part of the unknowns for the plate. It is my own observation today that these Hallen-type equations are unduly cumbersome and that one is better off to go to one of the recently developed finite difference formulations. Of those formulations there are again basically two that I am well aware of. One is the Wilton, Glisson and Butler formulation (Ref. [4]) which uses

---

[1] Y. Rahmat-Samii and R. Mittra, "Integral Equation Solution and RCS Computation of a Thin Rectangular Plate,' IEEE Trans. Ant. and Prop., v. AP-22, no. 4, July 1974, pp. 608-610.

[2] Y. Rahmat-Samii and R. Mittra, "Electromagnetic Coupling through Small Apertures in a Conducting Screen," IEEE Trans. Ant. and Prop., v. AP-25, no. 2, March 1977, pp. 180-187.

[3] O.C. Dunaway and D.R. Wilton, "A Numerical Solution for the Distribution of Time Harmonic Electromagnetic Fields in an Arbitrarily Shaped Aperture in a Ground Screen," EMP Interaction Note 214, Air Force Weapons Laboratory, Kirtland AFB, New Mexico.

[4] D.R. Wilton, A.W. Glisson and C.M. Butler, "Numerical Solutions for Scattering by Rectangular Bent Plate Structures," Final Report, Contract No. N00123-75-C-1372, Naval Electronics Laboratory (now Naval Ocean Systems Center), San Diego, California, October 1976. (Adaptable to wire attachment).

staggered zoning on the plates and finite differencing scheme to arrive at the derivative terms in the electric field integral equation. We shall see shortly what the staggered zoning implies. A code exists for this problem, but I do not know whether it is readily available.

Professor C. Butler - Such codes could be distributed easily but they are not designed for this as they are not user-oriented and would need a great deal of additional documentation before other people are able to use them.

Professor Pearson - They have coded their scheme for rectangular plates and for bent plate structures. The work on triangular patching of closed bodies is an extension of this zoning scheme that Dr Miller referred to in his lecture. Then a second method by Dr Rahmat-Samii, Professor Mittra and Dr Parhami (Ref. [5]) uses a simpler zoning scheme and finite differencing and their work has been coded for the rectangular plate problem with a wire attached normally. I note in passing that Pearson and Mittra (Ref. [6]) have derived an SEM* data to include the resonances through to the second resonance of rectangular apertures with aspect ratios from 0.1 to 1 and compounded into an SEM data base. I would be inclined to sub-title this data base: "all you wanted to know about the aperture but were afraid to ask" because the application of that data base would be almost as formidable a problem for the C.W. case as actually applying one of these codes. I cannot go into details but basically Wilton and his

---

[5] P. Parhami, Y. Rahmat-Samii and R. Mittra, "Technique for Calculating the Radiation and Scattering Characteristics of Antennas Mounted on Finite Ground Plane," Proc. IEE (London), v. 124, no. 11 November 1977, pp. 1009-1016.

[6] L.W. Pearson and R. Mittra, "The Singularity Expansion Representation of the Transient Coupling through a Rectangular Aperture," EMP Interaction Note 296, Air Force Weapons Laboratory, Kirtland AFB, New Mexico.

Data base available from L.W. Pearson, Department of Electrical Engineering, University of Kentucky, Lexington, Kentucky 40506.

---

* SEM - Singularity Expansion Method

AFWL-TR-74-192, January 1975.

colleagues stagger the zones for vertical (full line in Fig.1) patches versus the horizontal (dashed line) patches and then

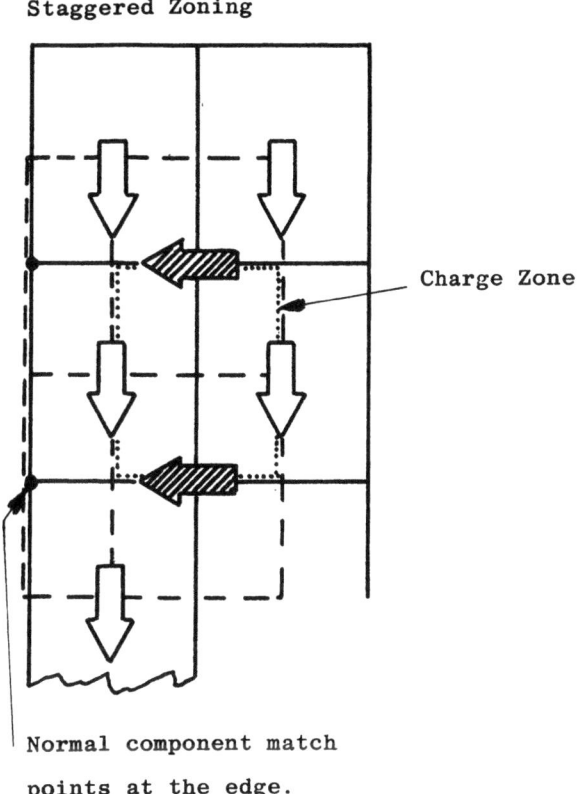

Staggered Zoning

Charge Zone

Normal component match points at the edge.

Fig. 1

explicitly represent the scalar potential and charge contributions in the integral equation, while the charge zones are staggered with respect to both of these. The nice thing about this is the charge at the centre of a charge zone (dotted line) defined by a first order central difference about the charge zone. It also gives the freedom of placing half zones at the edge for components normal to the edge. This gives you the prerogative of placing the edge, which does a better job for match points at representing the true extent of the plate. The simple zoning scheme terminology implies simply that the components of both currents are represented by the same set of patches (Fig. 2). The case for match points

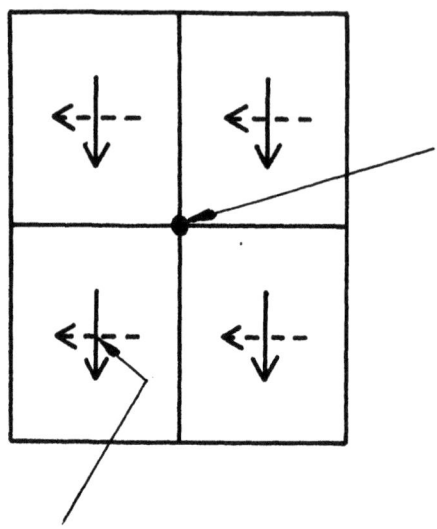

Fig. 2

right at the edge is perhaps more satisfying. However, if you consider the problem of wire attachment, then the trade-off is given in the other direction. The staggered zoning scheme allows you to place a wire at the intersection of two zone boundaries and so there are basically four components of current on the surface which can be matched to a wire current by explicit enforcement of Kirchhoff's current law. This is a four degree of freedom for attachment. The so called zoning scheme places a wire at the corner common to four zones (Fig. 3), each zone containing two components of currents and there are eight degrees of freedom available. In the implementation of this scheme, Kirchhoff's current law is not explicitly enforced. An electrically complete potential formulation is used so that Kirchhoff's current law is satisfied implicitly through the field equations.

I would like now to make just a couple of comments as a user of codes in some of our experimental work in SEM extraction work. We have been heavy users of other peoples programs and I would like to state a cardinal principle of program borrowing. If I may use an American colloquialism "there aint no free lunch", it

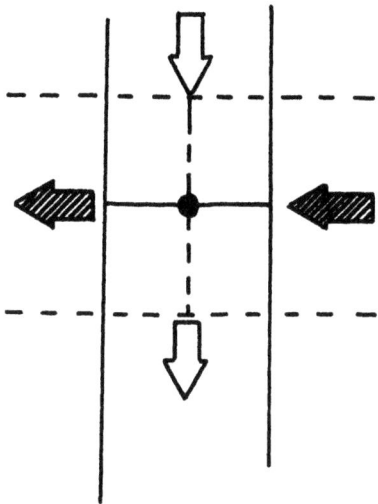

Wire attachment

4 degrees of freedom
Kirchoff's Law explicitly
enforced

Fig. 3

is very seldom that you can borrow someone's code to extend it to a new problem, or to a new class of problems, without at some point having to find the details in that coding, both at the level of understanding formulation methodology and perhaps even the algorithmics in the code. For instance, the algorithmics might matter, if you simply want to expand the size of the code to a larger problem. For example it may require a better Bessel function routine for your extended implementation. I have a brief example of that. A student brought me these results in the past few days. Actually, I have brought them here to discuss the results privately, but they make a point here quite nicely. The solid line in this (Fig. 4) represents an electric field which we have measured experimentally and which excites a cylindrical scatterer on which we are measuring currents. The dashed line represents a Gaussian approximation to the experimental waveform that we are using. We were trying to replicate our experimental currents using the thin wire time domain program available from Lawrence Livermore Laboratory. Let me show you comparative results for these two excitations. This is (Fig. 5a) the idealised source case and a description of the experimental configuration (Fig. 5b). The wave shape is the time history of the

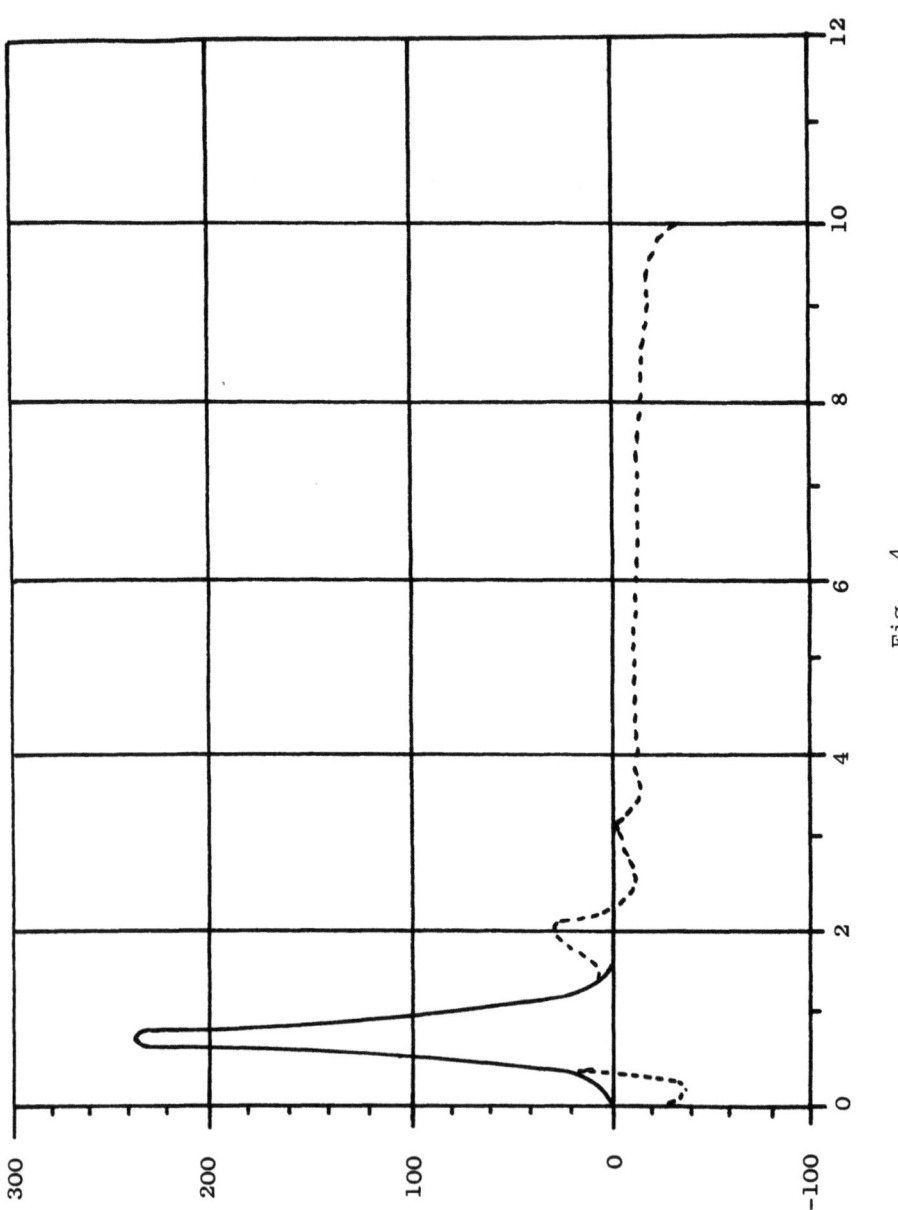

Fig. 4

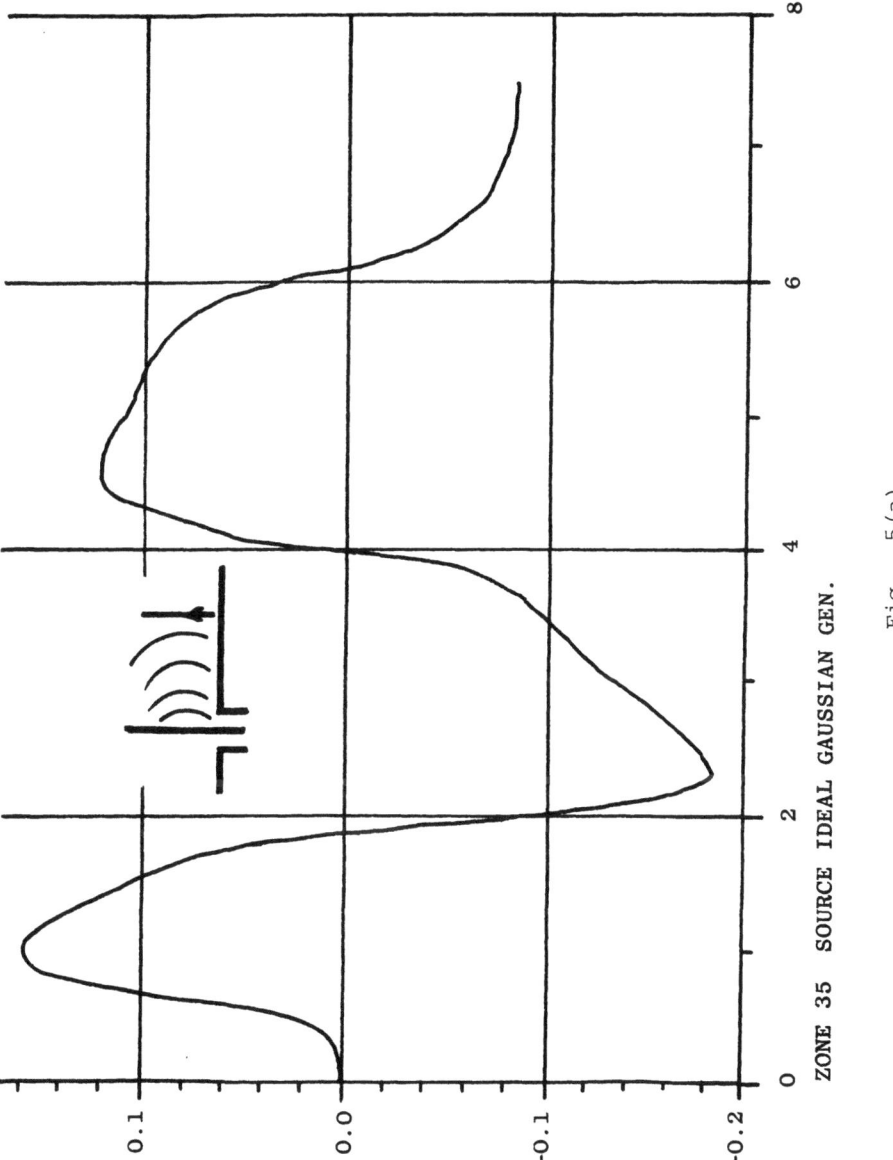

Fig. 5(a)

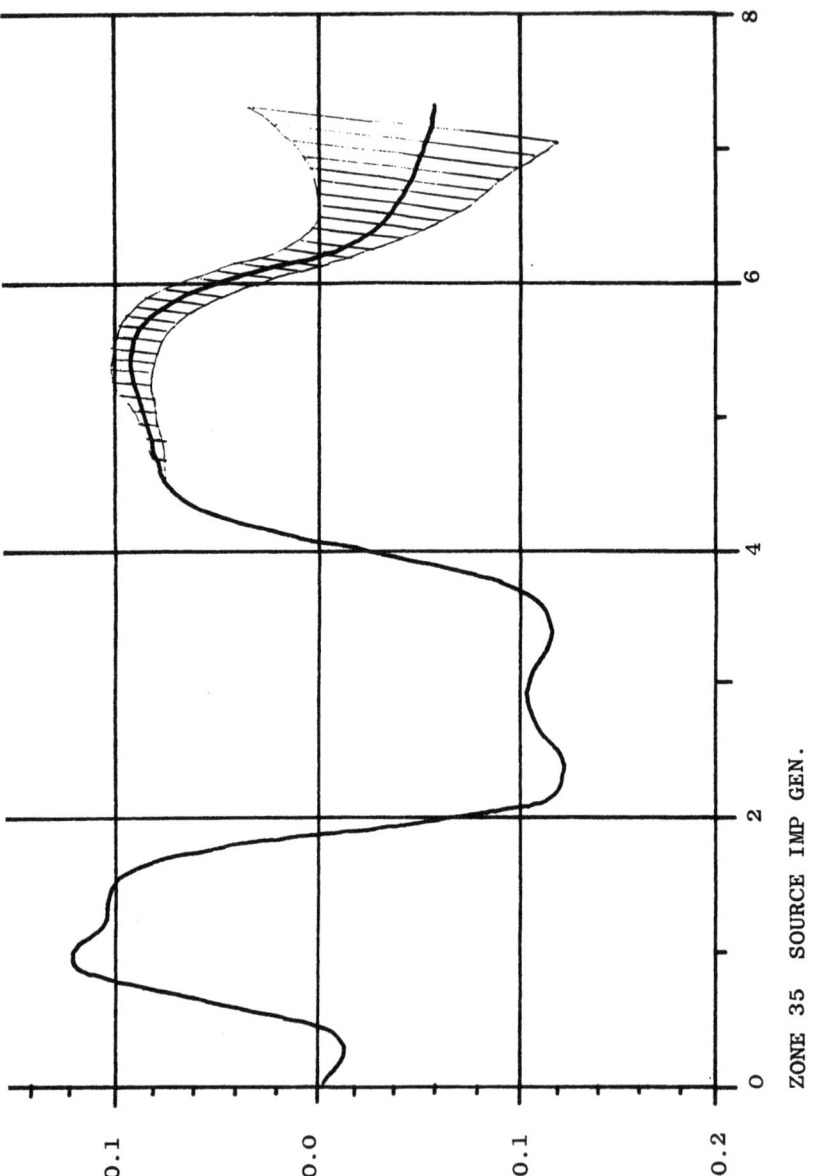

Fig. 5(b)

spherical wave emanating from the long wire radiator and so we have got quite a nice solution. The time scale is in nano-seconds and the electrical length of the scatterer is one nano-second long here. We have quite a nice solution over an 8 nano-second period for the idealised generator. Now the issue here is not the differences in the time histories. We understand then through the wave shape differences, it is in using the experimental data that some of the instabilities alluded to earlier (Dr E.K. Miller Time Domain Solutions via Integral Equations, in this volume) had a very rapid onset. The experimental waveform represents something like 16 to 32 signal averages of a measured response and so it is fairly smooth yet being experimental it contains noise and it is likely that through the finite differencing involved in the TWTD code the presence of noise in the experimental data brought about the early onset of the instability in the time domain code. Thus either we have to be willing to accept just 4 nano-seconds worth of data here or think about a way to patch the code to embrace the new situation.

Dr Miller - Thank you very much Wilson, are there any questions? Well, I think in summary then the number of existing computer codes that have been developed for plates is limited to perhaps 4 or so, none of which have been fully documented as at this date. Is that correct?

Dr Y. Rahmat-Samii - One is documented in a report.

Dr Miller - Which one was that?

Dr Rahmat-Samii - The last one. The Finite Differencing scheme. The paper appeared in '76 IEE and the report is available from the University of Illinois - I hope.

Dr Miller - But he is talking about the computer code. So if they want to obtain a copy of that report, who would they contact, Professor Mittra?

Professor R. Mittra - There may be some people here interested in obtaining a copy of those codes; at least one is available.

Dr Rahmat-Samii - I think the comment was made earlier that they aren't user oriented, the codes may be available but the documentation is limited or non existent or comment cards are not included throughout. The standards are very variable depending on who the sponsor was and how much money was available and who was doing the work.

Professor C. Butler - In the same sense the work of Glisson and Wilton is available in a US Navy report, which Professor Pearson referred to, the title of which I don't remember, the authors are Wilton, Glisson and way down low is the name of Butler and if you write to me and I can get it in the hands of Glisson and if he is in the right mood he will make a duplicate card deck. But I do want to emphasise that it's the sort of program that I think not everyone will be successful in trying to use because its not intended to be user-oriented.

Dr N. A. Adatia - I think it would be useful if the program operation actually indicated what is outputted from the program. Whether it only addresses the radiation problem, or works out input impedances or what.

Dr Miller - Here at this discussion you mean.

Dr Adatia - That is right, yes.

Professor Mittra - Wilson, were any of these programs intended for antenna radiation pattern calculations?

Professor Pearson - To my knowledge all the outputs are limited to currents. But they have been tested in the kind of near field calculation involving antenna radiation as opposed to scattering for example.

Dr Miller - Right, we did do some input impedance computation but it wasn't that successful, whereas for the scattering problem or far fields it appeared to be working well.

Yes one more question.

Dr R.J. Chignell - We have in fact implemented the code that Rahmat-Samii is talking about and in fact we use the accurate current as input data to a variational technique and we get a much better input impedance in that way, by going through the additional step of the variational calculations.

Dr. Miller - Thank you. I must mention that we at Lawrence Livermore Laboratory, for three or four years while our sponsors permitted us to do so, published a Newsletter. There are some twenty or so computer codes that have been documented fairly thoroughly and made available through our Laboratory to recipients of this Newsletter. Unfortunately the rules of the game changed during the course of this work and one of the effects was that we couldn't really send computer codes anywhere in the world as we had once done, so not all of the codes are available to anyone who might request them.

<u>Dr E.K. Miller</u> - The next speaker will be Dr Bennett of Sperry Research Center who will summarize the state of the art of time-domain computations.

<u>Dr C.L. Bennett</u> - Time Domain Integral Equations have been developed and solved for the target classes shown in Figure 6. Wire

Smooth Closed Bodies - Rotational, Plane, Asymetric

Closed bodies with edges - Rotational Plane, Asymetric

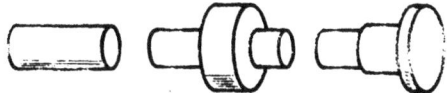

Thin wires on closed surfaces Plane Symmetric

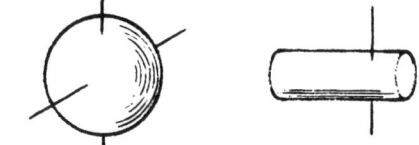

Open thin surfaces - Plane Symmetric

Closed bodies with fins - Plane Symmetric

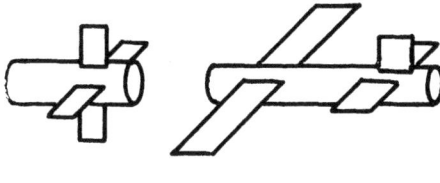

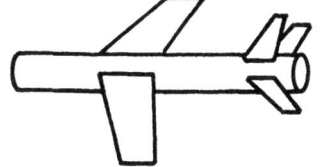

Fig. 6

problems have been solved for thin straight wires with arbitrary
resistive loading, operating either as scatterers or antennas.
That code is available and written up [1,2]. Dr Miller and his
colleagues at Lawrence Livermore have developed a code that treats
thin wire structures in the time domain [3]. The development of
codes for all the remaining classes or problems displayed in
Figure 6 has been limited to perfectly conducting bodies operating
as scatterers. Time Domain smooth closed body programs have been
written for rotationally symmetric problems, for plane symmetric
problems, and for asymmetric problems. One of these codes has
been written up and is available [4]. Spheres, prolate spheroids,
sphere-capped cylinders, sphere-cone-spheres and discs as sketched
in Figure 6 are some of the simple shapes for which time domain
solutions have been obtained, as I discussed in my lecture (p   ).
Time domain techniques have been applied to closed bodies with
edges and run for right circular cylinders and stacks of right
circular cylinders. The problem class of thin wires on closed surfaces has been treated. The time domain integral equation solution for this problem has been coded for the plane symmetric case
and applied to spheres with wires attached and cylinders with wires
attached as sketched in Figure 6. Open thin surface programs have
been written for both the plane symmetric and the asymmetric cases.
They have been applied to square plates, rectangular plates, quad-
rilateral plates, triangular plates, and curved surfaces where the
contour is rectangular. Finally, codes have been written for
closed bodies with fins attached for the plane symmetric case and
applied to targets like those sketched in Figure 6. The above
programs are discussed in reports that are available and the codes
themselves may be available. You can write to me if you are interested and I will send you information and forward your request
to the appropriate people.

---

[1] A.M. Auckenthaler and C.L. Bennett, "Computer Solution of
Transient and Time Domain Thin-Wire Antenna Problems,"
IEEE Trans. Microwave Theory and Techniques, Vol. MTT-19,
No. 11, pp 892-93, November 1971.

[2] A.M. Auckenthaler and C.L. Bennett, "SWIRE," ASIS-NAPS
Document No. NAPS-01541, 1971. For program listing order
document NAPS-01541 from ASIS National Auxiliary Publications Service, c/o CCM Information Corporation, 909
Third Avenue, New York, N.Y. 10022.

[3] E.K. Miller, A.J. Poggio, G.J. Burke, J.A. Landt, "WT-MBA/
LLIB," See Electromagnetic Computer Code Newsletter, R.M.
Bevensee, (Editor), Vol. 1, No. 2, pp 18-24, October 1974.
Information on this program can be obtained from

R.M. Bevensee, Electromagnetic and Systems Research
Group, Lawrence Livermore Laboratory, Livermore, California 94550.

[4] C.L. Bennett, "S3T-CLB," See Electromagnetic Computer
Code Newsletter, R.M. Bevensee (Editor), Vol. 3, No. 1,
pp 15-18, April 1976. Information on this program can
be obtained from R.M. Bevensee, Electromagnetic and
Systems Research Group, Lawrence Livermore Laboratory,
Livermore, California 94550.

[5] D.R. Wilton, S.S.M. Rao, A.W. Glisson, "Triangular Patch
Modelling of Arbitrary Bodies - An Electric Field Integral Equation Approach for Both Open and Closed Bodies,"
1979 International IEEE Antennas and Propagation Symposium,
18-22 June. Seattle, Washington, pp 151-154.

[6] C.L. Bennett and H. Mieras, "Time Domain Calculation of
scattering Center Response," 1979 URSI Meeting, Boulder
Colorado, 5-8 November 1979.

Dr Miller also asked me to discuss the time domain problems that
have not been solved. This is a summary of some of the ones that
come to mind.

> Curved plates with arbitrary contours
> E-field integral equation for non planar surfaces
> Wires on surfaces as an antenna problem
> Wires and Fins on surfaces
> Dielectric bodies
> Lossy bodies
> Inhomogeneous problems

An E-field integral equation formulation is better suited than an
H-field integral equation formulation for curved plates with
arbitrary contours. However, the E-field integral equation normally
requires that space derivatives be taken on the surface and from
a numerical implementation standpoint, this causes difficulties.
The work that Wilton and his colleagues have been doing with triangular patch modelling for plates and surfaces may be a way to
handle this problem [5]. It was noted in my lecture (p.   )
that the wires on surfaces problem has only been done for the
scattering problem. The application to the antenna problem should
be a "simple extension", however, the experience of some workers
has been that the antenna problem is much more difficult than the
scattering problem. Putting wires and fins on the same surface

should also be do-**able**. The remaining problems on the list are scattering by dielectric bodies, by lossy bodies, and by inhomogeneous bodies. There may be workers who are addressing some or all of these problems. We at Sperry Research Center are currently working on the problem of scattering by dielectric bodies.

That concludes my comments on the status of Time Domain Integral Equation solutions.

Dr Miller - Thank you Leonard, are there any questions?

Mr J.E. Summers - Can we assume that all the American programs are only available in Fortran?

Dr Bennett - These time-domain codes are all written in Fortran.

Dr Miller - For what kind of computer?

Dr Bennett - These programs have been written in Fortran IV, which is compatable with most computers, and run on Univac, IBM, CDC, and Honeywell computers.

Dr Miller - I should mention, however, the second line on Leonard's list, closed 3-dimensional smooth bodies, is one of the codes that is in the Lawrence Livermore library [4]. There were a couple of questions up here.

Dr L.W. Pearson - I was going to make essentially the same comment that codes are critically dependent on word length and exponent normalization algorithms from one machine to another and changing that may well introduce instabilities that are very subtle because it involves understanding floating point algorithms of a given machine.

Dr Miller - These are generally the kind of differences that 'drive people up the wall', because the thing will run but it doesn't give the right results and that is really hard to trouble-shoot. I think there was another question?

Dr T.K. Sarkar - Even on the same computer you may have different results depending on what compiler you use. Even if you write your program in Fortran IV, if you use a G compiler you get one result, if you use an H compiler you get another result.

Dr Miller - Yes there are a list of horror stories one could compile. Are there any other questions or comments?

Dr Bennett - I would like to make an observation. Yesterday we heard that the space time domain may be one of the ways you could get some good insight into problems involving ray tracing and the homogeneous lossy media problems. One of the things we did recently is an application of time domain techniques which will be described shortly at a conference presentation [6]. With time domain techniques, events are localized in space time at the leading edge. It's the leading edge part of the response which translates to the high frequency part of the response in the frequency domain and it's the leading edge part of the response that gives you the specular return. In impulse response terminology, it gives you the impulse in the impulse response. The leading edge part of the response will also give you the returns from edges. It's the same thing that you get from GTD. Therefore, the time domain approach may provide an efficient way of calculating the diffraction coefficients of GTD, for arbitrary shaped discontinuities in the target [6].

<u>Professor R. Mittra</u> - For the lossy medium you will have diffusion, not just dispersion and the pulse is going all over the place. It's not going to remain confined, so to speak.

<u>Dr Bennett</u> - For lossy medium the problem is much more difficult. There may be some people looking at that now. We are not.

<u>Dr Miller</u> - Thank you Dr Bennett.

The third speaker will be Professor Butler from the University of Mississippi who will discuss the general status of computations involving apertures.

<u>Professor C. Butler</u> - I am going to give you a brief introduction to where you can find results and where you can find what has been done in the past in the aperture literature. To my knowledge the first extensive review of what is available on apertures was done by Bouwkamp [1]. Then in [2] 1978 in the IEE Transactions on Antennas and Propagation special issue devoted to EMP, Rahmat-Samii, Mittra and I, attempted to provide a classified bibliography which essentially served to up-date the review of Bouwkamp. This is a review paper and, in the interest of saving space, we did not list the references that Bouwkamp had given. We tried to include those references that have become available since 1954

---

[1] C.J. Bouwkamp, "Diffraction Theory", <u>Rep. Progr. Phys.</u> Vol. 17, 35-100, 1954.

[2] C.M. Butler, Y. Rahmat-Samii, and R. Mittra, "Electromagnetic Penetration through Apertures in Conducting Screens" <u>IEEE</u> Trans. Ant. and Prop. Vol. AP-16, pp. 82-93, 1978 (Also EMC).

with emphasis on the material that is applicable to EMP problems. There is a vast amount of literature that pertains to apertures in waveguides which was not addressed to any extent in this review paper. Then Professor van Bladel and I attempted to up-date this one more time [3]. There is an extensive list of references in the NATO notes pertaining to apertures in cavities and in waveguides.

<u>Dr Miller</u> - Thank you Chalmers. Any questions concerning Apertures? It seems as though if you get those three papers you will have a pretty good start on what the latest literature is. I think Dr Brittingham at Lawrence Laboratory put together a list but I don't recall whether that includes any you might not have.

<u>Professor Butler</u> - I have Brittingham's list. We exchanged lists during the preparations of this paper, and I don't think that he has anything on his list that we didn't include except items that were not available in the open literature.

<u>Dr Miller</u> - OK - Thank you Chalmers. I think the status of computer codes on these various areas, the degree of development, reflects to some extent the amount of time spent over the years on these various areas. Numerical computations for apertures is more recent than e.g. some of the time-domain and thin-wire work. Thus the codes in these other areas have matured a bit more and as a result are more widely available.

We will next hear about quite a different type of modelling procedure based on the T-matrix approach. Peter Barber of the University of Utah will summarise the status of it.

<u>Dr P.W. Barber</u> - I would like to mention the availability of a computer program that is our best and most used program - one which we have made available to quite a few people. The program is based on the extended boundary condition or T-matrix approach and it is specifically written and optimised to calculate scattering and absorption for nonspherical dielectric obstacles. The obstacles must be axisymmetric, for example, finite cylinders, spheroids, or sphere-cone-spheres. Although the calculation of the T-matrix can be quite lengthy, our procedure is to calculate the T-matrix for an object only once, and then use the matrix as input to a family of programs that easily make calculations for different arbitrary orientations or for random orientation. The latter calculation is often needed in atmospheric or biological problems.

---

[3]   J van Bladel and C.M. Butler, "Aperture Problems" (NATO Notes).

The input to the program is the complex dielectric constant, the size parameter ka, the appropriate indices N and M of the spherical harmonic functions, and the number of integration steps which are required for the Gaussian integration scheme. The particle shape is input in a subroutine which is very easily written - we have a large number of them for different shapes. Basically you just have to specify the distance from an internal origin to the surface. The primary outputs that we have been interested in are scattering parameters. The main output is the differential scattering cross-section, which is proportional to intensity. We calculate the scattering from the forward direction, which is the direction of the incident wave, around to the backscatter direction. Generally, the calculation is done for angles from $0°$ to $360°$, because the pattern is not symmetric about $180°$ for arbitrarily oriented nonspherical objects. The program also prints out the extinction, scattering, and absorption cross-sections. The run time is on the order of seconds for the problems we have looked at. For example, using a CDC 7600, a run time of 12 seconds is required for a 4:1 prolate spheroid with a dielectric constant of 2 and ka equal to 10. The run time goes up as the matrix size, so the run time can be considerable for large problems.

As for the limitations of the program, there has been a lot of discussion about ill-conditioning problems. We have encountered the problem only once and that was for a very high dielectric constant lossy object. As far as running on different computers is concerned, we have two versions of the program - one is optimised for CDC machines and the other for IBM and Univac machines. However, either program will run without difficulty on any machine with Fortran IV. The version for CDC is written in single precision. The version for IBM and Univac is written in double precision for those machines. The program is completely self contained - the Bessel and Legendre functions are calculated in the program as are the Gaussian integration points.

We have given copies of the program to about 12 to 15 different users in the United States. The National Centre for Atmospheric Research is using it to make scattering calculations for ice crystals. A group at Indiana University is using it to make scattering calculations for two micrometer barium sulfate particles which they use in a flow system. The State University of New York at Stony Brook is also using it for atmospheric problems as is UCLA. Harry Diamond Laboratory is using the program to develop some new methods for sizing particles, and of course we use it at the University of Utah.

As far as program availability is concerned, we have been providing card decks with test cases and some simple documentation. We have never sent programs outside of the United States and we might have to negotiate a postage charge for that. In the United States, we ship card decks via United Parcel Service and it takes about a week.

I can make one comment though, along the same line that Wilson Pearson commented on. We would prefer to give the program only to people who are interested in learning something about the method and how the program works, because if you are trying to solve problems different from the exact problems that we solve, you should learn something about convergence testing.

I think my biggest fear is that someone will use the program to generate incorrect results and not realise it. If these results are then used in a paper or compared with other results from other methods, it could generate a lot of concern. We try to help people that actually have our programs if they do have questions and we are willing to run some cases to check results of others.

<u>Dr Miller</u> - Thank you Peter. You raise a very interesting point and that is, is there any degree of liability on those who supply programs when they are misused by people who get them. I know some of the researchers I have talked to, expressed a real concern about this. They are essentially exporting their reputations and when the programs do not provide the kind of result that was expected on the part of the user, for whatever reason, he tends to blame the person who gave it to him. It's a hard game to win at sometimes. I think that somehow or other we will have to provide some sort of protection for the developers if these things are to be readily transferred around. Are there any questions for Professor Barber?

<u>Mr D.J. Brain</u> - Can your technique be extended to polarisation or depolarisation effects?

<u>Dr Barber</u> - Well yes, you can input any polarisation and it will bring out the same polarisation as well as the cross polarisation components.

<u>Dr Miller</u> - OK any other questions? Just to remind people, Prof. Barber did talk the other day as well about an extension which is being considered to combine the T-matrix approach with the unimoment approach for inhomogeneous dielectrics and that is a very interesting problem.

Well we come now to a gentleman that not very many of you people know I am sure, so in order to help identify him for you I have a small limerick here. A knight Sir Roger of Syracuse, at a banquet his inhibitions did lose, so he climbed on a chair and amazed everyone there, with his talents to amuse. Sir Roger!

Professor R.F. Harrington - Well, we have so many programs we
don't know what to do with them all. Essentially every report
and every paper we at Syracuse University publish gives the program, and the papers usually tell if they are available, so if
you are interested in any of these programs you can write. We
do like to cover our cost. Right now it is $20 a card-deck plus
postage if you want overseas airmail. Otherwise we shall ship
it. - Printed matter takes a few weeks - airmail costs at least
$10 for overseas. Well, I wrote a few of them down but the list
got too long so I stopped. We did a lot of work on bodies of
revolution, which was Joe Mautz's thesis originally. His first
work was on the E-field equation for both scattering and radiation from apertures, and also with loading on the body. Primarily resistive loading was chosen so that one could compute the
effects of losses if required. Then more recently we did the H
field, the combined field and the combined source. As I pointed
out earlier, the E field and H field solutions will break down in
very narrow bands, and on an odd shaped body you don't know where
they are. So it is probably better to go to the combined field
or the combined source solution. All of these solutions are for
conducting bodies and, as I mentioned earlier in the week, if you
just take the E field and H field solutions, combine them in the
proper way, and use them twice, you have the dielectric body
program which is also available. Then we have a number of programs
involving wire objects for both radiation problems and scattering
problems, both loaded and unloaded, with junctions and bent wires.
These use triangular bases and triangular testing functions with
certain mathematical approximations. They can be easily changed
to piecewise sinusoids, just changing a few subroutines. That was
done originally by Chao and Strait [1]. We did a number of two
dimensional problems, both radiation and scattering again, E field,
H field solutions and also the GTD solution for an arbitrary polygon. The latter problem was also done at Ohio State, but I don't
know what condition their programs are in. More recently we have
worked a number of aperture problems. We treat two dimensional
apertures in planes, both zero thickness planes and thick planes.
We consider apertures in thick planes, which are formed by cascaded
parallel-plate sections. We consider apertures in thick planes
which have an arbitrary cross section between the two half-space
regions. It's a trivial modification to do it for cavities behind
ground planes. Most of those programs are available, again they
may have to be modified for your particular problem. We also have

---

[1] Kuo et. al., "Analysis of Radiation and Scattering
by Arbitrary Configurations of Thin Wires", IEEE Trans.,
Vol. AP-20, p.814, Nov. 1972.

rectangular apertures in conducting planes, we have cavity-fed rectangular apertures in conducting planes (this is where the apertures are different in size from waveguides) and we have arrays of rectangular apertures (again different in size from the waveguides). So we have a number of programs. Most of the theory has appeared in the literature. It tells in the papers what the report number is for the program and we also have a relatively complete list of reports. If you write to me, I will send you what we have. You can look at it and see if anything is of interest.

<u>Dr Miller</u> - OK, thank you very much. That is quite a long list of programs, are there any questions concerning...

<u>Professor Harrington</u> - I might relate a little story here about Joe Mautz. When he first started reviewing papers for the journals, he would get a paper and he would go through the mathematics and then he would re-run every curve in the paper. Woe be to any authors whose curves were wrong and even worse, woe be to any author for which he couldn't re-run the curve. That paper went back to the author, and he had to put more detail in his paper so that Joe could run every curve.

<u>Dr Miller</u> - Formidable work in all those programs.

<u>Professor Harrington</u> - Yes, then Joe wondered how anybody got time to review many papers.

<u>Dr Miller</u> - I think, Roger, I didn't hear you mention apertures and three dimensional bodies but I believe you have also done some work there on that problem as well as with wires in three dimensional cavities coupled with apertures.

<u>Professor Harrington</u> - Yes, I think this was probably done by Schumann.

<u>Dr Miller</u> - Ah! yes that is right. So that is another problem to which those programs apply. Well Professor Harrington has indicated that if you want to obtain any of those codes you can write to him. For a modest fee he will send them to you. It strikes me that it is unfortunate we don't have someone here from Ohio State University, since they have developed quite a large collection of computer codes there as well. I know that Professor Richmond (E.E. Dept., Ohio State University,) has made some of these codes available. We have at least one in the Lawrence Lab. library but there are quite a number of others there, specialising also in GTD.

If there are no further questions we will now hear about some problems more germane to practical application. First from Tom Armour of AWRE.

Mr T.W. Armour

Just a few further words about our modelling codes at Aldermaston. These include a simple in-house program using pulse functions and collocation, then the AMP program, again using collocation but with sinusoidal basis functions, and more recently our CHAOS program. This uses piece-wise linear functions and a Galerkin type solution of the Electric Field Integral Equation rather than collocation or point matching. The program originated at Syracuse University. We have made a number of modifications, for example, to include the use of finite resistance. Also dielectric wires and rods. A number of (wire) junction routines have been introduced including one due to Sayre. This was first used with pulses and was quite difficult to program for piece-wise linear basis functions. Although satisfactory we have found the most successful method to be that of Chao and Strait who used a complex of overlapping wires at a junction. Some junction routines use the integrated form of the continuity equation, others the differential form.

Our main interest is with surface currents and the EMP problem. Experimental checks are carried out in a travelling-wave simulator. Comparisons between theory and practice may be made in either frequency or time-domain. CHAOS is also used in radar scattering studies and again experimental checks are available. In the case of EMP studies additional comparisons have been made with the results of a finite difference code and with patch programs.

We would certainly like to have a general purpose patch program (not restricted to bodies of revolution and preferably using triangular patches) and the work of Don Wilton at the University of Mississippi is of very great interest.

Some additional points. We make use of rotational symmetry (in both the AMP and CHAOS programs) to help reduce computer run-time. Bodies of revolution may thus be modelled in very considerable detail. (CHAOS is in fact capable of inverting large dense matrices of order 750 x 750. This is in the most general case, without regard to symmetry.) Also left-right symmetry and the effects of a perfect ground. (So far we have not included the effects of a lossy earth.) As regards accuracy we usually quote a figure of some 30% for surface currents on large structures such as aircraft. Figures for surface currents on cylinders are usually rather less but depend on shape factor and the number and distributions of wires. In the case of an elliptic cylinder we achieved an accuracy of about 15%. Errors for total (axial) current on a cylinder (or body of revolution) are usually very small - some 2-3%. Similar results are obtained for a string of wires or simple arrangements of crossed-wires. Patch programs are certainly more accurate than wire-grid but lack the ability to model undercarriages and pylons, for example.

The occurrence of cavity resonances does not trouble us unduly since the frequencies are well above the range of interest (in EMP interaction studies) but difficulties may occur with radar scattering. We are very grateful for the work of Professor Harrington in highlighting this problem. His use of a combined Electric and Magnetic Field Integral Equation is of considerable value.

We are certainly fortunate in being concerned with both EMP and radar studies. Performance may be assessed over a very wide range of frequencies and at any orientation. Difficulties have occurred, however, at very low frequencies (mainly with collocation programs). Amplitudes should tend uniformly either to zero (indicating a normal resonance response) or to a finite value (indicating the presence of a 'magnetostatic' contribution). In EMP studies we have found that CHAOS works surprisingly well over a wide range of frequencies and it has not been necessary to use separate 'high' and 'low' frequency models. This would appear to be an advantage of a Galerkin type solution.

We very much appreciate the excellent theoretical work which has been done on apertures, cavities and wires behind apertures. Such detailed analysis is essential as a check on our programs since experimental data appears to be so limited. Fields calculated at the University of Mississippi (using patch programs and a novel variation of Schelkunoff's principle) in open-ended and 'choked' cylinders are of particular interest. They are not untypical of the kind of apertures and cavities occurring in aircraft and represent quite a challenge to any modelling program.

Finally I must draw attention to the work of Don Williams at one of our research establishments (SRDE) who is very much concerned with antennae on many types of structure and their input-impedance. He has used wire-grid, patch and hybrid programs. Changes in resonant frequency when wire-grid modelling are very evident in these studies. Similar results for antennae have been obtained at Ohio State University.

<u>Dr Miller</u> - I take it that most of your calculations are for scattering problems rather than those concerned with antennae.

<u>Mr T.W. Armour</u> - That is correct.

Dr Miller - Have you ever seen resonances, similar to those described by Professor Harrington, at frequencies other than the natural internal resonances, when using a wire-grid modelling program?

Mr T W Armour - We have seen no evidence of such resonances.

Professor Harrington - I would agree that these resonances are not seen very often.

Mr T W Armour - It is as well to be warned, however, about their occurrence. Of course internal cavity resonance frequencies occur on very many occasions. In modelling a closed cylindrical or spherical surface, cavity resonances appear but their amplitude may not mean very much since the matrix has (possibly) become ill-conditioned. If a large aperture is introduced then, interestingly, the amplitudes are very close to the expected theoretical values for apertures in a closed metallic surface.

Dr Miller - The 30% figure you quote - does this refer to surface current density?

Mr T W Armour - Yes indeed - the figure seems to be fairly general and holds for a number of structures. Exceptions may occur at the intersection of, say, wing and fuselage, or along the edge of a wing. Experimental measurements are also difficult in these situations.

In the case of radar cross-sections the results are usually very good but occasionally disappointing. The explanation for this is not known.

Professor C. Butler - Concerning the internal resonance problem. When Professor Harrington published his report on this matter I considered a situation that could be solved analytically in order to try and determine the precision required to detect the internal resonance. For this particular problem I could see analytically the difficulty. A term in the denominator went to zero. I found I could not see the internal resonance unless the computation was done to at least 3 significant figures. The computer just stepped right over the resonance frequency.

Dr Miller - I agree with your comments, that essentially good precision is required to see these resonances. Noise in the computation is somewhat similar to loss of precision, it can tend to make resonances disappear. An example comes to mind in the application of matrix techniques to solve for complex exponentials (Prony's Method). If the problem is over-determined the matrix becomes singular but noise keeps it from doing so. At least that is my understanding of the case.

Professor Butler - If one does a patch model of a closed surface then this really is a 'leaky' body unless the patches have currents on them which exactly reproduce the right boundary conditions. There is, of course, greater penetration through a wire-grid model.

Dr Miller - I agree with your remarks about a 'leaky' surface. Actually Professor Harrington may not have made the point very emphatically the other day in connection with 'zoning' of the problem. If we have a resonance region that is numerical (due to the internal resonances) and the problem is zoned more and more finely the resonance will become narrower and narrower so that as the solution becomes more and more accurate, the region over which the resonance has an influence tends to be reduced to a very small region about the actual resonance point.

Well, I feel we might be running a little bit behind and we have two more speakers to hear from. The first, John Owen of RAE (Royal Aircraft Establishment).

Mr J.I.R. Owen - The computer programme that will be described was developed at the RAE about 5 years ago for the analysis and prediction of the performance of HF aircraft aerials. A wire grid method is used to model the airframe, mutual impedances between the segments of the grid being determined by using the short dipole approximation and assuming constant currents on the segments. The vector component of the electric field is computed at the centre of the observation segment, but instead of calculating the scalar component of the electric field, the potential difference induced by changes at the ends of the source segment is computed across a proportion of the observation segment, governed by its wire radius. The whole of the observation segment is not used as this would lead to unrealistic values for self-impedances and mutual impedances between adjacent segments.

Once the scheme for current interpolation etc, have been programmed, the optimum values of the parameters for segment distribution and wire radius must be determined if accurate results are to be achieved. Consistency in the accuracy of the results for a range of aircraft/aerial combinations is vital when the programme is being used for aerial design work. A type of HF aerial commonly used is a notch or slot Fig. 1; aerials of this type are an integral part of the airframe with their performance depending to a large degree on their position in the structure. Being part of the airframe, however, makes these aerials expensive to install, and therefore a high, degree of confidence must be attained in any method used for predicting their performance.

FIG 1   H F NOTCH AERIAL

Constant checks are carried out, comparing the results of scale model measurement, full-scale mock ups and flight trials with those from the programme. Overall accuracy may only be determined by the use of more than one method, as not all of the results can be obtained from one method due to physical and measurement constraints.

Comparisons were made using a full-scale mock up of a tail section containing a notch with the wire grid scheme of Fig. 2, for a number of different segment radii, although in each calculation the radius was constant throughout the model. The most accurate agreement with the measured values for radiation resistance occurred when a radius of 4.5 ins was used. Provided that the radii and lengths of segments in the immediate vicinity of the aerial were approximately the same length, but shorter than a quarter of a wavelength, and the complete surface area of the structure was modelled, there were no significant variations in the results. The RAE programme has a constraint that prohibits segments to be placed closer than their diameter at their mid points, thus maintaining a physical realisable grid structure. Hence the minimum segment spacing and maximum wire radius are the same. Away from the aerial site in regions of comparatively low current density the grid structure can be made very sparse, without degrading the accuracy of the calculations for radiation resistance. However, if the surface of the airframe is not modelled accurately there is a degradation in the radiation patterns especially in the ratio between horizontal and vertical polarisation.

The results obtained from using these values for grid parameters had given good results when used for fixed wing aircraft as shown in Fig. 3. When used for a helicopter model, Fig.4, the results produced had some large discrepancies in the ratio of horizontal to vertical polarisation when compared to scale model measurements. Although the graphs had the same basic shape, Fig.5, differences of 8 to 10 dB occurred at frequencies of 6 to 8 MHz. As this was considered outside the limits of acceptable accuracy, usually 3 dB, the cause of the deviation was investigated.

An inaccuracy of 3 dB is considered to be the average value for the radiation measured from 1/30th scale models on the range or from aircraft of the same type due to construction tolerances.

For this type of helicopter the fundamental rotor blade resonance is a 7 MHz, which is also the frequency at which the largest discrepancy occurred between the results, and the rotors were therefore, considered as the most likely source of modelling error. A decrease in the H/V ratio was required in the wire grid models results, which may be achieved by increasing the mutual impedance between the rotor blades and the fuselage. This is most easily

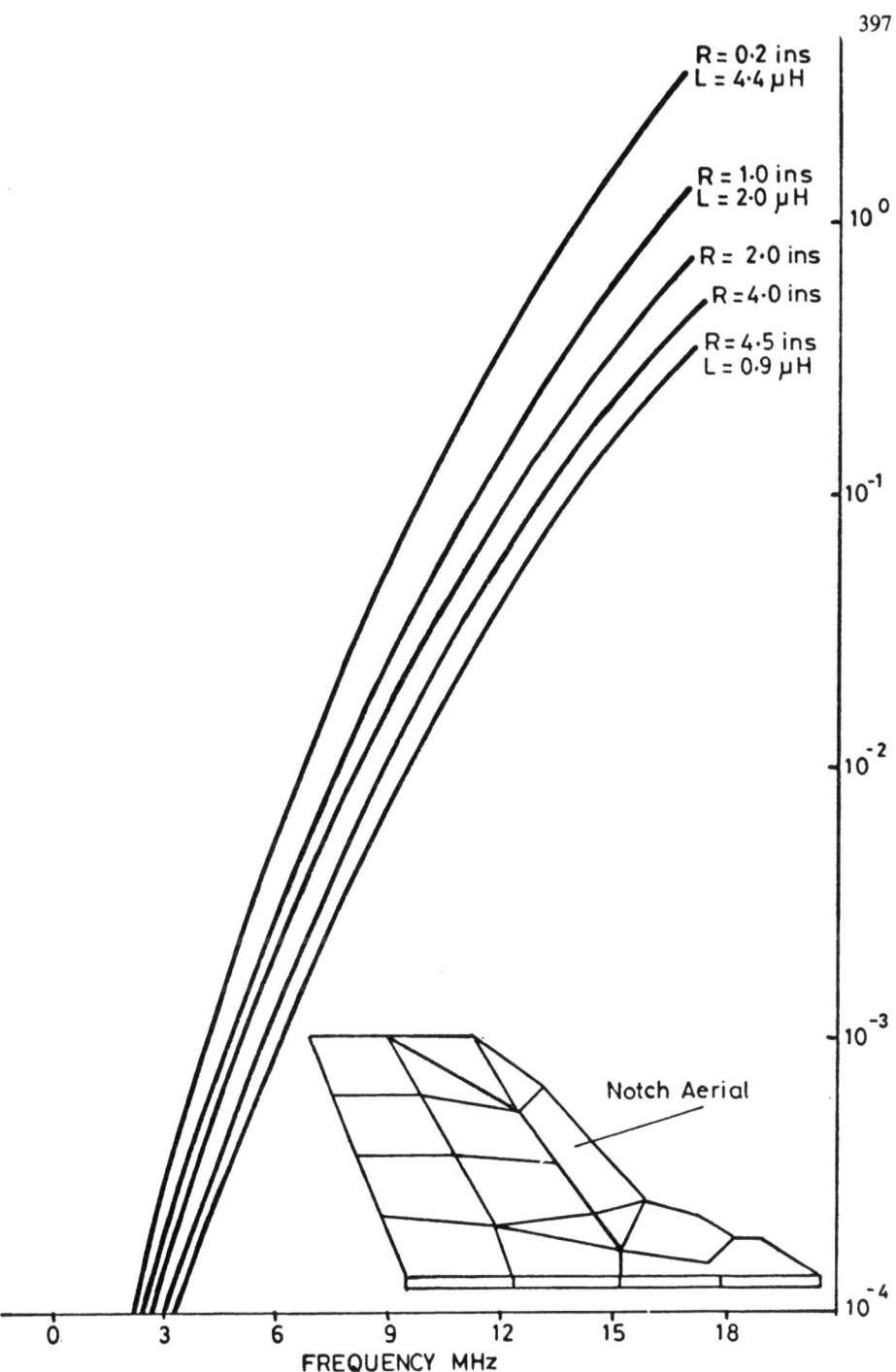

COMPARISON OF WIRE RADII AND RADIATION RESISTANCE FOR NOTCH AERIALS FIG. 2

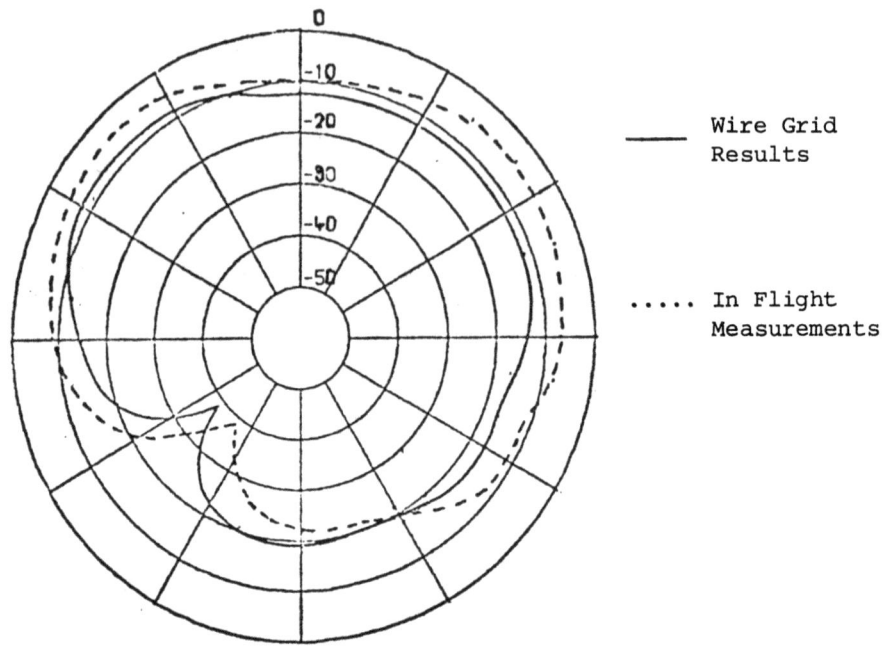

Fig. 3 Comparison of Wire Grid Results and In Flight Measurements for Concorde Notch at 5.7 MHz.

accomplished by increasing the diameter of the wires used in the rotor blades. After many attempts the radii which produced the closest correlation between the results were tapers of 4.5 ins to 8.0 ins along the short upright stem and 8.0 ins to 10 ins along the blades. This met the requirements to maintain segment radii constant at junctions and to keep the dimensions of the wire grid the same as those for the full size airframe.

Significant improvements were also obtained in the radiation patterns calculated from the grid especially those around the 7 MHz resonance region which has been a difficult frequency to model accurately, as illustrated in Fig.6.

Dr E.K. Miller - What code did you use for this?

Mr. J.I.R. Owen - The code and methods of deriving the grid have been developed at the RAE over the last 7 years. The main part of

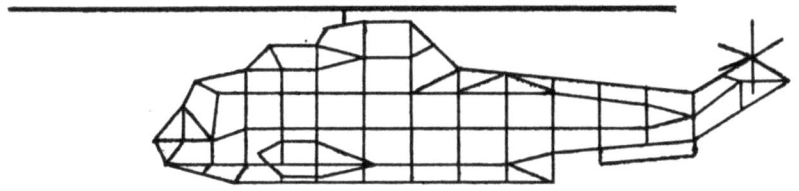
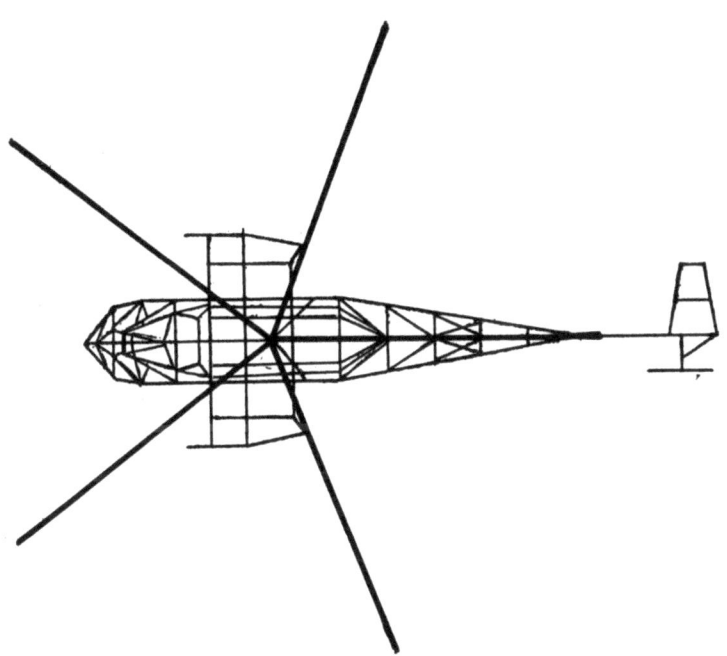

Fig.4 Helicopter Wire Grid

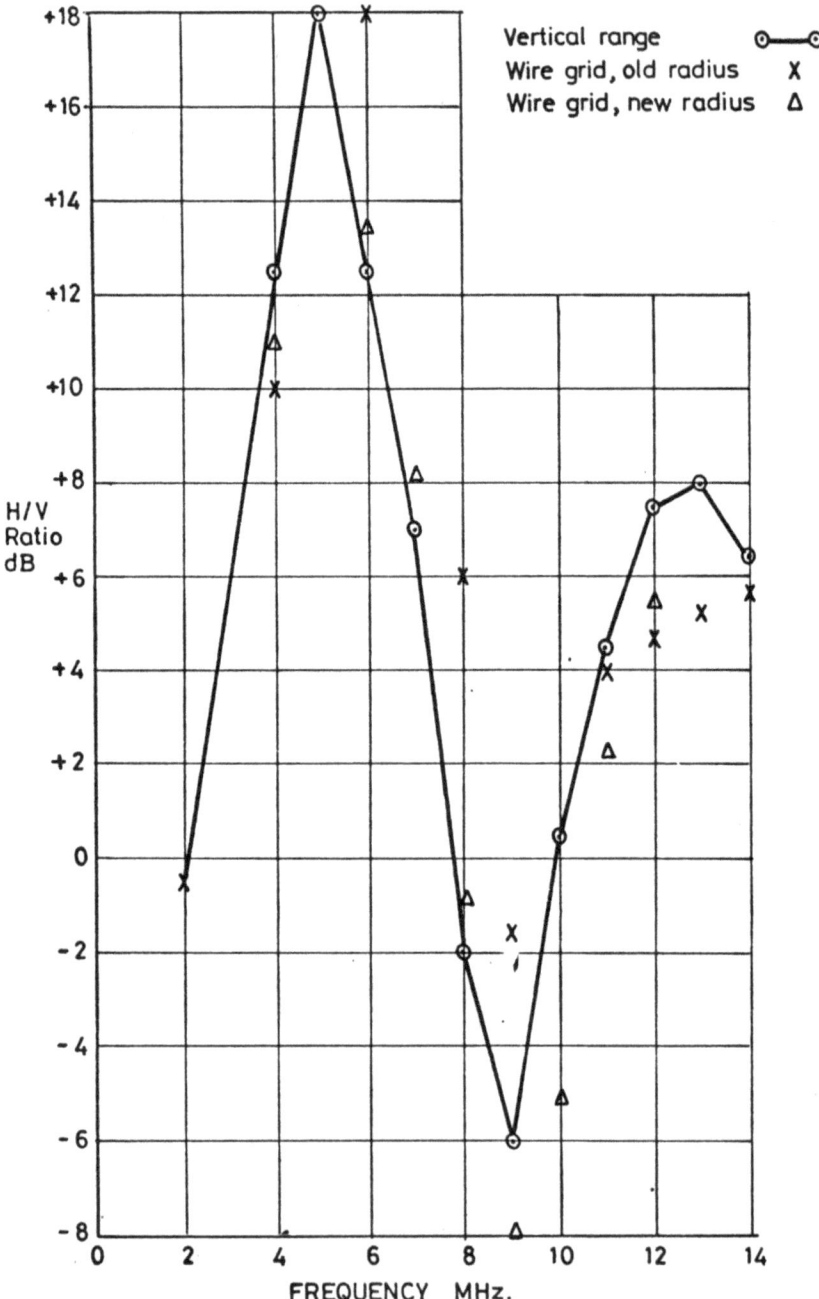

H/V RATIO HELICOPTER WIRE GRID   FIG. 5

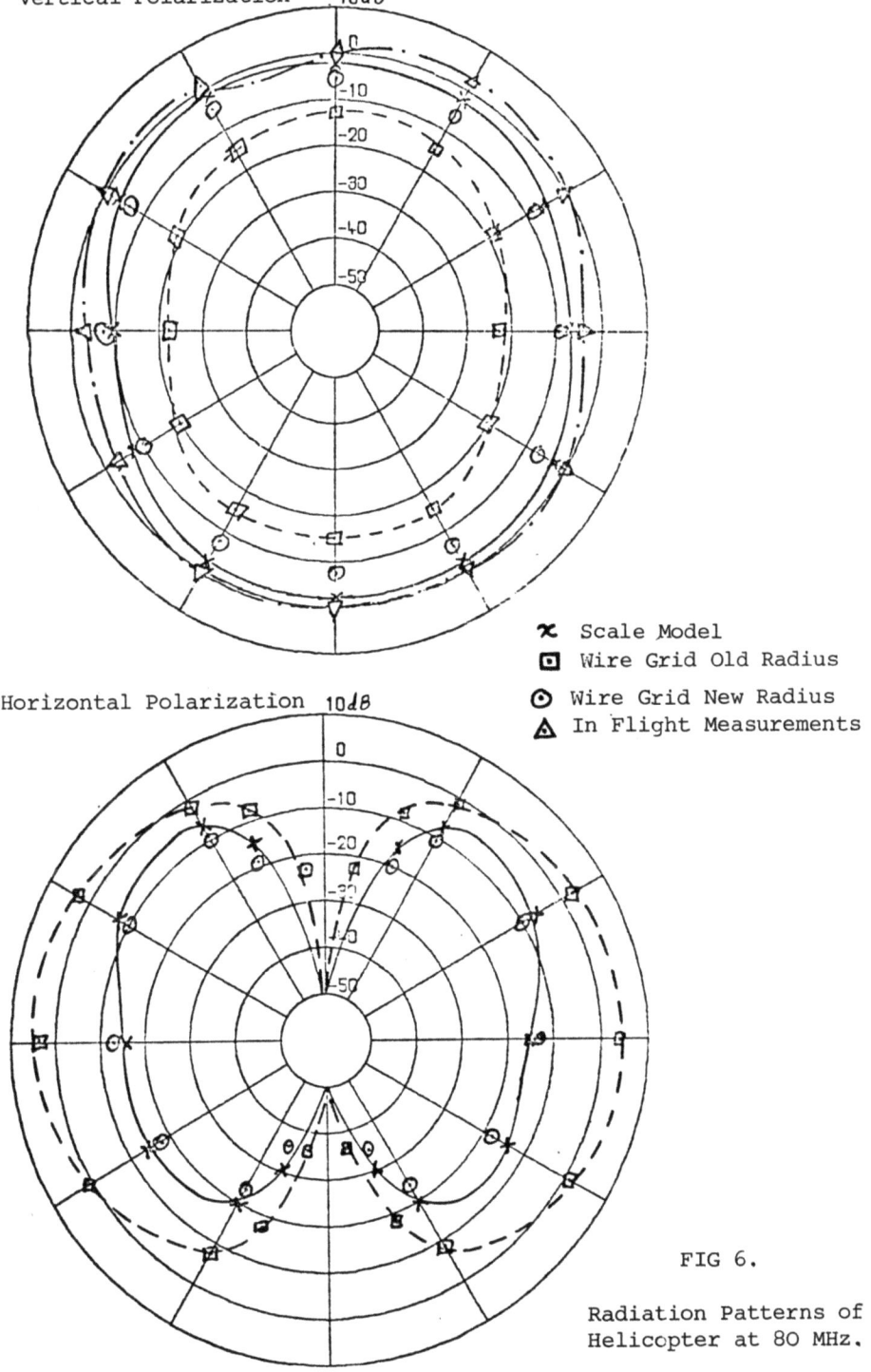

FIG 6.

Radiation Patterns of Helicopter at 80 MHz.

the code was written by Dr D. Forgan of RAE after he had looked at
AMP 1, work done by Albertson at TUD and Thiele at Ohio. The
programme does not have as many facilities as AMP 1, eg there are
no facilities for modelling transmission lines, ground planes,
near field radiation patterns and output normalisation. It does,
however, use far less store than AMP 1 and is typically 5 times
as quick to run.

Dr Miller - I think you have indicated one really important point
here, and that is that the correspondence between the model and
the problem intended to be modelled has to be first of all established, and it has been our experience too that people frequently
do not give you the proper dimensions.

Mr W.L. Jones - How does your optimum radius choice correspond to
the rule of thumb that the wire surface area should equal the area
of the surface being represented.

Mr Owen - There is not really very much correspondence. Around
the aerial, the wire segments are placed very close together, and
the surface area of the wires is greater than that of the surface
being modelled; along the fuselage and wings, however, the surface
area of the wires is probably only a tenth of that being modelled.
As I have shown the results using this method agree with those
from experimental measurements.

Dr Miller - I am not sure what to think about that, except to say
that it is my understanding that the condition of wire radius is
very dependent on the numerical and analytical treatment used.
However, I still refer back to some experiments which seem to
demonstrate that equal area ratios, is the one that is most appropriate, however the experiment reality obviously knows nothing
about linear versus piecewise sinusoidal point matching and all
that. So I suspect some of these numerical treatments require
certain interpretation in their application and it is learning how
to use them that is important.

Mr Owen - I would make one point. It is radiation patterns from
an aerial, which may be considered as a point or line source that
is being calculated. If radar cross section is required, then I
suspect, the whole structure has to be modelled with uniform
segment density as reflections may occur anywhere on the structure.

Dr Y. Rahmat-Samii - An announcement for those who have been working
on computer codes. It would be useful if they could send them to
the Newsletter so that whoever wishes can use them. We would also
appreciate some idea of the problems that they have been using
them for, with the values they have used for various parameters and
the accuracy of the results.

Dr E.K. Miller - Thank you Mr Owen and lastly we hear from Dr P.A. Ramsdale.

Dr P.A. Ramsdale - You have heard about some colossal problems which have been tackled successfully by moment methods. I have managed to get wrong solutions for many simpler problems, using these techniques!

This reference [1] includes some guidelines which I think will help explain what the various routines do and the differences between using different schemes for fairly straightforward wire antennas; HF antennas, VHF antennas, loaded antennas, active antennas and things of this nature - not enormous grid structures though. The other thing which is a little bit different from most of the work that has been talked about so far is that most of the solutions have been of the Pocklington equation.

$$\left[\frac{\partial^2}{\partial z^2} + k^2\right] \int I_z(z') \, G(z \, z') \, dz' = -j\omega\varepsilon \, E_z^i \quad (1)$$

and Roger Harrington's original method [2] can be regarded as a specific solution to this. We have tended to use the Hallen equation

$$\int I_z(z') \, G(z \, z') \, dz' = \frac{-j}{\eta} (B \sin kz + C \cos kz + \frac{V}{2} \sin k|z|) \quad (2)$$

because it has various advantages. These are that usually a sub-domain solution will converge with fewer segments, the matrix will tend to be better conditioned and the integrand has a lower order singularity because it is not differentiated twice. You heard Chalmers Butler earlier [3] giving an alternative scheme for

---

[1] P.A. Ramsdale 'Wire Antennas', Advances in Electronics & Electron Physics (Ed L Marten) Vol 47, pp123-196 (Academic Press; 1978).

[2] R.F. Harrington, 'Matrix Methods for Field Problems', Proc. IEEE Vol 55, No 2, pp136-149, February 1967.

[3] C.M. Butler, 'Introduction to Moment Methods with Simple Applications' Nato ASI - Norwich, pp   July 1979.

getting away from some of these problems. A big disadvantage of the Hallen equation is that your boundary conditions have to be put in separately, whereas the Pocklington equation contains its own boundary conditions. This can be an advantage in some cases; if you are considering an antenna loaded with various circuit components it is handy to be able to put the boundary conditions in at a fairly late stage. Regarding the conditioning of this equation, it is worth noting that the unknown constants B and C do need to be kept on the right-hand side, Raj Mittra and Klein [4] considered this. If you move the constants over to the left-hand side the matrix is just as bad as it was before. There are several solutions about where this has not been taken into account. If you integrate the Pocklington equation twice it can be shown that the general solution is the Hallen equation and Chalmers Butler and Don Wilton have developed a solution bearing that in mind [5]. Basically I will be talking about solutions where we use the Hallen equation - generalised somewhat, so that instead of using a delta function generator you can use any source.

In equation (2) $\frac{V}{2} \sin k \ |Z| \rightarrow V \int_0^s f(s) \sin k \ (z-s) \ ds$

Popovic et al used a delta function generator, and other approximations for a practical driving region such as a magnetic frill, can be used. There was also some mention the other day of entire domain solutions which most American workers have rejected on the grounds that they are not very stable. However if you are talking about a low order solution, which is quite adequate for many low frequency problems the entire domain solution enables non specialists to write programs which are likely to work (rather than trying to use techniques which they do not understand too well.) Richmond [6] investigated various functions and simple functions were considered by Popovic [7]. These were mainly collocation

---

[4] R. Mittra & A. Klein, 'Stability and Convergence of Moment Method Solutions' Numerical and Asymptotic Techniques in Electromagnetics, (Ed R Mittra) pp129-163 (Springer-Verlag, New York 1975).

[5] D.R. Wilton & C M Butler, 'Efficient Numerical Techniques for Solving Pocklington's Equation & Their Relationships to Other Methods'. Trans. Antennas Propag., Vol 24 pp83-86, January 1976.

[6] J.H. Richmond, 'Digital Computer Solutions of the Rigorous Equations for Scattering Problems', Proc. IEEE, Vol 53, pp796-804, August 1965.

[7] B.D. Popovic, 'Polynomial approximation of current along thin symmetrical cylindrical dipoles', Proc. IEE, Vol 117, pp873-878, May 1970.

solutions. Galerkin methods have been used by Silvester and Chan [8].

From a simple point of view, the reason entire domain solutions go wrong when many matching points are used can be seen by considering the lines of the matrix individually. Instead of understanding the matrix behaviour in terms of condition numbers and singularities if you look at the matrix it becomes clear why the entire domain solution tends to go wrong. By putting the matching points closer and closer together you find that the consecutive lines get more nearly identical the closer they get, hence the matrix becomes ill-conditioned. Whereas if you are using a sub-domain solution the self terms are located on the diagonal and as the self terms tend to be different from the mutual terms, the matrix does not go off quite so rapidly. The trouble you have with the entire domain solution is that you tend to put more matching points where you think you are in trouble. When they are too close together the solution explodes but if you are not too greedy you can get a good result. As an aside, I think this explains why a reduced kernel can go wrong in a sub-domain solution because as soon as you use the reduced kernel the self term and the mutual term in the next segment, tend to get a bit closer than they really are, when using an exact kernel. Hence the matrix tends to be ill-conditioned. This is an engineers view of what goes wrong rather than Raj Mittra's more rigorous study [4].

Let us look at some specific cases which have been tackled. The inductance loaded antenna; Fig. 1 if you put an inductor into an

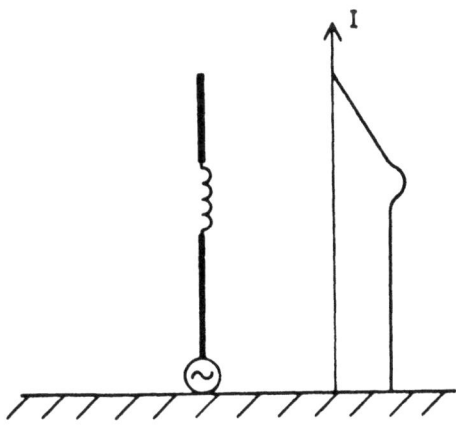

Fig.1 Inductance loaded antenna –
vary current moment by optimising the choice of load point.

antenna you can increase the current moment on the structure and a very simple approach might suggest putting it near to the top to maximise the current moment. However, there are several factors and you find that the optimum point for efficiency or signal to noise ratio, in the receiving case, works out at about 30 to 40% of the way up. Whether you get 30 or 40% depends on how well the current distribution is modelled and it is interesting that there is a little blip in the current distribution just above the inductor which is a good guide on how well you are doing with a low-order solution. Bob Hansen [9] used piecewise sinusoids and I used an entire-domain Popovic type solution [10] and also Roger Harrington's original moment method. All of these give approximately the right answer for the current moment even when the detail around the inductor is lacking. Another interesting problem is the active antenna- Fig. 2, where you run into

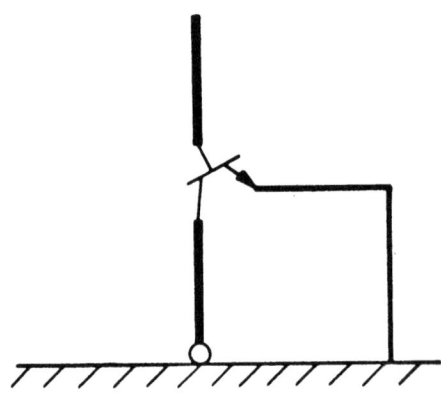

Fig. 2 Active antenna -
boundary conditions at transistor require an injected current.

---

[8] P Silvester, and K.K Chan 'Bubnov-Galerkin Solutions to Wire-Antenna Problems' Proc IEE, Vol 119, pp1095-1099 August 1972.

[9] R.C. Hansen 'Efficiency and Matching Tradeoffs for Inductively Loaded Short Antennas' IEEE Trans Commun., Vol 23, pp 430-435, April 1975.

[10] P.A. Ramsdale 'Signal/Noise Ratio of Inductive-Loaded Antennas' Electronics Letters, Vol 11, No 24, pp540-591, November 1975.

some interesting boundary conditions because you are capable of injecting current at the active device. If you use a sub-domain solution with overlapping segments which satisfy current continuity you have the problem of injecting a current somehow or other and that current is dependent on the current somewhere else on the structure. For these sort of problems it is very convenient to have a separate matrix and put boundary conditions in afterwards. This is easy to do if you use the Hallen equation. There have been sub-domain solutions [11, 12]. I cannot remember the exact details of these; I think piece-wise sinusoids were used but I am not too sure of the validity of the boundary conditions at the active device. I think it is quite a problem to get them in. With an entire domain solution though you can solve for the wires and put the boundary conditions in at the end in order to satisfy the transistor or more advanced active load [13].

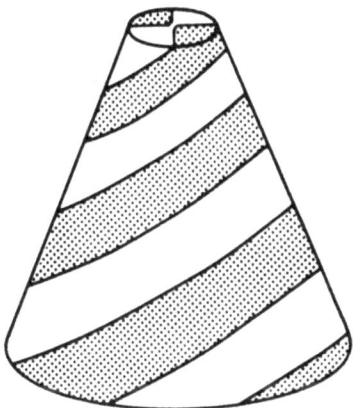

Fig. 3 Equiangular logarithmic spiral antenna.

---

[11] M. Pelletier, J.A. Cummins & S. Sanzgiri, 'Analyses of Active Antennas by the Method of Moments' IEEE G-AP, Int Symp, pp77-79, December 1972.

[12] W.C. Wong, 'Signal and Noise Analysis of a Loop-Monopole Active Antenna' IEEE Trans. Antennas Propag., Vol 24, pp574-580, July 1974.

[13] T.S.M. Maclean, and P.A. Ramsdale, 'Signal/Noise Ratio for Short Integrated Antennas,' Electronics Letters, Vol 11, No 3 pp62-63, Feb 1975.

As well as these very low order problems we have also been tackling the equiangular logarithmic spiral antenna.

Our initial studies were experimental but then we decided we needed an accurate analysis and carried out a literature search. Miller, Burke and Selden [14] show that this is a very slowly converging problem, so that even though it is a fairly straightforward looking structure it is quite difficult to solve. The first moment method is due to Yeh and Mei [15] and is the first use of the three term basis functions which Ed Miller now uses. They used a collocation technique for Hallen's equation. Because of the singularities between the sections you would expect the Hallen equation to converge faster than Pocklington. We have a listing of the original program and have run that and get the same results. The listing of Atia and Mei's extension [16] was not available but it is a straightforward extension to go from a two arm antenna up to an N arm antenna. The basic equation that we are solving uses Mei's extension [17] for a completely general wire rather than the straight wires considered before

$$\int I_s(s') \left\{ G(s,s') - \int_0^s \frac{\partial G(t,s')}{\partial s'} \bar{t} \cdot \bar{s}' + \frac{\partial}{\partial t} G(t,s') + G(t,s') \frac{\partial \bar{t}}{\partial t} \cdot \bar{s}' \right\} \cos(k(s-t)) \, dt = B \cos ks - j \frac{V_o}{2Z_o} \sin ks \quad (3)$$

and you see we have various differentials on the kernel all of which I am assured are well-behaved. We also have step discontinuities to feed this in a realistic fashion at the top. Although we have Yeh and Mei's program we have written our own in order to understand what we are doing. We apply the end condition separately because we think this keeps the matrix better conditioned,

---

[14] E.K. Miller, G.J. Burke & E.S. Selden 'Accuracy-Modelling Guidelines for Integral-Equation Evaluation of Thin-Wire Scattering Structures'. IEEE Trans Antennas Propag, Vol 19, pp534-537, July 1971.

[15] Y.S. Yeh & K.K Mei, 'Theory of conical equiangular spiral antennas - Part I - numerical technique', IEEE Trans. Antennas Propag, Vol 15 pp634-639, Sept 1967.

[16] A.E. Atia & K.K. Mei, 'Analysis of Multiple-Arm Conical Log-Spiral Antennas' IEEE Trans. Antennas Propag, Vol 19, pp320-331, May 1971.

[17] K.K. Mei, 'On the integral equations of thin wire antennas' IEEE Trans. Antennas Propag, Vol 13, pp374-378, May 1965.

as I mentioned before, and recently we have been trying to change the approximate kernel to the exact kernel. Both the unknown constant, B condition applied on the left-hand side and the use of the approximate kernel introduce ripple into the answer and you can get situations where the ripple starts to build up before the solutions start to converge. I think probably the exact kernel is the way out of this. Before I came to this Institute I was using different approximations for different regions to represent the exact kernel. This is very difficult because you find that the regions are not very well defined. I was talking to Chalmers Butler and I think I now have a far better idea of how this should be tackled. To someone who hasn't run many programs I think they should realise that it is a very good idea to get the exact kernel in at the start rather than try to put it in by modifying the program afterwards.

We rarely have numerical trouble with our solutions and we have very good experimental results with the double check of both a near and far field measurement. As with Mr. Owen just now we get good agreement if we use the correct equivalent arm width. There did not seem any particular justification for the wire radius he used. Similarly there seems no particular reason why we cannot use various values to get good answers. I raised this problem at one of the other panel discussions you may recall. Should you unravel the wire to model a tape, what about the effects of the singularities on the edge, etc. and we seem to be in a situation between surfaces and isolated wires. We have used the programs and they work to some extent, we don't get miraculous agreement with experiments but they are giving us some sort of guidance on what we can do to optimise our antennas. If any of the theoreticians here want any guides as to what we would like, our experimental antennas now use zig-zag structures within the original envelope and they also contain loading resistors on the bottom, so that you have extra boundary problems. We are also using dielectric on the inside and the outside so perhaps you could use the Sommerfeld equation in there as well. If anyone can give us a solution to this configuration, we will be quite happy to use it.

<u>Professor R.F. Harrington</u> - Anybody want to make any comments?

<u>Dr Y. Rahmat-Samii</u> - For an active antenna. If you use the integral equation as you describe what was the procedure to obtain the unknown coefficients B and C. Hallen's Integral equation for this particular geometry; you said you can impose directly the condition at the active element point. How was that done?

<u>Dr Ramsdale</u> - Well we are using an entire domain on each section of the antenna.

Dr Rahmat-Samii - What is the current at the edge, do you force it to be zero just because of the way you use your basis functions? Do you impose that condition beforehand?

Dr Ramsdale - No, we impose all the boundary conditions at the end of the solution. We are really using super-position to say that this antenna is excited by a voltage source at the centre together with some other generators which are in effect the boundary conditions. So we are using a connection matrix to satisfy all the different boundary conditions.

Dr Rahmat-Samii - In that case what are your basis functions for the current at the edge in your expansion?

Dr Ramsdale - Well these don't go to zero, because if you set up an equation which goes to zero you are in fact putting the unknown constants on the left-hand side of the equation resulting in poorer matrix conditioning. So we keep everything outside the matrix and incorporate it afterwards. This is very handy for such problems as synthesis which I think Dr Djordjevic wants to talk about later on. By keeping all the wires completely disconnected and then using a continuity matrix afterwards we can juggle around with different transistors, different loadings, etc.

Professor Chalmers Butler - Two or three questions and a comment. Concerning that integral equation, the Hallen integral equation where you enforce the boundary conditions at the end, we have done this and we have had difficulty in doing it. The reason we have difficulty is that if one looks at the left-hand side of the equation; the integral of I against the Green's function with the forcing function, for example sin (kz), then the solutions to that are infinite at the edge. How do you add these infinities so they all add up to zero? That is a problem that we finally left.

Dr Ramsdale - That is why I talked about the low order solutions, where the series of the I's is incapable of taking any energy from the infinity. If you take a delta function generator Wu [18] showed that the energy contained close to it was very small and short lived. This is one trouble with the entire domain solution, the great temptation is to move your matching points closer and closer together and so start representing one of these infinities.

---

[18] T.T. Wu & R.W.P. King, 'Driving-Point and Input Admittance of Linear Antennas', J. Apply Phys. Vol 30, pp74- January, 1959.

Professor Butler - My real fear is the satisfying of the boundary conditions at the end, because of the fact that the solutions to that problem, broken down into three parts, are infinite at the end. The kernel has a large singularity and that is what is forcing it.

The other comment is that we have re-concluded a few years ago that with piece-wise sine testing of the Pocklington equation the convergence will always be as fast as the convergence of Hallen's equation even if you used Hallen's equation with spline functions, but you have to be a little bit careful about that because if you do an antenna problem you might find a different convergence trend from what you would find if you do a scattering problem. So we concluded something quite different from what you concluded. Furthermore, it was shown in the paper that you referred to, [5] and this was shown analytically not by some numerical comparison, it was shown by solving a difference equation, that the piece-wise sine testing of Pocklington's equation yielded precisely the same result that one obtains by starting with Hallen's equation, so we couldn't see any way that one would converge faster than the other.

Except may be in the way people write programs.

Dr Ramsdale - Well, on those engineering examples we have considered we have found that Hallen's equation gives rapid convergence most of the time. We have mainly been using the reduced kernel and Hallen is probably better for that. If you use the exact kernel I think what you say is probably true.

Professor Harrington - Thank you. We seem to have some further time left and I would like to ask Dr T.K. Sarkar to speak on the solution of operator equations.

Dr T.K. Sarkar - The subject that I am trying to cover is the general solution of operator equations. We will talk of some of the advantages of Method of Moments and of the other conventional methods. We shall also make an attempt to establish some of the conditions of convergence of the various methods. Finally, we shall illustrate what we mean by ill-posed problems.

The problem of interest is the solution of the operator equation $Au = f$ where A is a mapping from elements in the space E (called the domain of the operator) to the elements in the space F (called the range of the operator). This is explained in Figure 1. Also A^*, $N(A)$ and $N(A^*)$ are defined as the adjoint, null space of the operator A and the null space of the adjoint, respectively. Thus $N(A)$ is that mapping which takes elements in E and maps it to the zero element in F. Here A is a known operator and f is a given

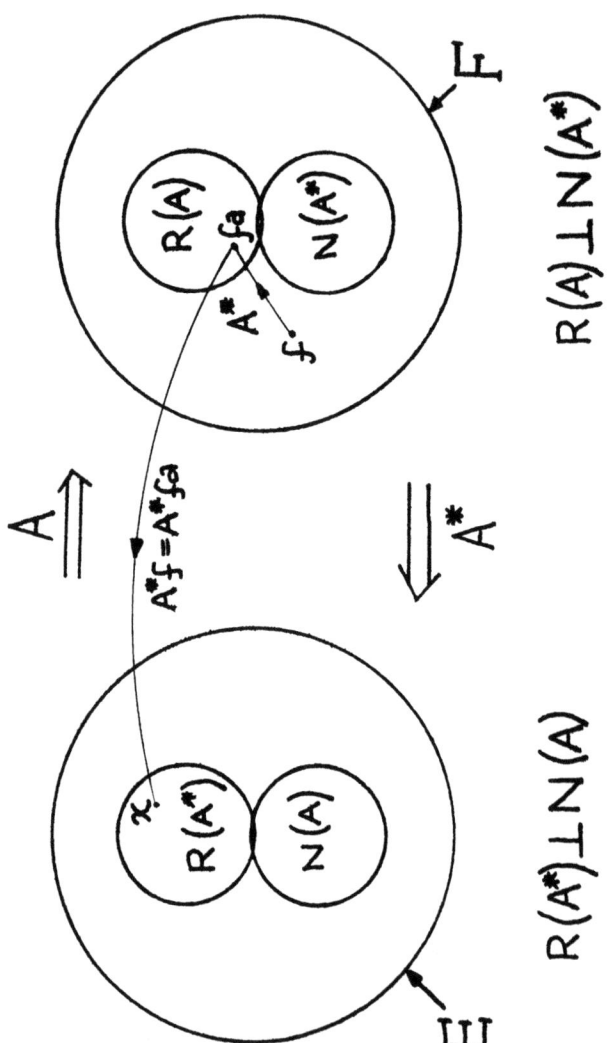

Fig 1
General Operator Mapping

element of F. The problem is to find u. This problem naturally raises the following four questions:

(1) For what elements of F is a solution possible? The answer to this question is given by the compatibility or the consistency conditions. By definition of R(A), an equation Au = f is compatible (has a solution) if and only if f ε R(A). So if A is onto F then f must be in R(A) and a solution must exist. But A is onto F if and only if N(A*) = θ, where A* is the adjoint of A, θ is the empty set and N(·) denotes the null space of an operator. In turn, N(A*) = θ, if and only if A* is one-to-one. These three conditions on A and A* are equivalent ways of stating that the equation Au = f is compatible for all f ε F. However, if N(A*) ≠ θ, then we have the more general condition:

R(A) = $[N(A^*)]^\perp$, where $[N(A^*)]^\perp$ constitutes the space of all vectors which are orthogonal to every vector belonging to N(A*). This is called the Fredholm alternative. It is interesting to note that N(A*) = $[R(A)]^\perp$ is valid for any arbitrary operator, whereas the Fredholm alternative may fail. This is because the Fredholm alternative is valid only if R(A) is closed. When R(A) is closed the above two conditions are equivalent.

<u>Dr Y. Rahmat-Samii</u> - What do you mean exactly saying that the range is closed?

<u>Dr Sarkar</u> - I mean that the range is a closed set. A closed set contains all the limits of all sequences which can be constructed from the set.

(2) Is the solution u unique? We say that the solution is unique if A is one-to-one. Moreover A is one-to-one if and only if N(A) = θ. And N(A) = θ if and only if A* is onto E. Thus we have three equivalent ways of establishing the uniqueness condition for Au = f.

In most practical problems of mathematical physics we have A as one-to-one but not onto F. In other words N(A) = θ, but N(A*) ≠ θ. Equivalently, we satisfy the uniqueness condition but not the compatibility condition. These problems are solved by the commonly called least squares method. Note that if f ε R(A) the equation has a unique solution and the least square error is zero. However if f ∉ R(A) we seek a solution such that $||f - Au||_F$ is minimized. Under the condition f ∉ R(A) a Galerkin method may fail, whereas a least squares method would always yield meaningful solutions.

(3) Do small perturbations in f result in small perturbations in the solution u without the need to impose additional

constraints?

This implies that A^{-1} should be continuous. If A is a linear operator then mathematically continuity and boundedness are synonymous. We show in the last section that A^{-1} is bounded if R(A) is closed or A is positive/negative definite or A is a bounded below operator.

If any of the above three conditions are violated, then the mathematicians call the problem ill-posed. Thus condition (3) implies that most differential equations are better posed than integral equations because for most differential equations R(A) is often closed, so we have a bounded inverse. Whereas for many integral operators of the first kind the inverse may not be bounded.

An example of an ill-posed problem where conditions 1) and 3) are violated is the linear least-squares problem. Why is this problem ill-posed? This is because we may have a system of equations Au = f such that the excitation element f is not on the range of the operator A. In many radiation problems when we model the system we are not sure whether $f \in R(A)$ or not. This is typical when we model some kinds of physical situations or have an experimentally determined f. The characteristic of this problem is that, this problem has a unique least-squares solution. If you have a matrix equation and you know that a least-square approximation to the solution can be obtained, then that solution is unique. Now our equation doesn't satisfy the compatibility condition because f is not in the range of the operator. How do we solve this equation? Well if we project f into range of the operator plus the null-space of the adjoint we have:

$$f = f_{R(A)} + f_{N(A^*)}$$

and if we apply A* on both sides we have $A^*f = A^* f_{R(A)} = A^* Ax$, since $A^* f_{N(A^*)} = \phi$. The solution is unique and is given by $x = |A^*A|^{-1} A^*f$. Now you see small perturbations in f would produce small perturbations in x, but if you produce small perturbations in the elements of the matrix, it wouldn't produce small perturbations in the solution. The characteristics of solutions for these type of ill-posed problems is expressed by the following theorem:

<u>Theorem</u>: If in Au = f, we have $f \notin R(A)$, then one can always find a solution u_k such that $\lim_{k \to \infty} ||Au_k - f|| \to 0$ whereas $\lim_{k \to \infty} ||u_k|| \to \infty$. This implies that in a problem where $f \notin R(A)$, one could always find a solution u_k for a k th order approximation such that the residuals $|Au_k - f| \to 0$ whereas the solution

diverges. This is typical of all ill-posed problems.

Then how do we make an ill-posed problem well-posed? I was told by some gentleman from Delft, that if they have an ill-posed operator the way they make it well is by giving it some aspirin. However, the mathematicians do it by imposing certain restrictions on the solutions possible, thereby restricting u_k to a certain admissible class of solutions. For example in antenna pattern synthesis, one could apply the Tikhonov regularization scheme by making $||u_k|| \leq K$, a constant independent of M. This procedure has been applied successfully by Deschamps and Cabyan in antenna pattern synthesis problem. It is worthwhile to mention that the regularization scheme of Tikhonov is not the only regularization scheme that exists in the literature. There are other regularization schemes, namely the statistical regularization schemes such as maximum entropy, maximum likelihood schemes which Dr Miller and Dr Bennett mentioned in their presentations.

<u>Dr E.K. Miller</u> - Could you give an example in solving such a simple problem in wire antenna? When using your terminology is the excitation not in the range of the operator?

<u>Mr T.K. Sarkar</u> - If you take Hallen's integral equation for the plane wave scattering by a thin wire of length L then we have:

$$\int_0^L I(z')G(z,z')\, dz' = A \cos kz + B \sin kz - \frac{j\omega\varepsilon_o E^i}{k^2}$$

Now if you break the currents up into three parts. Namely, I_1 being the solution to the excitation of $\cos kz$, I_2 the corresponding excitation $\sin kz$ and I_p for the constant incident field, then we believe the equation:

$$\int_0^L I_p(z')\, G(z,z')\, dz' = \text{a constant for } 0 < z < L$$

has no solution as

$$\int_0^L I_p(z') \frac{\partial G(z,z')}{\partial z}\, dz' = 0$$

So for the scattering problem we believe the broadside constant plane wave excitation is not in the range of the operator. Now we haven't done an exhaustive mathematical proof to show that it isn't, we are in the process of doing that.

<u>Dr E.K. Miller</u> - Excuse me, is that the thin wire kernel?

<u>Mr T.K. Sarkar</u> - You could have a thin wire kernel or you could have the exact kernel. I might say you probably ought to put the exact kernel in there. I think it's well known that if you put a delta function as an excitation for the radiation problem then it meets the condition of the theorem presented. But again I haven't done an exhaustive mathematical verification to show that the delta function is not in the range of the operator.

(4) The fourth question, which is really of importance to us as engineers is how do we construct a solution. This is apparently the most difficult question to answer and I shall try to explain why it is so difficult. Most of the methods that we have heard today are mainly these methods:
1) Rayleigh - Ritz variational method. 2) Galerkin method.
3) Least-squares - all of which are classified primarily by the name of Method of Moments. Then you have the other methods of solution like the Neumann series or the Fredholm series which apply to second kind of integral equations. In the latter case you write the solution in the form of infinite series: However the application of Neumann series is valid only if $||A|| < 1$ and A is a Hilbert-Schmidt (square integrable) kernel. Then one could solve the Fredholm equation of the second-kind

$$(I + A) u = f$$

by $u = (I + A)^{-1} f = \left[I + A + A^2 + \ldots \right] f$

If these conditions are not satisfied then you can't use the Neumann series to solve the problem but then one could use the Fredholm series to solve these type of second-kind integral equations if the operator belongs to the trace-class. Now what I mean by that is even if we have a singular kernel it must be integrable and most of the kernels in electromagnetic theory belong to that trace-class. Therefore if you get a magnetic field integral equation and if you do not have any internal resonance, hopefully you could write the exact solution to the problem in terms of the infinite series. The solution is unique, converges and is exact, but the question is why this is not done? This is because you run into a little bit of computational problem in that each term of the series is the determinant of an N x N matrix, and each entry of the matrix, is an n-fold integral. For a more general class of problem, methods which have been very popular to antenna engineers are primarily 1, 2 and 3. The variational method was first used by Lord Rayleigh (1871) in the solution of a boundary value problem. After 30 years Ritz wrote a paper which was published as variational method. As Finlayson points out Rayleigh himself (1911) commented on Ritz paper (1908) saying he was surprised that Ritz thought his method was new. As Courant (1943) remarked, it was probably the tragic circumstances of Ritz's work that caught the general interest. Ritz wrote his papers (1908, 1909) while aware that he soon was to die of tuberculosis. The variational method is also clearly presented in Ritz papers.

The next method that was developed was Galerkin's method. I would like to point out that Galerkin's method is the only method that I know of, was founded by an engineer. It is quite ironic that Galerkin wrote his first technical paper on the buckling of bars, columns, and systems of bars, while he himself was behind bars due to his anti-Tzarist views.

Galerkin's method is less restrictive than the variational method. In fact one could use the Galerkin's method to all sorts of problems but one may not be able to use the variational method in this way. As an example suppose we have a self-adjoint operator A and $Au = f$. What we do in a variational formulation is that we set up a potential function $F(u) = <Au, u> - 2<f, u>$. Then we take the functional derivative which is referred to as the Gateaux derivative and equate it to zero. $F'(u) = 2<Au - f, u'> = 0$. We make the expansion $u \sum \alpha_k u_k$, and we come up with an equation which is identical to Galerkin's method $\sum_k \alpha_k <Au_k - f, u_j> = 0$ for $j = 1,\ldots$ There is a pitfall over here that one has to be careful about. In many cases the functional F may never exist. As an example, suppose we have a second-order differential equation $u'' = -1/2\ x^{-3/2} = f$. If you impose this boundary condition $u(0) = 0$, $u(1) = -2$, the solution is $u = 2\sqrt{x}$. Now this functional is unbounded as $<f, u_0> \triangleq \int_0^1 f u_0\ dx = \int_0^1 -\frac{1}{x} dx = \infty$.

This implies that variational method is not as broadly applied as the Galerkin's method. That difference comes about physically because the variational method comes from the principle of conservation of energy, whereas Galerkin's method comes from the principle of virtual work. Those of you who are familiar with statics and dynamics know that the virtual work is a virtual displacement, it is not an actual displacement. Hence you don't have to worry about the application of the Galerkin's method. Another way to explain why the functional exists is to note that the potential function could be written as the line integral of an electric field and you just cannot substitute any arbitrary function for the potential to obtain an electric field. Only those potential functions are possible candidates in $\left[-\int \vec{E} \cdot \vec{dl} = v\right]$ which satisfy the curl equation ($\vec{\nabla} \times \vec{E} = 0$). This is one reason why one may not be able to define a variational method. Thus Galerkin's method is very versatile because we don't have to worry in certain cases about the creation of a fictitious function. Of course, variational method is advantageous when the functional F(u) has some physical meaning. Also a variational method guarantees the convergence of $U_n \to u$ exactly but does not guarantee the convergence of $||Au_n - f|| \to 0$.

Next is the method of moments. Historically when Galerkin's method is applied to differential equations it is called Bubnov-Galerkin method and when it is applied to integral equations it is called the method of moments. The name was coined by Yamada (1952) when he applied Galerkin's method to integral equations and used $1, x, x^2 \ldots$ as expansion functions. Naturally the integral of function multiplied by x is its first moment, by x^2 is its second moment and so on. Later on when Krylov started developing his theory on the convergence of Galerkin's method and the method of least-squares he used the generic name method of moments which subsequently has been used by Harrington. In the method of moments we choose the weighting functions in D(A*) because for any arbitrary operator the domain of the adjoint is always closed or could always be made closed. So in method of moments we make the residuals of the k th approximation $(Au_k - f)$ equal to zero in a certain functional space with respect to some weighting functions, i.e. $<Au_k - f, \omega_i> = 0$, where $\omega_i \in D(A^*)$ and $u_k = \sum_k \alpha_k x_k$. Mathematically this is equivalent to weak convergence of the residuals, i.e. $\underset{k \to \infty}{\text{Lim}} Au_k \overset{\omega}{\to} f$. This is satisfied as long as $||Au_k|| \leq M$, where M is a certain constant M independent of k. One would have strong convergence of the residuals, i.e. $\underset{k \to \infty}{\text{Lim}} ||Au_k - f|| \to 0$ if and only if A is bounded. This can be easily proven.

Now, so far we haven't said anything about the convergence of the solution, what we are talking about now is the convergence of the residuals. If you were to talk in terms of the weak or the strong convergence of the solution, then you have to have your inverse operator (A^{-1}) bounded, i.e. ($||A^{-1}|| \leq M$). If the inverse operator is not bounded then mathematically you cannot say (or rather it is difficult to say) that there is convergence of the solutions

u_k to u_o. I think it is important also to point out the method of least-square, even though it uses more work than the Galerkin's method may yield strong convergence for residuals whereas Galerkin's method may diverge. When both Galerkin's and that of the least-squares exists, one could show mathematically that the Galerkin's method converges faster than the method of least-squares. I was told by Professor Slivnik from Yugoslavia that the way some improve the rate of the convergence of an operator equation is to solve the problem on a faster computer. Thank you very much.

(Several reports on this subject may be available from the author).

<u>Dr. E.K. Miller</u>. Thank you Dr. Sarkar. That brings to a close this panel discussion and I would like to thank the members of the panel and the contributors in the audience for their participation.

Part II

HIGH FREQUENCY SCATTERING METHODS AND APPLICATIONS

Organised by Professor H. Bach
Technical University of Denmark, Denmark

INTRODUCTION TO GEOMETRICAL OPTICS

S. Cornbleet

Department of Physics, University of Surrey,
Guildford, Surrey, GU2 5XH, England.

1. INTRODUCTION

Geometrical Optics has an extremely long history and is probably among the oldest of sciences. From its origins it has been bound up with the history of mathematics and even today the conic sections play a fundamental rôle in the design of optical reflectors and the electromagnetic theory of waves in the theory of differential equations. The geometrical nature is undoubtedly due to the intuitive concepts of rays as straight lines and the visual experience of foci and caustics. The more recent investigations, say, from Snell and Descartes (1637) through to Hamilton and Faraday (ca 1850) saw the introduction of an increasing number of optical effects, refraction, double refraction, finite velocity of light, Newton's decomposition of white light, interference, diffraction, the theory of colour, polarization, total internal reflection, gratings, spectroscopy, Doppler effect and many more not explicable by the simple laws of rectilinear rays. For these a wave theory was developed by, among others, Huygens and Fraunhofer. The line above drawn at 1850 was done deliberately. It comes after the major work of Hamilton on geometrical optics [1] but precedes Maxwell's treatise on the electromagnetic theory of light [2] in 1873. The point to be made is that an enormous volume of work existed on the pure geometry of rectilinear rays, including the design of nearly all optical instruments and the work of Huygens (1690) and with full knowledge of the phenomena of interference and colour, before the advent of the only theory capable of explaining all of the above mentioned concepts. This dichotomy exists to the present day. It is easily possible to select extensive literature on the theory of optics and optical instruments that contains no reference to Maxwell's theory of

light. In general such literature will concern itself with
optical design and as such deals more with the geometrical part of
the theory. This will have its own postulations, theorems and
methods and will exist purely on the hypothesis of straight lines
(in homogeneous media) and the laws (postulated if not derived) of
reflection and refraction. That such straight lines can be replaced by light rays seems almost incidental. If, on the other hand,
the subsidiary effects need to be included then the more complete
electromagnetic problem has to be resolved. Herein lies another
dilemma. The complete solution of a problem involving the interaction of the electromagnetic field with material bodies, has only
been obtained exactly for a very few fundamental problems, whereas some extremely complicated optical systems can be derived with
comparative simplicity by geometrical methods. The procedure of
recent times has become one of stretching both of the underlying
concepts towards meeting and even of overlapping. This means
including in the term "geometrical optics" some concepts conventionally approached by E.M. theory and to evaluate some of the
more intractable E.M. problems by methods adopted from geometrical
optics. This diffuses the boundary of what one calls geometrical
optics considerably. One can have on one hand a geometrical
optics theory of diffraction and obtain on the other hand an
asymptotic E.M. solution in the geometrical optics limit.

2. FERMAT'S PRINCIPLE

Fermat's principle is the statement that the true optical path is
the path for which the optical length, defined as the product of
the physical length and the refractive index of the medium, is an
extremum. That is, between points P_1 and P_2

$$\delta \int_{P_1}^{P_2} \eta ds = 0 \qquad (1)$$

where η is the refractive index of the medium. We choose this as
our starting point for the reasons that this law is completely
general and is obeyed in very diverse and complex situations,
subject only to the conditions, normally required by the calculus
of variations, on the behaviour of η. It is thus applicable to
non-homogeneous media [3] crystalline media [4] and absorbing
media [5] and from it the laws of reflection and refraction can
easily be obtained. These are the usual postulates governing
geometrical optics. It is thus natural to use it, and such
advances as can be derived from it, in any future extensions of
the theory.

For a pure reflector, Fig. 1a, we have for an infinitesimal
variation of the path of the ray from P_1 to P_2

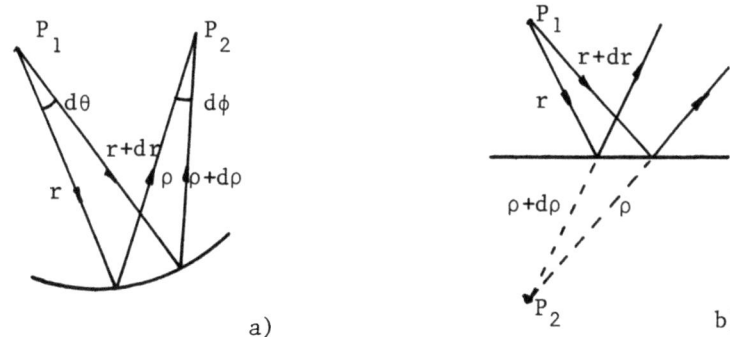

FIG. 1. Fermat's principle - reflection
a) Real image   b) Virtual image

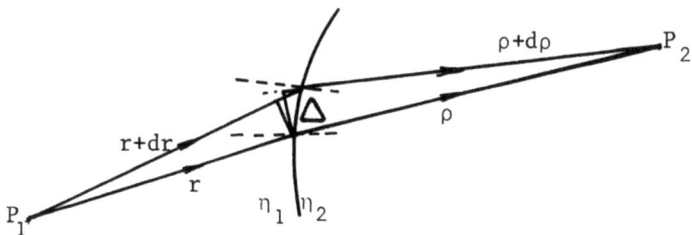

FIG. 2. Fermat's principle - refraction

$$r + (\rho + d\rho) = (r + dr) + \rho \quad \text{or} \quad d\rho = \mp dr \tag{2}$$

with sign depending on whether P_2 is a real (upper sign), or virtual point of intersection, Fig. 1b.

Expressed in bi-polar co-ordinates equation 2 is

$$r\, d\theta = \pm \rho\, d\phi \tag{3}$$

In a refraction, Fig. 2, we have by the two adjacent paths from P_1 to P_2

$$n_1(r + dr) + n_2 \rho = n_1 r + n_2(\rho + d\rho)$$

$$\text{or} \quad n_1 dr = \pm n_2 d\rho \tag{4}$$

with allocation of sign as before. With $n_2/n_1 = -1$ for a reflector this expression includes that of equation 2. These laws are by themselves sufficient to design most optical systems. They incorporate Snell's laws, for, dividing equation 4 by the common increment Δ

$$\eta_1 \frac{dr}{\Delta} = \eta_2 \frac{d\rho}{\Delta} \quad \text{or} \quad \eta_1 \sin \theta_1 = \eta_2 \sin \theta_2 \tag{5}$$

If performed vectorially [6] the rays and the surface normal can be shown to be coplanar. It has to be noted that this derivation is strictly local about the region of the true ray path. It applies to all surfaces for which an increment Δ exists, that is, for surfaces with continuous first derivatives in the describing variables.

By contrast the electromagnetic problem deals with the interaction of an infinite plane wave obliquely incident on an infinite plane interface and completely solves the boundary value problem, that is the continuity of electric and magnetic field components. This proves that the respective Poynting vectors and surface normal are coplanar, derives Snell's laws but includes many other major effects such as the existence of a reflected as well as refracted wave, polarization effects and Brewster's angle, the complex field arising in the situation of total internal reflection and the application to absorbing and highly conducting media [7]. It also retains full information regarding the amplitudes and phases of all the scattered components, the Fresnel formulae. However one has to go to a much deeper analysis [8,9] using the transport of discontinuities of the E.M. field, before the application to non-planar discontinuities or non-homogeneous media can be justified.

We continue hereafter with ray analysis in which as we have seen the operative element is the product of η with the increment ds of the optical path. If $\hat{s}$ is a unit vector tangential to the ray at any point, then we define the "directed ray vector" $\underline{t} = \eta \hat{s}$ and the integral in equation (1) for simplicity

$$S = \int \eta \, ds \tag{6}$$

Snell's laws of equation 4 vectorially are

$$\eta_1 \hat{n} \times \hat{s}_1 = \eta_2 \hat{n} \times \hat{s}_2 \quad \text{or} \quad (\underline{t}_1 - \underline{t}_2) \times \hat{n} = 0$$

so that $(\underline{t}_1 - \underline{t}_2)$ is parallel to $\hat{n}$, or

$$(\underline{t}_1 - \underline{t}_2) = \gamma \hat{n} \tag{7}$$

γ is dependent solely upon the incident ray, the surface normal and the relative refractive index of the interface media [10].

Let this surface have equation $\phi(x,y,z) = 0$ then $\hat{n}$ is proportional to $\nabla\phi$, so taking the curl of equation 7 (with γ constant)

$$\nabla \times \underset{\sim}{t}_1 = \nabla \times \underset{\sim}{t}_2$$

or $\nabla \times \underset{\sim}{t}$ is a constant along a true ray (8)

Consequently

$$\frac{d}{ds}(\nabla \times \underset{\sim}{t}) = 0 \tag{9}$$

and since for any ray with tangent vector $\hat{s}$ at a point $P(\underset{\sim}{r})$

$$\frac{d\underset{\sim}{r}}{ds} = \hat{s}$$

equation 9 becomes

$$\nabla \times \left| \frac{d}{ds}\left(\eta \frac{d\underset{\sim}{r}}{ds}\right) \right| = 0, \quad \text{and thus} \quad \frac{d}{ds}\left(\eta \frac{d\underset{\sim}{r}}{ds}\right)$$

is the gradient of a space function, and originally the normal to the surface of discontinuity. Then if there is a continuous variation in the refractive index we have for a true ray path

$$\frac{d}{ds}\left(\eta \frac{d\underset{\sim}{r}}{ds}\right) = \nabla \eta \tag{10}$$

This derivation has been simplified greatly to give a physical appreciation of the result. The complete analysis can be found in many places in the literature of optics [11,12].

Equation 10 gives the geometrical optics of rays in non-uniform media and can be used to design non-homogeneous lenses [13] and fibre optical guiding media [14]. The electromagnetic analysis of such structures shows the same effects as does the refraction described above [15] but without the a priori ability to design lenses and guiding structures.

From equation 8, if we take the arbitrary constant to be zero, then $\eta\hat{s}$ is also the gradient of the function S given by equation 6 as can be seen by formally integrating equation 10. That is

$$\frac{d\underset{\sim}{r}}{ds} \equiv \eta\hat{s} = \int \nabla\eta ds = \nabla \int \eta ds = \nabla S \tag{11}$$

There remains the ambiguity of sign as in equations 2 and 4 and thus strictly

$$\nabla S = \pm \eta \hat{s} \quad \text{or} \quad |\text{grad } S|^2 = \eta^2 \tag{12}$$

commonly known as the "eikonal" equation. If the end points P_1 and P_2 of a ray are varied subject to the ray remaining an extremum by Fermat's principle then we have at the ends only (since $\delta S = 0$ along each ray proper)

$$\delta S = (\nabla \eta \cdot \delta \underset{\sim}{r}) ds \Big|_{P_1}^{P_2} = \eta \frac{d\underset{\sim}{r}}{ds} \cdot \delta \underset{\sim}{r} \Big|_{P_1}^{P_2} \tag{13}$$

Consider now the ray from the point P_0 on a surface $F(x,y,z) = 0$ and normal to the surface that eventually passes through a point P. Along this ray the optical path will be P.

$$T = \int_{P_0}^{P} \eta \, ds$$

then

$$\delta T = \int_{P_0}^{P} \left\{ \frac{\partial \eta}{\partial \underset{\sim}{r}} \delta \underset{\sim}{r} - \frac{d}{ds}\left(\eta \frac{d\underset{\sim}{r}}{ds} \right) \right\} \cdot \delta \underset{\sim}{r} \, ds + \eta \frac{d\underset{\sim}{r}}{ds} \cdot \delta \underset{\sim}{r} \Big|_{P_0}^{P} \tag{14}$$

after integrating by parts in the classical manner [16]. The first term is zero since the path was taken along a ray and hence along an extremum. The second term is zero at the end point P because of the condition that the ray is normal to F there. Consequently as P varies

$$\delta T = \eta \frac{d\underset{\sim}{r}}{ds} \cdot \delta \underset{\sim}{r} \text{ at } P_0 \quad \text{or} \quad \eta \frac{d\underset{\sim}{r}}{ds} = \nabla_P T \tag{15}$$

This is the law of Malus and Dupin which states that if the rays are at any time normal to a surface F, then there exist surfaces T = constant which are normal to all the rays.

The geometrical relationship between S and T is Huygens' principle [17] namely "The surface T = constant is the envelope of surfaces $S(P_0,P)$ = constant associated with each point P_0 of F". For if we let u, v be parameters defining the surfaces, the equation of the envelope is obtained by differentiation and elimination that is

$$\frac{\partial S}{\partial u} = 0 \; ; \quad \frac{\partial S}{\partial v} = 0 \quad \text{so that} \quad \text{grad}_{P_0} S \cdot \frac{\partial \underset{\sim}{r}}{\partial u} = \text{grad}_{P_0} S \cdot \frac{\partial \underset{\sim}{r}}{\partial v} = 0 \tag{16}$$

From which it follows that just those rays which are perpendicular to the given surfaces meet the envelope. More extensive proofs can be found in the other references [18] or alternatively the converse could be proved. That is, that two surfaces with common ray normals of equal length ("parallel" surfaces in a homogeneous medium) can be shown to satisfy equation 10.

We thus arrive at a statement of Huygens' principle from purely geometrical considerations. Normally the good behaviour of the phase surface S = constant is assumed and the propagation of wave fronts and rays are synonymous. A divergence between the two concepts occurs immediately the surface S becomes irregular and in particular when it becomes discontinuous we shall show later which of the two approaches is more in keeping with experience.

3. CONSERVATION LAWS

The second requirement for the application of Huygens' principle is to allocate an amplitude at each point of the surface considered. In geometrical optics this attribution is usually a scalar function of position. Where the situation allows it, a vector distribution can be created by the appropriate scalar components. The propagation rules for amplitude depend on the previous definition of geometrical rays. A system of rays will enclose a tube and energy conservation within the tube will give the variation in amplitude as the rays of the tube are reflected, refracted or *scattered*. The analysis has only been applied to smooth surfaces S = constant with definable principle radii of curvature. Interference effects between adjacent tubes or intersecting tubes are not considered. Then if the principle radii of curvature on S at P are R_1 and R_2, Fig. 3, after the surface has progressed to S' a "parallel" distance P from S the principle radii of curvature at P' are $R_1' = (R_1 + p)$ and $R_2' = (R_2 + p)$. If W is the power density at p then in an element dA conservation will give

$$WdA = W'dA' \quad \text{or}$$

$$W' = \frac{WdA}{dA'} = W \frac{R_1 R_2}{(R_1+p)(R_2+p)} \tag{17}$$

The Gaussian curvature of a surface with principle radii of curvature $R_1 R_2$ is defined as $K = \frac{1}{R_1 R_2}$.

We can likewise define $K(p) = \frac{1}{(R+p)(R+p)}$

and thus equation 17 becomes

$$W' = W \frac{K(p)}{K(0)} \tag{18}$$

The relation is between power density, so for scalar field amplitude the square root of equation 18 is required [19].

Some quite complex scattering problems can be dealt with on this basis alone [20] and with no reference to phase or application of Huygens' principle, and it is a standard method for the design of reflectors with complicated shaped radiation patterns [21]. It constitutes the zeroth degree of approximation to the geometrical optical field, where shadows are sharp geometrical projections of the scatterer and a qualitative result is given for the distribution near to a focus or a caustic.

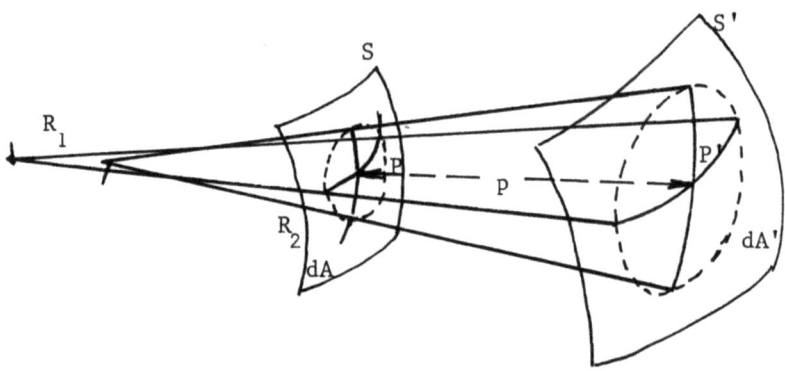

FIG. 3. Conservation of flux

4. THE RADIATED FIELD

The problem requires that we obtain the scattered field at a large distance from the scattering object. This cannot be done by ray methods *alone*. Indeed it is worth noting in this respect, that even in geometrical optics *design*, where ray methods predominate, the image field has to be of a special type such as a focus or a caustic, where *all* the rays can be accounted for. Attempts to treat non-exact focusing and aberrations by ray methods, for instance Herzberger's "spot diagrams" [22] are of the same qualitative nature as the conservation procedure. Alternatively, each ray direction can be considered as its associated infinite plane wave and the totality of rays as the "infinite spectrum" of plane waves. Complex problems with edges then require this spectrum to be continued into the region of complex angular directions [23].

If the radiated field is obtained directly from the induced sources on the scatterer, we have the "physical optics" approximation [24]. This method has to be extended for more complex problems, say, involving edges or penetrable media, by proposing additional forms of induced sources.

The method to which we are led by Huygens' concept can also be obtained from it by the addition of extensions that continue the simple solution in a way that can include the more complex ones. In this formulation, a modification of the Huygens'-Green formulation, the radiated field is obtained by applying Green's theorem

$$U(P) = \frac{1}{4\pi} \iint_S \left\{ U(Q) \frac{\partial}{\partial n} \left(\frac{e^{-iks}}{s} \right) - \frac{e^{-iks}}{s} \frac{\partial}{\partial n} U(Q) \right\} dS \qquad (19)$$

to the surface of a volume contained by an infinite plane and boundary hemisphere of infinite radius. We assume that all sources and scatterers lie in a localised region near to the origin on the outside of the volume. ∂/∂n represents differentiation in the direction of the inward pointing normal. The result is exact if U(Q) is known exactly and contains the elements of interference and, by components, polarization. In a concise survey, we will take a proven that the field obeys "the radiation condition at infinity" [25], which states that the field decays at large distances in such a way that the integration carried out over the infinite hemisphere tends to zero. S can then be taken to be the surface of the (arbitrary chosen) infinite plane only.

The functions to be used in this application of Green's theorem are:- a) $v = e^{-iks}/s$ the asymptotic solution of the time independent wave equation

$$(\nabla^2 + k^2)v = 0 \quad ; \quad k = \omega/c = 2\pi/\lambda \qquad (20)$$

s is the distance from the source point Q in the plane to the far-field point P.

[Note: the wave equation was known in physics long before Maxwell's formulation. It was when the latter showed wave behaviour in agreement with equation 20 that the electromagnetic wave nature of light was postulated].

The function v is also referred to as the free space Green's function. It represents an outgoing spherical wave from a point source and is the wavelet source required in Huygens' construction of subsidiary waves. It is asymptotically the radiation from a harmonic oscillator, the primary electromagnetic radiating source; b) U(Q), the complex distribution over the plane derived in the geometrical optics approximation by the principles of the previous section.

We vary the standard procedure by fixing P on the normal axis to S through the origin and sampling the far field by varying the direction of the infinite plane. Customarily the plane is fixed and the position of P varied, but this gives only a confined range over which the approximations required are valid.

The field $U(Q)$ will consist of two main contributions, the direct radiation from the source which is the basic geometrical field, and the field scattered from the object, thus

$$U(Q) = U_i(Q) + U_s(Q)$$

The scattered field $U_S(Q)$ is "carried" from the scatterer to the plane under consideration by a variety of processes which increase in complexity with increasing complexity of the problem posed. Those that are given by "ray" processes form a natural extension to geometrical optics.

The zeroth approximation is to regard the field to be $U_i(Q)$ alone, that of the incident field *as it would have occurred* in the absence of the scattering object. In the case of an aperture in a conducting plane, the incident field in the aperture, and in the case of a solid object the field outside the geometrical shadow. This is the Kirchhoff approximation and, as can be seen, is *not* an inherent requirement of Huygens' principle in spite of its geometrical basis and an eminent statement to that effect [26]. Even with this approximation and in a scalar representation, the interference effects resulting in diffraction and axial field variations, notably the Airy pattern [27] at the focus of a circular objective, are obtained.

Approximation methods for evaluating the integral of equation 19 can appropriately be called a first geometrical optics approximation, and a correction to the Kirchhoff solution, in that it is based upon the geometrically obtained Huygens' solution. In this we take, instead of the uniform phase surface S = constant, the arbitrary infinite plane surface of Green's theorem. To each point of this we have accordingly to attribute a value of phase, creating a "phase surface" over the *plane*. This phase surface being modulo 2π can always be confined between surfaces $\pm \pi$ and can have a finite number of discontinuities as a consequence. It too will consist of two main regions, the unperturbed region created by $U_i(Q)$ and a perturbed one created by $U_S(Q)$, the latter possibly extending well beyond the geometrical "shadow" of the scatterer. Methods are available to evaluate equation 19 with complex $U(Q)$. A statistical evaluation can be made even for highly disrupted phase surfaces such as would occur when a plane wave passes through a random space and time dependent phase changing screen [28].

We propose an approximation procedure which presents a more physical picture. This is based on the diffraction grating well known in optics and which is unique in that its action is obtained by a direct application of Huygens' principle and cannot be derived by ray methods.

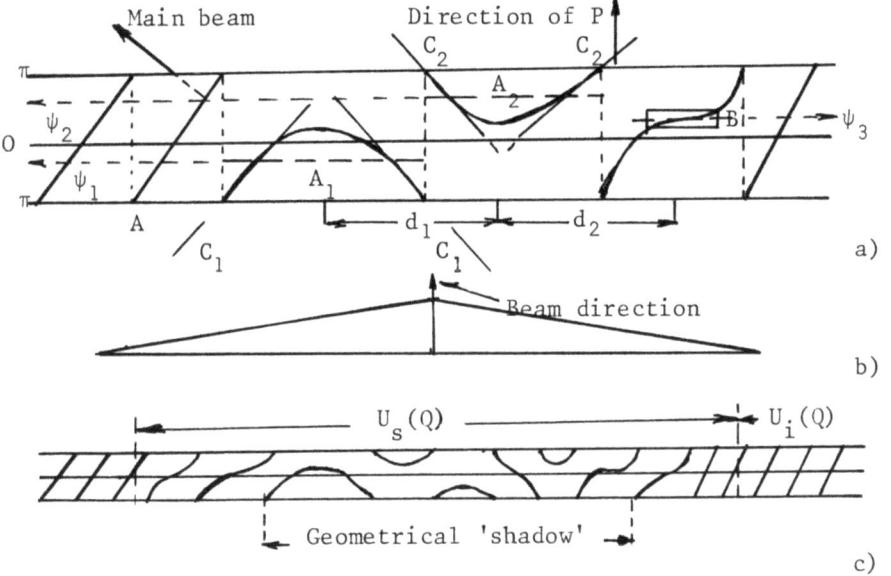

FIG. 4. Analysis of phase surface
a) Grating construction   b) Roof-top phase surface   c) Perturbed region $U_s(Q)$.

Parallel to the chosen plane there are regions about the turning values through which mean linear surfaces A_i and B_i can be taken, Fig. 4. Each will be even about its central value. The first approximation is to regard the radiated pattern, in the *locality* of the normal, as the result of these regions combining as a diffraction grating with phase distribution ψ_i and centres distance d_i as shown. This process is comparable to the "method of stationary phase" [29]. The amplitude distributions, or ray *density* over A and B have to be included in the evaluation, and the fundamental result for the unperturbed radiation from a slot of given width is required [30].

The next approximation is either to include in the foregoing the effect of the curvature of phase over the regions A_i and B_i, or to obtain separately the effects as a grating of *even* distributions such as C_i. This requires the inclusion of an obliquity factor. In certain cases where a "roof-top" surface of phase occurs these distributions may be the only type which arise. They are created mainly in conical systems such as the axicon [31]. *Odd* distributions can be ignored as having no effect in the locality of the normal.

The same procedure can be applied to the more complex phase fronts which occur in any scattering process. In some directions grating elements can occur with very wide spacing compared with the dimensions of the scatterer, giving rise to the very narrow low amplitude lobes which occur in scattering processes. There is no other way that such high resolution peaks can be physically explained.

The process is then repeated with other chosen directions of source plane, building up the angular spectrum. Once a method of deriving U(Q) is available it can usually be employed for all orientations of the Huygens' plane of interest. The principle that ray directions *alone* do not predict the far field behaviour has been experimentally verified in the case of an antenna design based solely upon Huygens' principle and actually contravening the laws of reflection [32].

Many scattering techniques currently used for the more complex situations, are based on "ray" methods. These methods propose additional rays, creeping rays, edge diffracted rays the elements of the "Geometrical theory of diffraction" [33]. It may well be that the use of rays alone has given rise to some of the anomalies that occur in these processes such as at caustics and foci, whereas the Huygens-Green formulation has been shown to be accurate in those areas.

More important still are the processes for which real rays transporting energy to the far field cannot be obtained, such as the inhomogeneous fields occurring at a total internal reflection, the inductive field close to impedance surfaces, whispering gallery modes and others [34]. This includes even the fundamental solutions for the scattering by an infinite straight edge [35]. Such complex rays or their analogue in the complex region of the angular spectrum of plane waves, cannot have any direct effect in the far radiation zone, but they can and do influence the local Huygens' source plane and thence via Green's formulation, the far field. This is instanced in the treatment of the illuminated aperture by a "boundary wave" solution [36]. This is, in principle, a standing wave *in the aperture* and its effect is transverse to the unperturbed wave direction, but its addition to the *aperture* field, gives corrected radiated patterns. Most interestingly it is shown that this same result derives from the separation of equation 19 into two integrals with "overlapping" boundary conditions, the Rayleigh-Sommerfeld integrals [37].

It has still to be proved, but can readily be demonstrated, that the exact location of the arbitrary plane considered "near" to the scatterer does not affect the heuristic result as quoted.

With this proviso, we can state that the predominantly Huygens' concept, used in the way illustrated, forms the major extension to geometrical optics in a variety of scattering situations.

Finally we note the very different definition of geometrical optics obtained by the assumption of a solution of equation 20 of the form [38]

$$V = A(x,y,z)e^{ik_0 S(x,y,z)} \qquad (21)$$

Substitution into the time independent wave equation results in a series, the first two terms of which are

$$-k_0^2 v \left\{ \left(\frac{\partial S}{\partial x}\right)^2 + \left(\frac{\partial S}{\partial y}\right)^2 + \left(\frac{\partial S}{\partial z}\right)^2 - \frac{k^2}{k_0^2} \right\} \qquad (22)$$

$$2ik_0 v \left\{ \frac{1}{2} \nabla^2 S + \frac{1}{A} \text{grad } A \cdot \text{grad } S \right\} \qquad (23)$$

The remaining terms remain finite in the limit $k_0 \to \infty$ thus those terms have to be equated to zero. Equation 22 then becomes equation 12 and after some manipulation equation 23 agrees with equation 18. This limit is therefore termed the "geometrical optics limit". However the essential elements that lead to the ray congruences and Huygens' principle have been bypassed and consequently other means have to be devised such as asymptotic solutions to include the effects noted above.

REFERENCES

References to SO are in: A. Sommerfeld Lectures on Theoretical Physics Vol. IV Optics Academic Press 1964.
References to Sl are in: S. Silver Microwave Antenna Theory and Design M.I.T. Radiation Laboratory series Vol. 12 1949.
References to DSJ are in: D.S. Jones The Theory of Electromagnetism Pergamon Press 1964.
References to RKL are in: R.K. Luneburg The Mathematical Theory of Optics University of California Press 1964.

1. A.W. Conway and J.L. Synge (Eds.) The collected papers of W.R. Hamilton, 1, Geometrical Optics, Cambridge University Press, 1931.
2. The Scientific Papers of J.C. Maxwell (Ed. W.D. Niven) Dover Publications, N.Y., 1965.
3. a) RKL 86; SO 355.
   b) O.N. Stavroudis, The optics of rays wavefronts and caustics, Academic Press, 1972, 14.

4. a) W. Pauli, Lectures on Physics (ed. C.P. Enz) $\underline{2}$, Optics and the Theory of Electrons, M.I.T., 90 and 99, 1973.
   b) L.D. Landau and E.M. Lifshitz, Electrodynamics of Continuous Media, Pergamon Press, 1960, 317.
5. a) D. Censor, Fermat's principle and real space-time rays in absorbing media, Jour. Phys. A: Math. Gen. No. 10, $\underline{10}$, 1977, 1781-90.
   b) J.J. Brandstatter, Waves Rays and Radiation in Plasma Media, McGraw Hill, 1963, 345.
6. S1 123; DSJ 340.
7. a) DSJ 321; SO 13-40.
   b) M. Born and E. Wolf, Principles of Optics, Pergamon Press, 1959, 46.
8. RKL 18-30.
9. M. Kline and I.W. Kay, Electromagnetic Theory and Geometrical Optics, Interscience, 1956, Chaps. 1-6.
10. H.P. Brueggemann, Conic Mirrors, Focal Press, 1968, Chap. 1.
11. DSJ 343.
12. SO 338.
13. a) RKL 182.
    b) S. Cornbleet, Microwave Optics, Academic Press, 1976, 108-155.
14. S. Cornbleet, Optical guiding in a general non-homogeneous cylindrical medium (to be published).
15. C.T. Tai, Dyadic Green's functions in electromagnetic theory, Intext Press, 1971, Chap. 12.
16. a) DSJ 343; ref. 3b, 17.
    b) H. Goldstein, Classical Mechanics, Addison-Wesley, 1964, 33.
17. ref. 4a, 9; DSJ 678; SO 148 and 180.
18. RKL 82; ref. 7b, 131; ref. 9, 138; ref. 4a, 88.
19. a) S1 139.
    b) R.E. Collin and F.J. Zucker, Antenna Theory, Part II, McGraw Hill, 1969, 31.
20. a) D.G. Burkhard and D.L. Shealy, Flux density for ray propagation in geometrical optics, Jour. Opt. Soc. Amer. $\underline{63}$, No. 3, 299.
    b) H.J. Riblet and C.B. Barker, A general divergence formula, Jour. App. Phys., Jan. 1948, 63.
21. a) S1 465, 506, 391.
    b) B.S. Westcott and A.P. Norris, Reflector synthesis for generalized far-fields, Jour. Phys. A: Math. Gen. No. 4, $\underline{8}$, 1975, 521 (also 1976, 113).
22. M. Herzberger, Modern Geometrical Optics, Wiley Interscience, 1956, Chap. 24.
23. H.G. Booker and P.C. Clemmow, The concept of an angular spectrum of plane waves and its relation to that of polar diagram and aperture distributions, Proc. I.E.E., No. 3, $\underline{97}$, 1950, 11.
24. W.V.T. Rusch and P.D. Potter, Analysis of Reflector Antennas, Academic Press, 1970.

25. DSJ 563; SO 197 and 250.
26. SO 201.
27. Ref. 7b, 395; ref. 4a, 57.
28. a) M.V. Berry, Disruption of wavefronts: statistics of dislocations in incoherent Gaussian random waves, Jour. Phys. A: Math. Gen. No. 1, $\underline{11}$, 1978, 27.
   b) P. Beckmann and A. Spizzichino, The Scattering of Electromagnetic Waves from Rough Surfaces, Pergamon Press, 1963.
   c) A. Ishimaru, Wave Propagation and Scattering in Random Media, 2 Vols., Academic Press, 1978.
29. DSJ 449; SO 317.
30. Ref. 7b, 266; S1 183; SO 237; ref. 4a, Chap. 2.
31. J.H. McLeod, Axicons and their uses, Jour. Opt. Soc. Amer., No. 2, $\underline{50}$, 1960, 166.
32. S. Cornbleet and B.J. Gunney, Corrected plane reflectors, I.E.E. Conference on Antennas and Propagation, Nov. 1978, Conference Publication No. 169, 201.
33. a) J.B. Keller, Geometrical theory of diffraction, Jour. Opt. Soc. Amer. $\underline{52}$, 1962, 116.
   b) J.B. Keller and B.R. Levy, Decay exponents and diffraction coefficients for surface waves, Trans. I.R.E., AP7, 1959, S52-S61.
34. a) S. Choudhary and L.B. Felsen, Asymptotic theory for inhomogeneous waves, I.E.E.E. Trans., AP21, No. 6, Nov. 1971, 827.
   b) Articles in the special issue on Rays and Beams, Proc. IEEE, Nov. 1974 with numerous references.
35. a) B.B. Baker and E.T. Copson, The mathematical theory of Huygens' principle, Oxford Univ. Press, 1969.
   b) DSJ, Chap. 9; SO 249-272.
   c) V.I. Smirnov, Course of Higher Mathematics, Vol. 3, Pt. II, 120-123 and 202-217. Pergamon, 1964.
   d) H. Lamb, On Sommerfeld's diffraction problem and on reflection by a parabolic mirror, Proc. London Math. Soc., $\underline{4}$, 1907, 190-203.
   e) E.T. Hanson, Diffraction, Phil. Trans. Roy. Soc. A: $\underline{229}$, 1930, 87.
   f) H.S. Carslaw, Some multiform solutions of the partial differential equations of Physical Mathematics and their applications, Proc. Lond. Math. Soc., Nov. 10, 1898, 121.
36. a) K. Miyamoto and E. Wolf, Generalization of the Maggi-Rubinowicz theory of the boundary diffraction wave, Jour. Opt. Soc. Amer., $\underline{52}$, 1962, 615-637.
   b) E. Wolf, Some recent research on diffraction of light, Proceding of Symposium on Modern Optics, 1967. Polytechnic Press, Brooklyn, N.Y., 433.
37. E. Wolf and E.W. Marchand, Comparison of the Kirchhoff and the Rayleigh-Sommerfled theories of diffraction at an aperture, Jour. Opt. Soc. Amer., May 1964, $\underline{54}$, 587.
38. a) DSJ 352; ref. 9b, 4.
   b) J.C. Heurtley, Scalar Rayleigh-Sommerfeld and Kirchhoff diffraction integrals, Jour. Opt. Soc. Amer., $\underline{63}$, 1973, 1003.

FOUNDATIONS OF THE GEOMETRICAL THEORY OF DIFFRACTION

Leopold B. Felsen

Polytechnic Institute of New York, Farmingdale, New York 11735, U.S.A.

1. Ray Method

At high frequencies, the electromagnetic fields away from source regions (whether actual, as on an antenna, or induced, as on a passive scatterer) can be approximated by equations that are simpler than the Maxwell vector field equations. The simplification arises from the fact that in the far zone of source distributions, the evanescent storage field can be neglected. The simplified equations are based on the assumption that $kL \gg 1$, where $k = 2\pi/\lambda$ (with λ representing the wavelength) is the wavenumber in the medium and L is a scale length describing the characteristic observation distances, obstacle dimensions or medium inhomogeneities. The precise nature of L depends on the problem under consideration, and it is therefore convenient in the mathematical treatment to regard k by itself as the large parameter, keeping in mind that normalization in the combined form kL is intended eventually. The simplified equations are found to characterize the high-frequency field in terms of local propagation phenomena involving propagation paths called rays. Thus, the field from an initial reference surface A to an observation point P can be tracked along a ray that passes through P and originates at a point P' on A. The field at P is affected only by the initial field values in the vicinity of P' and by the medium properties in the vicinity of the ray. It is possible that several rays pass through P, either because of the nature of the field distribution on A or because the presence of obstacles or scatterers gives rise to other local fields (Fig. 1). In that event, the total field at P is synthesized by the sum of the fields reaching P along the various rays. The localization arises from the fact that high-frequency source distributions generate a spectrum of plane waves that interfere constructively along certain preferred directions, the rays, and thereby give rise to a strong field; along other directions that deviate from the rays, these waves interfere destructively and cause weak effects that can be ignored.[1,2,3]

Along the ray trajectories, the constructively interfering plane waves can be characterized as a <u>local</u> plane wave. The local plane wave concept is central to the tracking of high-frequency fields. To understand what is involved, we consider the distinction between true and local plane waves. A true plane wave in a homogeneous medium has a plane phase front (equiphase surface) $A(\bar{r})$ = constant ≡ A_o and a constant amplitude $u(\bar{r}) = u_o$, where $\bar{r}$ is the position vector. The field propagates in the direction perpendicular to the phase front. These perpendicular trajectories, along which the phase front advances, point in the direction of $\Delta A(\bar{r})|_{A=A_o}$ and are the rays. For an observation point P along a ray, the field differs from that at P' on the initial surface A_o = const. only by the phase change $\exp(-jkd)$ where d is the distance between P' and P (Fig. 2).* A local plane wave describes a field characterized by non-planar phase fronts. The simplext example is that of a spherical wave. Ignoring the vectorial properties, the scalar field in the far zone is given by

$$u(\bar{r}) \sim C \frac{e^{-jkr}}{kr} \qquad (1)$$

where C is a constant. If $r = r_o$ denotes the initial wave front, then the rays point in the rays point in the direction $\Delta r|_{r=r_o} = \hat{r}$, where $\hat{r}$ is a radial unit vector. The field at P along the ray originating at P' differs from the field at P' not only by the phase change kd as in a true plane wave but also by the amplitude change $r_o/(r-r_o)$. This amplitude change identifies the field along the ray as a local plane wave field which is synthesized by constructive interference of a bundle of packet of true plane waves whose propagation directions are close to $\hat{r}$ (Fig. 3(a)).

Since energy in the high-frequency field, as carried by the local plane waves, flows along the direction of the rays, energy is conserved in a tube of rays. Thus, the energy density is inversely proportional to the ray tube cross section dA. The field amplitude is proportional to the square root of the energy density and therefore varies inversely with $\sqrt{dA}$. For the spherical wavefront in Fig. 3(a), the ray tube is conic and the cross section varies as r^2; the amplitude therefore decreases as $(1/r)$ as in (1). When the wavefront has a more general shape

---

* In this chapter, the assumed time dependence is $\exp(j\omega t)$.

the various rays. The localization arises from the fact that high-frequency source distributions generate a spectrum of plane waves that interfere constructively along certain preferred directions, the rays, and thereby give rise to a strong field; along other directions that deviate from the rays, these waves interfere destructively and cause weak effects that can be ignored.[1,2,3]

Along the ray trajectories, the constructively interfering plane waves can be characterized as a <u>local</u> plane wave. The local plane wave concept is central to the tracking of high-frequency fields. To understand what is involved, we consider the distinction between true and local plane waves. A true plane wave in a homogeneous medium has a plane phase front (equiphase surface) $A(\bar{r}) = \text{constant} \equiv A_o$ and a constant amplitude $u(\bar{r}) = u_o$, where $\bar{r}$ is the position vector. The field propagates in the direction perpendicular to the phase front. These perpendicular trajectories, along which the phase front advances, point in the direction of $\nabla A(\bar{r})|_{A=A_o}$ and are the rays. For an observation point P along a ray, the field differs from that at P' on the initial surface $A_o = \text{const.}$ only by the phase change $\exp(-jkd)$ where d is the distance between P' and P (Fig. 2). A local plane wave describes a field characterized by non-planar phase fronts. The simplest example is that of a spherical wave. Ignoring the vectorial properties, the scalar field in the far zone is given by

$$u(\bar{r}) \sim C \frac{e^{-jkr}}{kr} \qquad (1)$$

where C is a constant. If $r = r_o$ denotes the initial wave front, then the rays point in the direction $\nabla r|_{r=r_o} = \hat{r}$, where $\hat{r}$ is a radial unit vector. The field at P along the ray originating at P' differs from the field at P' not only by the phase change kd as in a true plane wave but also by the amplitude change $r_o/(r-r_o)$. This amplitude change identifies the field along the ray as a local plane wave field which is synthesized by constructive interference of a bundle or packet of true plane waves whose propagation directions are close to $\hat{r}$ (Fig. 3(a)).

Since energy in the high-frequency field, as carried by the local plane waves, flows along the direction of the rays, energy is conserved in a tube of rays. Thus, the energy density is inversely proportional to the ray tube cross section dA. The field amplitude is proportional to the square root of the energy density and therefore varies inversely with $\sqrt{dA}$. For the spherical wavefront in Fig. 3(a). the ray tube is conical and the cross section varies as r^2; the amplitude therefore decreases as $(1/r)$ as in (1). When the wavefront has a more general shape

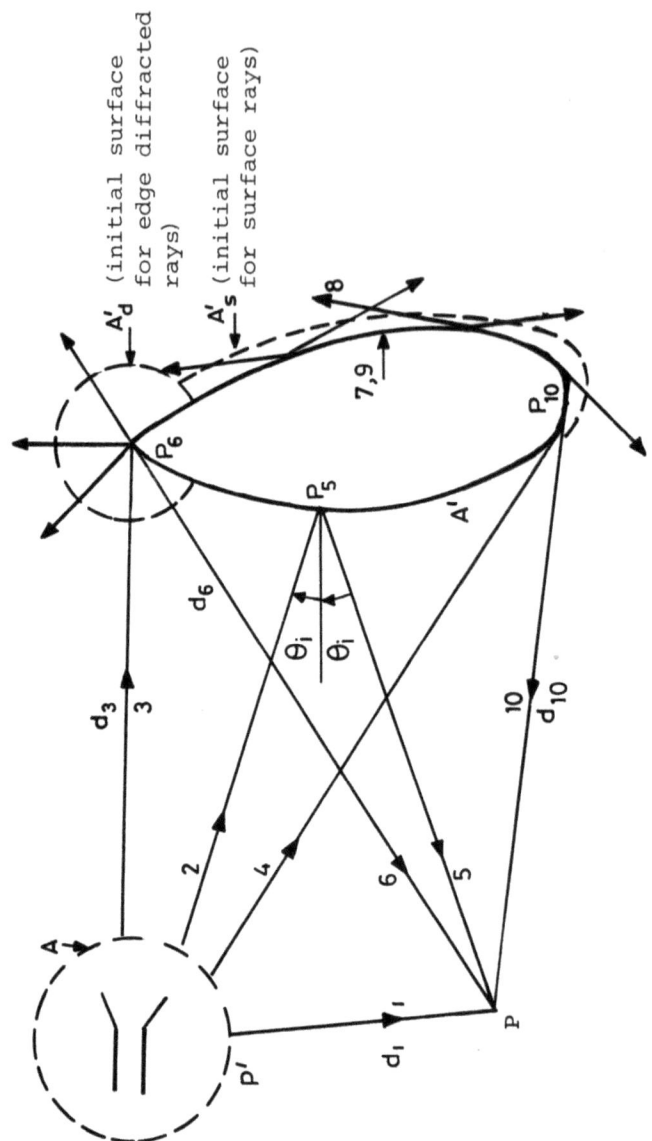

Figure 1. Scattering by a composite obstacle with surface contour A'. Various ray species have been identified by numbers explained in the text. The following are initial surfaces: A for incident rays, A' for reflected rays, A'_d for tip diffracted rays, and A'_s for surface (creeping) rays.

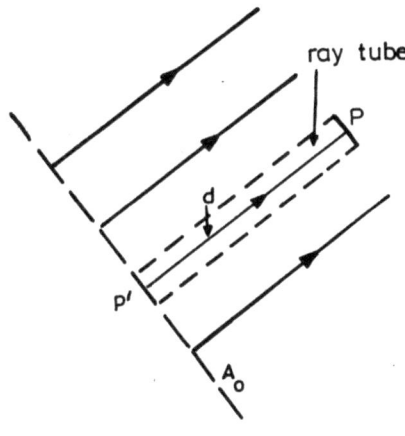

Figure 2. True plane wave

as in Fig. 3(b)), the ray tube cross section dA may have two principal radii of curvature R_1 and R_2, each centered on a surface called a caustic (Fig. 3(c)).

The preceding considerations lead to the following formulation of the local plane wave field carried along a ray (Sec. 2.2c):[2]

$$u(\bar{r}) \sim u_A \left(\frac{dA_A}{dA}\right)^{\frac{1}{2}} e^{-jkd},$$

$$\frac{dA_A}{dA} = \frac{R_1 R_2}{(R_1+d)(R_2+d)} \qquad (2)$$

where u_A is the initial field at P' on A (Fig. 3(b)), dA_A is the initial ray tube cross section, and d is the distance from P' to P. The expression for (dA_A/dA) in terms of R_1 and R_2 is inferred from Figs. 3(b) and (c).

By (2), the local plane wave field along a ray can be calculated from a knowledge of the initial field value and the ray geometry. The initial field value must be determined independently by solution of canonical problems. In the canonical problem, the incident local plane wave field may be replaced by a true plane wave field, and the obstacle configuration by a simpler geometry that nevertheless retains the correct local scattering properties. For the incident field in Fig. 1, the initial values on A can be calculated from the known source distribution. The reflected field is caused by incident rays (ray 2 in Fig. 1) that are reflected at the obstacle surface. The reflection laws for local plane waves on a curved surface are the same as for true plane waves on an infinite plane surface tangent to the curved surface at the point of impact of the incident ray (canonical problem). Thus, the incidence and reflection angles θ_i are equal. The initial value of the reflected field at P_5 is given by the incident field at P_5 multiplied by the plane wave reflection coefficient $\Gamma(\theta_i)$ descriptive of the reflecting properties of the boundary surface. The amplitude variation along ray 5, as determined by the ray tube cross section dA_5 (see (2)), involves the surface curvature at P_5, the curvature parameters for the incident wavefront, and the angle coordinates X and X_i specifying the directions of the reflected and incident rays, respectively. Therefore, the

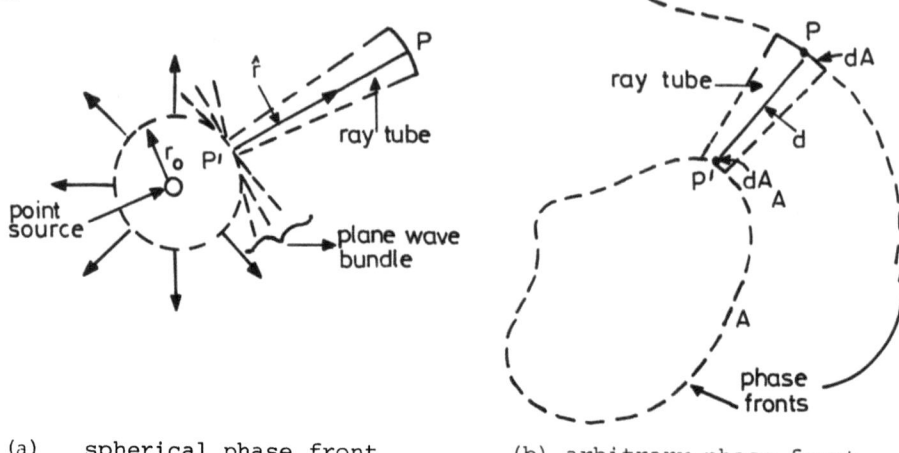

(a) spherical phase front   (b) arbitrary phase front

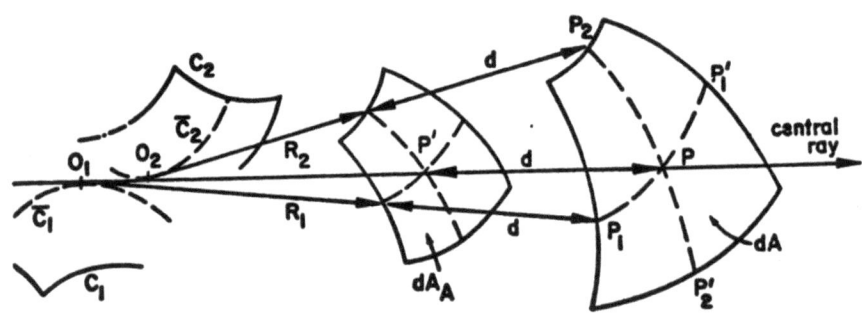

(c) Ray tube cross sections dA_A and dA for arbitrary phase front. The principal coordinate directions are shown dashed. The principal radii of curvature R_1 and R_2 are centered at the virtual foci O_1 and O_2, respectively, from which emanate the rays passing through the points (P_1, P_1') and (P_2, P_2'). These rays are tangent to caustic surfaces C_1 and C_2 whereon lie the virtual foci descriptive of ray tubes cut out elsewhere on the phase front. The curves $\bar{C}_1$ and $\bar{C}_2$ on the caustic surfaces correspond to rays passing through extensions of the principal coordinate curves P_1P_1' and P_2P_2'.

Figure 3. Local plane waves, rays and ray tubes.

reflected contribution to the field at P is (see ray 5 in Fig.1):

$$u_5 \sim u_2 \, \Gamma(\theta_i) \, M(5,2) \, e^{-jkd_5}, \quad u_2 \sim u_{A2} \, \frac{e^{-jkd_2}}{kd_2} \tag{2}$$

where u_{A2} is the initial field for ray 2 on surface A. To simplify the notation here and subsequently, we have written

$$(dA_{A5}/dA)^{\frac{1}{2}} \equiv M(\chi_5, \chi_2) = M(5,2) \tag{3a}$$

In this and subsequent formulas, d_i represents the path length along ray i.

The incident local plane wave field along ray 3 striking the conical tip of the obstacle in Fig. 1 excites a spherical wave front and therefore a family of rays centered at the tip. The canonical problem for the tip diffracted field is that of a plane wave incident on an infinite conical obstacle. That solution provides the <u>diffraction coefficient</u> $D(\chi, \chi_i)$*, by which the incident ray field is modified upon emerging from the conical tip. χ and χ_i denote, respectively, the angle coordinates specifying the directions of the diffracted and incident rays. Thus, the contribution at P due to the local plane wave along ray 6 is

$$u_6 \sim D(6,3) \, \frac{e^{-jkd_6}}{kd_6} \, u_3, \quad u_3 \sim u_{A3} \, \frac{e^{-jkd_3}}{kd_3} \tag{4}$$

where u_{A2} is the initial field for ray 3 on surface A. To simplify the notation here and subsequently, we have written

$$D(6,3) \equiv D(\chi_6, \chi_3) . \tag{4a}$$

Although the canonical problem yields the diffracted field on the initial surface A'_d in Fig. 1, the result can be expressed more conveniently as in (4) where relevant length parameters are measured from the tip.

When an incident ray grazes a smooth segment of a scatterer (ray 4 in Fig. 1), it excites a diffracted <u>surface ray</u> (also called creeping ray) that travels along a geodesic on the shadowed surface (ray 7) and sheds energy continuously (ray 8). The launching amplitude $D_\ell(\chi, \chi_i)$ of a surface ray field and its

---

* While the problem of plane wave diffraction by a conical obstacle can be solved, no convenient form for the diffraction coefficient has as yet been developed.

amplitude variation $M_s(d_7)$ along its geodesic path d_7 are determined from the canonical problem of plane wave diffraction by a smooth obstacle and have been found for special configurations. When a surface ray strikes the conical tip, it also gives rise to a spherical diffracted wave that adds a field u_6' to the contribution at P arriving along ray 6. The canonical problem of surface ray field diffraction by the tip of a conical obstacle is still under study; it furnishes the diffraction coefficient $D_s(\chi,\chi_i)$. The field u_6' is then given by:

$$u_6' \sim u_4 D_\ell(7,4) M_s(d_7) e^{-jkd_7} D_s(6,7) \frac{e^{-jkd_6}}{kd_6} , \quad u_4 \sim u_{A4} \frac{e^{-jkd_4}}{kd_4} , \quad (5)$$

where d_7 is the surface ray path (geodesic) between the point of tangency of the incident ray 4 and the tip.

Tip diffraction due to incident ray 3 also gives rise to a surface ray (ray 9) that emerges at P_{10} to contribute to the field at P along ray 10. The initial surface A_s for shed surface rays is displaced from the obstacle surface but in the formulation of the diffracted field, it is again convenient to measure distances from the obstacle surface. Thus,

$$u_{10} \sim u_3 D_s(9,3) M_s(d_9) e^{-jkd_9} D_\ell(10,9) \frac{e^{-jkd_{10}}}{kd_{10}} \quad (6)$$

where u_3 is given in (4), d_9 denotes the geodesic length along ray 9 from the tip to the shedding point P_{10}, and $D_\ell(10,9)$ gives the shedding amplitude, which can be normalized so that it equals in form the launching amplitude. Although the surface rays 7 and 9 in Fig. 1 appear to coincide, they describe different trajectories since the points of tangency of ray 4 and departure of ray 10 are not identical.

The total ray-optical field at point P in Fig.1 is now given by

$$u \sim u_1 + u_5 + u_6 + u_6' + u_{10} , \quad (7)$$

where

$$u_1 \sim u_{A1} \frac{e^{-jkd_1}}{kd_1} \quad (8)$$

denotes the field along the direct ray from the source, and the remaining contributions, given in (3) - (6), are due to the presence of the obstacle.

The result in (7) contains only the dominant contribution from each of the ray fields. Generally, each ray field has, in addition to this leading term, a series of higher order terms that decay inversely with k. The field is therefore given by an asymptotic expansion in inverse powers of k. Validity of the leading term alone implies that the higher order terms, in particular the second term, are small in comparison with the first. Some estimates on the accuracy of the asymptotically expanded field can be made on general grounds but for scattering problems encountered in practice, it is usually too difficult to apply them. Therefore, the range of parameters, for which formulas as in (7) are useful and valid, has been ascertained by comparison with canonical analytic solutions or with numerical solutions that can be generated independently. For example, in transition regions near shadow boundaries, caustics and foci, where relevant ray field amplitudes tend to infinity, the use of uniform asymptotic methods provides a valid description.

By the very construction of the field in (7), it is evident that the ray method decomposes a complicated composite scattering problem into a sequence of simpler (canonical) problems via the following steps:

1. Determination of the incident field over an initial surface A.

2. Determination of the reflected and diffracted ray fields that contribute at an observation point P.

3. Identification of canonical problems that treat separately each of the ray reflection and diffraction problems. The solutions of these problems furnish the initial amplitudes along various species of reflected and diffracted rays.

4. Synthesis of composite scattering problem by interaction (along rays) between canonical constituents.

The various ray species arising in the diffraction problem of Fig. 1 have been associated with a perfectly conducting obstacle. When the obstacle has other surface features (for example, edges) or when it is penetrable (lossless dielectric), additional diffraction mechanisms and corresponding ray fields may arise.[1,2,3]

2. Equations for the Ray-Optical Field, and their Solution[1,2,3]

2a. Derivation of the Equations

The scalar field $u(\bar{r})$ in an inhomogeneous medium with refractive index $n(\bar{r})$ satisfies the scalar wave equation

$$[\nabla^2 + k^2(\bar{r})] u(\bar{r}) = 0, \qquad k(\bar{r}) = k_o n(\bar{r}) \tag{9}$$

where k_o is the wavenumber in vacuum and $k(\bar{r})$ is the wavenumber in the medium. By assuming that $u(\bar{r})$ at high frequencies (large k_o) behaves like a local plane wave field, the wave equation can be simplified. In order to admit of corrections to the local plane wave assumption, we write $u(\bar{r})$ in the form of an asymptotic expansion (see Sec. 3a) in inverse power of k_o, wherein the local plane wave field represents the dominant (m = 0) term:

$$u(\bar{r}) \sim u_o(\bar{r}) \exp[-jk_o \psi(\bar{r})] + \left(\sum_{m=1}^{\infty} \frac{u_m(\bar{r})}{(-jk_o)^m} \right) \exp[-jk_o \psi(\bar{r})] \tag{10}$$

The amplitude functions u_m, m = 0, 1, 2, ..., and the phase function ψ, are independent of k_o; they are to be determined from equations obtained by substituting (10) into (9). In the resulting series

$$\sum_{m=0}^{\infty} Q_m(\bar{r}) (-jk_o)^{2-m} = 0, \tag{10a}$$

which must be satisfied for arbitrary (though large) k_o, one equates to zero the coefficients $Q_m(\bar{r})$ to obtain from $Q_o = 0$:

$$[\nabla \psi(\bar{r})]^2 = n^2(\bar{r}) \quad - - \text{ eikonal equation} \tag{11}$$

A unit vector

$$\hat{s} = \frac{\nabla \psi(\bar{r})}{|\nabla \psi(\bar{r})|} = \frac{\nabla \psi(\bar{r})}{n(\bar{r})} \tag{12}$$

compatible with (11), then points in the direction of the normal to the equiphase surfaces $\psi(\bar{r})$ = constant and may therefore be identified as the <u>ray vector</u> tangent to the ray trajectories. Since

$$\hat{s} = \frac{d\bar{r}}{ds} \tag{13}$$

if $\bar{r} = \bar{r}(s)$ denotes points on the ray trajectory and s measures distance along a ray (Fig. 4), one may write (12) as

$$\frac{d}{ds}\left[n(s) \frac{d\bar{r}(s)}{ds} \right] = \nabla n(s) \tag{14}$$

to obtain the **ray equation**. The notation n(s) implies that the observation points $\bar{r}$ are constrained to lie along a ray.

From $Q_1 = 0$ in (3):

$$\left[\nabla^2 \psi(\bar{r}) + 2\nabla\psi(\bar{r})\cdot\nabla\right] u_o(\bar{r}) = 0 \quad \text{- - - transport equation for geometric-optical field} \tag{15a}$$

This equation is equivalent to

$$\nabla \cdot \left[\,|u_o^2(\bar{r})|\, n(\bar{r})\, \hat{s}\right] = 0, \tag{15b}$$

with the quantity inside the square brackets being proportional to the energy flux density in the local plane wave field, which is seen to flow along the rays. (15b) therefore represents an energy flux conservation theorem, which has already been applied in Sec. 1 on physical grounds (within a tube of rays) for the determination of $u_o(\bar{r})$ along a ray.

From $Q_m(\bar{r}) = 0$, $m \geq 2$, in (10a), one obtains the transport equations for the higher order amplitude coefficients that correct the local plane wave field:

$$\left[\nabla^2 \psi(\bar{r}) + 2\nabla\psi(\bar{r})\cdot\nabla\right] u_{m-1}(\bar{r}) = -\nabla^2 u_{m-2}(\bar{r}), \quad m \geq 2 \tag{16}$$

These recursive equations are more complicated than that for the dominant term $u_o(\bar{r})$, and it usually impractical to effect a solution for the higher order coefficients $u_m(\bar{r})$, $m \geq 2$. However, these equations are useful for providing estimates on the range of validity of the local plane wave field assumption, which requires that $u_1(\bar{r}) k_o^{-1} \ll |u_o(\bar{r})|$. From this requirement, one may deduce the restriction

$$\frac{|\nabla n(\bar{r})|}{n(\bar{r})} \frac{1}{k_o n(\bar{r})} \ll 1 \tag{17}$$

which states that the relative change in the refractive index over an interval of the local wavelength $\lambda(\bar{r})$ in the medium must be small (note that $\lambda(\bar{r}) = 2\pi/k(\bar{r})$, where $k(\bar{r}) = k_o n(\bar{r})$). Thus the medium must be "slowly varying" on the local wavelength scale.

2b. **Ray Trajectories**

The ray trajectories are obtained by solution of (14) subject to prescribed initial conditions. In a homogeneous medium with $n(s) = $ constant, (14) reduces to $d\bar{r}/ds = $ constant, or $\bar{r}(s) = \bar{A}s + \bar{B}$, where $\bar{A}$ and $\bar{B}$ are constant vectors. Therefore, the rays are straight lines.

In an inhomogeneous medium, the rays are generally smoothly curved. It may be shown that the curvature K of the ray is given by

$$K(s) = \frac{1}{n(s)} \frac{dn(s)}{d\ell} \qquad (18)$$

where ℓ is the coordinate perpendicular to s and hence lies on a wavefront. Therefore, the ray is curved whenever the refractive index n varies along a wavefront, and the ray bends toward the direction of increasing n. This behaviour is in accord with the application of Snell's law of refraction when the continuously varying medium is approximated by a sequence of locally homogeneous layers. In a plane stratified medium where the refractive index $n(\bar{r}) = n(z)$ varies along the rectilinear coordinate z only, the rays may be shown to be plane curves. A typical ray $y = y(z)$ lying in the $x = 0$ plane has the functional form

$$y - y_o = \xi \int_{z_o}^{z} \frac{d\zeta}{[n^2(\zeta) - \xi^2]^{\frac{1}{2}}} \qquad (19)$$

where (y_o, z_o) is the initial point on the ray, and $\xi \equiv n(z)\sin\theta(z)$ = constant, with θ defined in Fig. 4, is the ray parameter. By choosing different values of ξ according to different initial conditions, one may generate the entire ray family. The condition ξ = constant is an analytic statement of Snell's refraction law when applied to a continuously varying medium.

2c.   **Phase and amplitude**

The local plane wave phase $\psi(\bar{r})$ is determined from (12) by integration along a ray:

$$\psi(\bar{r}) - \psi(\bar{r}_o) = \int_{\bar{r}_o}^{\bar{r}} n(s)\, ds, \qquad (20)$$

where $\bar{r}_o$ and $\bar{r}$ are the initial point and observation point, respectively, along the same ray, and $\psi(\bar{r}_o)$ is the initial phase.

In a homogeneous medium with constant $n(\bar{r}) = n_o$, where the rays are straight lines,

$$\psi(\bar{r}) - \psi(\bar{r}_o) = n_o |\bar{r} - \bar{r}_o| = n_o d \qquad (21)$$

as noted already in Sec. 1 from physical considerations pertaining to the local plane wave field. In a plane stratified inhomogeneous medium with $n(\bar{r}) = n(z)$, (20) yields for the ray in (19):

$$\psi(\overline{r}) - \psi(\overline{r}_o) = \xi(y - y_o) + \int_{z_o}^{z} \left[n^2(\zeta) - \xi^2\right]^{\frac{1}{2}} d\zeta \qquad (22)$$

The local plane wave amplitude $u_o(\overline{r})$ is determined by (15a) or (15b). By applying Gauss' divergence theorem to a volume contained within a tube of rays, (15b) may be reduced to:

$$|u_o(\overline{r})| \sim |u_o(\overline{r}_o)| \left[\frac{n(\overline{r}_o) \, dA(\overline{r}_o)}{n(\overline{r}) \, dA(\overline{r})}\right]^{\frac{1}{2}} \qquad (23)$$

where $dA(\overline{r}_o)$ and $dA(\overline{r})$ are, respectively, the ray tube cross section containing the initial point and the observation point along the same ray. For a spherical wavefront in a homogeneous medium, (23) reduces to the result in (1).

2d. Ray-Optical Field

The ray-optical (local plane wave) field $u_o(\overline{r}) \exp[-jk_o\psi(r)]$ in an inhomogeneous medium can be constructed by combining the results from (20) and (23):

$$u(\overline{r}) \sim u(\overline{r}_o) \left[\frac{n(\overline{r}_o) \, dA(\overline{r}_o)}{n(\overline{r}) \, dA(\overline{r})}\right]^{\frac{1}{2}} \exp\left[-jk_o \int_{r_o}^{r} n(s) ds\right] \qquad (24)$$

where $u(\overline{r}_o)$ is the prescribed initial field, with $\overline{r}$ and $\overline{r}_o$ denoting the observation points and initial points, respectively, along the same ray. $n(\overline{r})$ is the local refractive index and $dA(\overline{r})$ the ray tube cross section. The validity of (24) is confined to slowly varying media (see (17)). It is also restricted by the condition

$$dA(r) \not\approx 0 \qquad (25)$$

i.e., to the exterior of focusing regions where ray crossings occur. Ignoring (25) would lead to the incorrect conclusion of infinite fields at caustics and foci. To accommodate focusing regions where the ray-optical field is invalid, it is necessary to employ transition functions that are derived by a more sophisticated set of equations than those in (11) or (15).

3. Asymptotic Expansions

Approximate evaluations of fields in the high frequency regime commonly generate expansions in inverse powers of the wavenumber k, based on the assumption that k is large. An example is provided by the field representation in (10), wherein the leading term describes the local plane wave field. Expansions of this type, called "asymptotic expansions",

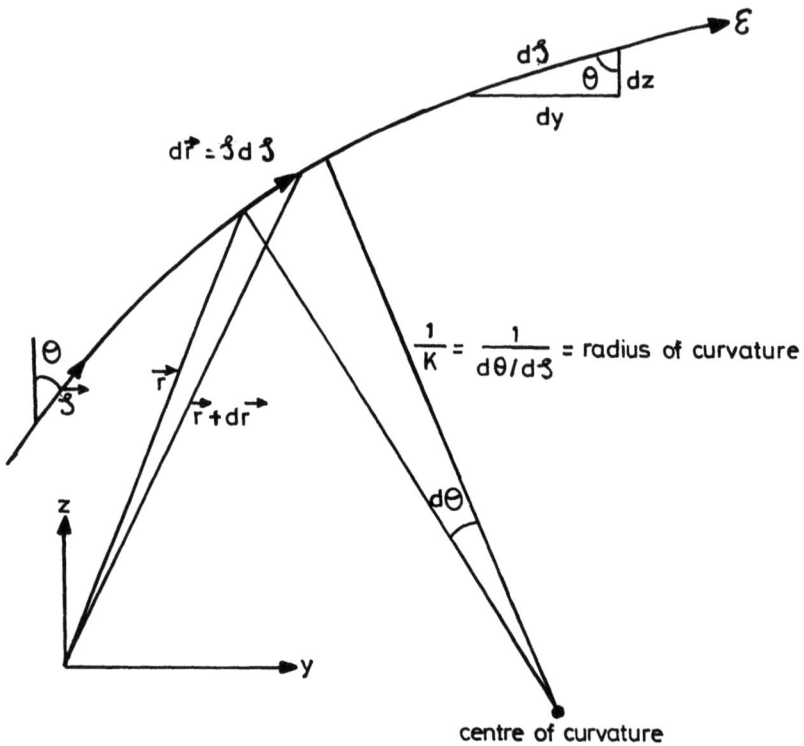

Figure 4. Ray parameters.

generally do not converge; i.e., for a fixed value of the parameters (k_o and $\bar{r}$ in (10)), the series diverges because the expansion coefficients ($u_m(\bar{r})$ in (10)) grow as $m \to \infty$. Nevertheless, these series, although divergent in the strict mathematical sense, are useful because they can be employed to approximate the field provided that only a finite number of terms is included. Loosely speaking, calculation involving the first M terms in the expansion is legitimate provided that the (M+1)th term is smaller in magnitude than the Mth term. For a given M, this can always be accomplished by choosing the asymptotic expansion parameter (1/k in (10)) large enough. A good indication of the quality of the approximation can be had from the recognition that the error between the asymptotic representation involving M terms and the true value of the field is roughly equal to the magnitude of the (M+1)th term.[4] Based on these considerations, one can deduce the necessary restrictions on the parameters in a function represented by an asymptotic series. For example, validation of the local plane wave (ray-optical) field in (10), as given by the first term, requires at the very least that $|u_1(\bar{r})/k_o| \ll |u_o(\bar{r})|$. Improvement of the local plane wave field by inclusion of the m=1 term requires, at the very least, that $|u_2(\bar{r})/k_o| \ll |u_1(\bar{r})|$, etc. In view of the error criterion cited above, it is evidently dangerous to push an asymptotic expansion containing a few terms, and especially if only the leading term is retained, near the limit where the first omitted term equals in magnitude the last term retained.

The preceding considerations can be phrased mathematically. Regarding k_o as the large parameter, a function $F(\bar{r}, k_o)$ (for example, the field in (10)) can be represented rigorously as follows:

$$F(\bar{r}, k_o) = \sum_{m=0}^{M} F_m(\bar{r}) k_o^{-m} + R_M(\bar{r}, k_o) \qquad (26)$$

where R_M is a remainder term. Making k_o large enough, one can reduce R_M to as small a value as desired (a good estimate of $|R_M|$ is given by $|F_{M+1}(\bar{r}) k_o^{-M-1}|$). Thus asymptotic series are not usually written in the form (26); instead, letting $k_o \to \infty$, the summation is extended over all values of m:

$$F(\bar{r}, k_o) \sim \sum_{m=0}^{\infty} F_m(\bar{r}) k_o^{-m}, \qquad k_o \to \infty \qquad (27)$$

and it is implied that for any m,

$$F_{m+1}(\bar{r})/k_o F_m(\bar{r}) \to 0 \quad \text{as } k_o \to \infty \qquad (28)$$

However, for practical calculations involving <u>fixed</u> k_o and $\bar{r}$, the series in (27) is to be understood in the sense of (26).

References

1. G.A. Deschamps, "Ray Techniques in Electromagnetics," Proc. IEEE, 60, pp. 1022-1035, 1972.

2. J.B. Keller, "Geometrical Theory of Diffraction," JOSA, 52, pp. 116-130, 1962.

3. R.G. Kouyoumjian and P.H. Pathak, "A Uniform Theory of Diffraction for an Edge in a Perfectly Conducting Surface," Proc. IEEE, 62, pp. 1448-1461, 1974.

4. H. Jeffreys, Asymptotic Approximations, Clarendon Press, Oxford, 1962, Chapter 7.

CANONICAL PROBLEMS AND DIFFRACTION COEFFICIENTS

Peter L. Christiansen

Laboratory of Applied Mathematical Physics
The Technical University of Denmark, Lyngby,
DK-2800 Denmark.

1. Introduction

   1.1 On the theory of geometrical optics (GO).

   The propagation problem. The rays are curves along which the field energy propagates. The wave fronts are surfaces in space to which the energy has reached along the rays after the elapse of a given time from an initial moment with an initial distribution. The amplitude of the field varies along the rays. The rays are determined as extremal curves to Fermat's variational principle (ray tracing). An envelope curve for a family of rays is called a caustic. A common point for a family of rays is called a focal point. Also the wave fronts can be determined by means of variational calculus methods. The amplitude variation can be found by means of energy conservation considerations.

   The scattering problem. The GO description of the scattering of the field energy by an obstacle is reflection or refraction of the rays. The GO scattering process takes place at a point at the surface of the scatterer. This process is described by a GO reflection or refraction coefficient which is the ratio between the scattered field and the incident field at the scattering point. The reflection law or refraction law for the rays is derived from Fermat's principle while the GO scattering coefficient may be extracted from the solution to a canonical problem as we shall see.

   Time-harmonic waves and GO in isotropic media. Time-harmonic fields are of the form $u(\bar{r})\exp(-i\omega t)$. In isotropic media $u(\bar{r})$ satisfies the reduced wave equation

$$(\Delta + k^2)u = 0 \qquad (1.1.1)$$

with the propagation constant $k = n\omega/c_o$ where ω is the angular frequency, $n = n(\bar{r})$ is the so-called ray index for the medium, and c_o is the vacuum velocity.

Introduction of Debye's assumption

$$u(\bar{r}) \sim A(\bar{r})e^{ik_o \psi(\bar{r})} \qquad (1.1.2)$$

where $k_o = \omega/c_o$, $\psi(\bar{r})$ is a phase function, and $A(\bar{r})$ is an amplitude function yields the eiconal equation

$$(\nabla \psi)^2 = n^2(\bar{r}) \qquad (1.1.3)$$

and the transport equation

$$2\nabla\psi \cdot \nabla A + A\Delta\psi = 0 \qquad (1.1.4)$$

in the short wavelength limit $k_o a \to \infty$ (a being a characteristic length). If the assumption (1.1.2) is replaced by the Luneburg series

$$u(\bar{r}) \sim \left(\sum_{m=0}^{M} A_m(\bar{r})(ik_o)^{-m}\right) e^{ik_o \psi(\bar{r})} \qquad (1.1.5)$$

the transport equation (1.1.4) is replaced by a system of transport equations for the expansion coefficients A_m.

The solutions $\psi(\bar{r})$ = constant to the eiconal equation (1.1.3) may be called wave surfaces. In the short wavelength limit the wave surfaces are identical with the wave fronts of GO. In the isotropic medium curves these orthogonal to the wave surfaces satisfy the same equations as rays in GO and may therefore be identified with these. Integration of the eiconal equation (1.1.3) along the rays yields the phase behaviour along the rays. Integration of the transport equation along the rays yields the amplitude behaviour along the rays. In a homogeneous medium e.g., the rays are straight lines. Here the phase varies linearly with the distance, r, measured along the rays. For diverging rays the amplitude varies asymptotically as $r^{-\frac{1}{2}}$ in two dimensions and as r^{-1} in three dimensions. These factors are called divergence factors.

1.2 On the geometrical theory of diffraction (GTD).

In GTD diffracted rays are added to the GO rays. The rays are determined as extremal curves to the extended Fermat's principle. The diffracted rays are named after the nature of the scattering process by which the rays are started. Examples are shown in Fig. 1.1. The phase and amplitude behaviour along the diffracted rays is similar to the behaviour along the GO rays. A new feature is an exponential damping along the curved surface rays in

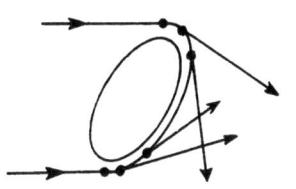

a) Surface diffraction

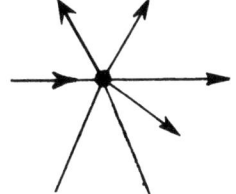

b) Edge diffraction

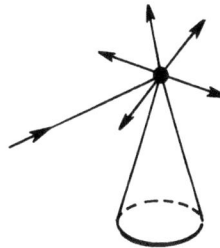

c) Vertex diffraction

Figure 1.1. Types of diffracted rays.

the case of surface diffraction. This effect is due to the energy loss to the tangentially diffracted rays which are continuously launched by the surface ray (or creeping wave). As in GO the GTD scattering process takes place at the scattering point and is described by a GTD diffraction coefficient. Diffraction laws for the diffracted rays are derived from the extended Fermat's principle and GTD scattering coefficients may be determined from a canonical problem as we shall see.

Like in GO we may derive eiconal and transport equations for GTD by insertion of the proper assumptions into the reduced wave equation. Again this approach leads to identical rays and yields the phase and amplitude behaviour along the rays.

2. Solution of some canonical problems and extraction of GO and GTD scattering coefficients.

2.1 Plane wave incidence on plane surface and interface.

Let an incident plane wave

$$u_i = e^{ik(-x \sin\chi_i + y \cos\chi_i)} \qquad (2.1.1)$$

strike a plane surface at $x = 0$ with surface impedance Z under the glancing angle χ_i (see Fig. 2.1).

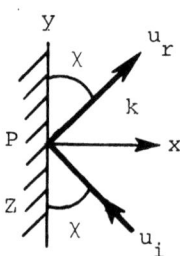

Figure 2.1. Reflection at plane surface at $x = 0$.

The field u_i is a solution to the reduced wave equation

$$(\Delta + k^2)u = 0 \qquad (2.1.2)$$

where k is the propagation constant. However u_i does not satisfy the boundary condition

$$\partial_n u + ikZu = 0 \quad \text{at } x = 0 \qquad (2.1.3)$$

where ∂_n denotes the outward normal derivative (here in the positive x-direction). By adding another solution to (2.1.2), the reflected field

$$u_r = R\, e^{ik(x \sin\chi_r + y \cos\chi_r)}, \qquad (2.1.4)$$

to u_i we get the total field

$$u = u_i + u_r, \qquad (2.1.5)$$

which satisfies the boundary condition (2.1.3) provided

$$\chi_i = \chi_r = \chi \quad \text{(the reflection law)} \quad (2.1.6)$$

and the constant R is given by

$$R = \frac{\sin\chi - Z}{\sin\chi + Z} . \quad (2.1.7)$$

At $x = 0$

$$u_r = Ru_i . \quad (2.1.8)$$

Therefore R is the GO scattering coefficient that describes the scattering by the surface at $x = 0$. Since this process is a reflection R is called the reflection coefficient. Fig. 2.1 illustrates the reflection of the particular incident ray through the point P with rectangular coordinates $(0,0)$. Note also

$$u = (1 + R)u_i \quad \text{at } x = 0 . \quad (2.1.9)$$

If $x = 0$ is an interface between two media with propagation constants k_1 and k_2 there will be a transmitted or refracted field, u_t, in addition to u_i and u_r. In the case where the incident wave

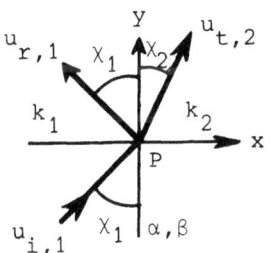
a) Incident k_1-wave

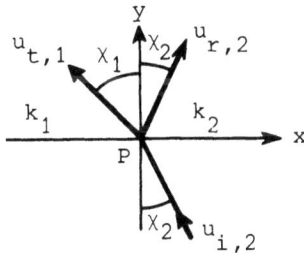
b) Incident k_2-wave

Figure 2.2. Reflection and refraction at plane interface at $x = 0$.

is present in the medium with propagation constant k_1 (see Fig. 2.2) we get

$$\left. \begin{array}{l} u_{r,1} = R_{11}\, u_{i,1} \\ \\ u_{t,2} = T_{12}\, u_{i,1} \end{array} \right\} \quad \text{at } x = 0 \,. \qquad (2.1.10)$$

In the opposite case (incident wave of k_2-type) we get

$$\left. \begin{array}{l} u_{r,2} = R_{22}\, u_{i,2} \\ \\ u_{t,1} = T_{21}\, u_{i,2} \end{array} \right\} \quad \text{at } x = 0 \,. \qquad (2.1.11)$$

The angles χ_1 and χ_2 in Fig. 2.2 are related by

$$k_1 \cos\chi_1 = k_2 \cos\chi_2 \qquad \text{(Snell's law)} \,. \qquad (2.1.12)$$

The GO scattering coefficients for reflection (R_{11} and R_{22}) and transmission (T_{12} and T_{21}) become

$$\left. \begin{array}{l} R_{11} = -R_{22} = \dfrac{\beta \sin\chi_1 \cos\chi_2 - \alpha \sin\chi_2 \cos\chi_1}{\beta \sin\chi_1 \cos\chi_2 + \alpha \sin\chi_2 \cos\chi_1} \\ \\ T_{12} = \dfrac{2\alpha\beta \cos\chi_2 \sin\chi_1}{\beta \sin\chi_1 \cos\chi_2 + \alpha \sin\chi_2 \cos\chi_1} \\ \\ T_{21} = \dfrac{2 \cos\chi_1 \sin\chi_2}{\beta \sin\chi_1 \cos\chi_2 + \alpha \sin\chi_2 \cos\chi_1} \end{array} \right\} \qquad (2.1.13)$$

when the boundary conditions at $x = 0$ are

$$\left. \begin{array}{l} u_2 = \alpha\, u_1 \\ \\ \partial_x u_2 = \beta\, \partial_x u_1 \end{array} \right\} \,. \qquad (2.1.14)$$

2.2 Cylindrical wave incidence on circular cylinder.

We now place a unit source at the point P_o with polar coordinates (r_o, θ_o) and a circular cylinder of radius a with axis through the origin (see Fig. 2.3). The canonical problem can then be formulated as the boundary value problem

$$(\Delta + k^2)u = - \frac{\delta(r-r_o)\delta(\theta-\theta_o)}{r} \qquad 0 \le \theta - \theta_o < 2\pi, \qquad (2.2.1)$$

$$\partial_r u + ikZu = 0 \qquad r = a, \qquad (2.2.2)$$

and

$$\lim_{r \to \infty} \sqrt{r}[\partial_r u - iku] = 0. \qquad (2.2.3)$$

Here (2.2.2) is the boundary condition (2.1.3) in polar coordinates, Z being the surface impedance of the cylinder and (2.2.3)

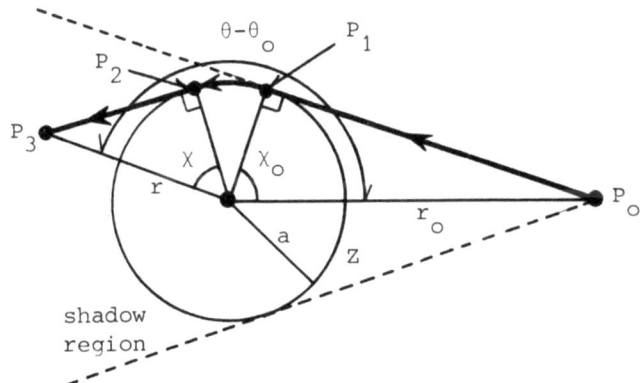

Figure 2.3. Diffraction by a circular cylinder.

is the radiation condition expressing that only outgoing waves may exist at infinity. By separation of variables we find the exact solution to the problem

$$u = \frac{i}{8} \sum_{n=-\infty}^{\infty} H_n^{(1)}(kr_>) \{ H_n^{(2)}(kr_<) - \frac{H_n^{(2)'}(ka) + iZH_n^{(2)}(ka)}{H_n^{(1)'}(ka) + iZH_n^{(1)}(ka)}$$

$$\times H_n^{(1)}(kr_<) \} e^{in(\theta-\theta_o)} \qquad a \leq r_< \qquad (2.2.4)$$

where $H^{(1)}$ and $H^{(2)}$ are the Hankel functions of first and second kind, respectively. Furthermore, we have used the convention

$$r_\gtrless = \begin{array}{c} \max \\ \min \end{array} \{ r_o, r \} . \qquad (2.2.5)$$

Note that the reciprocity theorem holds.

The series (2.2.4) is slowly convergent in the short wavelength limit $ka \to \infty$. In order to obtain a suitable representation in this case we use the Watson transformation which consists of three steps:

1. Write the expansion (2.2.4) as a contour integral in the complex order plane for the Hankel functions, ν. Using the calculus of residues we get

$$u = \frac{1}{8} \int_C \frac{\cos\nu(\theta-\theta_o-\pi)}{\sin\nu\pi} H_\nu^{(1)}(kr_o) \{ H_\nu^{(2)}(ka)$$

$$- \frac{H_\nu^{(2)'}(ka) + iZH_\nu^{(2)}(ka)}{H_\nu^{(1)'}(ka) + iZH_\nu^{(1)}(ka)} \times H_\nu^{(1)}(kr) \} d\nu \quad \text{for } r < r_o . \quad (2.2.6)$$

(The case $r_o < r$ is treated analogously). Here C is the contour shown in Fig. 2.4.

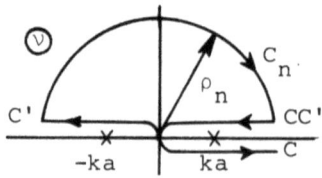

Figure 2.4. Integration contours in the ν-plane.

2. Deform the contour C into C' by replacing the integral
$\int_{0+i0-}^{\infty+i0-}$ by $\int_{0+i0+}^{-\infty+i0+}$. This is possible because the integrand in
(2.2.6) is an odd function of ν. Close the contour C' along a large half circle, C_n, in the upper half plane with center at $\nu=0$ and radius ρ_n. It can be shown that $\int_{C_n} \to 0$ for $\rho_n \to \infty$ as $n \to \infty$ when the sequence of half circles is chosen such that the poles of the integrand are avoided. This is true when the field point P_3 lies in the shadow region with respect to the source point P_0 behind the cylinder (see Fig. 2.3). If P_3 lies in the illuminated region a term is extracted in (2.2.6) which evaluated by the method of steepest descent yields the primary field and the reflected GO field from the source. (The reflection process at the cylinder turns out to be described by R (2.1.7) even though the surface is curved). For the remaining part of (2.2.6) the contour C' can be closed as described above.

3. Evaluate the resulting contour integral by means of calculus of residues. The poles of the integrand in the upper half plane are determined by

$$H_\nu^{(1)'}(ka) + iZH_\nu^{(1)}(ka) = 0 \ . \tag{2.2.7}$$

For large values of ka this equation possesses solutions in the vicinity of $\nu = ka$. These are shown in Table 2.1 for $Z = z_o(ka)^0 + O((ka)^{-\frac{1}{3}})$ and for $Z = 0$. In the Table x_p and x'_p denote the p'th solution to the equations $Ai(-x) = 0$ and $Ai'(-x) = 0$. Here Ai is the Airy function of first kind given by

$$Ai(y) = \frac{1}{\pi} \int_0^\infty \cos(\frac{1}{3} t^3 + yt) dt \tag{2.2.8}$$

and Ai' its derivative. A solution to (2.2.7), ν_o, away from the vicinity of ka is found when $\text{Re}\{Z\} \geqslant 0$ and $\text{Im}\{Z\} < 0$ with $Z \sim z_o(ka)^0$.

Application of the calculus of residues then finally yields

$$u_d = \frac{\pi i}{4} \sum_{p=0}^\infty \frac{\cos\nu_p(\theta - \theta_o - \pi)}{\sin\nu_p \pi} H_{\nu_p}^{(1)}(kr_o) H_{\nu_p}^{(1)}(kr)$$

$$\times \frac{H_{\nu_p}^{(2)'}(ka) + iZH_{\nu_p}^{(2)}(ka)}{\partial_\nu [H_\nu^{(1)'}(ka) + iZH_\nu^{(1)}(ka)]} \bigg|_{\nu = \nu_p} \tag{2.2.9}$$

Surface impedance Z	Poles ν_p	Decay exponents $\alpha_p(a)$
$z_0(ka)^0 + O((ka)^{-\frac{1}{3}})$	$ka\sqrt{1-z_0^2}$, $p=0$ $ka + 2^{-\frac{1}{3}} e^{i\frac{\pi}{3}} x_p (ka)^{\frac{1}{3}} + \frac{1}{iz_0} + O((ka)^{-\frac{1}{3}})$, $p>0$	$\operatorname{Im}\{ka\sqrt{1-z_0^2}\}$, $p=0$ $[2^{-\frac{1}{3}} e^{-i\frac{\pi}{6}} x_p (ka)^{\frac{1}{3}} - \frac{1}{z_0} + O((ka)^{-\frac{1}{3}})]a^{-1}$, $p>0$
0	$ka + 2^{-\frac{1}{3}} e^{i\frac{\pi}{3}} x_p'(ka)^{\frac{1}{3}} + O((ka)^{-\frac{1}{3}})$, $p>0$	$[2^{-\frac{1}{3}} e^{-i\frac{\pi}{6}} x_p'(ka)^{\frac{1}{3}} + O((ka)^{-\frac{1}{3}})]a^{-1}$, $p>0$

Table 2.1. Poles of the integrand in (2.2.6) and decay exponents for the surface waves.

as a result of the Watson transformation.

Expansion of $(\sin\nu_p\pi)^{-1}$ into a geometric series and insertion of the relevant asymptotic representations of the Hankel functions into (2.2.9) yield the result

$$u_d(P_2) \sim \sum_{p=0}^{\infty} d_p(P_1) (e^{ika(\theta-\theta_0) - \int_0^{a(\theta-\theta_0)} \alpha_p(t)dt}$$

$$+ e^{ika(2\pi-\theta+\theta_0) - \int_0^{a(2\pi-\theta+\theta_0)} \alpha_p(t)dt}) \sum_{q=0}^{\infty} e^{ika2\pi q - \int_0^{2\pi aq} \alpha_p(t)dt} d_p(P_2)$$

(2.2.10)

for the source point lying at P_1 and the field point lying at P_2 (see Fig. 2.3). The factor $\exp(ika(\theta-\theta_0) - \int_0^{a(\theta-\theta_0)} \alpha_p(t)dt)$ accounts for the phase along the creeping wave from P_1 to P_2 in the counterclockwise direction, the distance from P_1 to P_2 being $\rho_{P_1P_2} = a(\theta-\theta_0)$, and for the damping along the creeping wave due to the loss of energy for the continuously tangentially diffracted rays, $\alpha_p(t)$ being the decay exponent given in Table 2.1. The creeping wave possesses a mode structure wherefore there is a summation of the contributions from each mode. There is also a creeping wave from P_1 to P_2, circulating clockwise, of length $\rho_{P_1P_2} = a(2\pi-\theta+\theta_0)$. This wave gives rise to the contributions in the second term in the parenthesis. Furthermore there are multiple circulations accounted for in the last summation in (2.2.10). The diffraction coefficient $d_p(P_1) = d_p(P_2)$ describing the processes at P_1 and P_2 is given in Table 2.2. In the following formulae (2.2.11-13) we omit terms corresponding to circulation in clockwise direction and multiple circulations.

For source or/and field point away from the surface (see Fig. 2.3) we get asymptotic representations of the form

$$u_d(P_2) \sim \sum_{p=0}^{\infty} S(P_0) \frac{e^{ikr_{P_0P_1}}}{\sqrt{r_{P_0P_1}}} D_p(P_1) d_p(P_1) e^{ik\rho_{P_1P_2} - \int_{P_1}^{P_2} \alpha_p(t)dt} d_p(P_2),$$

(2.2.11)

$$u_d(P_3) \sim \sum_{p=0}^{\infty} d_p(P_1) e^{ik\rho_{P_1P_2} - \int_{P_1}^{P_2} \alpha_p(t)dt} d_p(P_2) D_p(P_2) S(P_2)$$

$$\times \frac{e^{ikr_{P_2P_3}}}{\sqrt{r_{P_2P_3}}}, \qquad (2.2.12)$$

and

$$u_d(P_3) \sim \sum_{p=0}^{\infty} S(P_0) \frac{e^{ikr_{P_0P_1}}}{\sqrt{r_{P_0P_1}}} D_p(P_1) d_p(P_1) e^{ik\rho_{P_1P_2} - \int_{P_1}^{P_2} \alpha_p(t)dt}$$

$$\times d_p(P_2) D_p(P_2) S(P_2) \frac{e^{ikr_{P_2P_3}}}{\sqrt{r_{P_2P_3}}}. \qquad (2.2.13)$$

The phases on the incident ray and the diffracted ray are given by $\exp\{ikr_{P_0P_1}\}$ and $\exp\{ikr_{P_2P_3}\}$ with $r_{P_0P_1} = r_0 \sin\chi_0$ and $r_{P_2P_3} = r\sin\chi$, (see Fig. 2.3), respectively. The angles χ_0 and χ are given by

$$\cos\chi_0 = a/r_0 \quad \text{and} \quad \cos\chi = a/r. \qquad (2.2.14)$$

The divergence factors for the incident and diffracted rays are given by $(r_{P_0P_1})^{-\frac{1}{2}} = (r_0 \sin\chi_0)^{-\frac{1}{2}}$ and $(r_{P_2P_3})^{-\frac{1}{2}} = (r \sin\chi)^{-\frac{1}{2}}$, respectively. The source factor for the isotropic unit source in two dimensional space is S given by

$$S = \frac{e^{i\frac{\pi}{4}}}{\sqrt{8\pi k}}. \qquad (2.2.15)$$

Note that S is used both at the source point P_0 and the diffraction point P_2. The diffraction coefficient, D_p, that describes additional diffraction processes at P_1 and P_2 is given in Table 2.2. (The case $Z \simeq z_0(ka)^{\frac{1}{2}}$ is usually referred to as the impedance boundary condition case. The case $Z=0$ is the Neumann boundary condition while $Z=\infty$ yields the Dirichlet boundary condition. In the

Surface impedance Z	d_p		D_p	
$z_0(ka)^0 + O((ka)^{-\frac{1}{3}})$	$i\sqrt{z_0}/(1-z^2)^{\frac{1}{4}}$	$p=0$	1	$p=0$
	$e^{i\frac{\pi}{4}} \frac{1}{z_0} (ka)^{-\frac{1}{2}}$	$p>0$	$2^{\frac{1}{3}} e^{-i\frac{\pi}{3}} z_0 [Ai'(-x_p)]^{-1} (ka)^{\frac{2}{3}}$	$p>0$
0	$-2^{-\frac{1}{3}} e^{i\frac{\pi}{12}} (x'_p)^{-\frac{1}{2}} (ka)^{-\frac{1}{6}}$	$p>0$	$2^{\frac{2}{3}} e^{-i\frac{\pi}{6}} [Ai(-x'_p)]^{-1} (ka)^{\frac{1}{3}}$	$p>0$

Table 2.2. Diffraction coefficients for surface diffraction processes.

latter case d_p must be replaced by d_p' which is a diffraction coefficient for the surface current proportional to $\partial_n \phi^1$).
Thus the short wavelength asymptotic representation of the exact solution to the canonical problem of a circular cylinder essentially confirms the expectations based on GO and GTD (apart from the mode structure of the creeping wave) and yields a determination of the canonical quantities α_p, d_p, and D_p.

2.3 Spherical wave incidence on sphere.

We next look at the three-dimensional problem of spherical wave incidence on a sphere of radius a (see Fig. 2.5) given by

$$(\Delta + k^2)u = -\frac{i}{2k} \delta(x)\delta(y)\delta(z-r_0) , \qquad (2.3.1)$$

$$\partial_r u + ikZu = 0 \qquad r = a , \qquad (2.3.2)$$

and

$$\lim_{r \to \infty} r[\partial_r u - iku] = 0 . \qquad (2.3.3)$$

Here the source is placed at $(x,y,z) = (0,0,r_0)$ and the normalization of the source is chosen such that the source factor for the three dimensional isotropic unit source becomes S^2 where S is given by (2.2.15). (In this manner u becomes dimensionless).

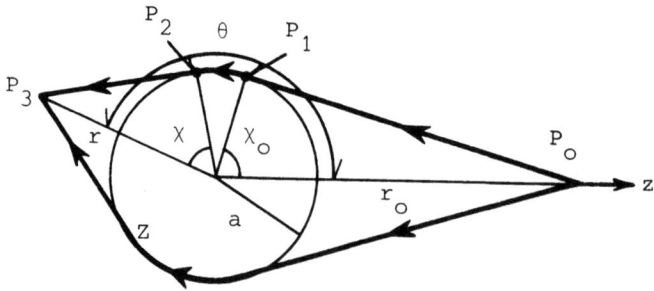

Figure 2.5. Point source at sphere.

Separation of variables in spherical coordinates (r, θ, ϕ) leads to the formula for the total field (corresponding to (2.2.4))

$$u = \frac{-1}{16\pi} \sum_{n=0}^{\infty} (2n+1) P_n(-\cos\theta) e^{in\pi} h_n^{(1)}(kr_>)$$

$$\times \left[h_n^{(2)}(kr_<) - \frac{h_n^{(2)'}(ka) + iZh_n^{(2)}(ka)}{h_n^{(1)'}(ka) + iZh_n^{(1)}(ka)} h_n^{(1)}(kr_<) \right]$$

$$a \leq r_< . \qquad (2.3.4)$$

Here P_n is the Legendre polynomial of order n given by

$$P_n(\cos\theta) = \frac{1}{2^n n!} \frac{d^n(-\sin\theta)^{2n}}{d(\cos\theta)^n} \qquad n = 0, 1, \cdots, \qquad (2.3.5)$$

while $h_n^{(1)}$, $h_n^{(2)}$ are the spherical Hankel functions given by

$$h_n^{(1)}{}_{(2)}(ka) = \sqrt{\frac{\pi}{2ka}} H_{n+\frac{1}{2}}^{(1)}{}^{(2)}(ka) . \qquad (2.3.6)$$

The diffracted field can be extracted from (2.3.4) by means of Poisson's summation formula

$$\sum_{n=0}^{\infty} f(n+\tfrac{1}{2}) = \sum_{q=-\infty}^{\infty} \int_0^{\infty} f(\nu) e^{-2\pi q\nu - i\pi q} d\nu \qquad (2.3.7)$$

which is actually equivalent to the Watson transformation described in the previous section. Application of (2.3.4) and (2.3.7) as well as the calculus of residues yields

$$u_d = \frac{1}{2ik^2 a\sqrt{r_o r}} \sum_{p=0}^{\infty} \nu_p \frac{H_{\nu_p}^{(1)}(kr_o) H_{\nu_p}^{(1)}(kr)}{H_{\nu_p}^{(1)}(ka)}$$

$$\times \frac{P_{\nu_p - \frac{1}{2}}(-\cos\theta) e^{i\nu_p \pi}}{\partial_\nu [H_\nu^{(1)'}(ka) + iZ H_\nu^{(1)}(ka)]\big|_{\nu=\nu_p}} \sum_{q=0}^{\infty} e^{i2\pi\nu_p q} e^{i\pi q} \qquad (2.3.8)$$

corresponding to (2.2.9). For $\theta \neq \pi$ the asymptotic representation of this result becomes

$$u_d \sim S^2 \frac{e^{ikr_{P_0P_1}}}{r_{P_0P_1}} \sum_{p=0}^{\infty} D_p(P_1) d_p(P_1) [e^{ika(\theta-\chi_o-\chi)-\alpha_p(a)a(\theta-\chi_o-\chi)}$$

$$+ e^{ika(2\pi-\theta+\chi_o+\chi)-\alpha_p(a)a(2\pi-\theta+\chi_o+\chi)-i\frac{\pi}{2}}] \sum_{q=0}^{\infty} e^{ika2\pi q - \alpha_p(a)a2\pi q}$$

$$\times (e^{-i\frac{\pi}{2}})^{2q} D_p(P_2) d_p(P_2) \sqrt{\frac{a \sin\chi_o}{r \sin\theta}} S \frac{e^{ikr_{P_2P_3}}}{\sqrt{r_{P_2P_3}}} \qquad (2.3.9)$$

where S is given by (2.2.15) and D_p and d_p are found in Table 2.2. The last result can be understood completely geometrically when it is noted i) that the creeping waves are great circles on the sphere, ii) that the creeping waves diverge in the plane normal to the drawing plane such that the resulting divergence factor from P_1 to P_3 becomes $(a \sin\chi_o/r \sin\theta)^{\frac{1}{2}}$, and iii) that every passage of the focal points on the z-axis gives rise to a phase jump of $-\frac{\pi}{2}$. Special results can be obtained for $\theta = \pi$ and source and observation points lying at P_1 and P_2, respectively. Thus the canonical quantities α_p, d_p, and D_p derived from a two-dimensional problem also apply in three-dimensional situations. It has been shown in a number of cases that the quantities also apply to problems with varying radius of curvature, a, and propagation constants, k.

2.4 Wedge and half-plane.

Consider the two-dimensional problem of a source and a wedge with exterior opening angle α (see Fig. 2.6)

$$(\Delta + k^2)u = -\frac{\delta(r-r_o)\delta(\theta-\theta_o)}{r} \qquad 0 < \theta_o < \frac{\alpha}{2}, \qquad (2.4.1)$$

$$u = 0 \text{ for } \theta = \begin{cases} 0 \\ \alpha \end{cases} \text{ (Dirichlet problem)}$$

or
$$\pm \frac{1}{r} \partial_\theta u = 0 \text{ for } \theta = \begin{cases} 0 \\ \alpha \end{cases} \text{ (Neumann problem)} \qquad (2.4.2)$$

and

$$\lim_{r \to \infty} \sqrt{r}\,[\partial_r u - iku] = 0 \,. \quad (2.4.3)$$

Again separation of variables yields an infinite series as exact solution. For the far field (kr >> 1) this series can be summed away from shadow and reflection boundaries ($\theta \neq \theta_o$, $\theta \neq \pi-\theta_o$ and $\theta \neq \pi+\theta_o$). The result can be written

$$u_d \sim D_e \frac{e^{ikr}}{\sqrt{r}} \qquad kr \gg 1 \quad (2.4.4)$$

where exp (ikr) is the phase factor and $r^{-\frac{1}{2}}$ is the divergence factor for the diffracted rays emitted from the edge of the wedge. The diffraction coefficient for the edge diffraction process at this point is

$$D_e = \frac{2\pi}{\alpha} S \left[(\cos\frac{\pi^2}{\alpha} - \cos\frac{\pi}{\alpha}(\theta_o - \theta))^{-1} \mp (\cos\frac{\pi^2}{\alpha} - \cos\frac{\pi}{\alpha}(\theta_o + \theta))^{-1} \right]$$

$$\times \sin\frac{\pi^2}{\alpha} \quad (2.4.5)$$

where S is given by (2.2.15). The upper and lower sign holds for the Dirichlet problem and the Neumann problem, respectively.

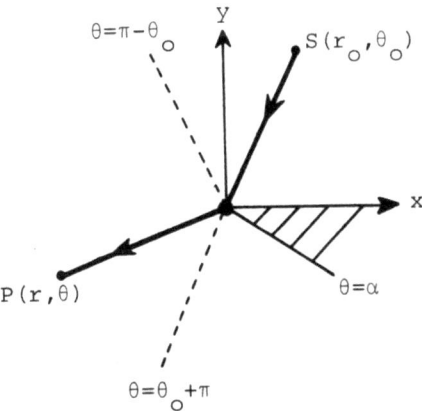

Figure 2.6. Point source at wedge.

When the conditions (2.4.2) are replaced by the impedance conditions

$$\frac{1}{r} \partial_\theta u \pm ikZ_\pm u = 0 \qquad \theta = \begin{cases} 0 \\ \alpha \end{cases} \qquad (2.4.6)$$

the method does not work.

Instead the problem may be solved by means of Wiener-Hopf technique or more general methods [2].

2.5 Circular cone.

The diffraction coefficient, D_v, for vertex diffraction can be obtained from the asymptotic representation of the exact solution of the three-dimensional problem of plane wave incidence from the direction (θ_o, ϕ_o) on a circular cone with semivertex angle $\delta = \pi - \theta_1$ (see Fig. 2.7) with Dirichlet or Neumann boundary condi-

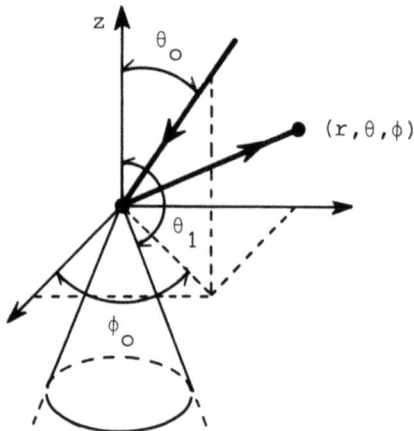

Figure 2.7. Plane wave incidence at circular cone.

tions. For $\theta + \theta_o$ not to close to $2\theta_1 - \pi$ the diffracted field becomes

$$u_d \sim D_v \frac{e^{ikr}}{r} \qquad (2.5.1)$$

with

$$D_v = -\frac{i\pi}{k} \sum_{m=0} \varepsilon_m \cos m(\phi - \phi_o) \int_0^\infty dx \, \frac{x \tanh \pi x}{\cosh \pi x}$$

$$\times \frac{K_x^m(\cos\theta) K_x^m(\cos\theta_o) \left\{\begin{matrix} 1 \\ \partial_{\theta_1} \end{matrix}\right\} K_x^m(-\cos\theta_1)}{\Gamma(\tfrac{1}{2}+m+ix)\Gamma(\tfrac{1}{2}+m-ix) \left\{\begin{matrix} 1 \\ \partial_{\theta_1} \end{matrix}\right\} K_x^m(\cos\theta_1)} \qquad (2.5.2)$$

with upper (lower) symbols for the Dirichlet (Neumann) problem. The symbol ε_m is given by

$$\varepsilon_m = \begin{cases} 1 & \text{for } m = 0 \\ 2 & \text{for } m = 1,2,\cdots \end{cases} \qquad (2.5.3)$$

and

$$K_x^m(\cos\theta) = P_{ix-\tfrac{1}{2}}^m(\cos\theta), \qquad (2.5.4)$$

P_ν^μ being the Legendre function of first kind. For $\mu = 0,1,2,\cdots$

$$P_\nu^\mu = (1-x^2)^{\tfrac{m}{2}} \frac{d^m}{dx^m} P_\nu(x),$$

where $P_\nu(x)$ is given by (2.3.6) for $\nu = 0,1,\cdots$. For $\delta \sim 0$ it can be shown [3] that the complicated diffraction coefficient, D_v (2.5.2), reduces considerably.

3. Application of GO and GTD scattering coefficients for construction of hybrid coefficients.

The scattering coefficients derived in the previous sections may be combined into hybrid scattering coefficients for complex processes.

3.1 Refraction of creeping waves.

As the first example of this application we consider the refraction of a creeping wave propagating along an interface with boundary conditions (2.1.14) separating two media with propagation constants k_1 and k_2 (see Fig. 3.1). If $k_2 < k_1$ and the creeping wave

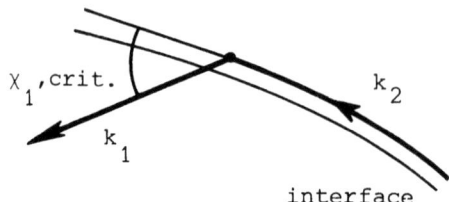

Figure 3.1. Refraction of creeping wave.

propagates with propagation constant k_2 refracted rays with propagation constant k_1 are refracted under the critical angle $\chi_{1,crit}$ given by $\cos\chi_{1,crit} = k_2/k_1$ (follows from (2.1.12) with $\chi_2 = 0$).
The complex process of diffraction and refraction is then described by [4]

$$\mathbb{D}_p = d_p \frac{T_{21}}{1+R_{22}} \qquad (3.1.1)$$

where d_p is given in Table 2.2 and T_{21} and R_{22} are given by (2.1.13). In d_p we let $k = k_2$ and $z \sim z_o (ka)^o$ where z_o is the equivalent surface for the ray. This quantity is found by comparison between R (2.1.7) and R_{22} (2.1.13) in the limit $\chi_2 = 0$. In the case of a convex interface seen from the k_2-side the radius of curvature at the diffraction point is positive while it is negative (with argument $-\pi$) in the case of a concave interface. In (3.1.1) the factor d_p gives the (total) field of k_2-mode at the diffraction point and the factor $T_{21}(1+R_{22})^{-1}$ gives the field of k_1-mode at the diffraction point on the refracted ray. This result can be verified from the canonical problem of a circular cylinder interface.

3.2 Edge diffraction of creeping waves.

When a creeping wave reaches an edge of a curved surface edge diffracted rays are emitted (see Fig. 3.2). This complex process of diffraction is then described by [1]

$$\mathbb{D}_e = d_p \frac{D_e}{1+R} \qquad (3.2.1)$$

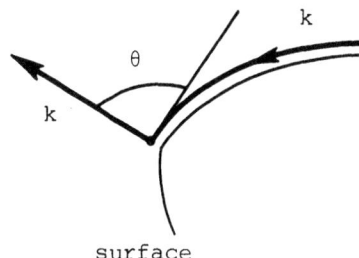

Figure 3.2. Edge diffraction of creeping waves.

where d_p is given in Table 2.2, R is given by (2.1.7), and D_e is given by (2.4.5) or in the case of a finite surface impedance by the relevant expression for D_e. Both R and D_e should be taken for vanishing angle of incidence. In the Dirichlet case ($z = \infty$) (3.2.1) must be modified [1]. In (3.2.1) the factor d_p gives the (total) field at the diffraction point, the factor $(1 + R)^{-1}$ reduces the field to the incident field on a tangential ray, and the factor D_e gives the diffracted field on the diffracted ray at the diffraction point. This result can be verified from solutions to canonical problems in special cases.

References.

1. Albertsen, N.C. and P.L. Christiansen, "Hybrid Diffraction Coefficients for First and Second Order Discontinuities of Two-Dimensional Scatterers", SIAM J. Appl. Math. 34, pp.398-414 (1978).

2. Christiansen, P.L., "Comparison Between Edge Diffraction Processes", Proc. IEEE 62, pp.1462-1468 (1974).

3. Bowman, J.J., T.B.A. Senior and P.L.E. Uslenghi, "Electromagnetic and Acoustic Scattering by Simple Shapes", North-Holland, Amsterdam (1969).

4. Christiansen, P.L., "Comparative Studies of Diffraction Processes in the Geometrical Theory of Diffraction", Polyteknisk Forlag, Lyngby, Denmark (1975).

UNIFORM THEORIES OF DIFFRACTION BY EDGES

G. A. Deschamps

Department of Electrical Engineering, University of
Illinois, Urbana, Illinois 61801, USA.

ABSTRACT. The geometrical theory of diffraction GTD has well
known shortcomings when applied in some transition regions such
as the vicinity of shadows boundaries or caustic surfaces.
Uniform theories provide field descriptions valid in these
regions. This article considers more specifically diffraction
by the edge of a screen and describes and compares two competing
theories that give the field near to shadow boundaries.

1. INTRODUCTION

Other authors in this series of lectures, devoted to high-
frequency scattering and diffraction, will have discussed the
need for special methods, in particular approximate ones, for
treating problems where characteristics dimensions of the
objects involved are large in term of wavelengths. Exact analytic solutions exist only for a few very special geometries.
Numerical methods based on solving integral equations are
impractical because of excessive memory requirements. On the
other hand methods inspired by the optical limit, where the
field may be thought of as propagating along rays, become acceptable. Not only do they provide good approximations but also a
useful viewpoint that help understand the mechanism of scattering and often suggests synthesis methods.

Classical geometrical optics considers rays that propagate
freely (along straight lines in homogeneous media or geodesics
in non homogeneous media) or are reflected by smooth surfaces.
It predicts zero fields in the regions not reached by either
type of rays.

The geometrical theory of diffraction (GTD), introduced by J. B. Keller and co-workers constituted a major progress. By enlarging the concept of ray to include diffracted and surface (creeping) rays a field in shadow regions can be computed and a more accurate field in illuminated regions is obtained.

The geometrical theory of diffraction however fails when applied in some transition regions, such as the vicinity of shadow boundaries or of caustic surfaces. The goal of the so-called "Uniform theories" is to provide field descriptions valid in those regions.

In this paper we shall address the problem of diffraction through an aperture in a screen and consider mainly representations of the field in transition regions about the shadow boundaries. Several solutions have been proposed that have their roots in the consideration of diffraction by a half-plane (Sommerfeld problem). Some solutions have been justified by searching for series representations that satisfy the wave equations (Helmoltz or Maxwell) asymptotically. Others are related to integral representations that are evaluated by stationary phase or saddle-point methods. In this latter process it is recognized that the transition regions occur when two critical points (saddle-point and pole of the integrand) come close to one another. These justifications are valid only for special problems, for instance, for an arbitrary field incident on a half plane. For a curved edge, or for the edge of a curved surface there are no such justifications and one must be contented with heuristic arguments based on replacing the exact situation by a simpler one for which the process can be justified. In a sense this is an extension of the concept of canonical problem used with such success by Keller. We shall follow this approach starting with the Sommerfeld problem, and briefly reviewing the GTD diffracted field since it plays a role in the uniform expressions.

2. THE SOMMERFELD PROBLEM

The Sommerfeld solution to the problem of diffraction of a field $F_1 = (E_1, H_1)$ by a conducting half plane is composed of two terms, each one of which can be interpreted as the field diffracted by an absorbing screen. The first term results from the diffraction of the incident field F_1, the second results from the diffraction of a field F_2 image of F_1 as described in appendix C. The geometrical optics field, the exact diffracted part of the field, and its approximations according to GTD or according to uniform theories, all have that same structure. It is therefore sufficient to describe how these fields are deduced from F_1, in presence of the absorbing screen. The second term resulting from the same procedure applied to F_2 will be denoted by (1→2) for whatever part of the field is being considered: total, diffracted, GTD, or uniform. The absorbing screen is precisely defined by considering the physical space P (complement of the screen) continued through the screen into a replica P' of P. The space Q thus obtained is made up of two sheets P and P' joined along the screen. The incident plane wave with wave vector k_1

$$F_1(r) = (E_1(0), H_1(0))e^{ik_1 \cdot r} \quad (2.1)$$

where $E_1(0)$, $H_1(0)$ are perpendicular to k_1 and related by $ZH_1(0) = \hat{k}_1 \times E_1(0)$ is an optical field with rays parallel to k_1. The points reached by these rays without being interrupted by the screen define the illuminated region. Introducing a function $\theta_1(r)$ equal to 1 when r is illuminated and 0 otherwise, the geometrical optic field corresponding to F_1 is $\theta_1(r) F_1(r)$. The total geometrical optic field is

$$\theta_1(r) F_1(r) + (1 \rightarrow 2). \quad (2.2)$$

The function $\theta_2(r)$ is defined by tracing the rays parallel to $k_2 = R(k_1)$, reflection of k_1 in the screen. If the incident field is a scalar field $u_1(r)$, the exact diffracted field is

$$F(r,k_1) u_1(r) + (1 \rightarrow 2) \quad (2.3)$$

where $F(r,k_1)$ is short for $F(\tau(r,k_1))$, F is the Fresnel integral (eq. A.3), and τ is the detour parameter (eq. D.2). The image field $u_2(r)$ takes into account the reflection coefficient of the screen as discussed in Appendix C. The result (2.3) can also be expressed as a sum

$$\theta_1 u_1(r) + e^{ik|r-p|} V(r,k_1) u_1(p) \quad (2.4)$$

The first term being the GO field, the second term is the exact diffracted field U_{D1}, corresponding to the incident field. The function $V(r,k_1)$ is $V(\tau(r,k_1))$ with $V(\tau)$ defined in eq. A-7. The point p on the edge is such that pr is a diffracted ray i.e., the ray from p to r make the same angle β with the edge as k_1. The length $R = |r-p|$ is the oblique distance from r to the edge. The total diffracted field is

$$u_D(r) = e^{ikR} u_1(p) V(r, k_1) + (1 \rightarrow 2) \quad (2.5)$$

$$= e^{ikR} u_1(p) [V(r, k_1) \pm V(r, k_2)]$$

since $u_2(p) = \pm u_1(p)$ according to the screen reflection coefficient. For an electromagnetic field such as (2.2) the exact solution is more complex. Part of it is still the incident field multiplied by the Fresnel function of the detour parameter

$$F(r, k_1) F_1(r) + (1 \to 2), \qquad (2.6)$$

but there are added terms that come from field components normal to the edge. It has been shown in [31] that those terms can be represented by a compact and coordinate-free expression

$$g(kR) X(r,k_1) (t^\psi - t^0) F_1(p) + (1 \to 2) \qquad (2.7)$$

In this formula and all of those that follow we shall use ψ and τ without subscript to mean ψ_1, τ_i - (This is to simplify the typing). When computing (1→2) one must remember to replace and τ by ψ_2, τ_2.

The various symbols in 2.7 are defined as follows:

$g(x) = (8\pi x)^{-1/2} \exp i(x + \pi/4)$ Green's function in R^2 for $x \gg 1$.

$R = |r-p|$, oblique distance to the edge.

t^ψ rotation operator through the angle ψ about t such that
$t^\psi: k_1 \to k$

k wave vector along the diffracted ray p to r

t vector along the edge. Later, for a curved edge Γ, t is a vector tangent to Γ at point p

$$X(r, k_1) = X(\psi) / \sin \beta = \operatorname{cosec} \psi/2 \qquad (2.9)$$

$$X(\psi) = \operatorname{cosec} \psi/2 \qquad (2.10)$$

in 2.9 and 2.10, ψ is a particular determination of the rotation angle as discussed in Appendix D. It is defined modulo 4π so that $\sin(\psi/2)$ and τ have unambiguously defined values.

t^0 rotation through angle 0, i.e., identity operator. It is expressed in this manner for more symmetry.

$t^\psi - t^0$ gives 0 when operating on vectors parallel to t. Thus these vectors are diffracted as scalar fields.
When applied to vector U perpendicular to t

$$(t^\psi - t^0) U = 2 \sin \psi/2 \; t^{(\psi+\pi)/2} U \qquad (2.11)$$

Introducing the operator $t_\perp = -t \times t \times \qquad (2.12)$
which projects vectors on the plane perpendicular to t, we have the general formula

$$t^\psi - t^0 = 2 \sin \psi/2 \cdot t^{(\psi+\pi)/2} \, t_\perp \tag{2.13}$$

which gives the small angle approximation

$$(t^\psi - t^0) \, U \simeq \psi t \times U \tag{2.14}$$

The first term in (2.6) can be decomposed into the GO field $\theta_1(r) \, F_1(r)$ plus

$$e^{i\tau^2} V(\tau) \, F_1(r) = e^{i(\tau^2 + k_1 \cdot r)} V(\tau) \, F_1(p) = e^{ikR} V(\tau) \, F_1(p) \tag{2.15}$$

Adding (2.7) gives the exact diffracted field:

$$F_D(r) = \left\{ e^{ikR} V(\tau) + g(kR) \, X(r,k_1) \, (t^\psi - t^0) \right\} F_1(p) + (1 \to 2) \tag{2.16}$$

Asymptotically, replacing V by $\tilde{V}$ and using (A.10), F_D reduces to the GTD diffracted field

$$F_d(r) = g(kR) \, X(r,k_1) \, t^\psi \, F_1(p) + (1 \to 2) \tag{2.17}$$

This field becomes infinite with $X(r,k_1)$ when ψ, and therefore $\tau(r,k_1)$, approaches 0. Using (2.12), (2.9) and (2.10), 2.14) can be written

$$F_D = e^{ikR} V(\tau) \, F_1(p) + g(kR) \, \mathrm{cosec}\,\beta \, 2t^{(\psi+\pi)/2} \, t_\perp \, F_1(0) + (1 \to 2) \tag{2.18}$$

We note that the field (2.16) is discontinuous accross the shadow boundary ($\tau=0$) and compensates exactly for the discontinuity of the GO field $\theta_1 F_1$. One term in (2.7) is the GTD field F_d (eq. 2.15) the other is the negative of

$$g(kR) \, X(r,k_1) \, F_1(p) = \tilde{F}(\tau) \, F_1(r) \tag{2.19}$$

Thus the total field at r can be written

$$F(r) = F_1(r) \, F(r,k_1) + F_d(r) - F_1(r) \, \tilde{F}(r,k_1) + (1 \to 2) \tag{2.20}$$

which will be recognized as the UAT expression presented in section 4. For small values of τ the last two terms become large and it is preferable to use

$$F(r) = \theta_1(r) F_1(r) + e^{ikR} V(\tau) +$$
$$[g(kR) \, \mathrm{cosec}\, \beta \cdot 2 \, t^{(\psi+\pi)/2} \, t_\perp] \, F_1(p) + (1 \to 2). \tag{2.21}$$

where the underterminacy $\infty - \infty$ has been removed.

3. THE GTD FIELD DIFFRACTED BY A CURVED EDGE

The GTD Solution F_d for the field diffracted by a curved

edge Γ is only defined when the incident field F_1 is also a ray optical field i.e., is represented by

$$F_1(r) = \exp(ik\, S_1(r)) \cdot \Sigma\, (ik)^{-m}\, F_{1m}(r) \tag{3.1}$$

The diffracted field is constructed by first finding its eikonal $S(r)$, which is done by tracing the diffracted rays. Then transport equations along these rays can be written and initial values of field amplitudes (vectorial for an electromagnetic field) are deduced from the Sommerfeld problem. Locally about each point p on Γ the incident field is replaced by a plane wave, and the screen by a half plane tangent to it and having a straight edge along t.

3.1 Diffracted Rays and Diffracted Pencils

The eikonal $S(r)$ of the diffracted field matches that of the incident field at each point p of Γ

$$S_1(p) = S(p), \quad p \in \Gamma \tag{3.2}$$

Alernatively the diffracted ray emanating from p toward r has a wavevector k that makes the same angle β with the tangent vector t to Γ at p as the incident wave vector $k_1(p)$. The vector k can be deduced from k_1 by a rotation about t through some angle, ψ_1 (denoted ψ according to an earlier remark)

$$t^\psi k_1 = k \tag{3.3}$$

The eikonal at point r is

$$S(r) = S_1(p) + pr \tag{3.4}$$

To apply the transport equations to the diffracted field one needs to relate the characteristics of the diffracted pencil in direction of k to those of the incident pencil in direction k_1. This problem is solved by the following formulas. The incident pencil is characterized by the location of its foci at $p - f_1 \hat{k}_1$, $p - f_2 \hat{k}_1$, and by the twist angle α between the plane of incidence $t\, k_1$ and the direction of focal line corresponding to focus 1. The amplitude variation in the incident pencil is

$$a_1(r) = a_1(0)\, (1 + r/f_1)^{-1/2}\, (1 + r/f_2)^{-1/2} \tag{3.5}$$

The radius of curvature at p of the incident phase front in the plane of incidence is f_o given by

$$\frac{1}{f_o} = \frac{\sin^2\alpha}{f_1} + \frac{\cos^2\alpha}{f_2} \tag{3.6}$$

The two foci of the forward diffracted pencil ($\psi = 0$) are the point p and $p - k_1 f_0$ with focal lines respectively perpendicular and parallel to the plane of incidence.

In the direction of $k = t^\psi k_1$ the diffracted pencil has foci at p and p-kf with

$$1/f_\psi - 1/f_0 = [(\hat{k}_1 - \hat{k}) \cdot n] / \rho \sin^2\beta \qquad (3.7)$$

where $p-\rho\hat{n}$ is the center of curvature of the edge Γ at p [Ref. 10]

3.2 Transport Equations

The variation of the field amplitude is according to the inverse of the square root of the pencil cross section hence follows the function

$$b_\psi(r) = r^{-1/2} (1 + r/f_\psi)^{-1/2} \qquad (3.8)$$

Note that f_0 and f_ψ are infinite for a plane wave incident on a straight edge.

3.3 The GTD Diffracted Field

To complete the description of the field F_d one needs some initial values of amplitudes (more precisely field vectors) and these are provided by the canonical Sommerfeld problem. The straight edge GTD solution (2.15) is simply modified by the appropriate divergence factor. The final result is

$$F_d(r) = (1 + r/f_\psi)^{-1/2} g(kr) \mathcal{X}(r,k_1) t^\psi F_1(0) + (1 \to 2) \qquad (3.6)$$

4. UNIFORM ASYMPTOTIC THEORY OF DIFFRACTION

The total field resulting from the diffraction of a scalar plane wave u_1 by an absorbing half-plane is obtained by multiplying the incident field $F_1(r)$ by the Fresnel function $F(\tau)$, where is the detour parameter (r, k_1). This suggest trying a solution of the same form for a general ray optical field (3.1), having eikonal $S_1(r)$, and being diffracted by a curved edge. A system of diffracted rays and a corresponding eikonal $S(r)$ can be constructed as described in section 3. A detour parameter is defined by

$$\tau = \pm \sqrt{k} \, (S-S_1)^{1/2} \qquad (4.1)$$

and a field

$$F_1(r)\ F(\tau) \tag{4.2}$$

can be considered. Using the asymptotic expansion of ( ) in A.4. This field can be shown to have S for eikonal and known amplitudes of various order. It can be observed however that these amplitudes (vectorial for an electromagnetic field) do not satisfy the recursive relations, or transport equations, resulting from the wave equation (Helmoltz or Maxwell). The solution proposed on [12, 13, 14, 15] consist of adding a complementary field $F_B(r)$ having also S for its eikonal and amplitudes determined by the condition that the resulting field

$$F(\tau)\ F_1(r) + F_B(r) \tag{4.3}$$

satisfies the appropriate transport equations. The expansion of F_B is of the form $\nu^{-1/2} \Sigma\ \nu^{-m} F_m(r)\ e^{\nu S}$. This leaves undetermined some initial values that result from an edge condition. Writing the solution

$$\theta_1(r)\ F_1(r) + F_D(r) \tag{4.4}$$

the expansion of the diffracted field can be evaluated up to any order.

In practical application one rarely needs the higher order terms. They are in general extremely complicated to evaluate and because of the asymptotic nature of the expansions one cannot be sure that they constitute an improvement of the solution.

In the following we shall discuss only the first term, which we still designate by F_B. It is found that

$$F_B = F_d - \tilde{F}\ F_1 \tag{4.5}$$

where F_d is the GTD diffracted field [eq. 3.6]. This is the central result, it was verified in section 2 that for Sommerfeld problem (Electromagnetic field) this happens to give the <u>exact solution</u> including longitudinal field components.

The total field can be written

$$F_t = (F-\tilde{F})\ F_1 + F_d \tag{4.6}$$

[Of course for a reflecting screen one must add a term (1→2) derived from the image field. See Appendix C].

This same assumption has been used to derive a uniform solution for the diffraction by a curved wedge. [17]

5. UNIFORM THEORY OF DIFFRACTION (UTD)

A uniform theory proposed in [18] expresses the field as the sum of a geometrical optic field $\theta_1 F_1$ and of the GTD diffracted field F_d multiplied by a transition function which reduces to unity when the detour parameter τ is large, while it compensates for the infinity of the GTD field as τ approaches 0.

The transition function with the notation (A-7) and (A-10) is $V(\tau)/\tilde{V}(\tau)$. Thus the total field, for an absorbing screen, is

$$F_k(r) = \theta_1(r) F_1(r) + (V/\tilde{V})(\tau) F_d(r) \tag{5.1}$$

When the screen is reflecting one must add a term (1→2).

In this formula as described in [18] the parameter τ (meaning τ_1, then τ_2) is not the actual detour parameter (eq. D.8) but its paraxical approximation (D.9). Using the exact τ does not change the result much because the factor $V/\tilde{V}$ differs significantly from 1 only when τ is small [Figure 2]. Formula (5.1) with the actual is in fact a variant on the UTD originally proposed. In a similar manner a variant to UAT results from replacing the actual by its paraxial value. This is in fact what is done to evaluate the UAT field on, or very near to, the shadow boundary.

Another variant of (5.1) can be constructed where the indeterminacy $0 \times \infty$ has been removed. It is based on the observation that F_d already contains a factor $V(\tau)$ and therefore the last term can be simplified. The diffracted field according to (5.1)

$$(1 + r/f_\psi)^{-1/2} \ g(kr) \ X(r,k_1) \ t^\psi F_1(0) \ [V(\tau)/\tilde{V}(\tau)] \tag{5.2}$$

But
$$g(kr) \cdot X (r,k_1) = e^{ikr} V(\tau_o) \tag{5.3}$$

where $\tau_o = \sqrt{2 \ kr} \ \sin \dfrac{\psi}{2} \sin \beta \tag{5.4}$

The ratio

$$\tilde{V}(\tau_o)/\tilde{V}(\tau) = \tau/\tau_o = (1 + r/f_o)^{1/2} \ (1 + r/f_1)^{-1/2} \ (1 + r/f_2)^{-1/2} \tag{5.5}$$

In the immediate vicinity of the shadow boundary $\psi \sim 0$ the above expression reduces to the incident field multiplied by $V(\tau)$. This agrees with the UAT solution $F(\tau) \ F_1(r)$, without the complementary field. Away from this immediate neighborhood of the edge but still close to $\psi = 0$ the following expression could be used

$$(1 + r/f_\psi)^{-1/2} \ (1 + r/f_o)^{1/2} \ t^\psi \ F(t^{-\psi} \vec{r}) \ V(\tau) \tag{5.6}$$

This equation shows how the radial variation of amplitude changes continuously from that along the incident shadow boundary field to the GTD field.

6. COMPARISON OF THE TWO METHODS

As mentioned in the introduction there is no general justification of either method. The only test is to apply the method to special problems with known solutions. The simplest special problem is the diffraction of a plane wave by a half plane. We have seen that UAT gives the exact solution while UTD fails near to the edge. Away from that region the two solutions agree. The uniform asymptotic solution has also been shown to apply to half plane diffraction for more general incident fields. [J. Boersma] UAT has the advantage of giving a complete expansion that satisfies the wave equation asymptotically. This however, is offset by the questionable value of such an expansion.

The UAT strays away from the principle of local interaction that characterizes high frequency fields. The field in the shadow region for instance depends on the incident field at points that it does not even reach. When generalized to a wedge, and particularly to a curved wedge, the UAT field also fails near the edge and its construction involves some arbitrariness as discussed in 4. In contrast UTD has a certain physical appeal: the field depends on what happens near the edge only and it can be interpreted as compensating for the GO discontinuities. The method is simpler to apply, in particular after the transformation discussed in 5.

Appendix A

FRESNEL INTEGRALS

The Fresnel integrals play a central role in expressing the field diffracted by a half plane and by an edge. The particular form used in this paper is

$$F(\tau) = \pi^{-1/2} \exp(-i\pi/4) \int_{\tau}^{\infty} e^{i\tau^2} d\tau \qquad (A-1)$$

It is represented graphically by a Cornu Spiral, locus of F in the complex plane as τ takes real values (Fig. A-1).

The following properties can be read on the figure

$$F(0) = 1/2, \qquad F(-\infty) = 1, \qquad F(+\infty) = 0,$$
$$F(\tau) + F(-\tau) = 1 \qquad (A-2)$$

As τ approaches $+\infty$ the spiral winds itself about the origin and can be represented approximately

$$\tilde{F}(\tau) = 1/2\pi^{-1/2}\,\tau^{-1}\,\exp\,i(\tau^2 + \pi/4) \qquad (A-3)$$

which is the first term of an asymptotic series

$$F(\tau) = \tilde{F}(\tau)\sum_n (1/2)_n\,(i\tau^2)^{-n}. \qquad (A-4)$$

Where

$$(1/2)_0 = 1,\ (1/2)_n = 1/2\,(1/2 + 1)\ \text{---}\ (1/2 + n-1) = \pi^{-1/2}\,\Gamma(1/2 + n)$$

As τ approaches $-\infty$

$$F(\tau) \simeq 1 - \tilde{F}(-\tau) \qquad (A-5)$$

Near $\tau = 0$ the function F is approximated by

$$F_0(\tau) = 1/2 - \pi^{-1/2}\,\exp(-i\pi/4)\,\tau \qquad (A-6)$$

Another form of the Fresnel integral is

$$V(\tau) = \varepsilon\,e^{-i\tau^2}\,F(\varepsilon\tau) = e^{-i\tau^2}(F(\tau) - \theta(-\tau)) \qquad (A-7)$$

where ε is the sign of τ and $\theta(x)$ the Heaviside unit step function (ε has been called the shadow indicator [17]). For a complex value of τ it is defined as sgn Re$[\tau\,\exp(-\pi/4)]$. The function V is represented in Fig. A-2. It is discontinuous at $\tau = 0$ as it jumps from $-1/2$ to $+1/2$ when τ goes from -0 to $+0$. The function is well represented by the following approximation [30].

$$\begin{aligned}
V &= V' + iV'' \\
V' &= 1/2(f+g) \\
V'' &= 1/2(f-g) \\
f &= (1 + 0.739\tau)/(2 + 1.430\tau + 1.976\tau^2) \\
g &= 1\,/\,(2 + 3.305\tau + 2.223\tau^2 + 3.388\tau^3).
\end{aligned} \qquad (A-8)$$

For large values of τ, positive or negative, the function V approaches

$$\tilde{V}(\tau) = \exp(i\pi/4)/2\tau\sqrt{\pi} = e^{-i\tau^2}\,\tilde{F}(\tau) \qquad (A-9)$$

The function V is useful in expressing the exact diffracted field in the Sommerfeld problem. Its asymptotic value (A-9) leads to the GTD solution by means of the identity.

$$\tilde{V}(\tau)\,e^{ikr} = \tilde{F}(\tau)\,e^{ik_1\cdot r} = X(r,k_1)g(kr) \qquad (A-10)$$

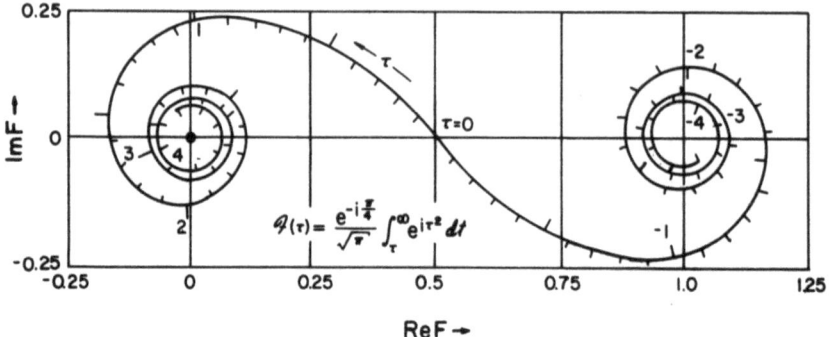

Fig. 1   Cornu Spiral for the Fresnel integral $F(\tau)$

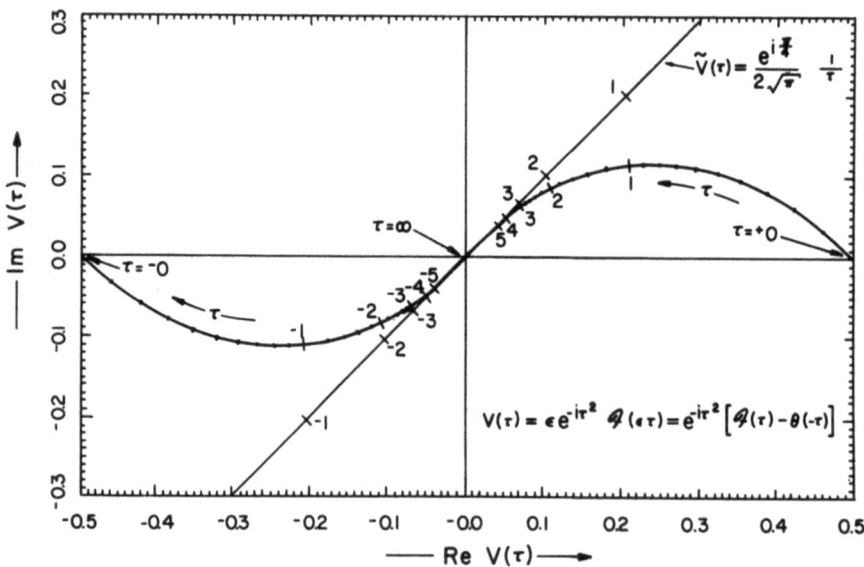

Fig. 2   Modified Fresnel Integral $V(\tau)$

where
$$g(x) = (8\pi x)^{-1/2} \exp i(x + \pi/4), \quad (A-11)$$

$$X(r, k_1) = \operatorname{cosec} \psi/2 \quad (A-12)$$

$$\tau = \sqrt{2\,kr} \sin(\psi/2) \quad (A-13)$$

Appendix B

ROTATION OPERATOR

Let t be a unit vector supported by the z axis, or tangent to a diffracting edge at the point of diffraction. We designate by t^ψ the operator that rotates about the axis supporting t, through the angle ψ, points in space or figures such as vectors considered as sets of points.

This operator has the following obvious properties that justify the exponential notation

Composition: $\quad t^\alpha t^\beta = t^{\alpha+\beta} \quad$ (B-1)
Identity: $\quad t^0 = \text{id} \quad$ (B-2)
Inverse: $\quad (t^\alpha)^{-1} = t^{-\alpha} \quad$ (B-3)

In [17] where this operator was first introduced in edge diffraction problems t^ψ was denoted by $\operatorname{rot}(t, \psi)$. The present notation is more compact and the exponential is justified by properties A-1, 2, 3.

To rotate a vector V about the unit vector t, one may use one of the following formulas, among others that may be better adapted to the specific manner in which V is given. A purely vectorial formula is:

$$t^\psi V = t(t \cdot V) + (t \times V)\sin\psi - t \times (t \times V)\cos\psi \quad (B-4)$$

If V is given by its components (X, Y, Z) in a rectangular coordinate system where $\hat{z}$ coincides with t

$$V = X\hat{x} + Y\hat{y} + Z\hat{z} \quad (B-5)$$

$$V' = t^\psi V = X t^\psi \hat{x} + Y t^\psi \hat{y} + Z t^\psi \hat{z} \quad (B-6)$$

with

$$\begin{aligned} t^\psi \hat{x} &= \cos\psi\,\hat{x} + \sin\psi\,\hat{y} \\ t^\psi \hat{y} &= -\sin\psi\,\hat{x} + \cos\psi\,\hat{y} \\ t^\psi \hat{z} &= \hat{z} \end{aligned} \quad (B-7)$$

$$V' = t^\psi V = (X \cos \psi - Y \sin \psi)\,\hat{x} + (X \sin \psi + Y \cos \psi)\,\hat{y} + Z\,\hat{z}. \quad (B-8)$$

If $\quad V' = X'\hat{x} + Y'\hat{y} + Z'\hat{z}, \quad\quad\quad\quad\quad\quad\quad\quad\quad\quad (B-9)$

a matrix representation follows immediately

$$\begin{bmatrix} X' \\ Y' \\ Z' \end{bmatrix} = \begin{bmatrix} \cos\psi & -\sin\psi & 0 \\ \sin\psi & \cos\psi & 0 \\ 0 & 0 & 1 \end{bmatrix} \begin{bmatrix} X \\ Y \\ Z \end{bmatrix} \quad (B-10)$$

Appendix C

IMAGE OF A FIELD UNDER REFLECTION

To construct the Sommerfeld solution, and also the GTD or the uniform solutions to the general problem of diffraction of a field F_1 by the curved edge of a curved screen, one needs to define an image field F_2. This field is then treated in the same manner as F_1 and the complete solution is the sum of the resulting two terms.

Starting with an arbitrary screen in a plane P let R designate the operator that transforms a geometrical figure A (point, line, vector, etc.) into its image R A. One may apply R to the vectors E_1 and and H_1 that describe the field F_1. However to obtain a field F_2 that satisfies Maxwell's equations and the boundary conditions on the screen one must reverse the sign of the image of E_1. The resulting field, called <u>odd image</u> of F_1, is then defined by

$$E_2(r) = -R\,E_1\,(Rr) \quad\quad\quad\quad (C-1)$$

$$H_2(r) = R\,H_1\,(Rr) \quad\quad\quad\quad (C-2)$$

If F_1 is a plane wave with wave vector k_1, F_2 has wave vector $k_2 = Rk_1$. The shading of k_2 must be indicated by the same sign ε as that of k_1 although we have the property $R(k_1^+) = k_2^-$.

For a scalar incident field $u_1(r)$ the image field $u_2(r)$ must be such that $u_1 + u_2$ satisfies the appropriate boundary condition ($u = 0$ or $\partial_n u = 0$). Correspondingly u_2 is defined by

$$u_2(r) = \mp u_1\,(Rr) \quad\quad\quad\quad (C-3)$$

When dealing with a curved screen the construction of the image field F_2 must be based on ray tracing and iterated transport equations [9] [11]. The rays must be continued backward through the surface. (This is why we prefer to call F_2 an image field rather than reflected field). There are two problems connected with that construction. (1) The screen must be continued past the edge into a smooth surface. This presents some arbitrariness. (2) The iteration of the transport equations should not be carried out too far because this may spoil rather than improve the solution. Anyhow even the second term is rather difficult to compute and is usually ignored.

Appendix D

DETOUR PARAMETERS

A parameter that plays an important role in the Sommerfeld problem as well as in the uniform theory of edge diffraction is the <u>detour parameter</u> τ. It is a function of the observation point r defined as a square root of the difference between the phase $kS(r)$ of the GTD diffracted field and that $kS_1(r)$ of the incident field. The square root is taken as positive when r is in the shadow region, i.e. when the incident ray through r is interrupted by the screen. It is negative otherwise and not defined if no ray goes through r. For an incident plane wave with wave vector k_1, the function will be denoted by $\tau(r,k_1)$, for a spherical wave with center at s by $\tau(r, s)$, more generally for a field F_1 by $\tau(r, 1)$ or $\tau_1(r)$. For a plane wave F_1, τ can be evaluated by first constructing the point p on the edge Γ from which emanates a diffracted ray going through r. This point is such that k_1 and pr make the same angle β with the tangent t to the edge Γ at point p. [There may be several such points and consequently τ may be multivalued]. The condition on p also implies the existence of a rotation t^ψ about t through angle ψ [see Appendix B] such that the vector k_1 is brought onto the vector k of the diffracted ray in direction from p to r. The detour parameter is then easily evaluated.

$$\tau^2 = k\mathcal{R} - k_1 \cdot (r-p) = k\mathcal{R}(1-\cos\psi)$$

where $\mathcal{R} = |r-p|$ is the oblique distance from the edge to point r and Ψ is the angle between k_1 and k. This angle is related to ψ and β by

$$\sin \frac{\Psi}{2} = \sin \frac{\psi}{2} \sin \beta \qquad (D.1)$$

Thus $\qquad \tau = \sqrt{2\ k\mathcal{R}}\ \sin \frac{\psi}{2} \qquad (D.2)$

with the sign chosen as explained above.

In this formula ψ is a particular determination of the rotation angle. Namely it is the angle of the continuous rotation that brings k_1 on k without crossing the screen. It is counted as positive if this rotation is from the illuminated to the shadow side of the vector k_1. The shadow side of k_1 may be indicated by shading half of its arrowhead or by adding a superscript $\varepsilon = \pm 1$ to k_1 according to whether this positive direction agrees or disagrees with a specified sense of rotation. We shall then designate this choice of angle by $(r, k_1\varepsilon)$. This angle may be modified by a multiple of without changing the value of (r, k_1^ε). This can also be written

$$\tau = \sqrt{2\ \kappa\rho}\ \sin \frac{\psi}{2} \tag{D.3}$$

where $\kappa = k \sin \beta$, $\rho = r \sin \beta$ are the length of projections of vectors k and pr on the plane perpendicular to Γ at p and ψ is the rotation angle defined in D

When F_1 is a spherical wave with center at s and r is the positive vector of the observation point the detour parameter is

$$\tau = \pm\sqrt{2\,kL}\ (r + s - R)^{1/2} \tag{D.4}$$

with $R = |r-s|$ and the appropriate sign. This may be written

$$t = \sqrt{2\,kL}\ \sin \frac{\psi}{2} \tag{D.5}$$

$$L = 2rs\ /(r+s-R) \tag{D.6}$$

with ψ the angle (specified as above) between k_1 and k at the diffraction point p. When ψ is small:

$$L \simeq rs\ /(r+s) \tag{D.7}$$

When F_1 is an arbitrary ray field with eikonal $S_1(r)$ the detour parameter

$$\tau = \sqrt{k}\ (S(r) - S_1(r))^{1/2} \tag{D.8}$$

can be approximated, close to a shadow boundary, by replacing S and S_1 by values corresponding respectively to the incident and forward diffracted pencils. These pencils have in common a central ray in direction $k_1 = k$ ($\psi=0$).

Near a point r on that ray, say at r + x where x is perpendicular to r and tangent to the diffraction cone the eikonals S_1 and S are quadratic functions of x defined by the curvature matrices at point p (taken as origin 0) of the incident and diffracted pencils. A somewhat lengthy derivation, that makes use of results in 3.2 and of the fact that $|x| \simeq r\psi \sin \beta$, $\sin \frac{\psi}{2} \simeq \frac{\psi}{2}$ leads to the following result

$$\tau = \sqrt{2\,kL}\ \sin (\psi/2)\ \sin \beta. \tag{D.9}$$

with the distance parameter introduced in [18] (here deprived of sin β factor)

$$L(r) = r\,(1 + r/f_0)\,(1 + r/f_1)^{-1}\,(1 + r/f_2)^{-1}. \qquad (D.10)$$

From this one recovers (D.2) for $f_0 = f_1 = f_2 = \infty$ and (D.7) for $f_0 = f_1 = f_2 = s$.

SELECTED REFERENCES

Some Basic Texts about Electromagnetics and Asymptotics
1. D. S. Jones, The Theory of Electromagnetism. New York: MacMillan, 1964.
2. L. B. Felsen and N. Marcuvitz, Radiation and Scattering of Waves. Englewood Cliffs, New Jersey: Prentice-Hall, 1973.
3. N. Bleistein and R. A. Handelsman, Asymptotic Expansions of Integrals. New York: Holt, Rinehart, and Winston, 1975.

Geometric Theory of Diffraction (GTD)
4. J. B. Keller, R. M. Lewis, and B. D. Seckler, "Asymptotic Solution of Some Diffraction Problem," Comm. Pure Appl. Math., Vol. 9, pp. 207-265, 1956.
5. J. B. Keller, "Diffraction by an aperture," J. Appl. Phys., Vol. 28, pp. 426-444, 1957.
6. J. B. Keller, R. M. Lewis, and B. D. Seckler, "Diffraction by an Aperture, II," J. Appl. Phys., Vol. 28, pp. 570-579, 1957.
7. J. B. Keller, "Geometrical Theory of Diffraction," J. Opt. Soc. Amer., Vol. 52, pp. 116-130, 1962.
8. R. M. Lewis and J. B. Keller, "Asymptotic Methods for Partial Differential Equations: The Reduced Wave Equation and Maxwell's Equations," Res. Rep. EM-194, Courant Institute of Mathematical Sciences, New York University, New York, 1964.

Ray Techniques
9. C. Schensted, "Electromagnetic and Acoustic Scattering by a Semi-Infinite Body of Revolution," J. Appl. Phys., Vol. 26, pp. 306-308.
10. G. A. Deschamps, "Ray Techniques in Electromagnetics," Proc. IEEE, Vol. 60, pp. 1022-1035, 1972.
11. S. W. Lee, "Electromagnetic Reflection from a Conducting Surface: Geometrical Optics Solution," IEEE Trans. Antennas Propagat., Vol. AP-23, pp. 184-191, 1975.

Uniform Asymptotic Theory (UAT) (ALB)
12. J. Boersma and P. H. M. Kersten, "Uniform Asymptotic Theory of Electromagnetic Diffraction by a Plane Screen," (in Dutch), Tech. Report, Department of Math., Tech. University of Eindhoven, Eindhoven, 1967.

13. D. S. Ahluwalia, R. M. Lewis and J. Boersma, "Uniform Asymptotic Theory of Diffraction by a Plane Screen," SIAM J. Appl. Math., Vol. 16, pp. 783-807, 1968.
14. D. S. Ahluwalia and R. M. Lewis, "Diffraction of Progressing Waves by a Screen," J. Inst. Math. Its Appl., Vol. 5, pp. 113-139, 1969.
15. R. M. Lewis and J. Boersma, "Uniform Asymptotic Theory of Edge Diffraction," J. Math. Phys., Vol. 10, pp. 2291-2305, 1969.
16. D. S. Ahluwalia, "Uniform Asymptotic Theory of Diffraction by the Edge of a Three-Dimensional Body," SIAM J. Appl. Math., Vol. 18, pp. 287-301, 1970.
17. S. W. Lee and G. A. Deschamps, "A Uniform Asymptotic Theory of Electromagnetic Diffraction by a Curved Wedge," IEEE Trans. Antennas Propagat., Vol. AP-24, pp. 25-34, 1976.

Uniform Theory of Diffraction (UTD) (KP)
18. R. G. Kouyoumjian and P. H. Pathak, "A Uniform Geometrical Theory of Diffraction for an Edge in a Perfectly Conducting Surface," Proc. IEEE, Vol. 62, pp. 1448-1461, 1974.
19. C. A. Mentzer, L. Peters, Jr., and R. C. Rudduck, "Slope Diffraction and Its Application to Horns," IEEE Trans. Antennas Propagat., Vol. AP-23, pp. 153-159, 1975.
20. C. E. Ryan, Jr., and L. Peters, Jr., "Evaluation of Edge-Diffracted Fields Including Equivalent Currents for the Caustic Regions," IEEE Trans. Antennas Propagat., Vol. AP-17, pp. 292-299, 1969. (See also correction: Vol. AP-18, p. 275, 1970).
21. R. G. Kouyoumjian, "The Geometrical Theory of Diffraction and its Application," Numerical and Asymptotic Techniques in Electromagnetics, R. Mittra, ed., New York: Springer-Verlag, pp. 165-215, 1975.
22. Cashman, J. D. R. G. Kouyoumjian, and P. H. Pathak, "Comments on a Uniform Theory of Diffraction for an Edge in a Perfectly Conducting Surface," IEEE Trans. Antennas Propagat., AP-25, pp. 447-451, 1977.

Fringe Current Method (Braunbek, Ufimtsev) or Physicol Theory (PTD) of Diffraction
23. E. F. Knott and T. B. A. Senior, "Comparison of Three High-Frequency Diffraction Techniques," Proc. IEEE, Vol. 62, pp. 1468-1474, 1974.
24. S. W. Lee, "Comparison of Uniform Asymptotic Theory of Ufimtsev's Theory in Electromagnetic Edge Diffraction," IEEE Trans. Antennas Propagat., Vol. AP-25, March 1977.
25. P. Ya, Ufimtsev, "Method of Edge Waves in the Physical Theory of Diffraction," Air Force System Command, Foreign Tech. Div. Document ID No. FTD-HC-23-259-71, 1971 (translation from the Fussian version published by Soviet Radio Publication House, Moscow, in 1962).

26. I. Kay and J. B. Keller, "Asymptotic Evaluation of the Field at a Caustic," J. Appl. Phys., Vol. 25, pp. 876-883, 1954.

High Order Wedges
27. L. Kaminetzky and J. B. Keller, "Diffraction Coefficients for Higher Order Edges and Vertices," SIAM J. Appl. Math., Vol. 22, p. 109, 1972.
28. T. B. A. Senior, "The Diffraction Matrix for a Discontinuity in Curvature," IEEE Trans. Antennas Propagation, Vol. AP-20, p. 326, 1972.
29. T. P. Fontana, G. A. Deschamps, "Uniform Solution for the Field Diffracted by Curvature Discontinuity in a Cylinder," EM Tech. Report, No. 77-14, June 1977, University of Illinois.

Other References
30. Abramowitz and Stegun, Handbook of Mathematical Functions, p. 302, Dover 1965.
31. G. A. Deschamps and S. W. Lee, Half-plane diffraction at oblique incidence of an electromagnetic plane wave of arbitrary polarization, 1979, IEEE/AP-S Symposium.

A UNIFORM GTD FOR THE DIFFRACTION BY EDGES, VERTICES AND CONVEX
SURFACES

R.G. Kouyoumjian, P.H. Pathak, and W.D. Burnside

ElectroScience Laboratory, Department of Electrical
Engineering, The Ohio State University, Columbus,
Ohio 43212, U.S.A.

I. INTRODUCTION
When a radiating object is large in terms of wavelength, the
scattering and diffraction are found to be essentially a local
phenomenon associated with specific parts of the object, e.g.,
specular reflection, shadow boundaries, and edges. The
high-frequency approach to be described in these notes employs
rays in a systematic way to describe this phenomenon. It was
originally developed by Keller and his associates at the Courant
Institute of Mathematical Sciences [1,2,3] and is referred to as
the geometrical theory of diffraction (GTD). The GTD is an
extension of geometrical optics in which rays diffracted from
edges, vertices and convex surfaces are introduced through a
generalization of Fermat's Principle. In its original form, the
GTD fails in the transition regions adjacent to shadow and
reflection boundaries. About fifteen years ago work began at
The Ohio State University [4] to overcome this and other limitations in the GTD. The result today is the new uniform GTD (UTD)
which provides expressions for the diffracted field so that the
total high-frequency field is continuous at shadow and reflection
boundaries. Uniform expression have been found for electromagnetic fields diffracted from edges and vertices in
perfectly-conducting surfaces and for the diffracted fields due
to sources either on or off perfectly-conducting convex surfaces.
These solutions have greatly enhanced the accuracy and utility
of the method, which can be readily demonstrated in the treatment
of a number of simple shapes. It is found that although the
method is approximate in nature it yields the leading terms in
the asymptotic high frequency solution. Moreover, in many cases
it works surprisingly well on radiating objects as small as a
wavelength or so in extent. Thus, if a solution is desired over

a wide spectral range, this high-frequency method nicely complements low-frequency methods. Finally, it is sufficiently flexible so that it can be combined with the moment method or the T-Matrix method, thereby extending the class of solutions for either method used separately.

The importance of the UTD, however, lies in its application to the complex shapes involved in problems encountered in practice, such as
1. Antennas
   a) antennas on aircraft, spacecraft, satellites, ships and in the presence of ground structures
   b) reflector antennas
   c) horn and waveguide antennas
2. Radar cross-section calculations
   a) aircraft
   b) ships
   c) ground targets
3. EMP (early-time response)
4. Underwater acoustics
   a) conformal arrays
   b) target strength calculations

II. FORMULATION OF THE THEORY

The treatment of high-frequency radiation to follow is for the most part restricted to perfectly-conducting objects located in isotropic, homogeneous media. The methods which will be presented, however, can be extended to objects with impedance surfaces or penetrable surfaces in inhomogeneous and anisotropic media.

A. Geometrical Optics

The geometrical optics field is the leading term in the asymptotic high-frequency solution of Maxwell's equations, and so it not unexpectedly provides the leading term in the GTD (and UTD) high-frequency approximations. Let us begin by examining this field and how it is used in these theories. The geometrical optics field is the sum of the leading terms in the incident and reflected fields. These fields can be expanded in Luneberg-Kline series for large ω of the form

$$\overline{E} \sim e^{-jk\psi} \sum_{m=0}^{\infty} \frac{\overline{E}_m}{(j\omega)^m} \qquad (2.1)$$

where an $\exp(j\omega t)$ time dependence is assumed and k is the wavenumber of the medium. Substituting the preceding expansion into the vector wave equation for the electric field and integrating the resulting transport equation for m=0 [5], [6], the leading term in (2.1) is found to be

$$\overline{E}(s) \sim e^{-jk\psi(s)}\overline{E}_0(s) = \overline{E}_0(0)e^{-jk\psi(0)}\sqrt{\frac{\rho_1\rho_2}{(\rho_1+s)(\rho_2+s)}}\, e^{-jks} \qquad (2.2)$$

in which s=0 is taken as a reference point on the ray path, and ρ_1, ρ_2 are the principal radii of curvature of the wavefront at s=0. In Fig. 1, ρ_1 and ρ_2 are shown in relationship to the rays and wavefronts.

Equation (2.2) is commonly referred to as the geometrical-optics field, because it could have been determined in part from classical geometrical optics. Specifically, the quantity under the square root, the divergence factor, follows from conservation of power in a tube of rays; in addition, we note that the eikonal equation could have been deduced from Fermat's principle, a fundamental postulate of classical geometrical optics. As is well known, classical geometrical optics ignores the polarization and wave nature of the electromagnetic field; however, the leading term in the Luneberg-Kline asymptotic expansion is seen to contain this missing information.

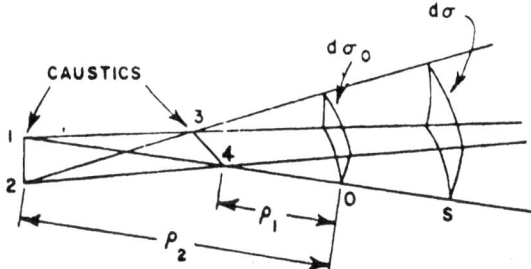

Fig. 1. Astigmatic tube of rays.

It is apparent that when $s=-\rho_1$ or $-\rho_2$ (2.2) becomes infinite so that it is no longer a valid approximation. The intersection of the rays at the lines 1-2 and 3-4 of the astigmatic tube of rays is called a caustic. As we pass through a caustic in the direction of propagation, the sign of $\rho+s$ changes and the correct phase shift of $+\pi/2$ is introduced naturally. Equation (2.2) is a valid high-frequency approximation on either side of the caustic; the field at a caustic must be found from separate considerations [7], [8], or by the use of equivalent edge currents, which will be described later.

From $\nabla \cdot \overline{E}=0$, one obtains

$$\hat{s} \cdot \overline{E}_0 = 0, \qquad (2.3)$$

so that the electric vector of the geometrical optics field is perpendicular to the ray path. Next, employing the Maxwell curl equation $\nabla \times \bar{E} = -j\omega\mu\bar{H}$, it follows from (2.1) that the leading term in the asymptotic approximation for the magnetic field is

$$\bar{H} \sim Y_c \hat{s} \times \bar{E} \qquad (2.4)$$

where $Y_c = \sqrt{\varepsilon/\mu}$ is the characteristic admittance of the medium, $\hat{s}$ is a unit vector in the direction of the ray path, and $\bar{E}$ is given by (2.2).

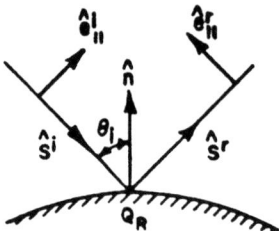

Fig. 2. Reflection at a curved surface.

Let a high-frequency electromagnetic wave be incident on a smooth curved perfectly-conducting surface S as depicted in Fig. 2. The geometrical-optics electric field reflected at Q_R on S (see Fig. 2) has the form given by (2.2). Choosing Q_R to be the reference point, it follows from the boundary condition for the total electric field on S that

$$\bar{E}_0^r(0) e^{-jk\psi^r(0)} = \bar{E}^i(Q_R) \cdot \bar{\bar{R}} = \bar{E}^i(Q_R) \cdot [\hat{e}_\parallel^i \hat{e}_\parallel^r - \hat{e}_\perp \hat{e}_\perp] \qquad (2.5)$$

in which $\bar{E}^i(Q_R)$ is the electric field incident at Q_R and $\bar{\bar{R}}$ is the dyadic reflection coefficient with $\hat{e}_\perp$ the unit vector perpendicular to the plane of incidence and $\hat{e}_\parallel^i$, $\hat{e}_\parallel^r$ the unit vectors parallel to the plane of incidence as shown in Fig. 2. In matrix notation

$$R = \begin{bmatrix} 1 & 0 \\ 0 & -1 \end{bmatrix} \qquad (2.6)$$

From (2.2) and (2.5)

$$\bar{E}^r(s) = \bar{E}^i(Q_R) \cdot \bar{\bar{R}} \sqrt{\frac{\rho_1^r \rho_2^r}{(\rho_1^r + s)(\rho_2^r + s)}} e^{-jks} \qquad (2.7)$$

in which ρ_1^r, ρ_2^r are the principal radii of curvature of the reflected wavefront at Q_R. It can be shown that

$$\frac{1}{\rho_1^r} = \frac{1}{2}\left(\frac{1}{\rho_1^i} + \frac{1}{\rho_2^i}\right) + \frac{1}{f_1} \qquad (2.8)$$

$$\frac{1}{\rho_2^r} = \frac{1}{2}\left(\frac{1}{\rho_1^i} + \frac{1}{\rho_2^i}\right) + \frac{1}{f_2}. \qquad (2.9)$$

The above equations are reminiscent of the simple lens and mirror formulas of elementary physics; this is particulary true of an incident spherical wave, where $\rho_1^i = \rho_2^i = s'$. Expressions for f_1 and f_2 are given in [9]. For an incident spherical wave,

$$\frac{1}{f_{\frac{1}{2}}} = \frac{1}{\cos\theta_i}\left(\frac{\sin^2\theta_2}{R_1} + \frac{\sin^2\theta_1}{R_2}\right) \pm \sqrt{\frac{1}{\cos^2\theta_i}\left(\frac{\sin^2\theta_2}{R_1} + \frac{\sin^2\theta_1}{R_2}\right)^2 - \frac{4}{R_1 R_2}} \qquad (2.10)$$

in which θ_1 and θ_2 are the angles between $\hat{s}^i$ and the principal directions associated with the principal radii of curvature of the surface R_1 and R_2, respectively. In the case of plane wave illumination, it follows from (2.8 to 2.10),

$$\sqrt{\rho_1^r \rho_2^r} = \sqrt{R_1 R_2}/2 \qquad (2.11)$$

which is useful in calculating the far-zone reflected field. If R_1 or R_2 become infinite, as in the case of a flat plate, cylinder or cone, it is evident that geometrical optics fails in the far zone for plane wave illumination.

Geometrical optics approximates the scattered field only in the direction of specular reflection, as determined by the law of reflection. In principle, the geometrical-optics approximation can be improved by finding the higher order terms $\bar{E}_1^r(R)$, $\bar{E}_2^r(R),\ldots$, in the reflected field, but in general it is not easy to obtain these from the higher order transport equations. Furthermore, these terms do not correct the following two serious defects of geometrical optics:
    a) the zero field in the shadow region,
    b) the discontinuity in the field at the shadow and reflection boundaries.
At the present time additional postulates are required to remove these limitations. They are given in the next subsection.

B. *Postulates*.
To overcome limitation (a) of the geometrical-optics field pointed out at the end of the last subsection, it is necessary to

introduce an additional field, the diffracted field. Keller [1,2,3] has shown how the diffracted field may be included in the high-frequency solution as an extension of geometrical optics. The postulates of Keller's theory, commonly referred to as the geometrical theory of diffraction (GTD), are summarized as follows.

(1) The diffracted field propagates along rays which are determined by a <u>generalization of Fermat's principle</u> to include points on the boundary surface in the ray trajectory.

(2) Diffraction like reflection and transmission is a <u>local phenomenon</u> at high frequencies, i.e., it depends only on the nature of the boundary surface and the incident field in the immediate neighborhood of the point of diffraction.

(3) The diffracted wave propagates along its ray so that
  (a) power is conserved in a tube (or strip of rays),
  (b) the phase delay along the ray path equals the product of the wave number of the medium and the distance.

The rays diffracted from an opaque object such as that depicted in Fig. 3 and which pass through the field point P or P' are found from the generalized Fermat's principle. The notion that points on the boundary surface may be included in the ray trajectory is not new. Imposing the condition that a point Q_R on a smooth curved surface be included in the ray path between the source and observation point is a time-honored method for deducing the reflected ray and the law of reflection. It seems reasonable to extend the class of such points as Keller has done. Diffracted rays are initiated at points on the boundary surface where the incident geometrical-optics field is discontinuous,[1] i.e., at points on the surface where there is a shadow or reflection boundary of the incident field. Examples of such points are edges Q_E, vertices and points Q_1 at which the ray incident from O is tangent to the curved surface. Note that a surface ray also may be excited at the edge Q_E. These diffracted rays like the geometrical-optics rays follow paths which make the optical distance between the source point and the field point an extremum, usually a minimum. Thus the portion of the ray path which traverses a homogeneous medium is a straight line, and if a segment of the ray path lies on a smooth surface, it is a surface extremum or geodesic.

---

[1] The incident field may be a diffracted field; a discontinuity of the diffracted field on the boundary surface initiates a higher-order diffracted ray.

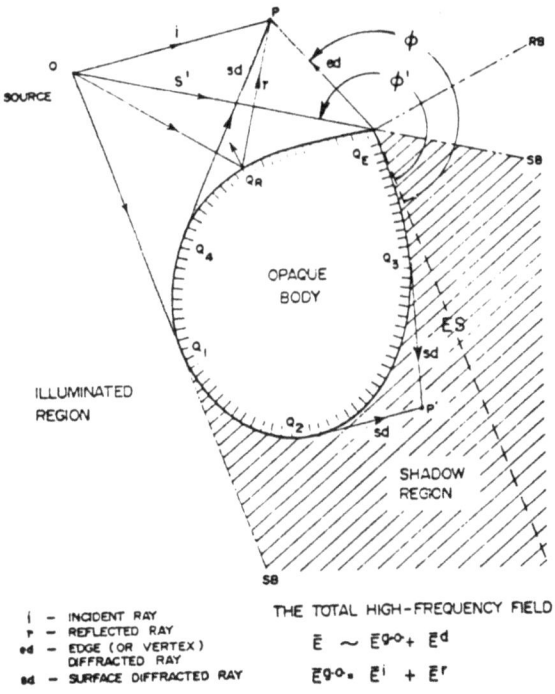

Fig. 3. Rays reflected and diffracted from an opaque object.

It is apparent that the rays provide a natural coordinate system for calculating the field; furthermore, the total high-frequency field at an observation point is synthesized from the fields of all the rays passing through that point.

Postulate 2 is essential for the method to be valid, i.e., the frequency must be sufficiently high and the illumination such that the complex radiation problem breaks down into a number of simple problems associated with specific parts of the object.

The uniform GTD (UTD) unlike the GTD requires that the diffracted field compensate the discontinuity in the geometrical optics field at a shadow or reflection boundary so that the total high-frequency field is everywhere continuous away from the radiating body. Thus the UTD diffracted fields assume their largest values at and near these boundaries, where their strength is comparable with the geometrical optics field.

As will be pointed out in the beginning of the next subsection, postulate 3 of Keller's theory actually follows from his first two postulates.

C. <u>Diffracted Fields</u>

Consider a normal congruence of rays emanating from a point of diffraction, such as Q_E, Q_2 or Q_4 in Fig. 3. The high-frequency diffracted field may be found by asymptotically approximating its integral representation, where the integral is taken over a wavefront associated with the diffracted ray [4]. Using the stationary phase method to approximate the integral one obtains

$$\vec{E}^d(s) \sim \vec{E}^d(0)\sqrt{\frac{\rho_1\rho_2}{(\rho_1+s)(\rho_2+s)}}\, e^{-jks} \qquad (2.12)$$

which has the same form as (2.2). This is not an unexpected result, because the procedure also can be applied to the incident and reflected fields of geometrical optics; however, it does not yield higher-order terms as does the Luneberg-Kline expansion.

In calculating the diffracted field it is convenient to choose the point of diffraction on the boundary surface as the reference point 0. However, this point of diffraction is also a caustic of the diffracted ray. First consider the case where the caustic is at an edge, forms a line on a smooth convex surface from which rays shed tangentially, or is a line of discontinuity in the electrical properties of the surface. Either ρ_1 or ρ_2 denoted by ρ' vanishes; however, $\vec{E}^d(s)$ must be independent of the location of the reference point; hence it follows from (2.12) that

$$\lim_{\rho' \to 0} \vec{E}^d(0)\sqrt{\rho'} = \vec{A} \qquad (2.13)$$

exists, so that

$$\vec{E}^d(s) = \vec{A}\sqrt{\frac{\rho}{s(\rho+s)}}\, e^{-jks} \qquad (2.14)$$

in which ρ is the distance between the caustic on the boundary surface (the point of diffraction) and the second caustic of the diffracted ray which is away from this surface. Thus the diffracted rays, like the geometrical-optics rays form an astigmatic tube, as shown in Fig. 1 with either the caustic 1-2 or 3-4 at the point of diffraction on the boundary surface. The caustic distance ρ may be determined by differential geometry; an expression for ρ will be given later.

For a two-dimensional problem we note that $\rho=\infty$, so (2.14) reduces to

$$\bar{E}^d(s) = \bar{A} \frac{e^{-jks}}{\sqrt{s}} \qquad . \qquad (2.15)$$

Next, let us consider the diffraction from a vertex or corner, where the diffracted rays emanate from a point caustic at the tip. In this case $\rho_1 = \rho_2 = \rho'$, and again since $\bar{E}^d(s)$ must be independent of the reference point s=0, it follows from (2.12) that

$$\lim_{\rho' \to 0} \bar{E}^d(0)\rho' = \bar{B} \qquad (2.16)$$

exists, and so for vertex or corner diffraction

$$\bar{E}^d(s) = \bar{B} \frac{e^{-jks}}{s} \qquad . \qquad (2.17)$$

Diffraction is a local effect according to Postulate 2, and since we are dealing with a linear phenomenon, $\bar{A}$ and $\bar{B}$ must be proportional to the incident field at the point where diffraction is initiated, if the incident field is not rapidly varying there.[2] Thus in the case of an edge Q_E or a more general line of diffraction or scattering Q_0, as shown in Fig. 4,

$$\bar{A} = \bar{E}^i(Q_E) \cdot \bar{\bar{D}} \qquad (2.18)$$

The constant of proportionality $\bar{\bar{D}}$ is referred to as a dyadic diffraction coefficient.

---

[2] If the incident field is rapidly varying at the point of diffraction, it may be possible to separate it into slowly-varying components for the purpose of calculating the diffracted field. Alternatively it may be desirable to introduce terms which are proportional to the spatial derivatives of the incident field; this will be considered later.

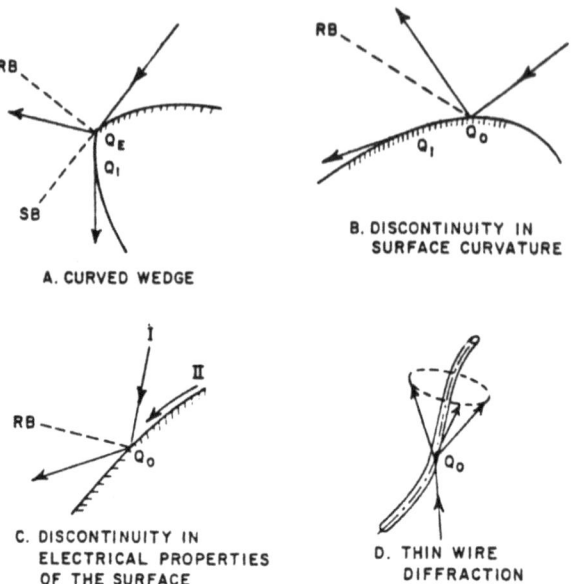

Fig. 4. Diffraction from lines of discontinuity.

In the case of diffraction at a convex surface as depicted in Fig. 3

$$\bar{A} = \bar{E}^i(Q_E) \cdot \bar{\bar{D}}(Q_1,\ldots,Q_2) \qquad (2.19)$$

in which the generalized dyadic diffraction coefficient $\bar{\bar{D}}$ is a function of the surface geometry at all points along the surface ray path between Q_1, where the surface ray is initiated, and the point of diffraction Q_2, where the diffracted ray is shed toward P'.

A second case of interest occurs if the source of the surface ray is on the surface S at Q_1. This configuration is relevant to the radiation from an aperture in S where the equivalent source is an infinitesimal magnetic current moment (magnetic dipole)

$$d\bar{p}_m(Q_1) = \bar{E}(Q_1) \times \hat{n}_1 da' \qquad (2.20a)$$

in which $\bar{E}$ is the aperture electric field and da' is an area element of the aperture. Another type of source which may be positioned at Q_1 is the infinitesimal normally-directed electric current moment

$$d\bar{p}_e(\ell') = \hat{n}_1 I(\ell') d\ell' \qquad (2.20b)$$

in which $\hat{n}_1$ is a unit normal vector to the surface at Q_1. This source represents a thin monopole of length $d\ell'$ and electric current $I(\ell')$. For these sources

$$\overline{A} = d\overline{p}_{m_e}(Q_1) \cdot \overline{\overline{L}}(Q_1,\ldots,Q_2) \tag{2.21}$$

in which the generalized dyadic launching coefficient is again a function of the surface curvature along the surface ray path between Q_1 and Q_2.

As will be seen in Section IV, $\overline{\overline{D}}$ and $\overline{\overline{L}}$ are expressed uniformly in terms of the scalar soft and hard Fock functions. These expressions as well as the scalar diffraction coefficients for the edge geometries are found from the asymptotic solution of the simplest boundary value problem (or problems) having the same local geometry as that near the point (or points) of diffraction. This procedure is justified by postulate 2 of the theory.

III. EDGE DIFFRACTION
A. Curved Wedge

Let us consider the field radiated from a point source at O and observed at P in the presence of a perfectly-conducting curved wedge, as shown in Figs. 3 and 5. Applying the generalized Fermat's principle, the distance along the ray path OQ_EP between O and P, which includes the edge point Q_E on its path, is a minimum and the law of edge diffraction

$$\hat{s}' \cdot \hat{e} = \hat{s} \cdot \hat{e} \tag{3.1}$$

results. Here $\hat{e}$ is a unit vector directed along the edge, and $\hat{s}'$ and $\hat{s}$ are unit vectors in the directions of incidence and diffraction, respectively. The above equation also follows from the requirement that the incident and diffracted fields be phase matched along the edge. If the incident ray strikes the edge obliquely, making an angle β_o with the edge, as shown in Fig. 5a the diffracted rays lie on the surface of a cone whose half angle is equal to β_o. The position of the diffracted ray on this conical surface is given by the angle ϕ, and the direction of the ray incident on the edge, by the angles ϕ' and β_o'; these angles are defined in Fig. 5a and b. Equation (3.1) may be used to develop a computer search program to locate the points of edge diffraction.

From (2.14, 2.18) the expression for the electric field of the edge-diffracted ray is

$$\overline{E}^d(s) = \overline{E}^i(Q_E) \cdot \overline{\overline{D}}(\phi,\phi';\beta_o') \sqrt{\frac{\rho}{s(\rho+s)}} \ e^{-jks} \tag{3.2}$$

Since the pertinent dimension in wedge diffraction is wavelength, it follows from dimensional considerations that the diffraction coefficient must vary as $k^{-1/2}$. The dyadic diffraction coefficient for a perfectly-conducting wedge has been obtained by Kouyoumjian and Pathak; their work is described in [9,10] and will only be summarized here. As noted before, the dyadic diffraction coefficient can be found from the asymptotic solution of canonical problems, which in this case involve the illumination of the wedge by plane, cylindrical, conical and spherical waves. The solution of these canonical problems serves as a basis for deducing the dyadic diffraction coefficient for arbitrary wavefront illumination and for the more general case where there are curved edges and curved surfaces.

The calculation of the caustic distance ρ in (3.2) is not a trivial matter for curved edge diffraction. Employing differential geometry, it is shown in [9,10] that

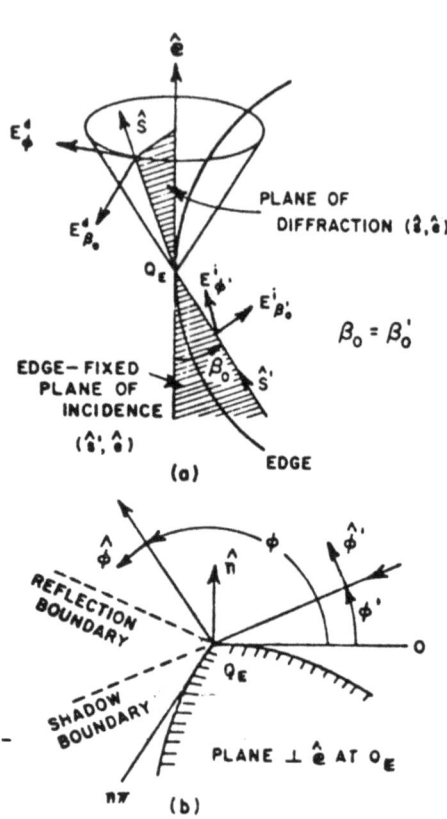

Fig. 5. Diffraction at a curved edge.

$$\frac{1}{\rho} = \frac{1}{\rho_e^i} - \frac{\hat{n}_e \cdot (\hat{s}' - \hat{s})}{a \sin^2 \beta_0'} \quad (3.3)$$

in which ρ_e^i is the radius of curvature of the incident wavefront in the plane which contains $\hat{s}'$ and $\hat{e}$ the unit vector tangent to the edge at Q_e; $\hat{n}_e$ is the unit vector normal to the edge at Q_E and directed away from the center of curvature; a is the radius of curvature of the edge at Q_e, a>0.

Let us introduce an edge-fixed plane of incidence containing the incident ray and $\hat{e}$ at Q_e and a plane of diffraction containing the diffracted ray and $\hat{e}$ at Q_e. The unit vectors $\hat{\phi}'$ and $\hat{\phi}$ are perpendicular to the edge-fixed plane of incidence and the plane of diffraction, respectively. The unit vectors $\hat{\beta}_0'$ and $\hat{\beta}_0$ are parallel to the edge-fixed plane of incidence and the plane of diffraction, respectively, and

$$\hat{\beta}'_0 = \hat{s}' \times \hat{\phi}', \quad \hat{\beta}_0 = \hat{s} \times \hat{\phi}.$$

Thus the coordinates of the diffracted ray (s, β_0, ϕ) are spherical coordinates and so are the coordinates of the incident ray (s', β'_0, ϕ'), except that the incident (radial) unit vector points toward the origin Q_F. These ray-fixed coordinates and their unit vectors are shown in Fig. 5.

For each type of edge illumination mentioned previously, it is shown in [10] that the dyadic diffraction coefficient can be represented simply as the sum of two dyads, if the ray-fixed coordinates mentioned in the preceding paragraph are used.

$$\bar{\bar{D}}(\phi,\phi';\beta'_0) = \hat{\beta}'_0\hat{\beta}_0 D_s(\phi,\phi';\beta'_0) - \hat{\phi}'\hat{\phi} D_h(\phi,\phi';\beta'_0) \quad , \quad (3.4)$$

where D_s is the scalar diffraction coefficient for the acoustically soft (Dirichlet) boundary condition at the surface of the wedge, and D_h is the scalar diffraction coefficient for the acoustically hard (Neumann) boundary condition. This result shows the close connection between electromagnetics and acoustics at high frequencies. If the dyadic diffraction coefficient is expressed in the edge fixed coordinate system, it is found to be the sum of seven dyads instead of the two dyads in (3.4). In this sense, the ray-fixed coordinate system is the natural coordinate system of the problem.

If the field point is not close to a shadow or reflection boundary and $\phi' \neq 0$ or $n\pi$, the scalar diffraction coefficients

$$D_{s \atop h}(\phi,\phi';\beta'_0) = \frac{e^{-j\frac{\pi}{4}} \sin\frac{\pi}{n}}{n\sqrt{2\pi k}\sin\beta'_0}$$

$$\left[\frac{1}{\cos\frac{\pi}{n} - \cos\left(\frac{\phi-\phi'}{n}\right)} \mp \frac{1}{\cos\frac{\pi}{n} - \cos\left(\frac{\phi+\phi'}{n}\right)} \right] \quad (3.5)$$

for all four types of illumination, which is important because the diffraction coefficient should be independent of the edge illumination away from shadow and reflection boundaries. The wedge angle is $(2-n)\pi$, where the plane surfaces forming the wedge are $\phi=0$ and $\phi=n\pi$, see Fig. 5b. This expression becomes singular as a shadow boundary (SB) or a reflection boundary (RB) is approached, which further aggravates the difficulties at these boundaries resulting from the discontinuities in the incident or reflected fields. The above scalar diffraction coefficients also have been given by Keller [3]. The case of grazing incidence $\phi'=0$ or $n\pi$ will be considered later.

Expressions for the scalar diffraction coefficients which are valid at all points away from the edge (again excluding $\phi'=0$ or $n\pi$) are

$$D_s(\phi, \phi'; \beta_0) = \frac{-e^{-j\frac{\pi}{4}}}{2n\sqrt{2\pi k}\sin\beta_0'}$$

$$\times \left[\cot\left(\frac{\pi + (\phi - \phi')}{2n}\right) F[k L^i a^+(\phi - \phi')]\right.$$

$$+ \cot\left(\frac{\pi - (\phi - \phi')}{2n}\right) F[k L^i a^-(\phi - \phi')]$$

$$\mp \left\{\cot\left(\frac{\pi + (\phi + \phi')}{2n}\right) F[k L^{ro} a^+(\phi + \phi')]\right.$$

$$\left.\left. + \cot\left(\frac{\pi - (\phi + \phi')}{2n}\right) F[k L^{rn} a^-(\phi + \phi')]\right\}\right], \tag{3.6}$$

where

$$F(X) = 2j\sqrt{X}\, e^{jX} \int_{\sqrt{X}}^{\infty} e^{-j\tau^2}\, d\tau. \tag{3.7}$$

When X is small

$$F(X) \simeq \left[\sqrt{\pi X} - 2X\, e^{j\frac{\pi}{4}} - \frac{2}{3}X^2\, e^{-j\frac{\pi}{4}}\right] e^{j\left(\frac{\pi}{4} + X\right)}, \tag{3.8}$$

and when X is large

$$F(X) \sim \left(1 + j\frac{1}{2X} - \frac{3}{4}\frac{1}{X^2} - j\frac{15}{8}\frac{1}{X^3} + \frac{75}{16}\frac{1}{X^4}\right). \tag{3.9}$$

If the arguments of the four transition functions in (3.6) exceed 10, then transition functions are approximately equal to one, and (3.6) reduces to (3.5).

Let $\phi \pm \phi' = \beta$, then

$$a^{\pm}(\beta) = 2\cos^2\left(\frac{2n\pi N^{\pm} - \beta}{2}\right) \qquad (3.10)$$

in which $N^{\pm}$ are the integers which most nearly satisfy the equations

$$2\pi n N^+ - \beta = \pi \qquad (3.11)$$

$$2\pi n N^- - \beta = -\pi \qquad (3.12)$$

Note that N^+ and N^- each have two values. For exterior edge diffraction (1<n<2), the value of N^+ or N^- at each boundary is included in Table 1 for convenience; these values can be used throughout their respective transition regions.

Table I

	The cotangent is singular when	value of N at the boundary
$\cot\left(\dfrac{\pi + (\phi - \phi')}{2n}\right)$	$\phi = \phi' - \pi$, a SB surface $\phi = 0$ is shadowed	$N^+ = 0$
$\cot\left(\dfrac{\pi - (\phi - \phi')}{2n}\right)$	$\phi = \phi' + \pi$, a SB surface $\phi = n\pi$ is shadowed	$N^- = 0$
$\cot\left(\dfrac{\pi + (\phi + \phi')}{2n}\right)$	$\phi = (2n-1)\pi - \phi'$, a RB reflection from surface $\phi = n\pi$	$N^+ = 1$
$\cot\left(\dfrac{\pi - (\phi + \phi')}{2n}\right)$	$\phi = \pi - \phi'$, a RB reflection from surface $\phi = 0$	$N^- = 0$

At a shadow or reflection boundary one of the cotangent functions in the expression for D given by (3.6) becomes singular; the other three remain bounded. Even though the cotangent becomes singular, its product with the transition function can be shown to be bounded. The location of the boundary at which each cotangent becomes singular is given in Table 1. The distance parameters

$$L^i = \frac{s(\rho_e^i+s)\rho_2^i\rho_1^i \sin^2\beta_0'}{\rho_e^i(\rho_1^i+s)(\rho_2^i+s)} \quad , \tag{3.13}$$

$$L^r = \frac{s(\rho^r+s)\rho_2^r\rho_1^r \sin^2\beta_0'}{\rho^r(\rho_1^r+s)(\rho_2^r+s)} \quad , \tag{3.14}$$

where ρ_1^r and ρ_2^r are the principal radii of curvature of the reflected wavefront at Q_E, and ρ^r is the distance between the caustics of the diffracted ray in the direction of reflection. It may be found from (3.3) with $\hat{s}=\hat{s}' - 2(\hat{n}\cdot\hat{s}')\hat{n}$. The additional superscripts o and n on L in (3.6) denote that the radii of curvature (and caustic distance ρ) are calculated at the reflection boundaries $\pi-\phi'$ and $(2n-1)\pi-\phi'$, respectively. In the far-zone where s>>ρ and the principal radii of curvature ρ_1 and ρ_2 of the incident and reflected wavefronts at Q_E, (3.13) and (3.14) simplify to

$$L = \frac{\rho_1\rho_2 \sin^2\beta_0'}{\rho} \quad ; \tag{3.15}$$

the appropriate superscripts are omitted here for the sake of simplicity.

In the case of the ordinary wedge where the edge is formed by intersecting plane surfaces

$$L^{ro} = L^{rn} = L^i \tag{3.16}$$

with L^i given by (3.13).

Grazing incidence, where $\phi'=0$ or $n\pi$ must be treated separately. Only the ordinary wedge is considered here. In this case $D_s=0$, and the expression for D_h given by (3.5) or (3.6) is multiplied by a factor of 1/2. The need for the factor of 1/2 may be seen by considering grazing incidence to be the limit of oblique incidence. At grazing incidence the incident and reflected fields merge, so that one half the total field propagating along

the face of the wedge toward the edge is the incident field
and the other half is the reflected field. Nevertheless, in
this case, it is clearly more convenient to regard the total
field as the "incident" field.

The present treatment does not include the modification of
the edge diffracted field which occurs when either the incident
or diffracted ray grazes a curved surface which forms the edge.
The angle between these rays and the surface should exceed
$(ka_i)^{-1/3}$, where a_i is the radius of curvature of the surface
at Q_E in the direction of the incident (or diffracted) ray. Also
the present treatment does not include the effect of surface rays
excited at the edge; this is considered in Subsection IVD.

Usually no more than one of the four transition functions
is significantly different from one; furthermore for the curved
wedge the nature of the approximation is such that the first two
terms within the brackets of (3.6) can be combined to give

$$\frac{-2 \sin \frac{\pi}{n} F\left[2kL^i \cos^2\left(\frac{\phi-\phi'}{2}\right)\right]}{\cos \frac{\pi}{n} - \cos\left(\frac{\phi-\phi'}{n}\right)} ,$$

which considerably simplifies the calculation of the scalar
diffraction coefficients.

In summary to calculate edge diffraction the scalar diffraction coefficients from (3.6) are substituted into (3.4) and the
resulting dyadic diffraction coefficient is substituted into
(3.2). The caustic distance ρ is calculated from (3.3).

B. Higher Order Edges

The preceding discussion has been restricted to ordinary
edges where the unit normal vector to the surface is discontinuous. However, in the case of higher-order edges, where some
j-th derivative of the surface has a jump discontinuity (while
all lower derivatives are continuous), it has been shown [9]
that the dyadic diffraction coefficient has the same form as in
(3.4). Also the dyadic diffraction coefficient for the
scattering from thin, curved wires has this form too. Geometries
of this type are shown in Figs. 4b,d.

Kaminetzky and Keller [11] and Senior [12] have obtained
expressions for the scalar diffraction coefficients in the case
of diffraction by an edge formed by a discontinuity in surface
curvature and Senior has given the dyadic (or matrix) diffraction
coefficient in an edge-fixed coordinate system. When transformed
to the ray-fixed coordinate system, Senior's expression for
the diffracted field reduces to the form in (3.2) with $\bar{D}$ given by
(3.4). Keller and Kaminetzky [11] also have given expressions
for the scalar diffraction coefficients in the case of higher-
order edges.

C. Slope Diffraction

It is assumed in (3.2) that the incident field $\bar{E}^i(Q_E)$ has a slow spatial variation (except for the phase along the incident ray). If this is not the case, Kouyoumjian and Hwang [13,14] have shown that a higher-order term must be included in the UTD expression for the edge diffracted field. Employing matrix notation the expression for the edge diffracted field then becomes

$$\begin{bmatrix} E^d_{\beta_0} \\ E^d_{\phi} \end{bmatrix} = \left\{ \begin{bmatrix} -D_s & 0 \\ 0 & -D_h \end{bmatrix} \begin{bmatrix} E^i_{\beta_0'} \\ E^i_{\phi'} \end{bmatrix} + \begin{bmatrix} -d_s & 0 \\ 0 & -d_h \end{bmatrix} \begin{bmatrix} \frac{\partial}{\partial n'} E^i_{\beta_0'} \\ \frac{\partial}{\partial n'} E^i_{\phi'} \end{bmatrix} \right\} \sqrt{\frac{\rho}{s(\rho+s)}} e^{-jks} \quad (3.17)$$

where

$$d_{s,h} = \frac{1}{jk\sin\beta_0} \frac{\partial}{\partial \phi'} D_{s,h} \quad . \quad (3.18)$$

The partial derivative with respect to the distance n' is taken in the direction ϕ' normal to the edge fixed plane of incidence. The first term in (3.17) is just (3.4) combined with (3.2) in matrix form; it makes the total high-frequency field continuous at shadow and reflection boundaries. The second term, referred to as the slope diffraction term, makes the first derivatives of the total high-frequency field continous (or nearly so) at these boundaries. Thus if this term is omitted, a "kink" may appear in the calculated pattern at a shadow or reflection boundary. A discontinuity of this type is evidence that the slope diffraction term should be included in the UTD solution.

As an example of the application of slope diffraction, let us consider the case of grazing incidence on the plane surface of a perfectly-conducting wedge. $E^i_{\beta_0'}(Q_E)=0$, so from (3.17)

$$E^d_{\beta_0}(s) = -\frac{1}{2} \frac{\partial}{\partial n'} E^i_{\beta_0'}(Q_E) \frac{1}{jk\sin\beta_0} \frac{\partial}{\partial \phi'} D_s \bigg|_{\phi'=0} \sqrt{\frac{\rho}{s(\rho+s)}} e^{-jks} \quad (3.19)$$

The above formula has been used to calculate the diffracted field of an infinitesimal slot (magnetic dipole) in the plane surface of a perfectly-conducting wedge. The far-zone pattern of a slot directed perpendicular to the edge is shown in Fig. 6; in this case $\rho=s'=d$, the distance of the slot from the edge. It is seen that the pattern calculated from the UTD solution is in excellent agreement with that calculated from the eigenfunction solution, even though the slot is only $\lambda/8$ from the edge. For comparison, in Fig. 7 the far-zone pattern is shown for the case where the axis of the slot is parallel to the edge, and again the

UTD pattern is seen to be in excellent agreement with the eigenfunction pattern, even though d=λ/8. Using (3.17) the diffracted field

$$E_\phi^d(s) = -\frac{1}{2} E_{\phi'}^d(Q_E) D_h \sqrt{\frac{d}{s(d+s)}} e^{-jks} \qquad (3.20)$$

in this case.

Equation (3.17) is a valid high-frequency approximation, if the incident field has a ray-optical behavior, i.e., it varies with distance s along its path as in (2.2). If this is not the case, then it may be possible to decompose the incident field into ray-optical components. For example, the near field of an array can be represented in terms of the ray-optical fields of each of its elements. Alternatively, one may be able to decompose the incident field in the spectral domain. Mittra and Rahmat-Samii [15,16] have employed plane-wave spectra to solve scattering and diffraction problems where the incident field does not exhibit a ray-optical behavior. Tiberio and Kouyoumjian [17,18] have extended this approach to cylindrical and spherical wave representations of the incident field in cases where the transition region field of one edge illuminates another edge.

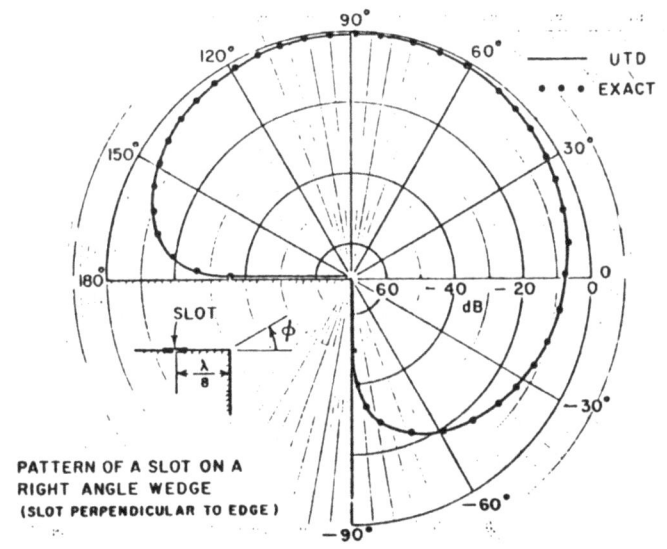

Fig. 6. Diffraction of the field of an infinitesimal slot by a right-angle wedge.

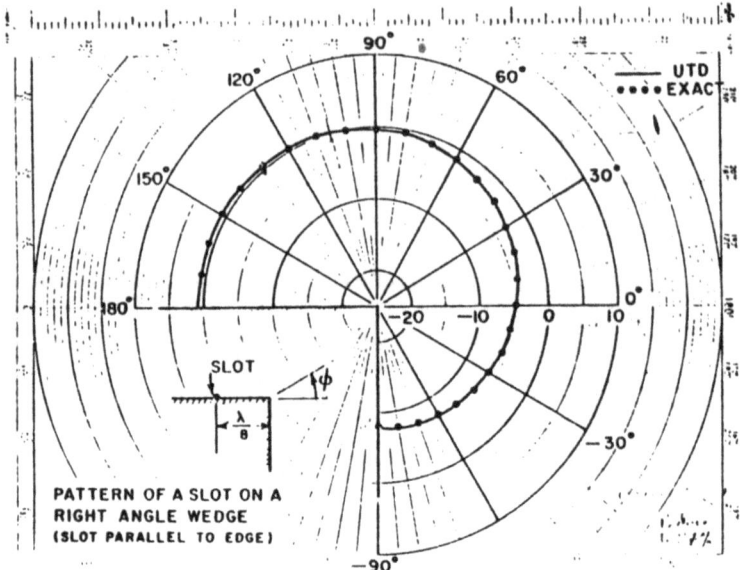

Fig. 7. Diffraction of the field of an infinitesimal slot slot by a right-angle wedge.

D. Equivalent Edge Currents

It was pointed out in Section II that ray-optical fields fail at caustics such as 1-2, 3-4 in Fig. 1. In the case of caustics resulting from edge diffraction (or diffraction at the lines of discontinuity shown in Fig. 4) equivalent electric and magnetic line currents may be used to make caustic corrections not close to a shadow or reflection boundary. These line currents are found from the edge diffracted field away from the caustic. The method is analogous to physical optics, which can be used to correct the caustics of geometrical optics [20]. In this case the surface currents obtained from the geometrical optics field are employed in a rigorous integral representation of the caustic field.

The equivalent line currents are an electric current

$$I(\ell) = \frac{-\hat{e} \cdot \bar{E}^i(Q_E)}{Z_c \sin\beta_o} D_s(\phi,\phi';\beta_o) \sqrt{\frac{8\pi}{k}} e^{-j\frac{\pi}{4}} \qquad (3.21)$$

and a magnetic current

$$M(\ell) = -\frac{(\hat{e} \times \hat{s}')}{\sin\beta_0} \cdot \bar{E}^i(Q_E) D_h(\phi,\phi';\beta_0) \sqrt{\frac{8\pi}{k}} \; e^{-j\frac{\pi}{4}} \quad (3.22)$$

flowing along the edge (see Fig. 5). Here ℓ is a measure of the distance along the edge; it is clear that $\hat{e}$, $\hat{s}'$ as well as the angles ϕ, ϕ' and β_0 are functions of ℓ. If these currents are substituted in the radiation integral and the integral is evaluated by the stationary phase method for a field point away from the caustic, one obtains the UTD expression for the edge-diffracted field. This makes it possible to blend the caustic field correction smoothly with the UTD field from (3.2). If the field incident at Q_E has a rapid spatial variation, a slope diffraction contribution obtained from (3.17) can be added to the equivalent edge currents.

Equations (3.21) (3.22) have been used to correct axial caustics arising from diffraction at a circular edge [4,19,20]; corrections for off-axial caustics are given in [21]. Equivalent edge currents also have been used to obtain the approximate vertex diffraction coefficient described in the next subsection and to derive the expressions for the edge-excited surface rays given in subsection IVD. In conclusion, it should be pointed out that the equivalent line current concept can be applied to correct caustics which may occur in the case of diffraction by convex surfaces.

E. Vertex Diffraction

The diffraction at a vertex formed by two intersecting edges is shown in Fig. 8; it will be referred to as the corner diffraction problem. It is included in the section on edge diffraction because the corner is treated as a terminated edge here (just as an edge can be regarded as a surface termination).

A diffraction coefficient is needed that is numerically efficient in order for the corner effect to be of any practical use in complicated modeling problems. A solution has been proposed which is based on the asymptotic evaluation of the radiation integral which employs the equivalent edge currents that would exist in the absence of the corners [22]. The corner diffraction term is then found by appropriately (but at present empirically) modifying the asymptotic result for the radiation integral which is characterized by a saddle point near an end point. This diffraction coefficient is still in the initial stages of its development. However it has been shown to very successfully predict the corner effect for numerous plate structures. For this reason, it is discussed here as a good engineering approximation to the problem.

The corner diffracted fields associated with one corner and one edge in the near field with spherical wave incidence are given by

$$\begin{Bmatrix} E^c_{\beta_o} \\ E^c_{\phi} \end{Bmatrix} \sim \begin{Bmatrix} IZ_o \\ MY_o \end{Bmatrix} \sqrt{\frac{\sin\beta_c \sin\beta_{oc}}{\cos\beta_{oc}-\cos\beta_c}} \; F[kL_c a(\pi+\beta_{oc}-\beta_c)] \; \frac{e^{-jks}}{4\pi s} \qquad (3.23)$$

where

$$\begin{Bmatrix} I \\ M \end{Bmatrix} = - \begin{Bmatrix} E^i_{\beta_o'}(Q_C) \\ E^i_{\phi'}(Q_C) \end{Bmatrix} \begin{Bmatrix} C_s(Q_E)Y_o \\ C_h(Q_E)Z_o \end{Bmatrix} \sqrt{\frac{8\pi}{k}} \; e^{-j\frac{\pi}{4}} \qquad (3.24)$$

and

$$C_{s,h}(Q_E) = \frac{-e^{-j\frac{\pi}{4}}}{2\sqrt{2\pi k}\sin\beta_o} \left\{ \frac{F[kLa(\beta^-)]}{\cos\frac{\beta^-}{2}} \left| F\left[\frac{La(\beta^-)/\lambda}{kL_c a(\pi+\beta_{oc}-\beta_c)}\right] \right| \right.$$
$$\left. \mp \frac{F[kLa(\beta^+)]}{\cos\frac{\beta^+}{2}} \left| F\left[\frac{La(\beta^+)/\lambda}{kL_c a(\pi+\beta_{oc}-\beta_c)}\right] \right| \right\} \qquad (3.25)$$

The function $F(x)$ was defined earlier, $a(\beta)=2\cos^2(\beta/2)$ where $\beta^{\pm}=\phi\mp\phi'$ and $L=s's''\sin^2\beta_o/(s'+s'')$ and $L_c=s_c s/(s_c+s)$ for spherical wave incidence. The function $C_{s,h}(Q_E)$ is a modified version of the diffraction coefficient for the half-plane case (n=2). The modification factor,

$$\left| F\left[\frac{La(\beta)/\lambda}{kL_c a(\pi+\beta_{oc}-\beta_c)}\right] \right| ,$$

is an empirically derived function that insures that the diffraction coefficient will not change sign abruptly when it passes through the shadow boundaries of the edge. There is also a corner diffraction term associated with the other edge forming the corner and is found in a similar manner.

If the vertex involves three intersecting edges, as at the corner of a cube, a third term associated with the third edge must be added. In addition the preceding expressions for the half-plane edge must be generalized to the wedge. The utility of the present result is evident in Figs. 9a,b,

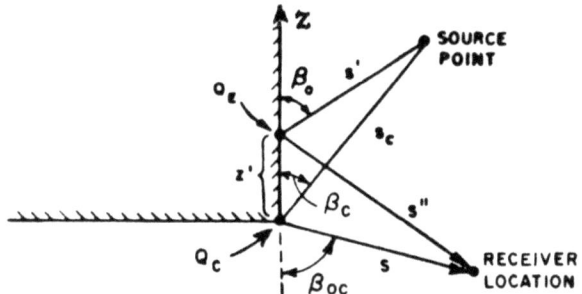

Figure 8. Geometry for corner diffraction problem.

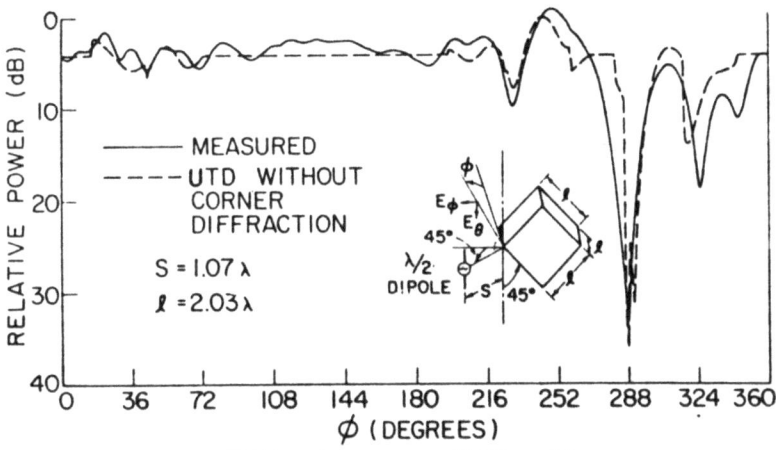

a. Without corner diffraction.

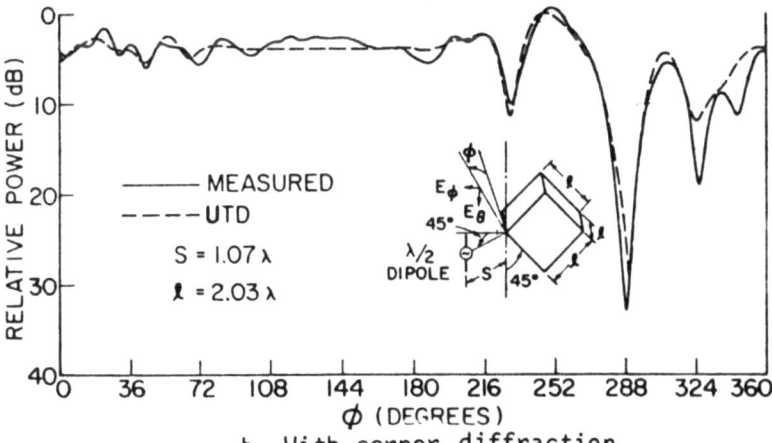

b. With corner diffraction.

Fig. 9. Comparison of measured and calculated E_θ radiation pattern for a dipole near a box in the indicated plane.

where the pattern of an electric dipole radiating in the presence of a cube is presented. In Fig. 9a, the vertex diffraction contribution is omitted, whereas in Fig. 9b it is included.

Very little work has been done on the high-frequency diffraction by vertices in smooth curved surfaces; it remains a complicated difficult subject. Some results for blunt vertices are given in [11], and the diffraction coefficient for the vertex of a cone in some special case may be obtained from Chapter 18 of [32] and in [23,24].

IV. SURFACE DIFFRACTION

When an incident ray system emanating from a source strikes a smooth, perfectly-conducting convex surface as shown in Fig. 10, it produces a system of rays reflected from that surface; at grazing, i.e., at Q_1, the incident ray merges with the reflected ray giving rise to a surface ray which propagates into the shadow region along a geodesic on the convex surface according to the generalized Fermat's principle described in Part II. This principle requires that the continuation of the incident ray path beyond the point of grazing on the convex surface to a field point be an extremum (generally a minimum); thus, the surface ray is a geodesic of the surface. The field associated with the surface ray exhibits an exponential decay in the deep shadow region due to the continual leakage of energy along the surface ray resulting from diffracted rays shed tangentially from the surface ray. One notes that surface rays can be excited by sources which are located either on or off a smooth convex surface; they can also be excited by the illumination of an edge, or other geometrical or electrical discontinuity in an otherwise smooth convex surface.

This section will focus on recently developed UTD solutions for calculating the electromagnetic fields excited by sources which lie either off or on a smooth convex surface. A UTD solution for calculating the electromagnetic fields associated with edge excited surface rays will also be presented. In all of the UTD solutions for convex surfaces presented below, it will be assumed that the principal radii of curvature of the surface R_1 and R_2 are large in terms of a wavelength so that the parameter

$$m = \left(\frac{k\rho_g}{2}\right)^{1/3}$$

is large at all points on the surface ray; in addition it is assumed that m is slowly varying along the ray path. Here ρ_g is the surface radius of curvature along the surface ray. These UTD solutions for the arbitrary convex surface all employ GTD rays, and they have been deduced from the asymptotic solutions to appropriate canonical problems as employed in the GTD procedure.

A. Both Source and Observation Points are off the Smooth Convex Surface.

The geometrical configuration of this problem is as shown in Fig. 10. In these notes, a uniform GTD (UTD) solution which is convenient and accurate for engineering applications is presented for this problem such that it overcomes the limitations of the GTD within the SB transition region;* furthermore, this UTD solution also automatically reduces to the GTD solution [3,4,25,26] exterior to the SB transition region where the latter solution is indeed valid. A study of this problem is useful for analyzing the blockage or shadowing effects of smooth convex structures as for example by a ship mast illuminated by a nearby shipboard antenna, or by the fuselage of an aircraft or missile which may be illuminated by a nearby wing, tail, or fin mounted antenna. Referring to Fig. 10, one notes that an extension of the incident ray beyond the point of grazing at Q_1 on the convex surface defines the SB which divides the space exterior to the surface into the lit and shadow regions. According to the GTD, the ray optical field exterior to the surface is associated with the usual geometrical optics (GO) incident and reflected rays in the deep lit region, i.e., region I of Fig. 10; whereas, in the deep shadow region, i.e., region III of Fig. 10, it is associated entirely with the surface diffracted rays. The shaded region II on either side of the SB is the SB transition region in which the ray optical field description changes rapidly but continuously from the GO field in I to the surface diffracted field in III.

The angular extent of II is of the order $\left[2 / k\rho_g(Q_1) \right]^{1/3}$ radians where $\rho_g(Q_1)$ is the surface radius of curvature along the incident ray direction at Q_1. Regions IV, V, and VI lie in the close vicinity of the shadowed part of the surface which is also a caustic of the surface diffracted rays; these regions constitute the caustic (or surface) boundary layer. It is noted that region IV is in the lit zone; whereas, V and VI are in the shadow zone. Specifically, the GTD solution fails in region II and it also fails in the surface boundary layer; whereas, the present UTD solution fails only within the surface boundary layer. A separate solution will be reported in the future to describe the fields within the surface boundary layer in a form suitable for applications.

The ansatz or formulation of the present UTD solution which employs GTD rays is based on an asymptotic solution given recently by Pathak [27] for the canonical problem of plane wave diffraction by a perfectly-conducting circular cylinder. The work in [27] extends and improves our earlier work [28], and it draws upon the pioneering work of Fock [29]. In the present UTD solution, it

---

* The SB refers to the shadow boundary.

is assumed that the incident high frequency electromagnetic field can be described according to ray optics as in (2.2). It is also assumed that the field point and the caustics of the incident ray system are not in the close vicinity of the surface, and that the amplitude of the incident field does not exhibit a rapid spatial variation in the vicinity of Q_1.

In this development, it is noted that the functional form of the GTD solution for the problem in Fig. 10 does not depend to first order on whether the surface is cylindrical or spherical [3,4,25,26]. Hence, the GTD solution also does not depend to first order on whether the surface ray is torsional or not; it is reasonable to conjecture that this property holds to first order for the UTD solution as well. Thus, the UTD solution for the arbitrary convex surface can be deduced to first order from just the solution to the canonical circular cylinder problem in [27]. The details of the construction of this solution together with some experimental verification are discussed in [30]. The final results for the total electric field, $\bar{E}$ which exists in the presence of a convex surface illuminated by an incident ray optical field $\bar{E}^i$ are as follows:

$$\bar{E}(P_L) \sim \bar{E}^i(P_L) + \bar{E}^i(Q_R) \cdot [R_s \hat{e}_\perp \hat{e}_\perp + R_h \hat{e}_\parallel^i \hat{e}_\parallel^r] \sqrt{\frac{\rho_1^r \rho_2^r}{(\rho_1^r + s^r)(\rho_2^r + s^r)}} \, e^{-jks^r} ; \tag{4.1}$$

for P_L in the <u>lit</u> region, and

$$\bar{E}(P_S) \sim \bar{E}^i(Q_1) \cdot [D_s \hat{b}_1 \hat{b}_2 + D_h \hat{n}_1 \hat{n}_2] \sqrt{\frac{\rho_2^d}{s^d(\rho_2^d + s^d)}} \, e^{-jks^d} ; \text{ for } P_S \text{ in the}$$

<u>shadow</u> region.

(4.2)

The quantities within brackets involving $R_s\atop h$ in equation (4.1) and $D_s\atop h$ in (4.2) may be viewed as generalized, dyadic coefficients for surface reflection[+] and diffraction, respectively. It is noted that (4.1) and (4.2) are expressed invariantly in terms of the unit vectors fixed in the reflected and surface diffracted ray coordinates. The unit vectors $\hat{e}_\parallel^i$, $\hat{e}_\parallel^r$, and $\hat{e}_\perp$ in (4.1) have

---

[+]Actually, cross terms exist in the above generalized dyadic reflection coefficient; but, in general their effect is seen to be weak within the SB transition region. Also these terms vanish in the deep lit region and on the SB, hence they have been ignored in (4.1).

been defined in Section IIA. At Q_1 let $\hat{t}_1$ be the unit vector in the direction of incidence, $\hat{n}_1$ be the unit outward normal vector to the surface, and $\hat{b}_1 = \hat{t}_1 \times \hat{n}_1$; at Q_2 let a similar set of unit vectors be defined with $\hat{t}_2$ in the direction of the diffracted ray as in Fig. 11. In the case of surface rays with zero torsion, $\hat{b}_1 = \hat{b}_2$. The distance s^r and s^d along the reflected and surface diffracted ray paths are shown in Fig. 10. It is noted that ρ_1^r and ρ_2^r have been defined in Section IIA and ρ_2^d is the wavefront radius of curvature of the surface diffracted ray evaluated in the $\hat{b}_2$ direction at Q_2. The $R_{\substack{s \\ h}}$ and $D_{\substack{s \\ h}}$ in (4.1) and (4.2) are given by

$$R_{\substack{s \\ h}} = -\left[\sqrt{\frac{-4}{\xi^L}} e^{-j(\xi^L)^3/12} \left\{\frac{e^{-j\frac{\pi}{4}}}{2\sqrt{\pi}\xi^L}[1-F(X^L)] + \tilde{P}_{\substack{s \\ h}}(\xi^L)\right\}\right], \quad (4.3)$$

for the <u>lit</u> region

and

$$D_{\substack{s \\ h}} = -\left[\sqrt{m(Q_1)m(Q_2)}\sqrt{\frac{2}{k}}\left\{\frac{e^{-j\frac{\pi}{4}}}{2\sqrt{\pi}\xi^d}[1-F(X^d)] + \tilde{P}_{\substack{s \\ h}}(\xi^d)\right\}\right]\sqrt{\frac{d\eta(Q_1)}{d\eta(Q_2)}} e^{-jkt} \quad (4.4)$$

for the <u>shadow</u> region

The function F appearing above has been introduced in Section IIIA dealing with edge diffraction (see (3.7)). The Fock type surface reflection function $\tilde{P}_{\substack{s \\ h}}$ is related to the $\binom{\text{soft}}{\text{hard}}$ Pekeris function $\binom{p^*}{q^*}$ by [31,32]

$$\tilde{P}_{\substack{s \\ h}}(\delta) = \begin{Bmatrix} p^*(\delta) \\ q^*(\delta) \end{Bmatrix} e^{-j\frac{\pi}{4}} - \frac{e^{-j\frac{\pi}{4}}}{2\sqrt{\pi}\delta}, \quad \text{(Note that } \delta=0 \text{ at SB)} \quad \begin{matrix}(4.5a)\\(4.5b)\end{matrix}$$

where p^* and q^* are finite and well behaved even when $\delta=0$; these universal functions are tabulated and plotted in [31,32]. Also,

$$\tilde{P}_{\substack{s \\ h}}(\delta) = \frac{e^{-j\frac{\pi}{4}}}{\sqrt{\pi}}\int_{-\infty}^{\infty}d\tau \frac{\tilde{Q}V(\tau)}{\tilde{Q}W_2(\tau)}e^{-j\delta\tau}; \quad \tilde{Q} = \begin{cases} 1 &, \text{ soft case} \quad (4.6a)\\ \frac{\partial}{\partial\tau} &, \text{ hard case} \quad (4.6b)\end{cases}$$

in which the Fock type Airy function $V(\tau)$ and $W_2(\tau)$ are [31,32]:

$$2jV(\tau) = W_1(\tau) - W_2(\tau); \quad W_1(\tau) = \frac{1}{\sqrt{\pi}} \int_{\infty}^{\infty} \frac{dt}{e^{-j2\pi/3}} e^{\tau t - t^3/3};$$
(4.7a)
(4.7b)

$$W_2(\tau) = \frac{1}{\sqrt{\pi}} \int_{\infty}^{\infty} \frac{dt}{e^{j2\pi/3}} e^{\tau t - t^3/3} \cdot$$
(4.7c)

The remaining quantities occurring in (4.1)-(4.4) are[+]:

$$\xi^L = -2m(Q_R)\cos\theta^i \quad ; \quad \xi^d = \int_{Q_1}^{Q_2} dt' \frac{m(t')}{\rho_g(t')} \quad ; \quad m(\cdot) = \left[\frac{k\rho_g(\cdot)}{2}\right]^{1/3}$$
(4.8a;4.8b;4.9)

$$t = \int_{Q_1}^{Q_2} dt' \quad ; \quad x^L = 2kL\cos^2\theta^i \quad ; \quad x^d = \frac{kL^d(\xi^d)^2}{2m(Q_1)m(Q_2)} \cdot$$

(4.10;4.11a;4.11b)

The $\rho_g(Q_R)$ in $m(Q_R)$ is the surface radius of curvature at Q_R in the plane of incidence; whereas $\rho_g(Q_{1,2})$ is the surface radius of curvature at $Q_{1,2}$ in the $\hat{t}_{1,2}$ direction. Here, dt' is an incremental arc length along the surface ray path. The angle of incidence θ^i is shown in Fig. 2. The $d_n(Q_1)$ and $d_n(Q_2)$ denote the widths of the surface ray tube at Q_1 and Q_2, respectively; the surface ray tube is formed by considering a pair of rays adjacent to the central ray as in Fig. 11. Let s' denote the distance along the

---

[+] In [30], ξ^L is more precisely given by $\xi^L = -2m(Q_R)f^{-1/3}\cos\theta^i$ where $f^{-1/3}$ depends upon the principal surface radii of curvatures at Q_R and θ^i in a complicated manner. However, it appears that replacing $f^{-1/3}$ by unity as is done in (4.8a) for all θ^i does not impair the accuracy of the solution. It is noted that f=1 when $\theta^i = \pi/2$ (i.e., at SB), and it differs from unity as $\theta^i \to 0$; however, as $\theta^i \to 0$, $\xi^L \ll 0$ and $P_s \to \sqrt{-\xi^L/4} \, e^{j(\xi^L)^3/12}$ so that $R_s \to \mp 1$ as it should with either definition for ξ^L.

incident ray which is measured from some reference point $\bar{r}_0$ along the ray, and let s_0 and ℓ denote its values at Q_1 and Q_R, respectively as shown in Fig. 10 for the special case of a point source; then for a diverging incident wavefront with principal radii of curvatures ρ_1^i and ρ_2^i at the reference point $\bar{r}_0$, one obtains:

$$L^L\bigg|_{SB} = L^d\bigg|_{SB} = \frac{(\rho_1^i+s_0)(\rho_2^i+s_0)}{(\rho_1^i+[s_0+s])(\rho_2^i+[s_0+s])} \frac{s(\rho_2^r+s)}{\rho_2^r} \text{, at SB} \quad (4.11c)$$

in which s denotes the identical values taken on by s^r and s^d at the SB, i.e., $(s^r=s^d)\equiv s$ at SB. The $L^{L,d}$ are slowly varying near SB, and (1-F) in (4.3) and (4.4) vanishes sufficiently rapidly as the field point moves far from SB; it is therefore convenient to use the form in (4.11c) for L^L and L^d even away from SB. Alternatively, away from SB, only the values of s and ρ_2^r in L^d of (4.11c) may be replaced by s^d and ρ_2^d for field points in the shadow zone, with everything else in that equation still being evaluated at SB; whereas, in the lit zone away from SB all quantities in L^L of (4.11c) may be evaluated in terms of the point of reflection Q_R which changes as the field point moves, e.g., s_0 and s are replaced by ℓ and s^r, respectively. It is noted that in addition to the large parameter m , the $kL^{L,d}$ product must also be large in this UTD solution.† One notes that

$$L^L = \frac{\ell s^r}{\ell+s^r} \text{ ; } L^d = \frac{s_0 s^d}{s_0+s^d} \text{ , } \begin{array}{l}\text{for cylindrical or}\\ \text{spherical wave illumination.}\end{array} \quad (4.12)$$

In the case of plane wave illumination, $\ell\to\infty$ and $s_0\to\infty$ in (4.12) so that $L^L=s^r$ and $L^d=s^d$ in this case. With $L^{L,d}$ as in (4.12) it is easily verified that the UTD solution satisfies reciprocity.

If the incident wavefront is of the converging, or converging-diverging type, then the parameters $L^{L,d}$ in (4.11c) can become negative. It has not been investigated in detail how the general solution can be completed when $L^{L,d}$ becomes negative. However, if one of the principal directions of the incident wavefront coincides with one of the principal planes of the surface at grazing, then one can treat a converging or converging-diverging type wavefront for which $L^{L,d}<0$, by replacing $F(X^{L,d})$ with $F^*(|X^{L,d}|)$. Note the asterisk (*) on $F^*(|X^{L,d}|)$ denotes the

---

†In general, the present solution appears to be accurate even for $kL^{L,d}$ as small as three; in some special cases $kL^{L,d}$ can be made as small as unity.

complex conjugate operator. The use of $F^*(|X^{L,d}|)$ when $L^{L,d}<0$
leads to a continuous total field at SB in this case.

It is shown in [30] that exterior to the SB transition region, the F function approaches unity so that only $\hat{P}_\hbar$ in (4.3) and (4.4) dominates allowing (4.1) and (4.2) to reduce uniformly to the GO and Keller's surface diffracted fields, as described in references [3,4,5]. Near and at the SB, the F function dominates; it is entirely responsible for ensuring the continuity of (4.1) and (4.2) at the SB. Finally, this UTD result remains accurate outside the paraxial regions of cylindrical (or elongated type) surfaces; a different solution is required for handling the paraxial regions. Some typical examples showing the accuracy of this UTD solution are illustrated in Figs. 12a, 12b, 13b, and 13c. Figures 12a and 12b indicate the radiation patterns of electric and magnetic dipole sources in the presence of a perfectly-conducting circular cylinder based on the UTD calculations; these patterns are compared with the corresponding exact eigenfunction (modal) series solutions. In Figs. 13b and 13c, the near zone patterns of a vertical and horizontal electric dipole near a conducting conical structure of the type shown in Fig. 13a are obtained via UTD and shown to compare well with measurements.

B. **Source Point on but Observation Point off the Convex Surface**

In this configuration, the source is located on the surface whereas the observation point is located at least a few wavelengths from the surface even though it may be in the near zone of that surface. This radiation problem is directly related by reciprocity to the problem of calculating the fields induced on the convex surface by a source which is located off the surface. The radiation from sources on perfectly-conducting convex surfaces is of interest in the design of antennas mounted on aircraft and missiles, and in the design of conformal arrays.

In the following development the GTD is extended to obtain a uniform high-frequency solution for the radiation from apertures and monopoles which may excite torsional surface rays on a perfectly-conducting smooth, convex surface. This problem was treated in an earlier paper [33], but only for torsionless surface rays. Subsequently solutions were given for cylindrical and conical surfaces, [34-37] where torsional surface rays exist. Although the solution in [37] was generalized to convex surfaces, it differs from the one described here; furthermore it does not appear to be as complete or as easy to implement.

Consider an infinitesimal, magnetic current moment $d\bar{p}_m(Q')$ or an electric current-moment $d\bar{p}_e(Q')$ located on a perfectly-conducting convex surface as shown in Fig. 14; these sources $d\bar{p}_m(Q')=\bar{E}(Q')\times n'da'$ and $d\bar{p}_e(Q')=I(Q')d\ell'n'$ pertain to the aperture and monopole type excitations, respectively as described in (2.20a; 2.20b) of Section IIC. The high frequency fields generated by these sources may be expressed in the ray coordinates of the GTD. Thus in the shadow region the radiation from Q' follows a surface ray to Q where it sheds tangentially from the surface to the field point P_s; whereas in the lit region, the radiation follows the incident ray of geometrical optics to P_L in the direction of the unit vector $\hat{s}$. At Q the orthogonal unit vectors $\hat{t}$ and $\hat{n}$ (i.e., parallel to the ray and normal to the surface, respectively) are introduced along with the binormal unit vector $\hat{b}=\hat{t}\times\hat{n}$. At Q' these same unit vectors are primed. In the lit region $\hat{n}=\hat{b}\times\hat{s}$, where $\hat{b}'=\hat{b}$ is perpendicular to the plane of incidence. Let $d\bar{E}_m$ and $d\bar{E}_e$ denote the electric fields radiated by the sources $d\bar{p}_m(Q')$ and $d\bar{p}_e(Q')$, respectively.

The high-frequency electric field is given by $d\bar{E}_{m_e} = \hat{n} dE_{m_e}^n + \hat{b} dE_{m_e}^b$ for points away from the convex surface in both the shadow and lit regions. Expressions for the above $\hat{n}$ and $\hat{b}$ directed electric field components have been deduced from a careful study of the cylinder and sphere canonical problems in which higher order terms are retained in the asymptotic solutions; in addition, experimental results for sources on a spheroid were helpful in the generalization to the convex surface. An important step in the present generalization of the canonical solutions to treat the arbitrary convex surface is based on the observation that the effect of surface ray torsion is localized to the source region at least to first order, and its effect on the radiation fields in both the shadow and the shadow boundary transition regions can be described explicitly in terms of a torsion factor T_o at the source. The torsion factor thus affects only the launching of the surface ray field at Q'; whereas, the diffraction of the surface ray at Q is not affected by surface ray torsion to first order, as pointed out in Part A of this section. Expressions for the fields $d\bar{E}_m$ radiated by $d\bar{p}_m$ are developed in detail in [38]; the final results are presented below. It is noted that the form of the generalized dyadic launching coefficient $\bar{\bar{L}}$ of (2.21) can be immediately deduced from these expressions.

(a) $d\bar{p}_m(Q')$ case:

$$dE_m^n(P_L) = C[(d\bar{p}_m \cdot \hat{b}')(H^\ell + T_o^2 F \cos\theta^i) + (d\bar{p}_m \cdot \hat{t}') T_o F \cos\theta^i] \frac{e^{-jks}}{s} + O[m_\ell^{-2}],$$

(4.13)

for $P=P_L$ in the <u>lit region</u>

$$dE_m^b(P_L) = C[(d\bar{p}_m \cdot \hat{b}') T_o F + (d\bar{p}_m \cdot \hat{t}')(S^\ell - T_o^2 F \cos^2\theta^i)] \frac{e^{-jks}}{s} + O[m_\ell^{-2}, m_\ell^{-3}].$$

for $P=P_L$ in the <u>lit region</u> (4.14)

and

$$dE_m^n(P_S) = C(d\bar{p}_m \cdot \hat{b}')H\, e^{-jkt} \left[\frac{\rho_g(Q')}{\rho_g(Q)}\right]^{-1/6} \sqrt{\frac{d\psi_0}{d\eta(Q)}} \sqrt{\frac{\rho_2^d}{s(\rho_2^d+s)}}\, e^{-jks}$$
$$+ O[m^{-2}], \qquad (4.15)$$

for $P=P_S$ in the <u>shadow region</u>

$$dE_m^b(P_S) = C[(d\bar{p}_m \cdot \hat{b}')T_0 S + (d\bar{p}_m \cdot \hat{t}')S]e^{-jkt} \left[\frac{\rho_g(Q')}{\rho_g(Q)}\right]^{-1/6} \sqrt{\frac{d\psi_0}{d\eta(Q)}}$$
$$\cdot \sqrt{\frac{\rho_2^d}{s(\rho_2^d+s)}}\, e^{-jks} + O[m^{-2}, m^{-3}]. \qquad (4.16)$$

with $\qquad\qquad\qquad\qquad$ for $P=P_S$ in the <u>shadow region</u>

$$C = \frac{-jk}{4\pi}. \qquad (4.17)$$

The quantity t denotes the length of the surface ray path from Q' to Q. The torsion factor T_0 is defined by:

$$T_0 = T(Q')\rho_g(Q') \qquad (4.18)$$

in which $\rho_g(Q')$ is the surface radius of curvature in the $\hat{t}'$ direction at Q', and $T(Q')$ is the surface ray torsion at Q'. One may show via differential geometry that

$$T(Q') = \frac{\sin 2\alpha'}{2}\left(\frac{1}{R_2(Q')} - \frac{1}{R_1(Q')}\right); \quad \frac{1}{\rho_g(Q')} = \frac{\cos^2\alpha'}{R_1(Q')} + \frac{\sin^2\alpha'}{R_2(Q')},$$

in which $R_1(Q')$ and $R_2(Q')$ denote the principal surface radii of curvatures at Q', and α' is the angle between $\hat{t}'$ and $\hat{\tau}_1'$ as shown in Fig. 15. Here $\hat{\tau}_1'$ and $\hat{\tau}_2'$ are the principal surface directions at Q' and $\hat{\tau}_{1,2}'$ is associated with $R_{1,2}(Q')$. It is assumed for the sake of definiteness that $R_1(Q') \geq R_2(Q')$. One notes that $T(Q')<0$ if $\pi/2<\alpha'<\pi$, or if $3\pi/2<\alpha'<2\pi$. The quantities H, H^ℓ, S, and S^ℓ contain the soft and hard Fock functions $\tilde{g}$ and g which are tabulated in [31] and plotted in [31,32]; in particular,

$$\tilde{g}(\delta) = \frac{1}{\sqrt{\pi}} \int_\infty^\infty e^{-j2\pi/3} d\tau \frac{e^{-j\delta\tau}}{W_2(\tau)}; \quad g(\delta) = \frac{1}{\sqrt{\pi}} \int_\infty^\infty e^{-j2\pi/3} d\tau \frac{e^{-j\delta\tau}}{W_2'(\tau)},$$

$$(4.19); (4.20)$$

$$S = \frac{-j}{m(Q')} \tilde{g}(\xi) \quad ; \quad H = g(\xi) \quad ; \quad \xi = \int_{Q'}^{Q} dt' \frac{m(t')}{\rho_g(t')} ,$$

(4.21a;4.21b;4.21c)

$$S^{\ell} = \frac{-j}{m(Q')} \tilde{g}(\xi_\ell) e^{-j\xi_\ell^3/3} \quad ; \quad H^{\ell} = g(\xi_\ell) e^{-j\xi_\ell^3/3} ;$$

$$\xi_\ell = - m_\ell(Q') \cos\theta^i .$$

(4.22a;4.22b;4.22c)

Also:

$$m(Q') = \left[\frac{k\rho_g(Q')}{2}\right]^{1/3} \quad ; \quad m_\ell(Q') = m(Q')[1+T_0^2 \cos^2\theta^i]^{-1/3}$$

(4.23);(4.24)

and $\cos\theta^i = \hat{n}' \cdot \hat{s}$ as shown in Fig. 14. Finally, F is given by:

$$F = \frac{S^\ell - H^\ell \cos\theta^i}{1+T_0^2 \cos^2\theta^i} .$$

(4.25)

(b) $\overline{dp}_e(Q')$ case:

$$dE_e^n(P_L) = CZ_c dp_e \sin\theta^i [H^\ell + T_0^2 F \cos\theta^i] \frac{e^{-jks}}{s} + O[m_\ell^{-2}] , \quad (4.26)$$

for $P=P_L$ in the <u>lit region</u>

$$dE_e^b(P_L) = CZ_c dp_e(Q') \sin\theta^i T_0 F \frac{e^{-jks}}{s} + O[m_\ell^{-2}] . \quad (4.27)$$

for $P=P_L$ in the <u>lit region</u>.

and

$$dE_e^n(P_S) = CZ_c dp_e(Q') H e^{-jkt} \left[\frac{\rho_g(Q')}{\rho_g(Q)}\right]^{-1/6} \sqrt{\frac{d\psi_0}{d\eta(Q)}} \sqrt{\frac{\rho_2^d}{s(\rho_2^d+s)}} e^{-jks}$$

$$+ O[m^{-2}]$$

(4.28)

for $P=P_S$ in the <u>shadow region</u>

$$dE_e^b(P_s) = CZ_c dp_e(Q')T_o S\ e^{-jkt}\left[\frac{\rho_g(Q')}{\rho_g(Q)}\right]^{-1/6} \sqrt{\frac{d\psi_o}{d\eta(Q)}} \sqrt{\frac{\rho_2^d}{s(\rho_2^d+s)}}\ e^{-jks}$$

$$+\ O[m^{-2}]\ . \qquad\qquad (4.29)$$

for $P=P_s$ in the <u>shadow region.</u>

In the above equations $Z_c=(Y_c)^{-1}$ denotes the characteristic impedance of the medium exterior to the surface. The $d\eta(Q)=\rho_2^d d\psi$ and $d\psi$ is the angle between the adjacent surface rays at Q. Likewise, $d\psi_o$ is the angle between the same pair of adjacent surface rays at Q' as in Fig. 16; also ρ_2^d is the wavefront radius of curvature of the surface diffracted ray in the $\hat{b}$ direction at Q.

The previous expressions reduce to the geometrical optics field in the deep lit region and to the GTD field in the deep shadow region. The expressions for the lit and shadow regions join smoothly at the shadow boundary. As expected, they reduce to the asymptotic solutions for the circular cylinder and sphere cases, but the higher order terms in m must be retained to pass smoothly to these two limiting cases. As the radii of curvature of the surface become infinite $T_o=0$, and (4.13), (4.14), (4.26) and (4.27) simplify to the field of magnetic and electric current moment sources on a ground plane.

In addition, this solution has been tested by applying it to calculate the radiation from slots and monopoles on perfectly-conducting circular and elliptic cylinders, cones, and spheroids [38]. The patterns of a radial slot on a cone are shown in Fig. 17; they are seen to compare well with the results based on the eigenfunction solution. In Figs. 18a and 18b, the patterns of a circumferential slot and a monopole are calculated and measured in the plane tangent to the spheroid at the source location. Note that the E_b component is due to the spheroidal surface; it would vanish if the source were on a ground plane.

C. Both Source and Observation Points on the Convex Surface

In this problem, the observation point P_s of Fig. 14 is moved to the point Q on the perfectly-conducting convex surface, with the source still being positioned at Q' on that surface. This problem is of interest in the calculation of the mutual coupling between a pair of antennas on a convex surface. Mutual coupling plays an important role in the design of conformal antenna arrays and in determining the electromagnetic compatibility between antennas. For convex surfaces, the mutual coupling calculation reduces to finding the electromagnetic surface fields at Q which are excited by a source at Q' on the surface.

Several previous studies have dealt with the asymptotic calculations of surface fields of sources on perfectly-conducting convex cylinders and cones [39,40,41,42]. More recently, an approximate asymptotic solution for the surface fields excited by an aperture in a perfectly-conducting arbitrarily shaped convex surface has been obtained in [43] based on an earlier solution in [44]. In particular the solution in [43] is obtained heuristically by modifying just the solution to the canonical problem dealing with the calculation of the surface fields excited by a $d\bar{p}_m(Q')$ (as in (2.20a)) on a conducting sphere.

An approximate asymptotic solution which is different from that in [43] and more complete than the previous ones in [39,40, 41,42] is presented here for the surface fields of an aperture in or a short monopole on a perfectly-conducting convex surface; furthermore, the effects of surface ray torsion have been explicitly identified in this solution which is based on the solutions to appropriate canonical circular cylinder and sphere problems. The details of the construction of this solution are given in [45]; the final results are presented below for the surface magnetic and electric fields $d\bar{H}_m^e(Q)$ and $d\bar{E}_m^e(Q)$ at Q, respectively due to sources $d\bar{p}_m^e(Q')$ which have been described earlier in this section. It is noted that the $d\bar{H}_m^e(Q)$ and $d\bar{E}_m^e(Q)$ are expressed invariantly in terms of the unit vectors which are fixed in the surface ray coordinates at Q' and Q as in Fig. 14, and these fields remain uniformly valid along the ray including the immediate vicinity of the source.

(a) $d\bar{p}_m(Q')$ case:

$$d\bar{H}_m(Q) = CY_c d\bar{p}_m(Q') \cdot \left[\hat{b}'\hat{b} \left(\left[1 - \frac{j}{kt}\right] \tilde{V}(\xi) + D^2\left(\frac{j}{kt}\right)^2 [\Lambda_s \tilde{U}(\xi) + \Lambda_c \tilde{V}(\xi)] \right. \right.$$

$$\left. + \tilde{T}_o^2 \frac{j}{kt} [\tilde{U}(\xi) - \tilde{V}(\xi)] \right) + \hat{t}'\hat{t} \left(D^2 \frac{j}{kt} \tilde{V}(\xi) + \frac{j}{kt} \tilde{U}(\xi) - 2\left(\frac{j}{kt}\right)^2 \right.$$

$$\left. \cdot [\Lambda_s \tilde{U}(\xi) + \Lambda_c \tilde{V}(\xi)] \right) + (\hat{t}'\hat{b} + \hat{b}'\hat{t}) \left(\tilde{T}_o \frac{j}{kt} [\tilde{U}(\xi) - \tilde{V}(\xi)] \right) \Big] D\; G(kt).$$

(4.30)

$$d\bar{E}_m(Q) = C\, d\bar{p}_m(Q') \cdot \left[\hat{b}'\hat{n} \left(\left[1 - \frac{j}{kt}\right] \tilde{V}(\xi) + \tilde{T}_o^2 \frac{j}{kt} [\tilde{U}(\xi) - \tilde{V}(\xi)] \right) \right.$$

$$\left. + \hat{t}'\hat{n} \left(\tilde{T}_o \frac{j}{kt} [\tilde{U}(\xi) - \tilde{V}(\xi)] \right) \right] D\; G(kt) \;.$$

(4.31)

In the above equations, D and G(kt) are defined by

$$D = \sqrt{\frac{t \, d\psi_0}{dn(Q)}} \quad ; \qquad G(kt) = 2e^{-jkt}/t \qquad (4.32);(4.33)$$

where t, $d\psi_0$, $dn(Q) = \rho_2^d d\psi$, C and Z_c have the same meaning as in Section IVB. The $\hat{T}_0$ is defined by

$$\hat{T}_0 = \pm \left| \sqrt{T_0(Q')T_0(Q)} \right| \quad , \qquad (4.34)$$

where the minus (-) sign in (4.34) is chosen if $T_0(Q') < 0$ and/or $T_0(Q) < 0$; otherwise the positive (+) sign is chosen. The quantities $T_0(Q')$ and ξ are defined in (4.18) and (4.21c), respectively; furthermore, $T_0(Q)$ is similar to $T_0(Q')$ except that it is evaluated at Q instead of Q'. The generalized soft and hard Fock functions $\hat{U}(\xi)$ and $\hat{V}(\xi)$ are defined as

$$\hat{U}(\xi) = \left(\frac{kt}{2m(Q')m(Q)\xi}\right)^{3/2} U(\xi) \quad ; \qquad (4.35a)$$

$$U(\xi) = \xi^{3/2} \frac{e^{j\frac{3\pi}{4}}}{\sqrt{\pi}} \int_{-\infty}^{\infty} \frac{d\tau}{e^{-j2\pi/3}} \frac{W_2'(\tau)}{W_2(\tau)} e^{-j\xi\tau} \quad , \qquad (4.35b)$$

$$\hat{V}(\xi) = \left(\frac{kt}{2m(Q')m(Q)\xi}\right)^{1/2} V(\xi) \quad ; \qquad (4.36a)$$

$$V(\xi) = \xi^{1/2} \frac{e^{j\frac{\pi}{4}}}{2\sqrt{\pi}} \int_{-\infty}^{\infty} \frac{d\tau}{e^{-j2\pi/3}} \frac{W_2(\tau)}{W_2'(\tau)} e^{-j\xi\tau} \quad . \qquad (4.36b)$$

The Fock functions $U(\xi)$ and $V(\xi)$ are tabulated in [31]; a useful summary of the large and small argument approximations for these functions is found in [44]. In order to interpolate smoothly between the canonical cylinder and sphere solutions, the weight factors Λ_C and Λ_S have been introduced heuristically into the solution for the arbitrary convex surface. These weight factors must be such that $\Lambda_S = 1$ and $\Lambda_C = 0$ for a sphere; whereas $\Lambda_S = 0$ and $\Lambda_C = 1$ for a cylinder. A reasonable choice for Λ_S and Λ_C appears to be:

$$\Lambda_S = \sqrt{\frac{R_2(Q')R_2(Q)}{R_1(Q')R_1(Q)}} \; ; \qquad \Lambda_C = 1 - \Lambda_S \; , \qquad (4.37a);(4.37b)$$

where R_1 and R_2 are the principal surface radii of curvatures.

(b) $d\bar{p}_e(Q')$ case:

$$d\bar{H}_e(Q) = Cd\bar{p}_e(Q') \cdot \left[\hat{n}'\hat{b}\left(\left[1 - \frac{j}{kt}\right] \hat{V}(\xi) + \tilde{T}_o^2 \frac{j}{kt} [\hat{U}(\xi) - \dot{\hat{V}}(\xi)] \right) \right.$$
$$\left. + \hat{n}'\hat{t} \left(\tilde{T}_o \frac{j}{kt} [\hat{U}(\xi) - \hat{V}(\xi)] \right) \right] D \; G(kt) \quad . \quad (4.38)$$

$$d\bar{E}_e(Q) = CZ_c d\bar{p}_e(Q') \cdot \hat{n}'\hat{n} \left(\hat{V}(\xi) - \frac{j}{kt} \dot{\hat{V}}(\xi) \; + \left(\frac{j}{kt}\right)^2 [\Lambda_S \dot{\hat{V}}(\xi) + \Lambda_C \hat{U}(\xi)] \right.$$
$$\left. + \tilde{T}_o^2 \frac{j}{kt} [\hat{U}(\xi) - \hat{V}(\xi)] \right) D \; G(kt) \quad . \quad (4.39)$$

It may be easily verified from the property $\hat{V}(\xi) \to 1$ and $\hat{U}(\xi) \to 1$, as $\xi \to 0$, that the above results for $d\bar{H}_m^e$ and $d\bar{E}_m^e$ reduce in the limit of vanishing surface curvature to the known, exact results for the fields on a planar, perfectly-conducting surface. Some typical results indicating the accuracy of the above solution are presented in Table II for the circular cylinder, and in Fig. 19 for a cone geometry; it is seen from these calculations that the solutions indeed agree well with both the eigenfunction (modal) series solution and with measurements available in [46]. In these calculations, the present solution is referred to as the OSU solution, whereas, those in [43] and [46] are referred to as the UI and the G-S-PB solutions, respectively.

D. Edge-excited Surface Rays

Curved wedges occur as part of many practical antenna and scattering shapes, e.g., the edge of a reflector antenna, the base of conical and cylindrical structures, and the trailing edge of wings and stabilizers. A curved wedge may have a plane surface; however, the edge-excited surface rays do not have to be introduced separately at this surface. They occur as part of the space ray system. At a concave surface forming a curved wedge, multiply-reflected waves and whispering gallery modes are excited.

In the case of the convex surface, outside the region where there is a confluence of the edge and curved surface shadow boundaries, the excitation of the surface rays at Q_E in Fig. 3 may be determined from the UTD parameters which are presently available. When there is a confluence of the edge and surface shadow boundaries SB and ES, respectively (see Fig. 3), the excitation of the surface rays at Q_E is more complicated; the latter case is presently being investigated.

Returning to the case when there is no confluence of the boundaries SB and ES of Fig. 3, the excitation of the surface rays may be determined via the equivalent edge current concept introduced in Section IIID; in this procedure, the field diffracted from Q_E is produced by equivalent edge currents at Q_E which then radiate in the presence of the convex portions of the structure as described in Part B of this section. For the sake of simplicity, this calculation will be illustrated here for a two dimensional edge configuration excited by a uniform electric or magnetic line source at 0 in Fig. 3. The details of this calculation may be found in [47]; only the final results will be summarized below.

It is noted that the electric field generated by the electric line source in the presence of the 2-D structure is entirely $\hat{e}$-directed; likewise the magnetic field due to the magnetic line source excitation is also entirely $\hat{e}$-directed. Here $\hat{e}$ denotes the unit vector parallel to the edge Q_E. Let the ray optical electric and magnetic fields which are incident at Q_E from the electric and magnetic line sources, be denoted by $\hat{e} E^i(Q_E)$ and $\hat{e} H^i(Q_E)$ respectively. Then, from the equivalent current concept introduced in Section IIID, the diffraction of the field E^i by the edge at Q_E may be calculated in terms of an equivalent magnetic line dipole source of strength $M_d(\phi,\phi')$ at Q_E; whereas, for the field H^i it may be calculated in terms of an equivalent magnetic line source of strength $M(\phi,\phi')$ at Q_E. From [47],

$$M_d(\phi,\phi') = -\sqrt{\frac{8\pi}{jk}} \; \frac{E^i(Q_E)}{2} \; \frac{D_s(\phi,\phi')}{\sin\phi} \quad , \qquad (4.40a)$$

where

$$\lim_{\phi \to 0} \frac{D_s(\phi,\phi')}{\sin\phi} = \left.\frac{\partial D_s(\phi,\phi')}{\partial \phi}\right|_{\phi=0} \; ; \qquad (4.40b)$$

and
$$M(\phi,\phi') = -Z_c \sqrt{\frac{8\pi}{jk}} \frac{H^i(Q_E)}{2} D_h(\phi,\phi') \qquad (4.41)$$

where D_s and D_h are the soft and hard edge diffraction coefficients.

The fields radiated from the equivalent edge currents M_d and M at Q_E in the presence of the rest of the 2-D convex cylindrical surface may be obtained from the 3-D solutions in Part B of this section, after specializing the latter to the 2-D case as follows. The 3-D spread factors for the rays exterior to the convex surface, i.e., 1/s and

$$\sqrt{\frac{\rho_2^d}{s(\rho_2^d+s)}}$$

in (4.13)-(4.16) for the lit and shadow regions must be replaced by $1/\sqrt{s}$ for the 2-D case; also the surface ray divergence factor

$$\sqrt{\frac{d\psi_0}{\rho_2^d d\psi}}$$

in the 3-D case must be replaced by unity for the 2-D case. Furthermore, the quantity $C=-jk/4\pi$ in (4.13)-(4.16) must be replaced by $-\sqrt{jk/8\pi}$ for the 2-D case. Then the diffracted components of the electric and magnetic fields $\hat{e}E^d$ and $\hat{e}H^d$ generated by M_d and M of (4.40a;b) and (4.41), respectively are given separately for the lit and shadow sides of the surface shadow boundary ES of Fig. 3 as follows:

(a) <u>LIT SIDE OF ES</u>:

$$E^d(P) = \begin{cases} C\, M_d(\phi,\phi')S^\ell \dfrac{e^{-jks}}{\sqrt{s}} & ;\; 0 \le \phi \le \pi/2 , \quad (4.42a) \\[6pt] E^i(Q_E)D_s(\phi,\phi') \dfrac{e^{-jks}}{\sqrt{s}} & ;\; \phi > \pi/2 , \quad (4.42b) \end{cases}$$

and

$$H^d(P) = \begin{cases} C\, M(\phi,\phi')H^\ell \dfrac{e^{-jks}}{\sqrt{s}} & ;\; 0 \le \phi \le \pi/2 , \quad (4.43a) \\[6pt] H^i(Q_E)D_h(\phi,\phi') \dfrac{e^{-jks}}{\sqrt{s}} & ;\; \phi > \pi/2 . \quad (4.43b) \end{cases}$$

(b) SHADOW SIDE OF ES

$$E^d(P') = C\, M_d(0,\phi')S\, e^{-jkt} \left[\frac{\rho_g(Q_E)}{\rho_g(Q_3)}\right]^{-1/6} \frac{e^{-jks}}{\sqrt{s}} \qquad (4.44)$$

and

$$H^d(P') = C\, M(0,\phi')H\, e^{-jkt} \left[\frac{\rho_g(Q_E)}{\rho_g(Q_3)}\right]^{-1/6} \frac{e^{-jks}}{\sqrt{s}}\,. \qquad (4.45)$$

It is noted that $\phi=0$ in M_d and M in (4.44) and (4.45). In the lit region, the fields E^d and H^d of (4.42a;b) and (4.43a;b) propagate a distance s from Q_E to P as in Fig. 3; on the other hand, these fields propagate a distance t on the surface from Q_E to Q_3 and then a distance s from Q_3 to P' in the shadow region as is evident from (4.44) and (4.45).

The above UTD expressions have been employed to calculate the edge excited surface rays in the problem of the scattering of an incident plane wave by a perfectly-conducting curved strip as shown in Fig. 20. The various rays employed in the analysis of this problem are also shown in this figure. The UTD calculations for the total fields surrounding the curved strip are shown in Figs. 21a and 21b for the H_z^i (hard) case along with the independent moment method calculations for comparison. It is seen from these figures that the UTD calculations indeed agree very well with the moment method calculations. As seen from Fig. 20, a few multiply reflected rays on the concave side of the strip have been included in the UTD calculations. However, a more complete treatment of the scattering by the concave side of the surface would require one to also include the diffraction of whispering gallery (WG) modes. An interesting and accurate representation involving a judicious combination of multiply reflected rays and WG modes for the fields on a concave surface has recently been obtained by Felsen [48]. The inclusion of the diffraction of these concave surface fields of [48] from the edges of the curved strip should yield even more accurate results in the regions of space where these effects are important (see shaded region of

Fig. 20).

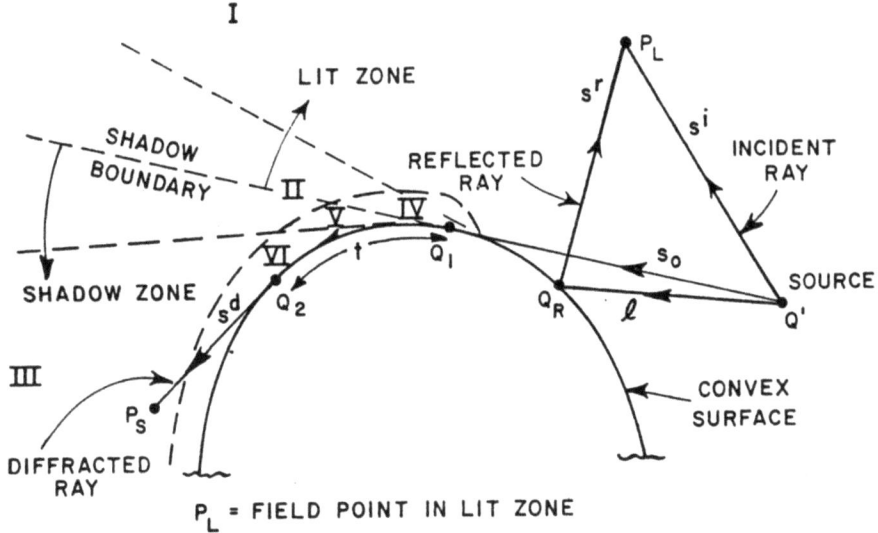

P_L = FIELD POINT IN LIT ZONE
P_S = FIELD POINT IN SHADOW ZONE

Figure 10. The rays and regions associated with scattering by a smooth convex surface.

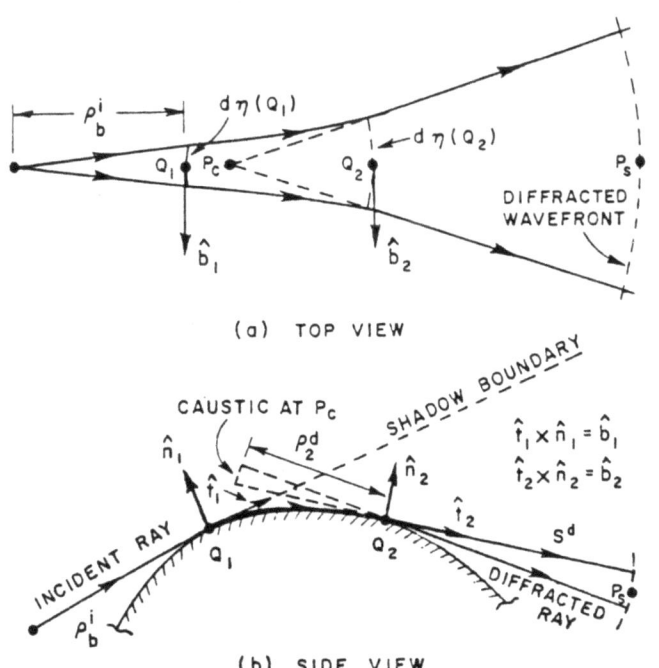

Figure 11. Diffraction by a smooth curved surface.

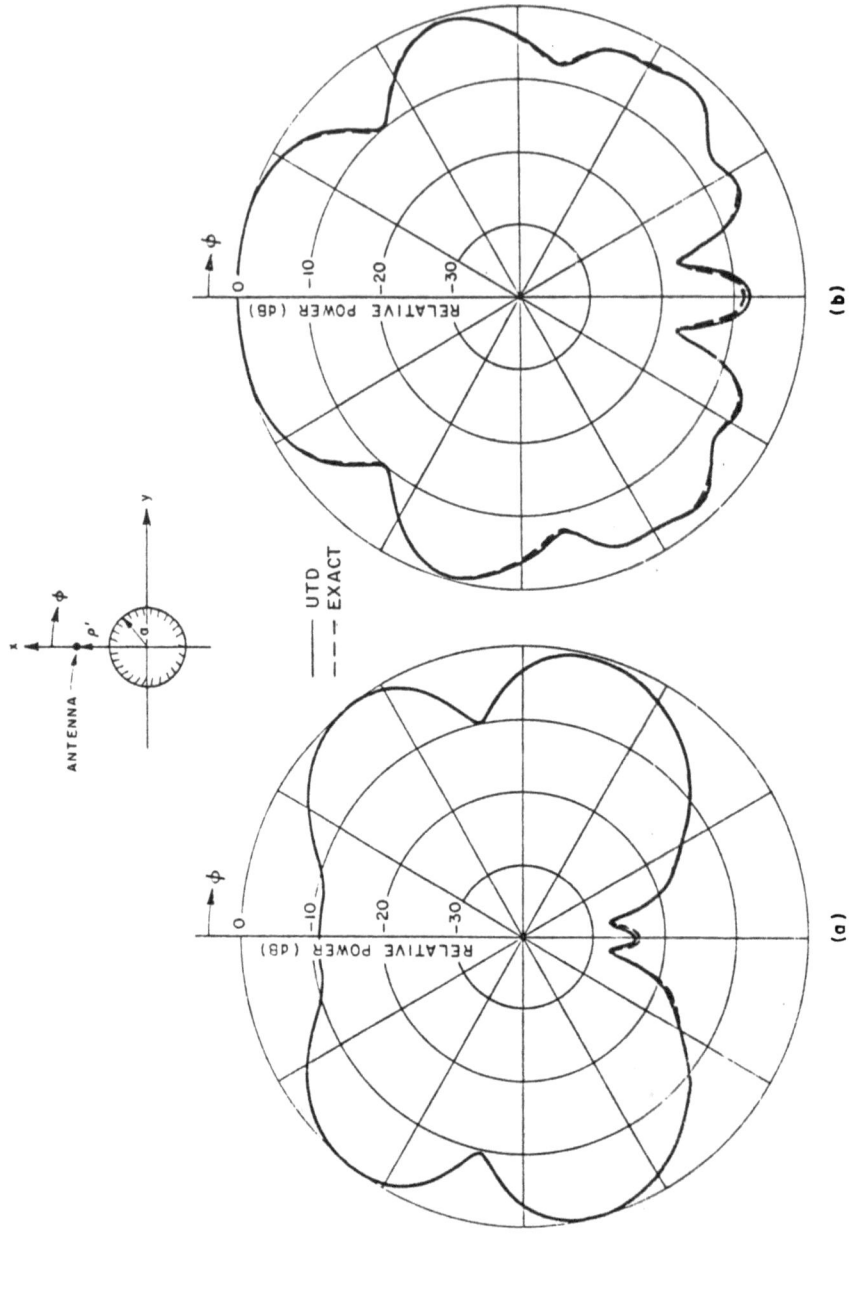

Figure 12. Comparison of UTD with exact modal solutions for the principal plane patterns of (a) electric, and (b) magnetic dipoles parallel to the cylinder axis.

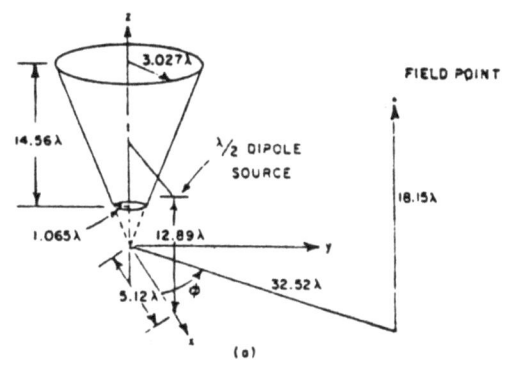

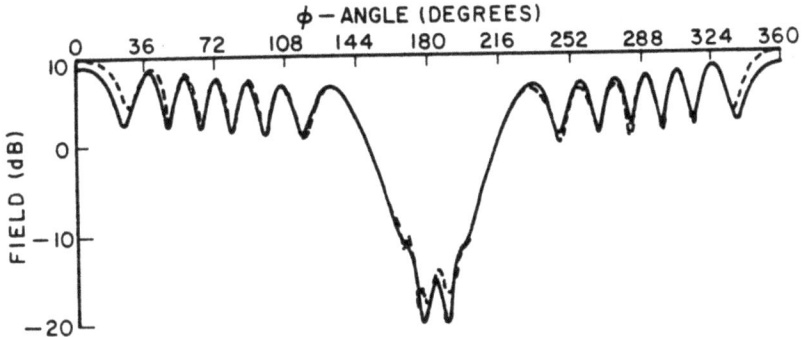

(b) VERTICAL POLARIZATION

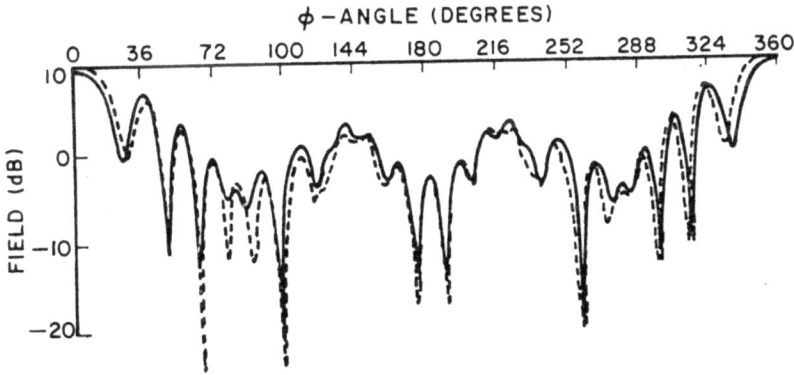

(c) HORIZONTAL POLARIZATION

Figure 13. radiation patterns of an electric dipole near the frustum of a cone.

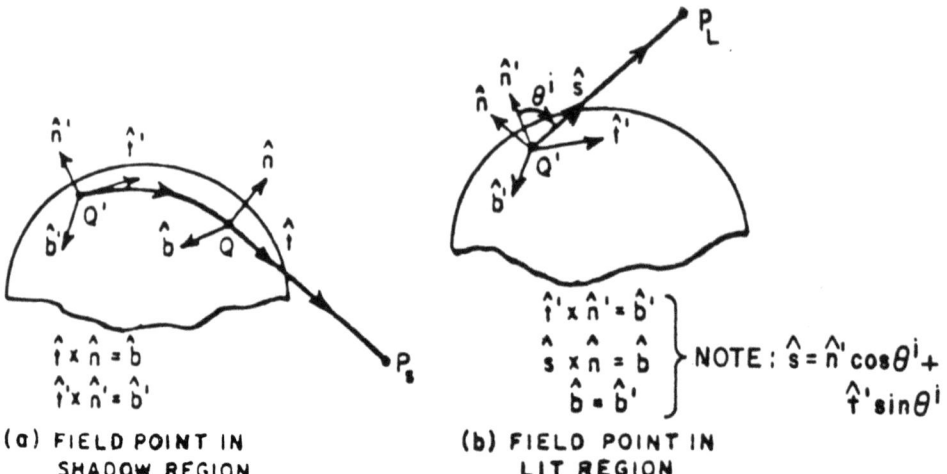

(a) FIELD POINT IN SHADOW REGION

(b) FIELD POINT IN LIT REGION

Figure 14. Rays emanating from a source on a convex surface.

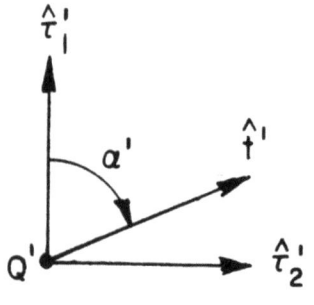

Figure 15. Principal directions $\hat{\tau}_1'$ and $\hat{\tau}_2'$ at Q'. Here, $\hat{\tau}_1' = \hat{\tau}_2' \times \hat{n}'$.

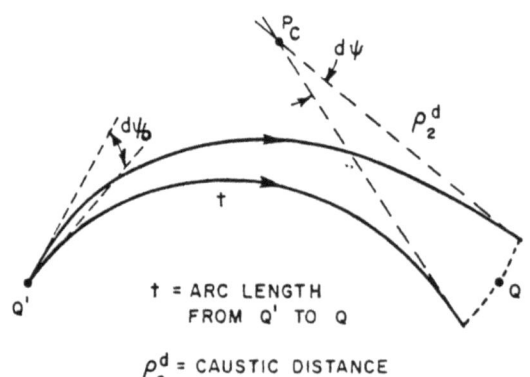

t = ARC LENGTH FROM Q' TO Q

ρ_2^d = CAUSTIC DISTANCE FROM P_C TO Q

Figure 16. Caustic distance associated with the spreading of the surface ray field.

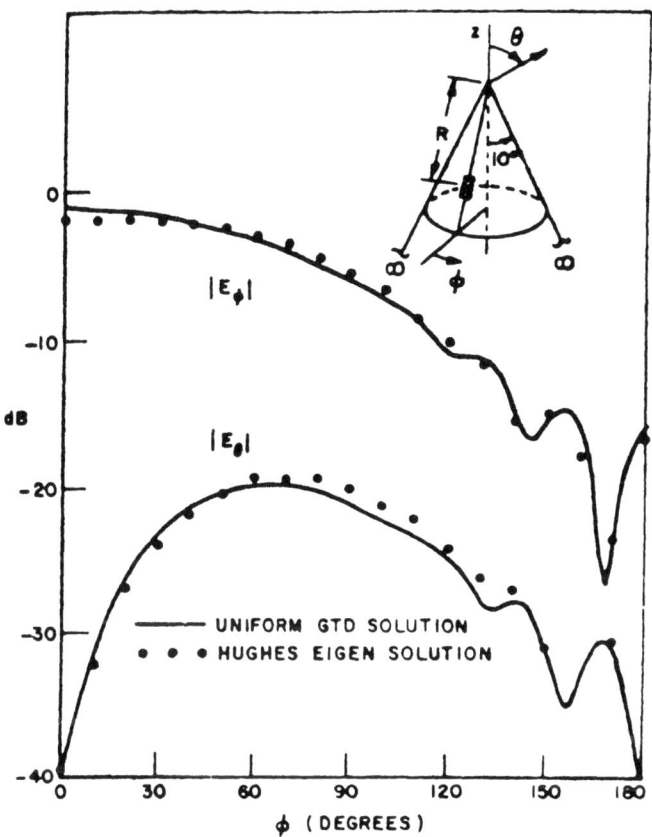

Figure 17. Radiation patterns of a half-wavelength radial slot in a cone calculated as a function of ϕ. $R = 6.22\lambda$ and $\theta = 80°$.

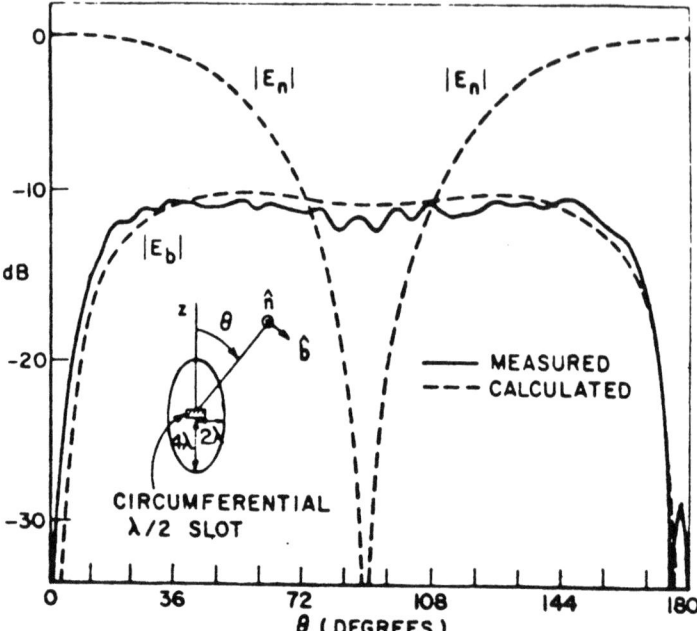

Figure 18a. Radiation patterns of a half-wavelength circumferential slot in a prolate spheroid calculated and measured in the shadow boundary plane.

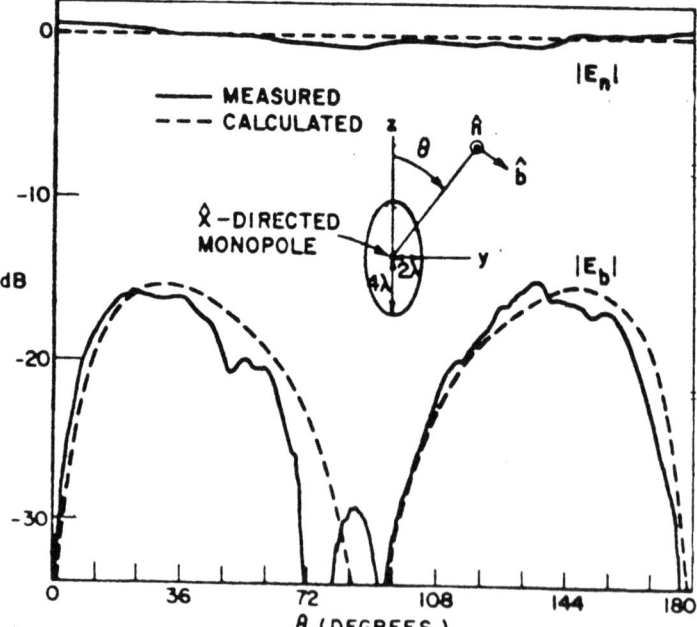

Figure 18b. Radiation patterns of a quarter-wavelength monopole on a prolate spheroid calculated and measured in the shadow boundary plane.

TABLE II

Y_{12} Between a Pair of Circumferential Slots in a Circular Cylinder of Radius a=1.991"; Frequency=9 GHz and for $\delta=\pi/2$; $z_0=0$ (H-Plane)

ϕ Degrees	Modal Hughes	Asymptotic OSU
30°	-25.98 dB -77°	-26.07 -75.73°
40°	-34.52 108°	-34.67 170.07
50°	-40.96 58°	-41.76 60.33°
60°	-46.62 -49°	-46.92 -47.55°

$\ell = 0.9"$
$d = 0.4"$

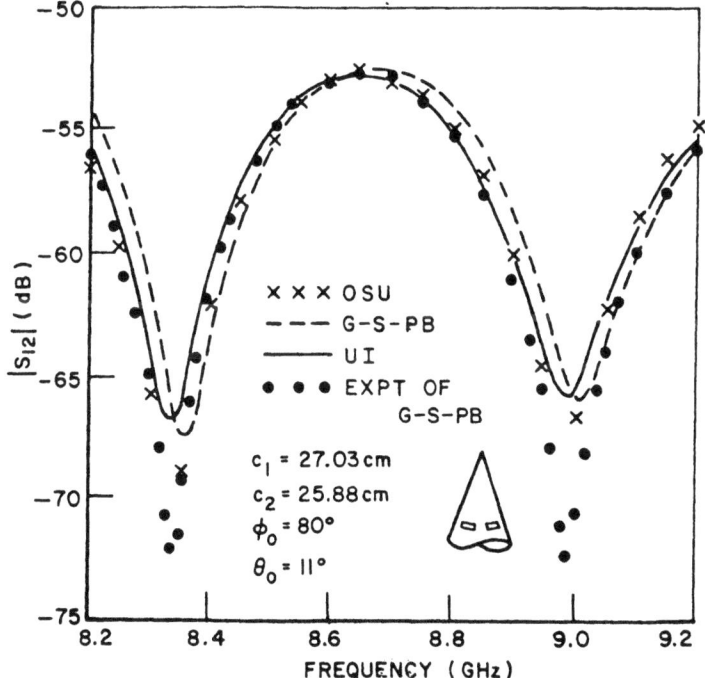

Figure 19. Coupling coefficient S_{12} between two circumferential slots on a cone vs. frequency. The radial separation between the slots is C_1-C_2 and angular separation is ϕ_0. The cone half angle is θ_0.

$c_1 = 27.03$ cm
$c_2 = 25.88$ cm
$\phi_0 = 80°$
$\theta_0 = 11°$

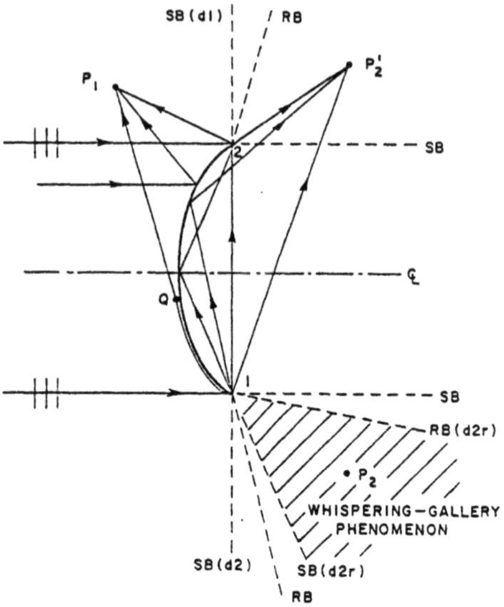

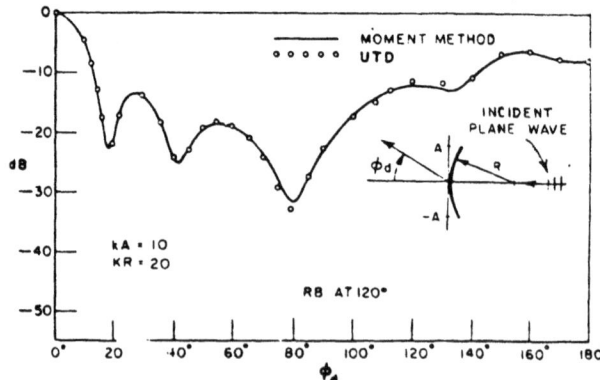

Figure 20. Rays diffracted and reflected by a curved strip.

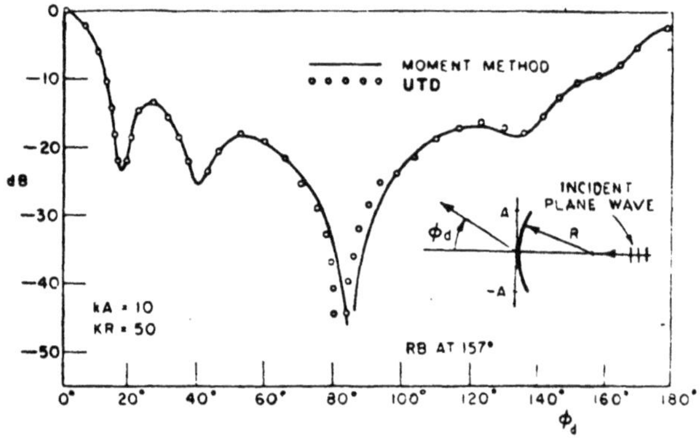

Figure 21a. Scattering from curved strip-hard boundary.

Figure 21b. Scattering from curved strip-hard boundary.

V. APPLICATIONS

In the previous sections, the UTD solutions for the general wedge and convex curved surface diffraction solutions were presented. These solutions provide the basic building blocks from which one can simulate much more complex structures. The concept applied here is that all scattered fields can be treated as ray optical. This being the case one can obtain the scattered field from a convex surface in terms of a ray optical field which in turn can diffract from a curved edge, for example. Recall that the UTD solutions for the wedge and curved surface are expressed in terms of general incident ray optical wavefronts. This concept allows one to develop high frequency numerical solutions for very complex electromagnetic problems. However, one must approach this panacea with caution. First, one cannot always assume that the scattered field from one structure which illuminates another is ray optical. Such a problem was addressed earlier in terms of the double diffraction problem where the source, receiver, and two edges align. Second, one must be aware of the limitations of the UTD and limited number of UTD diffraction solutions. This problem area manifests itself in terms of the UTD model used to simulate the actual structure. Since one does not have a UTD solution for all possible problems, he must attempt to approximate a given structure by a simpler UTD model which can be analyzed efficiently. This leads to two questions:
1) can one accurately analyze the simpler UTD model?
2) does the simpler UTD solution accurately model the actual situation?

From our point of view, the second question is all important and must normally be answered through an experimental verification process. In fact, the experimental verification actually answers both questions.

Based on our fifteen years experience in this area, one can approach the development of UTD solutions in a preferred manner. First, one should start with a UTD model which is known to be valid for a similar problem or with a simpler two-dimensional model in order to examine the significant features of the problem. One, then, proceeds to more complex configurations until he is satisfied that the UTD model resembles the actual situation. This is normally done in terms of critical comparisons with experimental results. One thing to keep in mind is that the UTD is a high frequency solution and will begin to fail as critical dimensions approach a wavelength. Thus, one can make critical tests of the solution at the lower frequency limit which simultaneously verifies the model representation and the UTD approximations.

In order to illustrate the above approach, let us consider the pattern analysis of airborne antennas realizing that most antennas are mounted along the fuselage center-line. With this in mind, a simple two-dimensional UTD model can be developed to study roll plane pattern (i.e., pattern in the x-y plane) as shown in Fig. 22. Using this model, the effect of the fuselage shape, engines, and wings were examined. As a result of that study it was ascertained that one must treat the fuselage cross-section in terms of an elliptic shape, the engine effects were minimal, and the three-dimensional outline of the wing was very significant. At this point, the UTD model was extended by simulating the fuselage with an elliptic cylinder and wings by finite flat plates. Thus, our UTD solution was extended to three dimensions as shown in Fig. 23. At this point, an efficient solution to describe the scattering from the wing was developed as will be presented later. With the development of this model, the UTD solution was compared with scale model aircraft measurements such as shown in Fig. 24. Based on numerous comparisons with various aircraft, antennas, and frequencies, the wing curvature did not play a significant role and the jet engine scattering was minimal for center-line antenna locations. This resulted in the basic roll plane aircraft model shown in Fig. 25a. A similar approach was used to analyze the elevation pattern with the resulting model shown in Fig. 25b. As presented in [49] these two models were used to develop a full volumetric pattern analysis for airborne antennas. Some of the results of that solution are presented in Fig. 24.

One should realize that he must study the simpler geometries and build to the more complex in that he cannot solve the problem exactly. If one starts with a complex model, he is most likely going to pay a premium for a given result, especially if a far simpler model can solve the same problem. The costs of complex solutions are rather obvious:
  1) large complex codes
  2) inefficient solutions.

Given that one has found a simple UTD model to represent a given situation, how can he develop an efficient solution to such a problem? The efficiency for most three-dimensional problems is dictated by the speed at which one computes the various ray paths. In order to illustrate how one might generate a very efficient ray path solution, let us consider the diffraction from a wing as shown in Fig. 26.

For the sake of brevity, only a straight edge will be treated here; however, the procedure used for the curved edge follows the same approach and is presented in [50]. It is also assumed that the source is positioned such that it does not illuminate any of the plate edges directly. In this case, the only signal transmitted by the source which illuminates the edges is that which first travels along a geodesic path on the cylinder surface, is subsequently diffracted from the curved surface, where it is finally edge diffracted to the receiver. The key to defining this ray path lies in determining the actual diffraction point along the edge.

As a first step in determining the diffraction point, it is known that the diffraction point must be located at a position along the edge such that the angle β between the incident ray and the edge equals the angle β' between the diffracted ray and the edge (see Fig. 26). The unit vectors $\hat{i}$ and $\hat{d}$ are in the direction of the incident and diffracted rays, respectively, and $\hat{e}$ is defined as a unit vector lying along the edge. The above noted basic law of edge diffraction (i.e., β=β') can be stated equivalently by the requirement that $\hat{i}\cdot\hat{e}=\hat{d}\cdot\hat{e}$. This requirement forms the basis for the technique used to compute the diffraction point. The edge unit vector $\hat{e}$ can readily be computed from the specified corner coordinates of the plate. For the discussion here it is sufficient to consider only the single edge bounded by points 1 and 2 in Fig. 26, and to express the edge unit vector simply as $\hat{e}=e_x\hat{x}+e_y\hat{y}+e_z\hat{z}$. With the notation of Fig. 26, the dot product $(\hat{i}\cdot\hat{e})$ can then be expressed as

$$\hat{i}\cdot\hat{e} = \frac{(x-x_{es})e_x+(y-y_{es})e_y+(z-z_{es})e_z}{[(x-x_{es})^2+(y-y_{es})^2+(z-z_{es})^2]^{1/2}} \quad (5.1)$$

where (x,y,z) are the coordinates of points along the edge and (x_{es},y_{es},z_{es}) locates the surface diffraction point. Similarly, the dot product $(\hat{d}\cdot\hat{e})$ can be expressed as

$$\hat{d}\cdot\hat{e} = \frac{(x_r-x)e_x+(y_r-y)e_y+(z_r-z)e_z}{[(x_r-x)^2+(y_r-y)^2+(z_r-z)^2]^{1/2}} \quad (5.2)$$

where (x_r,y_r,z_r) are the coordinates of the known (or specified) receiver location.

It appears that (5.1) and (5.2) could simply be equated and solved directly for the specific values of (x,y,z) yielding the unknown diffraction point coordinates (x_d,y_d,z_d). This is not the case, however, since the surface diffraction point is itself a complicated function of the coordinates (x,y,z) along the edge. Specifically, it can be shown that

$$x_{es} = \frac{a^2b^2x+a^2y\sqrt{a^2y^2+b^2x^2-a^2b^2}}{a^2y^2+b^2x^2}$$

$$y_{es} = \frac{a^2b^2y-b^2x\sqrt{a^2y^2+b^2x^2-a^2b^2}}{a^2y^2+b^2x^2} \quad (5.3)$$

$$z_{es} = \frac{bx_{es}z\,I_v-az_s(y-y_{es})I_v'}{bx_{es}I_v+a(y-y_{es})I_v'}$$

where

$$I_v = \int_{v_s}^{v_{es}} \sqrt{a^2\sin^2 v + b^2\cos^2 v}\, dv \quad \text{and} \quad I_v' = \sqrt{a^2\sin^2 v_{es} + b^2\cos^2 v_{es}}$$

with

$$v_{es} = \tan^{-1}\left(\frac{y_{es}/b}{x_{es}/a}\right), \quad v_s = \tan^{-1}\left(\frac{y_s/b}{x_s/a}\right).$$

In (5.3) the quantities a and b are the x and y cross-sectional dimensions, respectively, of the elliptic cylinder.

Since, as discussed above, the diffraction point coordinates (x_d, y_d, z_d) cannot be obtained directly, an iterative numerical search procedure is employed to determine the point at which $\hat{i}\cdot\hat{e} = \hat{d}\cdot\hat{e}$. As an initial step in the routine prior to initiating the actual search procedure, values of $\hat{i}\cdot\hat{e}$ are computed (using (5.1) and (5.3)) and stored for a sequence of n_p sample points along the plate edge. The coordinates of the sample points (i.e., specific values of the coordinates (x,y,z) in (5.1) and (5.3)), are selected and defined in the following manner. For an edge of length ℓ, the n_p sample points effectively subdivide the edge into (n_p-1) segments of length $\Delta\ell$, where $\Delta\ell = \ell/(n_p-1)$. The distance, t, between the end of the edge (i.e., point P_1 in Fig. 26) and the nth sample point along the edge is then given by $t(n) = (n-1)\Delta\ell$; $n=1,2,3,\ldots n_p$. These distances define the coordinates (x_n, y_n, z_n) for each of the n_p sample points.

After the $\hat{i}\cdot\hat{e}$ values are computed and stored at the n_p sample points along the edge, the routine initiates a search to determine which one of the (n_p-1) segments contains the diffraction point. The search begins by first determining whether (or not) the diffraction point does in fact lie anywhere along the physical edge. For a specified receiver location, this is done by computing the values of $\hat{d}\cdot\hat{e}$ at each end of the edge using (5.2) and comparing the values obtained with the respective stored values of $\hat{i}\cdot\hat{e}$ at these locations. If, for example, $\hat{i}\cdot\hat{e} < \hat{d}\cdot\hat{e}$ at one end of the edge and $\hat{i}\cdot\hat{e} > \hat{d}\cdot\hat{e}$ at the opposite end of the edge, then it follows that $\hat{i}\cdot\hat{e} = \hat{d}\cdot\hat{e}$ at some point along the physical edge. This point is, of course, the diffraction point. If, on the other hand, the sign of the inequality in the above example does not reverse, then the diffraction point does not lie on the physical edge. In this case the search is terminated, since there can be no diffraction from that edge for the receiver location specified.

Assuming that the diffraction point is found to lie along the physical edge, the routine proceeds with the search. The routine selects the midpoint of the edge as the next test point. A selected test point in the search routine always coincides with one of the n_p sample points in the list of previously computed $\hat{i}\cdot\hat{e}$ values. Again the sign of the inequality between $\hat{i}\cdot\hat{e}$ and $\hat{d}\cdot\hat{e}$ is noted, and the routine selects the half of the total edge which contains the diffraction point as the second search interval. The routine continues to halve the search interval in this manner until, after several cycles in the iteration, the length of the search interval is reduced to the segment length $\Delta\ell$ between successive sample points. This situation is depicted graphically in Fig. 27, which shows the diffraction point located within the segment bounded by distances $t(m)$ and $t(m+1)$. At this stage of the search, the routine uses linear interpolation to compute the distance h_d to the true diffraction point (see Fig. 27).

It should be noted that the search procedure described above, for determining the diffraction point along a given edge, must be repeated for each specified receiver location. Thus, if 360 receiver locations were used to compute a complete 360° pattern, then the above routine would be called 360 times for each edge of each plate used in the analytical model of the aircraft. For a typical aircraft model employing 6 plates with 4 edges each, the diffraction point search would be carried out 8640 (i.e., 360 x 6 x 4) times. Two primary features incorporated in this procedure result in vastly improved efficiency. One of these is the computation and storage of the list of $\hat{i}\cdot\hat{e}$ values at n_p sample points along each plate edge, prior to initiating the actual diffraction point search. Although the computation requires some time in itself, it is important to note that the same list of $\hat{i}\cdot\hat{e}$ values is used for all the receiver locations specified in computing a desired pattern. In computing a complete on-aircraft antenna pattern, storage of the list of $\hat{i}\cdot\hat{e}$ values for each edge results in a substantial net improvement in computational efficiency. The second feature incorporated in this routine which improves its computational efficiency is the use of linear interpolation to compute the exact diffraction point location after the search has bracketed the point within a small interval. The previous far-field routine [49] simply continued the iterative search until the search interval was reduced virtually to zero.

Another aspect of efficiency improvement has to do with the structure of the program. Considering the constraints of small, medium, and large computers, it is very advantageous to write computer codes which do not hop back-and-forth through the whole program. This is necessary in small computers because one must overlay various sections of the code into a small amount of memory. It is important in medium computers because they usually use paging algorithm which only alocate a certain number of pages per program. For big computers, they employ cache memory which is small but

extremely fast. This implies the following for large UTD numerical solutions:
1) define all fixed geometry aspects associated with the problem at the outset and store it in "COMMON",
2) define bounds on various terms such that complete ray paths do not have to be found before one decides to include a term or not,
3) use a single array to store the pattern data which is updated each time a new term is added, and
4) have the code compute each UTD term (such as plate reflected field) for the complete pattern before going on to the next term.

Using this structure, significant computations will be performed within smaller portions of the total code before proceeding to the next portion. This will reduce the amount of overlaying, decrease the amount of paging, and utilize cache memory to its fullest potential.

Let us consider a second major code development being done at the Ohio State University ElectroScience Laboratory. This code allows one to treat an antenna in the presence of a set of plates and a finite elliptic cylinder. This is a very general code such that one can model a wide variety of scattering problems. For example, it has been used to study wing mounted airborne antenna problems, a ship mounted antenna configuration, a communication van antenna farm, a radar antenna system mounted on a tank, etc.

The code is divided into three large sections. The first section contains the major scattered fields associated with the individual flat plates and the interactions between the different plates. These include the direct field, the singly reflected fields, doubly reflected fields, the single diffracted fields, the reflected-diffracted fields, and the diffracted-reflected fields. The diffracted fields include the normal diffracted fields as well as slope diffraction, a newly developed heuristic corner diffracted field and slope-corner diffracted field. The double diffracted fields are not included at present, but a warning is provided wherever this field component might be important. This is usually only a small angular section of space. This field may be included later whenever time and effort permit. The second section contains the major scattered fields associated with the elliptic cylinder. This includes the direct field, if not already computed in the plate section, the reflected field, the transition field, the deep shadow fields, the reflected field from the end caps, and the diffracted field from the end cap rims. The diffracted field from the end cap rim is not at present corrected in the pseudo caustic regions. This is where three diffraction points on the tim coalesce into one. This is only important in small angular regions in space and is not deemed appropriate to be included at the present time. An equivalent current method could be used for this small region but it is rather time consuming to use for the benefits derived from it for such a general code. The third section contains the major scattered fields associated with the interactions between the plates and cylinder. This includes, at present, the fields reflected from the plates then reflected or diffracted from the cylinder, the fields reflected from the cylinder then reflected from the plates, and the fields diffracted from the plates then reflected from the cylinder. These terms have been found to be sufficient for engineering purposes when analyzing wing-mounted aircraft antennas as well as many other structures.

The subroutines for each of the scattered field components are all structured in the same basic way. First, the ray path is traced backward from the chosen observation direction to a particular scatterer and subsequently to the source using either the laws of reflection or diffraction. Each ray path, assuming one is possible, is then checked to see if it is shadowed by any structure along the complete ray path. If it is shadowed the field is not computed and the code proceeds to the next scatterer or observation direction. If the path is not interrupted the scattered field is computed using the appropriate UTD solutions. The fields are then superimposed in the main program. This shadowing process is often speeded up by making various decisions based on bounds associated with the geometry of the structure. This type of knowledge is used wherever possible.

The shadowing of rays is a very important part of the UTD scattering code. It is obvious that this approach should lead to various discontinuities in the resulting pattern. However, the UTD diffraction coefficients are designed to smooth out the discontinuities in the fields such that a continuous field is obtained. When a scattered field is not included in the result, therefore, the lack of its presence is apparent. This can be used to advantage in analyzing complicated problems. Obviously in a complex problem not all the possible scattered fields can be included. In the UTD code the importance of the neglected terms are determined by the size of the so-called gliches or jumps in the pattern trace. If the gliches are small no additional terms are needed for a good engineering solution. If the gliches are large it may be necessary to include more terms in the solution. In any case the user has a gauge with which he can examine the accuracy of the results and is not falsely led into believing a result is correct when in fact there could be an error.

There have been many codes developed at the ElectroScience Laboratory over the past fifteen years using the UTD, two of which have been briefly described here. Some examples of the various problems that have been solved using UTD are presented in Figs. 28-30.

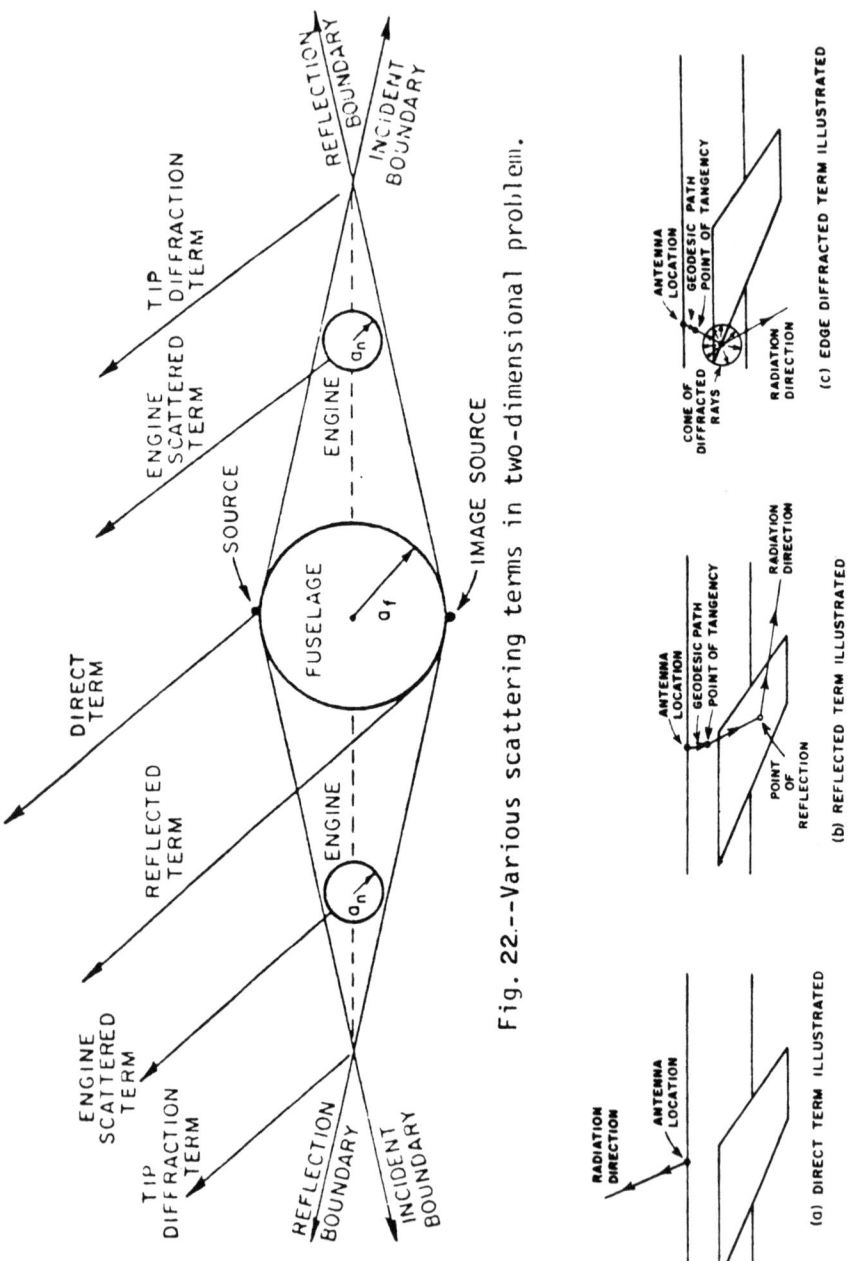

Fig. 22.--Various scattering terms in two-dimensional problem.

Figure 23. Dominant rays used in radiation pattern computation.

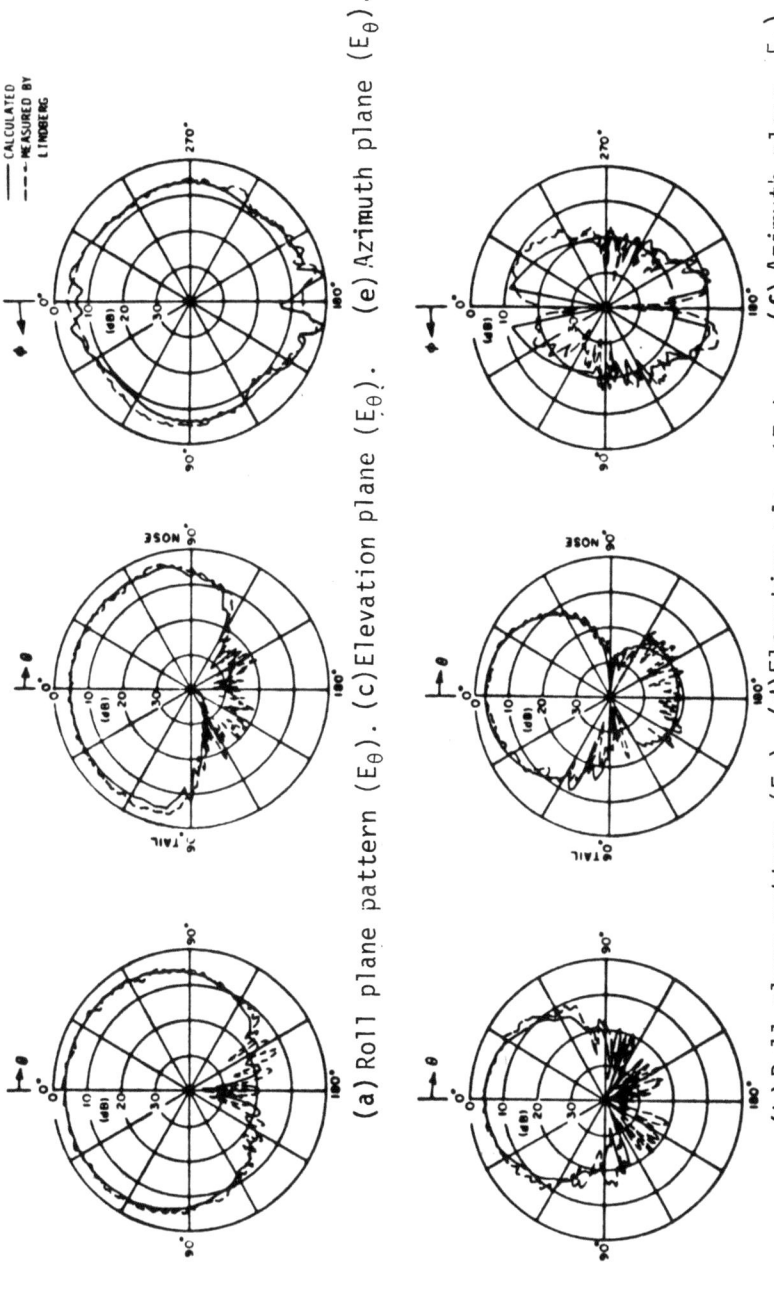

Figure 24. Radiation patterns of Lindberg crossed-slot antenna mounted at Station 470 on KC-135 aircraft.

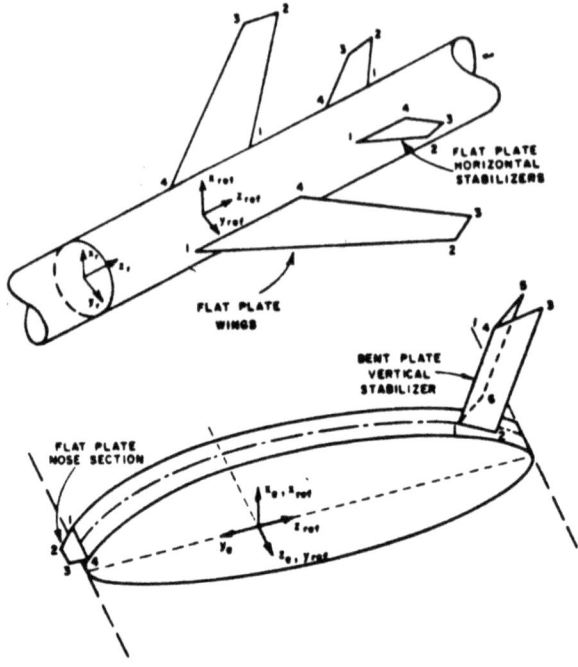

Figure 25. (a) Illustration of roll plane model.
(b) Illustration of elevation plane model.

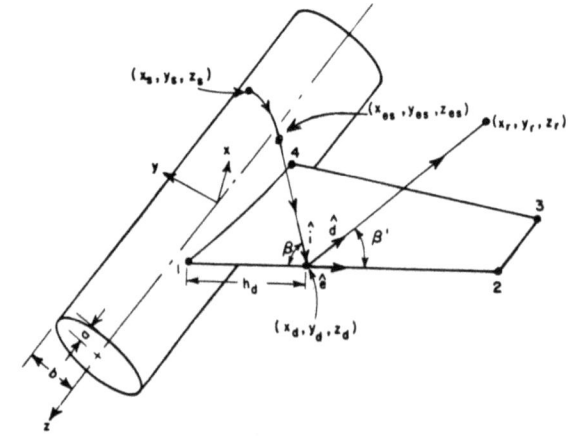

Figure 26. Geometry depicting diffraction from a straight edge of a finite flat plate.

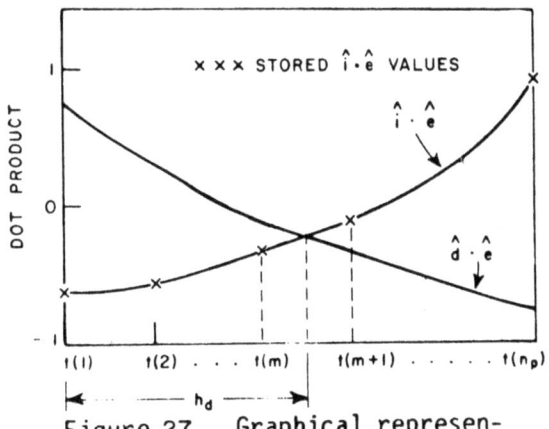

Figure 27. Graphical representation of how the diffraction point location is determined.

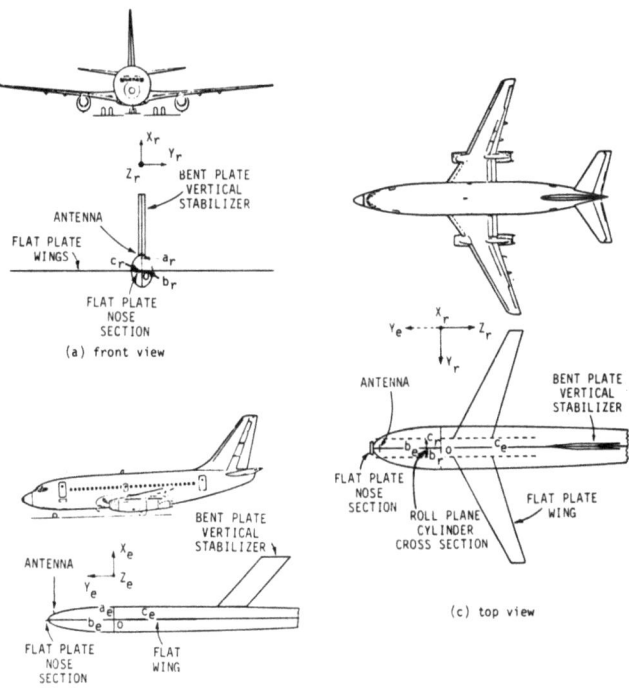

Figure 28. Computer simulation model of Boeing 737 aircraft with monopole located at Station 220 on top of fuselage.

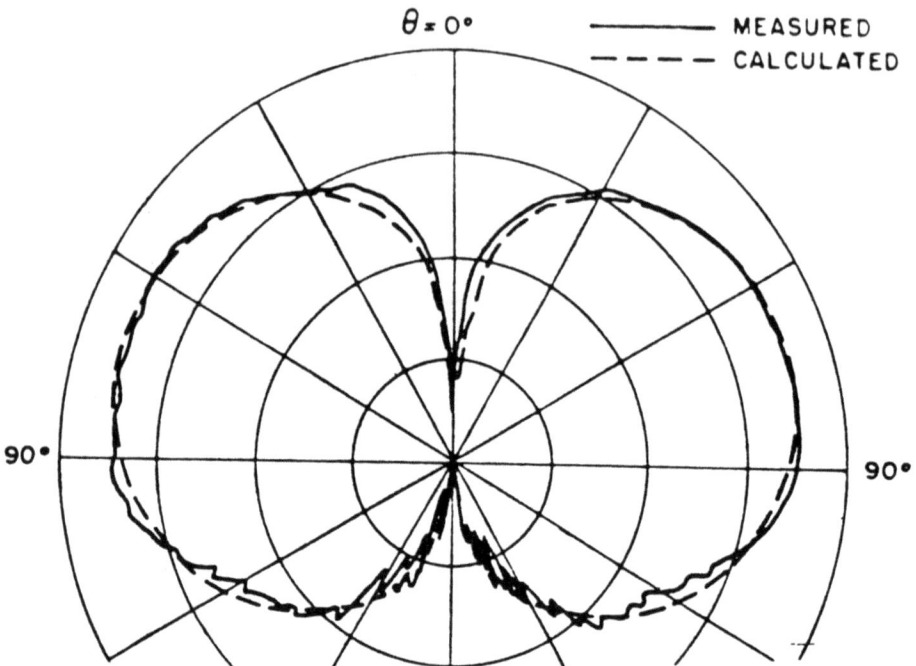

Fig. 29a - Roll plane pattern of a $\lambda/4$ monopole mounted at station 220 on top of a Boeing 737 aircraft.

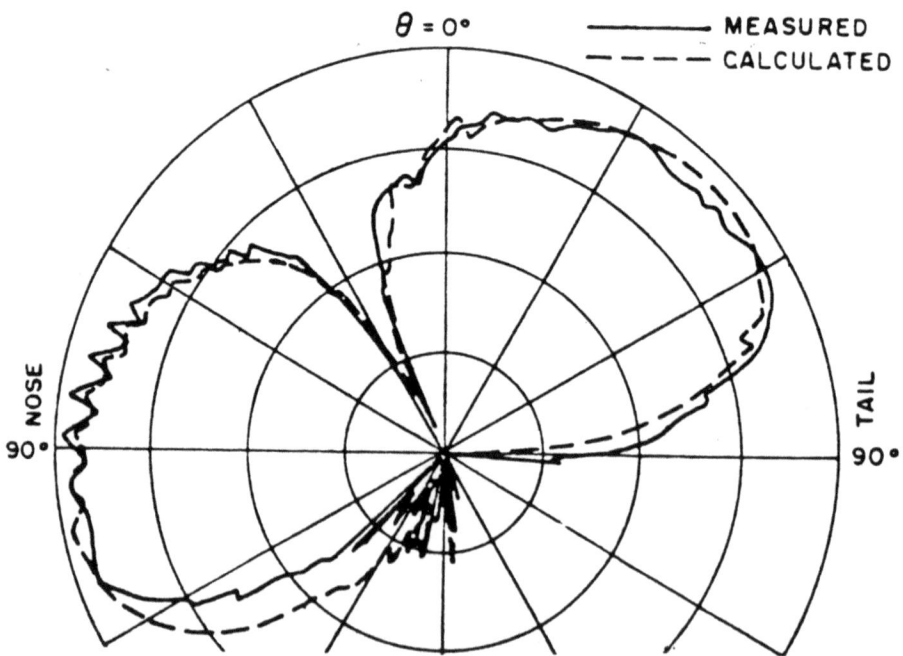

Fig. 29b - Elevation plane pattern of a $\lambda/4$ monopole mounted at station 220 on top of a Boeing 737 aircraft.

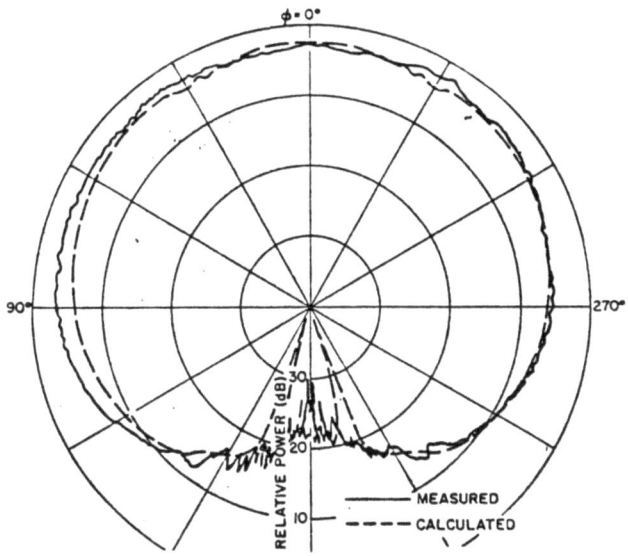

Fig. 29c - Azimuth plane pattern of a $\lambda/4$ monopole mounted at station 220 on top of a Boeing 737 aircraft.

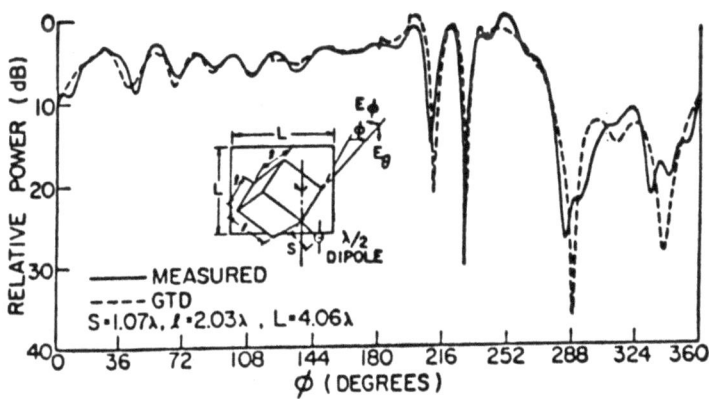

Figure 30. Comparison of measured and calculated E_θ radiation pattern for a dipole near a box on a finite ground plane in the indicated plane.

REFERENCES

[1] J.B. Keller, "The Geometric Optics Theory of Diffraction," presented at the 1953 McGill Symp. Microwave Optics, A.F. Cambridge Res. Cent., Rep. TR-59-118 (II), pp. 207-210. 1959.
[2] J.B. Keller, "A Geometrical Theory of Diffraction," in Calculus of Variations and Its Applications, L.M. Graves, Ed., New York: McGraw-Hill, 1958, pp. 27-52.
[3] J.B. Keller, "Geometrical Theory of Diffraction," J. Opt. Soc. Amer., Vol. 52, pp. 116-130, 1962.
[4] R.G. Kouyoumjian, "The Geometrical Theory of Diffraction and Its Applications" in Numerical and Asymptotic Techniques in Electromagnetics, R. Mittra, Ed., New York, Springer-Verlag, 1975.
[5] J.B. Keller, R.M. Lewis and B.D. Seckler, "Asymptotic Solution of Some Diffraction Problems," Commun. Pure Appl. Math., Vol. 9, pp. 207-256, 1956.
[6] R.G. Kouyoumjian, "Asymptotic High Frequency Methods," Proc. IEEE, Vol. 53, pp. 864-876, Aug. 1965.
[7] I. Kay and J.B. Keller, "Asymptotic Evaluation of the Field at a Caustic," J. Appl. Phys., Vol. 25, pp. 876-883, 1954.
[8] D. Ludwig, "Uniform Asymptotic Expansions at a Caustic," Commun. Pure Appl. Math., Vol. 19, pp. 215-250, 1966.
[9] R.G. Kouyoumjian and P.H. Pathak, "A Uniform Geometrical Theory of Diffraction for an Edge in a Perfectly Conducting Surface," Proc. IEEE, Vol. 62, pp. 1448-1461, 1974.
[10] P.H. Pathak and R.G. Kouyoumjian, "The Dyadic Diffraction Coefficient for a Perfectly Conducting Wedge," Electro-Science Laboratory, Dept. Elec. Eng., Ohio State Univ., Columbus, Ohio, Rep. 2183-4, June 5, 1970., prepared under Contract AF 19(628)-5929 for A.F. Cambridge Res. Labs. (AFCRL-69-0546), also ASTIA Doc. AD 707 827.
[11] L. Kaminetsky and J.B. Keller, "Diffraction Coefficients for Higher Order Edges and Vertices," SIAM J. Appl. Math., Vol. 22, pp. 109-134, 1972.
[12] T.B.A. Senior, "Diffraction Coefficients for a Discontinuity in Curvature," Electron. Lett., Vol. 7, No. 10, pp. 249-250, May 20, 1971.
[13] Y.M. Hwang and R.G. Kouyoumjian, "A Dyadic Diffraction Coefficient for an Electromagnetic Wave Which is Rapidly Varying at an Edge," USNC-URSI 1974 Annual Meeting, Boulder, CO., Oct. 1974.
[14] R.G. Kouyoumjian, Y.M. Hwang and R. Tiberio, "A Uniform Geometrical Theory of Diffraction for an Edge Illuminated by a Field with Rapid Spatial Variation," to appear.
[15] R. Mittra, Y. Rahmat-Samii and W.L. Ko, "Spectral Theory of Diffraction," Appl. Phys., Vol. 10, pp. 1-13, 1976.

[16] Y. Rahmat-Samii and R. Mittra, "On the Investigation of Diffracted Fields at the Shadow Boundaries of Staggered Parallel Plates-a Spectral Domain Approach," Radio Science, Vol. 12, pp. 659-670, 1977.

[17] R. Tiberio and R.G. Kouyoumjian, "A Uniform GTD Solution for the Diffraction by Strips Illuminated at Grazing Incidence," to appear in Radio Science.

[18] R. Tiberio and R.G. Kouyoumjian, "An Analysis of Diffraction at Edges Illuminated by Transition Region Fields," to appear.

[19] C.E. Ryan, Jr. and L. Peters, Jr., "Evaluation of Edge-diffracted Fields Including Equivalent Currents for the Caustic Regions," IEEE Trans. Antennas Propagat., Vol. AP-7, pp. 292-299, 1969.

[20] E.F. Knott and T.B.A. Senior, "Comparison of Three High-Frequency Diffraction Techniques," IEEE Proc., Vol. 62, pp. 1468-1474, 1974.

[21] N.C. Albertsen, P. Balling and N.E. Jensen, "Caustics and Caustic Corrections to the Field Diffracted by a Curved Edge," IEEE Trans., AP-25, pp. 297-303, May 1977.

[22] W.D. Burnside and P.H. Pathak, "A Corner Diffraction Coefficient," to appear.

[23] L.B. Felsen, "Asymptotic Expansion of the Diffracted Wave for a Semi-Infinite Cone," IRE Trans., AP-5, pp. 402-404, 1957.

[24] D.C. Pridmore-Brown, "Diffraction Coefficients for a Slot Excited Conical Antenna," IEEE Trans., AP-20, pp. 40-49, 1972.

[25] J.B. Keller, "Diffraction by a Convex Cylinder," Trans. IRE., Vol. AP-24, pp. 312-321, 1956.

[26] B.R. Levy and J.B. Keller, "Diffraction by a Smooth Object," Comm. Pure Appl. Math., Vol. 12, pp. 159-209, 1959.

[27] P.H. Pathak, "An Asymptotic Analysis of the Scattering of Plane Waves by a Smooth Convex Cylinder," paper to appear in J. Radio Science. (Also The Ohio State Univ. Electro-Science Lab. Tech. Rep. 784583-3, March 1978).

[28] P.H. Pathak, R.J. Marhefka and W.D. Burnside, "High Frequency Scattering by Curved Surfaces," The Ohio State University ElectroScience Lab. Tech. Rep. 3390-5, June 1974.

[29] V.A. Fock, "Diffraction, Refraction and Reflection of Waves: Thirteen Papers," A.F. Cambridge Res. Cent. Rep. AFCRC-TN-57-102, (AD 117 276), 1957. Also V.A. Fock, Electromagnetic Diffraction and Propagation Problems, Pergamon Press, 1965.

[30] P.H. Pathak, W.D. Burnside and R.J. Marhefka, "A Uniform GTD Analysis of the Diffraction of Electromagnetic Waves by a Smooth Convex Surface," submitted for publication to IEEE Trans. Antennas and Propagation.

[31] N.A. Logan, "General Research in Diffraction Theory," Vol. I, LMSD-288087; and Vol. II, LMSD-288-88, Missiles and Space Division, Lockheed Aircraft Corporation, 1959.

[32] J.J. Bowman, T.B.A. Senior and P.L.E. Uslenghi, Eds., Electromagnetic and Acoustic Scattering by Simple Shapes, Amsterdam, The Netherlands: North Holland Publ., 1969.

[33] P.H. Pathak and R.G. Kouyoumjian, "An Analysis of the Radiation from Apertures in Curved Surfaces by the Geometrical Theory of Diffraction," Proc. IEEE, Vol. 62, No. 11, pp. 1409-1447, Nov. 1974.

[34] P.H. Pathak and R.G. Kouyoumjian, "Effects of Torsional Surface Rays on the Radiation from Apertures in Convex Cylindrical Surfaces," USNC/URSI Annual Meeting, Boulder, CO., 1974.

[35] N. Wang, "Near Field Solutions for Antennas on Elliptic Cylinders," Rep. 784684-1, July 1977, The Ohio State Univ. ElectroScience Lab, Dept. Elec. Engr.; prepared under Contract N00019-77-C-0299 for Naval Air Systems Command.

[36] A. Hessel, J. Shmoys and Z.W. Chang, "Surface Ray Analysis of Conformal Arrays," Final Rep. for Phase 2, POLY-EE/EP-75-149, Dept. Elec. Engr. and Electro-physics, PINY, 1975.

[37] S. Safavi-Naini and R. Mittra, "Source Radiation in the Presence of Smooth Convex Bodies," Radio Science, Vol. 14, No. 2, pp. 217-237, Mar-Apr. 1979.

[38] P.H. Pathak, N. Wang, W.D. Burnside and R.G. Kouyoumjian, "A Uniform GTD Solution for the Radiation from Sources on a Perfectly-Conducting Convex Surface," to appear. (Also paper with above title was presented at the 1979 IEEE APS/URSI Meeting in Seattle, Wash., June 18-22, 1979).

[39] G. Hasserjian and A. Ishimaru, "Excitation of a Conducting Cylindrical Surface of Large Radius of Curvature," IRE Trans., Vol. AP-10, pp. 264-273, 1962.

[40] J.R. Wait, "Currents Excited on a Conducting Surface of Large Radius of Curvature," IRE Trans., Vol. MTT-4, No. 3, pp. 143-145, 1956.

[41] Y. Hwang and R.G. Kouyoumjian, "The Mutual Coupling Between Slots on an Arbitrary Convex Cylinder," Rep. 2902-21, March 1975, The Ohio State Univ. ElectroScience Lab., Dept. Elec. Engr.; prepared under Grant NGL 36-008-138 for National Aeronautics and Space Administration.

[42] Z.W. Chang, L.B. Felsen and A. Hessel, "Surface Ray Methods for Mutual Coupling in Conformal Arrays on Cylinder and Conical Surfaces," Polytechnic Institute of New York, Final Report (Sept. 1975-Feb. 1976), 1976; prepared under Contract N00123-76-C-0236. Also see K.K. Chan, L.B. Felsen, A. Hessel and J. Shmoys, "Creeping Waves on a Perfectly-Conducting Cone," IEEE Trans., Vol. AP-26, No. 5, pp. 661-670, Sept. 1970.

[43] S.W. Lee, "Mutual Admittance of Slots on a Cone; Solution by Ray Technique," IEE Trans. on AP-26, No. 6, pp. 768-773, Nov. 1978.

[44] S.W. Lee and S. Naini, "Approximate Asymptotic Solution of Surface Field Due to a Magnetic Dipole on a Cylinder," IEEE Trans., Vol. AP-26, No. 4, pp. 593-597, July 1978.

[45] P.H. Pathak and N. Wang, "Surface Fields of Sources on a Perfectly-Conducting Convex Surface," to appear. (Also see P.H. Pathak and N. Wang, "An Analysis of the Mutual Coupling Between Antennas on a Smooth Convex Surface," Final Rep. 784583-7, Oct. 1978, The Ohio State Univ. ElectroScience Lab., Dept. Elec. Engr.; prepared under Contract N62269-76-C-0554 for Naval Air Development Center, Warminster, PA.

[46] K.E. Golden, G.E. Stewart and D.C. Pridmore-Brown, "Approximation Techniques for the Mutual Admittance of Slot Antennas on Metallic Cones," IEEE Trans. on Antenna and Propaga., Vol. AP-22, pp. 43-48, 1974.

[47] P.H. Pathak and R.G. Kouyoumjian, "On the Diffraction of Edge Excited Surface Rays," paper presented at the 1977 USNC/URSI Meeting held at Stanford University, Stanford, CA., June 22-24, 1977.

[48] T. Ishihara, L.B. Felsen and A. Green, "High Frequency Fields Excited by a Line Source Located on a Perfectly-Conducting Concave Cylindrical Surface," IEEE Trans., AP-26, pp. 757-767, 1978.

[49] C.L. Yu, W.D. Burnside and M.C. Gilreath, "Volumetric Pattern Analysis of Airborne Antennas," IEEE Trans., AP-26, pp. 636-641, Sept. 1978.

[50] W.D. Burnside, N. Wang and E.L. Pelton, "Near Field Pattern Computations for Airborne Antennas," Rep. 784685-4, June 1978, The Ohio State Univ. ElectroScience Lab., Dept. Elec. Engr.; prepared under Contract N00019-77-C-0299 for Naval Air Systems Command.

INTRODUCTION TO GTD APPLICATIONS

H. Bach

Electromagnetics Institute, Technical University of Denmark

Introduction

   The application of the GTD to practical engineering problems is based on the principle of the local field which states that in the high frequency limit processes, such as reflection and diffraction, depend only on the electrical and geometrical properties of the scatterer in the immediate neighbourhood of the point of reflection and diffraction. The location of these points can be determined by the extended Fermat principle which states, that a ray follows a path for which the optical path length attains an extremum value. The application of the GTD to a given radiation problem requires that the given configuration may be decomposed into simple geometrical configurations for which the reflection and diffraction properties are known. All rays contributing significantly to the radiation intensity in the field point must be traced and the individual field contributions must be determined. The total field is then found as the sum of these contributions.
   It is thus clear that the application of the GTD to actual radiation problems in principle is rather straightforward. However, there are several complicating factors. One is that it is usually a laborious task to perform the ray tracing, since the number of rays increases very fast with the geometrical complexity of the scattering body. Another difficulty is that, even for a simple structure, all diffraction coefficients needed are usully not known. Furthermore, regions in space may exist, where the GTD does not describe the field with sufficient accuracy, such that special techniques must be applied. Finally, it may be mentioned that it is in general difficult to estimate the lower frequency limit above which the computed results are still suffi-

ciently accurate.

The fact that the GTD is based on the ray concept implies that the structure of GTD programs differs considerably from that of programs based on Moment Methods and Physical Optics. A typical GTD-program is conviently divided into two parts, namely a ray tracing part and a field computing part. The purpose of the ray tracing part is to provide the geometrical parameters to be used by the field computing part.

Ray tracing

The determination of a ray path is based on Fermat's principle. In its original form Fermat's principle states that a ray from a source point S to a field point F, both located in a medium with refractive index $n(\bar{r})$, is determined as the path of integration for which the integral

$$P = \int_S^F n(\bar{r}) ds$$

attains an extremum value i.e. either a minimum or a maximum. Thus in the general case rays are described as curves in space. Considering homogeneous media, for which $n(\bar{r})$ = constant, we find that the ray path is simply the straight line segment connecting the points S and F.

In GTD many different types of rays are introduced, even in the case where all media involved are homogeneous. From a tracing point of view this means that the integration path of the Fermat integral includes an additional point, which may be a fixed point (tip diffracted rays), a point restricted to some curve (edge diffracted rays) or a point restricted to some surface (reflected or refracted rays). Also the situation occur where a geodesic curve on a surface is part of the ray path (surface diffracted rays). Apart from these simple rays composite rays exist. These rays may undergo several processes such as reflections and diffractions, such that the resulting ray tracing problem may include by simultaneous determination of several points and/or curves. The primary purpose of the ray tracing is to determine such ray paths.

Since rays are described as curves in space they may be represented by parametric equations

$$\bar{r}(t) = (x(t), y(t), z(t)) \quad t_1 < t < t_2$$

where t is a parameter. The arc length s along the curve is then given by

$$s = \int_{t_1}^{t_2} |\bar{r}'(t)| dt$$

When the arc length is used as a parameter

$$\bar{r}(s) = (x(s), y(s), z(s)) \quad s_1 < s < s_2$$

the curve is given by its natural representation.

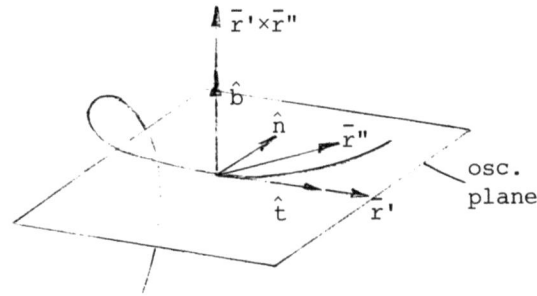

Fig. 1. Curve geometry.

In order to describe local properties of a curve we define, at each point of the curve, a local coordinate system, the moving trihedron. The basis vectors of the moving trihedron are the unit tangent vector $\hat{t}$, the unit normal vector $\hat{n}$ and the unit binormal vector $\hat{b}$. For the above representations we have

$$\hat{t} = \frac{\bar{r}'(t)}{|\bar{r}'(t)|} = \bar{r}'(s)$$

$$\hat{n} = \hat{b} \times \hat{t} = \frac{\bar{r}''(s)}{|\bar{r}''(s)|}$$

$$\hat{b} = \frac{\bar{r}'(t) \times \bar{r}''(t)}{|\bar{r}'(t) \times \bar{r}''(t)|} = \hat{t} \times \hat{n}$$

Each pair of unit vectors defines a coordinate plane of the moving trihedron. The osculating plane is defined by $\hat{t}$ and $\hat{n}$, the normal plane by $\hat{n}$ and $\hat{b}$ and the rectifying plane by $\hat{b}$ and $\hat{t}$.

The curvature κ describes the rate of turning and the torsion τ the rate of twisting of the curve and are given by

$$\kappa = \frac{|\bar{r}'(t) \times \bar{r}''(t)|}{|r'(t)|^3} = |\bar{r}''(s)|$$

$$\tau = \frac{|(\bar{r}'(t) \times \bar{r}''(t)) \cdot \bar{r}'''(t)|}{|\bar{r}'(t) \times \bar{r}''(t)|^2} = \frac{|\bar{r}'(s) \cdot (\bar{r}''(s) \times \bar{r}'''(s))|}{|r''(s)|^2}$$

The Frenet formulas express the relationship between these quantities and the unit vectors of the moving trihedron. Frenet's formulas may be written

$$\frac{d\hat{t}}{ds} = \kappa \hat{n}$$

$$\frac{d\hat{n}}{ds} = -\kappa \hat{t} + \tau \hat{b}$$

$$\frac{d\hat{b}}{ds} = -\tau \hat{n}$$

Since the torsion expresses the twisting of a curve, $\tau = 0$ for a plane curve, implying that $\hat{b}$ is a constant unit vector ortogonal to the plane of the curve, which is then the osculating plane. The radius of curvature $\rho = 1/\kappa$ and the center of curvature lies in the direction of the normal $\hat{n}$, at a distance ρ from the curve. In the immediate neighbourhood of a point of a non-planar curve it behaves like a plane curve. Accordingly the center of curvature also in this case lies on the normal to the curve i.e. in the osculating plane.

It is a remarkable fact that ray tracing problems can be solved analytically only in very few cases. Accordingly almost any GTD-program relies upon numerical procedures for the determination of ray paths. It is obvious that the efficiency of the program therefore depends strongly on the numerical methods used. Furthermore, there is no systematic approach to the ray-tracing problems although a few standard procedures exist. In the following such procedures will be discussed and applied to a few examples.

The second purpose of the ray tracing is to provide geometrical parameters such as path lengths, location of boundaries, transition regions, caustics and foci, radii of curvatures and all other geometrical information necessary for the field computation. The adequate tool for this task is differential geometry and the quantitites may generally be determined analytically once the ray path is known.

The simplest ray tracing problem is for the direct ray which is simply the straight line segment connecting the source point and the field point, such that the path length is the distance between these points. When the field point is at infinity the

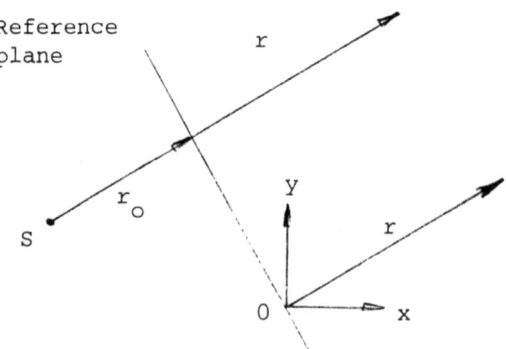

Fig. 2. Ray tracing for far-field point.

field is referenced to an isotropic point source at the origin. Assuming that the direct field is given by

$$\bar{E} = \bar{F}(\theta,\phi) \frac{e^{-jk(r_o+r)}}{r_o+r}$$

the normalized far field pattern is given by

$$\bar{P}(\theta,\phi) = \lim_{r \to \infty} \bar{F}(\theta,\phi) \cdot \frac{r}{r_o + r} e^{-jkr_o}$$

$$= \bar{F}(\theta,\phi) e^{-jkr_o}$$

The distance r_o is the distance from the field point to a orientated plane through the origin and perpendicular to the direction (θ,ϕ) to the field point. This distance is conviently found by the formula

$$r_o = -x_s \sin\theta\cos\phi - y_s \sin\theta\sin\phi - z_s \cos\theta$$

where (x_s, y_s, z_s) are coordinates to the field point. A similar technique obviously can be used for the determination of the far-field contributions of reflected and diffracted fields.

When a diffracting object is present shaded regions in space exist where the body obscures the field source. Thus no direct ray exists between the source point S and a field point F in the shadow region. The shadow region is bounded by the cone defined by the point S and the closed curve of the outer boundary of the object, when viewed from S. When the object is a polyhedron the boundary is a pyramidal surface, the shadow pyramid. A direct ray thus is possible only if the field point is located outside the shadow cone.

A singly reflected ray travels from the source point S to the field point F via a point of reflection R on a reflecting surface. The reflector may be concave, convex or plane and only in the last case the point of reflection can in general be found analytically. Plane and convex reflecting surfaces can support only one reflected ray, while a concave surface may give rise to several reflected and multiply reflected rays.

For any convex surface, the source point and the outer boundary of the body, when viewed from the source point, define a shadow boundary as mentioned above. For a smooth convex body this boundary is also the reflection boundary since a reflected ray passes through any point outside the shadow cone and since no reflected ray can reach a point within the shadow cone. However, if the body possesses edges, the reflection and shadow boundaries do not necessarily coincide. Thus in case the body is a polyhedron, and the source point is not located on a tangent plane, the shadow boundary and the reflected boundary have no points in common except edges. Since the geometrical optics field is composed by the direct and the reflected field, the geometrical optics field is discontinuous at the shadow and reflected boundaries.

Surfaces in spaces may be represented by the parametric equations

$$\bar{r}(u,v) = (x(u,v)\hat{x} + y(u,v)\hat{y} + z(u,v)\hat{z})$$

where u and v are parameters. Alternatively, the surface may be given in the form

$$z = f(x,y)$$

Using x and y as parameters we may write this in the parametric form

$$\bar{r}(x,y) = x\hat{x} + y\hat{y} + f(x,y)\hat{z}$$

which is analogous to the first form. At each point the surface has a normal given by

$$\bar{N} = \bar{r}'_u \times \bar{r}'_v = (-f'_x, -f'_y, 1)$$

provided the derivatives exist.

The curvature of surfaces plays an important role in GTD computations. In order to introduce this concept we consider the intersection curve formed by a surface and a plane containing the normal to the surface in one of its points. When the plane is rotated around the normal the curvature of the intersection curve varies between a minimum and a maximum value, which are obtained at orthogonal positions of the plane. The tangent vectors to the intersection curve at these positions define the principal directions $\hat{u}_1$ and $\hat{u}_2$. The corresponding curvatures of the intersection curve are the principal curvatures κ_1 and κ_2 while the principal radii of curvature are the reciprocals of the quantities. The directions of $\hat{u}_1$ and $\hat{u}_2$ may be such chosen that $\hat{u}_1$, $\hat{u}_2$ and $\bar{N}$ form a right-hand rectangular coordinate system. The curvature of the intersection curve, when the plane forms an angle θ with say $\hat{u}_1$ is then given by

$$\kappa = \kappa_1 \cos^2\theta + \kappa_2 \sin^2\theta$$

which is called Euler's formula.

For surfaces with rotational symmetry, which often occur in practical applications the determination of ρ_1 and ρ_2 is particularly simple.

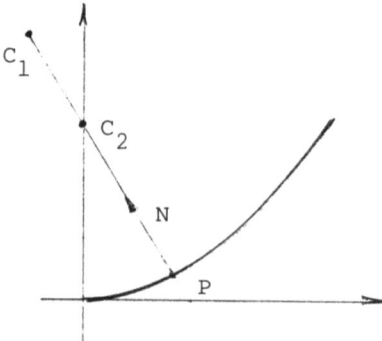

Fig. 3. Curvature of surface of revolution.

In fig. 3 a cut through the axis of symmetry is shown. At the point P, for symmetry reasons, the first principal curvature $\kappa_1 = 1/PC_1$, where PC_1 is the radius of curvature of the median, while $\kappa_2 = 1/PC_2$ where PC_2 is the distance from P to the axis measured along the normal. This is a consequence of Meusnier's formula, which will be dealt with later

We now consider a singly reflected ray travelling from S to F via one reflection point R on the surface. Let the surface be given by $z = f(x,y)$ and assume that S and F have coordinates (x_s, y_s, z_s) and (x_f, y_f, z_f), respectively. The distance from S to F via an arbitrary point $R(x,y,z)$ on the surface is then

$$s = \sqrt{(x-x_s)^2+(y-y_s)^2+(z-z_s)^2} + \sqrt{(x_f-x)^2+(y_f-y)^2+(z_f-z)^2}$$

Fermat's principle states that the point of reflection is determined by the requirement that this distance attains a stationary value. Accordingly we compute

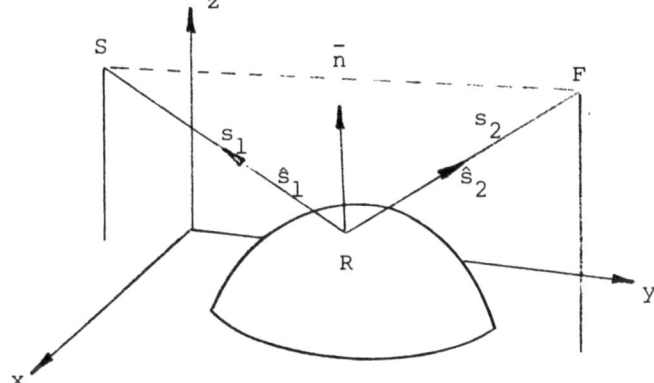

Fig. 4. Reflection by a curved surface.

$$\frac{\partial s}{\partial x} = \frac{x-x_s}{s_1} - \frac{x_f-x}{s_2} + \left(\frac{z-z_s}{s_1} - \frac{z_f-z}{s_2}\right)f'_x$$

$$\frac{\partial s}{\partial y} = \frac{y-y_s}{s_1} - \frac{y_f-y}{s_2} + \left(\frac{z-z_s}{s_1} - \frac{z_f-z}{s_2}\right)f'_y$$
(1)

The point of reflection $R(x,y,z)$ then, in principle, may be determined from the equations

$$z = f(x,y) \ ; \ \frac{\partial s}{\partial x} = \frac{\partial s}{\partial y} = 0$$

Although the above set of equations appears to be simple, they cannot in general be solved analytically except in the case where the reflecting surface is a plane. Among the special cases where an analytical solution is possible is the case of a source point located at the focus of a parabolic reflector. Points of reflection are thus generally determined numerically using the above set of equations. In general this is a time consuming task since a two-dimensional search is involved, but often the search may be accelerated by utilizing other properties of reflection.

We again consider reflection in a surface $z = f(x,y)$. Assuming that the point of reflection has been determined, we compute the vector $\bar{n}$ normal to the surface at the point reflection

$$\bar{n} = (-f'_x, -f'_y, 1) \quad \text{where} \quad f'_x = \frac{\partial z}{\partial x} \quad \text{and} \quad f'_y = \frac{\partial z}{\partial y}$$

The unit vectors $\hat{s}_1$ and $\hat{s}_2$, shown in fig. 4 are given by

$$\hat{s}_1 = \frac{1}{s_1} ((x_s-x)\hat{x} + (y_s-y)\hat{y} + (z_s-z)\hat{z})$$

$$\hat{s}_2 = \frac{1}{s_2} ((x_f-x)\hat{x} + (y_f-y)\hat{y} + (z_f-z)\hat{z})$$

and accordingly we can compute the vectors $\bar{n} \times \hat{s}_1$ and $\bar{n} \times \hat{s}_2$. The first of these vectors is normal to the plane of incidence determined by $\bar{n}$ and S, while the second vector is normal to the plane of reflection determined by $\bar{n}$ and F. Using eqs. (1) it is easily shown that

$$\bar{n} \times \hat{s}_1 + \bar{n} \times \hat{s}_2 = 0$$

which implies that the plane of incidence and the plane of reflection lie in the same plane. Furthermore, it follows that the angle of incidence and the angle of reflection are equal. We may therefore conclude that Snell's law of reflection is valid also for curved surfaces provided the surface at the point of reflection is replaced by its tangent plane. We also note that the normal $\bar{n}$ at the point of reflection intersects the line SF, which means that $\bar{n}$ is lying in the plane of the triangle SRF.

When the reflecting surface is a plane we can assume, without loss of generality, that $z = 0$ is the reflecting plane and that the source point has coordinates $(0,0,z_1)$. We then find the well-known results

$$(x,y,z) = (\frac{x_f s_1}{s_1+s_2}, \frac{y_f s_1}{s_1+s_2}, 0)$$

namely, that the point of reflection may be found by a simple geometrical construction using the image of the source point in the reflecting plane.

When the reflecting surface is a cylinder of arbitrary cross section any normal to the surface is parallel to a plane perpendicular to a generatrix of the cylinder. In this case projection is a useful technique. It is based on the fact that when a plane triangle is projected on a plane parallel to an angular bisector of the triangle then this bisector is projected in a line segment, which is an angular bisector in the projected triangle. Thus the fact that the angle of reflection equals the angle of incidence, which is expressed through the equation $\bar{n} \cdot \hat{s}_1 = \bar{n} \cdot \hat{s}_2$, involves the equation

$$\bar{n}' \cdot \bar{s}'_1 = \bar{n}' \cdot \bar{s}'_2$$

where $\bar{n}'$, $\bar{s}'_1$ and $\bar{s}'_2$ are the projections of $\bar{n}$, $\hat{s}_1$ and $\hat{s}_2$, respectively.

The point R' thus may be found by a one-dimensional search along a circle. Finally, the reflection point R may be found by projection of S and F on the plane determined by R' and the axis of the cylinder using simple image technique.

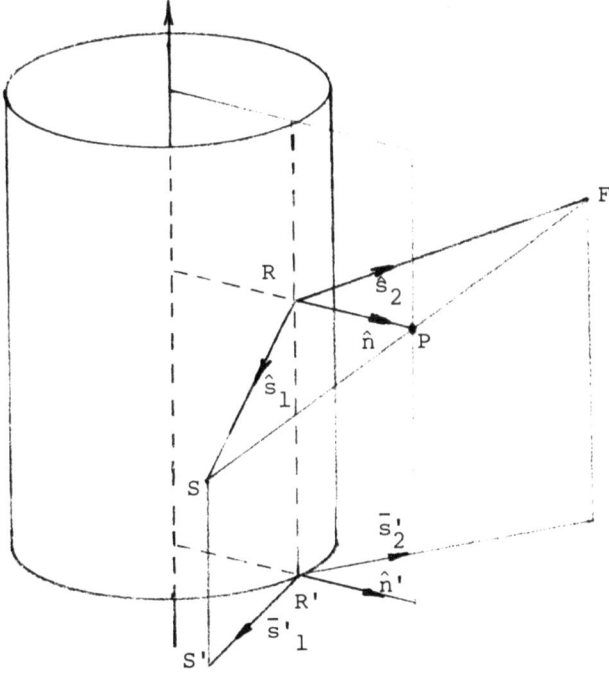

Fig. 5. Reflection by a circular cylinder.

For non-cylindrical surfaces the projection technique cannot in general be used, since no fixed plane parallel to the surface normals exists. In such cases a useful technique is to determine, on the reflecting surface, a curve, which contains the point of reflection. This point then may be determined by a one-dimensional search along the curve. As an example we consider reflection by a circular cone given by

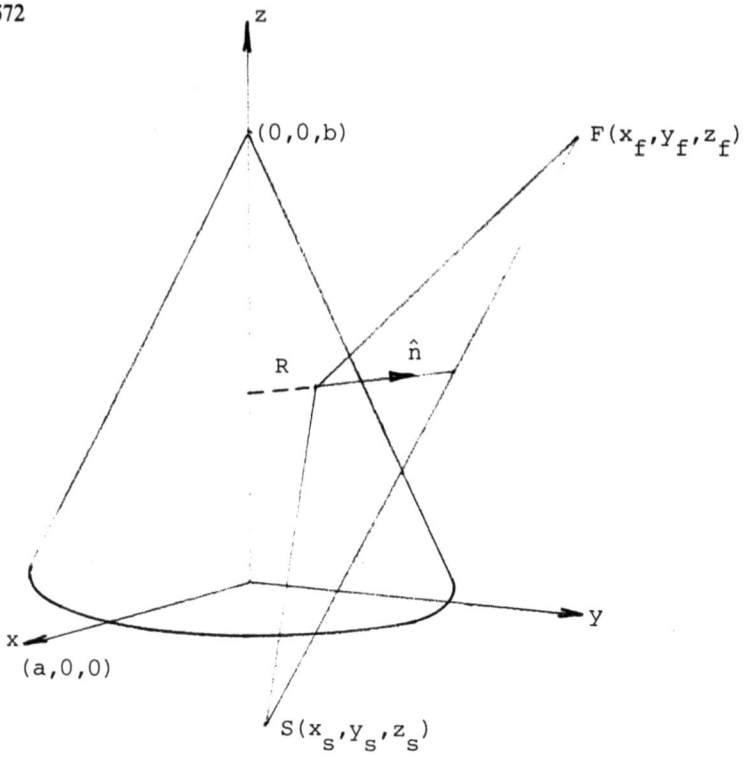

Fig. 6. Reflection by circular cone.

$$\bar{r} = (\hat{x}\rho\cos\phi + \hat{y}\rho\sin\phi + \hat{z}\, b(1 - \frac{\rho}{a})) \ ; \quad z \leq b$$

The line segment SF is given by

$$\bar{r} = \bar{r}_s + t(\bar{r}_f - \bar{r}_s) \qquad\qquad 0 \leq t \leq 1$$

and it is noted, that the normal at the point of reflection intersects the line SF. Now a line normal to the cone through a point P of the line segment is given by

$$\bar{r} = s\hat{n} + \bar{r}_p$$

When the point P traverses the line segment SF the above line generates part of a conoid, which surface obviously contains the normal at the point of reflection. Accordingly, this point must belong to the curve of intersection between the cone and the conoid. The conoid is given by

$$\bar{r} = \bar{r}_s + t(\bar{r}_f - \bar{r}_s) + s\hat{n}$$

and by elimination of the parameters s and t, using the equations for the cone, we find the equations

$$\rho = \frac{ab}{a^2+b^2}\left(b-z_s + \frac{\Delta z(x_s\sin\phi - y_s\cos\phi) + \frac{a}{b}(\Delta x y_s - \Delta y x_s)}{\Delta x \sin\phi - \Delta y \cos\phi}\right)$$

$$z = b\left(1 - \frac{\rho}{a}\right)$$

for the curve of intersection. Here $\Delta x = x_f - x_s$, $\Delta y = y_f - y_s$ and $\Delta z = z_f - z_s$.

Assuming that the reflector is a finite convex surface, the above curve intersects the boundary of the surface or the curve on the surface determining the shadow boundary for the reflected rays. The existence of a reflection point may in such cases be predicted prior to the actual determination of the reflection point [Rusch and Sørensen, 1979]. Defining

$$s = s_1 + s_2 \; ; \quad \bar{s}_1 = \overline{SR} \; ; \quad \bar{s}_2 = \overline{RF}$$

we may compute the gradient of s with respect to the coordinates of R

$$\nabla s = \frac{\bar{s}_1}{s_1} - \frac{\bar{s}_2}{s_2}$$

Since, as is easily shown,

$$\nabla s \cdot (\bar{s}_1 \times \bar{s}_2) = 0$$

the vector ∇s lies in the plane of the triangle SRF. We next consider the tangential component

$$(\nabla s)_t = \nabla s - (\nabla s \cdot \hat{n})\hat{n}$$

of ∇s and note that this vector will be directed away from the reflection point R_o since s is a minimum at R_o. Thus, since only one reflection point can exist on a convex surface, computation of $(\nabla s)_t$ at the end points of the reflection point curve allows an apriori determination of the existence of a reflection point.

Often object dealt with in practical applications are not convex, but concave. In such cases the determination of the ray path still may be based on Fermat's principle, but the function s now is the distance $SR_1R_2 \ldots R_nF$ where $R_1(x_1,y_1,z_1), \ldots, R_n(x_n,y_n,z_n)$ are successive reflection points. In consequence the ray path is determined by solving the equations

$$z = f(x,y), \quad \frac{\partial s}{\partial x_i} = \frac{\partial s}{\partial y_i} \; . \quad i = 1,2,\ldots,n$$

In most cases these equations can be solved only numerically, but often simplifications may be introduced by combinations of the techniques described above.

As an example we consider reflection from a circular cylinder and a plane through the axis of the cylinder. We use the techniques of projection mentioned earlier and project the configuration on a plane perpendicular to the axis of the cylinder and obtain the projection shown in fig. 7. Here S' and F' are projections of the source point and the field point, respectively, and R_1' and R_2' are the projected points of reflection, Now

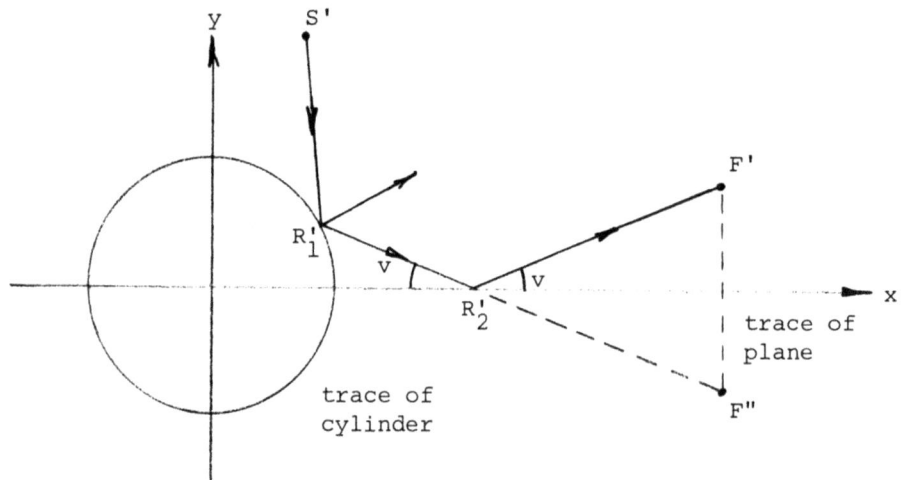

Fig. 7. Double reflection in a cylinder and a plane.

since the angles v are equal, the image point F" may be determined and the reflection point R_1' can be found (numerically) by considering F" as the field point. Once R_1' is found, R_2' may be determined by simple image theory. Finally, the z-coordinates of the points of reflection may be found from a projection on a vertical plane using the same techniques.

An edge diffracted ray travels from the source point S to the field point F via a point of diffraction on an edge. The edge may be a straight edge, a plane but curved edge, or an edge which is not a plane curve. The two most important cases are the straight edge and the circular edge.

We consider an edge given by the parametric equations

$$\bar{r}(t) = (f(t), g(t), h(t)) \qquad t_1 \leq t \leq t_2$$

where t is a parameter. The distance from S to F via an arbitrary point $D(x,y,z)$ on the curve then is given by

$$s = \sqrt{(x-x_s)^2+(y-y_s)^2+(z-z_s)} + \sqrt{(x_f-x)^2+(y_f-y)^2+(z_f-z)} \quad (2)$$

and the point of diffraction consequently may be determined from the equations for the curve and

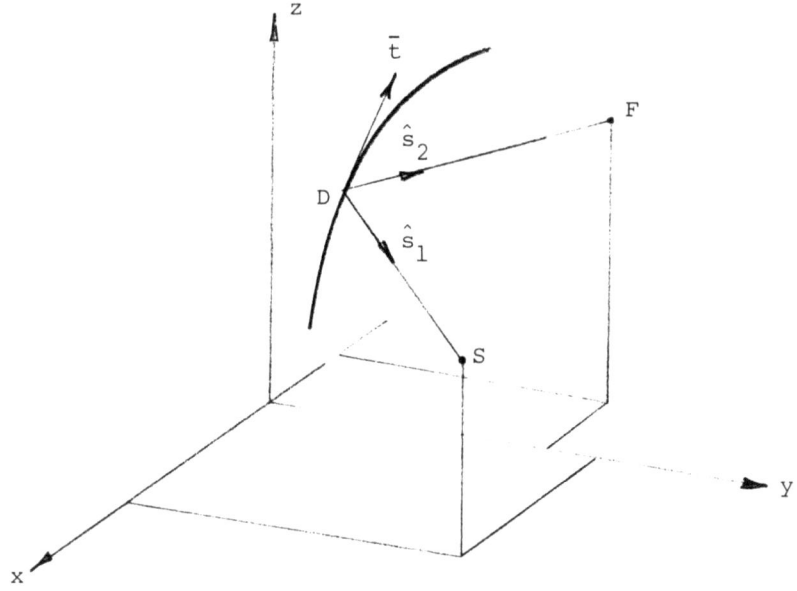

Fig. 8. Diffraction by a curved edge.

$$\frac{ds}{dt} = \frac{1}{s_1}((x-x_s)x' + (y-y_s)y' + (z-z_s)z')$$
$$- \frac{1}{s_2}((x_f-x)x' + (y_f-y)y' + (z_f-z)z') = 0 \qquad (3)$$

where a prime denotes differentiation with respect to t. Now,

$$\hat{s}_1 = \frac{1}{s_1}((x_s-x)\hat{x} + (y_s-y)\hat{y} + (z_s-z)\hat{z})$$
$$\hat{s}_2 = \frac{1}{s_2}((x_f-x)\hat{x} + (y_f-y)\hat{y} + (z_f-z)\hat{z}) \qquad (4)$$

while a vector $\bar{t}$, tangent to the curve, is given by $\bar{t} = (x',y',z')$. We then immediately obtain from eqs. (2) and (3)

$$\hat{s}_1 \cdot \bar{t} + \hat{s}_2 \cdot \bar{t} = 0 \qquad (5)$$

This equation implies that the angles between the tangent vector $\bar{t}$ and the vectors $-\hat{s}_1$ and $\hat{s}_2$ are equal. Furthermore, the plane of incidence, defined by $\bar{t}$ and S, and the plane of diffraction, defined by $\bar{t}$ and F, do not necessarily belong to the same plane. Here we have determined the diffracted ray through the point F, but it is obvious that eq. (5) is satisfied for all field points located on a circular cone with its apex at D, its axis along the tangent vector and with an apex angle β given by $\beta = 2\arccos(\hat{t} \cdot \hat{s}_1)$. This cone in the literature is often referred to as the

the Keller cone and the conception is adopted that diffracted rays emanate from the point of diffraction in all directions along the Keller cone.

Again it appears that the equations (3) and (5) can be solved analytically only in simple cases. The most important of these is the straight wedge, where the point of diffraction may be found by simple development technique. In fig. 9 the z-axis is along the straight edge of a wedge, and the source point S and the field point F are given in cylindrical coordinates (ρ, ϕ,z). Now the planes of incidence and diffraction may be developed into a plane figure by letting the angle $\phi_2-\phi_1$ between

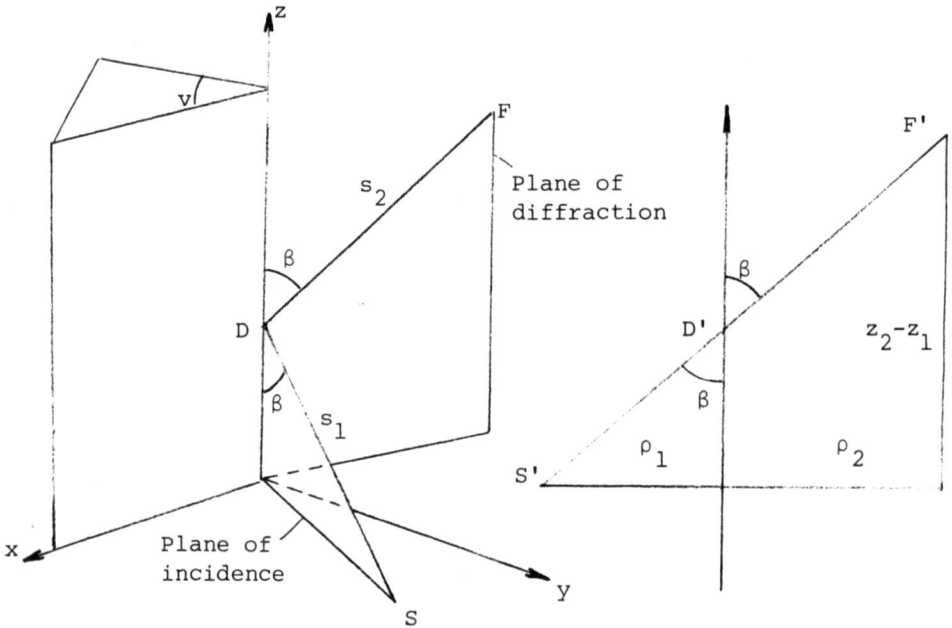

Fig. 9. Diffraction by a straight edge.

these plane go to $180°$. We then obtain the development shown to the right in fig. 9. Since eq. (5) holds also in the development the angles β are equal and the diffracted ray appears as a straight line in the development. We therefore immediately find the expressions

$$(x,y,z) = (0,0, \frac{\rho_1(z_f-z_s)}{\rho_2+\rho_1} + z_s)$$

for the coordinates of the point diffraction.

We next consider another example of large applicational importance, namely the ray tracing for the end of a circular cylinder as shown in fig. 10. Introducing the parametric equations

$$\bar{r} = (a\cos\phi, a\sin\phi, 0)$$

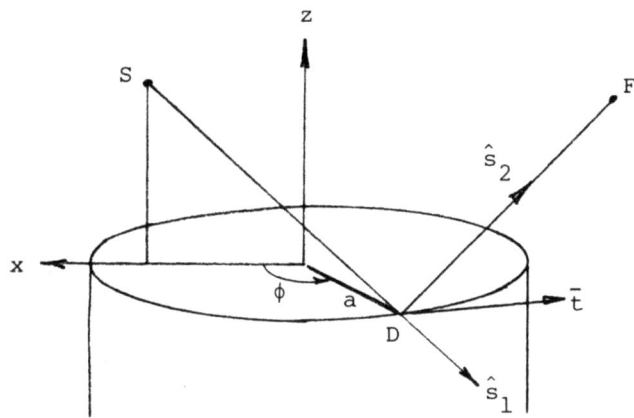

Fig. 10. Diffraction by a circular edge.

and using eq. (5), an equation in $\cos\phi$ for the determination of the point of diffraction D can be obtained. The derivation of this expression is left to the reader. It turns out that the equation is a sixth degree equation in $\cos\phi$ which cannot be solved in general. In Molinet [1973] this problem has been carefully investigated under the assumption that F is a far field point. Although the equation cannot be solved analytically, it has been shown that in general there exists only 1, 2, 3 or 4 distinct solutions to the equation. It is worthwhile to mention that when only two points of diffraction exist, one corresponds to a minimum optical path length and the other one to a maximum optical path pength. When two additional points of diffraction appear, one of the first points must change from a minimum point to a maximum point (or vice versa) since the total optical path via the edge must be a continuous function of ϕ. When both the source point and the field point are on the positive z-axis, the equation is singular and any point on the edge is a stationary point.

Since the problem of finding diffraction points on an arbitraryly curved edge may be reduced to solving an equation with only one unknown, namely a parameter along the curve, it is well-suited for numerical solution. The main problem then is to ensure that all statinary points are found, which often necessitates the computation of a test function, eq. (2) or eq. (5), in a number of closely spaced points.

A surface ray travels along a curved surface. Thus a ray of this type is curved in contrast to the rays considered so far. Now it can be shown, that between two points on a surface, within a suitable bounded region of the surface, only one curve along the surface exists, which minimizes the path length between the two points. This particular curve is called a geodesic and, accordingly, the path of a surface ray between two points A and B is the geodesic connecting A and B. In the applications, however, the problem is normally not to trace the ray between known points A and B, but rather to trace the total ray from the source

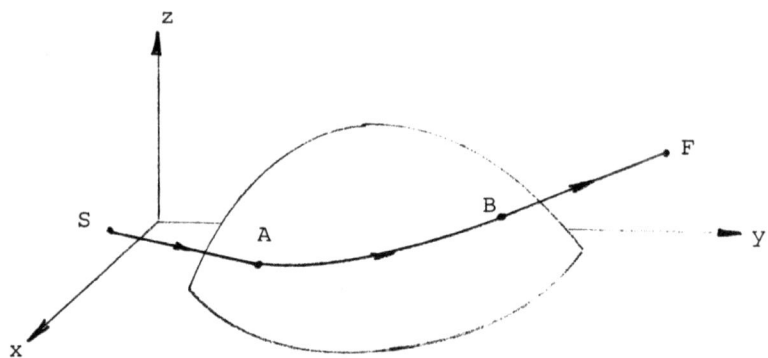

Fig. 11. Path of a surface ray.

point S to the point F via the surface of the body. The ray SA is tangential to the surface at A, and likewise BF is tangential to the surface at B as shown in fig. 11. Since the location of the points A and B is not given in advance, in general, tracing of surface rays is a problem, which is usually solved by means of tensor analysis. Of course, the problem may be solved numerically although this might be a time-consuming task due to the many free parameters. In many cases, however, the surfaces dealt with in practical applications are sufficiently simple that the problem may be solved more efficiently.

Since surface rays are described as curves on a surface they may be represented by the usual parametric equations

$$\bar{r}(s) = (x(s), y(s), z(s))$$

At an arbitrary point the surface normal $\bar{N}$ thus is ortogonal to the tangent vector $\hat{t}$ of the curve. In general the normal unit vector $\hat{n}$ to the curve is not coincident with the unit normal $\hat{v} = \bar{N}/N$ of the surface but forms an angle θ with this normal. We now define a unit vector $\hat{\sigma}$, sometimes called the lateral vector, by

$$\hat{\sigma} = \hat{t} \times \hat{v}$$

Obviously the vector $\kappa\hat{n}$ lies in the plane suspended by $\hat{v}$ and $\hat{\sigma}$ and we may write

$$\kappa \hat{n} = \kappa\cos\theta\hat{\nu} + \kappa\sin\theta\hat{\sigma} = \kappa_\nu \hat{\nu} + \kappa_g \hat{\sigma}$$

where κ_ν is called the normal curvature of the curve and κ_g the geodesic curvature. The formula

$$\kappa_\nu = \kappa\cos\theta$$

is called Meusnier's formula. Since κ_ν is the curvature of the surface in the plane determined by $\hat{t}$ and $\hat{\nu}$, Meusnier's formula relates the curvature of the surface to the curvature of the curve. In the special case of a geodesic curve it can be shown that $\kappa_g = 0$ and $\kappa = \kappa_\nu$. Thus $\hat{n} = \hat{\nu}$, which implies, that a surface ray follows a path where its normal is everywhere parallel to the surface normal and that its curvature is equal to the curvature along the surface.

We first consider the family of surfaces which are developable. This means that these surfaces can be rolled out on a flat surface without stretching. The class of developable surfaces comprises cylindrical surfaces, conical surfaces and the so-called tangent surfaces. The latter are generated by the tangents to a differentiable curve in space. Now it can be shown that a geodesic on a developable surface appears as a straight line, when the surface is developed. Thus the end points A and B can be simply determined from the development and transferred back to the actual surface. As an example we consider the circular cylinder $x^2 + y^2 = a^2$ assuming that the source point S and the field point F have coordinates (ρ_s, ϕ_s, z_s) and (ρ_f, ϕ_f, z_f), respectively. The entire surface, consisting of the tangent planes through S and F and the curved surface between A and B, is developable as shown in fig. 13. Assuming that A and B have coordi-

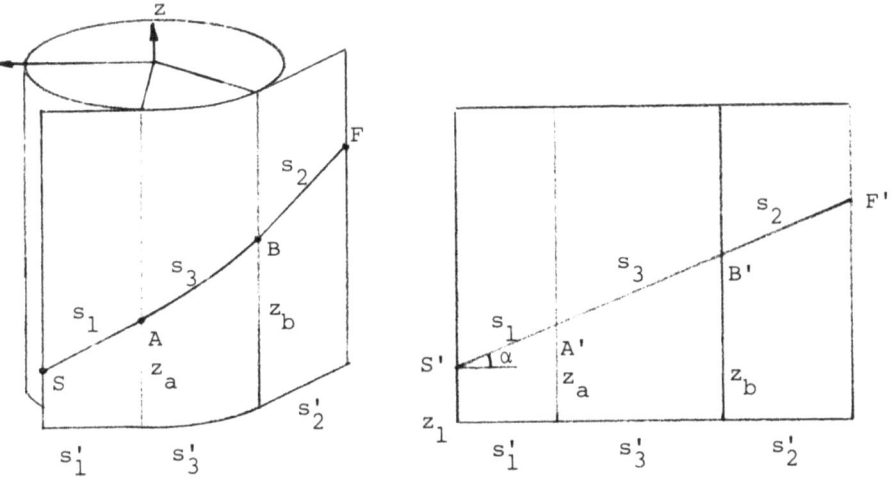

Fig. 13. Surface ray on circular cylinder.

nates (ρ_a, ϕ_a, z_a) and (ρ_b, ϕ_b, z_b), respectively, we have

$$s_1' = \sqrt{\rho_s^2 - a^2} \; ; \qquad s_2' = \sqrt{\rho_f^2 - a^2}$$

$$s_3' = a(\phi_b - \phi_a) = (\phi_f - \phi_s - \arccos\frac{a}{\rho_s} - \arccos\frac{a}{\rho_f})$$

We then immediately find

$$z_a = z_s + \frac{s_1'(z_f - z_s)}{s_1' + s_2' + s_3'} \qquad z_b = z_s + \frac{(s_1' + s_3')(z_f - z_s)}{s_1' + s_2' + s_3'}$$

and

$$s_1 = s_1' \sec\alpha$$
$$s_2 = s_2' \sec\alpha \qquad \text{where } \alpha = \arctg\frac{z_f - z_s}{s_1' + s_2' + s_3'}$$
$$s_3 = s_3' \sec\alpha$$

Among the varieties of surfaces which are not developable, the sphere is unique, since the geodesics on a sphere may be determined explicitly. They are simply parts of great circles, which are curves on the surface of the sphere also lying in a plane through the center of the sphere. Thus a ray on a spherical surface is a part of the great circle determined by a plane through S, F and the center of the sphere. The path of the surface ray then is simply that part of this great circle which connects the tangents to the sphere from S and F in the plane through the center. Apart from such special cases tracing of surface rays on non-developable surfaces must be performed numerically using the extended Fermat's principle.

From the preceding sections it may appear that the ray tracing on complex geometrical structures such, as spacecraft, aircraft and reflector antennas, may be a complicated task. However, many of these mechanical constructions are made up from simple geometrical structures for each of which the raytracing is simple. Furthermore, it is often possible to substitute for complex shapes simpler surfaces which resemble the original structure with an accuracy which is sufficient for engineering purposes. For example the fuselage of an aircraft may be substituted by two ellipsoids and the wings by segments of a plane.

First of all the complexity of actual scattering objects results in a large number of rays of different types. Since an accurate computation of the field often requires that composite rays must be included, such ray paths must be determined. As an example consider a cylindrical satellite body with a plane solar cell panel. A surface ray along the cylinder may be diffracted by an edge of the solar cell panel or a ray first diffracted in the circular edge of the body may undergo a new diffraction at an edge of the solar cell panel. Furthermore, the possibility exists that singly as well as doubly diffracted rays are reflected before or after diffraction, and also multiply reflected rays may exist. Thus it is obvious that the number of types of rays as well as the total number of rays increase drastically with the complexity of the scatterer.

Another major difficulty is the fact that on a composite structure parts of its surface may obstruct rays emanating from other parts of the structure. For example, using the idealized model of an aircraft mentioned above, a diffracted ray from the edge of a wing, for certain directions to the field point, may be obstructed by the fuselage and thus not contribute to the field in the direction considered. The ray may be reflected by the obstruction and contribute to the field in a completely different direction.

Field computations

The direct field is usually computed as the free space field from the source disregarding the presence of the diffracting object. It is obvious than an interaction takes place between the antenna and the object, resulting in a deformation of the current distribution existing when the antenna is in free space. However, when the body is in the far field of the antenna the interaction is weak and may be neglected in many engineering applications.

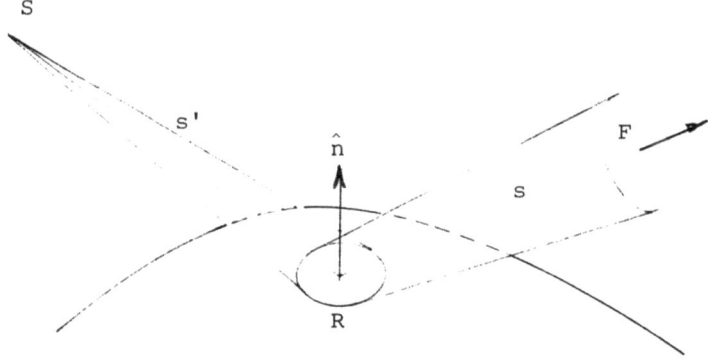

Fig. 14. Reflection in a convex surface.

The reflected fields, together with the direct field, form the geometrical optics field at the field point. In fig. 14 is shown a ray reflected at the point R. The distance SR is called s' and the distance RF is called s. The reflection process is expressed in the equation

$$\bar{E}^r(R) = \bar{\bar{R}} \cdot \bar{E}^i(R) \tag{10}$$

where the reflection coefficient $\overline{\overline{R}}$ is in general 3 × 3 matrix, and $\overline{E}^i(R)$ and $\overline{E}^r(R)$ are the incident and reflected fields at R. The above equation thus relates the incident field and the reflected field at the point of reflection.

The above equation may be simplified by the introduction of ray fixed coordinates as shown in fig. 15. Here $\hat{s}'$ is a unit vector pointing from S to R, and $\hat{s}$ a unit vector pointing from R to F. The unit vectors $\hat{\beta}'$ and $\hat{\beta}$ are associated with the angles β' and β, while

$$\hat{\phi}' = \hat{\beta}' \times \hat{s}' \quad \text{and} \quad \hat{\phi} = \hat{\beta} \times \hat{s}$$

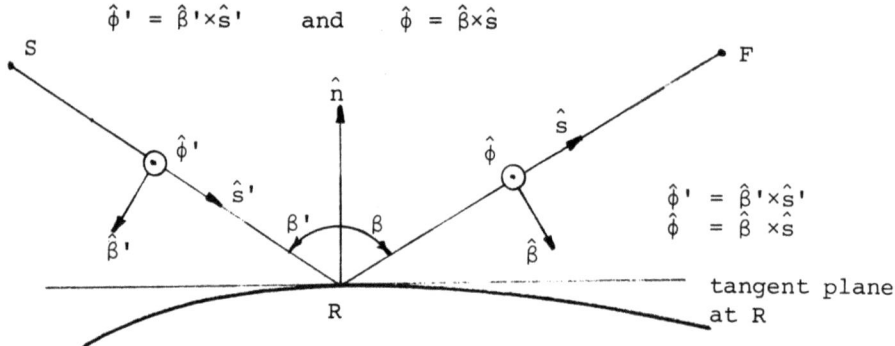

Fig. 15. Ray fixed coordinates for reflection.

The two coordinate systems (s,ϕ,β) are the ray-fixed coordinate systems. Expressing the fields in these coordinates we have

$$\overline{E}^i = E_\beta^i \hat{\beta}' + E_\phi^i \hat{\phi}' , \quad \overline{E}^r = E_\beta^r \hat{\beta} + E_\phi^r \hat{\phi}$$

With this specific choice of coordinates, which is a well-known result from classical electromagnetic theory, $\overline{\overline{R}}$ is given by

$$\overline{\overline{R}} = \begin{Bmatrix} R_p & 0 \\ 0 & R_v \end{Bmatrix} = \begin{Bmatrix} 1 & 0 \\ 0 & -1 \end{Bmatrix}$$

We find that eq. (10) can be replaced by the scalar equations

$$E_\beta^r = R_p E_\beta^i$$
$$E_\phi^r = R_v E_\phi^i$$

It thus results that the field components along the β- and ϕ-axes of the ray-fixed coordinate systems are reflected independently of each other.

We next turn to the problem of relating the reflected field $\bar{E}^r(R)$ at the point of reflection R to the reflected field $\bar{E}^r(F)$ at the field point F. This implies that eq. (10) must be modified to

$$\bar{E}^r(F) = \bar{\bar{R}} \cdot \bar{E}^i(R) \, A_r(s) e^{-jks}$$

where s is the distance from R to F. The divergence factor $A_r(s)$ describes the spatial attenuation, while the phase factor exp(-jks) describes the phase variation of the reflected field along the ray from R to F.

In the following we assume for simplicity that the incident field is a spherical wave. The divergence factor then is given by

$$A_r(s) = \sqrt{\frac{\rho_1^r \rho_2^r}{(\rho_1^r + s)(\rho_2^r + s)}}$$

where reference is made to fig. 16, and where ρ_1^r and ρ_2^r are the principal radii of curvature of the reflected wave front at the point of reflection. The determination of ρ_1^r and ρ_2^r is rather complicated and only the result is given here. Let R_1 and R_2 be

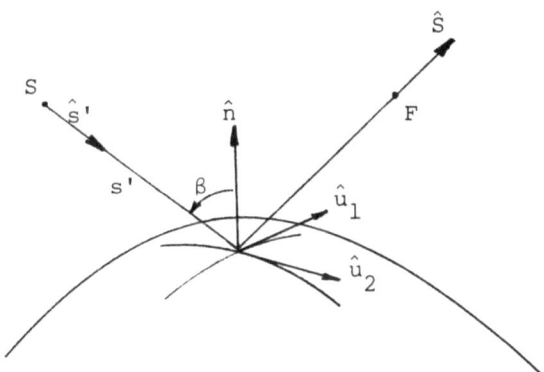

Fig. 16. Computation of wavefront curvature.

the principal radii of curvature of the reflecting surface at the point of reflection, and let $\hat{u}_1$ and $\hat{u}_2$ be the corresponding principal directions. Furthermore, let θ_1 be the angle between $\hat{s}'$ and $\hat{u}_1$ and let θ_2 be the angle between $\hat{s}'$ and $\hat{u}_2$. We then have that the principal radii of curvature of the reflected wave front are given by

$$\frac{1}{\rho_{1,2}^r} = \frac{1}{s'} + \frac{1}{R_o \cos\beta} \pm \sqrt{\frac{1}{R_o^2 \cos^2\beta} - \frac{4}{R_1 R_2}}$$

where

$$\frac{1}{R_o} = \frac{\sin^2\theta_1}{R_2} + \frac{\sin^2\theta_2}{R_1}$$

and β is the angle of incidence. A more general formula is presented in Kouyoumjian and Pathak [1974].

Next we consider diffracted fields emanating from a point of diffraction on an edge. The edge may be a curved edge, but we firstly consider the edge of a straight wedge with planar faces.

In fig. 17 is shown a straight wedge with an interior apex angle v. A rectangular coordinate system, with the z-axis along the edge of the wedge, is introduced. The x-axis lies on the one face of the wedge, and the y-axis is an outward normal to this face. The source point S may then be specified by the circular cylinder coordinates (ρ',ϕ',z') and the field point F by (ρ,ϕ,z). An edge-diffracted ray is assumed to pass from S to F via a point of diffraction D on the edge. The distance SD is denoted by s' and the distance DF by s, while the angles from the positive direction of the z-axis to the line segments DS and DF are called β' and β, respectively.

The diffraction process in D is expressed by the equation

$$\bar{F}^d(D) = \bar{\bar{D}}\,\bar{E}^i(D)$$

where $\bar{E}^i(D)$ is the incident field, $\bar{\bar{D}}$ the diffraction coefficient, and $\bar{F}^d$ an excitation factor. The diffraction coefficient $\bar{\bar{D}}$ is in general 3 × 3 matrix relating the incident field and the excitation factor at the point of diffraction.

We next relate the excitation factor at point D to the diffracted field at the field point F. The diffracted field is given by

$$\bar{E}^d(F) = \bar{F}^d(D)\,A_d(s)\,e^{-jks} \qquad (11)$$

where $A_d(s)$ is a divergence factor and $\exp(-jks)$ the phase factor. For a straight wedge the divergence factor is given by

$$A_d(s) = \sqrt{\frac{\rho}{s(\rho+s)}} \qquad (12)$$

where ρ is the radius of curvature of the incident wave front taken in the plane of incidence. If the incident wave is a spherical wave, $\rho = s'$, while for other shapes of the incident wave front ρ can be computed from the formula

$$\frac{1}{\rho} = \frac{\cos^2\theta}{\rho_1} + \frac{\sin^2\theta}{\rho_2}$$

Here ρ_1 and ρ_2 are the principal radii of curvature of the inci-

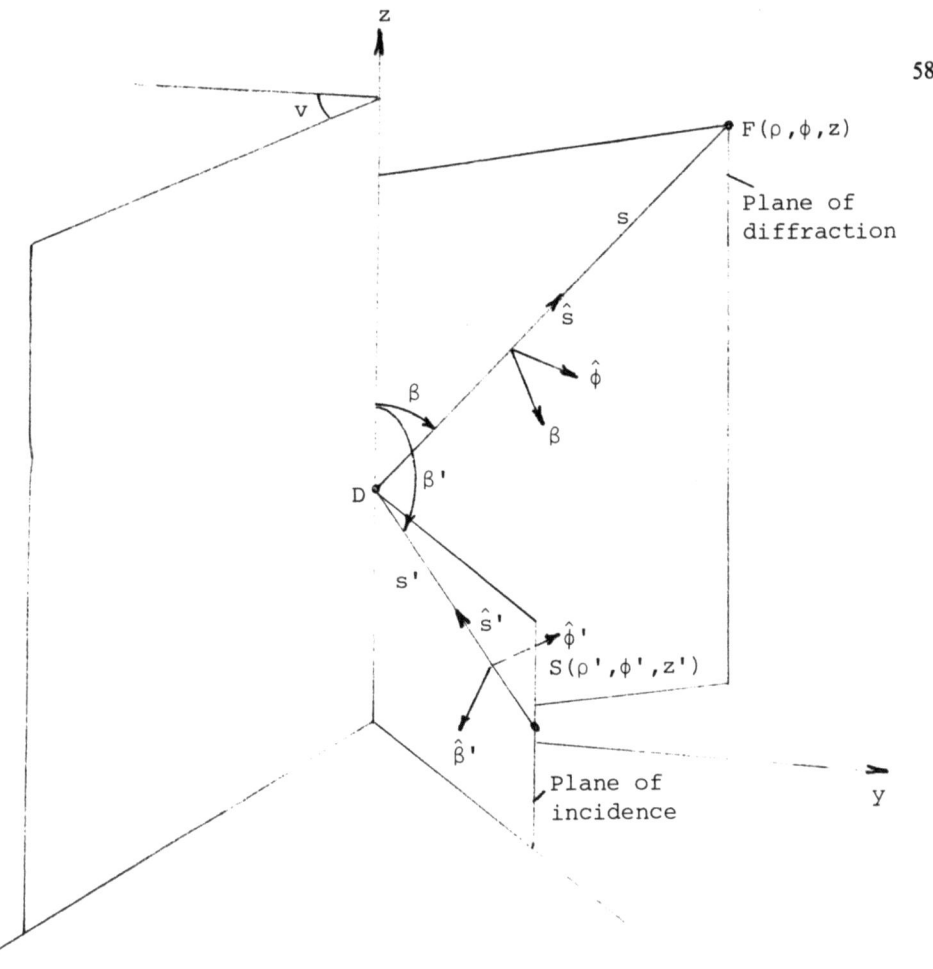

Fig. 17. Wedge geometry.

dent wave front and θ is the angle between the tangent vector to the incident wave front at D associated with ρ_1 and the tangent vector to the incident wave front at D lying in the plane of incidence.

Eq. (11) may be simplified by introduction of ray-fixed coordinates (s,ϕ,β). For the incident ray this coordinate system (s',ϕ',β') is defined by the unit vectors $\hat{s}'$ and $\hat{\beta}'$ associated with s' and β' and the equation $\hat{\phi}' = \hat{\beta}' \times \hat{s}'$. Similarly, for the diffracted ray the coordinate system (s,ϕ,β) is defined by the unit vectors $\hat{s}$ and $\hat{\beta}$ and the equation $\hat{\phi} = \hat{\beta} \times \hat{s}$. In fig. 18 is shown the development of the planes of incidence and diffraction and the ray-fixed coordinate systems. It is seen that this particular choice of coordinate systems implies that these are similarly orientated in the development. In the ray-fixed coordinate systems,

$$\bar{E}^i = E^i_\beta \hat{\beta}' + E^i_\phi \hat{\phi}' , \qquad \bar{E}^d = E^d_\beta \hat{\beta} + E^d_\phi \hat{\phi}$$

and

$$\bar{\bar{D}} = \begin{Bmatrix} D_s & 0 \\ 0 & D_h \end{Bmatrix}$$

where D_s and D_h are scalar diffraction coefficients. Accordingly we find

$$E_\beta^d(r) = D_s \, E_\beta^i \, A_d(s) e^{-jks}, \quad E_\phi^d(F) = D_h \, E_\phi^i \, A_d(s) e^{-jks} \quad (13)$$

We thus have the result that the field components along the β- and ϕ-axes of the ray-fixed coordinate system are diffracted independently of each other.

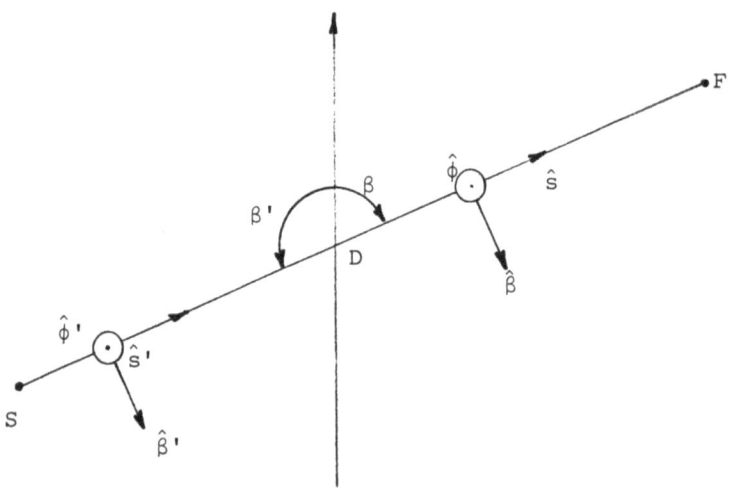

Fig. 18. Development and ray fixed coordinate systems.

The problem of deriving diffraction coefficients for the perfectly conducting wedge has been attacked by several authors. In Keller [1962] are presented the coefficients

$$D_{s,h}(\phi,\phi',\beta') = \frac{e^{-j\frac{\pi}{4}} \sin\frac{\pi}{n}}{n\sqrt{2\pi k} \sin\beta'} \left[\frac{1}{\cos\frac{\pi}{n} - \cos\frac{\beta^-}{n}} \mp \frac{1}{\cos\frac{\pi}{n} - \cos\frac{\beta^+}{n}} \right] \quad (14)$$

where n is given by $v = (2-n)\pi$ and where

$$\beta^- = \phi - \phi' \quad \text{and} \quad \beta^+ = \phi + \phi' \tag{15}$$

These coefficients are referred to as the Keller diffraction coefficients for the straight wedge. It is noted that at the shadow boundary for the direct field $\phi = \pi + \phi'$ and accordingly $\beta^- = \pi$. Similarly, at the shadow boundary for the reflected field $\phi = \pi - \phi'$ and accordingly $\beta^+ = \pi$. It is thus seen that the Keller diffraction coefficients diverge at these boundaries and therefore are useful in practical applications only for field points which are not close to any of the shadow boundaries.

This difficulty has been removed with the diffraction coefficients given by Pathak and Kouyoumjian [1970]. Here the scalar diffraction coefficients D_s and D_h are given by

$$D_{s,h} = \frac{1}{\sin\beta'} ((d^+(\beta^-)F(\kappa a^+(\beta^-)) + d^-(\beta^-)F(\kappa a^-(\beta^-))) \\ \mp (d^+(\beta^+)F(\kappa a^+(\beta^+)) + d^-(\beta^+)F(\kappa a^-(\beta^+)))) \tag{16}$$

Here

$$d^\pm(\beta) = -\frac{e^{-j\frac{\pi}{4}}}{2n\sqrt{2\pi k}} \cot\frac{\pi \pm \beta}{2n} \qquad F(\zeta) = 2j\sqrt{\zeta}\, e^{j\zeta} \int_{|\sqrt{\zeta}|}^{\infty} e^{-j\tau^2}\, d\tau$$

The function $F(\zeta)$ is the transition function, and it appears that for $F = 1$ the formula reduces to Keller's expression. The function a is defined by

$$a^\pm(\beta) = 1 + \cos(\beta - 2nN^\pm\pi); \quad N^\pm = \text{integer value } \left(\frac{\beta \pm \pi}{2n}\right)$$

while $\kappa = kL$ is called the largeness parameter. k is the wave number and L is a quantity dependent on the type of edge illumination, given by

$$L = \begin{cases} s \sin^2\beta' & \text{for plane wave incidence} \\ \dfrac{\rho'\rho}{\rho'+\rho} & \text{for cylindrical wave incidence} \\ \dfrac{s's \sin^2\beta'}{s'+s} & \text{for spherical wave incidence} \end{cases}$$

At a boundary where a d-function diverges, the corresponding F-function is zero, since the factor $a(\beta)$ in that case is zero. The product dF, however, attains a finite limiting value, which is different when the boundary is approached from the two sides. The resulting discontinuity has such a magnitude that the abrupt change of the primary field at the boundary is compensated for. The appearance of the quantity L in the diffraction coefficients implies that these are length-dependent, which is not in accordance with the principle of the local field as postulated by Keller. However, by introducing the largeness parameter, the discontinuities at the boundaries are removed, and theoretical results which are in accordance with results obtained by experiment may be found.

Until now we have considered only the straight wedge. In the general case of a curved wedge, formed by the intersection between two curved surfaces 1 and 2, $\overline{\overline{D}}$ is expressed by

$$D_{s,h} = \frac{1}{\sin\beta'}\left\{d^+(\beta^-)F(kL^ia^+(\beta^-))+d^-(\beta^-)F(kL^ia^-(\beta^-))\right.$$
$$\left.\mp\left[d^+(\beta^+)F(kL_1^r a^+(\beta^+))+d^-(\beta^+)F(kL_2^r a^-(\beta^+))\right]\right\}$$

where β' and the functions $d^{\pm}$, $\beta^{\pm}$, $a^{\pm}$ and F are defined as above. For the straight wedge the quantities L^i, L_1^r and L_2^r are all equal, whereas those for the curved wedge with curved surfaces are given by

$$L^i = \frac{s(\rho_e^i+s)\rho_1^i\rho_2^i}{\rho_e^i(\rho_1^i+s)(\rho_2^i+2)}(1-(\hat{s}'\cdot\hat{t})^2)$$

and

$$L_n^r = \frac{s(\rho_{en}^r+s)\rho_{1n}^r\rho_{2n}^r}{\rho_{en}^r(\rho_{1n}^r+s)(\rho_{2n}^r+s)}(1-(\hat{s}'\cdot\hat{t})^2) \quad n = 1, 2$$

ρ_1^i and ρ_2^i are the principal radii of curvature of the incident wavefront at the point of diffraction and ρ_e^i the corresponding radius of curvature in the plane containing $\hat{s}'$ and $\hat{t}$, a unit vector tangential to the edge at D. The radii of ρ_{1n}^r, ρ_{2n}^r and ρ_{en}^r for the reflected wavefront for surface 1 and 2 may be found by eq. (15) used at the point of diffraction. For a curved edge, the divergence factor is still given by eq. (12) but now with

$$\frac{1}{\rho} = \frac{1}{\rho_e^i} + \frac{1}{f} \; ; \quad \frac{1}{f} = \frac{\hat{n}\cdot(\hat{s}-\hat{s}')}{a^2\sin^2\beta'}$$

where ρ_e^i is the radius of curvature of the incident wavefront taken in the plane of incidence, a the radius of curvature of the edge, and $\hat{n}$ the principal normal to the edge at D. A detailed description of the diffraction by a curved edge is given in Kouyoumjian and Pathak [1974].

The diffraction coefficients are uniform in the sense that they yield fields which are continuous also in the transition regions. Special precautions must be taken in the case of grazing incidence ($\phi' = 0$ og $n\pi$). At grazing incidence the direct and reflected fields merge and the shadow and reflection boundaries coalesce. Since the incident field in this case is not naturally split into a direct and a reflected field, a factor of 0.5 must be included in eq. (13). In the case of polarization along $\hat{\beta}$, $D_s = 0$ and the diffracted field accordingly is zero. In this case the so-called slope diffraction coefficient must be applied [Kouyoumjian, 1975].

In this section we consider rays diffracted by a convex smooth surface. We restrict ourselves to developable surfaces and consider only the case where neither the source point nor the field point is located at the diffracting surface. In fig. 19 is shown a diffracting object, the source point S and the field point F. The ray connecting S and F is tangential to the surface of the object at points D_1 and D_2. The distance SD_1 is called s', the distance D_1D_2 along the surface s", and the distance D_2F is denoted by s.

The incident field at D_1 is supposed to excite a surface field, which propagates along the surface of the diffracting object. This surface field continuously sheds energy in tangential directions. The field propagating along the tangential ray through D_2 and F thus constitutes the diffracted field in F. Since we assume that the surface is an ideal conductor, the tangential component of $\bar{E}$ is always zero on the surface. On the other hand it is a fact that a field component along $\hat{b}$ is observed at F. It is therefore assumed that the surface field travels at a small, but finite, distance above the surface, such that the tangential component of $\bar{E}$ is zero, while the tangential $\bar{H}$ still can be calculated.

The surface field may be computed on the basis of an excitation function from which the diffracted rays are derived. The excitation function is often called the surface ray field although it is not a physically observable quantity and although it does not have the dimensions of either the electric or magnetic field. The excitation function is composed of infinitely many modes which propagate independetly of each other. We first consider the p'th mode $\bar{a}_p$. The mode may be written

$$\bar{a}_p = a_p^b \hat{b} + a_p^n \hat{n} \tag{20}$$

where $\hat{n}$ and $\hat{b}$ are the normal and the binormal to the ray. During its propagating along the surface the mode looses energy to the diffracted rays; furthermore, the amplitude is attenuated due to divergence along the surface. The attenuation of the surface mode is given by

$$A(s) = A(s_o) \sqrt{\frac{F(s_o)}{F(s)}} e^{-\alpha(s-s_o)}$$

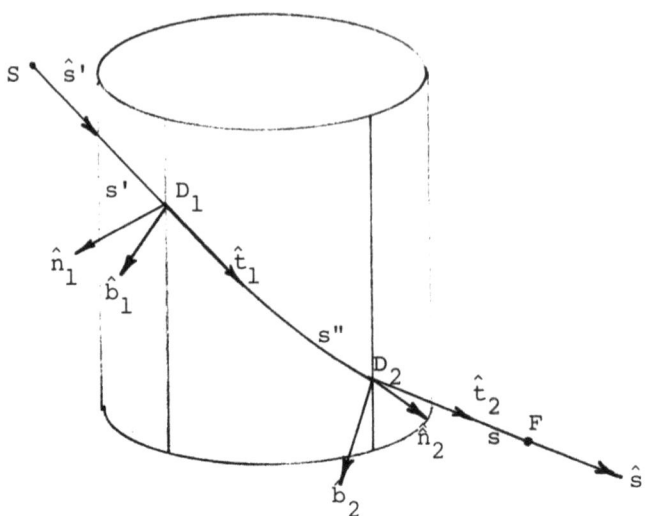

Fig. 19. Ray fixed coordinate systems on a cylinder.

where we have assumed that α is a constant and $s = s_o$ is an arbitrary point of reference. In the general case α is a function of the curvature of the ray and in such cases, the argument of the exponential is an integral along the ray from s_o to s. In the equation the square root may be interpreted as a divergence factor and α as an attenuation constant. Each component of eq. (20) then is of the form

$$a_p(s) = A_p(s_o) e^{j\phi_p^o} \sqrt{\frac{F(s_o)}{F(s)}} e^{-\alpha_p(s-s_o)} e^{-jk(s-s_o)} \qquad (21)$$

where ϕ_p^o is the phase of the amplitude at $s = s_o$. It is noted that the attenuation constants α_p^n and α_p^b for the two components of the excitation function mode are not in general equal to each other. The total excitation function $\bar{F}_s$ thus is given by

$$\bar{F}_s = \sum_{p=1}^{\infty} \bar{a}_p = \sum_{p=1}^{\infty} (a_p^b \hat{b} + a_p^n \hat{n})$$

where a_p^b and a_p^n are given by eq. (21).

We next relate the excitation function to the incident field $\bar{E}^i$ at D_1, where $\bar{E}^i$ is expressed in ray fixed coordinates system (t,n,b). At D_1 we have

$$\bar{a}_p(D_1) = \bar{\bar{D}}_p(D_1) \cdot \bar{E}^i(D_1)$$

Here $\bar{\bar{D}}_p$ is a diffraction coefficient for the p'th mode given by

$$\bar{\bar{D}}_p = \begin{Bmatrix} D_p^s & 0 \\ 0 & D_p^h \end{Bmatrix}.$$

where D_p^s and D_p^h are scalar diffraction coefficients. The value $\bar{a}_p(D_2)$ of the excitation function mode at D_2 now may be found from eq. (21), and the total excitation function at D_2 thus is given by

$$\bar{F}_s(D_2) = \sum_{p=1}^{\infty} \bar{a}_p(D_2) \tag{22}$$

Finally, we want to relate $\bar{F}_s(D_2)$ to the diffracted field in F. The diffracted field has one focal line at D_2 and another one at the distance ρ from D_2. The excitation factor $\bar{C}$ is linearly related to the excitation function (22) through

$$\bar{C} = \sum_{p=1}^{\infty} \bar{\bar{D}}_p(D_2) \cdot \bar{a}_p(D_2)$$

where $\bar{\bar{D}}(D_2)$ is a diffraction coefficient. The diffracted field in F thus is given by

$$\bar{E}^d(F) = \sum_{p=1}^{\infty} \bar{\bar{D}}_p(D_2) \cdot \bar{a}_p(\bar{D}_2) \cdot \sqrt{\frac{\rho}{s(s+\rho)}} \, e^{-jks}$$

where s is the distance D_2F.

As an example we consider diffraction by a circular cylinder in a plane perpendicular to the axis of the cylinder through a point source as shown in fig. 20. We assume that the radius a and the distance s" are large such that only the first mode of the excitation function contributes to the field. Assuming that the field is polarized in the direction of the axis of the cylinder, which is the z-axis, the excitation function is given by

$$\bar{F}_s(D_s) = \hat{z} \, a_1^b(D_1) \sqrt{\frac{s'}{s'+s''}} \, e^{-\alpha_p^b s''} e^{-jks''}$$

Assuming that the incident field is given by

$$\bar{E}^i = E^i(D_1)\hat{z}$$

we find

$$a_1^b(D_1) = D_1^s(D_1) \, E^i(D_1)$$

and the diffracted field in F thus is given by

$$\bar{E}^d(f) = \hat{z} \, E^i(D_1) D_1^s(D_1) \cdot \sqrt{\frac{s'}{s'+s''}} \, e^{-(\alpha_1^b+jk)s''} \cdot D_1^s(D_2) \cdot$$

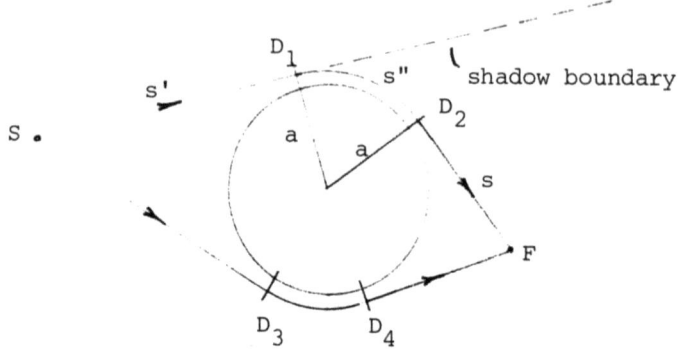

Fig. 20. Diffraction by a circular cylinder.

$$\sqrt{\frac{s'+s''}{s(s+s'+s'')}} \, e^{-jks}$$

$$= \hat{z} \, E^i(D_1)(D_1^2)^2 \cdot \sqrt{\frac{s'}{s(s+s'+s'')}} \, e^{-\alpha_1^b s''} \, e^{-jk)s+s'')}$$

where

$$D_1^s = D_1^s(D_1) = D_1^s(D_2)$$

It is noted that the total diffracted field in F includes also a diffracted field travelling via D_3 and D_4 in fig. 20.

We next present expressions for the attenuation constants and the diffraction coefficients in the case where the curvature of the surface ray is a constant a, in which case the diffraction coefficients at D_1 and D_2 are equal. Diffraction coefficients, which are more general, may be found in Kouyoumjian [1975]. We define x_p and x'_p by

$$Ai(-x_p) = 0, \quad x_p > x_{p-1} \quad \text{and} \quad Ai'(-x'_p) = 0, \quad x'_p > x'_{p-1}$$

The attenuation coefficients then are given by

$$\alpha_p^b = 2^{-\frac{1}{3}} k^{\frac{1}{3}} a^{-\frac{2}{3}} x_p \, e^{j\frac{\pi}{6}}, \quad \alpha_p^n = 2^{-\frac{1}{3}} k^{\frac{1}{3}} a^{-\frac{2}{3}} x'_p \, e^{j\frac{\pi}{6}}$$

while the diffraction coefficients are given by

$$(D_p^b)^2 = \pi \, 2^{-\frac{1}{2}} k^{-\frac{5}{6}} a^{-\frac{1}{6}} a^{\frac{1}{3}} \, e^{-j\frac{\pi}{12}} \cdot Ai'(x_p)^{-2}$$

$$(D_p^n)^2 = \pi \, 2^{-\frac{1}{2}} d^{-\frac{5}{6}} k^{-\frac{1}{6}} a^{\frac{1}{3}} \, e^{-j\frac{\pi}{12}} \, x'_p{}^{-1} \, Ai(x'_p)^{-2}$$

By means of the above expressions the diffracted field in the deep shadow region can be computed, while the field in the lit region is given by the GO-field. However, close to the shadow boundary, in the socalled penumbra region, the present field description fails, and special techniques must be applied. This topic will not be dealt with in the present notes, but interested readers are referred to Kouyoumjian [1975].

Conclusion

As we have seen the application of the GTD to an actual radiation problem in principle is simple. Since the high frequency approximation to the radiated field is determined as a sum of field contributions supplied by a number of rays, the problem is solved in two steps. First, all rays contributing to the field are traced, which is a purely geometrical task. Next, the field contributions corresponding to these rays are summed to yield the total field.

For a simple geometrical structure, as for example the straight wedge, the above procedure is easily carried out. However, in the case of actual configurations, the shapes of which are often of large geometrical complexity, a number of problems arise. The first problem is the ray tracing problem, which often turns out to be solvable only by numerical means. The difficulties increase when multiply diffracted rays are included and also the presence of curved surface and edges complicates the ray tracing. Another complicating effect is the shadowing problem, which arises because parts of the structure shadow rays propagating via other parts. Thus, although the general principles for performing the ray tracing are simple and wellknown, the possibilities of constructing programs which trace the rays for a general geometrical configuration are, with the present state of computer technology, very limited. On the other hand, for a given structure, the ray tracing problem can always be solved, at least numerically.

Assuming that the ray tracing has been performed the next step is the field computation. When the characteristic dimensions of the configuration are large in terms of wavelengths, it is possible, due to the principle of the local field, to decompose the structure into canonical configurations, and base the field computations on the fields diffracted by these simple geometrical structures. Therefore, when the diffraction coefficients for the canonical configuration are known, the field computation is in principle simple. In many practical cases, however, some diffraction coefficients are not known, in which case the associated field contribution cannot be computed. This results in inaccuracies which must be accepted or an approximate solution for example based on the equivalent current concept must be introduced. Another complication is the fact that GTD breaks down in certain regions. These difficulties cannot be remedied within the framework of the GTD but require a special treatment. In some cases general corrections may be worked out, as in caustic directions for example. A number of such corrections have been applied successfully, but in certain cases it appears to be difficult to introduce appropriate corrections. Generally, it can be stated that it is rather easy to obtain the pure GTD-pattern and that it requires a substantial effort to repair the deficiencies of the GTD.

The main advantage of a GTD solution is that it is valid for large scatterers (d $\gg \lambda$) where the Moment Methods often become unapplicable. Furthermore, GTD programs are fast programs and their speed of computation does not depend on size. Thus GTD is the natural tool, when high frequency antennas are placed in complex environments. The visual nature of the theory provides a clear physical insight into the effect of the scatterer upon the performance of the antenna and thus facilitates the design work in complicated practical cases.

List of references

Keller, J.B., "Geometrical theory of diffraction". J.Opt.Soc. Amer., Vol. 52, 116-130, 1962.

Kouyoumjian, R.G. and Pathak, P.H., "A univorm geometrical theory of diffraction for an edge in a perfectly conducting surface". Proc. IEEE, Vol. 62, 1448-1461, 1974.

Kouyoumjian, R.G., "The geometrical theory of diffraction and its application". Chap. 6 in: R. Mittra, Numerical and Asymptotic Techniques in Electromagnetics. Springer Verlag 1975.

Molinet, F. and Saltiel, L., "High frequency radiation pattern prediction for satellite antennas". Laboratoire Central de Telecommunication, Final report ESTEC Contract 1820/72 HP, 1973.

Pathak, P.H. and Kouyoumjian, R.G., "The dyadic diffraction coefficient for a perfectly conducting wedge". ElectroScience Lab., Ohio State Univ., Report No. 2184-4, 1970.

Rusch, W.V.T. and Sørensen, O., "On determining if a specular point exists". IEEE Trans. Antennas Propagation, Vol. AP-27, pp 99-101, Jan. 1979.

GTD APPLIED TO REFLECTOR ANTENNA ANALYSIS

Knud Pontoppidan

TICRA ApS, Kronprinsensgade 13, DK-1114 Copenhagen K, Denmark

1   Introduction

High gain antennas for satellite communication are often constructed as reflector antennas and this is likely to be the case also for future applications. However, in recent years still increasing requirements to the reflector antenna performance have become necessary. Due to the more stringent specifications a careful design is necessary and the availability of a fast and accurate method for the prediction of the radiation characteristics becomes essential.

Methods for the calculation of the scattered field from reflector antennas have been described by a number of authors, and most of these methods are variations of the physical optics (PO) method, such as the "aperture-field" method and the "current-distribution" method. PO methods involve a surface integration for each far-field direction and they are therefore expensive in computer time. The determination of the radiation pattern over the complete far-field sphere is entirely unrealistic in most cases.

The PO method gives good results near the main beam direction of a focusing reflector even with a fairly coarse integration grid. In the side-lobe region away from boresight, however, the accuracy decreases very rapidly. At a certain angular distance from boresight an asymptotic method, such as the geometrical theory of diffraction (GTD) will be superior to PO.

The PO method is applicable also to the subreflector scattered field in a dual reflector configuration. The effectively radiating part of the subreflector surface is a small region around the geometrical optics reflection point. In order to obtain an accurate PO solution a very fine integration grid must be applied and for large subreflectors this represents a very severe demand to computer time as well as core storage. The GTD is a very promising alternative for subreflector calculations. It is orders of magnitude faster and the caustic problems associated with focusing reflectors do not exist in the diverging subreflector fields. The main reflector is usually situated in the near-field region of a subreflector. This means that the PO-determined subreflector far-field pattern must be expanded into spherical modes in order to calculate the incident field on the main reflector. This additional step is not necessary if GTD is applied, since near fields and far fields are calculated with the same computational effort.

From the above it is seen that the PO method and the GTD method are complimentary to one another. PO must be used near the main beam of a focusing reflector but in all other directions and in the entire region around a subreflector GTD may be applied.

In the following sections the applicability of GTD will be demonstrated for a subreflector and for a focus-fed as well as an offset-fed main reflector.

All GTD calculations require three steps to be carried out:

- ray tracing, i.e. determination of the points of reflection and diffraction

- determination of the local geometrical and electrical parameters at the points of reflection and diffraction

- calculation of the field associated with each ray by means of general GTD programs

The solution to the ray tracing problem is illustrated in the following, whereas the calculation of the local parameters and the ray field may be found elsewhere, e.g. Kouyoumjian and Pathak, 1974.

2 Subreflector analysis

Figure 2.1 shows a dual offset reflector system. In order to calculate the incident field on the main reflector the most important ray is the reflected ray from the subreflector surface. This ray yields the geometrical optics approximation which is often used for simpler dual reflector antenna analyses. The direct ray from the feed does not contribute significantly to the main reflector illumination but it may play a role for the noise temperature of the antenna system. The rim of the main reflector is always lo-

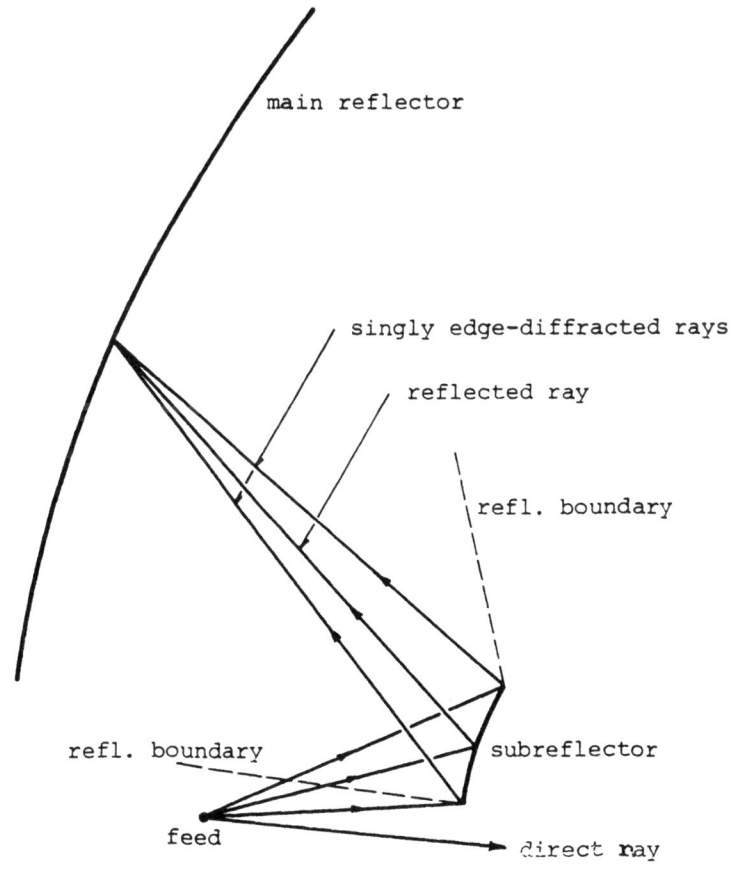

Figure 2.1 GTD rays for subreflector scattered field

cated close to the reflection boundary of the subreflector and this means that also the singly edge-diffracted rays from the subreflector rim must be included. These rays give rise to an additional tapering of the main reflector illumination (compared to the geometrical optics field alone) and they also account for the spill over of the subreflector field past the main reflector surface. All higher order diffracted rays, such as creeping rays along the curved surface and doubly diffracted rays, need not be taken into account. They will radiate primarily into regions of space away from both the main reflector surface itself as well as boresight direction.

2.1 Reflected ray

For the following it is assumed that the subreflector surface is expressed by

$$z = F(x,y) \tag{2.1}$$

which is defined within a circular area with center at O and radius r is illustrated in Figure 2.2.

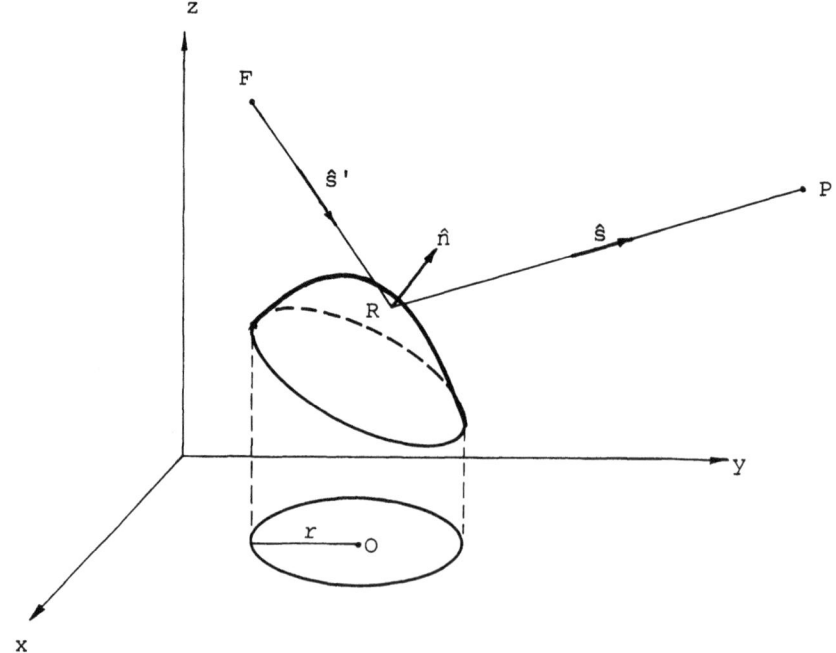

Figure 2.2 Ray tracing of reflection point

The reflection point is, according to Fermat's principle, determined as the minimum for the ray path length d = FR + RP, where F is the feed position, P is a field point and R is a point on the subreflector surface. The direction towards the minimum is

$$-\nabla d = \hat{s} - \hat{s}' \qquad (2.2)$$

where $\hat{s}$ and $\hat{s}'$ are unit vectors in the direction from F to R and from R to P, respectively. The projection of (2.2) on the surface tangent plane is

$$\hat{n} \times (-\nabla d \times \hat{n}) \qquad (2.3)$$

where $\hat{n}$ is the normal vector at R. The projection of (2.3) on the xy-plane is

$$\hat{c} = \hat{z} \times ((\hat{n} \times (-\nabla d \times \hat{n})) \times \hat{z}) \qquad (2.4)$$

and the unit vector $\hat{c}$ therefore represents the direction to the reflection point in the xy-plane.

A general reflection point search program may now proceed as follows. Starting with the center O of the def. area a unit vector $\hat{c}_1$ is calculated by (2.4) and a new approximation O_1 to the reflection point in the xy-plane is given by

$$\vec{OO}_1 = r\alpha\hat{c}_1 \qquad (2.5)$$

where $\tfrac{1}{2} < \alpha < 1$. At O_1 a new direction $\hat{c}_2$ is obtained by (2.4) and a new reflection point given by

$$\vec{OO}_2 = r(\alpha\hat{c}_1 + \alpha^2\hat{c}_2) . \qquad (2.5a)$$

After n iterations the xy-coordinates to the reflection point are determined by

$$\vec{OO}_n = r(\alpha\hat{c}_1 + \alpha^2\hat{c}_2 + \ldots + \alpha^n\hat{c}_n) \qquad (2.6)$$

and the process is stopped either because the reflection point tends to fall outside the definition area or because Snell's law is satisfied to a certain accuracy given by

$$|(\hat{s} - \hat{s}') \times \hat{n}| < \varepsilon \qquad (2.7)$$

where ε is a specified small number.

The total distance from O via O_1, O_2... to O_n is for n large

$$r' = r\frac{\alpha}{1 - \alpha} \quad . \tag{2.8}$$

The value of α should be small in order to obtain rapid convergence. However, the iteration points O, O_1, O_2... do not fall on a straight line in the xy-plane and therefore r' must be considerably larger than r. For practical applications $\alpha = 0.7$ (corresponding to r' = 2.3r) has been found adequate.

If the feed F and field point P are located on either side of the subreflector the above procedure will determine a point on the line connecting F and P. Therefore, the same method can be used to check whether the direct ray from F to P is shadowed by the subreflector surface.

It was assumed that the reflection point is found as a minimum for the ray path length FRP. If, instead, d is maximum at the reflection point, as in a gregorian system, the sign of $\hat{e}$ in (2.4) must be changed.

2.2 Diffracted rays

Assume that the subreflector rim is given by

$$\begin{aligned} x &= f(t) \\ y &= g(t) \end{aligned} \tag{2.9}$$

where $0 \leq t \leq 1$. Expression (2.9) is a projection of the rim on the xy-plane.

The ray tracing for the edge-diffracted rays is done as illustrated in Figure 2.3, where F and P are the position of the feed and the field point, respectively, and P_d is the diffraction point to be determined. $\hat{s}'$ and $\hat{s}$ are unit vectors in the direction from F to P_d and from P_d to P, respectively. $\bar{t}$ is a tangent vector to the edge given by

$$\bar{t} = \left(\frac{dx}{dt}, \frac{dy}{dt}, \frac{dz}{dt}\right) \tag{2.10}$$

where

$$z = F(x(t),y(t)) \qquad (2.11)$$

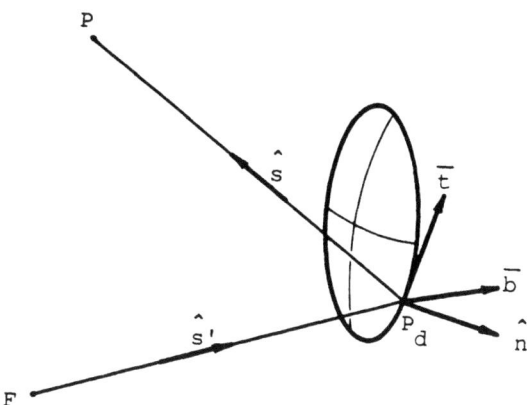

Figure 2.3 Ray tracing of edge-diffracted rays

P_d is a diffraction point if

$$\bar{t} \cdot (\hat{s}' - \hat{s}) = 0 \quad . \qquad (2.12)$$

The left hand side of (2.12) is a function of t and the equation is solved by a standard one-dimensional root finding procedure.

The number of diffraction points on the subreflector edge depends on the actual geometrical structure; it is usually 2 or 4.

The binormal to the edge at P_d is given by

$$\bar{b} = \bar{t} \times \bar{r} \qquad (2.13)$$

where

$$\bar{r} = \left(\frac{d^2 x}{dt^2}, \frac{d^2 y}{dt^2}, \frac{d^2 z}{dt^2} \right) \qquad (2.14)$$

from which the normal at P_d is

$$\bar{n} = \bar{b} \times \bar{t} \qquad (2.15)$$

The curvature of the edge is (Stavroudis, 1972),

$$\kappa = \frac{|\bar{t} \times \bar{r}|}{|\bar{t}|^3} \qquad (2.16)$$

The tangent vector (2.10), the normal vector (2.15) and the curvature (2.16) define, together with the surface function (2.1), the local geometry of the edge at the point of diffraction. The diffracted field in P is determined by an existing GTD program. This program calculates simultaneously the ordinarily diffracted ray (which is proportional to the incident field at P_d) and the slope diffracted ray (which is proportional to the slope of the incident field at P_d).

2.3 Subreflector geometry

The subreflector is completely described by the surface function (2.1) and the edge expression (2.9).

For practical purposes the following types of surfaces must be available:

- analytic surface expression such as hyperboloid, ellipsoid and paraboloid

- surface described by the z-coordinate at the nodes of a regular grid in the xy-plane

- rotational symmetric surface specified by the z-coordinate at intervals along a radius in the xy-plane

and the following types of subreflector edges:

- elliptical edge

- edge defined by a number of points in the xy-plane

The types of surfaces and edges may be arbitrarily combined. For example, a rotational symmetric surface may very well be combined with a non-rotational symmetric edge.

Where surfaces and edges are specified by discrete points, cubic interpolation is used for intermediate function values.

A typical offset subreflector is shown in Figure 2.4. The subreflector surface is a hyperboloid with focal points F_1 and F_2. The distance between the foci is $2c$ and the distance between the vertices of the hyperboloid is $2a$. The rim of the subreflector is defined by a circular cone with apex F_1 and half apex angle θ_c. An xyz-coordinate system is introduced with origin at F_1, as shown.

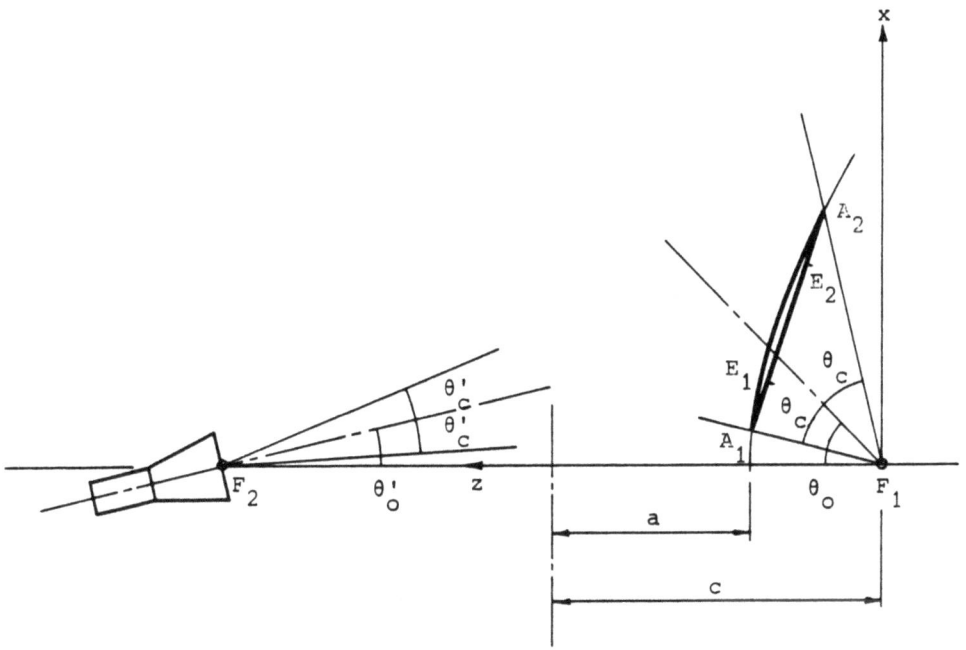

Figure 2.4 Typical offset subreflector

The subreflector edge curve exhibits several interesting geometrical properties which may be summarized as follows:

   a) The edge of the subreflector is plane. It is an ellipse with major axis in the xz-plane, denoted $A_1 A_2$ in Figure 2.4. The focal points of the ellipse are E_1 and E_2.

   b) The edge is, by definition, a circle when seen from F_1, but it also appears as a circle seen from F_2. This makes it possible to define a circular cone in F_2, similar to the one in F_2, with the offset angle θ_o' and the half cone angle θ_c'.

c) A hyperbola may be defined by A_1 and A_2 being the focal points and E_1 and E_2 the vertices. From any point on this hyperbola the subreflector will look as a circle. As a consequence of b) the hyperbola passes through F_1 and F_2. The path lengths from a point on the one hyperbola leg via any point on the subreflector edge to a point on the other leg are all equal. If a feed is located at F_2 this implies, that all points on the hyperbola leg through E_2 are caustic points where a correction to the divergence factor must be applied.

2.4 Computational example

The radiation patterns to be presented in the following are calculated for the specific parameter values of the offset subreflector in Figure 2.4:

$a = 15\lambda$

$c = 25\lambda$

$\theta_o = 45°$

$\theta_c = 31°$

The feed is a conical horn of diameter 5.6λ with a TE11-mode in the aperture and located at F_2 with the offset angle $\theta_o' = 12.7°$. The polarization is in the plane of symmetry. The E- and H-plane radiation patterns are shown in Figure 2.5.

Figure 2.6 shows the individual co-polar ray contributions in the plane of symmetry, $\phi = 0°$. The rays are numbered as shown inserted. Ray 1 is the reflected ray. The rays 2 and 3 (4 and 5) are the ordinarily diffracted ray and the slope diffracted ray, respectively, from the diffraction point at the bottom (top) of the subreflector. The two diffraction points do not move in this pattern plane. It is seen, that at the reflection boundaries $\theta = 14°$ and $\theta = 76°$ the slope diffracted rays are only about 5 dB below the ordinarily diffracted rays, but they decrease more rapidly away from the reflection boundaries. The direction $\theta \simeq 49°$ is a common caustic for all the dif-

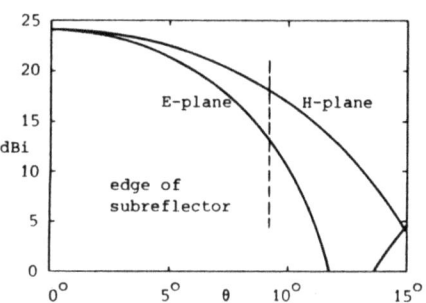

Figure 2.5   Radiation pattern of feed in Figure 2.4

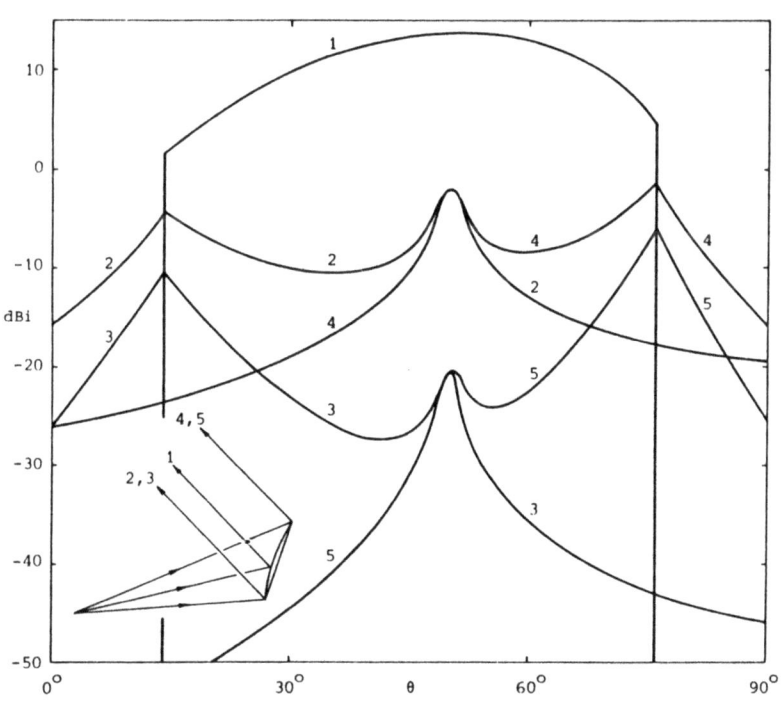

Figure 2.6   Ray contributions in plane of symmetry

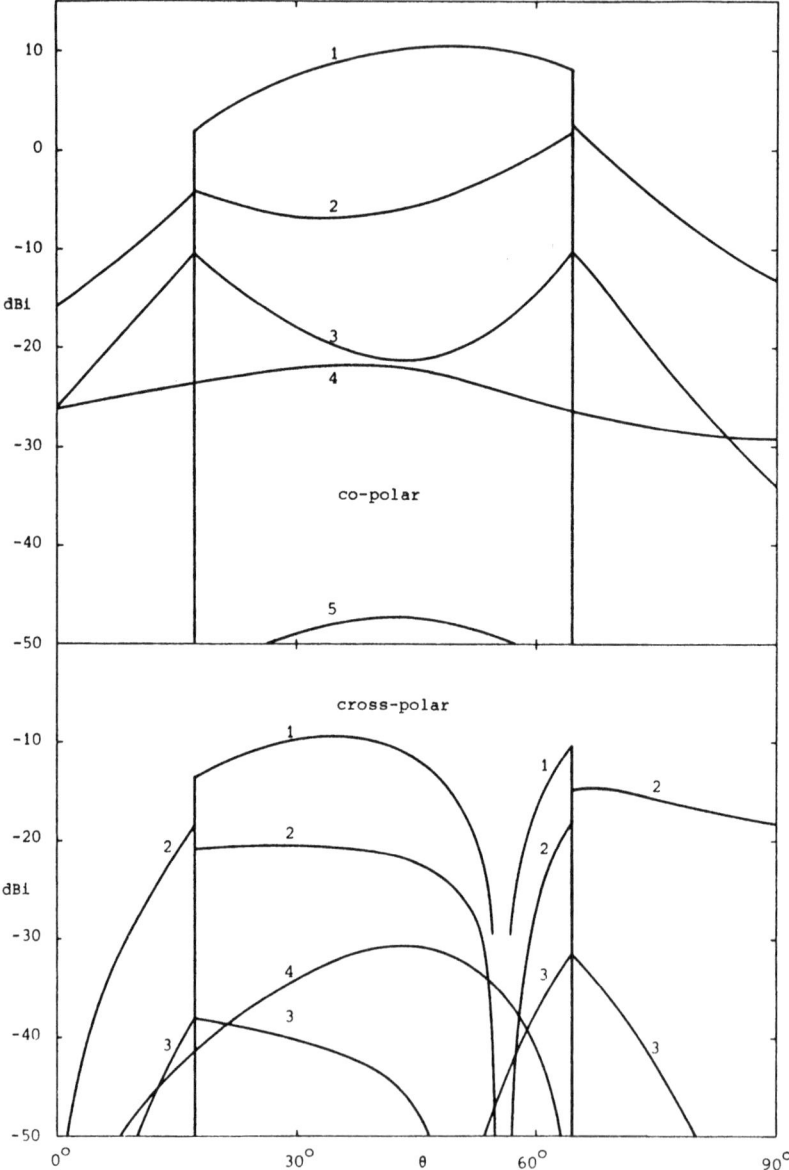

Figure 2.7 Co- and cross-polar ray contributions, $\phi = 30°$

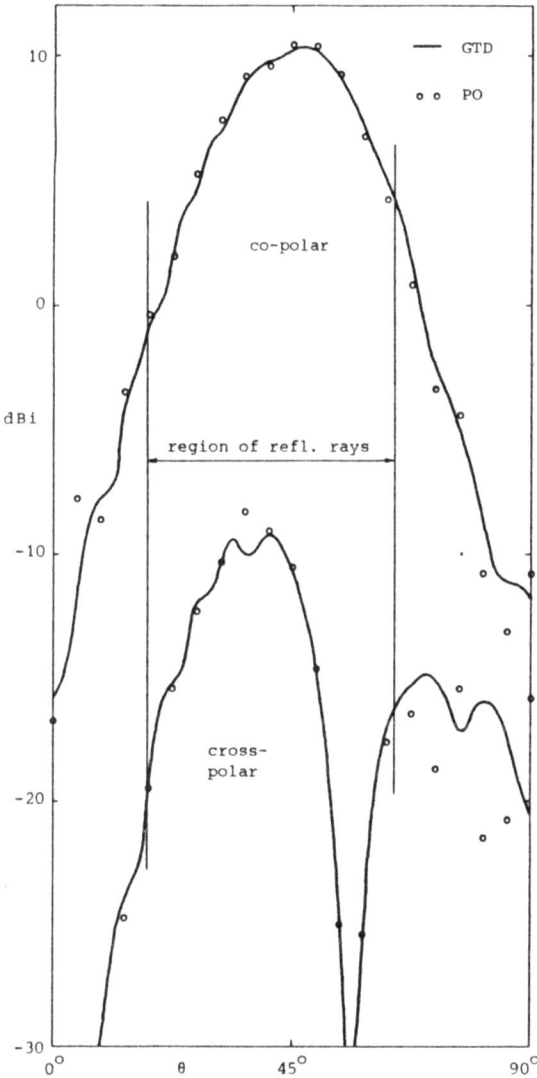

Figure 2.8 Comparison of amplitude between PO and GTD for offset subreflector, $\phi = 30°$

fracted rays and it appears only in the plane of symmetry. In this pattern plane all cross-polar components vanish.

Figure 2.7 shows the individual co- and cross-polar components in the plane $\phi = 30°$. The diffraction points are not stationary in this pattern plane, they move along the edge as the far-field direction varies. Ray 1 is the reflected ray. Ray 2 is an ordinary edge-diffracted ray for which the diffraction point coincides with the reflection point at the first reflection boundary, $\theta = 17.3°$. As the reflection point moves across the surface the diffraction point moves along the edge and they meet again at the other reflection boundary, $\theta = 64.5°$. Ray 3 is the slope diffraction corresponding to ray 2. Rays 4 and 5 are the ordinary and slope diffracted rays, respectively, from a diffraction point located on the opposite side of the subreflector. The cross-polar reflected ray has a minimum near $\theta = 56°$. The reason is that the cross polarization introduced by the tilt of the feed is counterbalanced by the cross-polar component of the feed pattern itself for this particular direction.

The total GTD solution for $\phi = 30°$ is presented in Figure 2.8 and compared to a physical optics result.

3   Focus-fed main reflector

The boresight direction of a focusing reflector is a caustic for both the reflected and for the diffracted rays and a GTD-solution is therefore not feasible near the main beam. The PO solution behaves the other way around, it is very attractive near and on the main beam whereas the accuracy decreases rapidly in the sidelobe region. For a full reflector antenna analysis, a combination of the two methods is necessary.

Previous results have shown that GTD is certainly applicable for focusing reflectors in off-axis directions. The problems that remain to be solved are merely to determine the accuracy of the results obtained and to establish criteria for where GTD should be used and where a PO-solution is more accurate. An answer to these questions is attempted in the following where GTD- as well as PO-results are compared to other solutions.

For focusing reflectors a reflected ray does not exist

except in one direction, the boresight direction. The rays diffracted from the edge of the reflector will exist in all directions and also for these rays the boresight direction is a caustic. The determination of the diffraction point and the field associated with each ray is carried out exactly as for the subreflector edge as described in section 2.2. Doubly diffracted rays and creeping rays are neglected due to their small amplitude.

3.1 Uniform illumination

A parabolic circular symmetric reflector antenna is illuminated by a balanced feed at the focus and with a radiation pattern of the form $1/(1+\cos\theta')$, where θ' is measured from the backward axial direction. This feed will yield an aperture distribution which is uniform in amplitude, direction and phase. The calculated radiation patterns are shown in Figure 3.1. The GTD-solution involves in this case only two diffracted rays from opposite points on the reflector edge. The individual ray contributions are also shown. The PO-pattern is calculated for two different values of the number of patches in the integration grid, $N_p = 97$ and $N_p = 181$. For this particular illumination the radiation pattern is given by (Silver, 1949)

$$A = 2(\pi D/\lambda)^2 J_1(x)/x \qquad (3.1)$$

where D is the diameter of the reflector and

$$x = k\, D/2\, \sin\theta \qquad (3.2)$$

The GTD-solution is poor within the region of the main beam but all side-lobes are accurately predicted. The two PO-solutions are almost identical on the main beam but in the side-lobe region they differ more and more both from each other and from the solution given by (3.1).

Figure 3.1 shows that PO is superior to GTD for the main beam. At the first and the second side lobe the two solutions are almost identical and from the third side lobe GTD is to be preferred.

The radiation pattern of the feed in Figure 3.1 is nearly isotropic and this means that the slope-diffraction contributions to the GTD-result are very small. In order to illustrate the influence of an in-

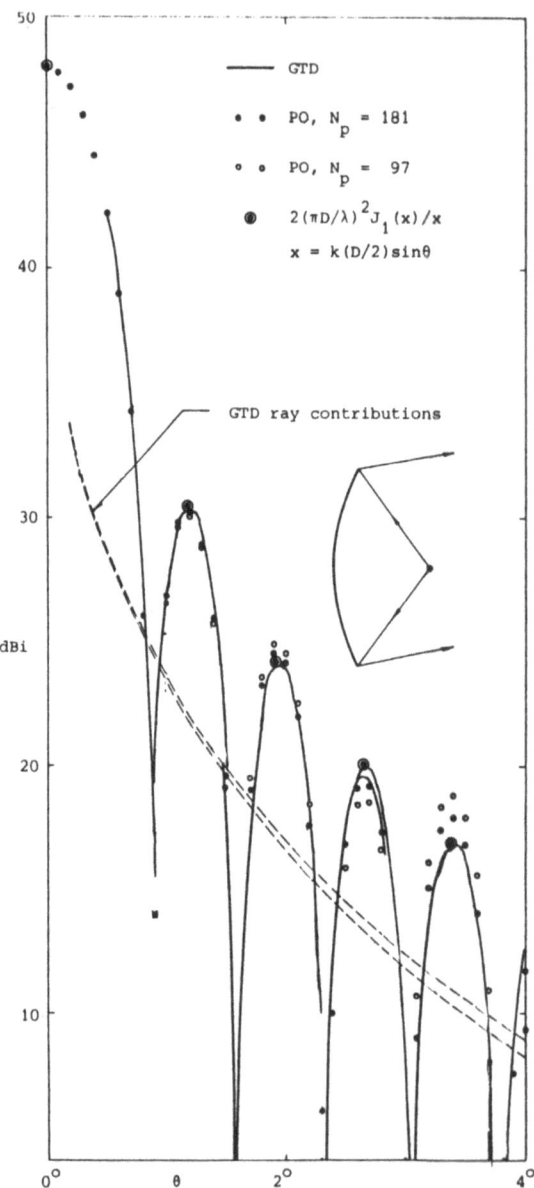

Figure 3.1 Radiation patterns for parabolic reflector with uniform aperture illumination, $D = 80\lambda$, $f/D = 0.5$

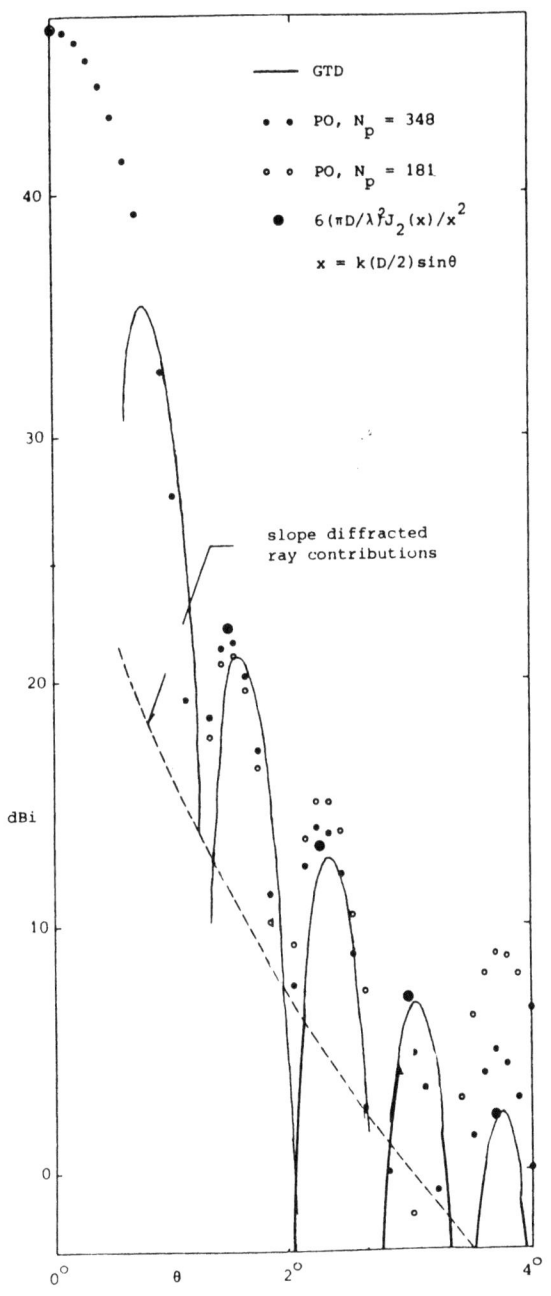

Figure 3.2 Radiation patterns for parabolic reflector with $1-r_n^2$ illumination, $D = 80\lambda$, $f/D = 0.5$

cident field slope the following example is investigated.

3.2 $1-r_n^2$ aperture illumination

The antenna configuration is again a parabolic focus-fed reflector but the illumination is such that the aperture distribution is uniform in direction and phase and the amplitude is $1-r_n^2$, where r_n is the normalized radius in the aperture. This is obtained with a feed pattern of the form

$$A = (1-4(f/D)tg(\theta/2)/(1+\cos\theta) \qquad (3.3)$$

The incident field on the reflector edge vanishes and so do the ordinarily diffracted rays. The slope of the incident field, however, is finite and the GTD-solution is therefore represented alone by two slope-diffracted rays.

The solution corresponding to (3.1) is

$$A = 6(\pi D/\lambda)^2 \, J_2(x)/x^2 \qquad (3.4)$$

The radiation patterns are shown in Figure 3.2. The comments from Figure 3.1 apply almost unchanged also for this highly tapered illumination. The side-lobe amplitudes are considerably lower and, accordingly, the errors are higher. It is interesting to note by comparing Figures 3.1 and 3.2 that the displacement of the side lobes, which is well known for tapered feeds, is accurately predicted by the GTD slope-diffracted rays.

A number of PO-calculations have been carried out for different values of the number of integration patches. The errors for the side lobe maxima are plotted versus θ in Figure 3.3. The figure shows that for $N_p = 181$ the transition point is at $\theta = 1.5$, which is near the first side lobe, and the maximum error for the combined solution is about 1.0 dB. With increased N_p the transition point moves slowly to the right and for $N_p = 5363$ it is near the 4th side lobe and a maximum error of 0.2 dB is obtained.

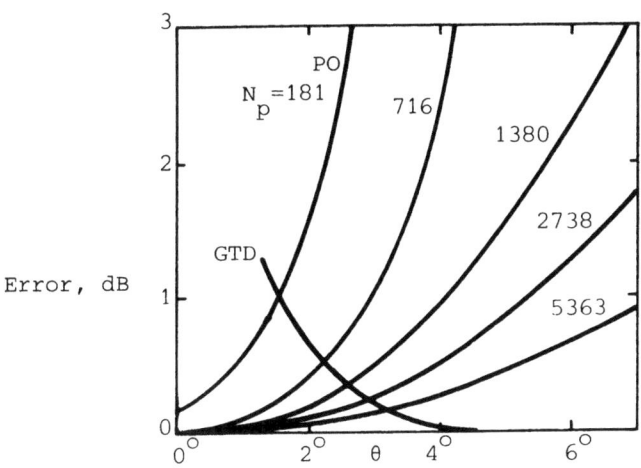

Figure 3.3 Errors for side-lobe maxima for PO and GTD for $1-r_n^2$ illumination

4 Main reflector with offset feed

For many applications, such as multibeam antennas, the feed can not be located at the focal point of the reflecting surface. This implies that reflected rays do exist in a region of far-field directions and they have to be included in a GTD solution.

The availability of an accurate GTD solution is especially important for multiple beam antennas. The prediction of the isolation between different beams requires an accurate knowledge of the individual beam radiation patterns up to ten side lobes away from boresight. This would be very difficult and expensive to accomplish with PO methods alone.

The following sections contain a preliminary study of the results that can be obtained by GTD for reflector antennas with offset feeds. The ray-tracing problems are described in Section 4.1 and the radiation pattern obtained by GTD is compared to an accurate PO solution in Section 4.2 and the agreement is discussed.

4.1 Ray tracing

For the determination of the reflection points R

Snell's law is used:

$$\hat{s} = \hat{s}_i - 2(\hat{n} \cdot \hat{s}_i)\hat{n} \tag{4.1}$$

where $\hat{s}_i$ and $\hat{s}$ are unit vectors in the direction of the incident and the reflected ray, respectively, and $\hat{n}$ is a unit normal vector to the surface at R. In general, the solutions to (4.1) must be found by a two-dimensional root-finding procedure. In many practical applications the surface is part of a paraboloid and in these cases equation (4.1) can be reduced to a one-dimensional expression of the form $f(x) = 0$.

In the following it is assumed that the reflector surface is a paraboloid with focal length f and that the feed is offset the angle v in the direction of the negative x-axis as illustrated in Figure 4.1.

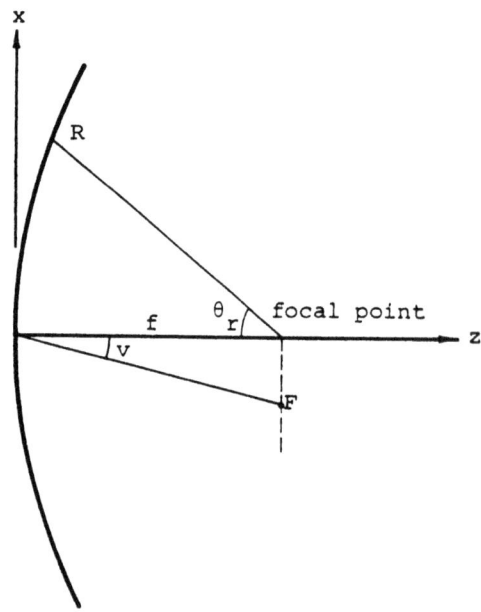

Figure 4.1 Paraboloidal reflector with offset feed

Furthermore, it is assumed that the far-field direction (θ,ϕ) is in the plane of symmetry, i.e. $\sin\phi = 0$. With these simplifying restrictions it is possible to determine the reflection points analytically, still maintaining the characteristic features of an offset feed.

Let a reflection point R be expressed by spherical coordinates (θ_r, ϕ_r) with origin at the focus. Calculating $\hat{n}$ and $\hat{s}_i^r$ and inserting into (4.1) one obtains for the y-component after some straightforward manipulations

$$\sin\phi_r \cos\phi_r = 0 \qquad (4.2)$$

which is satisfied for $\phi_r = p\pi/2$, $p = 0,1,2,3$. Thus the reflections are located in the plane of symmetry and in the plane of asymmetry.

The x-component of (4.1) gives for $\sin\phi_r = 0$

$$\text{tg}(\theta \pm \theta_r) = \pm\text{tg}\theta_r + \text{tg}v/(1-\text{tg}\theta_r^2/2) \qquad (4.3)$$

where the upper and lower sign corresponds to $\phi_r = 0°$ $\phi_r = 180°$, respectively. For $\cos\phi_r = 0$ one obtains

$$\text{tg }\theta = \text{tg}v \cos^2\theta_r/2 . \qquad (4.4)$$

The region of the far-field sphere which is occupied by reflected rays for a reflector with f/D = .5 and v = 6° is illustrated by the shaded area in Figure 4.2. The curve drawn in full line is the trace of the ray reflected from a point moving along the reflector rim. The dotted curve shows far-field directions where reflection points disappear without crossing the reflector edge. If a far-field pattern crosses this dotted curve a large discontinuity will show up. If a far-field pattern crosses the full curve the reflection discontinuities are counterbalanced by edge-diffracted rays.

The diffracted rays are calculated as described for the subreflector in Section 2.2.

4.2 Calculated radiation pattern

The radiation pattern obtained by GTD will be compared to an accurate PO solution. The reflector surface is a rotational symmetric paraboloid with diameter $D = 80\lambda$ (λ = wavelength) and with f/D = 0.5. The feed is offset 6° in the negative x-direction and polarized to the plane of symmetry. The feed pattern is balanced with an amplitude variation as $1/(1+\cos\theta)$. This feed would, if placed at the focus, yield a

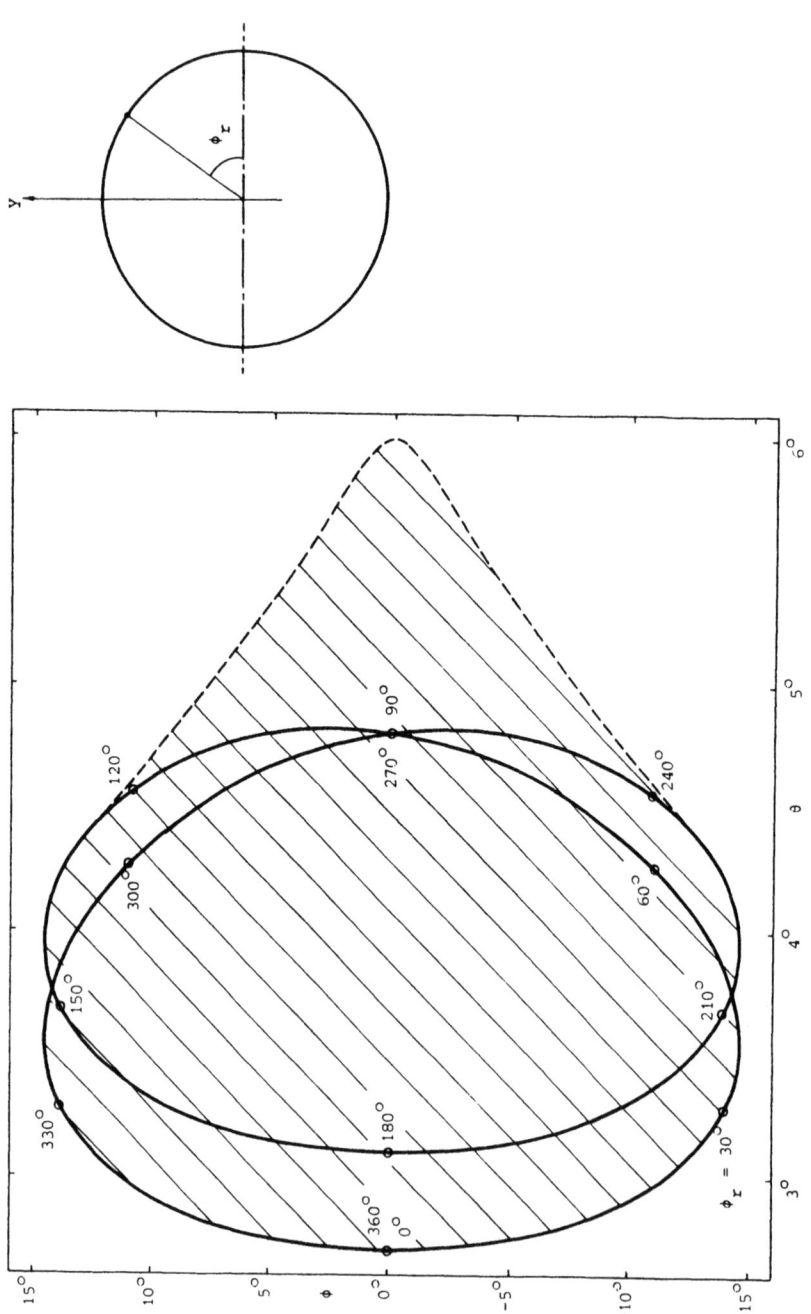

Figure 4.2 Region of reflected rays for circular symmetric paraboloid, f/D = 0.5, offset angle v = 6°

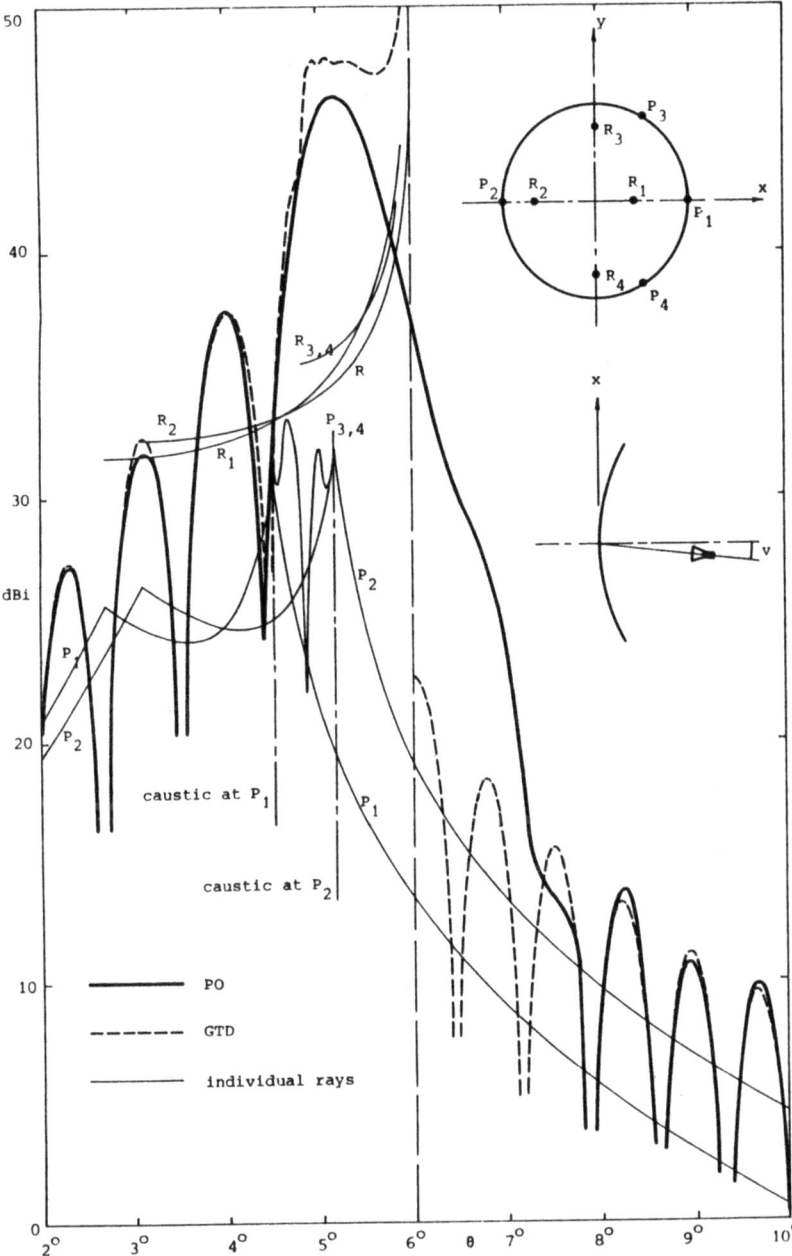

Figure 4.3 Radiation pattern of offset fed paraboloid, $D = 80\lambda$, $f/D = 0.5$, offset angle $v = 6°$

physical optics surface current which is uniform in amplitude, direction and phase. The uniform aperture distribution is of course slightly perturbed by the feed offset.

The focus-fed configuration was investigated previously and the results showed that with 716 surface patches (which will be used in the following), the PO-solution is considerably better than 1 dB up to the 4' side lobe. The accuracy of PO is only slightly affected by the offset at the feed.

Figure 4.3 shows the calculated radiation patterns. It is seeen that in the complete side-lobe region the agreement between PO and GTD is very good. Within the region of the main beam GTD fails and the reason is the caustic of the reflected rays in the specular direction, $\theta = 6°$. It is interesting to note that the coma effect is well described by GTD. The first two side lobes at the left of the main beam are primarily due to interference between the reflected rays from R_1 and R_2. All other side lobes are created by interference between the diffractions from P_1 and P_2.

The result presented shows that the applicability of GTD is very promising also for offset fed reflectors. Furthermore, the ray formulation of the GTD solution gives - apart from being accurate and fast - a clear and simple picture of the mechanisms involved, a property which is very valuable for design work.

5 References

Kouyoumjian, R.G. and Pathak, P.H. (1974)
"A uniform geometrical theory of diffraction for an edge in a perfectly conducting surface", Proc. IEEE, Vol. 62, No. 11, pp. 1448-61.

Silver, S. (1949)
"Microwave antenna theory and design", MIT Radiation Lab. Series, Vol. 12, McGraw-Hill, N.Y.

Stavroudis, O.N. (1972)
"The optics of rays, wavefronts and caustics", New York, Academic Press.

GTD APPLICATIONS AT LCT

Dr. F.A. Molinet

Laboratoire Central de Telecommunications,
78140 Velizy-Villacoublay,
France.

1. INTRODUCTION

Among the GTD applications which have been treated in the last years at LCT, three topics have been selected and will be presented with some details.

The first topic concerns the development of a new Hybrid Method which combines the Moment Method with asymptotic techniques. The principle of this technique is outlined and discussed in the special case of the wedge problem. Numerical results are shown and compared with those obtained by other authors.

The second topic concerns the application of GTD to bodies for which the following techniques are needed : asymptotic development of Physical Optics integrals, diffraction of creeping rays by edges, caustic corrections. The techniques used are briefly outlined and numerical results are presented and compared with measurements.

The third topic is devoted to the presentation of a general computer program for analyzing any kind of antenna mounted on an arbitrary structure consisting of plates, cylinders, spheres, cones. This analysis is accomplished by using GTD. Results obtained for antennas mounted on a satellite are given.

2. COMBINATION OF THE MOMENT METHOD WITH ASYMPTOTIC TECHNIQUES

The limitations of the Method of Moments (MM) and the Geometrical Theory of Diffraction (GTD) can in some cases be circumvented by combining these two methods. There are several ways to associate them depending on the nature of the problem.

In order to clarify the underlying concepts, we consider a scatterer consisting of two separated bodies A and B submitted to an incident field $\vec{E}_i$, $\vec{H}_i$ (Fig. 1).

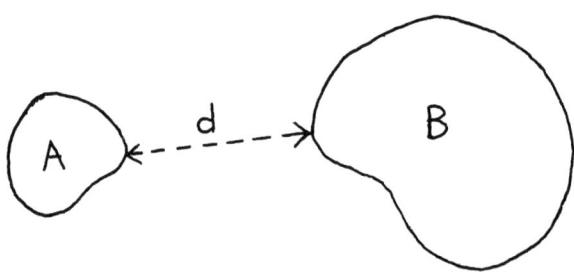

Fig. 1

We suppose that the field scattered by A can be calculated by the Moment Method. The size or the nature of the surface of A however do not permit the application of GTD. On the other hand, we suppose that the field scattered by B can be calculated by GTD, but that the large size of B dismisses the possibility of application of the Moment Method.

If A and B are separated by a distance d satisfying $kd \gtrsim 1$ where k is the wave number, it is possible to combine formally the Method of Moments and GTD in order to calculate the induced currents on the exterior surface of A as well as the diffracted field by the whole system. The method consists in restricting the integral equation satisfied by the currents on A and B to the surface of A and in calculating the interaction between the currents on A and the surface of B as well as the direct and indirect incident field on A by GTD. The resulting integral equation on A is then solved by the Method of Moments.

The application of this procedure requires the knowledge of the diffraction coefficients associated with the shape B.

A variant of this problem which can be solved in the same manner, is obtained by replacing A by a radiating element (wire or slot antenna), the incident field $\vec{E_i}$, $\vec{H_i}$ being absent.

This way to extend the use of the Moment Method by formally incorporating GTD into the generalized impedance matrix was first introduced by THIELE and NEWHOUSE (1) in 1975.

A second technique to combine the Geometrical Theory of Diffraction and Moment Method was published in 1975 by BURNSIDE, YU and MARHEFKA (2). It is a procedure for treating problems for which the required diffraction coefficients are unknown and consists in restricting the integral equation satisfied by currents on a body to the vicinity of the regions of diffraction (MM region). The interaction between the currents of these regions and the remaining surface of the body is evaluated with the aid of GTD in which the diffraction coefficients are treated as unknowns. They are determined by enforcing the integral equation to be satisfied at auxiliary points in the GTD region. The resulting set of simultaneous equations yields the current in the MM region and also the diffraction coefficients so that the current in the GTD region can be obtained.

A limitation of this method is that the form (spatial dependence) of the current away from the diffraction region must be known, which implies that the ray-path of the diffracted rays must be known at every point in the GTD region. For this reason, the method is particularly suited for the treatment of problems involving simple geometries and especially two-dimensional structures (3).

A third technique well adapted to three dimensional problems involving composite shapes formed by an assemblage of plates, has been proposed recently by MOLINET (4). The dominant feature of this approach is that it does not need an a priori knowledge of the asymptotic form of the current away from the diffracting MM region and therefore avoids the tedious problem of searching for diffracted ray-paths.

In order to introduce the principle of the method, a perfectly conducting infinite wedge illuminated by a plane wave is considered. The method is illustrated in the particular situation where only one of the walls of the wedge is illuminated and where the incident magnetic field $\vec{H_i}$ is parallel to the edge of the wedge. Under these conditions, the current $\vec{J}$ on the walls of the wedge satisfies the two-dimensional form of the H-integral equation :

$$\vec{J} = 2\hat{n} \times \vec{H}_i + 2\hat{n} \times \oint_c \vec{J} \times \vec{\nabla}_T' G \; dl' \qquad (1)$$

where

$$G = \frac{1}{4j} H_o^{(2)}\left(k_T |\vec{r}-\vec{r}'|\right)$$

and where the subscript T indicates that the transverse components are to be considered.

The surface of the wedge is divided into two regions as shown on figure 2. :

- The region in the vicinity of the edge called the MM region, defined by the contour C_{MM} which is divided into an illuminated part C_{MM}^e and a shadowed part C_{MM}^o.

- The region away from the region of diffraction called GTD or asymptotic region, defined by the contour C_A which is also divided into an illuminated part C_A^e and a shadowed part C_A^o.

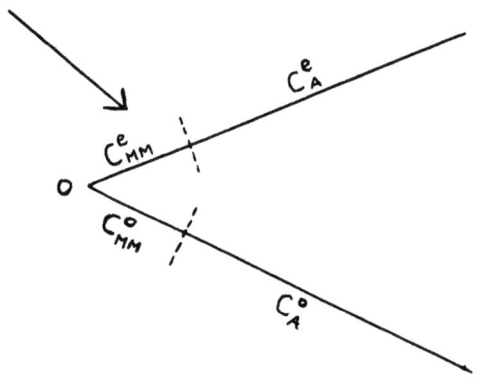

Fig. 2

The other symbols which are used are :

$\vec{J}^e$: current on the illuminated side
$\vec{J}^o$: current on the shadowed side
$\vec{J}_A^e$: current in region C_A^e
$\vec{J}_A^o$: current in region C_A^o
$\vec{J}_{MM}$: current in the MM region

Owing to these notations, the currents on the illuminated and shadowed side are given by :

$$\vec{J}^e = 2\hat{n} \times \vec{H}_i + 2\hat{n} \times \int_{C^o_{MM}} \vec{J}_{MM} \times \vec{V}'_T G \, dl' \qquad (2)$$

$$+ 2\hat{n} \times \int_{C^o_A} \vec{J}^o_A \times \vec{V}'_T G \, dl'$$

$$\vec{J}^o = 2\hat{n} \times \vec{H}_i + 2\hat{n} \times \int_{C^e_{MM}} \vec{J}_{MM} \times \vec{V}'_T G \, dl' \qquad (3)$$

$$+ 2\hat{n} \times \int_{C^e_A} \vec{J}^e_A \times \vec{V}'_T G \, dl'$$

$\vec{J}^e$ as given by equation (2) is the exact expression of the current at an arbitrary point M on the illuminated wall. When M is removed far away from the edge, the second and third terms on the right hand side of equation (2) tend to zero. The asymptotic value $\vec{J}^e_a$ of $\vec{J}^e$ is therefore $\vec{J}^e_a = 2\hat{n} \times \vec{H}_i$

Equation (3) gives the exact expression of the current $\vec{J}$ at an arbitrary point M' on the shadowed wall. When M' is removed far away from the edge, the second term on the right hand side of equation (3) tends to zero. The third term however does not tend to zero for grazing incidence on the shadowed wall and has the asymptotic value :
$$2\hat{n} \times \int_{C^e_A} \vec{J}^e_a \times \vec{V}'_T G \, dl'$$
The asymptotic value of $\vec{J}^o$ is therefore given by :

$$\vec{J}^o_a = 2\hat{n} \times \left(\vec{H}_i + \int_{C^e_A} \vec{J}^e_a \times \vec{V}'_T G \, dl' \right)$$

In (2) and (3), the unknown current $\vec{J}$ can be split into two parts $\vec{J} = \vec{J}_a + \vec{I}$ where $\vec{J}_a$ is the asymptotic value of $\vec{J}$ defined above and where $\vec{I}$ is the new unknown function. For an observation point lying in the MM region, this function satisfies the equations :

$$\vec{I}^o_{MM} = 2\hat{n} \times \int_{C^e_{MM}} \vec{I}^e_{MM} \times \vec{V}'_T G \, dl' + 2\hat{n} \times \int_{C^e_A} \vec{I}^e_{GTD} \times \vec{V}'_T G \, dl' \qquad (4)$$

$$\vec{I}^e_{MM} = 2\hat{n} \times \int_{C^o_{MM}} \vec{I}^o_{MM} \times \vec{V}'_T G \, dl' + 2\hat{n} \times \int_{C^o_A} \vec{J}^o_a \times \vec{V}'_T G \, dl' \qquad (5)$$

$$+ 2\hat{n} \times \int_{C^o_A} \vec{I}^o_{GTD} \times \vec{V}'_T G \, dl'$$

where C^o is the contour corresponding to the illuminated side and where we have stated :

$$\vec{J}^e_{MM} = \vec{J}^e_a + \vec{I}^e_{MM} \qquad \vec{J}^e_A = \vec{J}^e_a + \vec{I}^e_{GTD}$$

$$\vec{J}^o_{MM} = \vec{J}^o_a + \vec{I}^o_{MM} \qquad \vec{J}^o_A = \vec{J}^o_a + \vec{I}^o_{GTD}$$

The principle of the proposed method consists in treating the integrals containing $\vec{I}_{GTD}$ in equations (4) and (5) as perturbations.

If we neglect these integrals, we get a first order approximation for $\vec{I}^o_{MM}$ and $\vec{I}^e_{MM}$ by solving the system :

$$\vec{I}^o_{MM,1} = 2\hat{n} \times \int_{C^e_{MM}} \vec{I}^e_{MM,1} \times \vec{\nabla}'_T G \, dl' \qquad (6)$$

$$\vec{I}^e_{MM,1} = 2\hat{n} \times \int_{C^o_{MM}} \vec{I}^o_{MM,1} \times \vec{\nabla}'_T G \, dl' + 2\hat{n} \times \int_{C^o} \vec{J}^o_a \times \vec{\nabla}'_T G \, dl' \qquad (7)$$

For an observation point lying in the GTD region the function $\vec{I}$ satisfies the equations :

$$\vec{I}^o_{GTD} = 2\hat{n} \times \int_{C^e_{MM}} \vec{I}^e_{MM} \times \vec{\nabla}'_T G \, dl' + 2\hat{n} \times \int_{C^e_A} \vec{I}^e_{GTD} \times \vec{\nabla}'_T G \, dl' \qquad (8)$$

$$\vec{I}^e_{GTD} = 2\hat{n} \times \int_{C^o_{MM}} \vec{I}^o_{MM} \times \vec{\nabla}'_T G \, dl' + 2\hat{n} \times \int_{C^o} \vec{J}^o_a \times \vec{\nabla}'_T G \, dl' \qquad (9)$$

$$+ 2\hat{n} \times \int_{C^o_A} \vec{I}^o_{GTD} \times \vec{\nabla}'_T G \, dl'$$

If we neglect the integrals on the right hand sides of (8) and (9) which contain $\vec{I}^e_{GTD}$ and $\vec{I}^o_{GTD}$, we get a first approximation for $\vec{I}^o_{GTD}$ and $\vec{I}^e_{GTD}$ given by :

$$\vec{I}^o_{GTD,1} = 2\hat{n} \times \int_{C^e_{MM}} \vec{I}^e_{MM} \times \vec{\nabla}'_T G \, dl' \qquad (10)$$

$$\vec{I}^e_{GTD,1} = 2\hat{n} \times \int_{C^o_{MM}} \vec{I}^o_{MM} \times \vec{\nabla}'_T G \, dl' + 2\hat{n} \times \int_{C^o} \vec{J}^o_a \times \vec{\nabla}'_T G \, dl' \qquad (11)$$

In order to get a second order approximation for $\vec{I}^e_{MM}$ and $\vec{I}^o_{MM}$, we must replace $\vec{I}^e_{GTD}$ and $\vec{I}^o_{GTD}$ in (4) and (5) by their first order approximations given by (10) and (11).

The corresponding integral equations are :

$$\vec{I}^o_{MM,2} = 2\hat{n} \times \int_{C^e_{MM}} \vec{I}^e_{MM,2} \times \vec{V}'_T G \, dl' + 2\hat{n} \times \int_{C^e_A} \left(2\hat{n} \times \int_{C^o_{MM}} \vec{I}^o_{MM,2} \times \vec{V}''_T G \, dl''\right) \times \vec{V}'_T G \, dl' \quad (12)$$
$$+ 2\hat{n} \times \int_{C^e_A} \left(2\hat{n} \times \int_{C^o} \vec{J}^o_a \times \vec{V}''_T G \, dl''\right) \times \vec{V}'_T G \, dl'$$

$$\vec{I}^e_{MM,2} = 2\hat{n} \times \int_{C^e_{MM}} \vec{I}^o_{MM,2} \times \vec{V}'_T G \, dl' + 2\hat{n} \times \int_{C^o_A} \left(2\hat{n} \times \int_{C^e_{MM}} \vec{I}^e_{MM,2} \times \vec{V}''_T G \, dl''\right) \times \vec{V}'_T G \, dl' \quad (13)$$
$$+ 2\hat{n} \times \int_{C^o} \vec{J}^o_a \times \vec{V}'_T G \, dl'$$

Higher order approximations for $\vec{I}_{MM}$ can be obtained in the same way by continuing this procedure.

One can observe that the dimensions of the matrix which has to be inverted remain the same for all orders of approximation. Besides, at any order of approximation, the current in the GTD region is obtained by integrating known functions. The method does not need an a priori knowledge of the asymptotic form of the GTD current. This is an important advantage, especially when the method is applied to complex structures like ships for which the determination of the asymptotic form of the GTD current at a given point in the GTD region requires the determination of the diffracted ray-paths leading to this point.

The method has been applied to a 90° wedge for different configurations of the incident field. Pulse functions have been used as expansion functions in the MM region to solve the systems (6) (7) and (12) (13) corresponding to the first and second order approximations respectively.

The results obtained for the field configuration illustrated on figure 3 are given on figures 4 and 5. In the MM region, the crosses correspond to the first order approximation and the circles correspond to the second order approximation. One can observe that the procedure is very rapidly convergent and that the first order approximation gives very accurate results. The results in the GTD region, represented by dotted circles, have been calculated with the formulas (10) and (11) in which $\vec{I}^e_{MM}$ and $\vec{I}^o_{MM}$ have been replaced by their first order approximation. They fit very well the solid curve representing the exact solution.

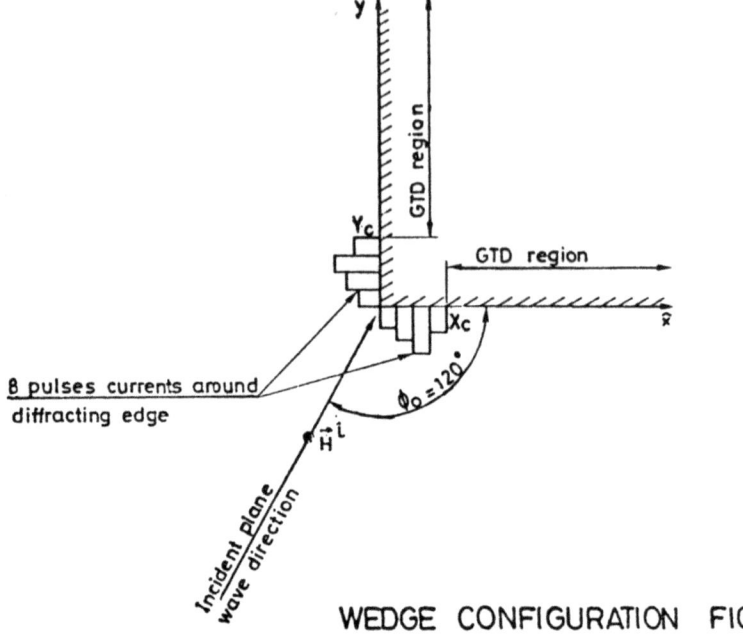

WEDGE CONFIGURATION FIG. 3

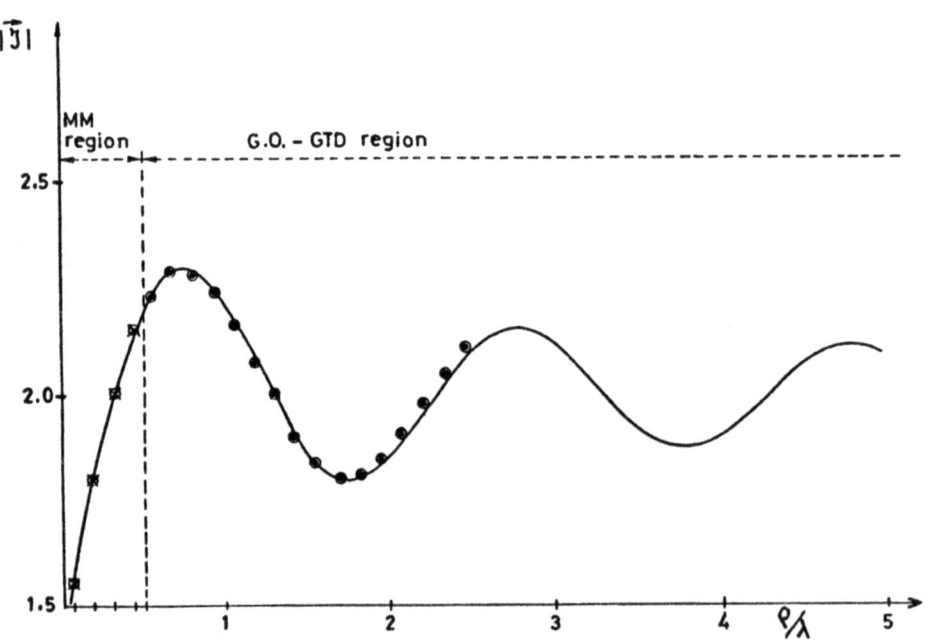

CURRENT ALONG X-WALL OF 90° WEDGE - $\phi = 120°$

FIG. 4

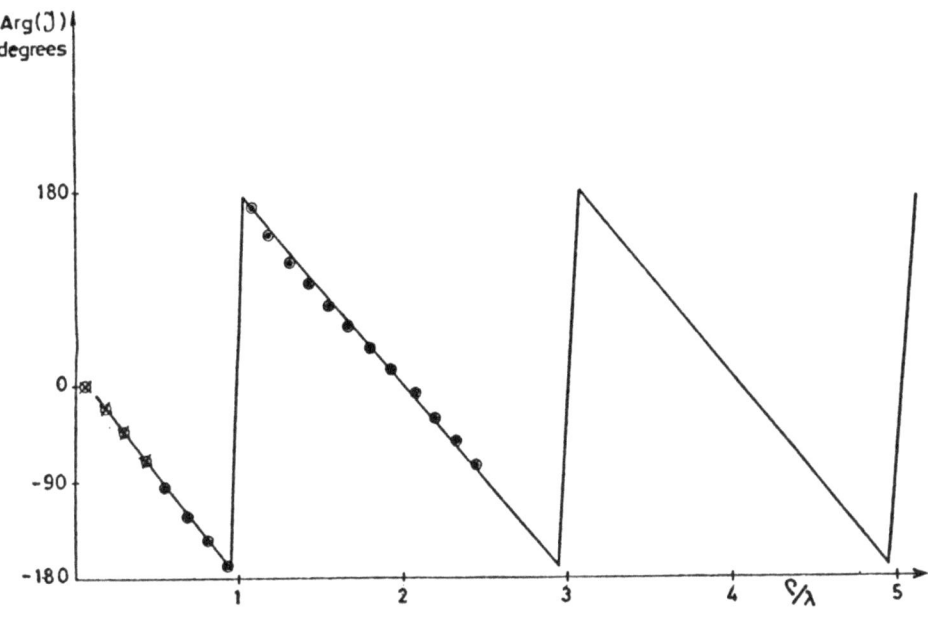

CURRENT ALONG X - WALL OF 90° WEDGE - $\phi = 120°$ FIG. 5

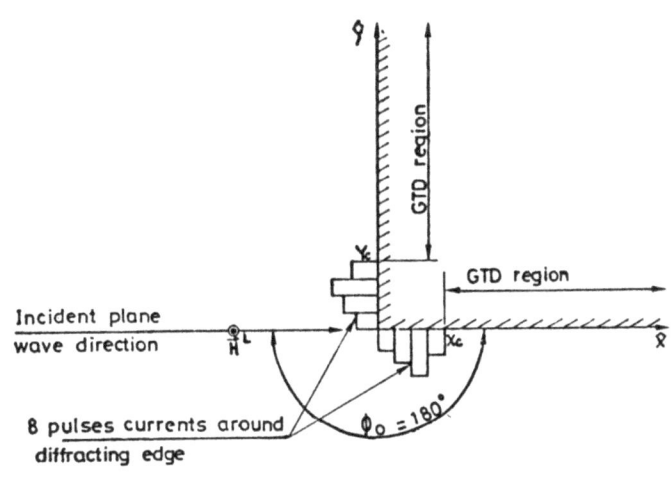

WEDGE CONFIGURATION   FIG. 6

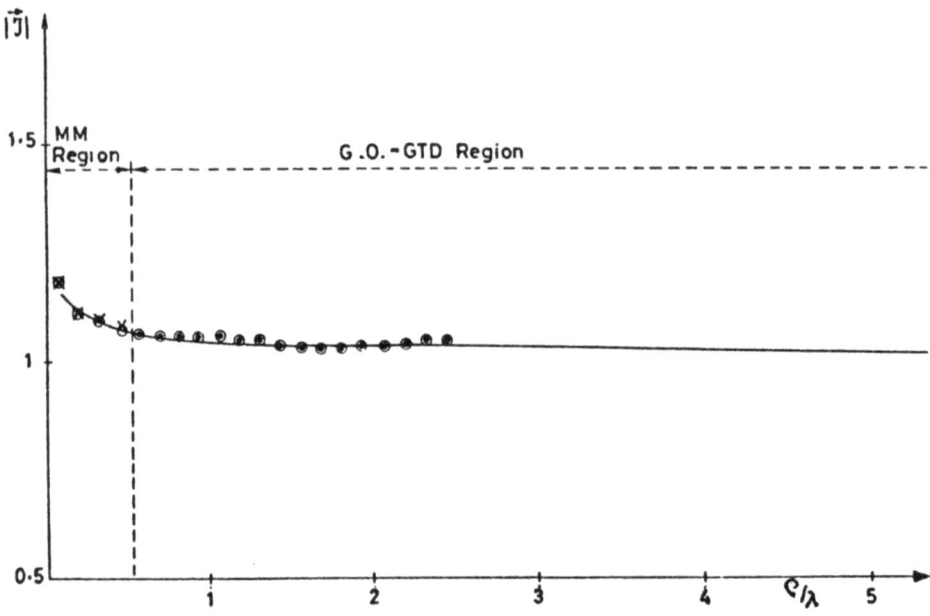

CURRENT ALONG X-WALL OF 90° WEDGE - $\phi = 180°$ FIG. 7

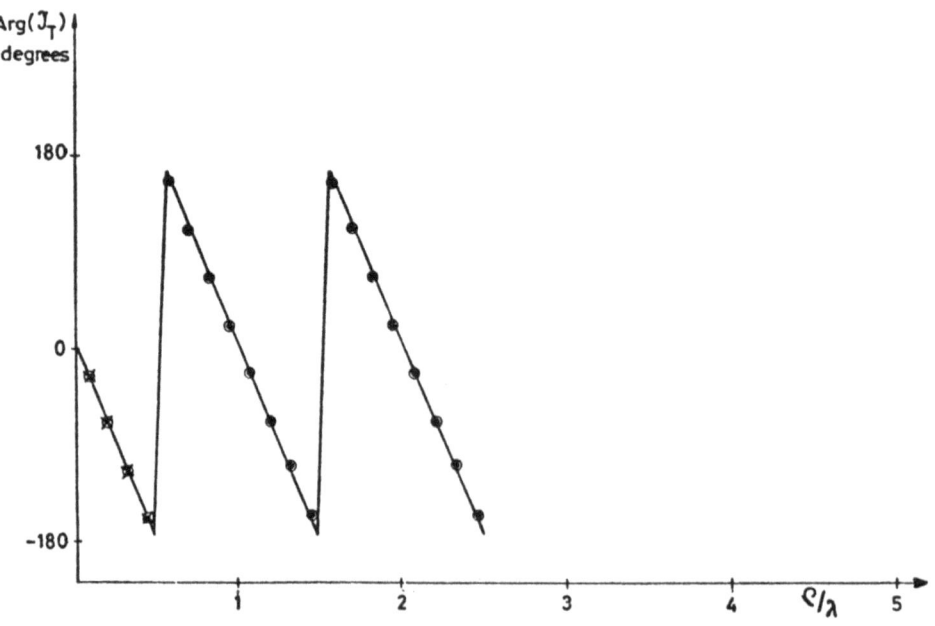

CURRENT ALONG X-WALL OF 90° WEDGE - $\phi = 180°$ FIG. 8

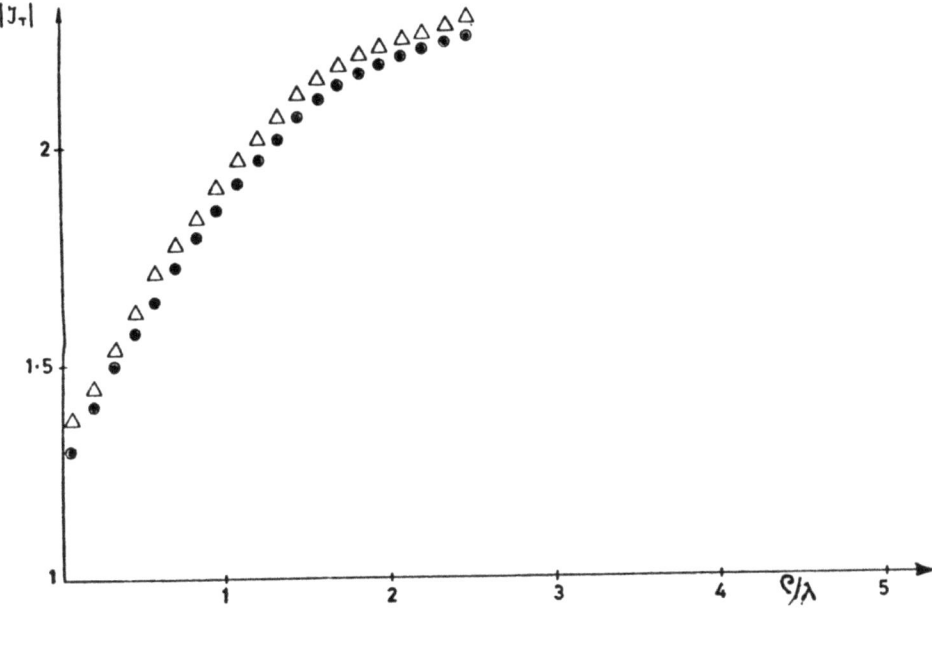

CURRENT ALONG Y-WALL OF 90° WEDGE - ɸ = 120°   FIG. 9

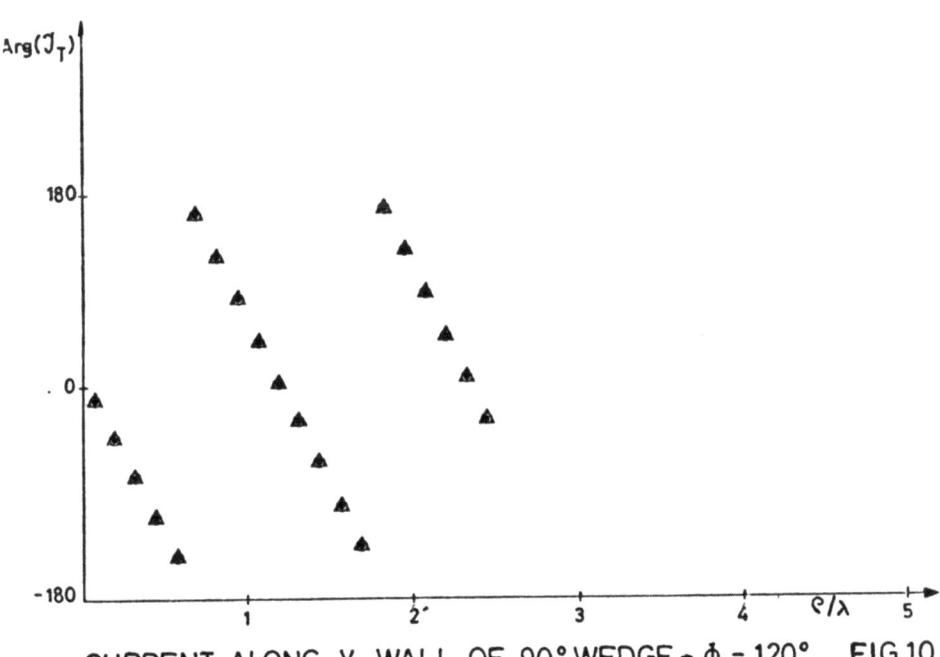

CURRENT ALONG Y-WALL OF 90° WEDGE - ɸ = 120°   FIG. 10

The results; (obtained for grazing incidence, Fig. 6) shown on figures 7 and 8 lead to the same conclusions.

When the walls of the wedge are both illuminated, the formulas (6), (7) and (10), (11) must be symmetrized. The numerical results shown on figures 9 and 10 have been obtained with unsymmetrized formulas. A slight difference appears in the values of the amplitude of the current (see figure 9), depending whether the x wall or the y wall is considered as illuminated in the formulas (6) to (11) above. No significant difference however appears between the values obtained for the argument of the current (see figure 10).

The asymptotic value of the current which has been introduced in the formalism can be easily recognized as being the Physical Optics contribution. The extension of the method to a square cylinder is therefore straightforward. It is also possible to extend this technique to three dimensional problems and especially to complex shapes composed of plates. When the size of the body is very large compared to the wave length, some Physical Optics integrals can be replaced by their asymptotic expansions.

3. APPLICATION OF GTD TO BODIES OF REVOLUTION

We consider bodies of revolution formed of combinations of elementary surfaces like parts of spheres, cones or cylinders. At the junction of two elementary surfaces, there is either a discontinuity of the tangent plane or a discontinuity of the curvature, the line of discontinuity being a circle having the same axis as the target.

When the radius of such a circular discontinuity is large compared to the wavelength, the corresponding scattered field can be calculated by GTD.

Some difficulties appear however when the observation point lies in the vicinity of a caustic. For backscattering of an incident plane wave, this situation arises when the direction of observation is either near broadside to a conical or a cylindrical part or close to the axis of the target. In the former case, the observation point lies in the vicinity of the caustic of the waves reflected by the lateral surface of the target, whereas in the latter case, the point of observation lies in the vicinity of the caustic of the rays diffracted by the circular discontinuities.

We present asymptotic formulas for the field scattered in each of these two situations. The considered geometries are a frustum near broadside incidence and a cone-sphere truncated on the spherical side by a plane perpendicular to the axis. In the last configuration, the back scattering near axial incidence of the circular wedge formed by the junction between the sphere and the plane, located in the shadow region of the incident field is studied.

3.1 High frequency scattering by the lateral surface of a frustum near broadside incidence.

Near broadside incidence, the field scattered by the lateral surface of a frustum can be calculated with the approximation of Physical Optics. Owing to this approximation, the scattered far field corresponding to an incident plane wave $\vec{E}^i = \vec{A}_i e^{-j\vec{k}.\vec{r}}$ is given by:

$$\vec{E}^d = -j \frac{k}{2\pi} \vec{A}_i \frac{e^{-jkr}}{r} \int_{S_e} (\hat{n}.\hat{i}') e^{-2j\vec{k}.\vec{r}'} ds' \qquad (1)$$

where the symbols used have the following significance:

$\hat{n}$: unit vector of the external normal to the surface S of the frustum

$\hat{i}'$: unit vector of the direction of observation

$\hat{r}'$: position vector of an arbitrary point on S

r : distance from the observation point to the phase origin O.

S_e : illuminated part of S.

The elements of the scattering matrix are given by:

$$S_{11} = S_{22} = -j \frac{k}{2\pi} I \quad , \quad S_{12} = S_{21} = 0 . \qquad (2)$$

with

$$I = \int_{S_e} (\hat{n}.\hat{i}') e^{-2j\vec{k}.\vec{r}'} ds' = \frac{2 e^{jkZ_i \cos\theta}}{\cos\theta_i} J \qquad (3)$$

and

$$J = \int_0^{\phi_m} \int_0^h (\cos\theta_i \sin\theta \cos\phi + \sin\theta_i \cos\theta) e^{2jk[(R_i - Z \, tg\,\theta_i)\sin\theta\cos\phi + Z\cos\theta]} \\ \times (R_i - Z \, tg\,\theta_i) \, dz \, d\phi \qquad (4)$$

The angle ϕ_m which delimits the illuminated part of the frustum is defined by:

$$\cos \varphi_m = \frac{- R_i \, tg\theta_i}{tg\theta \, (R_{i-1} + h \, tg\theta_i)} = - \frac{tg\theta_i}{tg\theta} \qquad (5)$$

where h is the height of the frustum and where the symbols are defined on the figure below.

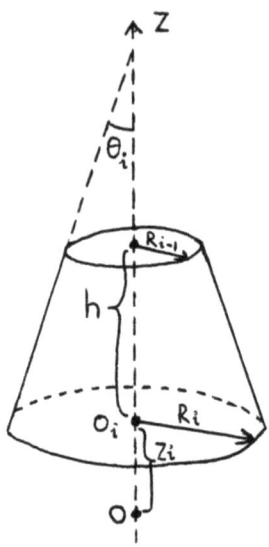

Integration of (4) with respect to Z gives :

$$J = R_i \, I_1 - tg\theta_i \, I_2 \qquad (6)$$

with

$$I_1 = \int_0^{\varphi_m} (\cos\theta_i \sin\theta \cos\varphi + \sin\theta_i \cos\theta) \, e^{2jkR_i \sin\theta \cos\varphi} \left(\frac{e^{jA_1 h} - 1}{jA_1} \right) d\varphi \qquad (7)$$

$$I_2 = \int_0^{\varphi_m} (\cos\theta_i \sin\theta \cos\varphi + \sin\theta_i \cos\theta) \, e^{2jkR_i \sin\theta \cos\varphi} \frac{1}{jA_1} \left[h \, e^{jA_1 h} - \frac{(e^{jA_1 h} - 1)}{jA_1} \right] d\varphi \qquad (8)$$

$$A_1 = 2k \, (\cos\theta - tg\theta_i \sin\theta \cos\varphi) \qquad (9)$$

The integrals (7) and (8) can be evaluated by the Method of Stationary Phase. However, in order to apply this method it is necessary to put the integrand in the form $F(\varphi) \, \exp[jkq(\varphi)]$, where $F(\varphi)$ is a slowly varying function of φ. For this reason we split each of the integrals I_1 and I_2 into two terms as follows :

$$I_1 = e^{2jkh\cos\theta} PF\, i_1 - PF\, i_2 \qquad (10)$$

$$I_2 = he^{2jkh\cos\theta} PF\, i_1 - e^{2jkh\cos\theta} PF\, i_3 + PF\, i_4 \qquad (11)$$

where the integrals i_n ($n = 1, 2, 3, 4$) have the following form:

$$i_n = \int_0^{\phi_m} F_n(\phi)\, e^{jkq_n(\phi)}\, d\phi \qquad (12)$$

and where $PF\, i_n$ is the finite part (Hadamard P.V.) of i_n. The integrand in (12) is given for the different values of n, by:

$$F_1(\phi) = F_2(\phi) = \frac{\cos\theta_i \sin\theta \cos\phi + \sin\theta_i \cos\theta}{2jk\,(\cos\theta - tg\,\theta_i \sin\theta\cos\phi)} \qquad (13)$$

$$F_3(\phi) = F_4(\phi) = \frac{\cos\theta_i \sin\theta\cos\phi + \sin\theta_i \cos\theta}{[2jk\,(\cos\theta - tg\,\theta_i \sin\theta\cos\phi)]^2} \qquad (14)$$

$$q_n(\phi) = 2R_n \sin\theta\cos\phi,\quad R_1 = R_3 = R_{i-1},\quad R_2 = R_4 = R_i \qquad (15)$$

Allowing for the change in variable $q_n(\phi) = q_n[\phi_s^{(n)}] - s^2$, with $\phi_s^{(n)}$ being the value of ϕ at the saddle-point, the integral (12) turns to:

$$i_n = e^{jkq_n(0)} \int_0^{S_n} G_n(s)\, e^{-jks^2}\, ds,\quad n = 1,2,3,4 \qquad (16)$$

with:

$$S_n = [2R_n \sin\theta\,(1-\cos\phi_m)]^{1/2},\quad n=1,2,3,4 \qquad (17)$$

$$G_n(s) = \frac{1}{jk\,tg\,\theta_i}\,\frac{2R_n \sin(\theta+\theta_i) - s^2\cos\theta_i}{(s^2 - s_{p_n}^2)\,(s_{q_n}^2 - s^2)^{1/2}},\quad n=1,2. \qquad (18)$$

$$G_n(s) = \frac{-R_n}{k^2 tg^2\theta_i}\,\frac{2R_n \sin(\theta+\theta_i) - s^2\cos\theta_i}{(s^2 - s_{p_n}^2)^2\,(s_{q_n}^2 - s^2)^{1/2}},\quad n=3,4. \qquad (19)$$

$$S^2_{P_n} = -\frac{2R_n}{\sin\theta_i} \cos(\theta+\theta_i) \qquad (20)$$

$$\left.\begin{array}{l}\\ \\ \end{array}\right\} n = 1, 2, 3, 4.$$

$$S^2_{q_n} = 4R_n \sin\theta \qquad (21)$$

By means of (17) and (21), S_{q_n} is equal to S_n for $\varphi_m = \pi$. Hence for directions of observation not close to the axis of the frustum ($\theta \neq 0$), S_{q_n} is quite different from the value $S = 0$ of the variable s at the saddle-point. We can therefore replace $(S^2_{q_n} - S^2)^{1/2}$ by S_{q_n} in the relations (18) and (19). The value of S_{P_n} however approaches zero at broadside incidence, $\theta = \frac{\pi}{2} - \theta_i$. The function $G_n(s)$ has therefore simple poles ($n = 1, 2$) or double poles ($n = 3, 4$) in the vicinity of the saddle point $S = 0$.

The locus of the poles in the complex S plane are related to the values of θ. According to (20) we have:

$$\begin{array}{lll} S^2_{P_n} > 0 & \text{for} & \theta > \frac{\pi}{2} - \theta_i \\ S^2_{P_n} < 0 & \text{for} & \theta < \frac{\pi}{2} - \theta_i \\ S^2_{P_n} = 0 & \text{for} & \theta = \frac{\pi}{2} - \theta_i \end{array} \qquad (22)$$

The asymptotic evaluation of the integrals (16) cannot be performed by the conventional method because of the presence in $G_n(s)$ of non-integrable singularities. In order to overcome this difficulty we consider extended forms of the integrals i_n in which the poles S_{P_n} are located in the first and third quadrant of the complex S plane as shown on the figure below.

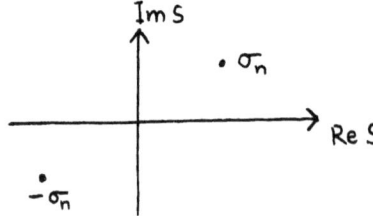

In a first stage we suppose that the poles are located away from the real axis ($\text{Im}\,\sigma_n \neq 0$). In the final formulas obtained, we then take the limit when $\text{Im}\,\sigma_n$ tends to zero.

Following FELSEN and MARCUWITZ (*), we separate the singularities from the function $G_n(s)$ by stating:

$$G_n(s) = \frac{a_n}{s-\sigma_n} + \frac{b_n}{s+\sigma_n} + T_n(s) \quad , n=1,2 \tag{24}$$

$$G_n(s) = \frac{\alpha_n}{(s-\sigma_n)^2} + \frac{\beta_n}{(s+\sigma_n)^2} + \frac{\gamma_n}{s-\sigma_n} + \frac{\delta_n}{s+\sigma_n} + U_n(s) \tag{25}$$

where $T_n(s)$ and $U_n(s)$ are regular functions.
According to (18) and (19):

$$a_n = -b_n = \frac{(R_n)^{1/2}\,\sigma_n^{-1}}{2jk\,tg\theta_i\,\sin\theta_i} \times \frac{\cos\theta}{(\sin\theta)^{1/2}}$$

$$T_n(s) = -a_n\,\frac{\sigma_n \sin 2\theta_i}{2R_n \cos\theta_i} \tag{26}$$

$$\alpha_n = \beta_n = \frac{C_n}{2\sigma_n^2} \quad , \quad \gamma_n = -\delta_n = -\frac{C_n}{2\sigma_n}\left[\frac{\cos\theta_i \sin\theta_i}{R_n \cos\theta} + \frac{1}{\sigma_n^2}\right]$$

$$U_n(s) = 0 \quad , \quad C_n = -\frac{R_n^{3/2}\,\cos\theta}{2k^2(\sin\theta)^{1/2}\sin\theta_i\,tg^2\theta_i}$$

The next step consists in extending the integration path in (16) from S_n to infinity. This is possible since the values of S far away from the saddle-point have a negligable contribution. Then, according to (24) and (25), the integrals to be evaluated can be written in the form:

$$i_n = e^{jkq_n(0)}\,a_n\int_{-\infty}^{+\infty}\frac{e^{-jks^2}}{s-\sigma_n}\,ds + e^{jkq_n(0)}\int_0^{+\infty}T_n(s)\,e^{-jks^2}\,ds, \tag{27}$$
$$n=1,2.$$

$$i_n = e^{jkq_n(0)}\,\alpha_n\int_{-\infty}^{+\infty}\frac{e^{-jks^2}}{(s-\sigma_n)^2}\,ds + e^{jkq_n(0)}\,\gamma_n\int_{-\infty}^{+\infty}\frac{e^{-jks^2}}{s-\sigma_n}\,ds, \tag{28}$$
$$n=3,4.$$

In order to evaluate the integrals on the right hand side of (27) and (28) we use the transformation:

$$\frac{1}{s-\sigma_n} = \frac{s}{s^2-\sigma_n^2} + \frac{\sigma_n}{s^2-\sigma_n^2} \tag{29}$$

(*) Radiation and Scattering of waves, Prentice-Hall, 1973.

This results in :

$$\int_{-\infty}^{+\infty} \frac{e^{-jks^2}}{s-\sigma_n} ds = \sigma_n \int_{-\infty}^{+\infty} \frac{e^{-jks^2}}{s^2-\sigma_n^2} ds \qquad (30)$$

It can be easily shown that the following identity holds for $|\mathrm{Re}\,\sigma_n| < |\mathrm{Im}\,\sigma_n|$ and for S located on the real axis :

$$\frac{1}{s^2-\sigma_n^2} e^{-jk(s^2-\sigma_n^2)} = \int_{C_1+C_2} e^{-z(s^2-\sigma_n^2)} dz \qquad (31)$$

where the contours C_1 and C_2 in the complex z plane are defined on the figure below :

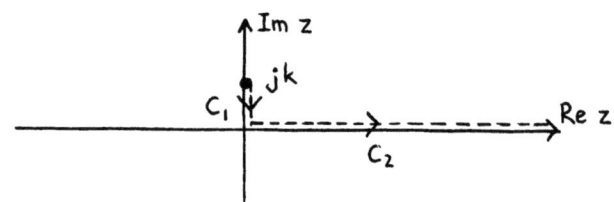

Applying this identity, we obtain :

$$\int_{-\infty}^{+\infty} \frac{e^{-jks^2}}{s-\sigma_n} ds = \sigma_n j \left(\frac{\pi}{j}\right)^{1/2} e^{-jk\sigma_n^2} \int_k^0 \frac{e^{jy\sigma_n^2}}{\sqrt{y}} dy + j\pi e^{-jk\sigma_n^2} \qquad (32)$$

For poles located on the imaginary axis away from the origin $(s_{p_n}^2 < 0)$ it follows that :

$$\int_{-\infty}^{\infty} \frac{e^{-jks^2}}{s-\sigma_n} ds = -2 \left(\frac{\pi}{j}\right)^{1/2} e^{-jk\sigma_n^2} \int_{-j\sigma_n\sqrt{k}}^{} e^{-ju^2} du + j\pi e^{-jk\sigma_n^2} \qquad (33)$$

Where the change in variable $u^2 = -y\sigma_n^2$ has been performed. The expressions on the right hand side of (32) and (33) are analytic-functions of σ_n. It is therefore possible to extend the validity of these relations by analytical continuation, to values of σ_n not restricted by the condition $|\mathrm{Re}\,\sigma_n|<|\mathrm{Im}\,\sigma_n|$, provided that $\mathrm{Im}\,\sigma_n \neq 0$. Consequently, relation (32) remains valid at the limit, when $\mathrm{Im}\,\sigma_n$ tends to zero. But, since the right hand side of (32) is a continuous function of σ_n it can be replaced by its value for $\mathrm{Im}\,\sigma_n = 0$. This gives :

$$\lim_{\mathrm{Im}\,\sigma_n \to 0} \int_{-\infty}^{+\infty} \frac{e^{-jks^2}}{s-\sigma_n} ds = 2j \left(\frac{\pi}{2}\right)^{1/2} e^{-jk\sigma_n^2} \int_{\sigma_n\sqrt{k}}^0 e^{ju^2} du + j\pi e^{-jk\sigma_n^2} \qquad (34)$$

Where the change in variable $u^2 = y\sigma_n$ has been performed. In order to evaluate the integral in the first term on the right hand side of (28), we use the relation:

$$\int_{-\infty}^{+\infty} \frac{e^{-jks^2}}{(s-\sigma_n)^2} ds = \frac{d}{d\sigma_n} \int_{-\infty}^{+\infty} \frac{e^{-jks^2}}{s-\sigma_n} ds \quad (35)$$

This gives:

$$\int_{-\infty}^{+\infty} \frac{e^{-jks^2}}{(s-\sigma_n)^2} ds = 4j \left(\frac{\pi}{j}\right)^{1/2} k\sigma_n e^{-jk\sigma_n^2} \int_{-j\sigma_n\sqrt{k}}^{0} e^{-ju^2} du \quad (36)$$

$$-2j \left(\frac{\pi k}{j}\right)^{1/2} + 2k\pi\sigma_n e^{-jk\sigma_n^2} \quad , \quad \sigma_n^2 < 0$$

$$\lim_{\text{Im } \sigma_n \to 0} \int_{-\infty}^{+\infty} \frac{e^{-jks^2}}{(s-\sigma_n)^2} ds = 4 \left(\frac{\pi}{j}\right)^{1/2} k\sigma_n e^{-jk\sigma_n^2} \int_{\sigma_n\sqrt{k}}^{0} e^{ju^2} du \quad (37)$$

$$-2j \left(\frac{\pi k}{j}\right)^{1/2} + 2k\pi\sigma_n e^{-jk\sigma_n^2}$$

Inserting the relations (33), (34), (36) and (37) into (27) and (28) and replacing the integral in (27) containing $T_n(s)$ by its asymptotic evaluation $\frac{1}{2} \left(\frac{\pi}{jk}\right)^{1/2} T_n(0)$, gives the final expressions for the integrals i_n. By using these expressions in (10), (11), (6) and (2) we obtain, after some pages of elementary calculus which does not involve any special difficulty, the following asymptotic expansions for the elements of the scattering matrix:

$$\theta < \frac{\pi}{2} - \theta_i ;$$
$$S_{11} = S_{22} = -\frac{A}{b}\sqrt{\frac{\pi}{jk}} e^{2jk(z_i\cos\theta + R_i\sin\theta)} \quad (38)$$
$$\times \left\{ \frac{1}{b}\left(\sqrt{R_i'} - \sqrt{R_{i-1}'}\right) e^{jk(R_i - R_{i-1})b^2} \right) \left(\frac{\sin 2\theta_i}{4\cos\theta} b^2 - 1\right)$$
$$+ \frac{1}{\sqrt{k}} e^{jkR_i b^2} \left[\bar{F}\left(\sqrt{kR_{i-1}'} b\right) - \bar{F}\left(\sqrt{kR_i'} b\right) \right] \times \left(\frac{1}{b^2} - \frac{1}{2}\frac{\sin 2\theta_i}{\cos\theta}\right) \right\}$$

$$\theta > \frac{\pi}{2} - \theta_i ;$$
$$S_{11} = S_{22} = -\frac{A}{b}\sqrt{\frac{\pi}{jk}} e^{2jk(z_i\cos\theta + R_i\sin\theta)} \quad (39)$$
$$\times \left\{ \frac{1}{b}\left(\sqrt{R_i'} - \sqrt{R_{i-1}'}\right) e^{-jk(R_i - R_{i-1})b^2} \left(\frac{\sin 2\theta_i}{4\cos\theta} b^2 + 1\right) \right.$$
$$\left. - \frac{1}{\sqrt{k}} e^{-jkR_i b^2} \left[\overset{+}{F}\left(\sqrt{kR_{i-1}'} b\right) - \overset{+}{F}\left(\sqrt{kR_i'} b\right) \right] \right.$$
$$\left. \times \left(\frac{1}{b^2} + \frac{1}{2}\frac{\sin 2\theta_i}{\cos\theta}\right) \right\}$$

where :

$$b = \left|\frac{\cos(\theta+\theta_i)}{\sin\theta_i}\right|$$

$$A = \frac{1}{2\pi} \frac{\cos\theta}{\sin^2\theta_i \sqrt{\sin\theta'}}$$

$$\overset{\pm}{F}(x) = \int_x^0 e^{\pm ju^2} du$$

For $\theta = \frac{\pi}{2} - \theta_i$ or $b = 0$, the expressions (38) and (39) tend to the finite limit :

$$S_{11} = S_{22} = -A_1 \sqrt{\frac{\pi}{jk}} e^{2jk(z_i \sin\theta_i + R_i \cos\theta_i)}$$
$$\left[-jk\tfrac{2}{3}\left(R_{i-1}^{3/2} - R_i^{3/2}\right) + \tfrac{\cos\theta_i}{2}\left(\sqrt{R_{i-1}} - \sqrt{R_i}\right)\right], A_1 = A\left(\tfrac{\pi}{2} - \theta_i\right)$$

For $\theta \sim \frac{\pi}{2} - \theta_i$

$$S_{11} = S_{22} = -A(\theta) \sqrt{\frac{\pi}{jk}} e^{2jk(z_i \cos\theta + R_i \sin\theta)}$$
$$\left[-jk\tfrac{2}{3}\left(R_{i-1}^{3/2} - R_i^{3/2}\right) + \tfrac{\cos\theta_i}{2}\left(\sqrt{R_{i-1}} - \sqrt{R_i}\right)\right]$$

Numerical values obtained with the asymptotic expansions (38) and (39) and with an exact method are given on figure 11.

3.2 <u>Backscattering near axial incidence of a circular wedge located in the shadow region of the incident field.</u>

We consider the back scattering of a plane wave by a curved wedge formed by the intersection between a sphere and a plane. The configuration studied as well as the notations used for the geometrical quantities are specified on the figure below.

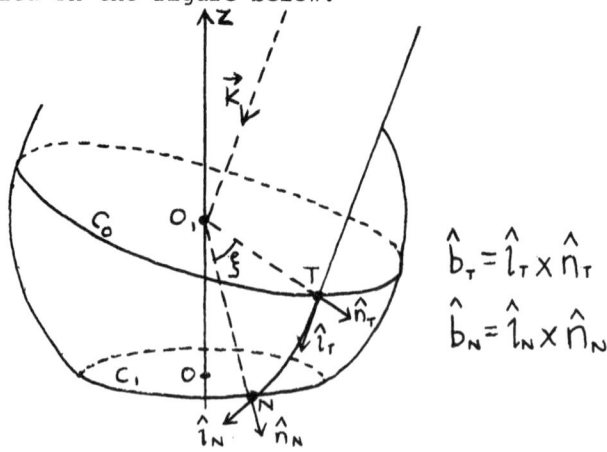

The wedge is located in the shadow region of the incident field and is therefore indirectly illuminated by creeping waves launched at the shadow boundary Co. These creeping waves are diffracted by the circular edge C_1 of the curved wedge. We suppose that the direction of propagation of the incident field is close to the axis of the structure.

The method used to treat this problem consists in replacing the diffraction by the curved wedge by the radiation of equivalent currents located on C_1.

From the definition of these currents, we can write :

$$d\vec{I}_e(N) = 0$$
$$d\vec{I}_m(N) = - \frac{2\hat{t}_N [\hat{t}_N . \vec{H}_i(N)]}{jKY_o \sin^2\beta} D_h e^{-j\frac{\pi}{4}} \left(\frac{8\pi}{K}\right)^{1/2} dl \quad (1)$$

where N is a point of C_1 and D_h the diffraction coefficient of the wedge for the hard boundary condition. All the symbols used in this text, are defined in the articles of KOUYOUMJIAN and PATHAK (5, 6).

Let $\vec{H}_v(N)$ be the total field which would exist at the point N on C_1 if the surface was continuous there. The asymptotic expansion of this field is given by :

$$\vec{H}_v(N) = \vec{H}_i(T).[\hat{b}_T \hat{b}_N F_1 - \hat{n}_T \hat{l}_N G_1] e^{-jk\widehat{TN}}$$
$$F_1 = \sum_P A_P^h D_P^h e^{-\alpha_P^h \widehat{TN}} \quad (2)$$
$$G_1 = \sum_P A_P^s D_P^s e^{-\alpha_P^s \widehat{TN}}$$

where $\vec{H}_i(T)$ is the incident field at the starting point T of the surface ray $\widehat{TN}$ and where $A_P^{s,h}$, $D_P^{s,h}$ and $\alpha_P^{s,h}$ are respectively the attachment coefficient, the surface diffraction coefficient and the attenuation coefficient of the pth mode.

It can be shown, by extending GTD to the diffraction of creeping waves by a line of discontinuity of the tangent plane (7, 8) that $\hat{t}_N . \vec{H}_i(N)$ is given by :

$$\hat{t}_N . \vec{H}_i(N) = \frac{1}{2} \hat{t}_N . \vec{H}_v(N) \quad (3)$$

Furthermore, in order to calculate the surface ray-field emanating from the magnetic current element $d\vec{I}_m$, it is necessary to separate the source from its image, which results in an effective magnetic dipole source of strength $\frac{1}{2}d\vec{I}_m$. The pth mode of the surface ray-field excited by this source at another point on the surface is therefore given by :

$$d\vec{E}_p(T) = -j\frac{K}{8\pi} e^{-jk\widehat{NT}} [\hat{n}_T(\hat{b}_N.\hat{t}_N)L_p^h(N)e^{-\alpha_p^h \widehat{NT}} + \hat{b}_T(\hat{t}'_N.\hat{t}_N)L_p^s(N)e^{-\alpha_p^s \widehat{NT}}]dI_m \left(\frac{d\Psi_N}{d\eta_T}\right)^{1/2} \times \exp(-jk\,\widehat{NT}) \quad (4)$$

where $L_p^{s,h}$ are the launching coefficients and where the divergence factor is given by :

$$\left(\frac{d\Psi_N}{d\eta_T}\right)^{1/2} = (R \sin \xi)^{-\frac{1}{2}} \quad (5)$$

The preceding relations allow us to write down without any special difficulty the expressions of the scattered field at an observation point P away from the body as well as the elements of the diffraction matrix.

For rays simply diffracted by the curved wedge C_1, the scattered field is given by :

$$\vec{E}_d(P) = \frac{1}{2\pi} \int_{C_1} \frac{e^{-2jk\widehat{NT}}}{\sin^2\beta} e^{-j\vec{k}.\vec{oT}} e^{-jk|\vec{TP}|} \left(\frac{d\sigma(T)}{d\sigma(N)}\right)^{1/2} \left(\frac{d\Psi_N}{d\eta_T}\right)^{1/2} X(N)\, \mathcal{A}(P) \quad (6)$$

$$[(\vec{A}.\hat{n}_T)(\hat{b}_N.\hat{t}_N)F_1(T,N) + (\vec{A}.\hat{b}_T)(\hat{t}_N.\hat{t}_N) G_1(T,N)]$$

$$[(\hat{b}'_N.\hat{t}_N)\hat{n}_T\, F_2(T,N) + \hat{b}'_T(\hat{t}'_N.\hat{t}_N) G_2(T,N)]\, d\ell$$

where
$$F_2(T,N) = \sum_{p=1}^{\infty} D_p^h(T)\, L_p^h(N)\, e^{-\alpha_p^h \widehat{TN}}$$
$$G_2(T,N) = \sum_{p=1}^{\infty} D_p^s(T)\, L_p^s(N)\, e^{-\alpha_p^s \widehat{TN}}$$
$$\mathcal{A}(P) = \frac{1}{|\overline{OP}|}(R\,tg\,\xi)^{1/2}, \quad X(N) = \frac{1}{K} \frac{\sin\frac{\pi}{n}}{\cos\frac{\pi}{n}-1}, \quad \hat{b}'_T = -\hat{b}_T,\; \hat{t}'_N = -\hat{t}_N$$

An example of application of these formulas is given on figure 12.

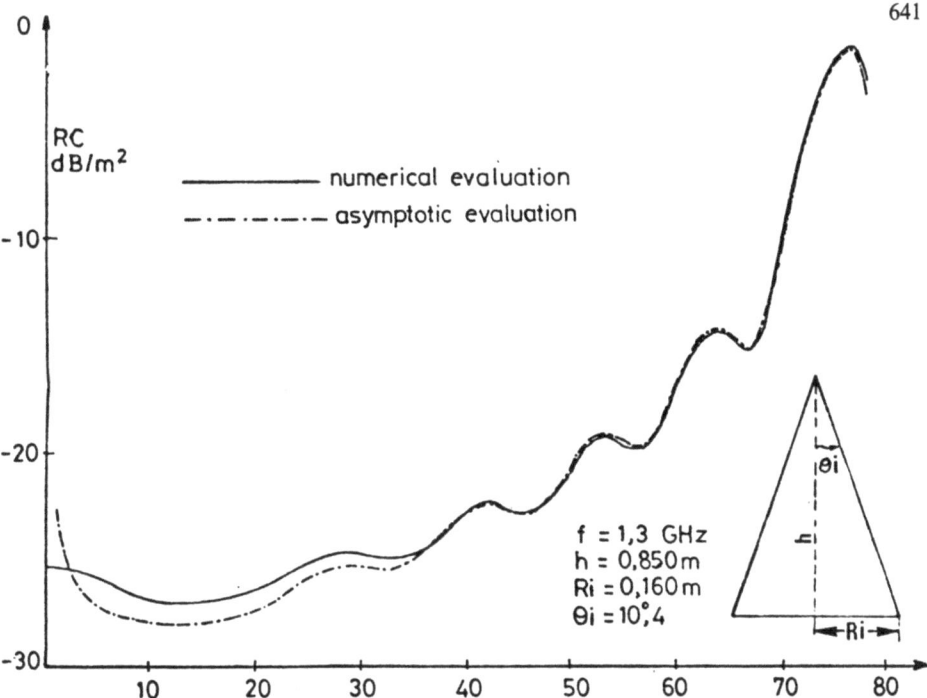

SCATTERING BY THE LATERAL SURFACE OF A CONE   FIG. 11

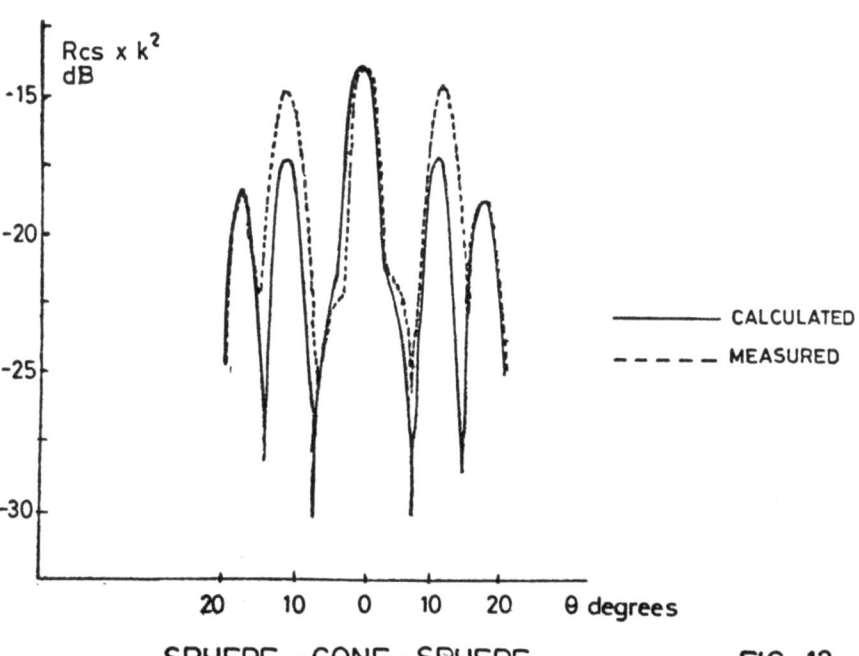

SPHERE - CONE - SPHERE   FIG. 12

4. PRESENTATION OF A GENERAL COMPUTER PROGRAM
(DR. L. BEAULIEU)

4.1 Introduction

The presented code is devoted to the computation of the radiation pattern of any point source located in the vicinity of a scattering body. It uses the high frequency method known as the ray technique or the geometrical theory of diffraction. It is possible to use it for any antenna type provided that it is not fitted in a structure. This code is not fully automatic and its use requires some basic knowledge about GTD.

Before computation the user has to perform a preliminary analysis to retain significant contributions. Then he builds a main program in order to add local fields.

A general coordinate system is defined, then a local coordinate system is associated with each geometrical structure. A third coordinate system is required to describe the radiation pattern of the isolated antenna and a fourth coordinate system is introduced to observe diffracted fields, whose phase depends on the position of a specific point chosen by the user. Later a more detailed description of each of these particular points is made. An example illustrates the use of this code.

4.2 Geometry specification

The modelling of a scattering body is accomplished using elementary structures such as finite plates, a cut elliptical cylinder, a cut cone, an ellipsoid of revolution and a sphere. As described above, a local coordinate system is associated with each elementary surface. Significant parameters describing surfaces (corner coordinates for a plate, wedge angle, sphere radius, etc.) are expressed in this system.

For each local system, coordinates of its origin and components of unit vectors are expressed in a general coordinate system.

4.3 Interactions taken into account

The user should have some elementary knowledge about GTD. The total diffracted field is computed from the local fields associated with rays. A ray is produced by discontinuities in the scattering body. Thus there are reflected, diffracted, reflected-reflected, reflected-diffracted and diffracted-reflected rays. Modules are available for the types of rays described below:

- Direct ray.
- Ray reflected by a surface such as a plate, a cut elliptic cylinder, a cone, an ellipsoid of revolution or a sphere.
- Ray reflected by a plate then reflected either by a plate, a cylinder or a cone.
- Ray reflected by a plate then diffracted by an edge.
- Ray diffracted by an edge then reflected either by a plate, a cylinder or a sphere.

For cylindrical and conical structures it is possible to include creeping rays. In the corresponding subroutines corrections are included for a finite length edge or a transition zone.

In some subroutines near and far field computations are available.

4.4 Antenna specification

The isolated antenna radiation pattern is described in a local coordinate system in terms of E_θ and E_ϕ (spherical coordinates). In this routine it is possible to include :

- Field computations from currents.
- Measured field in specific planes. Then for any direction (θ, ϕ) the field is computed using two dimensional interpolation.

4.5 Shadowing effects

Each ray is made up of a set of straight lines. For a typical segment the user takes into account only likely obstacles in order to reduce computer time. He then builds a shadowing routine for each ray. This procedure is not tedious because it is possible to write routines almost automatically, and it is easy to include new obstacles from a basic structure.

Elementary routines are available to compute shadowing effects of basic structures such as the cone, cut elliptical cylinder, cut ellipsoid and plates.

For any ray the related shadowing subroutine is built from a set of calls to the elementary routines associated with the geometrical elements which can intercept this ray.

4.6 Basic structure for any subroutine

For each subroutine computing the field associated with a given ray, inputs include geometry, observation direction, and routines for antenna type and shadowing effects.

Computations are performed as follows :

- Search for interaction points using Keller and Snell laws.
- If this ray exists, computation of shadowing effects.
- If this ray is not intercepted, computation of related local field (E_θ, E_ϕ).

4.7 User requirements

So a set of subroutines is available to compute the local fields associated with the ray types already described. There is also a set of subroutines to compute elementary shadowing effects.

Thus it is possible to compute the interaction between any structure and antenna provided that the corresponding modules are available. Of course it is possible to build new modules for other ray types.

For a scattering body the user analyzes the types of
rays to be retained, then he builds the main program
which is a series of calls to subroutines computing
ray fields.

Then, as previously indicated, he builds shadowing
routines related to each ray and routines computing
the fields of the isolated antenna. So we can see
that the procedure is not fully automatic because ray
tracing is impossible with a computer.

In order to use the code efficiently it is better to
have some basic knowledge about GTD.

4.8 Additional facilities

For the far field, computations are performed, according
to user requirements, in either the θ or φ cut. For
the near field, computations are performed on a sphere
whose center is specified by the user. Program outputs
include :

- Field components (linear) for the isolated antenna.

- Diffracted field components (linear, left and right circular, ellipticity ratio).

- Types of rays taken into account.

- Components of unit vectors related to antenna and observation systems expressed in the general coordinate system.

- Coordinate of the point source and of the phase reference point.

- Frequency.

Additional facilities such as printer plotting and pen
plotting are also available, including perturbed
and unperturbed fields. So for each run it is easy to
observe the influence of the structure.

An example.
This code has been used to compute the
overall radiation pattern of two s-band antennas mounted
on the French satellite SPOT. Core requirements are
very high and it is necessary to segment the program.
For a typical (θ, φ) direction, the run time is approximately 0.05 seconds on a CDC 7 600. For illustration an
example is presented below.

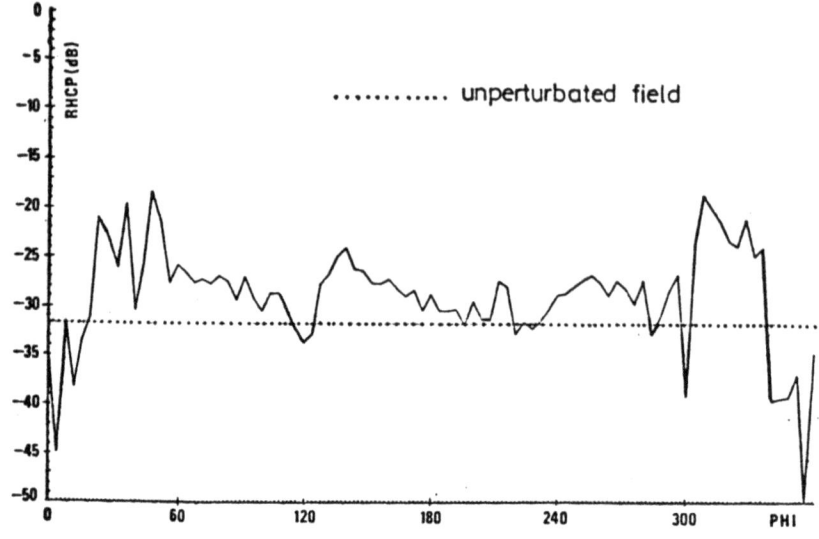

CONICAL SCAN (θ = 85°) TYPICAL PATTERN FIG 13

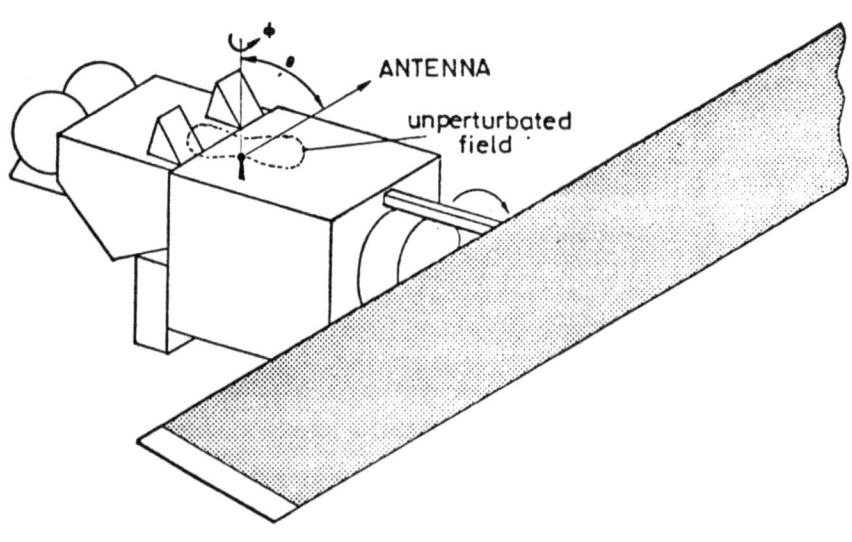

SPOT SATELLITE FIG. 14

4.9 Acknowledgment

Many routines described here were developed during the evaluation of the SPOT satellite antenna. So we thank CNES French Administration for their support.

5. REFERENCES

(1) G.A. THIELE and T.H. NEWHOUSE,

A Hybrid Technique for Combining Moment Methods with the Geometrical Theory of Diffraction.

IEEE Transactions on Antennas and Propagation, Vol. AP-23, pp. 62-69, January 1975.

(2) W.D. BURNSIDE, C.L. YU and R.J. MARHEFKA,

A Technique to Combine the Geometrical Theory of Diffraction and the Moment Method.

IEEE Transactions on Antennas and Propagation, Vol. AP-23, pp. 551-558, July 1975.

(3) J.A. AAS,

Control of electromagnetic scattering from Wing profiles by impedance loading.

Report 3424-4, August 1975, The Ohio State University, ESL, AD A019134.

(4) F.A. MOLINET,

On the Methods which Combine the Moment Method with Asymptotic Techniques.

URSI, XIX General Assembly, 1-8 August 1978, Helsinki, Finland.

(5) R.G. KOUYOUMJIAN, P.H. PATHAK,

A Uniform Geometrical Theory of Diffraction for an edge in a perfectly conducting surface.

Proceedings of the IEEE, Vol. 62, No. 11, pp. 1448-1461, November 1974.

(6)   P.H. PATHAK, R.G. KOUYOUMJIAN,

An Analysis of the Radiation from apertures in curved surfaces by the Geometrical Theory of Diffraction.

Proceeding of the IEEE, Vol. 62, No. 11, pp. 1438-1447, November 1974.

(7)   N.C. ALBERTSEN and P.L. CHRISTIANSEN

- Diffraction of Creeping waves by an edge, URSI Symposium on Electromagnetic Wave Theory, IEE London, pp. 114-116, 1974.

- Hybrid Diffraction Coefficients for first and second order discontinuities of two-dimensional scatterers.
SIAM J. Appl. Math., Vol. 34, No. 2, pp. 399-414, March 1978.

(8)   F.A. MOLINET

Diffraction d'une onde rampante par une ligne de discontinuité du plan tangent.

Annales des Télécommunications, tome 32, No. 5-6, pp. 197-202, Mai-Juin 1977.

TRANSFORM APPROACH TO ELECTROMAGNETIC SCATTERING[*]

Raj Mittra,[+] Yahya Rahmat-Samii,[++] and Wai Lee Ko[+]

[+]EM Lab, Dept. of Electrical Engineering
University of Illinois, Urbana, Illinois 61801

[++]Jet Propulsion Laboratory, California Institute
of Technology, Pasadena, California 91103

1. INTRODUCTION

Recently, the digital computer has opened up new vistas to researchers in electromagnetics and other disciplines, and has enabled them to attack a much wider class of problems than has been possible prior to the availability of high-speed computers. In employing the computer as a problem-solving tool, it is often tempting to let the computer do "all the work." In this event, one employs little or no analytical preparation prior to proceeding with the numerical steps on the computer. However, experience shows that the computer offers no panacea, at least when it is used in the manner indicated above. This is because for many practical problems, the requirements on storage, computational time, and cost can be prohibitively large if one uses a purely numerical approach. These limitations become particularly severe in the resonance region and above, where the geometrical dimensions of the object are on the order of two to three wavelengths or more. Below this region, the integral-equation formulation combined with matrix methods [1], [2] is eminently suited for deriving numerically rigorous solutions to radiation and scattering problems involving objects of arbitrary shape.

For frequencies above the resonance region, the transform technique has been found useful for deriving efficient numerical solutions to radiation and scattering problems by combining

---

[*]This work was supported in part by Joint Services Electronics Projects under Grant DAAG-29-78C-0016, and in part by the Office of Naval Research under Grant N000 1475C-0293.

integral-equation formulation with asymptotic techniques in the transform domain. The early application of the spectral technique had demonstrated its usefulness for a class of geometries involving open waveguides and scattering problems [3]-[9]. Recent studies have shown that the scope of application can be considerably enlarged while simultaneously improving the efficiency of computation by incorporating the existing high-frequency asymptotic solutions into the solution procedure. Consequently, this paper focuses mainly on this combination procedure which appears to offer the greatest potential for systematic and accurate solutions to radiation and scattering problems in the resonance region and above.

Turning now to a brief review of the asymptotic techniques, perhaps the most significant works on the subject are the two pioneering papers by Keller [10], [11] which introduced the concepts of the Geometrical Theory of Diffraction, or GTD, by augmenting the classical Geometrical Optics (GO) field expressions with the edge-diffracted fields from sharp wedges. Keller's GTD and some of its extensions have been described in a number of review papers and book chapters [12]-[15]; an exhaustive set of useful references can be found in [16]. Although in the last fifteen years Keller's GTD has played a very significant and unique role in solving high-frequency scattering problems, its original form is known to break down at shadow boundaries, reflection boundaries and caustics. Two uniform theories have recently been developed [17]-[19] for circumventing the difficulties associated with Keller's GTD in the neighborhood of shadow boundaries. Each one of these theories is based on its own Ansatz and yields different forms for the expressions of the field in the transition region between the lit and shadow zones. Another totally different approach is based on the modification of the Physical Optics (PO), called the Physical Theory of Diffraction (PTD) developed by Ufimtsev and others [20], [21]. A comparative study of the various asymptotic techniques [22], [23] can be useful for evaluating the different methods for computing the edge-diffracted fields in the transition regions.

All of the aforementioned asymptotic techniques based on GTD, be they uniform or nonuniform, have two important and fundamental limitations:

(a) It is difficult to estimate the accuracy of the solution;

(b) No systematic procedure is available for improving the solution.

The addition of higher-order, multiply-diffracted terms in the asymptotic series can and often does lead to a divergent result.

In contrast, the spectral approach to scattering, which was introduced in the literature [24] under the name of Spectral Theory of Diffraction (STD), not only provides a built-in test for the satisfaction of the boundary condition, but also is convenient for improving the GTD or other asymptotic solutions via Galerkin or iterative procedures [23]-[30].

The principal motivation of this paper is to present a comprehensive review of the spectral approach for solving electromagnetic scattering and diffraction problems. In contrast to the ray approach or GTD, which is essentially goemetric in nature, the scattered-field representation in the spectral approach is in terms of the Fourier transform of the induced surface current on the scatterer. Although the basic foundations of GTD and STD are different, the GTD solution, where valid, is indeed identifiable as the result obtained from an asymptotic evaluation of the transform-domain representation of the scattered field. In addition, the integral representation provides a uniform solution requiring no a posteriori correction [24]. A discussion of this topic is given in Section 3 of this paper.

The important subject of systematic improvement of high-frequency asymptotic solutions via a combination of GTD methods and integral-equation formulation is described in Sec. 4, and some unique numerical advantages that accrue from the use of this procedure are also demonstrated. Finally, a few concluding remarks on the spectral approach are given in Sec. 5. Possible extensions of this approach to a number of other important problems of practical interest are also mentioned in this section.

2. FORMALISM FOR TRANSFORM APPROACH TO HIGH FREQUENCY SOLUTION — STD

In this section we introduce the basic concepts of the spectral theory of diffraction (STD) via the half-plane problem. Since the canonical geometry of the half-plane is one of the cornerstones of the various non-uniform and uniform theories of high frequency diffraction, we also choose this geometry to build the foundations of STD.

The geometry of a perfectly conducting half-plane located at $y = 0$, $x \leq 0$ and illuminated by a plane wave is shown in Fig. 1a. We let the direction of propagation of the incident plane wave be normal to the edge, i.e., $\vec{k}^i \cdot \hat{z} = 0$. This assumption changes the vector nature of the three-dimensional problem to a two-dimensional scalar problem. Furthermore, the problem may be classified as in the cases of E-wave (nonzero field components E_z, H_x, H_y), or H-wave (nonzero field components H_z, E_x, E_y) by simply letting the incident E-field or H-field be directed alter-

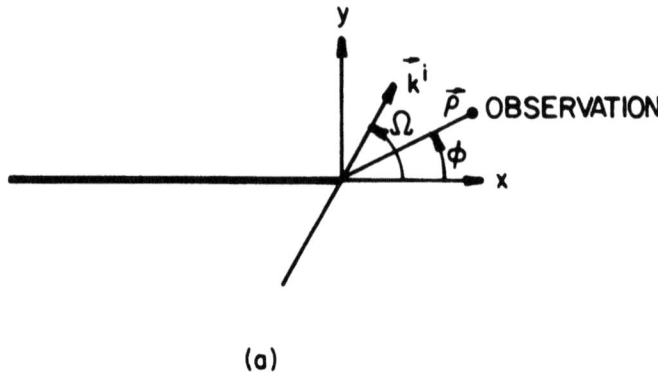

(a)

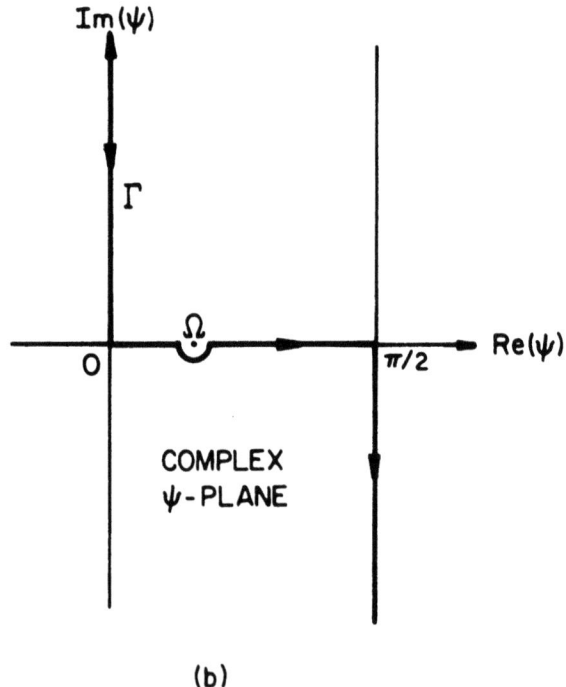

(b)

Figure 1a and b. (a) Diffraction of a plane wave by a half-plane, $0 \leq \Omega \leq \pi$ and $-\pi \leq \phi < \pi$. (b) integration path Γ for integral representation (11).

natively along the z-axis. Unless otherwise specified, the cases of E- and H-waves are treated simultaneously, with the help of two symbols u and τ such that

for E-waves: $u = E_z$, $\tau = -1$ (1a)

for H-waves: $u = H_z$, $\tau = +1$.   (1b)

The total field u^t may be split into the incident field u^i and the scattered field u^s to give

$$u^t = u^i + u^s . \qquad (2)$$

For a perfect electric conductor the total field u^t is subject to the boundary condition $u^t = 0$ or $\partial u^t/\partial y = 0$ for E-wave or H-wave cases, respectively, on the half-plane. If one defines the induced electric current on the half-plane as

for E-wave: $J_z = \left.\dfrac{\partial u^t}{\partial y}\right|_{0^+}^{0^-}$ (3a)

for H-wave: $J_x = -u^t\Big|_{0^+}^{0^-}$,   (3b)

and uses the time convention $e^{-i\omega_0 t}$, one can readily arrive at the following equations using Maxwell's equation

for E-wave: $u^s = E_z^s = i\omega_0\mu \displaystyle\int_{-\infty}^{0} J_z(x')g_0(k|\vec{\rho}-x'\hat{x}|)\,dx'$ (4a)

for H-wave: $u^s = H_z^s = \dfrac{\partial}{\partial y} \displaystyle\int_{-\infty}^{0} J_x(x')g_0(k|\vec{\rho}-x'\hat{x}|)\,dx'$ (4b)

where $k = \omega_0\sqrt{\mu\varepsilon}$ μ and ε are the permeability and permittivity of the medium, respectively, and $g_0(k\rho) = iH_0(k\rho)/4$ (H_0^1 is the Hankel function of first kind and zero order). The objective is to determine J and u^s for the half-plane illuminated by an incident plane wave. This is done by using the transform technique and employing the results given in [31].

2.1. Spectral Diffraction Coefficient and Total Field

Let us define the Fourier transform pair as

$$U(\alpha) = \int_{-\infty}^{\infty} u(x) e^{i\alpha x} dx = F[u(x)] \qquad (5a)$$

and

$$u(x) = \frac{1}{2\pi} \int_{-\infty+i\Delta}^{\infty+i\Delta} U(\alpha) e^{-i\alpha x} dx = F^{-1}[U(\alpha)] \qquad (5b)$$

where Δ is a small positive number. The incident plane wave can be written as

$$u^i = e^{i\vec{k}\cdot\vec{\rho}} = e^{i(k_x x + k_y y)} = e^{ik\rho\cos(\Omega-\phi)} \qquad (6)$$

where $k_x = k\cos\Omega$, $k_y = k\sin\Omega$ and $0 \leq \Omega \leq \pi$ is the incident angle shown in Fig. 1a. Transforming (4) into the spectral (Fourier) domain and applying the Wiener-Hopf technique, one arrives at the following:

for E-wave: $F[J_z] = (i\omega_0\mu)^{-1} X(k_x, \alpha) \qquad (7a)$

for H-wave: $F[J_x] = -\gamma^{-1} X(k_x, \alpha) \qquad (7b)$

where $X(k_x, \alpha)$ is

$$X(k_x, \alpha) = 2 \frac{\sqrt{k + \tau k_x} \sqrt{k + \tau\alpha}}{\alpha + k_x}, \qquad (8)$$

and $\gamma = \sqrt{\alpha^2 - k^2}$ such that $\text{Re } \gamma > 0$, and $\alpha = -k\cos\psi$. In this work, unless otherwise stated, $\sqrt{\cdot}$ and $(\cdot)^{1/2}$ are defined with their proper branch cut slightly below the negative real axis. Using the transform version of (4) and then incorporating (7), one finally obtains

$$U^s = \begin{Bmatrix} 1 \\ \text{sgn}(y) \end{Bmatrix} X(k_x, \alpha) \frac{e^{-\gamma|y|}}{2\gamma} \quad \text{for} \quad \begin{Bmatrix} \text{E-wave} \\ \text{H-wave} \end{Bmatrix}. \qquad (9)$$

Furthermore, one may notice that the following equation has been used in the construction of (9)

$$F[g_0(k\rho)] = \frac{e^{-\gamma|y|}}{2\gamma} . \tag{10}$$

Introducing the change of variables $x = \rho\cos\phi$, $y = \rho\sin\phi$, $k_x = k\cos\Omega$, $k_y = k\sin\Omega$, $\alpha = -k\cos\psi$ and $\gamma = -ik\sin\psi$ into (9) and substituting the result into (5b) one finally arrives at

$$u^s = \frac{i}{4\pi} \begin{Bmatrix} 1 \\ \text{sgn}(\phi) \end{Bmatrix} \int_\Gamma \chi(\Omega,\psi) \, e^{ik\rho\cos(\psi-|\phi|)} \, d\psi$$

$$\text{for} \begin{Bmatrix} \text{E-wave} \\ \text{H-wave} \end{Bmatrix} . \tag{11}$$

In the preceding equation ψ is the complex angle defined on the path Γ, shown in Fig. 1b, and $\chi(\Omega,\psi)$ is

$$\chi(\Omega,\psi) = K(k\cos\Omega, -k\cos\psi) = \chi_i(\Omega,\psi) + \tau\chi_r(\Omega,\psi) \tag{12}$$

where

$$\chi_{\substack{i \\ r}}(\Omega,\psi) = \mp\csc\frac{\Omega \mp \psi}{2} . \tag{13}$$

We may notice that $\chi_i(\cdot)$ and $\chi_r(\cdot)$ have the same functional form, i.e., $\csc(\cdot)$. This definition of χ_i and χ_r is closely related to the definition used by Deschamps in [12]. Clearly χ_i and χ_r are infinite at $\psi = \Omega$ and $\psi = -\Omega$, respectively. These two values of ω correspond to the incident and reflection shadow boundaries appearing in the classical GTD expressions which are obviously 'nonuniform' as a function of the observation angle. As a matter of fact, $\chi(\Omega,\psi)$ is precisely the angular part of Keller's diffraction coefficient, when ω is replaced by the observation angle ϕ. Although χ tends to infinity at the shadow boundaries, it does not mean that the field itself is also infinite as Keller's GTD predicts. Instead, the correct value of the field is obtained from (11), which is always bounded. To distinguish it from Keller's coefficient, which is associated with the diffracted field, we will refer to $\chi(\Omega,\psi)$ as the *spectral diffraction coefficient* for the half-plane. This terminology is chosen since $\chi(\Omega,\psi)$ is associated with the spectrum, or equivalently, the Fourier transform, of the induced current and appears only inside the kernel of the plane wave spectrum representation for the field and not directly in the form of a factor multiplying the incident field as in the case of Keller's representation.

We may further use (4) and (7) and introduce the spectral coefficient of the physical optics field X^{po} as the Fourier transform of the physical optics induced current to arrive at

$$\text{for E-wave:} \quad X^{po}(k_x,\alpha) = \frac{2k_y}{\alpha + k_x} \tag{14a}$$

$$\text{for H-Wave:} \quad X^{po}(k_x,\alpha) = \frac{2i\sqrt{\alpha^2 - k^2}}{\alpha + k_x}. \tag{14b}$$

The application of the change of variables used in (11) allows one to express (14) as

$$\chi^{po}(\Omega,\psi) = \chi_i^{po}(\Omega,\psi) + \tau \chi_r^{po}(\Omega,\psi) \tag{15}$$

where

$$\chi_{\substack{i\\r}}^{po}(\Omega,\psi) = \mp \operatorname{ctn} \frac{\Omega \mp \psi}{2}. \tag{16}$$

It is worthwhile to mention that χ^f, as defined in the following equation, is bounded at the shadow boundaries

$$\chi^f(\Omega,\psi) = \chi(\Omega,\psi) - \chi^{po}(\Omega,\psi). \tag{17}$$

$\chi^f(\Omega,\psi)$ could be called the fringe diffraction coefficient.

For the problem at hand, i.e., incident plane wave, the spectral integral (11) can be expressed exactly in terms of the Fresnel integral, viz.,

$$u^s = -e^{ik\rho\cos(\Omega-\phi)} F(-\xi_i) + \tau\, e^{ik\rho\cos(\Omega+\phi)} F(\xi_r) \qquad (18)$$

where the Fresnel integral F is defined as

$$F(\xi) = \frac{e^{-i\pi/4}}{\sqrt{\pi}} \int_\xi^\infty e^{it^2}\, dt \qquad (19)$$

and

$$\xi_{\substack{i\\r}} = \mp\sqrt{2k\rho}\,\sin\frac{\Omega \mp \phi}{2}. \qquad (20)$$

Using the analytic continuation argument, one can show that, for complex angles of incidence, (18) is still the proper solution of the diffraction problem. In this context Ω is replaced by the complex angle ω which follows the path Γ_i, $[(i\infty, 0) \cup (0, \pi) \cup (\pi, -i\infty)]$, in the complex ω-plane to cover the infinite spectrum of incidence angles.

3. FINITE BODIES WITH EDGES -- COMPARISON OF SPECTRAL AND UNIFORM THEORIES FOR A TYPICAL EXAMPLE -- THE STRIP

In the previous section we introduced the concept of spectral domain representation of the field scattered from a half-plane and showed how the difficulties with Keller's formulas in the transition regions could not only be interpreted, but circumvented as well. Though the above result may be interesting from a theoretical point of view, the diffraction formula would be even more useful if it were applied to practical structures, which are obviously of finite extent. The classical GTD formulas, when applied to such finite structures, still contain spurious infinities*, notwithstanding the fact that the spectrum of the induced current, and hence the scattered field, must be necessarily finite. Thus, in contrast to the half-plane problem for which it was necessary to express the scattered field in terms of the integral representation (11) because of the singularities in the diffraction coefficient, the scattered far field for the finite structure should be obtainable directly from the spectrum of the induced current by a straightforward application of the saddle point integration technique (which would merely require a substitution of variables) in the integral representation for the

---

* Except in special situations when the aggregate contributions of these infinities cancel.

scattered field. If the transform or the spectrum of the induced current is properly computed, the far-field expression derived from it should be uniformly valid for all observation angles, including the transition regions, and consequently, no a posteriori correction would be needed for formulas derived in this manner. This is in contrast to conventional GTD formulas which are typically "repaired," after the fact, by employing one of the uniform theories, e.g., the Uniform Theory of Diffraction (UTD) of Kouyoumjian and Pathak [18] or Uniform Asymptotic Theory of Lewis, Ahluwalia and Boersma [17]. With this background in mind, we proceed in the following section to compare the various uniform theory formulas with the one derived from the spectral approach by considering a simple illustrative example, viz., the strip. We show that not only are the spectral formulas uniform in their original format, and therefore require no a posteriori correction, they are actually simpler and more accurate as well. The increased accuracy results from the inclusion of a term which arises naturally in the spectral formulas from a physical interpretation of the diffraction phenomenon, but is shown to be missing in the other uniform formulas which are based on the assumption that diffraction is a "local" phenomenon. We will see that the absence of this term can create non-trivial and non-physical discontinuities in the scattered far-field.

3.1. Comparison of STD and Uniform Asymptotic Theory (UAT)

The uniform asymptotic theory (UAT) of Lewis, et. al. [17] has been introduced to overcome some of the difficulties in GTD that occur at the transition regions associated with shadow and reflection boundaries. In UAT, the fictitious infinities in Keller diffraction coefficients for edges are annihilated by the introduction of additional terms which themselves go to infinity at the shadow boundaries such that the singularities in the Keller coefficients are canceled out exactly.

We first present some uniform formulas based on UAT for some typical edge diffraction problems, and then show that the UAT ansatz cannot only be physically interpreted using the spectral approach, but can also be improved and generalized in a straightforward manner. To illustrate our point, we consider the problem of diffraction by a strip of width 2a which is illuminated by an H-polarized wave (see Figure 2). The incident angle is ϕ_0, and other variables such as ρ_1, ϕ_1, ρ_2, and ϕ_2 are shown in the diagram.

Following the UAT prescription [16], we write the total field as:

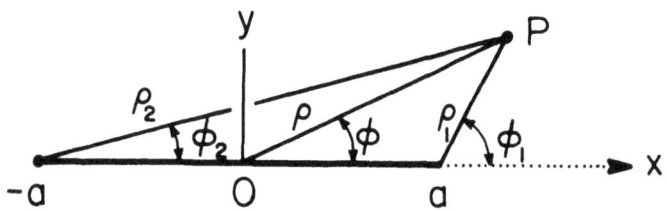

Figure 2. Geometry of the strip problem.

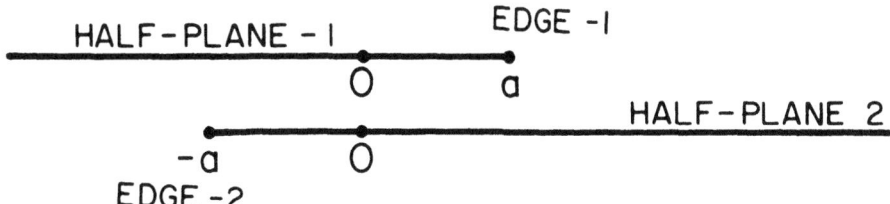

Figure 3. Two half-planes corresponding to two edges of the strip.

$$H^t = H^i[F(\xi_1^i) - \hat{F}(\xi_1^i)] + H^r[F(\xi_1^r)] - \hat{F}(\xi_1^r)] + H_1^d - H^i$$

$$+ H^i[F(\xi_2^i) - \hat{F}(\xi_2^i)] + H^r[F(\xi_2^r) - \hat{F}(\xi_2^r)] + H_2^d - H^r \qquad (21)$$

where H^i is the incident field given by

$$H^i = e^{-ik\rho\cos(\phi-\phi_0)}. \qquad (22)$$

H^r is the reflected field (from an infinite plane) and has the form:

$$H^r = e^{-ik\rho\cos(\phi+\phi_0)}. \qquad (23)$$

F is the Fresnel integral defined in (19)

and

$$\hat{F}(x) = \frac{1}{2x\sqrt{\pi}} e^{i(x^2+\pi/4)}. \qquad (24)$$

$\hat{F}$ is the asymptotic form of F for large positive argument. The subscripts 1 and 2 refer to the two edges which form the origins of the coordinate systems for the corresponding diffracted fields associated with these edges.

The diffracted fields H_1^d and H_2^d are derived from Keller's GTD formulas in a standard manner and are expressed as:

$$H_1^d = e^{ika\cos\phi_0} \frac{e^{i(k\rho_1 + \pi/4)}}{\sqrt{8\pi k\rho_1}} \left[\frac{1}{\cos\left[\frac{\phi_0+\phi_1}{2}\right]} - \frac{1}{\cos\left[\frac{\phi_1-\phi_0}{2}\right]} \right] \qquad (25a)$$

and

$$H_2^d = -e^{ikacos\phi_0} \frac{e^{i(k\rho_2 + \pi/4)}}{\sqrt{8\pi k\rho_2}} \left[\frac{1}{\cos\left[\frac{\phi_2-\phi_0}{2}\right]} + \frac{1}{\cos\left[\frac{\phi_2+\phi_0}{2}\right]} \right]. \quad (25b)$$

Finally, $\xi_1^{i,r}$, $\xi_2^{i,r}$, which are the so-called "detour functions," are given by

$$\xi_1^i = -\sqrt{2k\rho_1} \sin \frac{1}{2}(\pi + \phi_0 - \phi_1), \quad 0 \le \phi_1 \le \pi \quad (26a)$$

$$\xi_1^r = \sqrt{2k\rho_1} \sin \frac{1}{2}(\pi - \phi_0 - \phi_1), \quad 0 \le \phi_1 \le \pi \quad (26b)$$

and similarly for the edge 2.

Note that we have deliberately added the term $-(H^r+H^i)$ to the UAT-prescribed formula in order to obtain the proper GO limit, although no direct allusion for such a required modification is found in the UAT formulas.

Let us now concentrate on the upper hemisphere since the discussion pertaining to the lower hemisphere is expected to be very similar in view of the symmetry of the problem. For the region $y > 0$, $F(\xi_{1,2}^i) - \hat{F}(\xi_{1,2}^i) \approx 1$, and we can simplify (21) to obtain:

$$H^t = H^i + H^r[F(\xi_1^r) - \hat{F}(\xi_1^r) + F(\xi_2^r) - \hat{F}(\xi_2^r) - 1] + H_1^d + H_2^d. \quad (27)$$

We note from (27) that the geometrical optics field has been replaced in the UAT formulation by:

$$H^i + H^r[\{F(\xi_1^r) - \hat{F}(\xi_1^r)\} + \{F(\xi_2^r) - \hat{F}(\xi_2^r)\} - 1],$$

whereas the Keller GTD terms, viz., H_1^d and H_2^d, have remained unchanged. This is the essence of the UAT ansatz as interpreted by Lee [16].

We will now show that the UAT formula (27) cannot only be physically interpreted, but improved and generalized as well.

To this end, we rewrite (27) as

$$H^t = H^i + (H_1^d - H_1^p) + (H_2^d - H_2^p) + \{H^r[F(\xi_1^r) - \hat{F}(\xi_1^r)$$

$$+ F(\xi_2^r) - \hat{F}(\xi_2^r) - 1] + H_1^p + H_2^p\} \tag{28}$$

where H_1^p and H_2^p are proportional to the Fourier transforms (or spectrum) of the physical optics currents that would exist on *semi-infinite* half-planes erected at edges 1 and 2, respectively (see Figure 3). These quantities are given by:

$$H_1^p = \frac{e^{i(k\rho_1 + \pi/4)}}{\sqrt{8\pi k\rho_1}} (\sin\phi_1) \cdot 2 \frac{e^{ika\cos\phi_0}}{\cos\phi_0 + \cos\phi_1} \tag{29a}$$

$$H_2^p = \frac{-e^{i(k\rho_2 + \pi/4)}}{\sqrt{8\pi k\rho_2}} (\sin\phi_2) \cdot 2 \frac{e^{ika\cos\phi_0}}{\cos\phi_0 + \cos\phi_2} . \tag{29b}$$

Note that $H_{1,2}^p$ have exactly the same singularities at the shadow boundaries ($\phi_{1,2} = \pi - \phi_0$) as the Keller coefficients. This is not unexpected, however, since it was pointed out in [24] that the singularities in Keller diffraction coefficients can be identified as the contributions of the physical optics currents of semi-infinite extent. In fact, the two terms H_1^p and H_2^p have been deliberately introduced in the expression given in (28) so that we can provide a physical interpretation of the various terms. We can, for instance, identify $(H_1^d - H_1^p)$ as the spectrum or the far-field associated with the J^1 current on the semi-infinite half plane -1, J^1 being the difference between the total current and the physical optics current on the same half-plane. A similar interpretation can obviously be given to the companion term $(H_2^d - H_2^p)$. We now proceed to show that the terms appearing in the curly braces in (28) have a very simple physical interpretation, and can be replaced by a much simpler form that does not require the use of Fresnel integrals. Additionally, we find that the spectral domain approach also provides an insight into the generalization of the formulas for curved surfaces.

In order to develop a physical interpretation of the UAT formula, we turn now to the spectral domain interpretation of the scattered field. Unlike the ray optical representation, which expresses the total field as a superposition of geometrical optics (GO) and diffracted fields, we use the conventional representation of the total field as:

$$H^t = H^i + H^s \tag{30}$$

where H^s is the scattered field. The reason for choosing this form for the total field rather than its ray-optical counterpart

$$H^t = H^g + H^d \tag{31}$$

is that the geometrical optics field H^g is $O(k^0)$ and is obviously discontinuous at the shadow boundaries. Since the total field must be continuous everywhere in space, it follows that the exact H^d must also be discontinuous. Consequently, neither of these fields can be associated with physical induced currents on the surface of the scatterer, for the obvious reason that these currents produce continuous fields everywhere in space external to the scatterer.

Thus, we return to (30) and interpret the scattered field H^s as proportional to the spectrum or Fourier transform of the induced current J on the scatterer. The next step is to derive a high frequency approximation of these currents under the same assumptions as employed in GTD. Basically, we know that for a single half-plane erected at edge -1, the induced current J can be thought of as a superposition of a physical optics current J^0 and a higher-order ($k^{-1/2}$) component, which we call J^1. For the truncated half-plane, i.e., the strip, we can approximate J as

$$J = J^b + J_1^{tr} + J_2^{tr} \tag{32}$$

where J^b is the physical optics current on the strip and J_1^{tr}, J_2^{tr} are the $O(k^{-1/2})$ current components from edges 1 and 2, respectively, truncated over the strip. Thus, we can write

$$J_{1,2}^{tr} = J_{1,2}^1 - J_{1,2}^\ell \tag{33}$$

where J_1^1 is the $O(k^{-1/2})$ current induced on the semi-infinite half plane 1 and J_1^ℓ is the portion of that current for $x > a$. We can similarly interpret the quantities with subscript 2, which are associated with half-plane 2.

The final step is to invoke the Fourier transform relationship between the scattered field and the induced current, which reads:

$$H^s = -ik \frac{e^{i(k\rho+\pi/4)}}{\sqrt{8\pi k\rho}} \sin\phi \, \tilde{J}_x \, \hat{z} \tag{34}$$

where $\tilde{J}_x$ is the transform or the spectrum of J_x. Thus, from (32) and (33), we can write

$$H^t = H^i + H^s = H^i + (H_1^d - H_1^p) + (H_2^d - H_2^p) + H^b - H_1^\ell - H_2^\ell \tag{35}$$

where we have again made use of the spectral domain interpretation of the scattered field produced by J^1 and have expressed this field as the difference between the Keller diffracted field, H^d, and the field, H^p, which is proportional to the transform of the physical optics current on the half-plane. Also, we have used the symbol H^b for the scattered field due to the physical optics current on the strip. The expression for H^b is very straightforward, and is given by

$$H^b = -ik \frac{e^{i(k\rho+\pi/4)}}{\sqrt{8\pi k\rho}} (\sin\phi) \, 4a \, \frac{\sin[ka(\cos\phi_0+\cos\phi)]}{ka(\cos\phi_0+\cos\phi)} . \tag{36}$$

Let us now compare the UAT expression (28) and the STD expression (35). To facilitate this comparison, we have numerically evaluated H^b and H^u where

$$H^u = H^r[F(\xi_1^r) - \hat{F}(\xi_1^r) + F(\xi_2^r) - \hat{F}(\xi_2^r) - 1] + H_1^p + H_2^p . \tag{37}$$

Shown in Figure 4 are the plots of H^b and H^u for a 2λ wide strip and $\rho = 50\lambda$, which is the distance of the observation point from the origin of the coordinate system located at the center of the strip. It is evident that the two quantities are virtually indistinguishable. Thus, we have a simple expression for H^b which is equivalent to the collection of several involved expressions that comprise H^u.

Next, we note a conspicuous absence of the $H_{1,2}^\ell$ terms in (28).

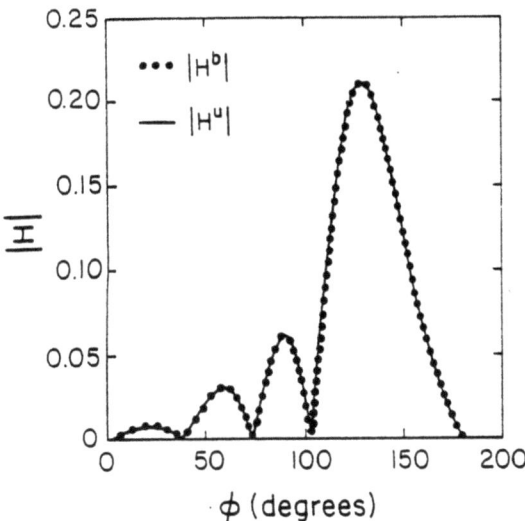

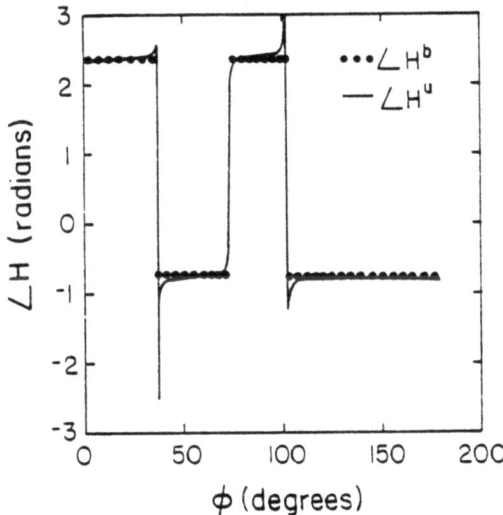

Figure 4. Comparison of STD and UAT for a 2λ strip: (a) Magnitude of H^b (STD; cf. Eq. 36) and magnitude of H^u (UAT; cf. Eq. 37); (b) Phase of H^b (STD; cf. Eq. 36) and phase of H^u (UAT; cf. Eq. 37).

Although these terms can make significant contributions to the scattered field, they are not present in any of the ray optical expressions derived to-date, perhaps because it is difficult to extract them from a ray-optical approach. We demonstrate a little later that the exclusion of these terms from the scattered field can introduce significant errors in some cases.

Returning to the comparison of (28) and (35), we draw the conclusion that the application of the STD concept provides us with a simpler and more accurate expression than the UAT. In particular, we note that the expressions of each of the terms H^d, H^p and H^b in (35) are trigonometric in nature, whereas (28) requires the evaluation of the Fresnel integrals and cancellation of numbers which are equal, opposite, and infinitely large in the vicinity of the shadow boundaries. Next, we demonstrate that the same physical interpretation can also be applied to UTD, the Kouyoumjian-Pathak uniform formula [14]. The simplification in the expression for the scattered field is equally dramatic.

3.2. Comparison of STD with Uniform Theory of Diffraction (UTD)

The uniform theory of diffraction (UTD) of Kouyoumjian-Pathak has also been introduced to overcome some of the difficulties in GTD that occur at the transition regions associated with shadow and reflection boundaries. However, in contrast to UAT, the fictitious infinities in Keller diffraction coefficients are eliminated in UTD by the introduction of multiplicative factors which go to zero at the shadow boundaries such that the product has a finite discontinuity that matches the discontinuity in the geometrical optics field. As a result, the total field computed by such a modified diffraction coefficient is continuous for all observation angles.

We first present the UTD solution of the problem of diffraction by a strip of width 2a illuminated by an H-polarized wave (see Figure 2), and then show that the UTD ansatz can also be physically interpreted using the spectral approach. Furthermore, as in UAT, the UTD formulas can be improved and generalized for complex geometries in a straightforward manner.

Following the UTD formulation [14], we write the total field for $y > 0$ as

$$H^t = H^i + [\theta(-\varepsilon_{r1}) + \theta(-\varepsilon_{r2}) - 1] H^r + H_1^D + H_2^D \tag{38}$$

where H^i and H^r are given in (22) and (23), respectively; $\theta(\cdot)$ is the step function; ε_{r1} and ε_{r2} are the shadow indicators associated with half planes 1 and 2, respectively. A shadow indicator takes the value 1 in the shadow region, and 0 in the lit region. The $\theta(-\varepsilon_{r1})$ and $\theta(-\varepsilon_{r2})$ for this example can be written explicitly as

$$\theta(-\varepsilon_{r1}) = \frac{1}{2} + \text{sgn}[-(\pi-\phi_0) + \phi_1] \cdot \frac{1}{2} \tag{39a}$$

$$\theta(-\varepsilon_{r2}) = \frac{1}{2} + \text{sgn}(\pi-\phi_0-\phi_2) \cdot \frac{1}{2} \tag{39b}$$

where

$$\text{sgn}(x) = \frac{x}{|x|} = \begin{cases} 1 & \text{for } x > 0 \\ -1 & \text{for } x < 0 \end{cases} \tag{39c}$$

is simply a sign function.

The H_1^D and H_2^D in (38) are the diffracted fields calculated by the UTD formulas. Using the modified diffraction coefficients, the Keller GTD diffracted fields $H_{1,2}^d$ as given in (25a) and (25b) can be made uniform according to UTD ansatz and become $H_{1,2}^D$, which are given below

$$H_1^D = e^{-ika\cos\phi_0} \left\{ -e^{ik\rho_1} \left[F(k\rho_1;\phi_0-\phi_1) e^{-i2k\rho_1\cos^2\{(\phi_0-\phi_1)/2\}} \cdot \right. \right.$$

$$\text{sgn}(\pi-\phi_0+\phi_1)$$

$$\left. \left. + F(k\rho_1;\phi_1+\phi_0) e^{-i2k\rho_1\cos^2\{(\phi_1+\phi_0)/2\}} \text{sgn}(\phi_0-\pi+\phi_1) \right] \right\} \tag{40a}$$

$$H_2^D = e^{ika\cos\phi_0} \left\{ -e^{ik\rho_2} \left[F(k\rho_2;\phi_2-\phi_0) e^{-i2k\rho_2\cos^2\{(\phi_2-\phi_0)/2\}} \cdot \right. \right.$$

$$\text{sgn}(\pi+\phi_0-\phi_2)$$

$$\left. \left. + F(k\rho_2;\phi_2+\phi_0) e^{-i2k\rho_2\cos^2\{(\phi_2+\phi_0)/2\}} \text{sgn}(\pi-\phi_0-\phi_2) \right] \right\} \tag{40b}$$

where sgn(x) is defined in (39c) and F(x) is defined in (19) with the argument given by

$$x = \sqrt{2k\rho} \left| \cos \frac{\phi \pm \phi_0}{2} \right| \quad (40c)$$

where $\rho = \rho_1$ or ρ_2, and $\phi = \phi_1$ or ϕ_2, accordingly.

In order to give a physical interpretation to the UTD formulation, we now rearrange the terms in (38) following the steps that are similar to the ones used in conjunction with (21). In UAT, these steps led to (28), which was then compared with the STD expression (35). To this end, we rewrite the total field given in (38) as

$$H^t = H^i + (H_1^d - H_1^P) + (H_2^d - H_2^P)$$

$$+ \{[\theta(-\varepsilon_{r1}) + \theta(-\varepsilon_{r2}) - 1] H^r + H_1^D - H_1^d + H_1^P + H_2^D - H_2^d + H_2^P\}. \quad (41)$$

Note that $H_{1,2}^d$ and $H_{1,2}^P$ have been added and subtracted from (38) in writing (41). H_1^d and H_2^d are given in (25a) and (25b), respectively. H_1^P and H_2^P are given in (29a) and (29b), respectively.

Following the same steps which led to H^u in (37), we obtain the H_u, which is the UTD counterpart of E^u:

$$H_u = [\theta(-\varepsilon_{r1}) + \theta(-\varepsilon_{r2}) - 1] H^r + H_1^D - H_1^d + H_1^P + H_2^D - H_2^d + H_2^P, \quad (42)$$

which may be identified as the collection of terms inside the braces { } in (41).

Shown in Figure 5 are the plots of H^b given in (36) and H_u given in (42) for a 2λ wide strip, and $\rho = 50\lambda$, which is the distance of the observation point from the origin of the coordinate systems located at the center of the strip. We have thus demonstrated that the Kouyoumjian-Pathak uniform theory (UTD) can also be physically interpreted using the spectral concept. The simplification in the expression for the scattered field is equally dramatic, as evidenced by a comparison of H^b with H_u, the former being trigonometric and the latter involving the Fresnel integrals. Moreover, H^b is continuous throughout the entire region of observation, whereas H_u

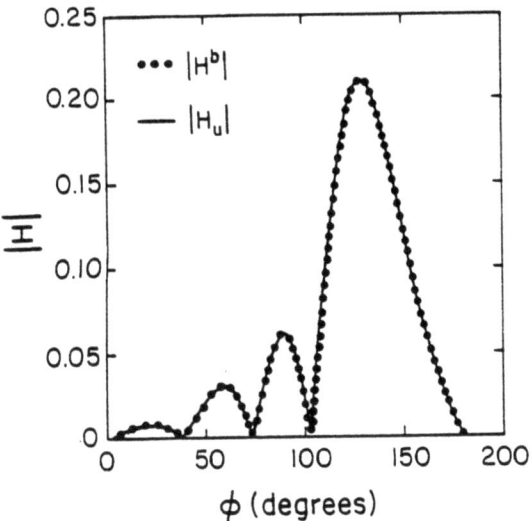

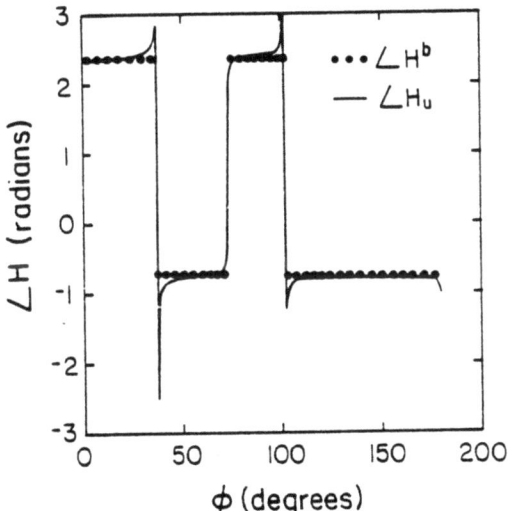

Figure 5. Comparison of STD and UTD for a 2λ strip: (a) Magnitude of H^b (STD; cf. Eq. 36) and magnitude of H_u (UTD; cf. Eq. 42); (b) Phase of H^b (STD; cf. Eq. 36) and phase of H_u (UTD; cf. Eq. 42).

has terms with discontinuities in both the reflected and the diffracted parts of the expression.

3.3. Improvement of UAT and UTD

Next, we go on to show that the inclusion of $-(H_1^\ell + H_2^\ell)$ in the expression for the scattered field diffracted from edges can be very important for accurate evaluation of the field. To illustrate our point, we consider the example of a rectangular cylinder illuminated by a plane wave (see Figure 6). Using conventional GTD uniform formulas, we obtain the scattering pattern shown in Figure 7. The STD result (whose details derivation can be found in [28]), which has been verified numerically by comparison with the moment method solution of the scattered field, is shown in the companion diagram, Figure 8. It is evident that the GTD results contain discontinuities (~3dB) in the vicinity of angles $\phi = 0$, $\pi/2$, π and $3\pi/2$, which, incidentally, are located far from the shadow boundaries. The investigation of the wedge problem also reveals that the generalization of the STD approach to wedge-type structures can be carried out in a very simple manner, and the resulting expressions are once again very simple, the Fresnel integrals being absent here, too.

In summary, we have demonstrated that the uniform GTD expressions for edge diffraction formulas can be simplified, physically interpreted, refined and generalized using the STD concept.

4. SYSTEMATIC IMPROVEMENTS OF ASYMPTOTIC SOLUTIONS

One of the most challenging problems in the solution of high-frequency scattering analyses is the establishment of the accuracy of the results and the refinement of the solution when the need for its improvement is clearly indicated. The difficulty in verifying whether the asymptotic expression, typically derived from the ray approach, does indeed solve the boundary value problem under consideration stems primarily from the fact that there is no obvious way to "build in" the boundary conditions in solution procedures based on ray methods. Another reason is that the high-frequency solutions are often constructed for the radiated far fields, whereas the application of the boundary conditions clearly requires the near-field information. In contrast, the integral-equation formulation for the scattering problem is based directly on the application of the boundary condition and, consequently, the boundary-condition check is redundant for this approach. However, the conventional moment-method solution of integral equations is limited strictly to the low-frequency and resonance regions as the matrix size becomes unmanageably large

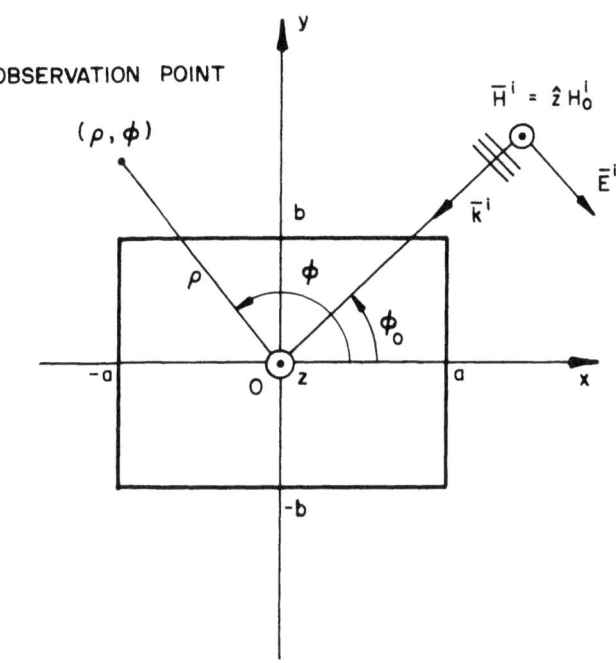

Figure 6. Diffraction by a rectangular cylinder illuminated by an H-polarized plane wave incident at an angle ϕ_0.

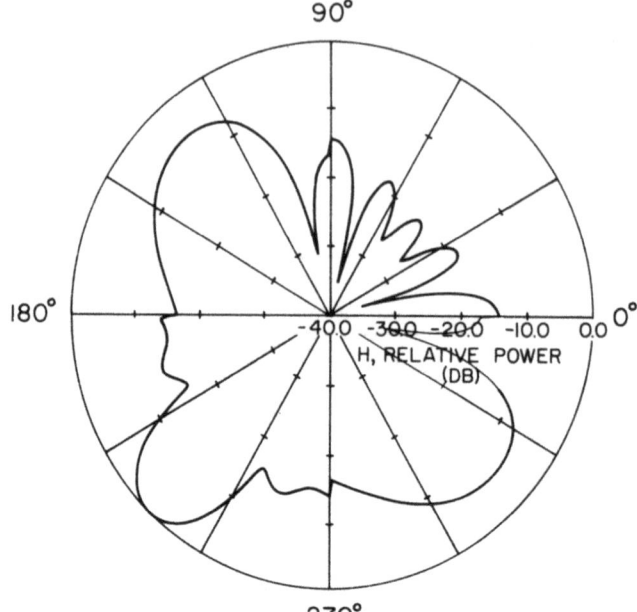

Figure 7. GTD diffracted far-field pattern of the rectangular cylinder; $\phi_0 = \pi/4$, $a = b = 1\lambda$. Note that the discontinuities in the scattered field are on the order of 3 dB.

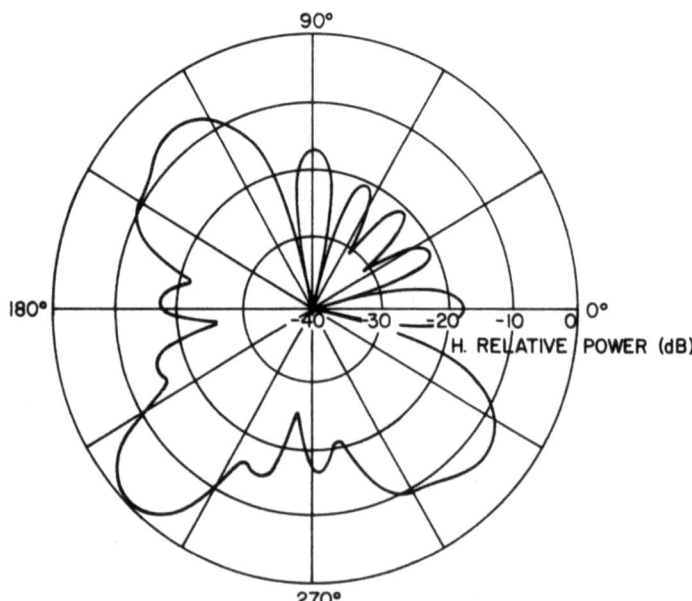

Figure 8. Scattered far-field pattern of the rectangular cylinder using STD approach; $\phi_0 = \pi/4$, $a = b = 1\lambda$.

beyond the resonance region.

In this section we will briefly outline a spectral-domain method for bridging the two approaches, viz., the integral-equation and asymptotic techniques. The hybrid method has the desirable feature that it not only verifies the accuracy of the ray solutions but provides a systematic means for improving the solution for a large class of problems of practical interest. This fact will be illustrated via a typical example, viz., plane-wave diffraction by a strip. Other cases have also been treated and may be found in [26, 27] [32,33].

4.1. Development of Spectral-Domain Formulation of the Integral Equation and Its Iterative Solution

The key to combining the asymptotic solution with the integral-equation formulation lies in recognizing the fact that the Fourier transform of the induced current on a scatterer is directly related to the scattered far field and that a good approximation to this scattered field is often available from any asymptotic method, e.g., GTD. To take advantage of these facts, we choose to work with the "Fourier-transformed" or "spectral-domain" version of the integral equation rather than with the conventional spatial-domain counterpart. We begin, however, with the conventional electric-field integral equation (E-equation) for a perfectly conducting scatterer:

$$(\bar{\bar{G}} * \bar{J})_t = -\bar{E}^i_t, \quad \bar{\bar{G}} = (\bar{\bar{I}} + \frac{1}{k^2}\nabla\nabla)e^{+ik|\bar{r}-\bar{r}'|}/(4\pi|\bar{r}-\bar{r}'|) \quad (43)$$

where $\bar{\bar{G}}$ is the Green's Dyadic, $\bar{\bar{I}}$ is the unit dyad and $\bar{J}(\bar{r}')$ is the unknown induced surface current density. The subscript t signifies the tangential component of the field on the surface S of the scatterer, $\bar{E}^i$ is the incident electric field on the scatterer, and * symbolizes the convolution operation.

As a preamble to Fourier transforming (43), we first extend it over all space. To this end, we define a truncation operator $\theta(\bar{A})$:

$$\theta(\bar{A}) = \int \bar{A}_t \delta(\bar{r}-\bar{r}_s) \, d\bar{r}, \quad r_s \in S \quad (44)$$

where δ is the Dirac delta function. Let $\hat{\theta}(\bar{A})$ be defined as the complementary operator.

$$\hat{\theta}(\bar{A}) = \bar{A} - \theta(\bar{A}). \quad (45)$$

We can then rewrite (43)

$$\bar{\bar{G}} * \bar{J} = \theta(-\bar{E}^i) + \hat{\theta}(\bar{\bar{G}} * (\theta\bar{J})), \quad \text{for all space.} \tag{46}$$

As indicated above, in contrast to (43), (46) is valid at all observation points whether on or off the surface S. Note that the integral equation (43) is embedded in (46) and that we have made use of the obvious identity $\theta\bar{J} = \bar{J}$. We have also dropped the subscript t in writing (46), because by referring to (44) we observe that the θ operator selects the tangential component of the function in its argument.

Next we Fourier transform (46) by introducing the transform relationships

$$\tilde{F}(\bar{k}) = \int_{-\infty}^{\infty} F(\bar{r}) \, e^{-i\bar{k}\cdot\bar{r}} \, d\bar{r} = F[F(\bar{r})] \tag{47a}$$

and

$$F(\bar{r}) = \left(\frac{1}{2\pi}\right)^3 \int_{\infty}^{\infty} \tilde{F}(\bar{k}) \, e^{i\bar{k}\cdot\bar{r}} \, d\bar{k} = F^{-1}[\tilde{F}(\bar{k})] \tag{47b}$$

with ~ on top denoting the transformed quantities.

The transformed version of (46) reads

$$\tilde{\bar{\bar{G}}}\tilde{\bar{J}} = -\tilde{\bar{E}}_I + \tilde{\bar{F}} \tag{48}$$

where $\tilde{\bar{F}} = F[\hat{\theta}(\bar{\bar{G}} * (\theta\bar{J}))]$ and $\tilde{\bar{E}}_I$ is the transform of the tangential component of the incident field truncated on S. Note that the convolution operations in (46) are transformed into an algebraic product upon Fourier transformation.

A formal solution to (48) can now be written as

$$\tilde{\bar{J}} = \tilde{\bar{\bar{G}}}^{-1}(-\tilde{\bar{E}}_I + \tilde{\bar{F}}). \tag{49}$$

Equation (49) indicates that if we had available the Fourier transform of the scattered electric field, we could construct the solution for the induced surface current density in the transform domain by adding it to $-\bar{E}_I$, which is known, and by performing an algebraic division represented by $\bar{\bar{G}}^{-1}$. In practice, of course, $\bar{F}$ is not known and must be solved for along with $\bar{J}$ if (49) is to be used in the form as shown. However, instead of using this form, we proceed to derive an iterated form of the equation as shown below:

$$\tilde{J}^{(n+1)} = \tilde{\bar{G}}^{-1}(-\tilde{\bar{E}}_I + \tilde{\bar{F}}^{(n)}) \tag{50}$$

which indicates that the $(n + 1)$th approximation of $\tilde{\bar{J}}$ can be derived from the nth approximation for $\bar{F}$. We next show how $\tilde{\bar{F}}^{(n)}$ itself can be derived from $\tilde{J}^{(n)}$. To this end, we use the identity

$$\tilde{\bar{F}} = F[F^{-1}[\tilde{\bar{G}}\tilde{\bar{J}}] - \theta(F^{-1}[\tilde{\bar{G}}\tilde{\bar{J}}])] \tag{51}$$

which may be verified by writing (51) as

$$\tilde{\bar{F}} = F[\bar{\bar{G}} * \bar{J} - \theta(-\bar{E}^i)] \tag{52}$$

and using (46) to get

$$\tilde{\bar{F}} = F[\hat{\theta}(\bar{\bar{G}} * (\theta\bar{J}))] \tag{53}$$

which, of course, is the definition of $\tilde{\bar{F}}$. We can now use (51) to derive the nth approximation $\tilde{\bar{F}}^{(n)}$ of $\tilde{\bar{F}}$ from the nth approximation of $\tilde{\bar{J}}$, i.e., $\tilde{J}^{(n)}$. The relationship is written as

$$\tilde{\bar{F}}^{(n)} = F[F^{-1}[\tilde{\bar{G}}\tilde{J}^{(n)}] - \theta(F^{-1}[\tilde{\bar{G}}\tilde{J}^{(n)}])]. \tag{54}$$

The desired iteration formula relating $\tilde{J}^{(n+1)}$ and $\tilde{J}^{(n)}$ may now be written by using (50) and (54). The expression is

$$\tilde{J}^{(n+1)} = \tilde{\bar{G}}^{-1}[-\tilde{\bar{E}}_I + F[F^{-1}[\tilde{\bar{G}}\tilde{J}^{(n)}] - \theta(F^{-1}[\tilde{\bar{G}}\tilde{J}^{(n)}])]]. \tag{55}$$

4.2. Procedure for Applying the Iterative Method

The step-by-step procedure for constructing the solution of the transformed surface current $\tilde{\tilde{J}}$ will now be given:

1. Begin with an estimate of $\tilde{\tilde{J}}^{(0)}$, which is the Fourier transform of the induced surface current, or equivalently, the scattered *far field* within a known multiplicative factor. That the scattered far field is directly related to the Fourier transform of the induced current is well-known in electromagnetics and has been derived in several standard texts (see for instance Eq. (2.23b) of [34]. Typically, the initial approximation for $\tilde{J}$, viz., $\tilde{\tilde{J}}^{(0)}$, can be obtained as follows:

    (a) Estimate $\tilde{\tilde{F}}$, the Fourier transform of the scattered field, $\bar{F}$, outside the scatterer, using GTD or other asymptotic solutions.

    (b) Subtract $\tilde{\tilde{E}}_I$, the Fourier transform of the tangential component of the incident electric field truncated to the surface of the scatterer.

    (c) Multiply the result of Step (b) by $\tilde{\tilde{G}}^{-1}$. Note that $\tilde{\tilde{G}}^{-1}$ is known and the operation is algebraic. The Fourier transform is typically done numerically using the Fast Fourier Transform (FFT).

    (d) Take the inverse Fourier transform of the result of Step (c) and truncate it to the surface of the scatterer to obtain $\tilde{J}^{(0)}$, the initial approximation for $\tilde{J}$.

2. Multiply $\tilde{\tilde{J}}^{(0)}$ by $\tilde{\tilde{G}}$, the known transform of the Green's Dyadic. Note this involves algebraic multiplication and not the usual time-consuming convolution operation.

3. Take the inverse Fourier transform of the product $\tilde{\tilde{G}}\tilde{\tilde{J}}^{(0)}$ using both *visible* and *invisible ranges*.

4. Apply the truncation-projection operator θ to $F^{-1}[\tilde{\tilde{G}}\tilde{\tilde{J}}^{(0)}]$, which gives the approximation to the tangential component of the scattered electric field $\bar{E}_t^S$ on the surface S. The accuracy of the solution can be conveniently checked at this point by verifying the satisfaction of the boundary condition by the tangential component of $\bar{E}^S$, viz., $\{\bar{E}_t^S = -\bar{E}_t^I\}$ on S. As mentioned in the introduction, this is an important feature of the method.

5. Subtract $\theta(F^{-1}[\tilde{\tilde{G}}\tilde{\tilde{J}}^{(0)}])$ from the total $F^{-1}[\tilde{\tilde{G}}\tilde{\tilde{J}}^{(0)}]$ already evaluated.

6. Take the Fourier transform of the difference obtained in Step 5.

7. Subtract $\tilde{\tilde{E}}_I$, the Fourier transform of the tangential component

of the incident electric field truncated on the surface, from the result in Step 6.

8. Multiply the result obtained in Step 7 by $\bar{\bar{G}}^{-1}$. Note that $\bar{\bar{G}}^{-1}$ is also known and the operation is again algebraic as in Step 2.

9. Take the inverse Fourier transform of $\tilde{\bar{J}}^{(1)}$ obtained in Step 8 and evaluate it on S to get the desired induced surface current on the scatterer. In other words, perform the operation $\theta(F^{-1}[\tilde{\bar{J}}^{(1)}])$. For an exact solution, this operation is redundant, since $\bar{J} = \theta\bar{J}$, and hence, $\theta(F^{-1}[F[\theta\bar{J}]]) = \theta\theta\bar{J} = \bar{J}$. However, the Fourier inversion of an nth approximate solution $\tilde{\bar{J}}^{(n)}$ will not give rise to a current distribution that is nonzero except on S. This step provides a test for the accuracy and for the convergence of the approximate solution by comparing the approximate $\bar{J}^{(0)}$ with $F[\theta(F^{-1}[\tilde{\bar{J}}^{(1)}])]$.

10. Take $F[\theta(F^{-1}[\tilde{\bar{J}}^{(1)}])]$ to derive an *improved* approximation for $\bar{J}^{(1)}$.

11. Repeat as necessary using, for instance, the *improved* $\tilde{\bar{J}}^{(1)}$ from Step 10 in the iteration Equation (55) to generate the next higher-order approximation $\tilde{\bar{J}}^{(2)}$.

We now show in some detail the application of the iteration procedure just described to a two-dimensional scattering problem, viz., the diffraction by a strip.

4.3. Diffraction by a Strip

The application of the general procedure outlined in the last section is illustrated in this section by using it to solve the two-dimensional problem of a plane-wave diffraction by a finite screen or a strip. This problem was chosen for the following reason: it is shown that when the angle of incidence is normal or near normal, the GTD solution accurately satisfies the boundary condition $E_{tan} = 0$ on the strip even when the multiple interaction between the two edges of the strip is neglected. However, it is found that when the angle of incidence is near grazing, the GTD solution is quite unsatisfactory, while the iterated solution generated by the hybrid technique does display the correct behavior.

The geometry of the electromagnetic scattering problem involving a perfectly conducting infinite strip of zero thickness illuminated by a uniform plane wave, whose electric intensity vector is oriented parallel to the edges of the strip, is depicted in Figure 9. For convenience of analysis, an arbitrary incident wave can always be decomposed into two components with respect to the z-axis, namely,

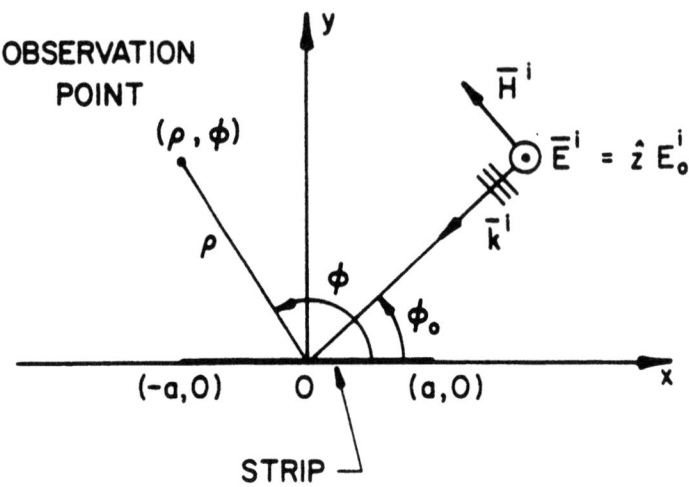

Figure 9. Diffraction by a strip illuminated by an E-wave.

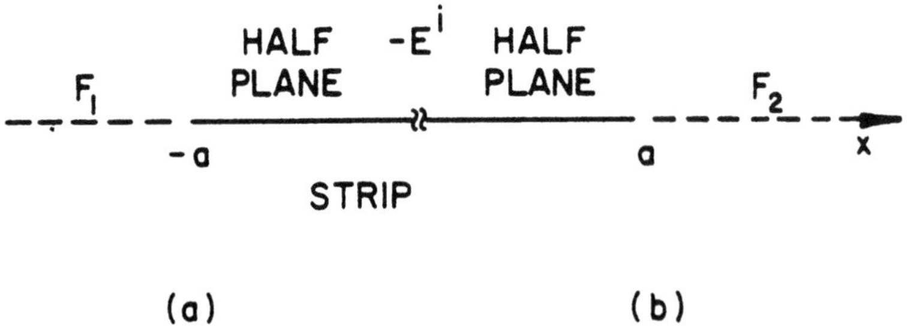

Figure 10. $F_1(x)$ can be approximated by the GTD solution to the half-plane problem (a) shown on the left-hand side; $F_2(x)$ can be approximated by the GTD solution to the half-plane problem (b) shown on the right-hand side.

TM_z (E-wave) and TE_z (H-wave). In the following discussion, we consider the E-wave case only; the H-wave case can be solved in a similar manner by considering $\bar{H}^i = \hat{z}H_0^i$.

The incident field is given by

$$E_z^i(\rho,\phi) = e^{-ik(x\cos\phi_0 + y\sin\phi_0)} \tag{56}$$

where the $e^{-i\omega t}$ time dependence is understood.

The integral equation formulation [35] for the problem at hand takes the form

$$-E_z^i(x) = \int_{-a}^{a} J_z(x')G(x-x')\,dx', \quad x \in [-a,a] \tag{57}$$

where $J_z(x')$ is the algebraic sum of the induced surface current densities on the top and the bottom surfaces of the thin strip. The kernel G is the two-dimensional free-space Green's function given by

$$G(x-x') = \frac{i}{4} H_0^{(1)}(k|x-x'|) \tag{58}$$

where $H_0^{(1)}$ is the Hankel function of the first kind of order zero, and $k = 2\pi/\lambda$ is the free-space propagation constant. Note that (57) is the conventional integral equation which equates the integral representation of the tangential component of the scattered E-field radiated by the induced surface current density to the negative of the tangential component of the incident E-field on the surface of the perfectly conducting scatterer as required by the satisfaction of the boundary condition. Hence, (57) is valid on the strip only.

An extended integral equation that is valid for all x can be obtained by including the scattered fields outside the strip as well. If the scattered field on the interval $(-\infty,-a)$ is designated by $F_1(x)$ and the scattered field on the interval (a,∞) is designated by $F_2(x)$, then the extended form of (57) becomes:

$$\int_{-a}^{a} J_z(x')G(x-x')\,dx' = \theta(-E_z^i(x)) + F_1(x) + F_2(x) \tag{59}$$

where θ is defined in (44).

Since the Fourier transform of the induced surface current density can be related to the far field, (59) is Fourier transformed to give

$$\tilde{J}_z(\alpha)\tilde{G}(\alpha) = \widetilde{\theta(-E_z^i)}(\alpha) + \tilde{F}_1(\alpha) + \tilde{F}_2(\alpha) \tag{60}$$

where ~ on top indicates the Fourier transform pair defined in (47) which simplifies in the present one-dimensional problem to

$$\tilde{F}(\alpha) = \int_{-\infty}^{\infty} F(x) e^{-i\alpha x} dx \tag{61a}$$

and

$$F(x) = \frac{1}{2\pi} \int_{-\infty}^{\infty} \tilde{F}(\alpha) e^{i\alpha x} d\alpha . \tag{61b}$$

The Fourier transform of the two-dimensional Green's function in (60) takes the form

$$\tilde{G}(\alpha) = \frac{i}{2\sqrt{k^2-\alpha^2}} . \tag{62}$$

Note that (60) is an algebraic equation in the spectral domain in contrast to the convolution form of the integral equation (59) in the spatial domain. The reason for working in the spectral domain will become clear when the method of solution for (60) is developed. Following the procedure discussed in the last section and in terms of the notations introduced in the present problem, we proceed as follows:

1. Obtain $\tilde{J}_z^{(0)}(\alpha)$, the initial approximation of the Fourier transform of the induced surface current density, or equivalently, the scattered *far field* within a known multiplicative factor, as follows:

   (a) Find the expressions for the first estimate of $\tilde{F}_1^{(0)}(\alpha) + \tilde{F}_2^{(0)}(\alpha)$. Note that GTD may be used to get closed-form expressions for $\tilde{F}_1^{(0)}(\alpha)$ and $\tilde{F}_2^{(0)}(\alpha)$ since these can be obtained from the GTD solutions to the two half-plane problems as shown in Figure 10. The expressions for $\tilde{F}_1^{(0)}$ and $\tilde{F}_2^{(0)}$ as obtained from GTD read

$$\tilde{F}_1^{(0)}(\alpha) = \frac{i}{2}\left[e^{ika\cos\phi_0}\sqrt{8k}\sin\frac{\phi_0}{2}\right]\frac{e^{i\alpha a}}{(\alpha + k\cos\phi_0)\sqrt{\alpha+k}}$$

$$- \frac{ie^{ia(\alpha+k\cos\phi_0)}}{(\alpha + k\cos\phi_0)} \tag{63}$$

and

$$\tilde{F}_2^{(0)}(\alpha) = -\frac{i}{2}\left[e^{ika\cos\phi_0}\sqrt{8k}\cos\frac{\phi_0}{2}\right]\frac{e^{i\alpha a}}{(\alpha + k\cos\phi_0)\sqrt{k-\alpha}}$$

$$+ \frac{ie^{-ia(\alpha+k\cos\phi_0)}}{(\alpha + k\cos\phi_0)} \tag{64}$$

Note that these expressions are free of singularities for all α.

(b) Solve for the initial approximation of $\tilde{J}_z(\alpha), \tilde{J}_z^{(0)}(\alpha)$, by carrying out the operations shown below:

$$\tilde{J}_z^{(0)}(\alpha) = F\left\{\theta\left(F^{-1}\left[\frac{\widetilde{\theta(-E^i)}(\alpha) + \tilde{F}_1^{(0)}(\alpha) + \tilde{F}_2^{(0)}(\alpha)}{\tilde{G}(\alpha)}\right]\right)\right\} .$$

2. Use (55) to further improve the solution as necessary. The check for satisfaction of the integral equation can be applied very simply by computing $\tilde{J}(\alpha)\tilde{G}(\alpha)$, taking its inverse Fourier transform, and verifying how well it approaches $-E^i$ on the surface of the scatterer.

Figure 11 shows the calculated induced surface current density distribution on the strip with $ka = 4$ (1.3λ wide) for normal incidence. Note that the current density becomes large at the edges, as it should for E-wave incidence, although no specific condition was enforced at the edges, nor any special care exercised. Note also that the approximate current is confined essentially on the surface of the strip and extends very little outside of this surface. Thus, the solution in this case is very close to the true solution and this is easily verified by truncating the current density, computing the scattered field it radiates on the strip, and verifying that the scattered field is indeed very nearly equal to $-E^i$.

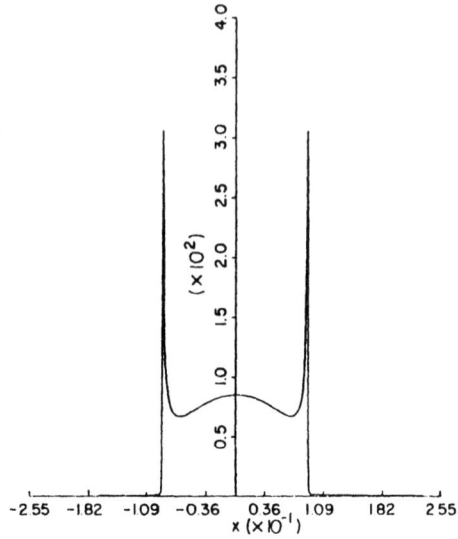

Figure 11. Magnitude of the induced surface current density distribution normalized to $(ikZ_0)^{-1}$ on the strip of $ka = 4$. (1.273λ wide), $\phi_0 = 90°$.

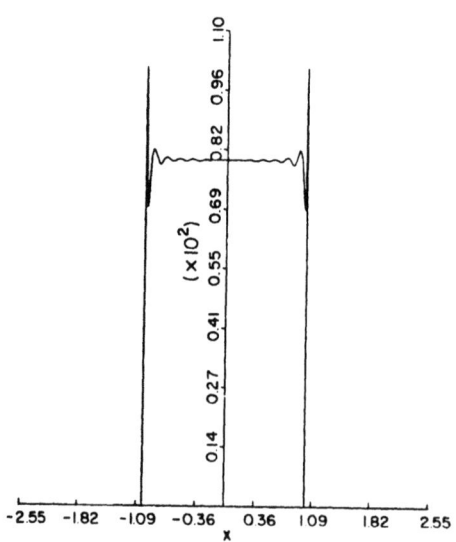

Figure 12. Magnitude of the induced surface current density distribution normalized to $(ikZ_0)^{-1}$ on the strip of $ka = 40$. (12.73λ wide), $\phi_0 = 90°$.

Figure 12 depicts the result for ka = 40, i.e., a 13λ strip. Note that the peak in the center is no longer present and the current there approaches that given by the physical optics approximation. There are now more oscillations, however, and the current density has a sharp dip before rising to infinity at the edges.

Figure 13 exhibits the satisfaction of the boundary condition after one iteration. As mentioned before, such a test is not available in the conventional GTD approach.

Let us next turn to the interesting case of a near grazing incidence where the zero-order current density has a long tail extending beyond the edge of the strip (see Figure 14). This result is to be expected since the two half-plane GTD solutions used in the zero-order approximation represent a poor approximation for the induced current for shallow incidence angles. If this tail is truncated, the remaining portion of the current density on the strip produces a scattered field on the surface of the strip which is significantly different from $-E^i$, where $|E^i| = 1$, as may be seen from Figure 15.

Figure 16 shows the effect of one iteration on the zero-order GTD solution shown in Figure 14. Note that the current density is significantly altered in the neighborhood of the shadowed edge demonstrating the fact that even with a relatively poor initial guess, the convergence is quite rapid in this case.

To see that this is indeed an improved solution, the truncated portion of it is used to calculate the scattered field. It is observed that the satisfaction of the boundary condition has been improved as shown in Figure 17.

To verify the convergence of the solution numerically, one more iteration is performed and the result is depicted in Figure 18. Note that the shape of the surface current density does not change much which indicates a settling down of the solution has occurred. Also, note that the tail extending outside of the strip has been reduced to an insignificant quantity, which, when truncated, will produce little effect on the scattered field on the surface of the strip.

To recapitulate, the strip problem has been solved by a combination of the integral-equation and asymptotic high-frequency techniques. Formulation of the integral equation in the Fourier transform domain allows one to conveniently obtain the zero-order approximation to the transformed unknown surface current density from the solution of two half-plane problems.

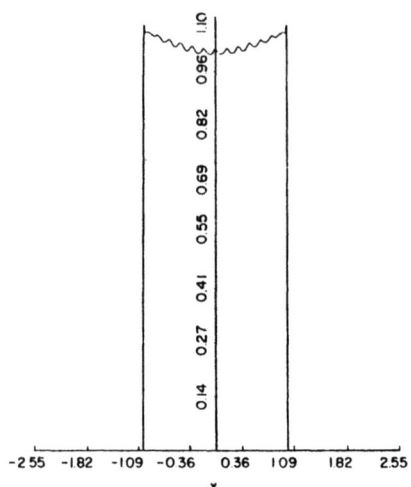

Figure 13. Magnitude of the scattered E-field evaluated on the strip of ka = 40., ϕ_0 = 90° (one iteration).

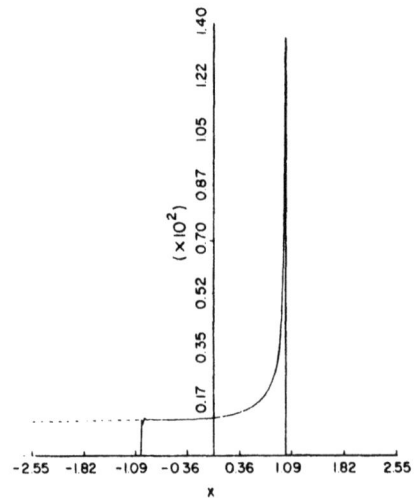

Figure 14. Magnitude of the induced surface current density distribution normalized to $(ikZ_0)^{-1}$ on the strip of ka = 40., ϕ_0 = 10° (no iteration).

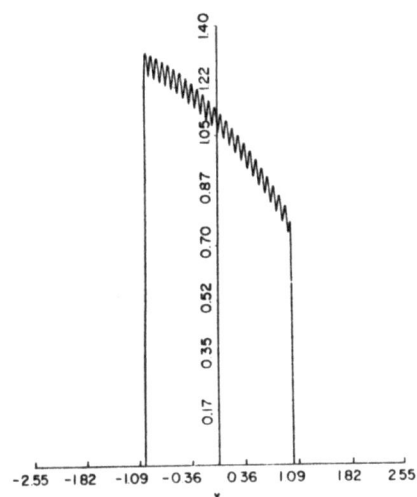

Figure 15. Magnitude of the scattered E-field evaluated on the strip of ka = 40., ϕ_0 = 10° (no iteration).

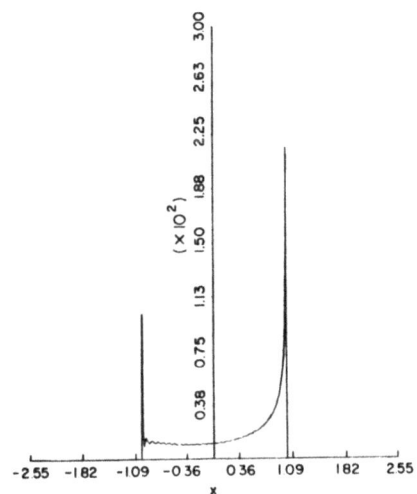

Figure 16. Magnitude of the induced surface current density distribution normalized to $(ikZ_0)^{-1}$ on the strip of ka = 40., ϕ_0 = 10° (one iteration).

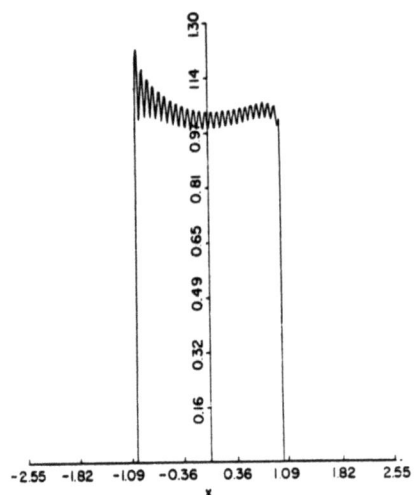

Figure 17. Magnitude of the scattered E-field evaluated on the strip of ka = 40., ϕ_0 = 10° (one iteration).

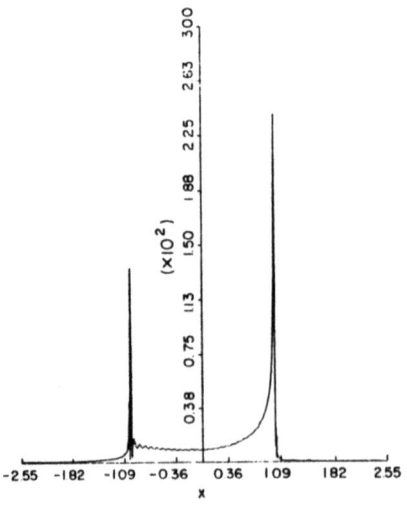

Figure 18. Magnitude of the induced surface current density distribution normalized to $(ikZ_0)^{-1}$ on the strip of ka = 40., ϕ_0 = 10° (two iterations).

Higher-order solutions have been obtained via the iteration steps outlined above and the numerical convergence has been demonstrated. The iteration process generates the proper edge singularities even when they are not present in the original approximation, e.g., physical optics. However, additional iterations are necessary in that case. Validity of the solution has been substantiated by numerically verifying the satisfaction of the boundary condition to within a close tolerance.

5. APPLICATION OF MOMENT METHOD IN THE TRANSFORM DOMAIN

A different approach to handling (48) would be to employ the Galerkin procedure in the transform domain [27]. One may write

$$\tilde{J} \simeq \tilde{J}^{(0)} + \sum_p C_p \tilde{J}_p \tag{65}$$

where $\tilde{J}^{(0)}$ is the approximate solution derived from a suitable asymptotic formula for the scattered field and $\tilde{J}_p$ represents a set of basis functions in the transform domain. Typically, there are certain angular regions in the far field where the asymptotic solutions require refinement. One may choose to concentrate the basis functions in these regions in the transform domain. Alternatively, the $\tilde{J}_p$'s could be chosen as the transforms of a suitable set of basis functions in the space domain, and the location (support) of these subdomain basis functions may be selected to coincide with transition regions or corners, etc., where the canonical solution of the asymptotic solution may require refinement.

In either case, the problem of determining $\tilde{J}$ may be reduced to that of finding the unknown coefficients C_p such that (65) satisfies (48). The Galerkin procedure provides a way for accomplishing this, as we will soon see. This technique also has the advantage that the other unknown in (48), viz., $\tilde{F}$, is conveniently eliminated from this equation upon application of Galerkin's method. We demonstrate this fact in the manipulations presented below.

Substituting (65) into (48) and taking a scalar product of the resulting equation with a set of suitable series of testing functions $\tilde{W}_q$, we arrive at

$$\sum_p C_p <\tilde{W}_q, \tilde{\tilde{G}} \tilde{J}_p> = -<\tilde{W}_q, \tilde{E}_I> + <\tilde{W}_q, \tilde{F}> \tag{66}$$

where $<\,,\,>$ is the scalar inner product. If we now choose $\tilde{W}_q$ to be transforms of functions which are nonzero only on the surface of the scatterer, then the scalar product $<\tilde{W}_q, \tilde{F}>$ can be shown to vanish. To show this, one uses Parseval's theorem and transforms the scalar product of $\tilde{W}_q$ and $\tilde{F}$ in terms of a similar product of their counterparts in the space domain. Since the inverse transforms of $\tilde{W}_q$ and $\bar{F}$ exist in complementary regions, viz., on the surface of the scatterer and in the region complementary to this surface, respectively, one finds that their scalar product is identically zero. One can now proceed in the usual manner to solve for the coefficients C_p by solving the matrix equation represented by (66) with the term $<\tilde{W}_q,\bar{F}>$ deleted. It is evident that the use of this method would be practical only when relatively few terms are needed in (65) to modify the available asymptotic solution; however, this is typically the situation for many problems. It should also be noted that (66) represents a direct check on the satisfaction of the boundary condition, in the sense of moments. The choice of $\tilde{W}_p$'s is governed by the locations on the surfaces of the scatterer where these boundary conditions are applied. Typically these will be the zones where the asymptotic solution might be inaccurate, e.g., the transition region between the lit and shadow regions.

To illustrate the procedure we consider a smooth convex surface with no wedges - a circular cylinder. One of the important attributes of this canonical geometry is that it permits convenient comparison with the exact series solution available for the representation of scattered fields from this structure. The geometry of the problem is shown in Figure 19. We consider the case of an E-polarized wave incident from $\phi = 180°$.

The first step in attacking the problem is to use a geometrical optical approach to derive the scattered far-field. When this is done, one obtains the dotted curve in Figure 20 which also exhibits the exact series solution for the scattered field as a solid curve. It is evident that in the range $-60° < \phi < 60°$ the GO solution is not adequate. This is not totally surprising since it is well-known that creeping-wave contributions need to be included in the scattered field expression in the shadow and transition regions. Rather than following this procedure, we will now show how the Galerkin method can be readily and conveniently applied to this problem to derive an accurate solution.

To this end we consider, as a first step, the behavior of the scattered field on a surface erected in juxtaposition to the cylinder at the point $x = a$, the farthest point away from the incident field. Referring to Figure 21, in the deep shadow region, say $|y| < 2$, we expect the scattered field $E_z^s = -E_z^i$ to be a very good approximation. On the other hand, when we go far onto the lit region on this surface, say for $|y| > 6$, we expect the E_z^s to be described adequately by the GO formulas. If we had a good estimate of the scattered field behavior in the transition region $2 < |y| < 6$, we would be able to get a good representation of the excess scattered field (over and above the GO field) on the entire surface at $x = a$. We should then be able

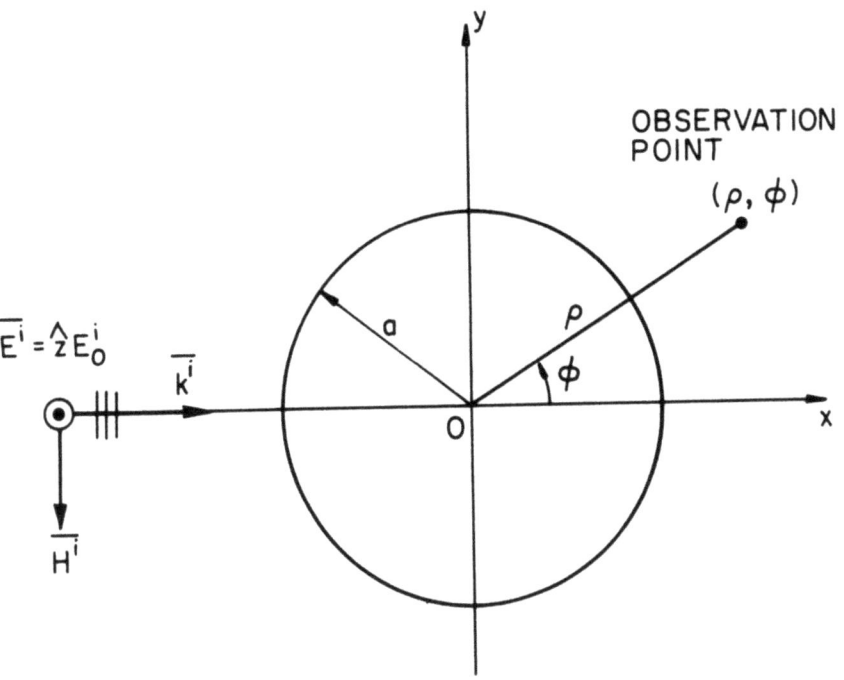

Figure 19. Diffraction by a circular cylinder illuminated by an E-wave incident along the x-axis.

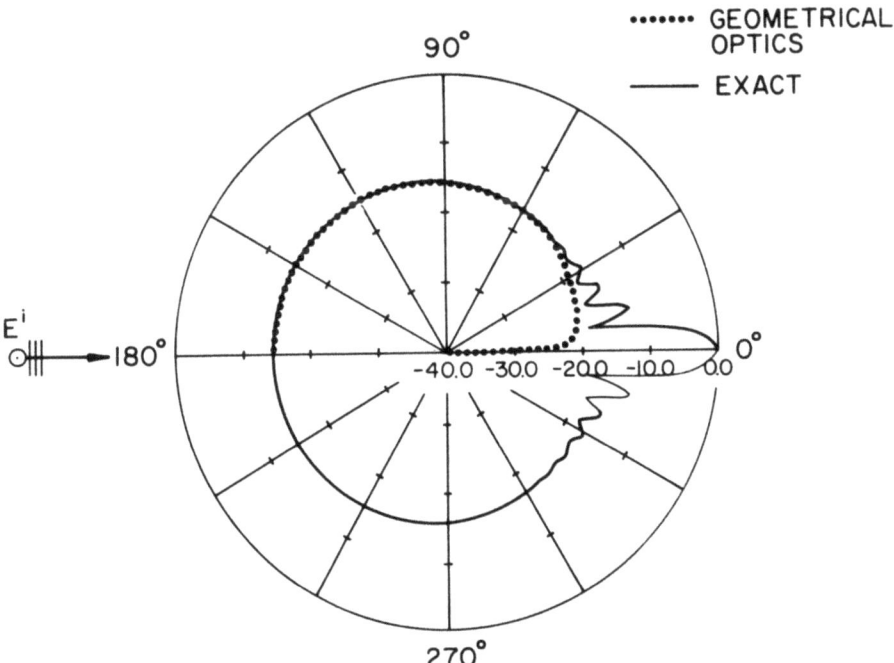

Figure 20. Geometrical optics scattered far-field pattern in dB of a circular cylinder with radius $a = 3\lambda$.

to compute the field radiated in the r.h.s. of the cylinder by this excess field using the concept of Huyguens' source and use this radiated field to fill in the gap between the GO pattern and the true pattern.

To derive the E_z^s field in the transition region, we first interpolate the magnitude of this field from E^i at $|y| = 2$, to 0 at $y = 6$ and the phase from π at $|y| = 2$, to the GO phase at $|y| = 6$. Next we introduce a set of basis functions, with undetermined coefficients, to describe the correction to the interpolated E_z^s field at the plane $x = a$, $2 < |y| < 6$. To determine these coefficients, we apply the concept of Galerkin's method in the spectral domain as briefly outlined in the last section. In the example being considered here, the transforms of the basis functions play the role of $\tilde{J}_p$ in (65), and the zero-order, scattered far field $\tilde{J}^{(0)}$ is obtained by adding the contributions of GO and the approximate excess E_z^s field derived from the interpolation procedure just described.

The choice of the testing functions, $\tilde{W}_q$ in (66), is suggested by the fact that the error in the high-frequency asymptotic solution is mostly concentrated around the transition region on the surface of the cylinder, i.e., in the neighborhood of the junction between the lit and shadow regions. Thus, a suitable choice for the testing functions would be to locate them at the transition region as shown in Figure 21, where the location of the basis functions is also shown. Note that we need not be restricted in our choice for the location of these functions by demanding that they have a common support, although this is almost always the case in the conventional moment or Galerkin methods. We may also note from Figure 21 that the shape of the basis and testing functions are both Gaussian. Since we are dealing with transforms, this choice is not only convenient for deriving the Fourier transformations $\tilde{J}_p$ and $\tilde{W}_q$, but is also desirable from a numerical point of view because the transforms are not oscillatory as they would be for a pulse or triangular basis. This feature is important when numerically computing the scalar products $<\tilde{J}_p, \tilde{W}_q>$, needed for the determination of the unknown coefficients C_p.

Only a few (3 to 7) unknowns C_p are needed to derive an accurate solution for both the radiated far field and the surface current on the cylinder. The accuracy itself can be verified by computing the tangential E-field on the surface via Fourier inversion of the hemispherical far-field pattern centered around the point to be tested. This procedure is also used to compute the surface current distribution from the knowledge of the scattered H-field at large distances. Of course an independent check is available for this problem via the exact series solution. A comparison of the Galerkin solution and the exact series solution is shown in Figures 22 and 23 to illustrate the highly accurate nature of the

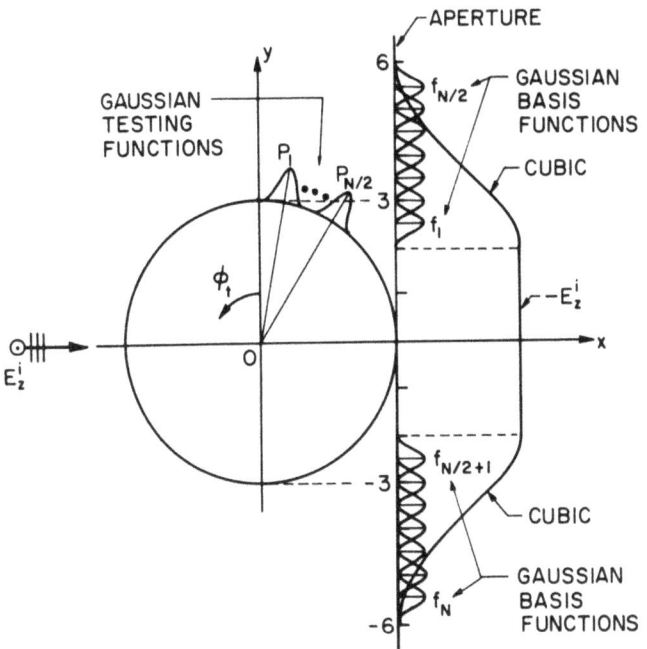

Figure 21. Locations of the basis functions on the aperture and the testing functions on the surface of the obstacle.

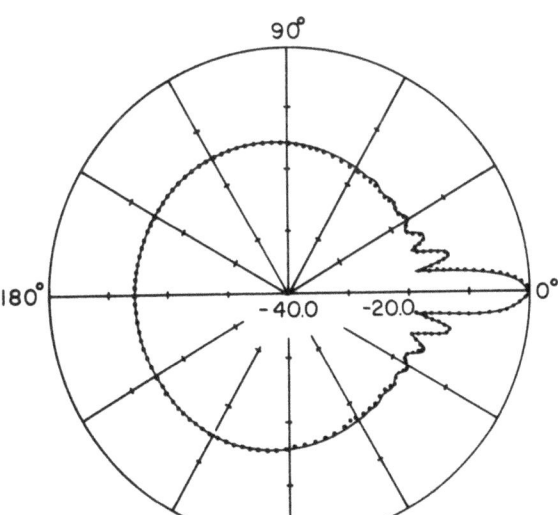

Figure 22. Scattered far-field pattern in dB of a circular cylinder with radius a = 3λ obtained by Galerkin's method.

692

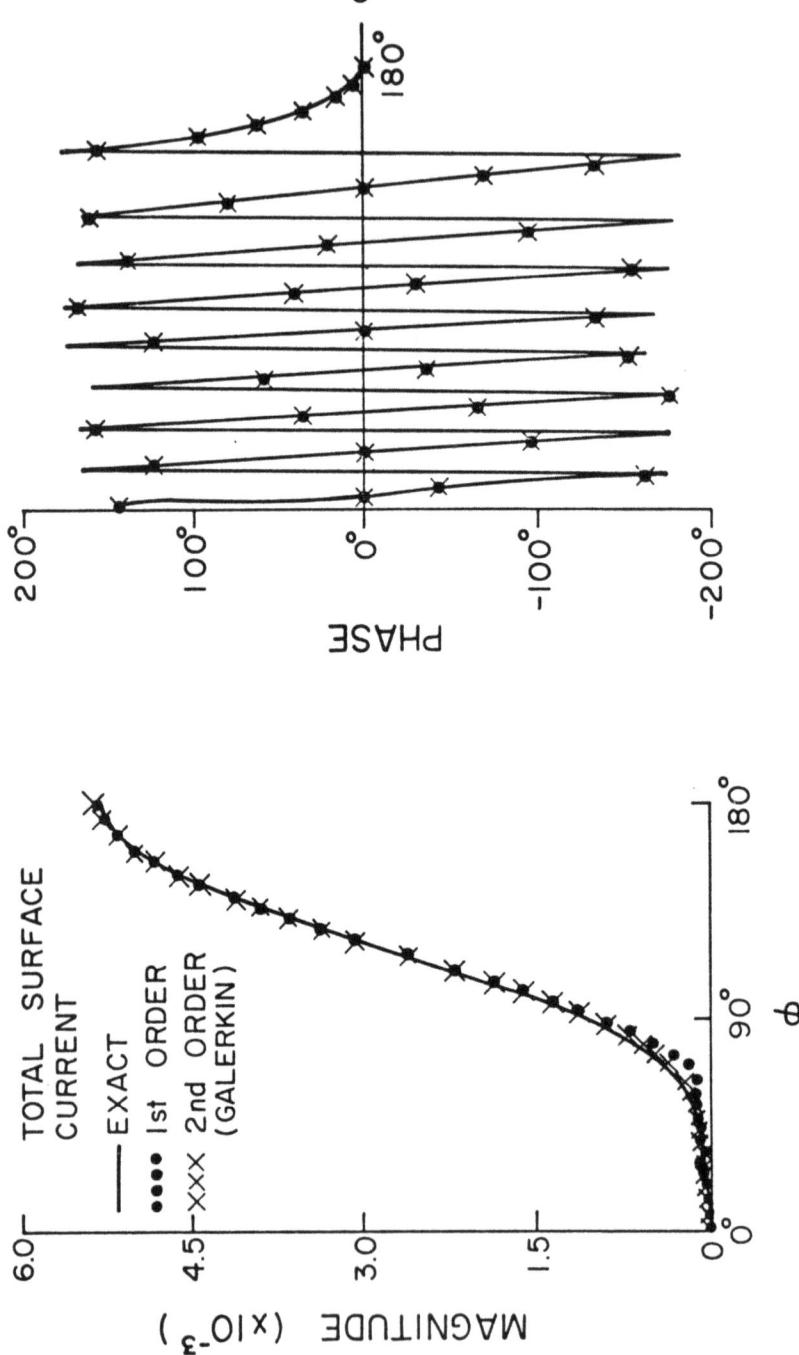

Figure 23. Total surface current on a perfectly conducting circular cylinder with radius $a = 3\lambda$.

Galerkin solution. In fact, the solution is almost identical to the exact solution except for a slight error in the transition regions, even without the Galerkin refinement, as evidenced by the dotted curve in Figure 23 which exhibits this case.

6. CONCLUSION

The purpose of this paper was to show how the GTD formulas can be interpreted and extended using the spectral domain concepts. The problems discussed herein are only representative in nature and the space does not permit us to include many other radiation and scattering problems of practical interest which have been successfully attacked via the transform approach. We wish to emphasize the fact that two important benefits of using the spectral approach are the availability of a convenient accuracy check and the possibility of deriving an improved solution via iteration or Galerkin techniques. The spectral domain approach also provides an avenue for bridging the gap between the low and high frequency regions by combining integral equaiton methods with asymptotic techniques.

REFERENCES

1. R. F. Harrington, Field Computation by Moment Methods, MacMillan, New York, 1968.
2. R. Mittra, Ed., Computer Techniques for Electromagnetics, Pergamon Press, Oxford, 1973.
3. Y. Rahmat-Samii, T. Itoh and R. Mittra, "A Spectral Domain Technique for Solving Microstrip Line Problems," AEU Electronics and Communications, Band 27, pp. 69-71, 1973.
4. Y. Rahmat-Samii, T. Itoh and R. Mittra, "A Spectral Domain Analysis for Solving Discontinuity Problems," IEEE Trans. Microwave Theory Tech., vol. MTT-22, pp. 372-378, April 1974.
5. T. Itoh and R. Mittra, "Spectral Domain Approach for Calculating the Dispersion Characteristics of Microstrip Lines," IEEE Trans. Microwave Theory Tech., vol. MTT-21, pp. 496-499, July 1973.
6. J. B. Davies and D. Mirshekar-Syahkal, "Spectral Domain Solution of Arbitrary Coplanar Transmission Line with Multilayer Substrate," IEEE Trans. Microwave Theory Tech., vol. MTT-25, pp. 143-146, February 1977.
7. R. Mittra and T. S. Li, "A Spectral Domain Approach to the Numerical Solution of Electromagnetic Scattering Problems," AEU, Electronics and Communications, Band 29, pp. 217-222, 1975.
8. T. S. Li and R. Mittra, "Applications of the Spectral Domain Approach for Numerical Solution of Electromagnetic Scattering from Corner Reflectors, Prisms and Rectangular Plates," AEU, Electronics and Communications, Band 29, pp. 323-333, 1975.

9. N. N. Bojarski, "K-Space Formulation of the Electromagnetic Scattering Problem," Technical Report AFAL-TR-71-75, March 1971.
10. J. B. Keller, "Diffraction by an Aperture," J. Appl. Phys., vol. 28, pp. 426-444, 1957.
11. J. B. Keller, "Geometrical Theory of Diffraction," J. Opt. Soc. Am., vol. 52, pp. 116-130, 1962.
12. G. A. Deschamps, "Ray Techniques in Electromagnetics," Proc. IEEE, vol. 60, pp. 1022-1035, 1972.
13. S. W. Lee, "Electromagnetic Reflection from a Conducting Surface: Geometrical Optics Solution," IEEE Trans. Antennas Propagat., vol. 23, pp. 184-191, 1975.
14. R. G. Kouyoumjian, "The Geometrical Theory of Diffraction and Its Application." Chapter 6 in: R. Mittra, Ed., Numerical and Asymptotic Techniques in Electromagnetics, Springer-Verlag, New York, 1975.
15. H. Bach, "Engineering Applications of the Geometrical Theory of Diffraction," Chapter 5 in: E. J. Maanders and R. Mittra, Ed., Modern Topics in Electromagnetics and Antennas, Peter Peregrinus, London, 1977.
16. S. W. Lee, "Uniform Asymptotic Theory of Electromagnetic Edge Diffraction: A Review," EM Report 77-1, Electromagnetics Lab., University of Illinois, Urbana, Ill. 1977. (This report includes 300 some references).
17. R. M. Lewis, D. S. Ahluwalia, and J. Boersma, "Uniform Asymptotic Theory of Diffraction by a Plane Screen," SIAM J. Appl. Math., vol. 16, pp. 783-807, 1968.
18. R. G. Kouyoumjian and P. H. Pathak, "A Uniform Geometrical Theory of Diffraction for an Edge in a Perfectly Conducting Surface," Proc. IEEE, vol. 62, pp. 1448-1461, 1974.
19. S. W. Lee and G. A. Deschamps, "A Uniform Asymptotic Theory of Electromagnetic Diffraction by a Curved Wedge," IEEE Trans. Antennas Propagat., vol. 24, pp. 25-34, January 1976.
20. P. Ya. Ufimtsev, Method of Edge Waves in the Physical Theory of Diffraction, Air Force Systems Command, Foreign Tech. Div. Document ID No. FTD-HC-23-259-71, 1971. (Translation from the Russian version published by Soviet Radio Publication House, Moscow, 1962).
21. K. M. Mitzner, "Studies in Physical Optics," Tech. Rept. AFAL-TR-74-22, Air Force Avionics Lab., Wright-Patterson Air Force Base, OH, 1974.
22. S. W. Lee, "Comparison of Uniform Asymptotic Theory and Ufimtsev's Theory of Electromagnetic Edge Diffraction," IEEE Trans. Antenna Propagat., vol. 25, pp. 162-170, March 1977.
23. Y. Rahmat-Samii and R. Mittra, "Spectral Analysis of High Frequency Diffraction of an Arbitrary Incident Field by a Half Plane--Comparison with Four Asymptotic Techniques," Radio Science, vol. 13, pp. 31-48, Jan./Feb., 1978.

24. R. Mittra, Y. Rahmat-Samii and W. L. Ko, "Spectral Theory of Diffraction," Appl. Phys., vol. 10, pp. 1-13, January 1976.
25. Y. Rahmat-Samii and R. Mittra, "A Spectral Domain Interpretation of High Frequency Diffraction Phenomena," IEEE Trans. Antennas Propagat., vol. 25, pp. 676-687, September 1977.
26. W. L. Ko and R. Mittra, "A New Approach Based on a Combination of Integral Equation and Asymptotic Techniques for Solving Electromagnetic Scattering Problems," IEEE Trans. Antennas Propagat., vol. 25, pp. 187-197, March 1977.
27. R. Mittra and W. L. Ko, "An Approach to High-Frequency Scattering from Smooth Convex Surfaces," IEEE Trans. Antennas Propagat., vol. 25, pp. 781-788, November 1977.
28. W. L. Ko and R. Mittra, "A New Look at the Scattering of a Plane Wave by a Rectangular Cylinder," AEU, Electron. & Comm., pp. 494-500, December 1977.
29. Y. Rahmat-Samii and R. Mittra, "On the Investigation of Diffracted Fields at the Shadow Boundaries of Staggered Parallel Plates--A Spectral Domain Approach," Radio Science, vol. 12, pp. 659-670, Sept./Oct. 1977.
30. R. Mittra and Y. Rahmat-Samii, "A Spectral Domain Approach for Solving High Frequency Scattering Problems," in Electromagnetic Scattering, P.L.E. Uslenghi, Ed., Academic Press, New York, 1978.
31. R. Mittra and S. W. Lee, Analytical Methods in the Theory of Guided Waves, McMillan and Company, 1971.
32. M. Tew and R. Mittra, "Accuracy Tests for Asymptotic ons Solutions to Radiation from a Cylinder," University of Illinois at Urbana-Champaign, EM Lab Tech. Report No. 77-22, October 1977; see "Accuracy Tests for High Frequency Asymptotic Solutions," by R. Mittra and M. Tew, IEEE Trans. Ant. and Propagation, Vol. AP-27, No. 1, January 1979, pp. 62-68.
33. M. Tew and R. Mittra, "Reciprocity Test for Asymptotic Solutions," National Radio Science Meeting; Boulder, Colorado, January 1978; see "An Integral E-Field Accuracy Test for Asymptotic Solutions," by R. Mittra and M. Tew, (to appear).
34. R. E. Collin and F. J. Zucker, Antenna Theory, McGraw-Hill Book Company, New York, 1969.
35. A. W. Maue, "Formulation of General Diffraction Problems Through an Integral Equation," Zeltschrift fur Physik, vol. 126, pp. 601-619, 1949.

SOME USEFUL ASYMPTOTIC AND NUMERICAL TECHNIQUES
FOR SOLVING HIGH-FREQUENCY SCATTERING PROBLEMS

Yahya Rahmat-Samii

Jet Propulsion Laboratory
California Institute of Technology
Pasadena, California 91103

ABSTRACT. The objectives of this paper are: (1) to conduct an analytical and numerical comparison among various widely used asymptotic theories in high-frequency electromagnetic diffraction and (2) to demonstrate the application of the Bessel-Jacobi series for the efficient and accurate evaluation of the radiation integral for offset reflector antennas. Only the basic concepts and final results are reported here and the interested reader is referred to references [1] - [8] for complete analytical and numerical details. Time conventions $\exp(-i\omega t)$ and $\exp(j\omega t)$ are used in Sections 1 and 2, respectively, but are suppressed in the formulations.

1. ANALYTICAL AND NUMERICAL COMPARISON OF HIGH-FREQUENCY DIFFRACTION TECHNIQUES

After Keller's pioneering paper [9] on the Geometrical Theory of Diffraction (GTD), numerous attempts have been made toward extending the domain of GTD to regions where it predicts an infinite field and is therefore invalid. Most of the efforts in this direction have been concentrated on constructing uniform formulations such as UTD, MSD, UAT and STD that overcome the difficulties of GTD at the incident and shadow boundaries, and at the transition regions. The Uniform Theory of Diffraction (UTD) and its Modified Slope Diffraction (MSD) version have been introduced by Kouyoumjian, Pathak and Hwang [10], [11]; the Uniform Asymptotic Theory (UAT) has been developed by Ahluwalia, Lewis and Boersma [12] and has been employed extensively by Boersma, Lee and Deschamps [13] - [15]; and, finally, the Spectral Theory of Diffraction (STD) has been applied by Mittra,

Rahmat-Samii and Ko [16], [1]. UTD, MSD and UAT are asymptotic techniques based on individual hypotheses or Ansatz. Typically, there is no general proof available for the validity or the completeness of these Ansatz and one has to resort to certain test examples to establish their accuracy. The formulations of these asymptotic theories are reduced to the classical Sommerfeld's result when the plane wave illumination on a half-plane is considered. In fact, Sommerfeld's formulation is the basic foundation of all the different Ansatz that have been proposed to date. For more complex situations, the validity of the various asymptotic theories is typically checked against numerical results or experimental data, since analytical results are often unavailable in a closed form. Even for the problem of diffraction of a halfplane illuminated by a line source possessing isotropic or nonisotropic patterns there is no substantial check available for establishing the accuracy of the various asymptotic theories. Specifically, at the transition regions it is not known how well these asymptotic theories compare with each other or with the exact solution.

The objective of this section is to compare the aforementioned asymptotic theories with the exact solution that can be constructed in an integral form using STD. The comparison has been carried out, both analytically and numerically, for a wide range of observation angles including the shadow boundaries, their vicinities, and at angles away from these boundaries. For completeness, different source locations and pattern functions have been investigated and extensive results have been derived. In the following, we begin by reviewing briefly the nature of the formulation of each of the uniform theories, and follow this with the presentation of the results and conclusions.

The geometry under consideration is a half-plane illuminated by an incident field of a nonisotropic (multimoment) line source. The space is oriented by the use of a set of Cartesian coordinates (x,y) and polar coordinates (ρ,ϕ), such that the origins of the coordinate systems are at the edge of the halfplane located at $x < 0$. All the angles are measured positively counterclockwise, and the range of angles is either $[0,\pi]$ or $[0,-\pi]$. Next, let a coordinate system (ρ_i,ϕ_i) be erected at the source location. In the following we first present the formulation of STD for this diffraction problem. The half-plane is illuminated by an arbitrary incident field with its plane wave spectral representation given by (see Fig. 1)

$$u^i(\rho_i,\phi_i) = i/4\pi \int_\Gamma P(\omega) \, e^{ik\rho_i \cos(\omega - |\phi_i|)} \, d\omega \qquad (1.1)$$

where Γ is the path $[(i\infty,0)U(0,\pi)U(\pi,-i\infty)]$ in the complex ω-plane. For the multimoment line source, $P(\omega)$ may take the form $P(\omega) = \Sigma\, I_n e^{in\pi/2}\, [\cos\omega + \text{sgn}(\phi_i)\sin\omega]^n$. Using Sommerfeld's solution and applying the STD construction, one can show [2] that the following expression for the total field diffracted by the half-plane illuminated by the incident field (1) holds,

$$u^t(\rho,\phi) = u^i(\rho_i,\phi_i) + u^s(\rho,\phi), \qquad (1.2a)$$

where the scattered field $u^s(\rho,\phi)$ is

$$u^s(\rho,\phi) = i/4\pi \int_\Gamma P(\omega)\exp[iks\cos(\omega-\Omega)]\{-\exp[ik\rho\cos(\omega-\phi)] \cdot F(-\xi_i) + R\exp[ik\rho\cos(\omega+\phi)]F(\xi_r)\}d\omega \quad (1.2b)$$

and s and $0 < \Omega < \pi$ are the distance and the angle of the source point with respect to the edge of the half-plane. In (1.2b) R is the reflection coefficient that takes the value of -1 or $+1$ for E or H waves, respectively. Furthermore, F, the Fresnel function, and $\xi_{\frac{i}{r}}$ are defined as

$$F(\xi) = e^{-i\pi/4}\,\pi^{-1/2} \int_\xi^\infty e^{it^2}\,; \quad \xi_{\frac{i}{r}} = \mp\sqrt{2k\rho}\,\sin\frac{\omega \mp \phi}{2} \quad . \quad (1.3)$$

Equation (1.2) is exact and hence it can be used as a basis for comparing the aforementioned asymptotic techniques. We have constructed the asymptotic solution of (1.2) up to and including term of order k^{-3} (dominant asymptotic term of u^i is $k^{-1/2}$) for observation points away from the transition regions and exactly at the shadow boundaries. Also, numerical schemes have been developed to handle (1.2) for all observation points and specifically attention is given to the determination of the field in the transition region for different locations of the source and its pattern function. In the following the final results are summarized.

In order to construct the asymptotic expansion of the total field, one must first determine the asymptotic expansion of the incident field (1.1). This can be done rather simply [2] and the final expressions are

$$u^i(\rho_i,\phi_i) = \frac{e^{ik\rho_i + i\pi/4}}{2\sqrt{2\pi k}} \sum_{m=0}^\infty (ik)^{-m} z_m^i(\rho_i,\phi_i) \qquad (1.4)$$

$$z_0^i(\rho_i,\phi_i) = P(|\phi_i|)\rho_i^{-1/2} \qquad (1.5a)$$

$$z_1^i(\rho_i,\phi_i) = [1/2\, P''(|\phi_i|) + 1/8\, P(|\phi_i|)]\,\rho_i^{-3/2} \qquad (1.5b)$$

$$z_2^i(\rho_i,\phi_i) = [1/8\, P''''(|\phi_i|) + 5/16\, P''(|\phi_i|) + 9/128\, P(|\phi_i|)]\,\rho^{-5/2}\,. \qquad (1.5c)$$

where prime denotes the differentiation operation. Expressions (1.4) and (1.5) complete the asymptotic expansion of u^i up to and including the order of $k^{-5/2}$, and they agree with the results given by Boersma and Lee [17] and Keller et al [18].

Next, we give the asymptotic expansions of u^s. For simplicity of notation (1.2) is rewritten as follows

$$u^s(\rho,\phi) = u^s_i(\rho,\phi) + Ru^s_r(\rho,\phi), \qquad (1.6)$$

and the following notations are used

$$\chi_{\substack{i \\ r}} = \mp \csc[(\Omega \mp \phi)/2], \quad g(k\rho) = e^{ik\rho + i\pi/4}/2\sqrt{2\pi k\rho} \,. \qquad (1.7)$$

Using the analytical developments of Rahmat-Samii and Mittra [2], the asymptotic expansions of u^s can be determined at observation points away from the transition regions and those exactly at the shadow boundaries.

For observation points away from the transition regions one obtains

$$u^s_{\substack{i \\ r}}(\rho,\phi) = u^g_{\substack{i \\ r}}(\rho,\phi) + u^d_{\substack{i \\ r}}(\rho,\phi) \,, \quad u^d_r(\rho,\phi) = u^d_i(\rho, 2\pi - \phi) \qquad (1.8)$$

where

$$u^g_i(\rho,\phi) = \begin{cases} -u^i(\rho_i,\phi_i), & \text{for } \Omega < \phi \notin T^i \\ 0, & \text{for } \Omega > \phi \notin T^i \end{cases} \qquad (1.9a)$$

$$u^g_r(\rho,\phi) = \begin{cases} 0, & \text{for } -\Omega < \phi \notin T^r \\ u^r(\rho_r,\phi_r), & \text{for } -\Omega > \phi \notin T^r \end{cases} \qquad (1.9b)$$

and u^r is the field radiating from the image point. Customarily, $u^i + u^g_i + Ru^g_r$ is called the Geometrical Optics (GO) field in the GTD construction of the total field. The diffracted field u^d_i may be expressed as

$$u^d_i(\rho,\phi) = u^d_{i1}(\rho,\phi) + u^d_{i2}(\rho,\phi) + u^d_{i3}(\rho,\phi) + O(k^{-4}) \qquad (1.10)$$

which, after performing all the necessary manipulations, can be asymptotically expanded as

$$u^d_{i1}(\rho,\phi) = g(k\rho)\chi_i g(ks)P(\Omega) \qquad (1.11a)$$

$$u_{i2}^d(\rho,\phi) = (1/4)g(k\rho)(ik\rho)^{-1}\chi_i^3 g(ks)P(\Omega) + (1/2)g(k\rho)[(1/2)\chi_i^3$$
$$\cdot P(\Omega) + \cos((\Omega-\phi)/2)\chi_i^2 P'(\Omega) + \chi_i P''(\Omega)](iks)^{-1}g(ks) \qquad (1.11b)$$

$$u_{i3}^d(\rho,\phi) = (3/16)g(k\rho)(ik\rho)^{-2}\chi_i^5 g(ks)P(\Omega) + (1/8)g(k\rho)$$
$$\cdot(ik\rho)^{-1}[(-2\chi_i^3+3\chi_i^5)P(\Omega) + 3\cos((\Omega-\phi)/2)\chi_i^4 P'(\Omega) + \chi_i^3 P''(\Omega)]$$
$$\cdot(iks)^{-1}g(ks) + (1/8)g(k\rho)[(3/2)\chi_i^5 P(\Omega) + \cos((\Omega-\phi)/2)$$
$$\cdot(2\chi_i^2 + 3\chi_i^4)P'(\Omega) + (\chi_i+3\chi_i^3)P''(\Omega) + 2\cos((\Omega-\phi)/2)\chi_i^2$$
$$\cdot P'''(\Omega) + \chi_i P''''(\Omega)](iks)^{-2}g(ks). \qquad (1.11c)$$

The diffracted field may then be simply constructed using the relation

$$u^d(\rho,\phi) = u_i^d(\rho,\phi) + Ru_r^d(\rho,\phi) = u_i^d(\rho,\phi) + Ru_i^d(\rho,2\pi-\phi). \qquad (1.12)$$

The expression of u_1^d is precisely Keller's GTD solution and is only valid away from the transition regions. In addition, the spectral analysis also provides higher-order terms whereas GTD is incapable of determining these terms. The expression of u_2^d agrees completely with the one obtained using the UAT formulation. As was shown by Boersma and Lee [17], the modified slope diffraction (MSD) failed to produce this higher-order term correctly. Furthermore, we have constructed u_3^d to demonstrate the ease in using spectral analysis for the systematic determination of higher-order terms. The contribution of the higher-order terms can be significant when one wants to correctly order the asymptotic solution of multiple-edge diffraction problems.

Having determined the asymptotic expansion of (1.2b) for observation points away from the transition regions, we now proceed to construct them for observation points exactly at the shadow boundaries, i.e., $\phi = \pm\Omega$. Along these directions, the solutions given in (1.11) diverge to infinity, because χ_i and χ_r take infinite values at $\phi = \Omega$ and $\phi = -\Omega$, respectively. To overcome this difficulty, one has to evaluate (1.2b) more carefully. Following the procedures outlined in [2], we arrive at

$$u_i^s(\rho,\Omega) = -u_r^s(\rho,-\Omega) = -(1/2)g[k(s+\rho)] \{ P(\Omega) + (ik)^{-1}$$
$$\cdot [(1/2)P''(\Omega) + (1/8)P(\Omega)](s+\rho)^{-1} + (ik)^{-2}[(1/8)P''''(\Omega)$$
$$+ (5/16)P''(\Omega) + (9/128)P(\Omega)](s+\rho)^{-2} \} -g(ks)g(k\rho)(2\rho)/(\rho+s)$$
$$\cdot P'(\Omega) - (ik)^{-1}g(ks)g(k\rho)(\rho^2+3\rho s)/[3s(s+\rho)^2] [P'(\Omega) + P'''(\Omega)]$$
$$- g(ks)g(k\rho)(ik)^{-2}(15\rho s^2+10s\rho^2+3\rho^3)/[60s^2(s+\rho)^3] [4P'(\Omega)$$
$$+ 5P'''(\Omega) + P'''''(\Omega)] + O(k^{-7/2}) \qquad (1.13)$$

which, of course, is bounded. The first term on the right-hand side of (1.13) can be identified with the negative of one-half of the incident field when it is compared with (1.4) and (1.5). The total field may then be obtained by determining $u_r^s(\rho,\Omega) = u_i^s(\rho, 2\pi - \Omega)$.

Let us now briefly focus our attention on the formulation of various uniform asymptotic theories. All the uniform asymptotic theories essentially approximate (1.2) in an asymptotic sense. It is interesting to note that as yet there is no proof available that justifies the use of these forms for arbitrary and complex situations. In all of these techniques the following is true for the total field

$$u^t(\rho,\phi) = v(\rho,\phi) + Rv(\rho, 2\pi-\phi) \qquad (1.14)$$

where $v(\rho,\phi)$ takes different forms for different theories. For convenience of interpretation, all of the asymptotic theories are schematically represented in Figure 2. The notations in this figure are explained below.

We first give the expression of $v(\rho,\phi)$ for the GTD theory, which reads as follows:

$$v(\rho,\phi) = u^g(\rho,\phi) + u^d(\rho,\phi) + O(k^{-3/2}) \qquad (1.15)$$

where

$$u^g(\rho,\phi) = \theta\{\sin((\Omega-\phi)/2)\}u^i(\rho_i,\phi_i) \qquad (1.16)$$

and

$$u^d(\rho,\phi) = -\csc((\Omega-\phi)/2)g(k\rho)u^i(s,\Omega) \ . \qquad (1.17)$$

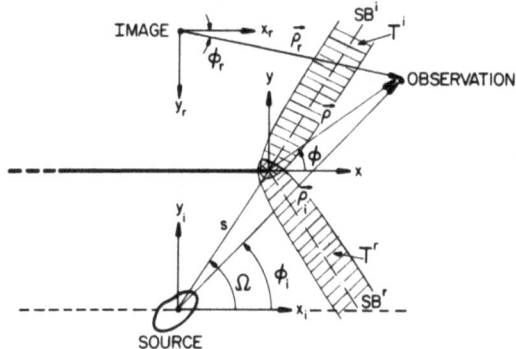

Figure 1. Diffraction of an arbitrary field by a half-plane.

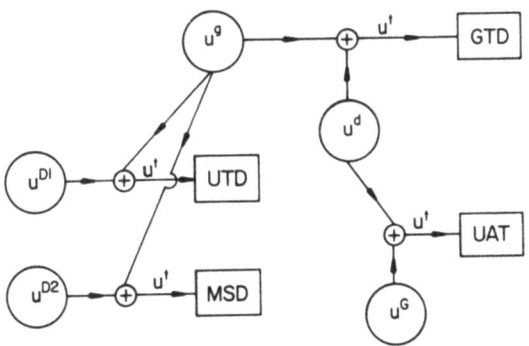

Figure 2. Schematic diagram of inherent relationships among different asymptotic techniques.

In the preceding equations, θ is the unit step function and

$$u^i(\rho_i, \phi_i) = g(k\rho_i)P(\phi_i) \quad (1.18)$$

that is, the first asymptotic term of (1.4). It is clear that the GTD formulation fails at the shadow boundaries, i.e., $\phi = \pm\Omega$, and their vicinities.

The UTD formulation overcomes this failure of GTD at the shadow boundaries by expressing $v(\rho, \phi)$ from (1.14) as

$$v(\rho, \phi) = u^g(\rho, \phi) + u^{D1}(\rho, \phi) + 0(k^{-3/2}) \quad (1.19)$$

where u^g is already defined in (1.16) and u^{D1} is

$$u^{D1}(\rho, \phi) = 2\sqrt{\pi}\, \text{sgn}[\sin((\phi-\Omega)/2)][2k\rho s/(\rho+s)]^{1/2}$$

$$\cdot g(k\rho)e^{-i(\pi/4+\xi^2)}\, F(\xi)u^i(s,\Omega). \quad (1.20)$$

In (1.21), F is the Fresnel integral (1.3) and

$$\xi = |\sin((\phi-\Omega)/2)|[2k\rho s/(\rho+s)]^{1/2} \quad (1.21)$$

which goes to zero at the shadow boundaries. The UTD formulation is often referred to in the literature as the Kouyoumjian-Pathak (KP) formulation.

Since UTD is only suitable for a plane wave or isotropic line source type of incident field, recently a modified version of UTD that employs the notion of the slope diffraction coefficient was introduced to generalize this procedure. This formulation, which is referred to here as MSD, modified (1.19) as

$$v(\rho, \phi) = u^g(\rho, \phi) + u^{D2}(\rho, \phi) + 0(k^{-3/2}) \quad (1.22)$$

where

$$u^{D2}(\rho, \phi) = u^{D1}(\rho, \phi) + \{2\sqrt{\pi}\, \text{sgn}[\sin((\phi-\Omega)/2)] \cdot \sqrt{2ks\rho}\, \rho$$

$$\cdot (s+\rho)^{-3/2} \sin(\phi-\Omega)e^{-i(\pi/4+\xi^2)}F(\xi) - 2\rho(\rho+s)^{-1}$$

$$\cdot \cos((\phi-\Omega)/2)\}\, g(k\rho)(\partial u^i/\partial \phi^i)\big|_{s,\Omega}. \quad (1.23)$$

In summary, UTD and its modification MSD have been introduced to modify u^d in the GTD formulation in such a manner as to cancel the discontinuity of u^g.

Finally, the Uniform Asymptotic Theory (UAT) has been invented to circumvent the shadow boundary difficulties of the GTD by writing v as

$$v(\rho,\phi) = u^G(\rho,\phi) + u^d(\rho,\phi) + O(k^{-3/2}) \tag{1.24}$$

where u^d is the same term used in the GTD formulation, i.e., (1.17). $u^G(\rho,\phi)$ has the form that exactly cancels the infinite value of u^d at the shadow boundaries. In this theory u^G takes the form

$$u^G(\rho,\phi) = [F(\tau) - \hat{F}(\tau)] u^i(\rho_i,\phi_i) \tag{1.25}$$

where $\hat{F}$ is the first term of the asymptotic expansion of F

$$\hat{F}(\tau) = e^{i\pi/4} + i\tau^2/2\sqrt{\pi}\,\tau \tag{1.26}$$

and τ, which is called the detour function, is

$$\tau = \text{sgn}\,[\sin\,(\phi-\Omega)/2]\,\{k[(\rho+s) - \rho_i]\}^{1/2} ;$$

$$\rho_i = [s^2+\rho^2+2s\rho\,\cos(\phi-\Omega)]^{1/2} . \tag{1.27}$$

Note that at the shadow boundaries the detour τ is zero. The higher-order asymptotic expression of UAT to the order of k^{-2} (the dominant asymptotic term of u^i is $k^{-1/2}$) for the problem at hand is given by Boersma and Lee [17]. It is worth mentioning that in all of the above asymptotic techniques the field incident on the diffracting edge must be a ray field (local plane wave). Some difficulties related to the application of incident fields that do not fulfill this criterion have been discussed by Rahmat-Samii and Mittra [3], Lee and Boersma [19], Jones [20], and Boersma [21]. Furthermore, one should realize that in the UTD and MSD construction of the total field, the higher-order derivatives of the field are discontinuous at the shadow boundaries, whereas in the UAT the total field is finite and regular with continuous derivatives.

Based on the analytical and asymptotic evaluation of (1.1) and (1.2), both away from the transition regions and exactly at the shadow boundaries, and the results given in [17], the following observations can be made from the various solutions of the problem of the diffraction of an anisotropic line source by a half-plane: (i) STD formulation can be used in a very systematic and straightforward fashion to determine arbitrary higher-order terms in the asymptotic expansion of the total field. In fact, by using this approach we have computed the asymptotic expression of the total field up to the order of k^{-3}; (ii) UAT formulation, in its general version, can also furnish the higher-order asymptotic expression of the total field. However, in this method the higher-order terms in the diffracted field u^d must be generated through an explicit application of the edge condition. Results up to the order of k^{-2} have been reported in [17] and they agree perfectly with the results of STD; (iii) UTD provides the correct asymptotic expression of the field up to the order of k^{-1}, only for the isotropic line source. Furthermore, UTD formulation does not provide higher-order asymptotic terms in a direct fashion; (iv) MSD, which is the modified version of UTD, allows one to determine the asymptotic values of the field for the anisotropic line source up to the order of k^{-1}. Again, this formulation does not give higher-order terms in a systematic way.

All of the above observations are made on the basis of a comparative study of the various results directly at the shadow boundaries ($\xi = \tau = 0$) and far away from the transition region ($\xi \gg 0$, $\tau \gg 0$) where the field can be expressed in analytic forms. For the sake of completeness we have also investigated the behavior of the fields predicted by the various theories in the neighborhood of the transition regions. For this purpose the field integral in STD has been evaluated using an efficient numerical algorithm which performs a numerical integration on the steepest descent path [2]. Note that the computed results have also been obtained, up to the range of higher-order terms available, using UTD, MSD, and UAT formulations, as described in the previous paragraphs.

Figure 3 shows the total field diffracted by a half-plane illuminated by a nonisotropic line source (E wave) with pattern function

$$P(\omega) = \cos\omega + \text{sgn}(\phi_i)\sin\omega. \qquad (1.28)$$

Results are shown for different techniques and the field is sampled in a 40° angular region about the incident shadow boundary (SB^i). The SB^i direction is shown in the figure, and as expected, the UTD formulation fails to provide the correct result due to the lack of proper slope discontinuity compensation. Since the incident field is well-behaved in the entire region, we have also compared the scattered field u^s for different techniques in Figure 4. Exactly at the shadow boundaries, we can employ our higher-order asymptotic results, derived in (1.11) and (1.13) to

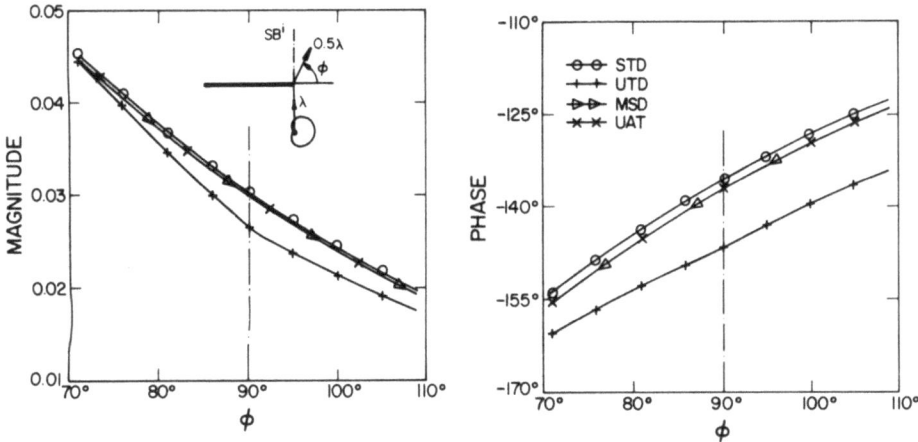

Figure 3. Total field u^t diffracted by a half-plane illuminated by a nonisotropic source (E wave) with pattern function of (28) and $|u^i(1,\pi/2)| = .0796$.

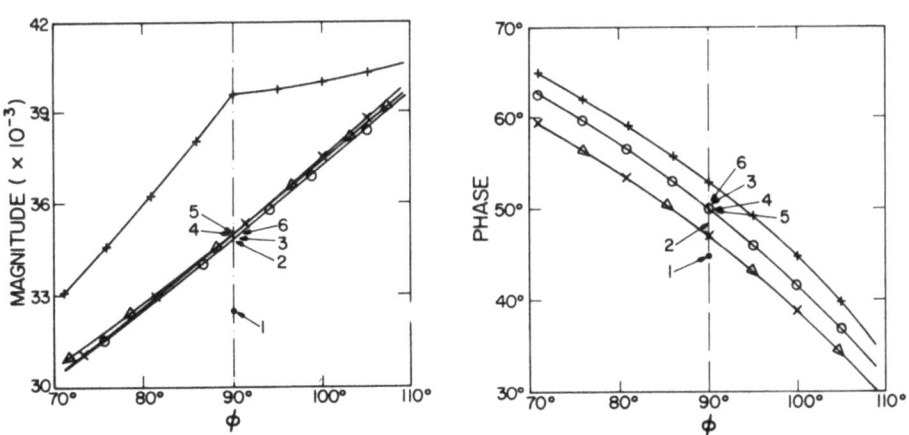

Figure 4. Scattered field u^s from a half plane illuminated by a nonisotropic source (geometry shown in Figure 3). Points 1-6 are the number of terms in asymptotic expansion of u^s.

determine the scattered field. The numerical values of these asymptotic terms are indicated in Figure 4 with a view to exhibiting the behavior of the asymptotic solution. The number of terms used in the asymptotic evaluation is indicated by numerals inserted alongside the points representing the evaluation of the diffracted field at the shadow boundary.

Figure 5 displays the behavior of the scattered field calculated using the aforementioned techniques in the transition region of a half-plane illuminated by an isotropic line source (E wave). As expected, UTD and MSD are the same and all the techniques are in very good agreement. For completeness, the asymptotic expressions have been calculated, and are shown in Figure 5. Figure 6 shows a plot of the scattered field from a half-plane illuminated by an isotropic line source (E wave) coplanar with the half-plane. For this geometry the incident and reflected shadow boundaries coincide with the half-plane itself. The scattered field was sampled in an angular region $-10° < \phi < 10°$ and results for different techniques including GTD are shown. For the sake of completeness the asymptotic results computed from (1.8) are also shown in Figure 6 for the observation angle $\phi = 0$, and the oscillatory nature of this asymptotic solution is evident from the figure.

Based on the extensive numerical results, several observations can be made regarding the different techniques. First, for a source location and an observation point satisfying the condition $s + \rho\cos(\Omega \mp \phi) > 0$, numerical evaluation of the STD integral is very rapid and thus this approach allows one to efficiently compute the scattered field. Second, the UTD formulation, i.e., (1.20), provides uniform numerical results for all observation points when an isotropic line source is considered. The results were checked with those of STD and the agreement is better than 0.6% for the amplitude, and the phase deviation is less than 2°, even in the transition regions. Third, the MSD formulation, i.e., (1.22), is numerically uniform and the results derived from it compare well with STD for the nonisotropic line source. And fourth, the UAT formulation, i.e., (1.24) also provides excellent agreement when compared with STD, and also agrees quite well with MSD. However, one must exercise some care when using UAT, as given in (1.24), for numerical computations. It involves a subtraction of two large numbers that go to infinity at the shadow boundaries. The results are subject to numerical error if one attempts to get too close to these boundaries. Exactly at the shadow boundaries one must employ an analytical limiting procedure or rearranged expression before the formula can be handled on the computer.

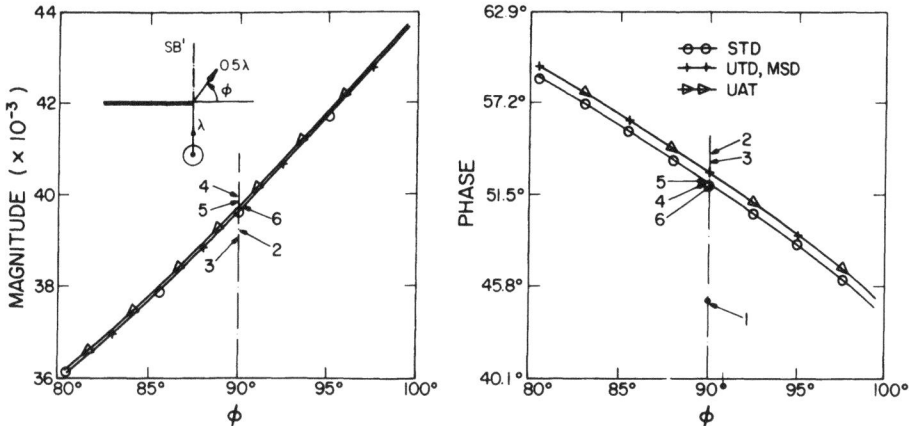

Figure 5. Scatterred field u^s from a half plane illuminated by an isotropic source (E wave) with $|u^i(1,\pi/2)| = .0796$. Points 1-6 are the number of terms in asymptotic expansion of u^s.

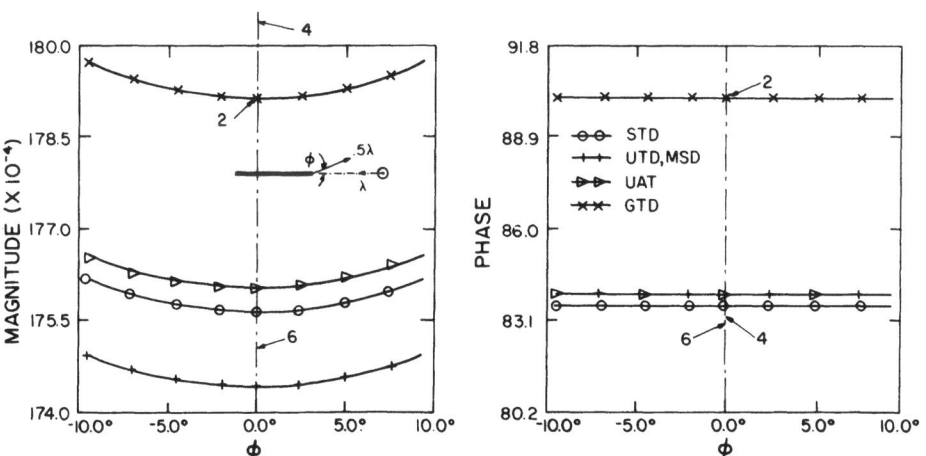

Figure 6. Scattered field u^s from a half plane illuminated by an isotropic source (E wave) with $|u^i(1,\pi/2)| = .0796$. Points 2, 4, and 6 are the asymptotic evaluation at $\phi = 0°$.

2. OFFSET REFLECTOR ANALYSIS USING THE BESSEL-JACOBI SERIES

Recent interest in dual offset reflector antennas has motivated many researchers to investigate the problem of scattering from subreflectors and main reflectors using various asymptotic and numerical techniques. Typically in dual reflector antenna systems, the subreflector has a broad pattern while the main reflector, generally concave, radiates a pencil beam pattern. A study of an arbitrarily shaped subreflector can be found in [22] where the Geometrical Theory of Diffraction (GTD), in a uniform format, is used as the basic analytical tool. In this paper, we consider only analysis of scattering from the main reflector. Due to the caustic difficulties at the main beam, integration techniques are generally preferred to ray optical techniques for the analysis of the main reflector. A comprehensive review of different techniques may be found in [23] - [25]. It has been found that the Physical Optics (PO) radiation integral provides a very accurate solution for predicting the far-field radiation characteristics of the reflector antennas in the main beam region and out to several sidelobes. Although the method is very useful, it has the drawback of being very time-consuming even on today's modern and fast digital computers. The difficulty stems from the fact that, for large reflectors and wide observation angles, the integrand of the radiation integral oscillates rapidly and is therefore hard to evaluate accurately. More importantly, one has to evaluate this integral each time as the observation angle changes. In this paper, the integrand is first rearranged so that it is possible to use a biconvergent series expansion [6]. This feature of expansion is related to the asymptotic evaluation of the integral. Finally, the integral is expressed by a Bessel-Jacobi series expansion which has many unique features for calculating the far-field of pencil beam patterns. In this presentation only the final results will be discussed and the reader is referred to [5] - [8] for details.

The geometry of the reflector surface Σ is depicted in Figure 7. The projection of Σ on x-y is the circular region σ with radius a and will be used to define the integration region. Various primed and unprimed Cartesian and spherical coordinates are defined in this figure. Reflector surface Σ may be described as

$$z' = f(x', y') = \tilde{f}(\rho', \phi') , \qquad (2.1)$$

and the induced physical optics current on Σ due to a source field $\vec{H}_s$ is defined by

$$\vec{J} = 2\hat{n} \times \vec{H}_s \qquad (2.2)$$

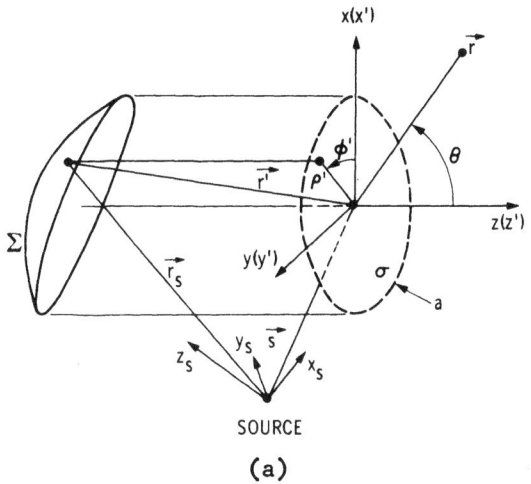

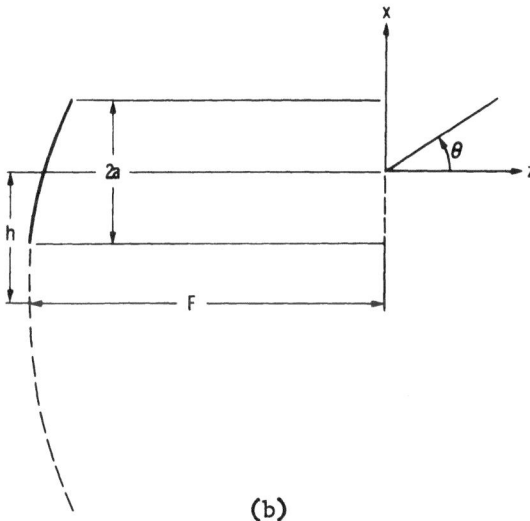

Figure 7. (a) Offset reflector antenna illuminated by an arbitrarily located source.
(b) Offset parabolic reflector.

where $\hat{n}$ is the unit normal to the surface. The far-field is then obtained from the well-known integral

$$\vec{E} = -jk\eta \, (e^{-jkr})/(4\pi r)(\hat{\hat{I}}-\hat{r}\hat{r}) \cdot \vec{T}(\theta,\phi) \tag{2.3}$$

where $\eta = \sqrt{\mu/\epsilon}$, $\hat{I}$ is the unit dyadic and

$$\vec{T}(\theta,\phi) = \iint_\Sigma \vec{J}(\vec{r}') \, e^{j k \vec{r}' \cdot \hat{r}} ds' . \tag{2.4}$$

Integration of (2.4) is performed on the reflector surface Σ. This integration can be transformed into an integration over the projected circular region σ, using the surface Jacobian transformation which can be expressed as

$$J_\Sigma = [1 + (\partial \tilde{f}/\partial \rho')^2 + (\rho')^{-2}(\partial \tilde{f}/\partial \phi')^2]^{\frac{1}{2}} \tag{2.5}$$

This leads to the following expression for (2.4)

$$\vec{T}(\theta,\phi) = \int_0^a \int_0^{2\pi} \vec{\tilde{J}}(\rho',\phi') \, [e^{jkr' \cos\theta' \cos\theta}]$$
$$\cdot \{e^{jk\rho' \sin\theta \cos(\phi'-\phi)}\} \, \rho' d\rho' d\phi' . \tag{2.6}$$

where

$$\vec{\tilde{J}}(\rho',\phi') = \vec{J}(\vec{r}') \, J_\Sigma . \tag{2.7}$$

$\vec{\tilde{J}}$ is referred to as the equivalent induced current and has its support on the reflector surface. It is worth mentioning that, although integration (2.6) is performed on the projected aperture, the current is evaluated on the reflector surface and (2.6), therefore, is identical to (2.4). This is why (2.6) is different from the integration obtained when using the aperture field method.

We now proceed to rearrange the integrand of (2.6) in a manner which will allow us to integrate on the curved reflector with little more effort than would be necessary for integration on a flat surface (Note that the term $\{\cdot\}$ defines a Fourier kernel.) The procedure follows [5] - [7] where the mathematical details are given. We focus our attention on the offset parabolic reflector for which (2.1) can be expressed as

$$z' = F[-1 + (\rho'^2)/(4F^2)] + h/(2F) \rho' \cos\phi' + h^2/(4F) \tag{2.8}$$

where F and h designate the focal length and offset height, respectively. A unique feature of (2.8) is the appearance of the linear term in ρ' and $\cos\phi'$ which allows one to combine it with the Fourier kernel. Substituting (2.8) into the term $[\cdot]$ of (2.6), we find

$$r'\cos\theta'\cos\theta = z'\cos\theta = F[-1 + (\rho'^2/(4F^2))]\cos\theta + h/(2F)$$
$$\cdot \rho'\cos\phi'\cos\theta + (h^2)/(4F)\cos\theta. \qquad (2.9)$$

As the starting point for rearrangement of the integrand we assume that the scattered field has its pencil beam directed at (θ_B, ϕ_B) and then try to rearrange the Fourier kernel $\{\cdot\}$ such that its center is in this direction. To this end we set

$$\begin{cases} B \sin\Phi = \sin\theta \sin\phi + C_v \\ B \cos\Phi = \cos\theta \sin\phi + C_u + h/(2F)\cos\theta \end{cases} \qquad (2.10)$$

where C_v and C_u are unknown constants yet to be determined. We now require that B becomes zero at the observation angle (θ_B, ϕ_B) which allows us to determine C_v and C_u as

$$\begin{cases} C_v = -\sin\theta_B \sin\phi_B \\ C_u = -\sin\theta_B \cos\phi_B - h/(2F)\cos\theta_B. \end{cases} \qquad (2.11)$$

Substituting (2.19) into (2.6) and using (2.10) we finally arrive at

$$e^{-jk(h^2/4F)\cos\theta} \vec{T}(\theta,\phi) = \int_0^a \int_0^{2\pi} \vec{\tilde{K}}(\rho',\phi') [e^{jkL}]$$
$$\cdot \{e^{jk\rho' B \cos(\phi'-\Phi)}\} \rho'd\rho'd\phi' \qquad (2.12)$$

where

$$\vec{\tilde{K}} = \vec{J} e^{-jk\rho'(C_u\cos\phi' + C_v\sin\phi')} \qquad (2.13)$$

and

$$L = F[-1 + (\rho'^2)/(4F)]\cos\theta. \qquad (2.14)$$

It is worthwhile to mention here that $\vec{\tilde{K}}$ is independent of the observation coordinates θ and ϕ.

We now focus our attention on the exp(jkL) term which is a function of both the observation and the source coordinates. From (2.14) the following functions can be defined as

$$\begin{cases} L_0 = F[-1 + \rho'^2/(4F)] \cos\theta_0 \\ L_a = F[-1 + a^2/(4F)] \cos\theta \\ L_{a0} = F[-1 + a^2/(4F)] \cos\theta_0 \end{cases} \quad (2.15)$$

where θ_0 is a fixed angle typically taken along the beam maximum and usually is set to be $\theta_0 = \theta_B$. Employing (2.15) and (2.14), we define Δ as

$$\Delta = L - L_0 - L_a + L_{a0} = 1/(4F)(\rho'^2 - a^2)(\cos\theta - \cos\theta_0) \quad (2.16)$$

which has the following property

$$\Delta \to 0 \text{ as } \theta \to \theta_0 \text{ and/or } \rho' \to a.$$

We now return to (2.12) and rewrite it as

$$e^{-jk(h^2/4F)\cos\theta_T} \vec{T} = e^{jk(L_a - L_{a0})} \vec{T} \quad (2.17a)$$

where

$$\vec{T}(\theta,\phi) = \int_0^a \int_0^{2\pi} \vec{K} e^{jkL_0} [e^{jk\Delta}] e^{jk\rho'B\cos(\phi'-\phi)} \rho' d\rho' d\phi'. \quad (2.17b)$$

It is helpful at this point to relate the motivation for the manipulations leading up to (2.17b). If we can demonstrate that the exponent Δ is small for certain ranges of observation angles to the extent that $\exp(jk\Delta) \approx 1$, we would obtain a Fourier transform for the radiation integral with an "effective aperture" distribution proportional to $\vec{K}$. Referring to the definition of Δ and various L's given in (2.15), we observe that $\Delta = 0$ at the angle $\theta = \theta_0$ (typically $\theta_0 = \theta_B$, i.e., the main beam direction).

Thus, $\exp(jk\Delta)$ is identically equal to 1 for this choice of observation angles. In addition, for ρ' in the neighborhood of the rim, i.e., $\rho' = a$, Δ is small. This is the region from which the dominant contribution of the radiation integral is derived in the wide-angle region (typically just a few sidelobes away from the beam maximum) where θ is not close to θ_0. This observation is consistant with the asymptotic evaluation of the integral where the contribution to the wide-angle region is predominantly from edge-diffracted rays. Thus, unusual as it may appear at first sight, what we have accomplished through manipulation described above is the derivation of an "effective aperture" distribution which is simultaneously valid not only at the beam maximum but at wide angles as well. In fact, numerical results have clearly demonstrated that only in a small intermediate angular region does the setting of $\exp(jk\Delta) = 1$ in (2.17b) introduce any significant error [5] - [7].

To fill in this gap, we expand $\exp(jk\Delta)$ in a Taylor series, obtaining

$$\vec{\tilde{T}} = \sum_{p=0}^{p\to\infty} 1/p! (jk)^p (1/4F)^p (\cos\theta - \cos\theta_0)^p \vec{\tilde{T}}_p \qquad (2.18)$$

where

$$\vec{\tilde{T}}_p = \int_0^a \int_0^{2\pi} \vec{\tilde{Q}}_p(\rho',\phi') e^{jk\rho'B\cos(\phi'-\Phi)} \rho' d\rho' d\phi' \qquad (2.19a)$$

and

$$\vec{\tilde{Q}}_p(\rho',\phi') = \vec{\tilde{K}} e^{jkL_0} (\rho'^2 - a^2)^p . \qquad (2.19b)$$

In its present form expansion (2.18) is <u>rapidly</u> convergent at the observation angles near the main beam and those few sidelobes away. For this reason we refer to (2.18) as a "biconvergent" expansion. Due to its Fourier transform nature, integration (2.19) can be numerically computed using a variety of methods including the FFT. In this work, however, we employ the Bessel-Jacobi series expansion as it has the most suitable feature for constructing beam shaped patterns from a circular projected aperture.

Introducing the change of variable

$$\rho' = as' \qquad (2.20)$$

and using the orthogonality properties of the Fourier series and modified Jacobi polynomial [26], we can then expand $\vec{Q}_p$ as

$$\vec{Q}_p(as', \phi') = \sum_{n=0}^{N \to \infty} \sum_{m=0}^{M \to \infty} [\,_p\vec{C}_{nm} \cos n\phi' + \,_p\vec{D}_{nm} \sin n\phi'] F_m^n(s') \quad (2.21)$$

where F_m^n is the modified Jacobi polynomial, $\vec{C}$ and $\vec{D}$ are obtained from

$$\begin{Bmatrix} _p\vec{C}_{nm} \\ _p\vec{D}_{nm} \end{Bmatrix} = \frac{\varepsilon_n}{2\pi} \int_0^{2\pi} \int_0^1 \vec{Q}_p(as', \phi') \begin{Bmatrix} \cos n\phi' \\ \sin n\phi' \end{Bmatrix} F_m^n(s') s' \, d\phi' \, ds' \quad (2.22)$$

and $\varepsilon_n = 1$ for $n = 0$ and $\varepsilon_n = 2$ for $n \neq 0$. Substituting (2.21) into (2.19a) and using some basic properties of the Bessel function and modified Jacobi polynomial [26], we finally can express $\vec{T}_p$ as

$$\vec{T}_p = 2\pi a^2 \sum_{n=0}^{N \to \infty} \sum_{m=0}^{M \to \infty} j^n [\,_p\vec{C}_{nm} \sin n\Phi + \,_p\vec{D}_{nm} \cos n\Phi]$$

$$\cdot \sqrt{2(n+2m+1)} \, \frac{J_{n+2m+1}(kaB)}{kaB} \, . \quad (2.23)$$

The above result can be sequentially used in (2.18) and (2.17a) to finally give the far-field from (2.3). We note that the radiation integral $\vec{T}$ can be readily calculated at any observation angle once the coefficients C and D are determined. This is in contrast to the direct evaluation of the original integral (2.4) which must be computed repeatedly for each observation angle. The numerical evaluation of (2.22) is much simpler than (2.4) as its integrand does not contain the highly oscillatory Fourier transform kernel. We also note from (2.23) that the leading term is the Airy disk function which is the radiation pattern of a circular aperture with uniform amplitude and phase distribution (except for a linear taper which shifts the beam maximum to θ_B, θ_B). All of the higher-order terms of the series in (2.23) become zero as $B \to 0$ and represent perturbations around the beam maximum. Furthermore, it can be shown [6] that the higher-order p series coefficient in (2.18) and (2.22) can be constructed from proper recursion relations which reduce the computation time.

A general computer program has been developed based on the expressions developed here and in [27] and extensive numerical results have been obtained for a variety of reflector dimensions and feed locations. Some of the representative results are reported in [5] - [8]. Even in the most difficult cases, the maximum number of terms used was under $P = 3$, $M = N = 10$.

ACKNOWLEDGEMENT

The author wishes to thank Prof. R. Mittra and Dr. V. Galindo-Israel who collaborated with the author in the development of most of the topics described herein and did much of the original work on the application of the Bessel-Jacobi series.

REFERENCES

1. Y. Rahmat-Samii and R. Mittra, "A Spectral Domain Interpretation of High Frequency Diffraction Phenomena," IEEE Trans. Antennas Propagat., vol. 25, pp. 676-687, September 1977.

2. Y. Rahmat-Samii and R. Mittra, "Spectral Analysis of High Frequency Diffraction of an Arbitrary Incident Field by a Half Plane--Comparison with Four Asymptotic Techniques," Radio Science, vol. 13, pp. 31-48, Jan./Feb., 1978.

3. Y. Rahmat-Samii and R. Mittra, "On the Investigation of Diffracted Fields at the Shadow Boundaries of Staggered Parallel Plates--A Spectral Domain Approach," Radio Science, vol. 12, pp. 659-670, Sept./Oct. 1977.

4. R. Mittra and Y. Rahmat-Samii, "A Spectral Domain Approach for Solving High Frequency Scattering Problems," in Electromagnetic Scattering, P.L.E. Uslenghi, Ed., Academic Press, New York, 1978.

5. V. Galindo-Israel and R. Mittra, "A New Series Representation for the Radiation Integral with Application to Reflector Antennas," IEEE Trans. Antennas Propagat., vol. 25, pp. 631-641, September 1977.

6. R. Mittra, Y. Rahmat-Samii, V. Galindo-Israel and R. Norman, "An Efficient Technique for the Computation of Vector Secondary Patterns of Offset Paraboloid Reflectors." IEEE Trans. Antennas Propagat., vol. 27, pp. 294-304, May 1979.

7. Y. Rahmat-Samii and V. Galindo-Israel, "Shaped Reflector Antenna Analysis Using the Bessel-Jacobi Series," to appear.

8. Y. Rahmat-Samii, V. Galindo-Israel and R. Mittra, "A Plane-Polar Approach for Far-Field Construction from Near-Field Measurements," to appear in IEEE Trans. Antennas Propagat., 1980.

9. J. B. Keller, "Geometrical Theory of Diffraction," J. Opt. Soc. Amer., vol. 52, pp. 116-130, 1962.

10. R. G. Kouyoumjian and P. H. Pathak, "A Uniform Geometrical Theory of Diffraction for an Edge in a Perfectly Conducting Surface," Proc. IEEE, vol. 62, pp. 1448-1461, 1974.

11. Y. M. Hwang and R. G. Kouyoumjian, "A Dyadic Coefficient for an Electromagnetic Wave which is Rapidly Varying at an Edge," URSI 1974 Annual Meeting, Boulder, Colorado.

12. D. S. Ahluwalia and R. M. Boersma, "Uniform Asymptotic Theory of Diffraction by a Plane Screen," SIAM J. Appl. Math., vol. 16, pp. 783-807, 1968.

13. S. W. Lee and J. Boersma, "Ray-optical Analysis of Fields on Shadow Boundaries of Two Parallel Plates," J. Math. Phys., vol. 16, pp. 1746-1764, 1975.

14. S. W. Lee and G. A. Deschamps, "A Uniform Asymptotic Theory of Electromagnetic Diffraction by a Curved Edge," IEEE Trans. Antennas Propagat., vol. AP-24, No. 1, 1976.

15. S. W. Lee, "Uniform Asymptotic Theory of Electromagnetic Edge Diffraction: A Review," EM Report 77-1, Electromagnetics Lab., University of Illinois, Urbana, Ill., 1977. (This report includes some 300 references).

16. R. Mittra, Y. Rahmat-Samii and W. L. Ko, "Spectral Theory of Diffraction," Appl. Phys., vol. 10, pp. 1-13, January 1976.

17. J. Boersma, and S. W. Lee, "High-Frequency Diffraction of a Line-Source Field by a Half-Plane, 1, Solutions by Ray Techniques," IEEE Trans. Antennas Propagat., vol. 25, pp. 171-179, 1977.

18. J. B. Keller, R. M. Lewis, and B. D. Seckler, "Asymptotic Solution of Some Diffraction Problems," Commun. Pure Appl. Math., vol. 9, pp. 207-265, 1956.

19. S. W. Lee, and J. Boersma, "Ray-Optical Analysis of Fields on Shadow Boundaries of Two Parallel Plates," J. Math. Phys., vol. 16, pp. 1746-1764, 1975.

20. D. S. Jones, "Double Knife-Edge Diffraction and Ray Theory," Quart. J. Mech. Appl. Math., vol. 26, pp. 1-18, 1973.

21. J. Boersma, "Diffraction by Two Parallel Half-Planes," Quart. J. Mech. Appl. Math., vol. 28, pp. 405-425, 1975.

22. S. W. Lee, P. Cramer, Jr., K. Woo and Y. Rahmat-Samii, "Diffraction by an Arbitrary Subreflector: GTD Solution," IEEE Trans. Antennas Propagat., vol. 27, pp. 305-316, May 1979.

23. A. W. Love, Reflector Antennas, New York, IEEE Press, 1978.

24. P. J. B. Clarricoats, and G. T. Poulton, "High-Efficiency Microwave Reflector Antennas: A Review," Proc. IEEE, vol. 65, pp. 1470-1504, Oct. 1977.

25. A. W. Rudge and N. A. Adatia, "Offset-Parabolic-Reflector Antennas: A Review" Proc. IEEE, vol. 66, pp. 1592-1618, 1978.

26. D. Slepian, "Prolate Spheroidal Wave Functions, Fourier Analysis and Uncertainty-IV: Extensions to Many Dimensions; Generalized Prolate Spheroidal Functions," Bell Syst. Tech. J., vol. 43, pp. 3009-3057, July - Dec. 1964.

27. Y. Rahmat-Samii, "Useful Coordinate Transformations for Antenna Applications," IEEE Trans. Antenna Propagat., vol. 27, pp. 571-574, July 1979.

HIGH-FREQUENCY FIELDS EXCITED BY A LINE SOURCE LOCATED ON A
PERFECTLY CONDUCTING CONCAVE CYLINDRICAL SURFACE. *

T. Ishihara, L.B. Felsen and A. Green

Department of Electrical Engineering, Polytechnic Institute of
New York, Farmingdale, NY 11735, U.S.A.

ABSTRACT - Alternative representations are obtained for the
high-frequency surface field excited on a perfectly conducting
concave circular cylinder by an axial magnetic line current
located on the surface. Included are ray-optical, canonical
integral, whispering gallery mode, and near-field formulations,
and various combinations of these. Asymptotic evaluations in
different parameter ranges lead to results with varying accuracy
and physical content. Their utility is assessed by extensive
numerical calculations and comparisons. Most intriguing is a
form of the asymptotic solution that involves only a number of
geometric optical rays and a number of whispering gallery modes.

I. INTRODUCTION

When high-frequency fields impinge on a concave perfectly conducting surface, the induced surface currents can be well
approximated by their physical optics values since the entire
surface is illuminated. Phrased alternatively, the method of
geometrical optics is adequate to describe the surface field
behaviour. However, when actual sources or induced equivalent
sources (for example, at an edge termination) are located very
near or on the surface, geometrical optics becomes inapplicable
because the multiply reflected ray fields (Fig. 1) have caustics
that lie near the boundary (Fig. 2). Therefore it is necessary
to account in some other manner for the fields that would be
contributed by rays having undergone many reflections.

---

* Reprinted (with small changes) from IEEE Transactions on
Antennas and Propagation, AP-26 (1978), p.757-767.

A systematic study of the problem can be undertaken on the prototype configuration of a circular cylinder. Since a thorough understanding of the axially independent two-dimensional case is essential for extension to three-dimensional fields, attention is focused here on the problem of excitation by a magnetic line current located on the concave surface and oriented parallel to the cylinder axis. Starting from a rigorous formulation of the Green's function problem, one may extract from the azimuthally periodic solution for the closed surface a portion [1] that describes only the propagation characteristics between points Q and P in Fig. 1. Results obtained therefrom are applicable to concave surface segments that are open, and they are in a form suitable for subsequent generalization to variable surface properties. This basic representation, given as a contour integral, is the starting point for the present study, which has as its goal the exploration of alternative field representations that are useful for calculation in various parameter ranges. Included are ray-optical, canonical integral, and whispering gallery mode (Fig. 3) solutions, and various combinations of these. Strong emphasis is placed on the physical interpretation of the results and on their relative accuracy. Most pleasing from a physical standpoint is a field solution that comprises a mixture of (N + 1) geometric optical rays and M whispering gallery modes, with criteria provided for the proper choice of N and M.

II. ALTERNATIVE FIELD PRESENTATIONS

A. Green's Function Formulation

We consider the prototype problem of line source radiation inside a perfectly conducting circular cylinder of radius a. To make the analysis relevant to the propagation along concave surfaces in general, it is necessary to remove the azimuthal periodicity imposed on the fields in the cylindrical $\rho = (\rho, \phi)$ geometry where ρ represents the radial and ϕ the angular (azimuthal) coordinate. This can be done by extending the range of the ϕ coordinate from its physical periodic domain $0 \leq \phi \leq 2\pi$ into an infinite domain $-\infty \leq \phi \leq \infty$ [7]. Such an extension implies that angularly propagating waves originating at the source point (ρ', ϕ') are outgoing toward $|\phi| = \infty$; this "angular radiation condition" can be realized by placing at some angular location away from the source angle ϕ' a perfect absorber for angularly propagating waves. Such an absorber has, however, the undesired property of generating diffraction at the radial coordinate origin $\rho = 0$. Therefore, when considering propagation phenomena ascribable only to the cylinder surface, it is desirable to remove the spurious diffraction effects from the total field solution.

The line source Green's function in the cylindrical domain $0 < \rho < a$, $-\infty < \phi < \infty$ can be constructed by the method of separation of variables and expressed in its most general form as a contour integral involving the two one-dimensional characteristic Green's functions g_ρ and g_ϕ for the radial and angular domains, respectively [7]. When the Green's function is represented in terms of angularly propagating waves, the corresponding eigenmode spectrum in the radial domain involves a discrete (whispering gallery mode) and continuous portion [3]. The latter accounts in part for propagation phenomena associated with the portion of the concave boundary lying between the source and observation points (this is the desired contribution) and in part for the spurious diffraction effects of the angular absorber. When the spurious effects are removed and cognizance is taken of the high-frequency nature of the analysis, one may show that the relevant propagation characteristics along the portion of the concave boundary lying between the source point Q and the observation point P in Fig. 1 are contained in the partial Green's function |1|

$$G(\tilde{\rho},\tilde{\rho}') = \frac{1}{i(\pi k a)^2} \int_C \frac{e^{i\nu|\phi - \phi'|}}{H_\nu'^{(2)}(ka) J_\nu'(ka)} d\nu \qquad (1)$$

where k is the free-space wavenumber, the prime on the cylinder functions denotes the derivative with respect to the argument, and a time factor $\exp(-i\omega t)$ is implied. It has been assumed in (1) that P and Q both lie on the boundary (i.e. $\rho = \rho' = a$) and that the boundary condition requires that the normal derivative of the Green's function vanishes at $\rho = a$; this makes G proportional to the axial component of magnetic field. The contour C and the singularities of the integrand in the complex ν-plane are shown in Fig. 4. Contributions from the pole singularities ν_m arising from

$$J_{\nu_m}'(ka) = 0, \qquad m = 1, 2, \ldots \qquad (2)$$

are found to describe whispering gallery modes. While (2) has an infinite number of real solutions as indicated in Fig. 4, only those with $\text{Re } \nu_m > 0$ represent spectral contributions in the angular transmission representation which includes also a continuous spectrum.

B. Whispering Gallery Mode and Continuous Spectrum Representation

The spectral representation comprising a discrete (whispering gallery mode) portion and a continuous portion may be obtained directly from (1) by deforming C in Fig. 4 into a contour extending along the imaginary ν axis:

$$G = \sum_{m=1}^{M_1} G_m + R_{M_1} \tag{3}$$

where M_1 is the total number of poles satisfying Re $\nu_m > 0$ and

$$R_{M_1} = \frac{1}{i(\pi ka)^2} \int_{-i\infty}^{+i\infty} \frac{e^{i\nu|\phi - \phi'|} d\nu}{H_\nu^{'(2)}(ka) J_\nu'(ka)}$$

$$\approx \frac{1}{\pi} \int_0^\infty e^{-ka|\phi - \phi'|\sinh \nu} d\nu, \quad ka \gg 1 \tag{4a}$$

is the continuous spectrum. Following Wasylkiwskyj [3], the first integral may be simplified to the second form shown and also to

$$\tfrac{1}{2}\left[H_0(ks) - Y_0(ks)\right], \quad s = a|\phi - \phi'| \tag{4b}$$

where H_0 is the Struve function and Y_0 the Neumann function. Each residue contribution G_m represents a whispering gallery mode of the form

$$G_m = \frac{i}{ka} e^{i\nu_m|\phi - \phi'|} J_{\nu_m}(ka) \left[\frac{\partial}{\partial \nu} J_\nu'(ka)\right]_{\nu_m}^{-1} \tag{5}$$

On use of the uniform asymptotic formulas given by

$$J_\nu(ka) \sim \sqrt{\frac{2}{ka \cos w}} \sigma^{\frac{1}{4}} \text{Ai}|-\sigma|, \quad \nu = ka \sin w$$

$$J_\nu'(ka) \sim -\sqrt{\frac{2 \cos w}{ka}} \sigma^{-\frac{1}{4}} \text{Ai}'[-\sigma]$$

$$\sigma^{3/2} = \frac{3}{2} ka \left[\cos w - \left(\frac{\pi}{2} - w\right) \sin w\right], \quad \text{Re } w > 0 \tag{6}$$

which are justified here, one finds

$$G_m \sim \frac{\exp(iks \sin w_m + i\pi/2)}{ka(\pi/2 - w_m) \cos w_m} \tag{7}$$

The roots w_m are determined by solution of the resonance equation

$$\cos w_m - (\pi/2 - w_m) \sin w_m = \frac{2}{3ka} \sigma_m^{3/2} \tag{7a}$$

where

$$Ai'(-\sigma_m) = 0, \quad m = 1, 2, \ldots M_1.$$

C. Ray-Optical Representation

To generate a ray-optical formulation, one replaces $H_\nu'^{(2)}(ka)$ and $J_\nu'(ka) = \{H_\nu'^{(1)}(ka) + H_\nu'^{(2)}(ka)\}/2$ by their Debye asymptotic forms

$$H_\nu'^{(1,2)}(ka) \sim \sqrt{\frac{2 \cos w}{\pi ka}} \exp\left\{\pm ika\left[\cos w - \left(\frac{\pi}{2} - w\right)\sin w\right] \pm i\frac{\pi}{4}\right\} \quad (8)$$

$$\operatorname{Re} w > 0, \left\{ka\left[\cos w - \left(\frac{\pi}{2} - w\right)\sin w\right]\right\}^{2/3} \gg 1 \quad (8a)$$

where $\nu = ka \sin w$. Utilizing the travelling wave expansion,

$$\frac{1}{J_\nu'(ka)} = \frac{2}{H_\nu'^{(2)}(ka)} \sum_{n=0}^{\infty} (-1)^n r^n, \quad r = \frac{H_\nu'^{(1)}(ka)}{H_\nu'^{(2)}(ka)}, \quad |r| < 1, \quad (9)$$

in conjunction with (8), one reduces (1) to a series of integrals

$$G = \sum_{n=0}^{\infty} G_n \quad (10)$$

where

$$G_n \sim \frac{1}{\pi}(-i)^n \int_{C_n'} \exp\left[ika\, q_n(w)\right] dw \quad (10a)$$

and

$$q_n(w) = |\phi - \phi'| \sin w + 2(n+1)\left[\cos w - \left(\frac{\pi}{2} - w\right)\sin w\right]. \quad (10b)$$

The saddle points of $q_n(w)$, as obtained from $dq_n/dw = 0$, are

$$w_{sn} = \frac{\pi}{2} - \frac{|\phi - \phi'|}{2(n+1)}, \quad \overline{w}_s = \frac{\pi}{2} \tag{11}$$

and thus lie on the real w-axis between $w = 0$ and $w = \pi/2$. A typical path C_n' through the nth saddle point is shown in Fig. 5. The original path mapped from the ν plane can be deformed into C_n'. The Debye approximations in (8) are valid in this relevant region of the complex w-plane. Use of the conventional saddle point formula for evaluation of the integral in (10a) yields

$$G_n \sim e^{i\pi/4} \sqrt{\frac{2}{\pi k}} (-i)^n \frac{e^{ikD_n}}{\sqrt{D_n}} \tag{12}$$

where

$$D_n = 2(n+1)a \sin\left[|\phi - \phi'|/2(n+1)\right]. \tag{12a}$$

This result corresponds to a direct or multiply reflected ray as depicted in Fig. 1 and could have been constructed directly by ray-optical techniques.

Although the series in (10) with (12) formally contains ray contributions with an arbitrarily large number of reflections, these are suspect since the saddle points, from which they are derived, all cluster about $w = \pi/2$. Thus the asymptotic method, whereby each saddle point is treated as isolated, is inadequate. Moreover, (8) becomes invalid as $w \to \pi/2$, thereby invalidating the simplification of the integrand on which the saddle point evaluation is based. It is therefore necessary to truncate the number of legitimate ray-optical terms at some $n = N$ such that w_{sN} is sufficiently less than $\pi/2$. Phrased alternatively, the caustics for rays with many reflections approach the boundary (see Fig. 2) and thus invalidate the ray-optical field formulation in (12).

D. Ray plus Canonical Integral Representation

In view of the above observations, one may employ instead of (9) the partial expansion

$$\frac{1}{J_\nu'(ka)} = \frac{2}{H_\nu^{(2)'}(ka)} \sum_{n=0}^{N} (-1)^n r^n + \frac{1}{J_\nu'(ka)} (-1)^{N+1} r^{N+1}.$$

$$\tag{13}$$

Substitution of (13) and (1) yields

$$G = \sum_{n=0}^{N} G_n + R_N \qquad (14)$$

where

$$R_N = \frac{1}{i(\pi ka)^2} \int_C \frac{e^{i\nu|\phi - \phi'|}}{H_\nu'^{(2)}(ka) J_\nu'(ka)} r^{N+1} (-1)^{N+1} d\nu. \qquad (15)$$

The sum in (14) evidently describes ray contributions having experienced up to N reflections. The remainder integral R_N incorporates the cumulative effects of ray fields having been reflected more than N times. It should be noted that when the observation point approaches the source point so that $|\phi - \phi'| \to 0$, one has $\nu_{sn} \to ka$ even for n = 0, and the ray representation fails altogether. When N = -1, no ray contributions are extracted, and (15) agrees with (1).

Several options are open for dealing with R_N. One possibility is to explore under what conditions the contributing range of the integrand is localized so that approximation methods may be used effectively. To this end, the cylinder functions are first replaced by their uniform asymptotic approximations in terms of Airy functions. From the behaviour of the integrand over the contour C in the ν-plane or the contour C' in the w-plane, it may then be shown that the principal contribution to the integral arises from the vicinity of $\nu = ka$ or $w = \pi/2$ provided that the arc length parameter

$$\gamma = \left(\frac{ka}{2}\right)^{1/3} \frac{s}{a} \qquad (16a)$$

and the number of rays are related by the criterion (note that we designate a ray having undergone N reflections as the (N + 1)th ray)

$$\frac{\gamma}{2(2^{1/3}\Delta)^{1/2}} - 1 < (N + 1) \lesssim \frac{\gamma}{2(2^{1/3}\Delta)^{1/2}} \qquad (16b)$$

where $\Delta = \Delta(ka) = O(1)$ is chosen appropriately. When the left hand side of the inequality is negative, i.e., in the region near the source, no ray-optical contributions are included. The first inequality in (16b) is required to assure that the integrand in (18a) decays away from t = 0. The second inequality in (16b) (and the inequality in (21a)) is required to validate use of the Debye approximations in the first term in (13), which yields the geometric optical ray contribution.

Our criterion in (16b), the utility of which is confirmed by the numerical calculations in Section III, differs somewhat from that obtained by Babich and Buldyrev [1]. The numerical factor Δ in the inequalities (16b) and also in (21a) is not rigidly fixed. This implies that the location of the point at which a geometric optical ray field splits off from the remainder field can be varied. In each specific problem, the factor Δ should be chosen such that the discontinuity in the total field is a minimum when the new geometric optical wave emerges. The estimated value of $\Delta \doteq O(1)$ can be determined from a knowledge of the point of emergence of the direct ray. This point can be found by comparing the numerical results of the direct ray plus canonical integral representation here, or of the direct ray plus whispering gallery mode representation (Section II-E), with that of the exact solution (whispering gallery mode and continuous spectrum representation). In this manner, one finds from (16b) or (21a) with N = 0 that $\gamma \approx 3.1$ whence $\Delta \cong 2$ (see Figs. 6-8). This value of Δ is then used in (16b) to determine the splitting off points of the various reflected rays. It turns out that the results are not very sensitive to the choice of Δ because of the overlap of the curves for various N. Therefore Δ can be determined from the first overlap region for N = -1, N = 0, without the need of performing the comparison with the exact solution.

Changing variables to t via $\nu = ka + (ka/2)^{1/3} t$ and using

$$H_\nu^{(2)\prime}(ka) \sim -\left(\frac{2}{ka}\right)^{2/3} \frac{w_1'(t)}{2}.$$

$$J_\nu'(ka) \sim -\left(\frac{2}{ka}\right)^{2/3} Ai'(t) \qquad (17)$$

where $w_1(t) = Ai(t) \mp iBi(t)$, one may write (15) as

$$R_N \sim \frac{(-1)^{N+1} e^{iks}}{2i\pi^2 ka} \left(\frac{ka}{2}\right)^{2/3} I_N(\gamma) \qquad (18)$$

where

$$I_N(\gamma) = \int_{C_t} \frac{e^{i\gamma t}}{Ai'(t) w_2'(t)} \left[\frac{w_1'(t)}{w_2'(t)}\right]^{N+1} dt. \qquad (18a)$$

The contour C_t in the t-plane is inferred from the mapping and Fig. 4 by observing that $\nu = ka$ corresponds to t = 0. The integral in (18a) is in a canonical form analogous to the Fock integral for convex surfaces. Tabulations exist for selected ranges of γ and N [10]. When N = -1, the resulting expression in (18) is a valid representation for G in the range of small γ. N = 0 defines the domain of the direct ray, N = 1 the domain of the direct plus singly reflected ray, etc.

E. Ray plus Whispering Gallery Mode Representation

An alternative possibility is to represent R_N as a whispering gallery mode sum plus an explicit remainder. If C is deformed into C_N in Fig. 4, M residues at the poles ν_m of the integrand lying between C and C_N must be extracted. One finds

$$R_N = \sum_{m=1}^{M} G_m + R_{MN} \tag{19}$$

where G_m is given in (5) or (7). The remainder integral R_{MN} is the same as R_N except for the replacement of C by C_N. It may be shown (see Appendix) that the remainder integral R_{MN} can be approximated by

$$R_{MN} \sim -\tfrac{1}{2} G_N, \tag{20}$$

provided that observation points are characterized by $w_{SN} \sim (w_M + w_{M+1})/2$, i.e. for saddle points w_{SN} well removed from the pole singularities (see Section III for a discussion on procedure when this condition is not satisfied). Thus

$$G \sim \sum_{n=0}^{N} G_n + \sum_{m=1}^{M} G_m - \tfrac{1}{2} G_N. \tag{21}$$

The number of rays, (N + 1), is picked according to the criterion

$$N + 1 \lesssim \frac{\gamma}{2(2^{1/3}\Delta)^{1/2}}. \tag{21a}$$

F. Near-Field and Infinite Plane Limits

Except for the whispering gallery mode plus continuous spectrum representation in Section II-B, which is not convenient for numerical computation, the other formulations cannot account for the limiting case $\gamma \to 0$ in (16a), which arises either when the observation point approaches the source point ($s \to 0$ with α fixed) or when the radius of curvature becomes arbitrarily large ($a \to \infty$ with s arbitrary but fixed). The latter limit traces the transition from a concave to a plane surface, and if a is allowed to become negative, from concave to convex. We have derived the transition functions for these cases.

We begin with G expressed by (18), with N = -1 and use the Wronskian relation for the Airy functions together with Cauchy's theorem applied to the upper half of the t-plane to infer that, equivalently,

$$G \sim - \frac{e^{iks}}{2\pi ka} \left(\frac{ka}{2}\right)^{2/3} \int_{-\infty - i\delta}^{+\infty - i\delta} e^{i\gamma t} \frac{Ai(t)}{Ai'(t)} dt. \tag{22}$$

Employing large argument expansions, one may show that

$$-\frac{Ai(t)}{Ai'(t)} \sim \sum_{j=0}^{10} \frac{\alpha_j}{t^{(1+3j)/2}} + O(t^{-34/2}), -\pi < \arg t < 0 \tag{23}$$

where $\alpha_0, \alpha_1, \ldots, \alpha_{10}$ are given by

$\alpha_0 = 1,$ $\alpha_1 = -\frac{1}{4}$ $\alpha_2 = 7/32$
$\alpha_3 = -21/64,$ $\alpha_4 = 0.7143531708,$
$\alpha_5 = -2.071282248,$ $\alpha_6 = 7.557254769,$
$\alpha_7 = -33.32008068,$ $\alpha_8 = 161.5948751,$
$\alpha_9 = -1019.868939,$ $\alpha_{10} = 6845.104932$ (23a)

Then by Laplace inversion of (22) [8]

$$G \sim \frac{i}{2} H_0^{(1)}(ks) \left\{ \sum_{j=0}^{10} b_j \gamma^{(3/2)j} + O(\gamma^{33/2}) \right\} \tag{24}$$

where $b_0, b_1, \ldots, b_{10}$ are given by

$b_0 = 1,$ $b_1 = \sqrt{\pi} e^{-i\pi/4}/4,$
$b_2 = -7i/60,$ $b_3 = -7\sqrt{\pi} e^{i\pi/4}/512,$
$b_4 = -0.4398134 \times 10^{-2},$
$b_5 = 0.4109687 \times 10^{-3} \sqrt{\pi} e^{-i\pi/4},$
$b_6 = 0.1122861 \times 10^{-3} i,$ $b_7 = 0.9182121 \times 10^{-5} \sqrt{\pi} e^{-i\pi/4},$
$b_8 = 0.2093046 \times 10^{-5},$ $b_9 = 0.1637812 \times 10^{-6} \sqrt{\pi} e^{-i\pi/4},$
$b_{10} = -0.3633427 \times 10^{-7} i.$ (24a)

All the terms shown have been used for subsequent numerical evaluation of G. Actually, the inversion yields for the factor outside the braces in (24) the large argument approximation of $(i/2) H_0^{(1)}(ks)$, the Green's function for an infinite perfectly conducting plane. By inserting the exact limiting value as

$a \to \infty$, the formula in (24) may be applied as well in the near field of the source.

To trace the transition from concave to convex curvature when the distance s along the surface remains fixed, we allow a to change continuously from positive to negative values via the complex excursion $0 \leq \arg a \leq \pi$. To keep $s = a|\phi - \phi'|$ positive, it is necessary simultaneously to have $\arg |\phi-\phi'| = -\arg$. Moreover, one must continuously deform the integration path in the complex t-plane to keep the integral convergent (exp (iγt) oscillatory) when γ is allowed to become complex according to the rules stated above. These considerations lead to a straight line path along which $\arg t = -\pi/3, 2\pi/3$, when $\arg \alpha = \pi$. Changing variables $t = \mu \exp(i2\pi/3)$, one finds

$$G \sim \frac{e^{iks}}{2\pi ka} \left(\frac{ka}{2}\right)^{2/3} \int_{C_\mu} \frac{w_1(\mu)}{w_1'(\mu)} e^{i\gamma\mu} d\mu \qquad (25)$$

where γ is again given by (16a) and the path C_μ proceeds along the real μ axis. The expression (25) is the known result for the field on the surface of a convex perfectly conducting cylinder. By contour deformation about the singularities μ_p at $w_1'(\mu_p) = 0$, one derives the creeping wave series, and by expansion analogous to (23) and Laplace inversion, the limiting transition as $\gamma \to 0$. Thus one may track the field continuously as the curvature changes from concave to convex between fixed source and observation points on the surface.

III. NUMERICAL RESULTS

Extensive numerical computations have been performed for $k\alpha$ = 10, 30, 50, 100, and 1000 to check the accuracy and range of validity of the various formulations in Section II. Results are presented in Figs. 6-12 for ka = 10, 100, and 1000 only since all of the relevant formulas apply to ka = 30 and 50 with accuracy similar to that obtained for the larger values ka = 100, 1000.

Except for the low (ka) values (ka = 10), the whispering gallery mode and continuous spectrum representation in (3), for arbitrary γ, has been taken as the reference solution, with which the other results are compared. The failure of this representation for low (ka), as evidenced by the oscillations in the curves in Fig. 6(a) for small γ, is attributed to difficulties in evaluating the eigenvalues of the whispering gallery modes with sufficiently high precision; for the larger (ka) values, the uniform approximations in (6) are adequate. A remarkable feature of this representation is the sensitivity of the field to the exact number M_1 of whispering gallery modes, as evidenced in Fig. 9 for ka = 100. Here M_1 = 32, and any deviation from this number causes oscillations about the true field magnitude.

It can also be seen from Fig. 9 that as the number of whispering gallery modes approaches $M_1 = 32$, the field magnitude also approaches the true value. The continuous spectrum is necessary to establish the near field for moderate γ. Analytically, this means that the Neumann function in (4b) together with the discrete spectrum (whispering gallery modes) produces the proper singularity near the source point. Omission of the continuous spectrum leads to errors that can be assessed in Fig.10.

The ray plus canonical integral representation in (14) is seen to provide an excellent approximation for all cases (i.e. for ka = 10, 100, 1000) as shown in Fig. 6(a) - (c) provided that the number of rays (N + 1) is chosen accordingly to the criterion in (16b). Because of the overlap of the curves for various N, no difficulty arises in switching from one formulation to the other as γ varies. Tabulations of the canonical integral in (18a) are given in [10] for the perfectly conducting boundary and for nonvanishing surface impedance. When ka is small, this representation may serve as a reference because it does not exhibit the sensitivity (see Fig. 6(a)) encountered with the whispering gallery mode and continuous spectrum formulation. As shown in Fig. 8(a), the ray plus canonical integral representation for small ka agrees well with the near-field expansion in (24).

The most intriguing and physically most appealing formulation is based on the mix of (N + 1) rays and M whispering gallery modes in (21). The whispering gallery modes plus the last term in (21) account for the field near the surface where geometrical optics becomes invalid. Because of the inequality (21a) that must be satisfied by the number of rays, N + 1, the inclusion of multiply reflected rays requires larger separation γ between source and observation points. For any given N, as γ increases in its applicable range, a new whispering gallery mode must be included whenever the N-times reflected ray touches the caustic of that mode. The process for N = 0 is depicted in Fig. 11(a). The concentric circles show the modal caustics of the whispering gallery modes, while the angle w_K represents the eigenvalue of the Kth mode. Within the ranges of P_1, P_2, and P_K, one must include the M = 1, M = 4, and M = K modes. Corresponding curves for $|G|$ are shown in Fig. 7. It should be noted that the direct ray plus whispering gallery mode representation yields a good approximation for $\gamma \geqslant 2(2^{1/3} \Delta)^{1/2}$. This representation, however, is not recommended for large values of γ since many modes must then be included. For example, N = 0, M = 24, at $\gamma = 10$.

Inclusion of reflected rays, when appropriate, reduces the number of required whispering gallery modes, as depicted in

Fig. 11(b). Thus at the observation point $P(a,\phi)$ inclusion of only the direct ray (N = 0) requires six whispering gallery modes, while inclusion of the direct and singly reflected rays (N = 1) requires only a single mode, etc. Fig. 11(b) also shows that the applicable range of the single mode plus ray representation becomes wider as the number of rays increases. As seen from Figs. 8(b) and (c), it is possible to represent the field accurately with only the lowest (most tightly bound) whispering gallery mode provided that a higher order reflected ray is included, when necessary, as γ increases. Accordingly, although (21) is a good representation for the ranges satisfying the inequality (21a), the simplest and physically most appealing expression will be (21) with the inequality (16b). Figs. 8(a)-(c) have been obtained with the inequality (16b). These results are surprisingly simple compared with the whispering gallery mode and continuous spectrum representation, for which 32 modes for k_a = 100 and 319 modes for k_a = 1000 must be included in order to obtain accurate field values. In Figs. 8(a)-(c), the number indicated along the curves should be read as follows: for example, M = 2 [N = 1] denotes the range wherein the direct and singly reflected rays plus two whispering gallery modes are applicable, while the circles indicate the starting points of the relevant intervals.

A word of caution must be appended in connection with Fig.11 when one seeks to ascertain the ranges wherein only a single whispering gallery mode (M = 1) is retained. It is suggestive that the M = 1, N = K (K = 0, 1, 2, ...) representation is adequate as long as the ray N = K traverses the space between the M = 1 and M = 2 mode caustics. However, the criterion (21a), with the equality, identifies initial γ values larger than those corresponding to Fig. 11, when the N = K ray touches the M = 1 caustic (see the circles for M = 1, N = K in Fig. 8 and the corresponding starting values for the N = K intervals in Fig. 6 as obtained from (21a)). Thus, referring again to Fig. 11, the N = K ray should be separated out only when that ray passes well within the inverval between the M = 1 and M = 2 caustics, preferably closer to the M = 2 caustic. With this proviso in mind, the detailed criteria (21a) or (16b) can be dispensed with, and the appropriate combination of rays and modes may be inferred from Fig. 11.

Analytically, the emergence of a whispering gallery mode (contact of ray and model caustic) is characterized by the coalescence of a pole and a saddle point in the remainder integral R_{MN}. Therefore the field in (21) is discontinuous at these points. The discontinuity can be avoided on use of a Fresnel integral transition function [9] in the vicinity of these γ values. However, we have found that continuity can be established by using the simple result when pole and saddle point coincide (this corresponds essentially to the midpoint between the two discontinuous endpoints) and then resorting to a perturbation expansion.

The ray and whispering gallery mode representation is invalid in the near field but it can there be joined to the near-field expansion in (24). Concerning the latter, it should be noted in Fig. 12 that inclusion of more perturbation terms does not lead to continuing improvement or extension of the range of validity; for $ka = 100$, the perturbation terms up to $\gamma^{21/2}$ seem to give the best result. The near-field forms with eight perturbation terms are shown in Fig. 8(a)-(c) for $ka = 10$, 100, and $1,000$, respectively. From each figure, one can see that the changeover from the near-field form to the direct ray plus lowest whispering gallery mode (i.e., $M = 1$) representation can be performed at the crossover with negligible error.

APPENDIX
THE REMAINDER INTEGRAL R_{MN}

As noted in the text, the remainder integral R_{MN} is the same as R_N in (15) except for the replacement of C by C_N in Fig. 4. With the aid of the Wronskian relation

$$\frac{1}{i\pi ka\, J_\nu'(ka)\, H_\nu^{(2)'}(ka)} = \tfrac{1}{2}\left(\frac{J_\nu(ka)}{J_\nu'(ka)} - \frac{H_\nu^{(2)}(ka)}{H_\nu^{(2)'}(ka)}\right), \qquad (26)$$

R_{MN} can be represented by

$$R_{MN} = \frac{(-1)^{N+1}}{2\pi ka}\int_{C_N}\left(\frac{J_\nu(ka)}{J_\nu'(ka)} - \frac{H_\nu^{(2)}(ka)}{H_\nu^{(2)'}(ka)}\right)r^{N+1}\, e^{i\nu|\phi-\phi'|}\, d\nu \qquad (27)$$

Along the contour C_N, one may use in the integrand the Debye asymptotic formulas (see (8) and similarly for the undifferentiated cylinder functions) to obtain

$$R_{MN} \sim \frac{(-i)^{N+1}}{2\pi}\int_{C_N'}\left(\tan\left\{ka\left[\cos w - \left(\frac{\pi}{2}-w\right)\sin w\right] + \frac{\pi}{4}\right\} - i\right)e^{ikaq_N(w)}\, dw \qquad (28)$$

where $q_N(w)$ and w_{sN} are given in (10b) and (11), respectively.

Proceeding as in [3], we note that if $w_{sN} \approx (w_{M+1} + w_M)/2$, the asymptotic evaluation involving the tangent function in the integrand of (28) yields a negligible contribution compared with the remaining term. In this case, referring to (10a) and (12)

$$R_{MN} \approx -\tfrac{1}{2} G_N. \tag{29}$$

REFERENCES

[1] V.M. Babich and V.S. Buldyrev, Asymptotic Methods of Short Wave Diffraction. Moscow, USSR; Nauka, Ch.11,Sec.4,1972.

[2] V.M. Buldyrev and A.I. Lanin, "Asymptotic formulas for a wave propagating along a concave surface; limits of their applicability," Radio Tekh.Elektron, vol. 20, Jan. 1975.

[3] W. Wasylkiwskyj, "Exact and quasi-optic diffraction within a concave cylinder," IEEE Trans. Antennas Propagt., vol. AP-23, no. 4, pp.480-492, July 1975.

[4] L.B. Felsen and A. Green, "Ground wave propagation in the presence of smooth hills and depressions", paper presented at the AGARD Sym. on EM Propagation Characteristics of Surface Materials and Interface Aspects, held in Istanbul, Turkey, Oct. 18-19, 1976, Published in the Sym. Proc.

[5] L.B. Felsen and T. Ishihara, "Effects of smooth elevations and depressions on ground wave propagation," paper presented at URSI Comm F Open Sym. on Propagation in Nonionized media, held in La Baule, France, April.18- May 6, 1977.

[6] L.B. Felsen and T. Ishihara, "High-frequency surface fields excited by a point source on a concave perfectly conducting cylindrical boundary," Radio Science, Vol. 14, pp. 205-216, 1979.

[7] L.B. Felsen and N. Marcuvitz, Radiation and Scattering of Waves. Englewood Cliffs, NJ: Prentice Hall, Sec. 2.3, 3.4 1973.

[8] G. Hasserjian and A. Ishimaru, "Currents induced on the surface of a conducting circular cylinder by a slot," J. Res. Nat. Bur.Stand., vol.66D, no. 3, pp.335-365, May-June 1962.

[9] L.B. Felsen and N. Marcuvitz, Ibid., Ch. 4.

[10] T. Ishihara and L.B. Felsen, "High-frequency fields excited by a line source located on a concave cylindrical impedance surface," IEEE Trans. Antenna Propagat., vol. AP-27, pp. 127-179, 1979.

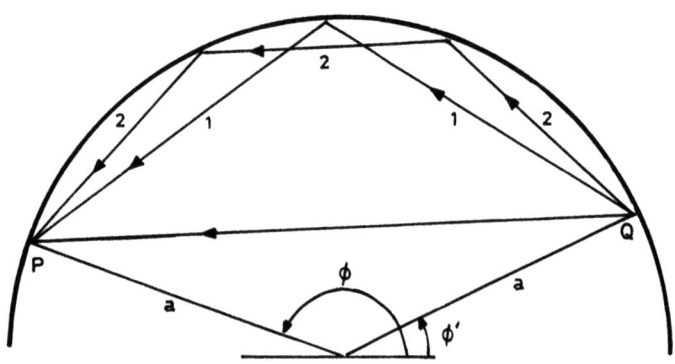

Fig. 1 Direct and multiply reflected rays for circular boundary.

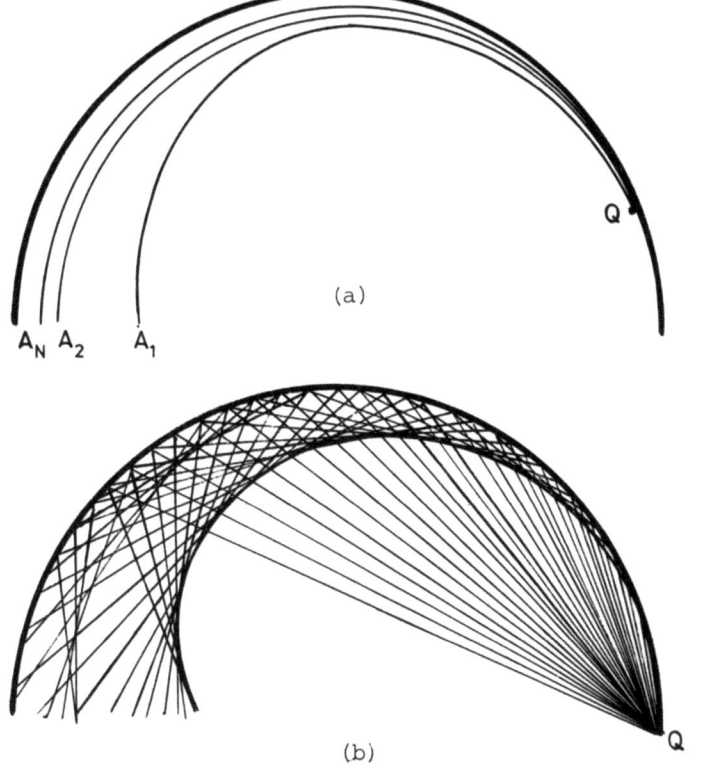

Fig. 2 Reflected ray caustics. (a) Caustics A_n for rays reflected n times; these ray contributions are confined to region between A_n and boundary. (b) Caustic formation due to singly and doubly reflected rays.

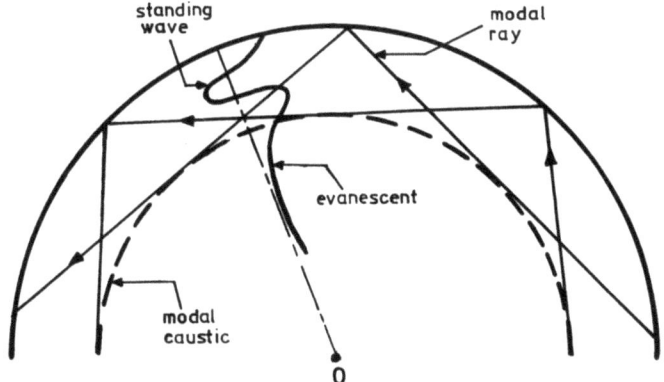

Fig. 3 Whispering gallery mode: modal field, modal rays, and caustic.

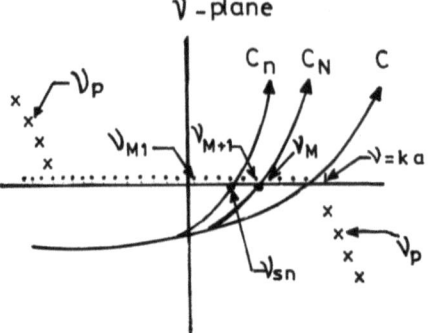

Fig. 4 Integration path and singularities in complex ν-plane. XXX zeros ν_p of $H_\nu^{(2)'}(ka)$. C_n is the steepest descent path through ν_{sn}.

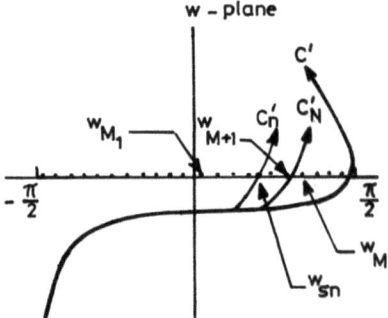

Fig. 5 Integration paths and singularities in complex w-plane ($\nu = ka \sin w$). C corresponds to C in Fig. 4 and can be deformed into the steepest descent path C' through w_{sn} .... zeros w_m of $J_\nu'(ka)$; the poles w_p are not shown.

Fig. 6 Ray plus canonical integral representation. Both the γ and $|\phi - \phi'|$ coordinates are indicated. Ranges in γ and $|\phi - \phi'|$ corresponding to $N = -1$ (no geometric-optical ray) in (14), $N = 0$ (one ray), $N = 1$ (two rays), and $N = 2$ (three rays) are as shown. — Whispering gallery modes and continuous spectrim. .—.—. $|G| = |R_{N=-1}|$. —.—.—.— $|G_{N=0} + R_{N=0}|$; —..—..— $|\Sigma_{n=0}^1 G_n + R_1|$. —...—...—...— $|\Sigma_{n=0}^2 G_n + R_2|$. (a) $ka = 10$. (b) $ka = 100$. (c) $ka = 1000$. (figure see next page).

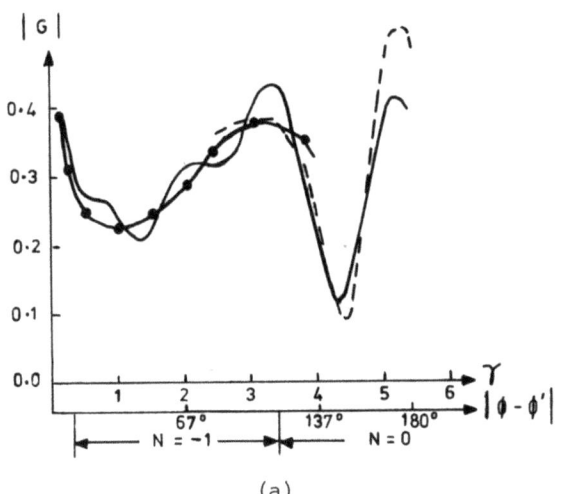

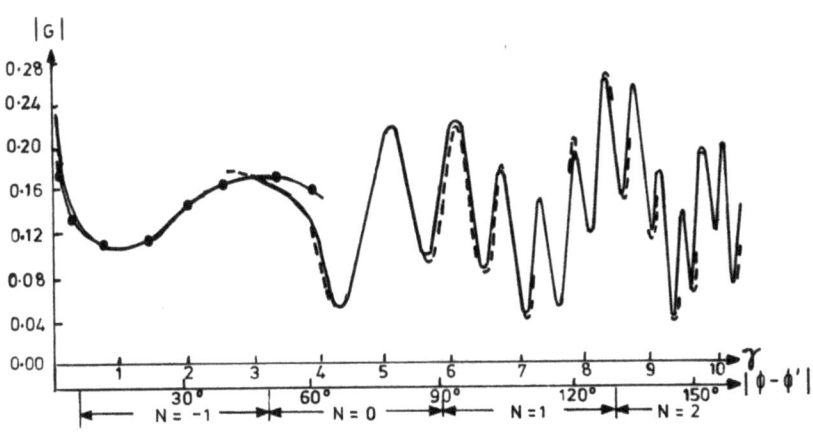

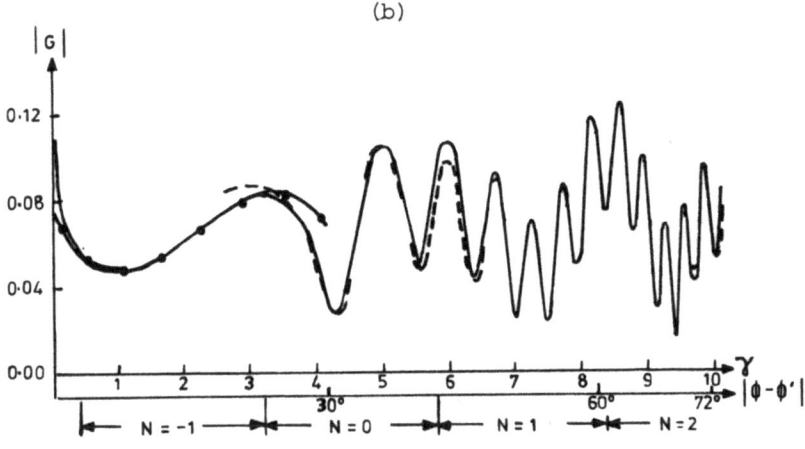

Fig. 6

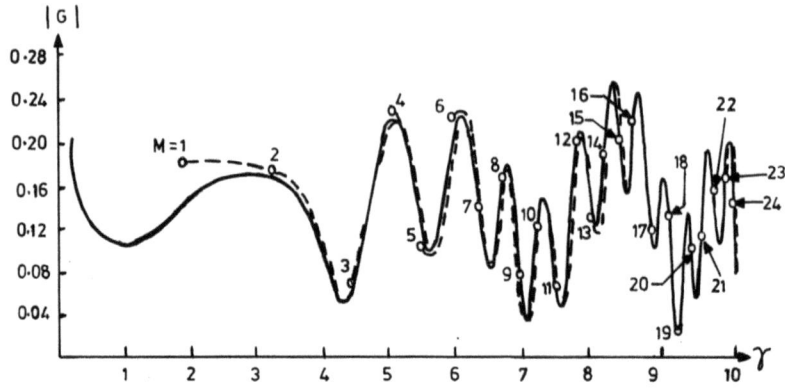

Fig. 7 Direct ray plus whispering gallery mode representation. M = K indicates point of emergency of Kth whispering gallery mode. Solid curve is calculated from (3).

Fig. 8 Ray plus whispering gallery mode representation. Solid curves serve as reference solutions. Here $|\phi - \phi'|$ coordinate is not shown since it has been depicted in Fig. 6. o-.-.-.- $M = K |N = 0|$, $|G| = 1\tfrac{1}{2} G^k_{N=0} + \Sigma^1_{m=-} G_m |$. o-..-..-..- $M = K |N = 1|$, $|G| = |\Sigma^1_{n=0} G_n + \Sigma^K_{m=1} G_m - \tfrac{1}{2} G_{N=1}|$. o-...-...- $M \neq 1 |N = 2|$, $|G| = |\Sigma^2_{n=0} G_n + G_m - \tfrac{1}{2} G_{N=2}|$. - - - Near-field form. (a) $ka = 10 (M_1 = 3)$. Solid curve is calculated from (14) and (18). (b) $ka = 100 (M_1 = 32)$. Solid curve is calculated from (3). (c) $ka = 1000 (M_1 = 319)$. Solid curve is calculated from (3). (figure see next page).

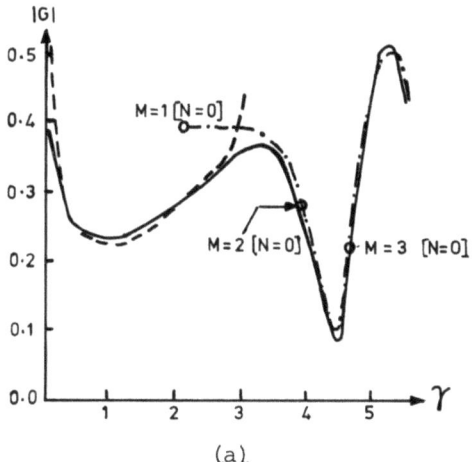

(a)

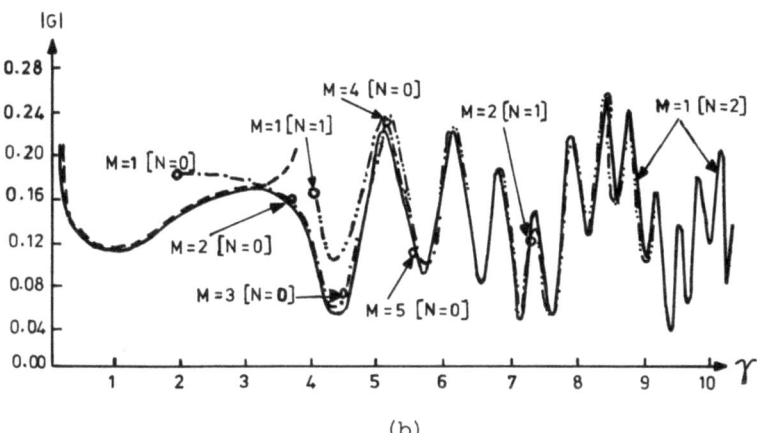

(b)

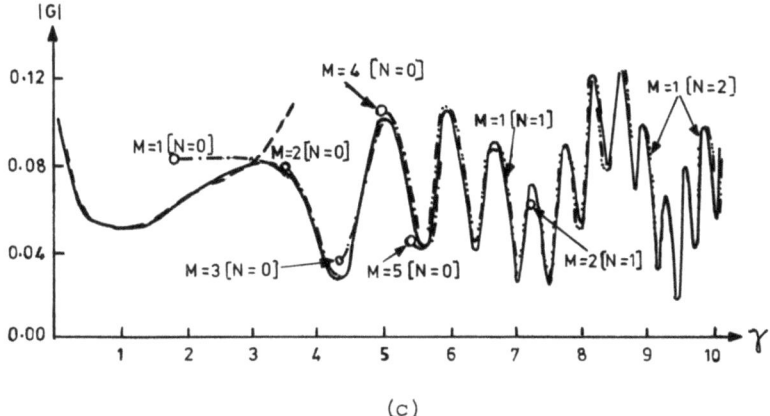

(c)

Fig. 8

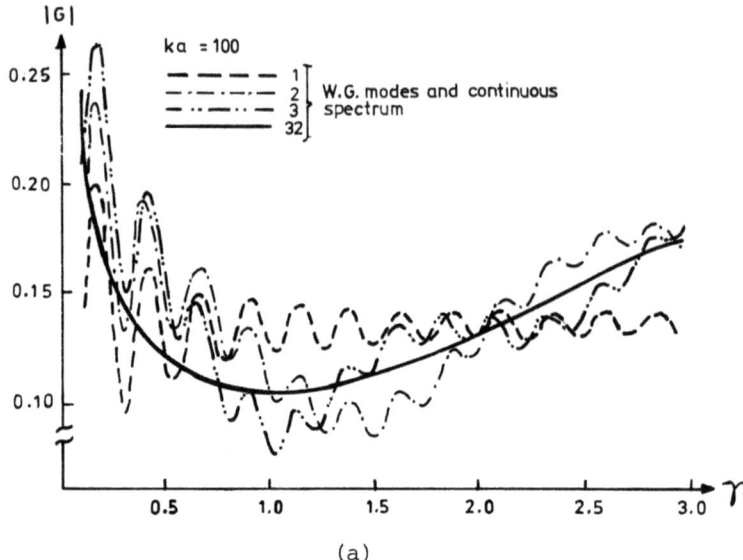

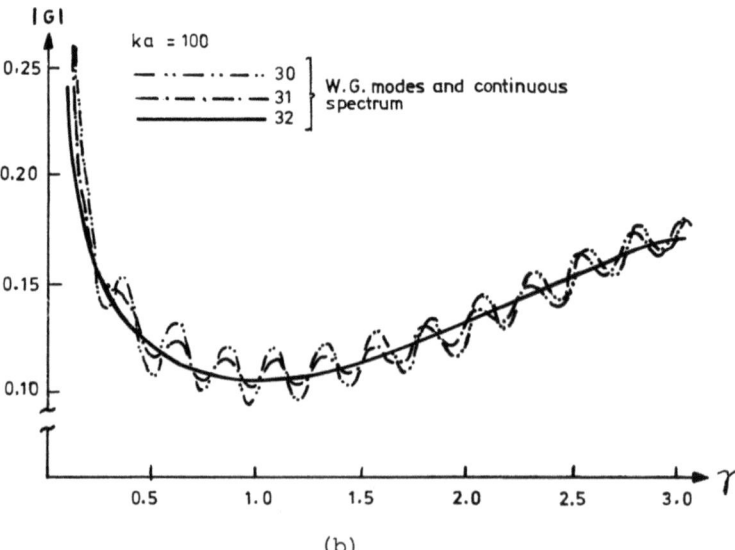

Fig. 9 Whispering gallery modes plus continuous spectrum. (a) Few modes. (b) Many modes.

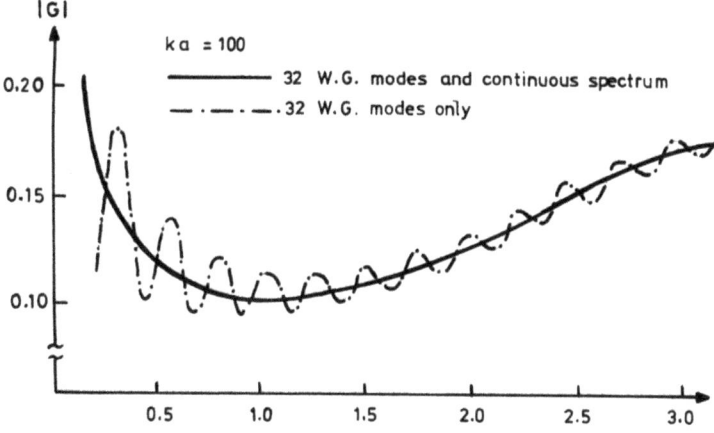

Fig. 10 Effect of continuous spectrum.

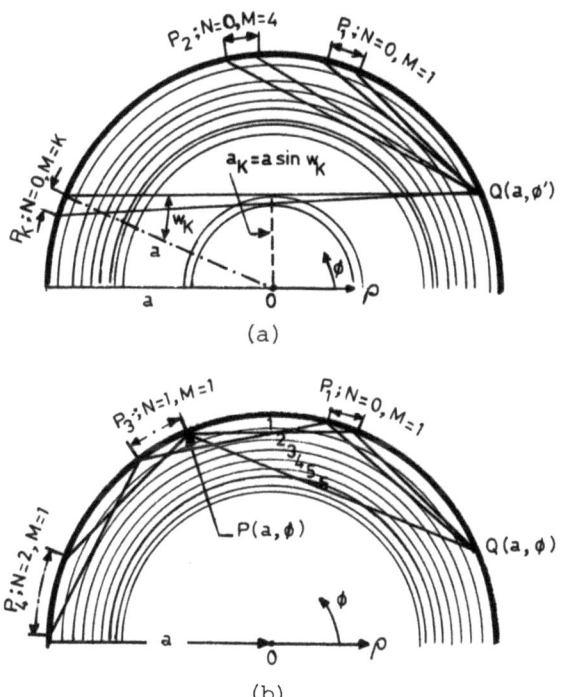

Fig. 11 Geometrical interpretation of direct ray plus whispering gallery mode representation. (a) $N = 0$ (direct ray only). (b) $N = 0, 1, 2$. The closest range for which the $M = 1$, $N = K$ representation may be employed is identified by an $N = K$ ray lying well within the interval between the $M = 1$ and $M = 2$ caustics, preferably closer to the $M = 2$ caustic. No such restriction applies to the intervals corresponding to $M > 1$. $N = K$ as identified in the figure.

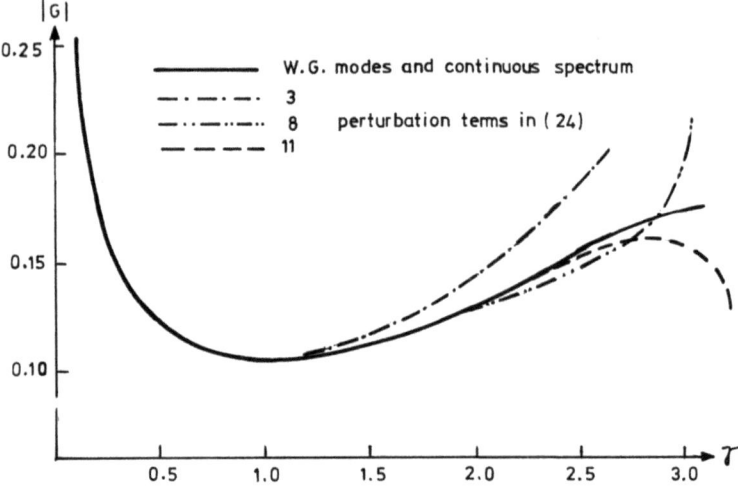

Fig. 12 Near-field form. ka = 100

GAUSSIAN BEAMS: PARAXIAL THEORY

Georges A. Deschamps

Department of Electrical Engineering
University of Illinois, Urbana, IL 61801, USA

ABSTRACT. A beam of electromagnetic radiation is a propagating field that remains substantially close to a central line called its axis. The transformation of such beams, either by propagation or reflection and refraction at some surfaces, will be analyzed by using the paraxial approximation or replacing the wave equation by a parabolic equation. A central result of the paper is that Gaussian beams can be represented by *pencils of complex rays* and that consequently their transformations through optical systems obey the laws of geometrical optics as they apply to pencils of real rays.

1. INTRODUCTION

A beam of electromagnetic radiation can be loosely defined as a field that essentially propagates close to a central line Oz which may be called its axis. An example is the shaft of light produced when a plane wave propagating in direction Oz is incident upon some aperture in the plane $z = 0$. Within the geometrical optic approximation, the field past the aperture would be represented by parallel rays and its transverse cross section would be independent of z. Observation and a more exact analysis show that the field spreads out and decreases in intensity as z increases. This is a diffraction effect which can be observed even with coherent light, and is more apparent when the beam initial cross section, for $z = 0$, is small in terms of wavelengths.

The practical interest in beams stems from their application to the guidance of electromagnetic energy and to the production of coherent radiation in lasers and masers. It was shown by Goubau [2] that one can compensate for the spreading of a beam by periodically correcting its phase by means of lenses or concave mirrors. At high frequency (wavelength of a few millimeters and shorter) this method of guiding the field presents some advantages over using conventional metallic or dielectric waveguides; losses are smaller and there is some flexibility in shaping propagating paths, for instance avoiding rotary joints in the feed system of a rotating antenna. In producing coherent radiation a common device is the Fabry-Perrot resonator where the space between two concave mirrors is filled with an active medium. Radiation bouncing back and forth between the reflectors is amplified and tends to form a beam of the type we shall analyze, one whose field in a cross section is represented by a Gaussian function. [3, 4, 5, 6].

In order to remain close to an axis Oz, the field in a beam must not only be concentrated near the origin 0 in the initial cross section $z = 0$, but also have a spectrum composed of plane waves close in direction to Oz. This property justifies replacing the exact wave equation by a parabolic equation which describes the evolution with z of the field in a cross section. The analysis presented in this paper will be exclusively concerned with this paraxial approximation. When applying the results of this analysis one should not forget the conditions under which they are valid.

2. FREE SPACE PROPAGATION OF BEAM

Let Oz be the axis of the beam oriented in the direction of propagation. The sources of the field are assumed to be in the region $z < 0$ and the field in the plane $z = 0$ is known and represented by the function $f_0(x)$ where $x \in R^n$, $n = 1$ or 2. The problem is to find the field $f_t(x)$ in the plane $z = t$, ($t > 0$). A solution, not restricted to a beam but applicable to any field having its sources in $z < 0$, consists of expressing the field $f(x, t)$ for $t > 0$ as a spectrum of plane waves

$$f_t(x) = \int e^{i(\xi \cdot x + \zeta t)} F_0(\xi) \, d\!\!\!/\xi \tag{1}$$

The slash $d\!\!\!/\xi$ means $(2\pi)^{-n} d\xi$. In this integral, ζ is the function of ξ

$$\zeta = \zeta_+(\xi) \tag{2}$$

defined as the square root of $k^2 - \xi^2$ that corresponds to waves crossing the plane $z = 0$ in the positive direction:

$\zeta = \sqrt{k^2 - \xi^2}$ if $\xi < k$ and $\zeta = i\sqrt{\xi^2 - k^2}$ if $\xi > k$.

In the plane $t = 0$ one obtains

$$f_0(x) = \int e^{i\xi x} F_0(\xi) \, d\xi \tag{3}$$

which shows that $F_0(\xi)$ is the inverse Fourier transform of $f_0(x)$

$$F_0(\xi) = \int e^{i\xi x} f_0(x) \, dx \tag{4}$$

The relations (1) (3) (4) are summarized in Fig. 1 which shows how the function f_t can be deduced from f_0.

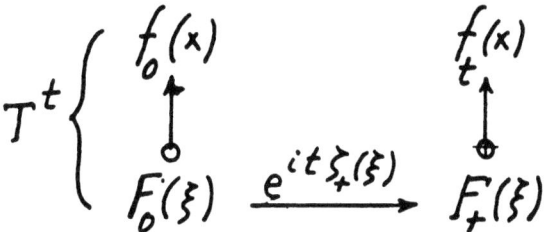

Fig. 1. Evolution of transverse field (f,F) from plane $z = 0$ to $z = t$.

In order to form a beam the field must satisfy two conditions. First the function $f_0(x)$ must be concentrated near to some point, say $x = 0$. Its *effective support*, i.e. the region where it is larger than some small fraction of its maximum, is near to 0, then the function $F_0(\xi)$ is also concentrated near $\xi = 0$. The second insures that the field is made up of plane waves whose directions are close to the axis. Consequently the effective support of $f_t(x)$ does not spread out too fast as t increases. It is well known that the effective supports of the functions $f_0(x)$ and $F_0(\xi)$ cannot be both arbitrarily small. When f_0 and F_0 are Gaussian functions, the product of their moment matrices, which indicates their spread, is the unit matrix (Appendix A). This is related to the uncertainty relation and is thoroughly discussed for instance in [23].

When only the second condition is satisfied the field is said to be *collimated* and can be though as a distribution of parallel beams.

Over the effective support of $F_0(\xi)$ where $\xi \ll k$ one can approximate the function $\zeta_+(\xi)$ by

$$\zeta_p(\xi) = k - \frac{1}{2k} \xi^2 \tag{5}$$

This amounts to replacing the circle Γ of radius k by a parabola Π having the same radius of curvature as shown in Fig. 2. (In R^3, Γ becomes a sphere, Π a paraboloid).

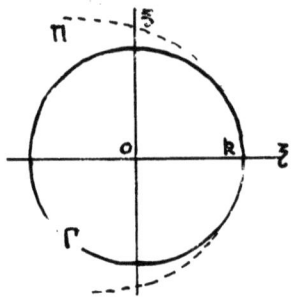

Fig. 2. Approximation of circle Γ by parabola Π (eq. 5)

This is known as the paraxial or parabolic approximation. If $\zeta_p(\xi)$ is substituted for $\zeta_+(\xi)$ the diagram in Fig. 1 becomes that in Fig. 3.

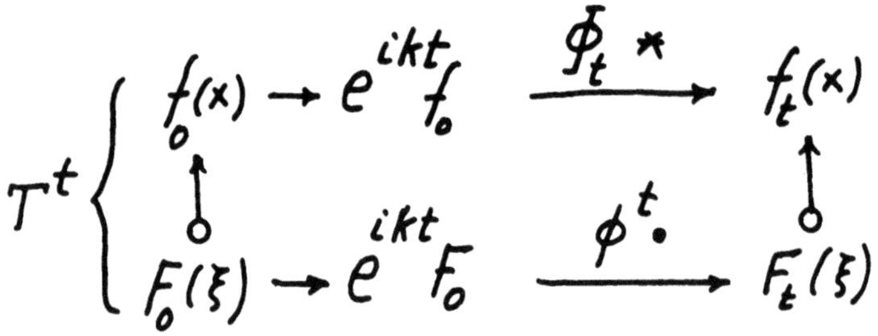

$$T^t \begin{cases} f_0(x) \to e^{ikt} f_0 \xrightarrow{\phi_t^*} f_t(x) \\ \updownarrow \qquad \qquad \qquad \updownarrow \\ F_0(\xi) \to e^{ikt} F_0 \xrightarrow{\phi_{\bullet}^t} F_t(\xi) \end{cases}$$

Fig. 3. Evolution of field according to parabolic approximation.

The field in the plane $z = 0$ being described by the Fourier pair,

$(f_0(x), F_0(\xi))$

its transformation into the field in the plane $z = t$ will be denoted by T^t and expressed by

$$T^t : (f_0, F_0) \to e^{\nu t}(\Phi_t * f_0, \phi^t \cdot F_0) \qquad (6)$$

where

$$\phi^t(\xi) = \exp(\frac{t}{\nu} \frac{\xi^2}{2}) \qquad (7)$$

and
$$\Phi_t(x) = \left(\det \frac{k}{2\pi i t}\right)^{1/2} \exp \frac{\nu}{t} \frac{x^2}{2} \tag{8}$$

also form a Fourier pair.

The group property

$$T^t T^s = T^{t+s} \tag{9}$$

follows from

$$\phi^t \phi^s = \phi^{t+s} \tag{10}$$

and the corresponding convolution

$$\Phi_t * \Phi_s = \Phi_{t+s} \quad . \tag{11}$$

If the field $f_0(x)$ is represented by a Gaussian function the same property will hold for its Fourier transform $F_0(\xi)$, for $F_t(\xi)$, and for $f_t(x)$. All cross sections of the field are Gaussian. This may be considered as a modal property characteristic of Gaussian beams.

Specifically assume that $f_0(x)$ equals the normal probability density function with *moment matrix* M

$$f_0(x) = (\det 2\pi M)^{1/2} \exp(-M^{-1}[x]) \quad . \tag{12}$$

Using notations in Appendix A, $f_0(x)$ is represented by

$$\phi^C(x) = \phi_{C^{-1}}(x) = \left(\det -\frac{\nu}{2\pi} C\right)^{1/2} \exp(\nu C[x]) \tag{13}$$

if we let $C = i \, k^{-1} M^{-1}$ or $C^{-1} = -\nu M$.

Let us first treat the general case where the beam cross section at $z = 0$ is described by the function

$$\phi^{C_o}(x) = \exp ik \, C_o[x] \tag{14}$$

with an arbitrary symmetric matrix C_o having dimension of length^{-1}. The effect of propagation to $z = t$ is to transform $\phi^{C_o}(x)$ into $\phi_t * \phi^{C_o}$. To perform this convolution we write

$$\phi^{C_o} = \left(\det\left(-\frac{\nu}{2\pi} C_o\right)\right)^{1/2} \phi_{C_o^{-1}} \tag{15}$$

(see Appendix for notations)

Then
$$f_t(x) = (\det \frac{-\nu C_o}{2\pi})^{1/2} \phi_{C_o^{-1}+t}(x) \tag{16}$$

Letting
$$C_0^{-1} + t = C_t^{-1} \tag{17}$$

this gives
$$f_t(x) = (\det C_t / \det C_0)^{1/2} \phi^{C_t}(x) \tag{18}$$

The phase of the output field is governed by the equation (17). The ratio of determinant, which gives the amplitude A, can be interpreted as a geometrical optic divergence factor. By rotation of axes we can diagonalize the matrix C_0, hence C_0^{-1}

$$C_0^{-1} = \begin{bmatrix} f_1 & 0 \\ 0 & f_2 \end{bmatrix} \tag{19}$$

Then
$$C_t^{-1} = \begin{bmatrix} f_1 + t & 0 \\ 0 & f_2 + t \end{bmatrix} \tag{20}$$

and
$$A = (1 + \frac{t}{f_1})^{-1/2} (1 + \frac{t}{f_2})^{-1/2} \tag{21}$$

Note that f_1 and f_2 as well as the angle of rotation will in general be complex numbers.

Let us apply this result to a real Gaussian distribution
$$f_0(x) = \exp(-M^{-1}[x]) \tag{22}$$

We can express it as ϕ^{C_o} where
$$ikC_0 = -M^{-1}$$

and apply the general formula.

Let us assume that the distribution has symmetry of revolution: $M = V\begin{bmatrix} 1 & 0 \\ 0 & 1 \end{bmatrix}$ where V is the variance, then

$$C_0^{-1} = -ikM = -ikV\begin{bmatrix} 1 & 0 \\ 0 & 1 \end{bmatrix} \tag{23}$$

We shall denote by a the "scaled" variance kV. The focal length of the beam is $f_1 = f_2 = -ia$. The focus is at $z = ia$. We have $C_t^{-1} = t - ia$ and therefore

$$f_t(x) = (\frac{t - ia}{0 - ia})^{n/2} \phi_{t-ia} \tag{24}$$

$$f_0(x) = \phi_{-ia} \tag{25}$$

3. TRANSFORMATION THROUGH AN IDEAL LENS

We define an ideal thin lens as a planar device that transforms the field u(x) incident at point x on one side of the plane into a field

$$v(x) = \phi^C(x)\, u(x) \qquad (x \in R^2) \tag{26}$$

at the same point x on the other side.

When C is real we have an ordinary lens that produces a phase shift $kC[x]$ quadratic function of x.

If the lens has symmetry of revolution, C is a multiple of the unit matrix

$$C = f^{-1} \begin{bmatrix} 1 & 0 \\ 0 & 1 \end{bmatrix} \tag{27}$$

and f is the lens focal distance (counted as positive for a divergent lens).

In general f may be a complex number. In particular if $f = -ia$, and a planewave represented by rays parallel to the axis is incident upon the lens, the field on the output side of the lens is

$$f_0(x) = \phi_{-ia} = \psi_{a/k} = \exp(-\frac{1}{2} k \frac{x^2}{a}) \tag{28}$$

which generates a Gaussian beam. Its field is represented by rays diverging from the focus at $z = ia$ (origin $z = 0$ at the center of the lens), hence it is naturally described by the function $G(\vec{r} - i\vec{a})$ as discussed in 7. When C is an arbitrary complex 2 x 2 matrix, it represents an astigmatic lens with quadratic attenuation. The output field for an incident planewave is a general Gaussian beam which can be quite naturally thought of as a pencil with two focal lines.

When the incident field is a Gaussian field $\phi^Q(x)$ on the input side of the lens, the field on the output side is

$$\phi^Q \phi^C = \phi^{Q+C} \tag{29}$$

The matrix transformation through the lens is

$$Q \rightarrow Q' = Q + C \tag{30}$$

The corresponding functions of ξ, or spectra, are related by

$$\Phi_Q * \Phi_C = \Phi_{Q+C} \tag{31}$$

4. REFRACTION AND REFLECTION OF A BEAM

Refraction of a beam at a curved interface Σ between two media of wavenumbers k and k', can be represented by a formula similar to that for an ideal lens. Paraxially the interface can be replaced by a parabolic surface tangent to the actual surface at point 0 where it meets the axial ray, and having the same curvature matrix.

Let us assume the axis 0z normal to the plane P tangent to Σ at point 0. The equation of Σ, near 0, will be $z = C[x[$ where C is a 2 x 2 matrix symmetric (and real). If the phase of the incident field at point $x \in P$ is $k\, Q[x]$, its phase at the point $(x, C[x])$ on Σ will be

$$k(Q + C)\,[x]. \tag{32}$$

The transmitted field will have a phase $k'Q'[x]$ in plane P and therefore, $k'(Q' + C)[x]$ on the surface Σ. Matching the two phase functions

$$k(Q + C) = k'(Q' + C) \tag{33}$$

hence

$$Q' = \frac{k}{k'} Q + (\frac{k}{k'} - 1)C. \tag{34}$$

This is a generalization of a familiar formula derived by ray tracing and relating the abscissa $-p$ of the center of curvature of the incident phase front and that of the transmitted phase front at p'

$$\frac{k'}{p'} - \frac{k}{p} = \frac{k - k'}{R} \qquad \text{where } R = \frac{1}{C} \tag{35}$$

This formula can be extended to the case of oblique incidence and to reflection from a smooth surface ([18], eq. 33 and 37 p. 1028). It applies to an astigmatic beam. In all cases it reduces to adding a quadratic phase shift $k\, L[x]$ to the input matrix Q multiplied by a factor $n = \frac{k}{k'}$ therefore the general transformation formula is

$$Q \rightarrow Q' = \nu Q + L \tag{36}$$

5. TRANSFORMATION THROUGH AN OPTICAL SYSTEM

An optical system consisting of homogeneous regions separated by smooth surfaces can be decomposed into a succession of elements considered in Sections 2, 3, 4.

For instance the system in Fig. 4

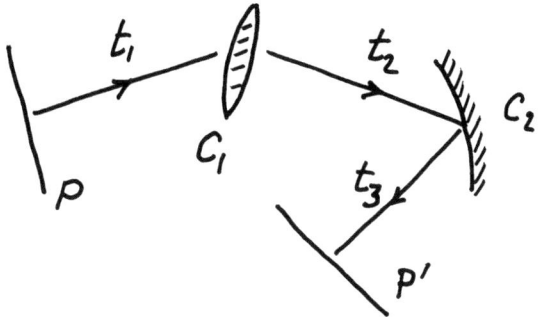

consisting of a lens (matrix C_1) and reflector (C_2) separated by propagation segments of length t_1, t_2, t_3 from the input plane P to the output plane P'. If I designates the operation "inverse of" and $\Phi^Q(x)$ is the field in plane P, the field in plane P' is proportional to $\Phi^{Q'}(x)$ where Q' is given by the formula

$$Q' = I(+t_3)IC_2I(+t_2)IC_1I(+t_1)IQ \tag{37}$$

This is a generalized continued fraction, to be read from right to left, $(+t_i)$ means the operation of adding the matrix $t_i \begin{bmatrix} 1 & 0 \\ 0 & 1 \end{bmatrix}$.

6. PARAXIAL RAYS AND PENCILS

Let us consider at point 0 on the axial ray Oz a plane P perpendicular to Oz. A paraxial ray may then be specified by two vectors: the *position vector* from point 0 to the intersection of the ray with plane P and the *direction vector* ξ projection on P of the wave vector $\vec{k}$ of the ray. The four coordinates of x and ξ are the *ray coordinates* and form a vector $\psi \in R^4$. For a paraxial ray both kx and ξ/k are small. (ξ/k indicates the slope with respect to the axial ray.) As the point 0 moves along the axial ray to a point of abscissa z, measured along the ray, the vector ψ is a function of z. Tracing the ray consists of finding this function. Because ψ is small, the transformation of ψ between two points such as the input and the output to a system is linear and represented by a matrix T called the ray transformation matrix

$$\psi' = T\psi \quad . \tag{38}$$

In a homogeneous region where the rays are straight lines, it is obvious that as z increases from 0 to z, ξ remains equal to itself and x increases by $(\xi/k)z$. Thus propagation along the axial ray from z = 0 to z is represented by

$$T^z: \begin{bmatrix} x \\ \xi \end{bmatrix} \to \begin{bmatrix} x' \\ \xi' \end{bmatrix} = \begin{bmatrix} 1 & z/k \\ 0 & 1 \end{bmatrix} \begin{bmatrix} x \\ \xi \end{bmatrix} \quad . \tag{39}$$

Now let us consider the transformation T_f through a lens of focal distance f (positive for divergent lens). Within the paraxial approximation the change of direction vector ξ is proportional to x hence

$$T_f: \begin{bmatrix} x \\ \xi \end{bmatrix} \to \begin{bmatrix} x' \\ \xi' \end{bmatrix} = \begin{bmatrix} 1 & 0 \\ k/f & 1 \end{bmatrix} \begin{bmatrix} x \\ \xi \end{bmatrix} \tag{40}$$

For an astigmatic lens k/f is replaced by a matrix kC. For transformation through a refracting surface at oblique incidence see [18].

The effect of an optical system such as the one on Fig. 4 can be evaluated by multiplying the matrices corresponding to each element. The resulting *ray transformation matrix* T describes the system. Here are some of its properties.

For any two paraxial rays that go through the system, there is a combination of their coordinates $\psi_1 = \begin{bmatrix} x_1 \\ \xi_1 \end{bmatrix}$ and $\psi_2 = \begin{bmatrix} x_2 \\ \xi_2 \end{bmatrix}$ that remains invariant. It is the expression

$$\psi_1^T J \psi_2 = x_1 \cdot \xi_2 - x_2 \cdot \xi_1 \tag{41}$$

called the Lagrange invariant. Here

$$J = \begin{bmatrix} 0 & 1 \\ -1 & 0 \end{bmatrix}$$

where 0 and 1 are the 2 x 2 zero and unit matrices. This implies for the matrix T that

$$\psi_1^T T^T JT \psi_2 = \psi_1^T J \psi_2$$

for any ψ_1, ψ_2, hence,

$$T^T JT = J \quad . \tag{42}$$

This property can be verified for the systems we have considered by noting that if it holds for T_1 and T_2 it also holds for their products, or their inverses, and then by checking that it applies to the matrices T for propagation in a homogeneous region and for transition through a discontinuity surface. It can be shown more generally to apply to curved rays in nonhomogeneous regions as well [5], [20], [21].

A matrix

$$T = \begin{bmatrix} A & B \\ C & D \end{bmatrix} \qquad (43)$$

that satisfies (42) is called *symplectic*. It can be deduced from (41) that

$$T^{-1} = \begin{bmatrix} D^T & -B^T \\ -C^T & A^T \end{bmatrix} \qquad (44)$$

that AB^T, CD^T are symmetric, and that $AD^T - BC^T = 1$ [5]. Also the determinant of T is unity and its four eigenvalues form two pairs of inverse numbers. The product of two symplectic matrices, the inverse, the transpose, and the complex conjugate of a symplectic matrix are also symplectic. A characteristic of two rays (ψ_1, ψ_2) that belong to the same pencil is that $\psi_1^T J \psi_2 = 0$. This follows from the symmetry of the matrix Q. Any other ray in the same pencil is a linear combination

$$\psi = a\psi_1 + b\psi_2 \qquad (45)$$

with coefficients (a, b) that may be complex.

7. FIELD OF A POINT SOURCE AT A COMPLEX LOCATION

It has been observed in [10], [19] and later in [21], [18] that the field of a point source at a complex location $i\vec{a}$ gives a simple representation of a Gaussian beam valid in the vicinity of its axis Oz. Let

$$G(\vec{r} - i\vec{a}) = e^{ikR}/4\pi R \qquad (46)$$

where R is the length of the vector $\vec{r} - i\vec{a}$. This function of $\vec{r}$ satisfies the wave equation except on a circle of radius a, axis Oz.
Let the vector $\vec{r} = z\hat{z} + \vec{x}$ where $\vec{x}$ is the component $\vec{r}$ transverse to the axis

$$R^2 = (z - ia)^2 + x^2$$

Hence

$$R \sim z - ia + \frac{1}{2}\frac{x^2}{z - ia}$$

provided $|x| \ll |z - ia|$.

Substituting in the expression for G

$$G \sim \frac{e^{\nu(z-ia)}}{z - ia} \exp(\frac{ik}{2} \frac{x^2}{z - ia}) \qquad (47)$$

which is proportional to the function Φ_{z-ia} representing a Gaussian beam.

It should be noted that outside of the paraxial region, G can differ appreciably from a Gaussian beam [21]. In particular, G is singular for R = 0 which occurs on a circle of radius a with center 0 and axis 0z.

8. PARABOLIC EQUATION APPROXIMATION

An alternative approach to the paraxial approximation is based on replacing the original wave equation

$$(\Delta + k^2)f(x,z) = 0 \qquad x \in R^2 \qquad (48)$$

by a parabolic equation. This is justified, as in Section 1, by the assumption that propagation occurs mainly in the z direction. This is taken into account by writing

$$f(x,z) = u(x,z) \; e^{\nu z} \qquad \nu = ik \qquad (49)$$

The equation for u becomes

$$(\partial_z^2 + \Delta_x + 2\nu\partial_z)u(x,z) = 0 \qquad (50)$$

Since the most significant part of the z dependence has been removed, u depends only weakly on z and $\partial_z^2 u$ can be neglected leaving the parabolic equation

$$\partial_{iz} u = \frac{1}{2k} \Delta_x u \qquad (51)$$

This could also be deduced as in [24] by first formally solving Eq. 48

$$f(x,z) = T^z f(x,0) \qquad (52)$$

with the operator

$$T^z = \exp\nu z (1 + \frac{\Delta_x}{k^2})^{1/2} \qquad (53)$$

The first term of its expansion is

$$T^z \simeq e^{\nu z} \exp(\frac{i}{2} \frac{z}{k} \Delta) \qquad (54)$$

which is equivalent to (4).

Plane wave solutions of Eq. 4 of the form

$$u(x,z) = \exp i(\zeta z + \xi \cdot x) \qquad (\xi \in R^2) \qquad (55)$$

are such that

$$\zeta = -\frac{1}{2k}\xi^2 \qquad (56)$$

For plane wave solutions (1) $f(x,z) = \exp i(\zeta z + \zeta \cdot x)$ factoring e^{ikz} and using (56) gives

$$\zeta = k - \frac{1}{2k}\xi^2 \qquad (57)$$

which is the parabolic approximation already obtained in Section 1.

The Schrödinger equation for a free particle if of the same form as Eq. 51 with z replaced by time t and $x \in R^3$

$$i\hbar \partial_t \psi = (-\frac{\hbar^2}{2m}\Delta + V)\psi \qquad (58)$$

Note that $V(x) = 0$ for a free particle and that the large parameter is $\nu = i/h$. Evolution of a wave function with time can be discussed in terms of beams [25]. To a particle acted upon by a potential corresponds the wave equation for a field in a medium where the refractive index is a function of position.

Returning to Eq. 51 let us seek a solution of the form

$$u = A_z \exp\nu C_z[x] \qquad (59)$$

(or with the notations of appendix A: $u = A \phi^C(x)$)

Substitution in Eq. 51 gives

$$k^2(C^2 + C')[x] + 2ik(\frac{A'}{A} + \frac{1}{2}\operatorname{tr} C) = 0 \qquad (60)$$

The z dependence of A and C is understood, the prime means a z derivative. Since (60) must hold for all z and all x we obtain

$$C' + C^2 = 0 \qquad (61)$$

$$\frac{A'}{A} = -\frac{1}{2}\operatorname{tr} C \qquad (62)$$

The first equation for the 2 x 2 matrix C gives

$$\partial_z C^{-1} = 1 \qquad (63)$$

hence

$$C_z^{-1} = C_0^{-1} + z\begin{bmatrix}1 & 0\\ 0 & 1\end{bmatrix} \tag{64}$$

By rotating the axes for x we can make C_0 diagonal and of the form

$$C_0 = \begin{bmatrix} 1/f_1 & 0 \\ 0 & 1/f_2 \end{bmatrix} \tag{65}$$

where f_1, f_2 are the radii of curvature of the phase front for $z = 0$. Because of (64)

$$C_z = \begin{bmatrix} \dfrac{1}{f_1+z} & 0 \\ 0 & \dfrac{1}{f_2+z} \end{bmatrix} \tag{66}$$

The second equation

$$\ln A = -\frac{1}{2}\int_0^z \text{trace } C_z\, dz$$

can be integrated. Since

$$\text{tr } C = \frac{1}{f_1 + z} + \frac{1}{f_2 + z} \tag{67}$$

$$\ln A_z - \ln A_o = \frac{1}{2}\ln\left(1 + \frac{z}{f_1}\right) + \frac{1}{2}\ln\left(1 + \frac{z}{f_2}\right)$$

$$A_z = A_o(\det C_z/\det C_o)^{1/2} \tag{68}$$

The interest of the parabolic equation approximation is that it can be applied to other geometries, where the role of z is played by some other privilege variable. For instance, when the propagation is mostly in the radial direction, one can factor out e^{ivp} and then neglect $\Delta_r = r^{-1}\partial_r r \partial_r$.

Furthermore, in a general non-uniform medium, one can trace the rays of the geometrical optics approximation and consider that propagation occurs mostly along these rays. One can take the eikonal S as the privileged variable, factor out e^{ivs} and neglect the longitudinal Laplacian Δ_S.

9. CIRCUIT ANALOGY

The formulas for the transformation of the ray coordinates $\psi = \begin{bmatrix}x\\\xi\end{bmatrix}$ through a system suggest interpreting x as a voltage, ξ as a current. The propagation through distance t represented by

$$T^t = \begin{bmatrix} 1 & t/k \\ 0 & 1 \end{bmatrix} \tag{69}$$

and the transformation through a symmetric lens of focal distance f

$$T_f = \begin{bmatrix} 1 & 0 \\ k/f & 1 \end{bmatrix} \tag{70}$$

suggest the analogy with series and parallel elements. A cascade of such elements is represented by a continued fraction (compare with eq. 37). When x and ξ are in R^2 they represent pairs of voltages and currents, but the formulation is essentially the same with 2 x 2 matrices taking the place of scalars. The invariance of the symplectic product $\psi_1^T J \psi_2$ which is the *Lagrange invariant* attached to two rays (when this invariant equal 0, the two rays belong to the same pencil) is interpreted in circuit theory as an expression of reciprocity.

10. SOME EXTENSIONS OF GAUSSIAN BEAMS

10.1 Anisotropic media

The present paraxial theory is easily extended to propagation of fields satisfying more general wave equations and in particular those relevant to a non isotropic medium. The simplest approach is via the dispersion surface Σ corresponding to the wave equation. A beam is then represented by a sum of plane waves whose wave vectors belong to a small patch of Σ around the wave vector k_1 such that the normal to Σ at that point indicates the direction Oz of the beam axis [Fig. 5]. The small patch is approximated by a

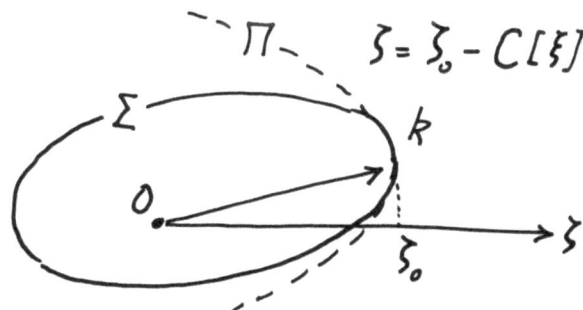

Fig. 5. Approximation of a patch of Σ by paraboloid Π

paraboloid Π having the same curvature as Σ at k_1 and having its axis parallel to oz. The equation of Π is of the form

$$\zeta = \zeta_0 - L[\xi] \tag{71}$$

with L a 2 x 2 symmetric matrix whose coefficients have dimensions of a length and

$$L[\xi] = \frac{1}{2} \xi^T L \xi \qquad (72)$$

The field of the beam for $z = 0$ being represented by the Fourier pair $(f_o(x), F_o(\xi))$, and for $z = t$ by $(f_t(x), F_t(\xi))$, we have the relation

$$F_t(\xi) = e^{i\zeta_o t} \phi^{tkL}(\xi) F_o(\xi) \qquad (73)$$

provided the effective support of $F_o(\xi)$ is contained in the projection of the patch on the $z = 0$ plane, and the corresponding convolution

$$f_t(x) = e^{i\zeta_o t} \Phi_{tkL}(x) * f_o(x) \qquad (74)$$

When f_o is a Gaussian function $f_o(x) = \phi^{C_o}(x)$ the same is true of $F_o(\xi)$, $F_t(\xi)$, $f_t(x)$. After propagation through the distance t the field is $f_t(x) = A(x)\phi^{C_t}(x)$ with

$$C_t^{-1} = C_0^{-1} + tkL \qquad (75)$$

which generalizes

$$C_t^{-1} = C_0^{-1} + t\begin{bmatrix} 1 & 0 \\ 0 & 1 \end{bmatrix} \qquad (76)$$

The amplitude

$$A = (\det C_t / \det C_o)^{\frac{1}{2}} \qquad (78)$$

as in Eq. 68.

10.2 Non-uniform media

We have assumed so far that the beams considered would propagate within homogeneous regions separated by smooth surfaces. Thus the rays were composed of segments of straight lines in each region, refracted (or reflected) at each transition surface according to the laws of geometrical optics.

One could also consider non-uniform media having characteristics that vary smoothly and slowly in terms of the local wavelength in the medium. Then it is still possible to apply geometrical optics. The rays become curves that are the geodesics with respect to the metric $nd\sigma = ds$ (n refractive index, σ Euclidean arc). The optical path, or the eikonal S, can be used as a parameter along each ray.

The rays in the vicinity of some central ray Γ_o form a 4 parameters family. Considering a plane P_0 orthogonal to Γ_o at the point 0 (s=0) a ray Γ in its vicinity is defined by a *position vector* $x = (x_1 x_2)$ (the intercept with the plane P_0) and a *direction vector* $\xi = (\xi_1, \xi_2)$ (the projection of the local wavevector $\vec{k}$ at 0 on plane P_0). The ray with coordinates $\psi = [\begin{smallmatrix}x\\\xi\end{smallmatrix}]$ will be denoted Γ_ψ. As s varies, the ray coordinates with respect to the plane P_s normal to Γ_o at point s will be a function $\psi(s)$. The ray itself is described by x(s). Strictly speaking x and ξ are infinitesimal and can be though of either as "small" or as finite velocity vectors with respect to a parameter describing the variation of the ray Γ away from Γ_o. The vector x(s) defines a Jacobi field on Γ_o [26].

10.3 Other problems

Other topics we have not discussed here are Gaussian Beams of higher order, described over a cross section by Gauss-Hermite functions and vectorial fields that have beam properties. Those may be considered as field of dipoles, more generally multipoles, at a complex location. The coupling between beam modes due to obstacles or simply to the finiteness of actual beam is also important. A general topic of great interest is the study of beams near to material boundaries.

11. CONCLUSION

The goal of this article has been to establish a correspondance between Gaussian beams and pencils of complex rays. The main consequence of this result is that one can trace the evolution of a Gaussian beam through an optical system, as long as the paraxial approximation remains valid, by using the formulas of "Gaussian optics" established for real rays and extended to astigmatic pencils. [Gauss's name enters this topic through more than one door!] Instead of a position along the central ray, described by a real number z, we have, for real rays, a 2 x 2 symmetric matrix Z that represents the focal lines of the pencil in position and direction. For a beam, the matrix Z becomes complex but obeys the same transformation rules as if it was real. Thus beam optics is an analytic continuation of real ray optics.

It is to be noted that the introduction of complex rays in this problem accounts for some diffraction effects - one may wonder whether this observation could be exploited in other circumstances.

APPENDIX A. GAUSSIAN FUNCTIONS

Functions that play a central role in the analysis of Gaussian beams and therefore deserve, on several counts, the name of *Gaussian* are the exponentials of quadratic functions of the variable. The most general function of this type, where the quadratic form can take complex values, describes the field in a Gaussian beam cross section and also its spectrum. Any cross section of the beam, under the paraxial approximation, enjoys this property. This is due to the protean properties of Gaussian functions: the Fourier transform of a Gaussian function is Gaussian and so are the product and the convolution of two Gaussian functions.

We have adopted the following notations. Given an $n \times n$ symmetric matrix in $GL(n, C)$, $A = A^T$ and a vector $u \in R^n$ (or C^n) we denote the associated quadratic form by

$$A[u] = \frac{1}{2} u^T A u \qquad (A.1)$$

and the associated Gaussian function by

$$\psi^A(u) = \exp(-A[u]) \qquad (A.2)$$

The matrix A in the exponent does not denote a power. However this notation enjoys the obvious property.

$$\psi^A \psi^B = \psi^{A+B} \qquad (A.3)$$

We shall denote by ψ_A the function $\psi^{A^{-1}}$.

When Re A > 0, i.e. (ReA)[u] > 0 for all real u, the function ψ^A can be integrated over real space. Multiplying it by a factor that makes its integral equal to 1 gives the normalized function that will be denoted by $\widehat{\psi}^A$

$$\widehat{\psi}^A(u) = (\det \frac{A}{2\pi})^{1/2} \psi^A(u) \qquad (A.4)$$

and similarly

$$\widehat{\psi}_A(u) = (\det 2\pi A)^{-1/2} \psi_A(u) = \psi^{A^{-1}} \qquad (A.5)$$

If the matrix A is the moment matrix

$$A = E(xx^T) \qquad (A.6)$$

of a real vector x normally distributed about 0, its probability density function is $\widehat{\psi}_A(x)$ and the Fourier transform, of ψ_A is the characteristic function, $\psi^A(\xi)$. Using the same notation for a Fourier pair as in Fig. 1 and 3 we write

$$\hat{\psi}_A(x) \leftrightarrow \psi^A(\xi) \qquad (A.7)$$

Replacing A by A^{-1} in A.7 we obtain

$$\hat{\psi}^A(x) \leftrightarrow \psi_A(\xi) \qquad (A.8)$$

and

$$\psi_A(x) \leftarrow \hat{\psi}^A(\xi) \qquad (A.9)$$

which results from applying the Fourier transformation twice:

$$f(x) \leftrightarrow F(\xi) \qquad F(x) \leftrightarrow (2\pi)^n f(-\xi) \qquad (A.10)$$

When A is complex the formulas (A.8-9-10) still hold but of course the probability interpretation does not apply.

To simplify further the notation when dealing with propagation problems in a medium of wave number k we introduce the following functions

$$\phi^Z(\xi) = \psi^{-Z/\nu}(\xi) \qquad (A.11)$$

where Z is an n x n symmetric matrix having the dimensions of a length, and $\nu = ik$. The Fourier transform of ϕ^Z is

$$\Phi_Z(x) = \psi_{-Z/\nu}(x) = (\det -\frac{2\pi}{\nu}Z)^{\frac{1}{2}} \exp(\nu y[x]) \qquad (A.12)$$

the matrix Y being the inverse of matrix Z.

Similarly the function

$$\phi^Y(x) = \psi^{-\nu Y}(x) = \exp(\nu Y[x]) \qquad (A.13)$$

where Y is a 2 x 2 matrix having dimensions of length^{-1}, has a Fourier transform

$$\Phi_Y(\xi) = (2\pi)^n \psi_{-\nu Y}(\xi) = (\det -\frac{\nu}{2\pi} Y)^{-\frac{1}{2}} \exp(\nu^{-1} Z[\xi]) \qquad (A.14)$$

Note that $\phi^Q(\cdot)$ is only defined when the variable x or ξ and the dimension of Q have been specified: the factor ν or ν^{-1} is inserted to make the exponent dimensionless.

Compensating for this inconvenience we have the simple rule that the Fourier transform, direct or inverse, is written by lowering or raising the subscript and replacing ϕ by Φ or vice versa.

The obvious rule for an ordinary product

$$\phi^A \phi^B = \phi^{A+B} \tag{A.15}$$

gives by Fourier transformation

$$\Phi_A * \Phi_B = \Phi_{A+B} \tag{A.16}$$

(For functions of ξ we understand convolution as resulting from integration with respect to $d\xi = (2\pi)^{-n} d\xi$) and with

$$C^{-1} = A^{-1} + B^{-1} \tag{A.17}$$

We have also

$$\phi^A + \phi^B = \phi^C$$

It can be verified that both functions ϕ and Φ satisfy:

$$\phi^Y = \phi_Z, \quad \Phi^Y = \Phi_Z \tag{A.18}$$

where $YZ = 1$.

The functions ϕ and Φ, when the matrix that specify them is real, could also be called Fresnel function as they play a role in Fresnel field computations. When the matrix is complex Gauss-Fresnel could be an appropriate name.

SELECTED REFERENCES

Early Work

1. Pierce, J. R., "Modes in Sequences of Lenses", Proc. Nat'l. Acad. Sci., 47, pp. 1808-13, Nov. 1961.

2. Goubau, G. and Schwering, F., "On the Guided Propagation of Electromagnetic Wave Beams", IRE Trans., AP-9, pp. 248-256, 1961.

3. Fox, A. G. and Li. T., "Resonant Modes in a Maser Interferometer", Bell Syst. Tech. J., 40, pp. 453-488, 1961.

4. Boyd, G. D. and Kogelnik, H., "Generalized Confocal Resonator Theory", 41, pp. 1347-1369, 1962, ibid.

5. Tien, P. K., Gordon, J. P., and Whinner, Y. J. R., "Focusing of a Light Beam of Gaussian Field Distribution in Continuous and Periodic Lens-Like Media", Proc. IEEE, 53, pp. 129-136, Feb. 1965.

6. Kogelnik, H. and Li, T., "Laser Beams and Resonators", Applied Optics, 5-10, pp. 1550-67, Oct. 1966.

Analysis: Geometrical Methods

7. Deschamps, G. A. and Mast, P. E., "Beam Tracing and Applications" in Proc. Symp. Quasioptics, New York: Polytechnic Press, pp. 379-395, 1964.

8. Collins, S. A., "Analysis of Optical Resonators involving Focusing Elements", Applied Optics, 3, pp. 1263-1275, 1964.

9. Gordon, J. P., "A Circle Diagram for Optical Resonators", B.S.T.J., 43, pp. 1826-7, July 1964.

Analysis Matrix Methods - Astigmatic Beams

10. Deschamps, G. A., "Matrix Methods in Geometrical Optics", Fall Meeting of URSI, University of Michigan, p. 84 of abstract, 1967.

11. Kahn, W. K. and Nemit, J., "Ray Theory of Astigmatic Resonators and Beam Waveguides", Proc. Symp. on Modern Optics, J. Fox, ed. Brooklyn, N.Y., Polytechnic Press, 1967.

12. Suematsu, Y. and Fukinuki, H., "Matrix Theory of Light Beam Wave-Guides", Bull. Tokyo Inst. Technology, 88, 33-47, March 1968.

13. Arnaud, J. A. and Kogelnik, H., "Gaussian Light-Beams with General Astigmatism" Appl. Opt., 8, pp. 1687-1963, 1969.

14. Arnaud, J. A., "Non-Orthogonal Optical Waveguides and Resonators", Bell Syst. Tech. J., 49, pp. 2311-2348, Nov. 1970.

15. Arnaud, J. A., "Mode Coupling in First Order Optics", J. Opt. Soc. Amer., 61, pp. 751-758, June 1971.

16. Kogelnik, H., "Imaging of Optical Modes - Resonators with Internal Lenses", Bell Syst. Tech. J., 44, pp. 455-494, March 1975.

17. Kogelnik, H., "On the Propagation of Gaussian Beams of Light Through Lenslike Media INcluding Those with a Loss or Gain Variation", Appl. Opt., 4, no. 12, pp. 1562-1569, 1965.

18. Deschamps, G. A., "Ray Techniques in Electromagnetics" Proc. IEEE, 60, no. 9, pp. 1022-1035, Sept. 1972.

Complex Rays Point of View

19. Deschamps, G. A., "Beam Optics and Complex Rays," URSI Symposium on Electromagnetic Waves, Stresa, June 1969.

20. Keller, J. B. and Streifer, W., "Complex Rays With an Application to Gaussian Beams," J. Opt. Soc. Amer., 61, no. 1, pp. 40-43, 1971.

21. Deschamps, G. A., "Gaussian Beam as a Bundle of Complex Rays," Electron. Lett., 7, no. 23, pp. 684-685, 1971.

Parabolic Approximation

22. Gloge, D. and D. Marcuse, "Formal Quantum Theory of Light Rays", J. Opt. Soc. Amer., 59, pp. 1629-1631, Dec. 1969.

Other References

23. Slepian, D., Pollak, H. O., and Landau, H. J. "Prolate spheroidal wave functions, Fourier analysis and uncertainty," Bell Telephone System, Monograph 3746, Jan. 1961.

24. Bremmer, H. "Series developments of diffraction integrals," McGill Symposium on Microwave Optics, P & II, pp. 226-34, (1959).

25. Lamb, Jr., Willis E., "Non relativistic quantum mechanics," Phys. Today, 22, no. 4, April 1969, pp. 23-28.

26. Milnor, J. "Morse Theory", Princeton University Press (1973).

EVANESCENT WAVES*

Leopold B. Felsen

Department of Electrical Engineering, Polytechnic Institute of
New York, Farmingdale, New York 11735, U.S.A.

I. INTRODUCTION

In the theory of wave propagation, evanescent fields have
played a secondary role since they are usually over-shadowed by
the dominant effects of nonevanescent fields. Although
evanescent waves must be included in complete representations of
arbitrary fields, they contribute generally in the near zone of
sources (actual, or induced as on an obstacle or scatterer), but
are negligible in the far zone. An exception occurs when the
evanescent fields represent the entire field contribution;
although exponentially small, they must then be explicitly
accounted for in the field description. Representative of this
circumstance, and most familiar to specialists in guided-wave
theory, are the evanescent fields exterior to slab or rod
dielectric surface wave structures, or the fields excited on the
optically thinner side of an interface between two media by a
totally reflected field incident from the optically denser side.
Evanescent fields also occur on the "dark" side of caustics
formed by a guided mode in an inhomogeneous duct or by a
focused incident field in a homogeneous or an inhomogeneous
medium.

With the development of laser sources and their application
to integrated and fiber optics, and also for applications in

---

* This material has been extracted from the paper by
L.B. Felsen, "Evanescent Waves", published in J.Opt.Soc.Am.,
Vol.66, No.8, August 1976.

microwave acoustics, Gaussian beams have assumed an important role as collimated fields that are coupled into transmission systems comprised of bulk media and guiding regions. Gaussian beams have usually been treated by plane-wave spectral analysis, which obscures the fact that they are actually evanescent fields. Stimulated primarily by the importance of Gaussian beam propagation, guiding and diffraction, evanescent waves have recently been studied from a more general viewpoint than heretofore. This general study has, in turn, clarified certain properties even of the more conventional evanescent wave fields.

The conventional view of evanescent waves is associated with modal fields that belong either to a discrete or continuous spectrum. For dielectric guiding structures, the discrete modes decay in a direction transverse to a preferred guiding axis; the guide axis is specified to follow a coordinate parallel to the boundaries. In plane-wave spectral or other eigenfunction representations of general fields, evanescent (below cutoff) modes decay along a selected guiding direction. In either event, the modal character ascribes to these fields global properties that pertain to the entire waveguide cross section or to the entire axial domain of evanescent decay. Consequently, the treatment of evanescent fields has conventionally taken place within the framework of guided mode theory.

The modal view of evanescent fields is unnecessarily restrictive. For nonevanescent fields it has long been recognized that modal representations are not the most useful formulations of certain field problems. Thus, in the range of short wavelengths in a free space or large-waveguide environment, ray-optical representations describe the field more compactly and with deeper physical content.[1,2] Such ray-optical formulations are based on the local character of short-wavelength propagation and permit the tracking of fields in terms of local plane waves along ray trajectories that may intercept boundaries, scatterers, and other perturbations in the local propagation environment. Such local tracking is not possible for modal fields because of their inherent global character.

The established versatility of ray (i.e., local wave tracking) methods for nonevanescent fields has motivated the study of similar methods for evanescent fields. These developments are reviewed in the present paper. For applications to surface wave propagation along curved layers, modal propagation in graded index slabs and fibers, evanescent plane wave and Gaussian beam scattering by cylinders and half-planes, see references 3-16.

II. LOCAL PLANE-WAVE FIELDS WITH COMPLEX PHASE

Guided by the conventional ray method, which involves a description of arbitrary nonevanescent high-frequency fields in terms of local ordinary plane waves, we seek a description of arbitrary evanescent high-frequency fields in terms of local evanescent plane waves. Such a formulation follows the route employed for the conventional ray method except that the local plane-wave fields are assumed to have a complex rather than a real phase. Thus beginning with the scalar wave equation (the theory for vector fields remains to be developed),

$$[\nabla^2 + k^2 n^2(r)] u(r) = 0, \tag{1}$$

where k is the free-space wave number, n(r) is the spatially dependent real refractive index and a time dependence $\exp(-i\omega t)$ is suppressed, one assumes a field representation in the form[3]

$$u(r) \sim \overline{A}(r) \exp[ikS(r)] + \exp[ikS(r)] \sum_{n=1}^{\infty} \frac{\overline{A}_n(r)}{(ik)^n}, \quad k \text{ large}, \tag{2}$$

where the phase function S and the amplitude functions $\overline{A}$, $\overline{A}_n$ are complex. The first term in (2), which represents the local evanescent plane-wave field, will be of principal concern. Substituting (2) into (1) and equating to zero the coefficients of each power of k leads to the eikonal equation

$$(\nabla S)^2 = n^2, \tag{3}$$

the transport equation for the local plane-wave amplitude $\overline{A}$,

$$\nabla^2 S + 2(\nabla S \cdot \nabla) \ln \overline{A} = 0, \tag{4}$$

and transport equations for the higher-order amplitude functions $\overline{A}_n$,

$$\overline{A}_n \nabla^2 S + 2(\nabla S \cdot \nabla) \overline{A}_n = -\nabla^2 \overline{A}_{n-1}, \quad n = 1, 2, 3, \ldots \tag{5}$$

with $\overline{A}_0 \equiv \overline{A}$. Writing

$$S(r) = R(r) + iI(r), \quad R \text{ and } I \text{ real} \tag{6}$$

and

$$\overline{A}(r) = \exp[w(r) + iv(r)], \quad w \text{ and } v \text{ real} \tag{7}$$

one obtains from (3), upon separation into real and imaginary parts,

$$(\nabla R)^2 - (\nabla I)^2 = n^2 \tag{8}$$

with $\nabla R \cdot \nabla I = 0$. Similarly, from (4),

$$\tfrac{1}{2}\nabla^2 R + \nabla R \cdot \nabla w - \nabla I \cdot \nabla v = 0, \tag{9}$$

$$\tfrac{1}{2}\nabla^2 I + \nabla R \cdot \nabla w + \nabla I \cdot \nabla v = 0, \tag{10}$$

and from (5),

$$\overline{A}'_n \nabla^2 R - \overline{A}''_n \nabla^2 I + 2(\nabla R \cdot \nabla \overline{A}'_n - \nabla I \cdot \nabla \overline{A}''_n) = -\nabla^2 \overline{A}'_{n-1}, \tag{11}$$

$$\overline{A}'_n \nabla^2 I + \overline{A}''_n \nabla^2 R + 2(\nabla I \cdot \nabla \overline{A}'_n + \nabla R \cdot \nabla \overline{A}''_n) = -\nabla^2 \overline{A}''_{n-1} \tag{12}$$

where $\overline{A}_n = \overline{A}'_n + i\overline{A}''_n$, $\overline{A}'_n$ and $\overline{A}''_n$ real.

To solve the eikonal and transport equations, it is convenient to introduce trajectories tangent to the unit vectors $\underset{\sim}{s}_0$ and $\underset{\sim}{t}_0$, where

$$\underset{\sim}{s}_0 = \nabla R/\beta, \quad \underset{\sim}{t}_0 = \nabla I/\alpha, \quad \text{with } \beta = |\nabla R|, \; \alpha = |\nabla I|. \tag{13}$$

Since $\nabla R \cdot \nabla I = 0$, these trajectories form an orthogonal grid (Fig. 1). The trajectories tangent to $\underset{\sim}{s}_0$, which are normal to the equiphase surfaces R = constant, are called "phase paths" because the equiphase surfaces advance along them; on these phase paths, the imaginary part I of the complex phase, and hence the exponential amplitude of the field u, remains constant. The trajectories tangent to $\underset{\sim}{t}_0$ lie on an equiphase surface and hence describe constant phase contours. If s and t are taken as length measures along the phase paths and equiphase contours, respectively, and referring to (13), one may rewrite the eikonal (dispersion) equation as

$$\beta^2 - \alpha^2 = n^2, \tag{14}$$

and derive the trajectory equations

$$\frac{d}{ds}(\beta \underset{\sim}{s}_0) = \nabla \beta, \quad \frac{d}{dt}(\alpha \underset{\sim}{t}_0) = \nabla \alpha. \tag{15}$$

Since $dI = \alpha dt$ remains constant on two neighbouring phase paths and $dR = \beta ds$ remains constant on two neighbouring equiphase contours, one may relate two values of α on a phase path by

$$\alpha_2 = \alpha_1 dt_1/dt_2 \tag{16}$$

and two values of β on an equiphase contour by

$$\beta_2 = \beta_1 ds_1/ds_2. \tag{17}$$

Thus α and β can be determined from known initial values α_1, β_1 and from the geometrical properties of the trajectory grid. The transport equations (9) and (10) become

$$\tfrac{1}{2}\nabla \cdot (\beta \underset{\sim}{s}_0) + \beta \frac{dw}{ds} - \alpha \frac{dv}{dt} = 0 \qquad (18)$$

$$\tfrac{1}{2}\nabla \cdot (\alpha \underset{\sim}{t}_0) + \beta \frac{dw}{ds} + \alpha \frac{dv}{dt} = 0, \qquad (19)$$

and similarly for the higher-order equations (11) and (12).

The calculation of the lowest-order field $u \sim \exp(ikR - kI + w + iv)$ proceeds by the following steps.

(i) Determination of the phase paths or equiphase contours from (15);

(ii) Calculation of α on a phase path from (16) or of β on an equiphase contour from (17), from given values on an initial surface. Knowledge of either α or β permits calculation of β or α, respectively, via (14);

(iii) Calculation of the complex phase $S = R + iI$ by (a) performing the integration $\int \beta ds = R$ along a phase path, with I remaining constant at its initial value, or (b) performing the integration $\int \alpha dt = I$ along an equiphase contour, with R remaining constant at its initial value; and

(iv) Calculation of the complex amplitude $\overline{A} = \exp(w + iv)$ by solution of (18) and (19).

A word of caution is in order. The structure of (14) and (15) is such that wave tracking along a phase path requires a <u>non-local</u> base of support on an initial surface. This is in contrast to geometric-optical field tracking, which can be accomplished locally along a ray. The degree of non-locality depends on how strongly evanescent are the local plane wave fields. For weakly evanescent fields, perturbation methods starting from the non-evanescent ray trajectories have been useful.

Even without constructing an explicit field solution, one may gain substantial insight into the propagation characteristics of evanescent fields from a study of the eikonal, trajectory, and transport equations. First, one observes from (15) that the phase paths and equi-phase contours are generally curved, even in a homogeneous medium with constant n. Defining the curvatures K and $\hat{K}$ of the phase paths and equiphase contours, respectively, as

$$\underset{\sim}{t}_0 K = d\underset{\sim}{s}_0/ds, \quad \underset{\sim}{s}_0 \hat{K} = d\underset{\sim}{t}_0/dt, \qquad (20)$$

one deduces from (15) that

$$K = d\beta/(\beta dt), \quad \hat{K} = -d\alpha/(\alpha ds). \qquad (21)$$

Thus K is nonzero whenever β varies along an equiphase contour, and $\hat{K}$ is nonzero whenever α varies along a phase path. This behaviour is plausible since a variation of β along an equiphase contour implies a variation in the propagation speed of different portions of the phase front. An initially plane phase front thereby becomes nonplanar.

Since the variations of α and β along these respective contours are determined from the trajectory grid [see (16) and (17)], it is evident that K and (or) $\hat{K}$ vanish only under very special circumstances. Thus $K = \hat{K} = 0$ only if the trajectory grid consists of parallel lines as in Fig. 2(a). The corresponding field $\sim \exp(i\beta_0 s + \alpha_0 t)$, with $\beta_0^2 - \alpha_0^2 = 1$ and β_0, α_0 constant, is an evanescent plane wave in a homogeneous medium; it can exist, for example, exterior to a straight dielectric layer whose boundaries are perpendicular to the t direction. $\hat{K} = 0$ and $K \neq 0$ may occur when the phase paths are circular as in Fig. 2(b), as is the case for weak evanescent fields exterior to a circularly curved di-electric layer with boundaries perpendicular to t. One observes from the contracting spacing ds that β, and therefore also $\alpha = (\beta^2 - 1)^{\frac{1}{2}}$, increases (i.e. the field becomes more strongly evanescent) with distance from the concave side of the layer, whereas the opposite is true for the convex side. In fact, for sufficiently great distances from the convex side, it is possible to have $\beta \to 1$, $\alpha \to 0$, with consequent transformation of the evanescent field into a nonevanescent field. Because of the energy leakage, which then takes place, α is not independent of s as implied in Fig. 2(b); however, Fig. 2(b) represents a lowest-order approximation. These aspects are discussed further in Sec. V.

When the phase paths are confocal hyperbolas and the equiphase contours ellipses, the corresponding field represents a two-dimensional Gaussian beam. Figure 2(c) reveals that the field becomes more evanescent as the observer moves along an equiphase contour away from the beam axis, but less evanescent as the observer moves along a phase path away from the beam waist. Since the hyperbolas have radial asymptotes, the evanescence of the field along a phase path decreases and $\alpha \to 0$, $\beta \to 1$ in the Fraunhofer region of the beam; thus the far-zone field of the beam is comprised of ordinary local plane waves although the radiation pattern decreases exponentially away from the beam axis.

It is appropriate to comment here on misleading terminologies that have been employed to describe Gaussian beams. The equi-exponential-amplitude contours (i.e. the phase paths) are often referred to as rays. This is a peculiar designation since conventional rays in a homogeneous medium are straight lines. Even their identification as complex rays is inappropriate.[17] The natural generalization of a straight ray with real direction cosines into a ray with complex direction cosines yields a straight line in complex coordinate space. An important attribute of a local plane-wave field transported along a straight ray, whether real or complex, is the constancy of its direction cosines (or the wave numbers along the various coordinate directions). Along a phase path, the wave numbers α and β are not constant so that the evanescent field changes continuously from one local evanescent plane wave to another. Complex rays do indeed play a role in the description of a Gaussian beam but not by their identification with the phase paths.

III. Complex source points and Gaussian beams

The two-dimensional Green's function for free space,

$$G_f(r, r') = \tfrac{1}{4}i\, H_0^{(1)}(kD), \qquad (22)$$

$$D = \left[(y - y')^2 + (z - z')^2\right]^{\tfrac{1}{2}}, \quad r = (y,z),$$

has for large kD the asymptotic approximation

$$G_f(r, r') \sim (-i8\pi k)^{-\tfrac{1}{2}} D^{-\tfrac{1}{2}} \exp(ikD). \qquad (23)$$

By analytic continuation, (22) and (23) can be extended into complex coordinate space, and the complex distance, $\hat{D}$, with $(\hat{y}', \hat{z}') = (0, ib)$, $b > 0$, and Re $\hat{D} > 0$, is then given in the paraxial region (Re $\hat{y})^2 \ll$ (Re $\hat{z})^2 + b^2$ as follows

$$\hat{D} \sim \text{Re } \hat{z} - ib + (\text{Re } \hat{y})^2 / 2(\text{Re } \hat{z} - ib), \qquad (24)$$

thereby reducing (23) to within a constant factor to the conventional paraxial Gaussian beam field. Actually, the solution in (22) with complex source point $(\hat{y}', \hat{z}')$ and real observation point (y, z) coordinates continues to satisfy the source free-wave equation exactly.

The preceding demonstration has important consequences. It implies that by the simple artifice of replacing the real source coordinates (y', z') of a line source field in free space by the complex values (0, ib), one may generate an exact solution of the wave equation, which behaves in the vicinity of the positive z axis like a Gaussian beam [note that the paraxial Gaussian beam field is only an approximate solution of the wave equation, invalid outside the paraxial region]. Since the complex-source-point substitution converts an incident cylindrical wave into an incident Gaussian beam, it follows that the same substitution in formulas for propagation and diffraction of line-source fields in the presence of obstacles or other medium perturbations will convert these into diffraction solutions for incident Gaussian beams (depending on the field representation employed, the analytic continuation from real to complex source points may involve certain minor restrictions[9]). Thus the whole arsenal of line-source diffraction results, either rigorous or high-frequency asymptotic, is available for the generation of corresponding results when the incident field is a two-dimensional Gaussian beam.

The same considerations apply in three dimensions, in which event the coordinates of a point source are assigned complex values. Thus, when $\hat{r}' = (\hat{x}', \hat{y}', \hat{z}') = (0, 0, ib)$, the three-dimensional free-space Green's function

$$G_f(\underline{r}, \hat{\underline{r}}') = (4\pi\hat{D})^{-1} \exp(ik\hat{D}), \tag{25}$$

$$\hat{D} = \left[\rho^2 + (z - ib)^2\right]^{1/2}, \quad \rho^2 = x^2 + y^2,$$

describes a rotationally symmetric field that satisfies the wave equation and behaves near the positive real z axis, where

$$\hat{D} \sim z - ib + \rho^2/2(z - ib), \quad \rho^2 \ll z^2 + b^2, \tag{26}$$

like a paraxial Gaussian beam. Moreover, by choosing $\hat{\underline{r}}' = (\hat{x}', \hat{y}', \hat{z}') = i\underline{b}, \underline{b} = (b_x, b_y, b_z)$, one may direct the beam axis along a preselected direction $\underline{b}$ instead of the z axis. Finally, by using the complex-source-point procedure for vector dipole fields, one may generate vector electromagnetic beam fields instead of the scalar fields considered so far. For example, when the complex distance $\hat{D}$ in (25) is substituted into the far-field formula for a y-directed magnetic dipole with moment m in free space, one obtains a three-dimensional electromagnetic beam field with magnetic field components [9]

$$H_{xf} = -km\left(\frac{\varepsilon}{\mu}\right)^{1/2} \frac{xy}{\hat{D}^2} G_f,$$

$$H_{yf} = km\left(\frac{\varepsilon}{\mu}\right)^{1/2} \frac{x^2 + (z-ib)^2}{\hat{D}^2} G_f, \tag{27}$$

$$H_{zf} = -km\left(\frac{\varepsilon}{\mu}\right)^{1/2} \frac{y(z-ib)}{\hat{D}^2} G_f,$$

and electric field components

$$E_{xf} = -km\frac{z-ib}{\hat{D}} G_f, \quad E_{yf} = 0, \quad E_{zf} = km\frac{x}{\hat{D}} G_f, \tag{28}$$

with ε and μ denoting the free-space permittivity and permeability, respectively, and G_f given by (25). This field is transverse electromagnetic on the beam axis due to the orientation of the dipole perpendicular to z. Other polarizations can be generated by tilting the dipole with respect to the z axis. Results dual to those in (27) and (28) are obtained for an electric dipole.

The complex-source-point method has the advantage of generating rigorous beam solutions from real line-source or point-source solutions without the intervention of evanescent wave tracking or field tracking along complex rays. However, the results obtained by the complex-coordinate substitution, while rigorous, are not easily interpreted in physical terms. Therefore, even when the beam field is formally constructed by the complex-source-point procedure, it is desirable to resort to the local evanescent plane-wave mechanism to gain an understanding of the wave processes involved. This combination of methods has been used to advantage in a variety of beam guiding and diffraction problems. The complex-source-point solutions are especially important for providing rigorous results against which asymptotic methods that accommodate a broader range of problems, such as the local evanescent wave tracking method or complex-ray tracing procedures, can be tested.

REFERENCES

[1]   J.B. Keller, Proceedings of the Symposium on Applied Mathematics (McGraw-Hill, New York, 1958), p. 27.

[2]   L.B. Felsen and N. Marcuvitz, Radiation and Scattering of Waves (Prentice Hall, Englewood Cliffs, N.J. 1973), Secs. 1.6 and 1.7.

[3]   S. Choudhary and L.B. Felsen, IEEE Trans. Antennas Propag. AP-21, 730, 1973.

[4]   S. Choudhary and L.B. Felsen, Proc. IEEE 62, 1530, 1974.

[5]   J.B. Keller and W. Streifer, J. Opt. Soc. Am. 61, 40, 1971.

[6]   L.B. Felsen, Philips Res. Rep. 30, 187, 1975.

[7]   G.A. Deschamps, Electron. Lett. 7, No. 23 (1971).

[8]   J.A. Arnaud, Appl. Opt. 8, 1909, 1969.

[9]   L.B. Felsen, "Complex-source-point solutions of the field equations and their relation to the propagation and scattering of Gaussian beams,". Symp. Mat. 18, (1976), p. 40-56.

[10] S.Y. Shin and L.B. Felsen, Appl. Phys. 5, 239, 1974.

[11] L.B. Felsen and S.Y. Shin, IEEE Trans. Microwave Theory Tech. MTT-23, 150, 1975.

[12] W-Y.D. Wang and D. Deschamps, Proc. IEEE 62, 1541, 1974.

[13] S. Choudhary and L.B. Felsen, J. Acoust.Soc.Am. 63 (1978), p.661-666.

[14] S. Choudhary and L.B. Felsen, J. Opt. Soc. Am. 67 (1977), p. 1192-1196.

[15] S.Y. Shin and L.B. Felsen, IEEE Trans. Microwave Theory and Techniques MTT-26 (1978), p.845-851.

[16] H.L. Bertoni, A.C. Green and L.B. Felsen, J. Opt.Soc. Am.68 (1978), p.983-989.

[17] G. Otis, J. Opt.Soc.Am.64 (1974), p.1545.

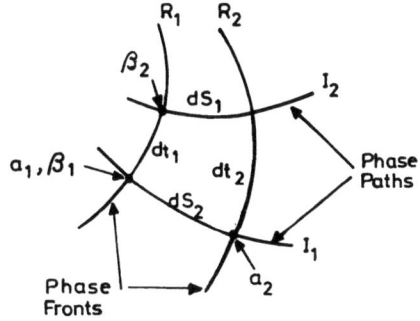

FIG. 1 Phase paths and paths of constant phase.

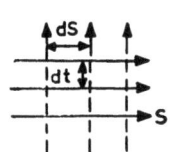

(a) evanescent field exterior to plane dielectric layer: $K = \hat{K} = 0$

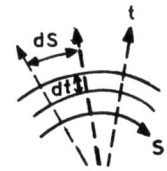

(b) weakly evanescent field exterior to curved dielectric layer: $\hat{K} = 0, K \neq 0$

(c) Gaussian beam

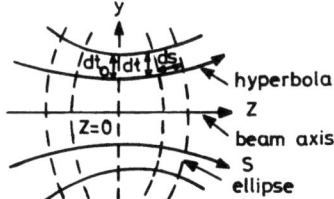

FIG. 2. Various evanescent fields in a homogeneous medium with n = 1. The phase paths are shown solid and the equiphase contours dashed. The incremental lengths ds and dt denote the spacing between adjacent equiphase contours and phase paths, respectively.

Part III

MECHANISMS OF BIOLOGICAL INTERACTION AND MICROWAVE BIOEFFECTS

Organised by Professor A.W. Guy
University of Washington, USA

HISTORY AND STATE OF THE ART ON THE QUANTITATION OF THE
INTERACTION OF ELECTROMAGNETIC FIELDS WITH BIOLOGICAL STRUCTURES

Arthur W. Guy

Bioelectromagnetics Research Laboratory
Department of Rehabilitation Medicine, RJ-30
University of Washington
Seattle, WA 98195

I. INTRODUCTION

The amount and the distribution of the absorbed electromagnetic energy in biological tissue exposed to electromagnetic (EM) fields is a function of many factors, including the magnitude of the electric field (E), magnitude of the magnetic field (H), the relative stored energy in the magnetic and electric fields, the source and tissue configurations, the tissue composition, frequency, environmental factors, and others. Therefore, in general, it is impossible to establish any meaningful relationships between a simple measurement of the external E and H fields and observable biological effects in a subject exposed to EM fields. We could expect, however, to establish a useful association between the measured internal electric field, E_t, and the effect because of its obvious relationship with the power loss, or specific absorption rate (SAR) of energy [W (per unit mass in the tissues)], induced current density (J), or other identifiable quantities that directly interact with the tissues. Furthermore, one would expect that once this association is established for a particular biologic tissue or system, it could be related much more meaningfully with other tissues or systems. Too often in the past, confusion has arisen and many erroneous conclusions have been drawn because of the lack of knowledge on the internal field distributions that produce the effect in the animal or the individual. Though it is quite proper and only practical to use the incident or free field magnitude in quantifying the hazards of EM radiation fields relating to man, it is not very quantitative to simply relate biologic effects in test animals to an incident or free field magnitude or power flux density in a test chamber.

The absorption, diffraction, and scattering effects of the test animal or specimen in the field are considerably different from the same effects pertaining to exposure of man. The greatest complicating factors result from interference patterns within the tissues which produce regions of intensifications of SAR or hot spots and regions of low SAR within a given animal. These SAR patterns will vary depending on the source, frequency, body size, geometry, and the environment of the subject. The only practical way to accurately quantify biologic damage or therapeutic benefits in terms of incident power density levels is to relate them to the internal field and SAR distributions. Much of the research pertaining to biological effects is done through animal experimentation or irradiation of biologic specimens in vitro. The question is, how do we relate the effects to the fields and extrapolate the results of these to the radiation effects in humans. In controlled experiments with animal or biologic tissues we have the option to set up any field configuration we desire, whether a radiation field or a near-zone field situation, where an effective power density can be determined with a survey meter. We cannot, however, assume that if a certain effective power density level produces quantifiable effects of damage or benefits in the animal or specimen, similar effects or damage can occur in the human with the same effective incident power density. With exposure of the animal or specimen, we also have scattering, absorption, and internal reflections that are uniquely associated with the animal's body and tissue characteristics or the specimen's geometry that result in SAR levels different from that for human exposure.

The most sensible approach is to quantify the actual fields, or absorbed energy in the tissues or specimen, and relate this finding to the biologic effects, damage, or benefits that may occur. Once this information is available, the next task is to determine what incident power density or outside field, whether predominantly radiation, electric, or magnetic, will produce the same effect in man. The essentials to know, then, are what rate of energy absorption per unit volume or mass in tissue of an animal or specimen under irradiation will yield an effect, damage, or benefit, and what level of incident power density or field as measured by a survey meter will produce the same rate of energy absorption in human tissues. These questions can be answered only through the development and application of appropriate theoretical and experimental techniques. Only then will there be a clear understanding of how EM fields interact with the tissue, whether the interaction is on a microscopic or macroscopic scale, involves the entire body structure, is thermal or non-thermal in origin or is merely an artifact to the nature of the experimental approach. By taking proper account of body size of the experimental animal, along with accurate in vivo dosimetry, the results from an investigator using rats can be related to those from another study on cats, monkeys, dogs, frogs, or a tissue sample in a test tube.

II. HISTORY

The historical evolution of the part microwaves have played in the health sciences is important in the elucidation of present day problems on the subject. Interest in the interaction of electromagnetic energy and biological media dates back to 1890 when d'Arsonval demonstrated that high frequency electrical currents of 10,000 Hz produced no muscular contractions but only heating of the tissues. This led to the use of radio frequency energy by physicians for heating human tissues for therapeutic purposes. The therapeutic benefits for certain musculoskeletal conditions arise from increased blood circulation and metabolic rate. The use of electromagnetic energy for achieving this became popular since the energy would penetrate deeply into the tissues producing heat by inducing current to flow through the resistive tissues. This form of therapy heats the muscle and joint tissue in contrast to the superficial heating of the cutaneous and subcutaneous tissue obtained by hot pads.

By the 1900's, physicians were using high-frequency currents between 0.5 and 3.0 MHz and by the year 1935 frequencies as high as 10 MHz were coupled directly to the body by electrodes to produce the required current density for therapeutic heating (1). During this time period, as early as 1928, radiations as high as 100 MHz (shortwave diathermy) were produced and used clinically (1). In 1937, it was reported by Williams (2) that EM waves with wavelengths of a few centimeters could be focused, and Southworth (3) pointed out that such radiation could be directed through hollow conducting tubes. The proposal to use microwaves for therapeutic purposes actually originated in Germany in 1938 and 1939 when Hollman (4,5) discussed the possible application of 25 cm long waves for therapeutics and predicted that the waves could be focused to produce heating of the deep tissues without excessive heating of the skin. Similar predictions were made by Hemingway and Stenstrom (6) in the United States. The lack of hardware at the time prevented clinical tests of the ideas and diathermy treatments continued at frequencies below 100 MHz.

It is interesting to note that arguments both for and against the existence of athermal effects developed with the early history of the therapeutic application of EM waves. Danilewsky and Worobjew (7) demonstrated that contractions in frog nerve muscle preparations increased in amplitude when high-frequency currents were applied along with minimum faradic stimuli. When high-frequency currents were removed, the excitability of the nerve rapidly returned to its original value. As increasing strength high-frequency current was applied (0.5 to 1.0 MHz exposure) a point was reached where a depression in excitability resulted. This same phenomenon of an increase or decrease of excitability was also obtained by irradiating the sciatic nerve of a warm-blooded animal. Audiat (8) asserted that since the excitability of the nerve muscle preparation diminished under the action of the EM waves, it had to be a "specific" effect, since heat would have an opposite effect. It was also claimed by Delharm and Fischgold (9) that high-frequency (HF) currents diminished excitability of the nerve muscle preparation in a manner similar to that produced by the anodic effect of a direct current. Later it was shown by Weissenberg (10) that interrupted HF current applied to a nerve muscle preparation of a frog showed stimulating effects similar to that obtained by a pulsating DC current. It was postulated that the nerve rectified a small portion of the applied HF current. Pflomm (11) stated that if a frog's heart preparation was placed in a short-wave field, the beat becomes slower and the excursions lessen with the diastolic beat finally ceasing. But if the field is switched off, the heart gradually resumes its activity. Hill and Taylor (12), on the other hand, replicated the work showing that weak high-frequency fields with wavelengths of 600, 22, and 6 meters would increase the excitability of a nerve muscle preparation, whereas stronger currents produced a depression of excitability. They showed that exactly similar effects could be produced by a hot wire placed near the nerve and concluded that the action of the high frequency current on the nerve muscle preparation was thermal. These researchers also demonstrated that the effects observed by Pflomm (11) on the frog's heart due to HF fields were identical to that obtained when the frog's heart was warmed.

During the period between 1931 and 1941 there were many basic problems in using shortwaves for the effective therapeutic heating of tissues mostly related to the fact that investigators were not able to quantify the actual energy absorbed in tissues during treatment. The results obtained were left entirely to chance and many quantitatively uncontrolled experiments resulted in contradictory statements appearing in the medical literature. The various shortwave generators, produced by different manufacturers, had variable outputs. It was implied through advertisements that the heating of deeper tissues would be enhanced with greater output of the machines. Since the heating of tissues seemed to vary considerably with frequency, even with the same apparent output of the various machines, many researchers jumped to the conclusion that there were selective and specific properties of the various wavelengths investigated.

Finally, in 1941, the power of the interdisciplinary approach, a lesson that has been learned and forgotten too many times, bore fruit. A research team consisting of engineers and physicians, E. Mittleman, S.L. Osborne, and J.S. Coulter (13), quantified the temperature rise in living tissues in terms of the absorbed energy and power instead of simply the output of the diathermy applicator or the tissue exposure level. The group instrumented a short-wave diathermy generator so that the actual power absorbed by the patient under treatment could be measured with an error limit of 5%. The absorbed power was correlated with the measured temperature in the tissues. The results illustrated in Table 1 relate the absorbed power, the temperature rise, and the watts per 1000 cc for 0.1° F temperature rise/min.

Table 1. Relationship of temperature rise per minute to power absorption (from Mittlemann, et al., ref. 13)

	No.	Power Absorbed Watts	Temperature Rise Per Min.	Watts Per 1000 cc for 0.1 F Temp. Rise Per Min.	Deviation Against Average Percent
Subject I	1	60	0.275	5.75	-1
thigh mass	2	72	0.310	6.15	+4.8
approximately	3	74	0.300	6.42	+8
3800 cc	4	85	0.384	5.86	0
	5	86	0.390	5.78	+0.7
	6	100	0.460	5.75	+0.7
	7	137	0.652	5.55	+5.2
Subject II	8	88	0.440	6.15	+4.8
thigh mass	9	88	0.440	6.15	+4.8
approximately	10	44	0.2600	5.20	-13
3280 cc	11	135	0.720	5.75	-0.7
		Average of all tests		5.86	

This work makes use, for the first time, of the concept of specific absorption rate (SAR) with units very close to the W/kg now widely accepted for RF dosimetry. The absorption results agreed well with theory. The amount of power absorption required per unit of tissue volume to raise the tissue temperature to a certain extent in a given period of time was the same in all of the tests. The group conducted another series of measurements in which the patient was exposed using different wavelengths and exposure techniques. These results shown in Figure 1 indicated that the temperature rise per minute was strictly proportional to the value of the actual power absorbed by the patient. The power absorption necessary to raise the temperature of 1000 cc of tissue volume by 0.1° F per minute was computed from the results of Figure 1 and recorded in Table 2, which is in excellent agreement with the results in Table 1. It is significant to note that the deviation from linearity due to blood cooling does not take place until the end of the 20 minute exposure period in cases where the values of power absorption exceed 100 W (calculated SAR = 48 W/kg) and also where the subject has been treated for a short time previously (curves with index 2 in Fig. 1). The work by Mittlemann et al. (13) clearly demonstrated that the degree of tissue heating was dependent solely on power absorption and not on wavelength for similar ratios of deep heating to superficial-heating.

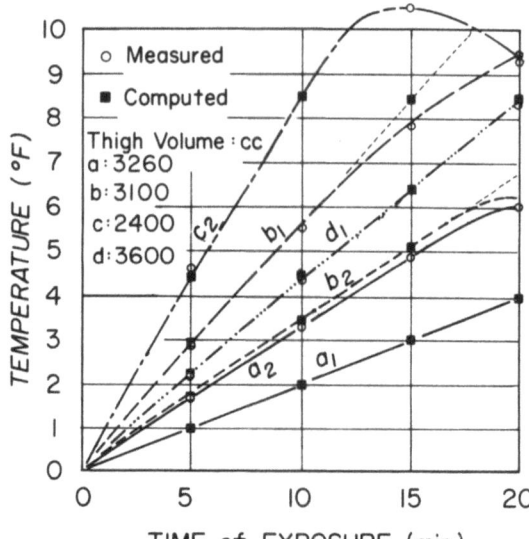

Figure 1. Heating curves showing temperature rise is proportional to power absorbed (from Mittlemann, et al., ref. 13) (see Table 2).

Table 2. Heating curves obtained on four different subjects using variable wattages, different technics, and various wavelengths (from Mittlemann, et al., ref. 13)

Vol. cc.	Curve	Wave-length Meters	Technic	Power Absorption Watts	Watts Per 1000 cc for 0.1 F Temp. Rise Per Min.	Deviation Against Average Percent
3280	a1	6	Air Sp.	35	5.35	9.
3280	a2	6	Air Sp.	63	5.83	0.
3100	b1	12	Air Sp.	110	6.25	6.2
3100	b2	12	Air Sp.	70	6.60	11.0
2400	c2	12	Pancake Coil	115	5.68	3.0
3600	d1	8	Db. cuffs	80	5.35	9.0
			Average of all tests		5.83	

It is also significant to note that there is a marked difference in the behavior of the temperature curves for high and low wattage. For observed powers <100 W (calculated SAR = 33-44 W/kg) temperature in the deep tissue rises along a straight line until near the termination of the 20 minute exposure period. If the power absorption exceeded these levels, then the final temperature is actually lower than the previous temperature, indicating an increase in blood circulation which rapidly dissipates the heat.

In the first ten years of the development and use of shortwave diathermy, research consisted of measuring the temperatures of superficial and deep tissues of both animal and human subjects exposed to capacitive and inductive type applicators with generator wavelengths varying from 6-24 meters. The only dosimetry at the time was the monitoring of the power output of each machine and relating it to temperature measurements in the tissues. A number of researchers felt that as a result of their observing temperature rises and heating in exposed tissues that varied as a function of wavelength, the absorption characteristics were wavelength specific. Others, however, felt that the variations were more a function of the electrode configuration and spacing and the geometry of the tissue being treated. It was also observed by these early researchers, for example, Osbourn and Coulter (14), that it was difficult to produce therapeutic temperature levels of 42-45° C in deep tissues, such as muscle and bone marrow, without coagulating and dilatareously affecting the skin and the superficial tissues.

In 1940, Johnson et al. (15) reported the action of short radio waves on transplantable tumors *in vivo* and *in vitro*. They developed a technique for measuring the temperatures in the tumors produced by means of short radio waves so that continuous temperature records could be made during the treatment. The group studied the thermal sensitivities of the Jenson rat sarcoma and the Walker rat carcinoma 256 over a range of temperatures from 47-43.5° C and quantified the exposure times necessary for various temperatures to produce a 50% regression. They found that the exposure times required to produce 50% regression of the Walker rat carcinoma 256 at 47, 45, and 43.5° C were 45 minutes, 1.5 hours, and 6 hours, respectively, whereas those for the Jenson rat sarcoma for the same range of temperatures were 25 minutes, 1 hour, and 3 hours. The group used thermocouples imbedded in hypodermic needles and tried to eliminate coupling of high-frequency currents through a tuning process. They had considerable problems with RF pickup under certain conditions, but felt that they were able to obtain reliable temperature measurements with proper tuning.

At the end of World War II, when security restrictions were lifted on magnetron microwave power sources, the first microwave diathermy was developed operating at a frequency of 3000 MHz. With the new equipment, the first work on the therapeutic applications of microwaves began at Mayo Clinic in 1946 by Krusen et al. (16) and Leden et al. (17) which involved the exposure of laboratory animals to 65 W of 3000 MHz radiation. The temperature distribution in the thigh of experimental dogs was measured with thermocouples before and after exposure to microwaves. In this work (with the thermal couples removed during the period of radiation) it was demonstrated that deep tissues could indeed be heating resulting in a number of physiological responses and increased blood flow to the area treated. But, as with shortwaves, it was again noted that the average temperature rise was greater in the skin and subcutaneous fat than the deeper muscle tissue, although the final temperature in the muscle tissue was higher.

Worden et al. (18) found that by using a hemispheric reflector backed monopole antenna energized with 30 W of 2450 MHz microwave energy placed 2.5 cm from the surface of the thighs of experimental dogs, that maximal heating was obtained after an exposure period of 20 minutes. But if the exposure was extended to 30 minutes, the temperatures dropped as a result of a sharp increase in blood flow. This is consistent with the temperature characteristics found earlier through the use of shortwaves by Mittlemann et al. (13). The same group also carried out comparative studies on the effects of exposing ischemic tissue and tissue with normal circulation to the same amount of microwave energy over a period of 15-20 minutes. They found that serious damage resulted in the ischemic area even though the average temperatures recorded were not any higher than those found in normal tissue with circulation that was not damaged.

In 1948, Siems et al. (19) made comparative studies of the effect of shortwave and microwave diathermy on blood flow. Their experiments clearly demonstrated that shortwave and microwave diathermy were equally effective in producing increased blood flow in the extremities of normal dogs. A number of such experiments were carried out, sometimes with conflicting results. Finally, in 1950, in order to resolve this problem, Richardson et al. (20) carried out research on the relationship between deep tissue temperature and blood flow during radiation with shortwave diathermy and microwaves. They found that with both modalities the blood flow in an area depended largely upon the degree of tissue hyperthermia developed. They found, in their experiments with dogs, that it was necessary to increase the tissue temperatures at 1 cm depth in the extremity of the dog to the level of 42-43° C before a consistent increase in blood flow occurred in the femoral artery. After reaching a critical temperature, they found that the increase in blood flow was sufficient to actually diminish the temperature. This again is consistent with the work of Mittlemann et al. (13) and has been observed by many researchers since that time.

After 1950 and up until 1965, research on the use of microwaves for diathermy expanded significantly. Clinical-type and experimental animal research was much more predominant than any quantitative work on dosimetry. Significant work on the engineering approaches in quantifying the various effects was done, however, by Schwan at the University of Pennsylvania. Schwan's work on the dielectric properties of biological media and wave propagation absorption by various tissue geometries deserves considerable attention (21-25). During this period it was demonstrated theoretically by Schwan that the microwave frequency of 2450 MHz was not a good choice for therapy as a result of major deficiencies, including: 1) excessive heating in the subcutaneous fat due to standing waves; 2) poor penetration into the muscle tissue due to small skin depth; and 3) poor control and knowledge of energy absorbed by patients due to the large variations in electrical thickness (compared to the wavelength of subcutaneous tissues). He recommended that frequencies be reduced to 900 MHz or less.

Between 1960 and 1966, Lehmann et al. (26-29) and Guy (30) experimentally verified Schwan's earlier theoretical prediction that 900 MHz or lower frequencies could produce better therapeutic patterns than obtained with 2450 MHz energy. Since 1966, Lehmann et al. (29), DeLateur (31), and Guy (32,33) have developed and clinically tested direct contact applicators operating at 915 MHz which are therapeutically more effective and safer in terms of leakage radiation than the existing 2450 MHz equipment.

In addition to medical applications, work was also directed to the understanding of potential biological hazards and methods for quantifying interaction of electromagnetic fields with biological tissues. In the late 1950's, the possibility of producing high field intensity, but nonthermal, "pearl chain" effects in biological fluids containing suspended small particles of biological material of differing dielectric properties was demonstrated (34-38). Studies began in the early 1960's on determining the relationship of exposure frequency, subject size, and subject shape with energy coupling (39-41). Analyses were performed with prolate spheroidal tissue models using static solutions to determine low-frequency quasi-static field coupling and spheres using the Mie theory to determine plane wave field coupling characteristics with the bodies of humans. The studies with the spheres indicated that the absorption cross section varied markedly with frequency with sharp minima and maxima.

After the mid-1960's, experimental phantom models of various tissues were developed and used for experimentally verifying the theoretical analyses and determining field coupling and absorption characteristics for more complex tissue structures not amenable to theoretical analysis (42,43). From the late 1960's until the present, thermography has played a powerful role in measuring the EM field-induced temperature changes in both phantom and actual biological tissues, allowing for a rapid and accurate quantitation of the absorbed energy and electric field distributions within the tissues (44-47). Both theoretical work and the development of new instrumentation increased substantially in the 1970's. Complex spherical models of the human head consisting of a core of brain tissue and spherical shells simulating the skull and the scalp indicated that hot spots or localized regions of high energy absorption could occur in the center of the brain with magnitudes much higher than observed at the surface of the head due to the focusing of energy by the high dielectric constant and spherical shape of the head (48,47). More extensive analyses using spherical, prolate spherical, and ellipsoidal models has recently created a much better understanding of the absorbed energy patterns in the bodies of man and animals exposed to EM fields (49-55). Theoretical work is being backed up by careful experiments utilizing special temperature sensing probes composed of microwave transparent materials such as fiber optics and miniature leads of low electrical conductivity (56-60). Finite difference techniques and other numerical methods are being used in computer programs for calculating EM field and associated heating patterns in arbitrarily shaped bodies more closely simulating man (61,62). Mathematical models have also been developed to include the effect of cooling mechanisms including blood flow for calculating steady state temperatures in various parts of the body including critical organs such as the eyes and the brain (63,64). Theoretical analyses coupled with animal experiments have indicated that the long known but unexplained microwave hearing effect where individuals exposed to pulse radars could hear clicks and buzzing sounds is due to the conversion of microwave pulses to heat in the tissues of the head (65-68). The effect previously was thought to be a nonthermal effect since the threshold energy of a microwave pulse required to elicit the effect is sufficient to produce a temperature rise of only 10^{-5}°C. Though there have been a number of theoretical analyses suggesting the possibility of low-level nonthermal effects, there has been no general agreement that such effects exist (69-71). The following sections describe in more detail our present understanding, in terms of theoretical approaches, of the interaction of electromagnetic fields with biological tissues.

III. DIELECTRIC PROPERTIES AND EM WAVE PROPAGATION THROUGH BIOLOGICAL TISSUES

Some of the basic characteristics of electromagnetic field interaction with biological materials can be characterized by the wave parameters tabulated in Tables 3 and 4. The first column lists selected frequencies between 1 MHz and 10 GHz. Frequencies of 27.12, 40.68, 433, 915, 2450, and 5800 MHz are significant since they are used for industrial, scientific, and medical heating processes. The frequencies of 27.12, 433, 915, and 2450 MHz are used for diathermy purposes. The second column tabulates the corresponding wavelength λ in air and the remaining columns pertain to the wave properties of a particular tissue group. Table 3 gives data for muscle, skin, or tissues of high water content, while Table 4 is for fat, bone, and tissues of low water content. Other tissues containing intermediate amounts of water, such as brain, lung, bone marrow, etc., will have properties that lie between the tabulated values for the two listed groups. The tables list the dielectric properties, the depth of penetration,

Table 3. Properties of microwaves in biological media*

Frequency (MHz)	Wavelength In Air (cm)	Muscle, Skin, and Tissues with High Water Content					Reflection Coefficient Air-Muscle Interface		Reflection Coefficient Muscle-Fat Interface	
		Dielectric Constant (ε_H)	Conductivity σ_H (S/meter)	Wavelength λ_H (cm)	Depth of Penetration (cm)	r	ϕ	r	ϕ
1	30,000	2,000	0.400	436	91.3	0.982	+179	--	--
10	3,000	160	0.625	118	21.6	0.956	+178	--	--
27.12	1,106	113	0.602	68.1	14.3	0.925	+177	0.651	-11.13
40.68	738	97.3	0.680	51.3	11.2	0.913	+176	0.652	-10.21
100	300	71.7	0.885	27	6.66	0.881	+175	0.650	-7.96
200	150	56.5	1.00	16.6	4.79	0.844	+175	0.612	-8.06
300	100	54	1.15	11.9	3.89	0.825	+175	0.592	-8.14
433	69.3	53	1.18	8.76	3.57	0.803	+175	0.562	-7.06
750	40	52	1.25	5.34	3.18	0.779	+176	0.532	-5.69
915	32.8	51	1.28	4.46	3.04	0.772	+177	0.519	-4.32
1500	20	49	1.56	2.81	2.42	0.761	+177	0.506	-3.66
2450	12.2	47	2.17	1.76	1.70	0.754	+177	0.500	-3.88
3000	10	46	2.27	1.45	1.61	0.751	+178	0.495	-3.20
5000	6	44	4.55	0.89	0.788	0.749	+177	0.502	-4.95
5800	5.17	43.3	4.93	0.775	0.720	0.746	+177	0.502	-4.29
8000	3.75	40	8.33	0.578	0.413	0.744	+176	0.513	-6.65
10000	3	39.9	10.00	0.464	0.343	0.743	+176	0.518	-5.95

*Based on data from Schwan and Piersal (21) for temperature of 37° C.

Table 4. Properties of microwaves in biological media*

Frequency (MHz)	Wavelength In Air (cm)	Fat, Bone, and Tissues with Low Water Content					Reflection Coefficient Air-Fat Interface		Reflection Coefficient Fat-Muscle Interface	
		Dielectric Constant (ε_L)	Conductivity σ_L (mS/m)	Wavelength λ_L (cm)	Depth of Penetration (cm)	r	ϕ	r	ϕ
1	30,000	--	--	--	--	--	--	--	--
10	3,000	--	--	--	--	--	--	--	--
27.12	1,106	20	10.9 - 43.2	241	159	0.660	+174	0.651	+169
40.68	738	14.6	12.6 - 52.8	187	118	0.617	+173	0.652	+170
100	300	7.45	19.1 - 75.9	106	60.4	0.511	+168	0.650	+172
200	150	5.95	25.8 - 94.2	59.7	39.2	0.458	+168	0.612	+172
300	100	5.7	31.6 - 107	41	32.1	0.438	+169	0.592	+172
433	69.3	5.6	37.9 - 118	28.8	26.2	0.427	+170	0.562	+173
750	40	5.6	49.8 - 138	16.8	23	0.415	+173	0.532	+174
915	32.8	5.6	55.6 - 147	13.7	17.7	0.417	+173	0.519	+176
1500	20	5.6	70.8 - 171	8.41	13.9	0.412	+174	0.506	+176
2450	12.2	5.5	96.4 - 213	5.21	11.2	0.406	+176	0.500	+176
3000	10	5.5	110 - 234	4.25	9.74	0.406	+176	0.495	+177
5000	6	5.5	162 - 309	2.63	6.67	0.393	+176	0.502	+175
5900	5.17	5.05	186 - 338	2.29	5.24	0.388	+176	0.502	+176
8000	3.75	4.7	255 - 431	1.73	4.61	0.371	+176	0.513	+173
10000	3	4.5	324 - 549	1.41	3.39	0.363	+175	0.518	+174

*Based on data from Schwan and Piersal (21) for temperature of 37° C.

and the reflection characteristics of various tissues exposed to
electromagnetic waves as a function of frequency.

The dielectric behavior of the two groups of biological tissues tabulated in Table 3 and 4 has been evaluated most thoroughly by Schwan and his associates (21-25) and by other researchers including Cook (72-74) and Cole (75). The interaction of electromagnetic wave fields with biological tissues is related to these dielectric characteristics. The tissues are composed of cells encapsulated by thin membranes containing an intracellular fluid composed of various salt ions, polar protein molecules, and polar water molecules. The extracellular fluid has similar concentrations of ions and polar molecules, though some of the elements are different.

The action of electromagnetic fields on the tissues produce two types of effects that control the dielectric behavior. One is the oscillation of the free charges or ions and the other the rotation of dipole molecules at the frequency of the applied electromagnetic energy. The first gives rise to conduction currents with an associated energy loss due to electrical resistance of the medium, and the other affects the displacement current through the medium with an associated dielectric loss due to viscosity. These effects control the behavior of the complex dielectric constant $\varepsilon^*/\varepsilon_o = (\varepsilon'-j\varepsilon'')$, where ε_o is the permittivity of free space, ε^* is the complex permittivity, ε' is the dielectric constant, and ε'' is the loss factor of the medium. The effective conductivity σ (due to both conduction currents and dielectric losses) of the medium is related to ε'' by $\varepsilon'' = \sigma/\omega\varepsilon_o$ and the loss tangent is given by $\tan \delta = \varepsilon''/\varepsilon' = \sigma/\omega\varepsilon'\varepsilon_o$. The quantity ε^* will be dispersive due to the various relaxation processes associated with polarization phenomena. These are indicated by the dielectric properties given in Tables 3 and 4.

Plane wave propagation characteristics in plane layered biological tissues may be examined to show how radiation is absorbed when the radius of curvature of the tissue surface is large compared to a wavelength. The propagation constant k for power transmission through biological tissues can be written in terms of the complex dielectric constants $\varepsilon^*_{H,L}$ and free space propagation constant k_o in the standard form:

$$k = k_o(\varepsilon^*\varepsilon_o^{-1})^{1/2} = \beta - j\alpha \tag{1}$$

where the wavelengths $\lambda = 2\pi/\beta$ are significantly reduced in the tissues due to the high dielectric constants. Tables 3 and 4 indicate that the factors of reduction are quite large, between 6.5 and 8.5, for tissues of high water content, and between 2 and 2.5 for tissues with low water content. In addition to the large reduction in wavelength, there will be a large absorption of energy in the tissue which can result in heating. The SAR, W, resulting from both ionic conduction and vibration of dipole molecules in the tissues is given by

$$W = (2\rho)^{-1} \sigma |E|^2 \tag{2}$$

where E is the magnitude of the electric field and ρ is the density of the tissue in kg/m^3. One may note from the conductivities listed in Tables 3 and 4 that absorption in tissue of higher water content may be as high as 60 times greater than in that of low water content for the same electric fields. The absorption of microwave power will result in a progressive reduction of wave power density as the waves penetrate into the tissues. We can quantify this by defining a depth of penetration $d = 1/\alpha$ or a distance that the propagating wave will travel before the power density decreases by a factor of e^{-2}. We can see from Tables 1 and 2 that the depth of penetration for tissues of low water content is as much as 10 times greater than the same parameter for tissues of high water content.

Since each tissue in a complex biological system such as man has different complex permittivity, there will in general be reflections of energy between the various tissue interfaces during exposure to microwaves. The complex reflection coefficient due to a wave transmitted from a medium of complex permittivity ε_1^* to a medium of permittivity ε_2^* and thickness greater than a depth of penetration is given by

$$\Gamma = re^{j\phi} = (\sqrt{\varepsilon_1^*} - \sqrt{\varepsilon_2^*})(\sqrt{\varepsilon_1^*} + \sqrt{\varepsilon_2^*})^{-1} \tag{3}$$

The values r and ϕ for various interfaces are tabulated in Tables 3 and 4. Note the large reflection coefficient for an air-muscle or a fat-muscle interface. When a wave in a tissue of low water content is incident on an interface with a tissue of high water content of sufficient thickness (greater than the depth of penetration), the reflected wave is nearly 180° out of phase with the incident wave, thereby producing a standing wave with an intensity minimum near the interface. If the wave is propagating in a tissue of high water content and is incident on a tissue of low water content, the amplitude of the reflected component is in phase with the incident wave, therby producing a standing wave with an intensity maximum near the interface. If there are several layers of different tissue media with thicknesses less than the depth of penetration for each medium, the reflected energy and standing wave pattern are influenced by the thickness of each layer and the various wave impedances. These effects may be obtained from the standard transmission line equations. The distribution of electric field strength E in a given layer is

$$E = E_o[e^{-jkz} + \Gamma e^{jkz}] \tag{4}$$

where E_o is the magnitude of the field. From (2), the equation for SAR in the tissue layer, we obtain

$$W = (2\rho)^{-1} \sigma E_o^2 [e^{-2\alpha z} + r^2 e^{2\alpha z} + 2r \cos(2\beta z + \phi)] \tag{5}$$

Schwan (21,38) and Johnson and Guy (45) have made extensive calculations of these absorption distributions in various tissues. Typical distributions are shown in Fig. 2 for a wave transmitted through a subcutaneous fat medium into a muscle medium. The absorption is normalized to unity in the muscle at the fat-muscle interface. The relative absorption curves shown remain unchanged for smaller fat thicknesses. The severe discontinuity between the SAR in the muscle and that in the fat is quite apparent. Also, it can be seen that the standing wave peaks become larger in the fat and the wave penetration into the muscle becomes less with increasing frequency. This illustrates clearly the desirability for using frequencies lower than the 2450 MHz for deep therapeutic heating of tissues. Subcutaneous fat may vary from less than a centimeter in thickness to as much as 2.5 cm in thickness for different individuals. Deep heating for diathermy requires the transmission of energy through this subcutaneous fat layer to the muscle layer. Optimum results are attained with maximum heating in the muscle. The absolute values of SAR in the tissue layers are dependent on incident power density, skin thickness, and fat thickness. Fig. 3 illustrates the SAR at the muscle interface and the peak SAR in a skin layer 2 mm thick as a function of fat thickness for an incident power density of 1 mW/cm^2. These values may be used to determine the SAR at other

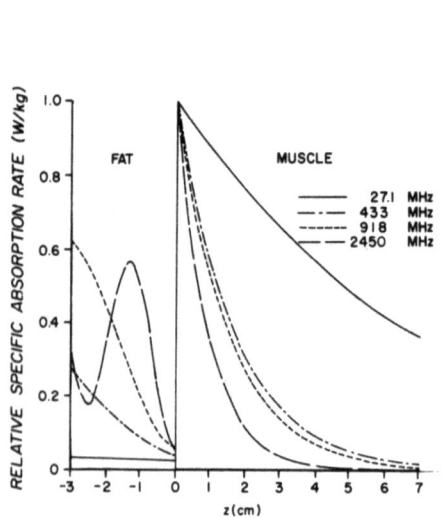
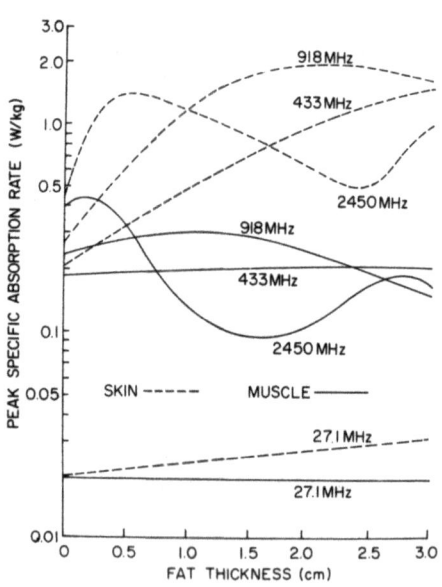

Figure 2. Relative specific absorption rate patterns (from Johnson & Guy, ref. 45).

Figure 3. Peak specific absorption rate in plane skin and muscle layers as a function of subcutaneous fat thickness (from Johnson & Guy, ref. 45).

locations in the muscle and fat by relating them to the curves in Fig. 2. The peak SAR is always maximum in the skin layer for the plane layered model. This is significant since the thermal receptors of the nervous system are located there and will indicate pain when the incident power density reaches levels that could thermally damage the tissue. With surface cooling of the skin, however, by natural environmental conditions or by controlled clinical procedures, the temperature increases may be higher in the fat or muscle. The peak absorption in the various tissues may vary over a wide range with fat thickness and frequency. It is apparent that frequencies below 918 MHz can penetrate more deeply into the tissue. The implications of this in terms of both radiation hazards and therapeutic applications are apparent. The first two figures clearly indicate the advantages of lower frequencies for therapeutic heating of deep tissues, including: 1) increased penetration into the muscle tissue, 2) less severe standing waves and resulting "hot spots" in the fat, and 3) better control and knowledge of the absorbed energy for a given incident power for a large variation of fat thicknesses between different patients.

There is a practical lower limit, however, of the frequency that can be used. As the frequency is decreased, the applicator needed becomes increasingly large until it is no longer possible to obtain desired selective heating patterns. If the applicator is not increased in size as frequency is lowered, only superficial heating will result. This has been discussed in detail by Guy and Lehmann (30) and Guy (32).

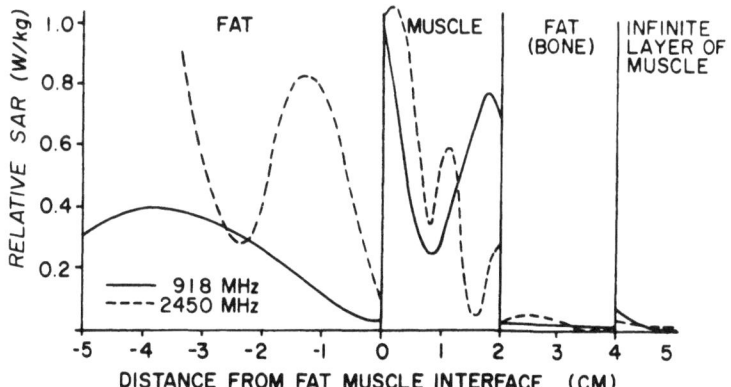

RELATIVE ABSORBED POWER DENSITY
IN
FOUR LAYERED TISSUE GEOMETRY

Figure 4. Relative specific absorption rate patterns in plane fat, muscle, and bone layers exposed to a plane wave source (from Johnson, et al., ref. 45).

A problem of interest in diathermy is the determination of how effective microwaves are in heating a layer of bone beneath a layer of subcutaneous fat and muscle. Fig. 4 illustrates heating patterns for this case using diathermy frequencies of 2450 MHz and 918 MHz for a 2 cm thick bone. The results clearly show that the absorption in the bone is very poor due to both a severe reflection and a low electrical conductivity. Since a standing wave peak at 918 MHz occurs in the muscle near the bone surface, we would expect significant bone heating due to thermal conduction from the muscle.

IV. THERMAL CONSIDERATIONS

The energy equation for the time rate of change of temperature (°C/sec) per unit volume of subcutaneous tissue in a subject exposed to EM fields is

$$dT/dt = 0.239 \ c^{-1} \ [W_a + W_m - W_c - W_b] \qquad (6)$$

where W_a is the SAR, W_m is the metabolic heating rate, W_c is the heat loss due to thermal conduction, W_b is the power dissipated by blood flow, all expressed in W/kg, and c is the thermal conductivity expressed in kcal/kg°C. The SAR for tissue exposed to an EM field is

$$W_a = (2\rho)^{-1} \ \sigma |E|^2 \qquad (7)$$

If it is assumed that blood enters the tissue at arterial temperature T_a and leaves at tissue temperature T, we may express blood cooling by

$$W_b = k_2 m \ c_b \ (\rho_b \Delta T')^{-1}$$

where $\Delta T' = T - T_a$, c_b is the specific heat of blood, ρ_b is the density of blood, m is the blood flow rate in ml/100 g/min, and the constant $k_2 = 0.698$. Prior to the time the tissue is exposed to fields, it is assumed that a steady state condition exists where $W_a = dT/dt = 0$ requiring $W_m = W_c + W_b$. According to the typical values of the physical and thermal properties of tissues given in Table 5, under normal conditions the metabolic rate W_m averages 1.3 W/kg for the total body, 11 W/kg for brain tissue, and 33 W/kg for heart tissue. According to the energy equation, we would expect to see some change in tissue temperature due to applied EM fields if the SAR, W_a, were of the same order of magnitude as W_m, or more. In fact, the safety guides in the United States that allow a maximum human exposure level of 10 mW/cm^2 of incident power are partially based on limiting the average W_a to the average resting value of W_m. Thus, SAR $W_a \gg W_m$ could be expected to produce marked thermal effects, whereas an SAR of $W_a \ll W_m$ would not be expected to produce any significant thermal effects.

Table 5. Thermal and physical properties of human tissues*

Tissue	Sub-script	Specific Heat c Kcal/kg	Density ρ gm/cc	Metabolic Rate (W_o) W/kg	Blood flow Rate (m) ml/100 gm·min	Thermal Conductivity k_c mW/cm·°C
skeletal muscle (excised)	m		1.07			4.4
skeletal muscle (living)	m	.83		0.7	2.7	6.42
fat	f	.54	0.937			2.1^1
bone(cortical)	bc	.3	1.79			14.6^1
bone(spongy)	bs	.71	1.25			
blood	bl	.93	1.06^2			5.06
heart muscle	m			33	84	
brain(excised)	br					5.0
brain(living)	br			11	54	8.05
kidney	k			20	420	
liver	l			6.7	57.7	
skin(excised)	s					2.5
skin(living)	s			1	12.8	4.42
whole body				1.3	8.6	

(1) For pig
(2) For human
*From Guy et al (46)

V. RELATIONS BETWEEN PLANE WAVE FIELDS AND ABSORBED POWER DENSITY IN EXPOSED BIOLOGICAL OBJECTS

A. Spherical Models

We have already discussed a case in the previous section for the SAR distributions in planar layered tissues exposed to plane wave fields. Considerable insight can be gained into the relationship between frequency, body size, and SAR by considering spherical tissue layers exposed to a plane electromagnetic wave. Figure 5 illustrates the relative SAR density patterns for various spherical tissue geometries exposed to a 1 mW/cm^2 plane wave source. The origin of the rectangular coordinate system used in the figures is located at the center of the sphere with wave propagation along the z axis and the E field polarized along the x axis. The maximum absorption and the average absorption density are tabulated on each plot. When the diameter of the exposed object is of the order of one wavelength as measured in the tissue, severe internal wave interference produces sharp maxima and minima in the SAR patterns as shown in Figure 5a for a sphere representing a human head exposed to 918 MHz radiation. The spherical model is composed of an inner core consisting of brain tissue surrounded by a layer of bone and skin. When the

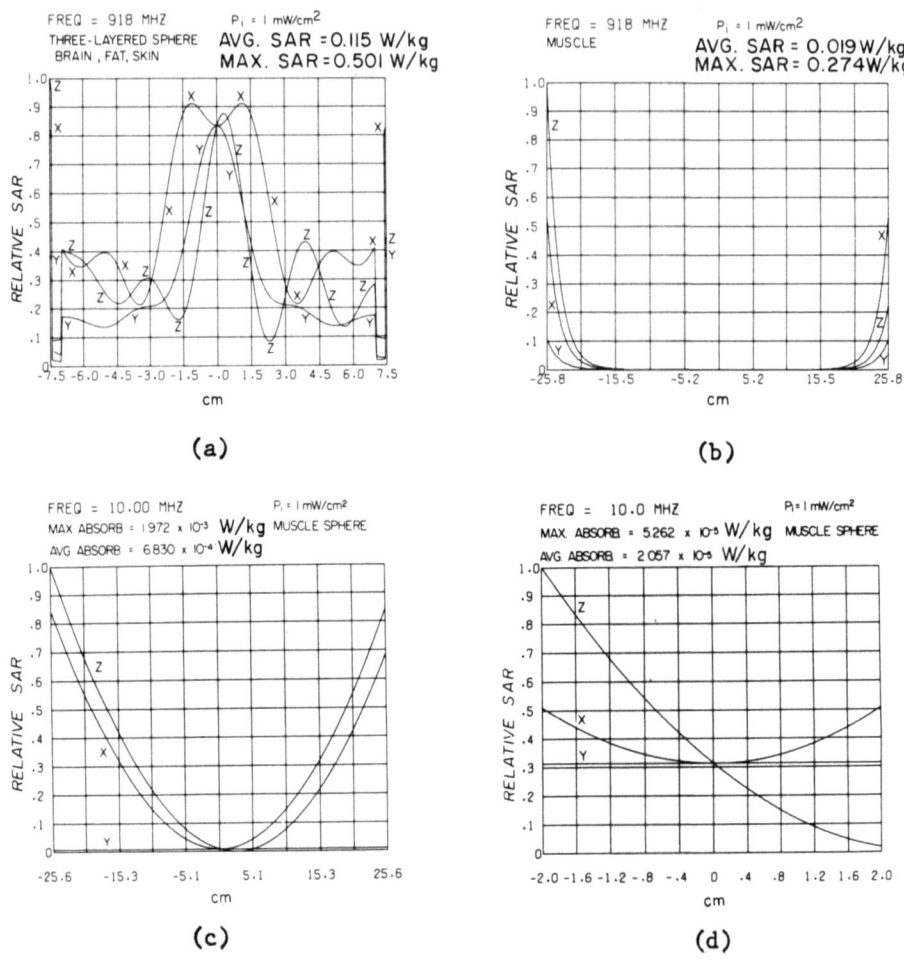

Figure 5. Specific absorption rate in spherical tissue layers exposed to 1 mW/cm² incident plane wave power density (from Guy, ref. 47).

object is large compared with a wavelength, as measured in the tissue, the maximum absorption occurs at the exposed surface, decaying nearly exponentially with depth as shown in Figure 5b for a homogeneous muscle sphere with the same mass as a 70 kg man exposed to 918 MHz radiation. When the exposed subject is very small compared with a wavelength, but of a mass approximating that of man, the SAR density varies nearly as the square of distance from the y axis (direction of magnetic field vector) as shown in Figure 5c. On the other hand, if the object is very small compared with a wavelength, but with a mass very small compared to that of man, the SAR is uniform along the y axis but increases with distance toward the exposed surface and decreases with distance toward the opposite surface, as shown in Figure 5d. The latter two absorption patterns for objects small compared with a wavelength can be explained from simple quasi-static field theory (76). The electric field component of the incident plane wave couples to the object in the same manner as a static electric field giving rise to a constant internal electric field which is $3|\epsilon^*|^{-1}$ times smaller, and in the same direction as the applied field where $|\epsilon^*|>>1$ is the dielectric constant of the tissue. Superimposed on the constant internal electric field is another magnetically induced electric field component encircling the y axis, as shown in Figure 6. The magnitude of the latter field, which varies directly with radial distance r from the axis, and directly with frequency f, is given by $E = \pi frH$, where H is the magnetic field. The H-induced E field component in a sphere with the same mass as man is much greater than the E-induced component, whereas, for a small object with the mass of a small rodent, both components are significant. The variation of the maximum and average SAR with frequency for an exposed homogeneous muscle sphere with the same volume is shown in Figure 7. Also shown in the figure is the average SAR per unit total surface area of the sphere. In the frequency range of 1 MHz to 20 MHz, the absorption characterized by the pattern in Figure 4c varies as the square of the frequency. This is due primarily to

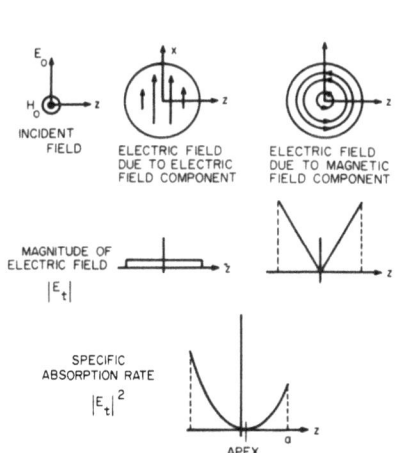

Figure 6. Sketch depicting how electric and magnetically-induced electric fields add in an exposed tissue sphere to produce the specific absorption rate pattern (from Guy, ref. 47).

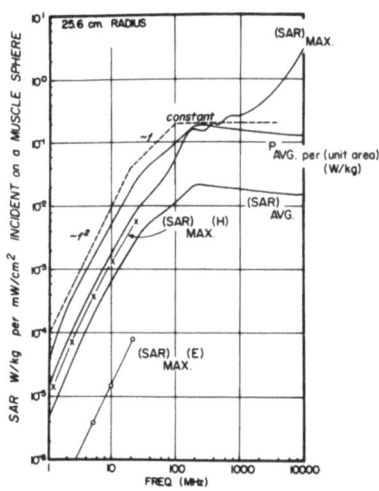

Figure 7. Specific absorption rate patterns vs. frequency in spherical muscle model of 70 kg man exposed to plane wave 1 mW/cm² source (from Guy, ref. 47).

the magnetically-induced fields. The maximum SAR induced by the incident H field is denoted by the curve marked with crosses, and that due to the incident E field is denoted by the curve with zeros in the range where the quasi-static coupling approximations apply. Note that in this range, the maximum SAR is only 10^{-5} to 10^{-2} W/kg per mW/cm^2 of incident power. In the frequency range 100-1000 MHz, internal reflections are significant for the man-size sphere and the average absorption attains a maximum of 2×10^{-2} W/kg per mW/cm^2 of incident power at 200 MHz, which remains relatively constant with frequency up to 10 GHz. The maximum absorption density increases with frequency above 1000 MHz, approaching that produced by non-penetrating radiation. The dashed lines illustrate roughly the frequency dependence of the average SAR and how safety standards might be relaxed as a function of frequency if the absorption characteristics in man were the same as for the sphere. The wide variation of absorption characteristics with body size is illustrated in Figure 8 for a sphere consisting of an inner muscle core surrounded by concentric layers of subcutaneous fat and skin exposed to 2450 MHz plane wave 1 mW/cm^2 radiation. The total radius, fat thickness, and skin thickness are noted on the figure. It is significant to note that based on the spherical models, the peak SAR could be as high as 4.2 W/kg in the body or head of a small bird or animal but as low as 0.27 W/kg at the surface, and 0.05 W/kg 215 cm deep in the human body exposed to 1 mW/cm^2, 918 MHz source (Fig. 5b).

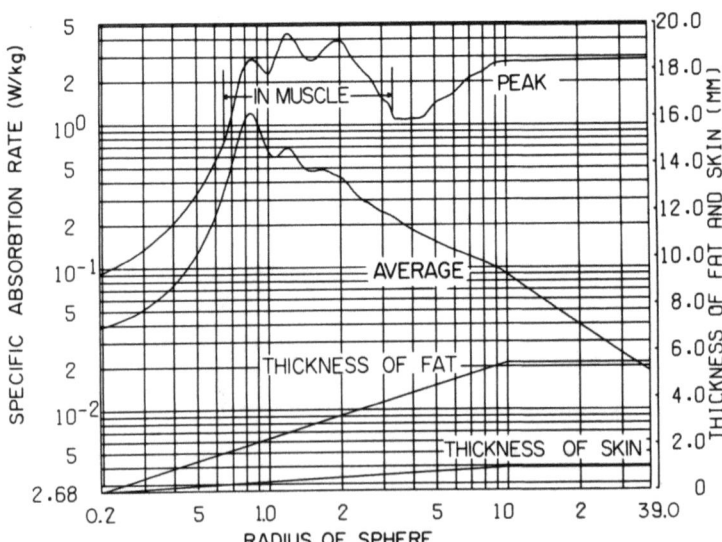

Figure 8. Specific absorption rate vs. outside radius in spherical tissue layer model of animal exposed to 2450 MHz 1 mW/cm^2 plane wave (from Guy, ref. 47).

Thus, 10 mW/cm^2 could be of extreme and 0.5 mW/cm^2 could be of mild thermal significance to the smaller animal in comparison with metabolic rate. For the human model, on the other hand, 10 mW/cm^2 would appear to be of mild thermal significance and 0.5 mW/cm^2 would have negligible thermal significance.

B. Prolate Spheroidal, Ellipsoidal and Cylindrical Models

Although the work with spherical models is useful for allowing a simple understanding of the coupling characteristics of plane waves to biological subjects, it has limitations. More useful models include the prolate spheroidal, and ellipsoidal geometries that have been analyzed in the frequency range where the wavelength is long compared to the major axis by Durney, et al. (50), Johnson, et al. (51), and Massoudi (52,53). In contrast with the analysis for the spherical model, the results from these analyses show that the average SAR and SAR distributions are strong functions of the orientation of the spheroids and ellipsoids with respect to the polarization of the incident plane wave fields. Three standard polarizations for exposing spheroids defined by which vector, $\bar{E}$, $\bar{H}$, or $\bar{K}$, of the incident plane wave lies along the major axis of the spheroid are denoted by E, H, or K, polarization, respectively. Polarization definitions for an ellipsoid are similar to those for a spheroid but, since the for-

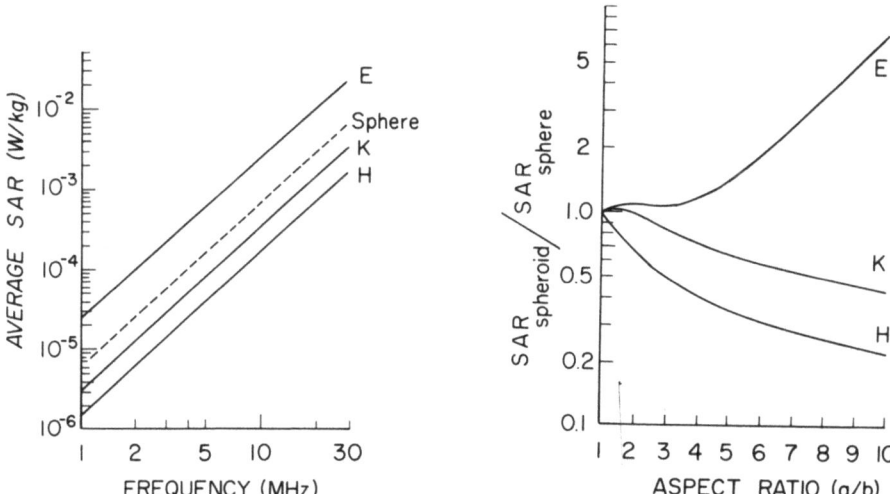

Figure 9. Average specific absorption rate within a muscle prolate spheroid for each of the three polarizations, E, H, and K, resulting from an incident electromagnetic plane wave with power density = 10 W/cm^2, volume = 0.07 m^3. Right values for a = 1 m and a/b = 7.73, and left values for 10 MHz as function of aspect ratio a/b (from Durney, et al., ref. 50).

mer has an elliptical cross section there are six standard polarizations for it. The three semi-axes of the ellipsoid are defined as a, b, and c, which lie along the x, y, and z coordinates, with a > b > c. The polarization is defined by the field vectors which lie parallel to the coordinates x, y, and z. For example, EKH polarization responds to the case where $\overline{E}$ lies along x, $\overline{H}$ lies along y, and $\overline{K}$ lies along z. Typical SAR characteristics of prolate spheroidal and ellipsoidal models exposed to a plane wave calculated by Durney (50) are shown in Figures 9 and 10. The results show that for a prolate spheroid the average SAR is maximum for E polarization and minimum for H polarization and somewhere between the two for K polarization. Figure 9 illustrates the SAR as a function of frequency and aspect ratio, a/b, for a prolate spheroid model exposed to a plane wave for three standard polarizations. The figure also illustrates the SAR data for a spherical model with the same volume as the spheroid. The calculations indicate that the SAR increases approximately as the square of the frequency and with a difference in SAR of more than an order of magnitude between the E and H polarizations. The marked change in SAR with polarization was verified in experimental studies by Gandhi (54) and Allen, et al. (55). Figure 10 shows the calculated average SAR for an exposed ellipsoidal model

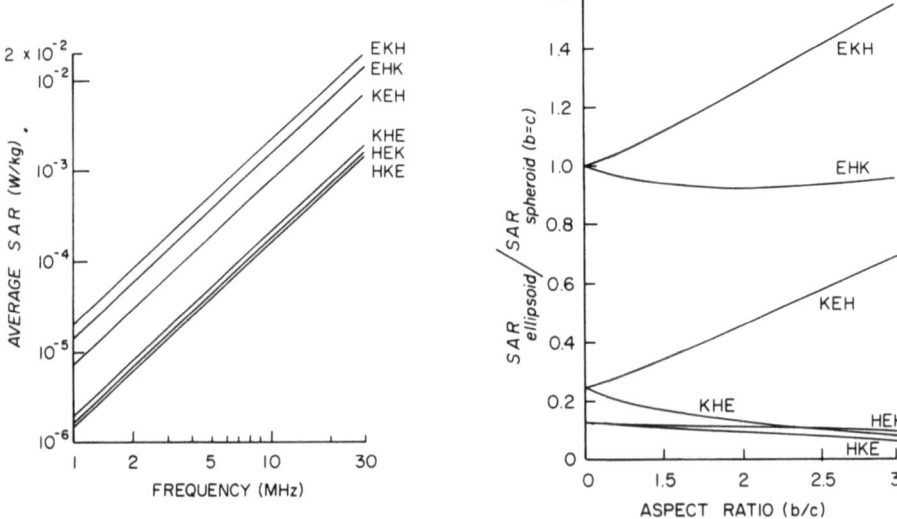

Figure 10. Average specific absorption rate within an ellipsoidal model of man for each of the six standard polarizations as a function of frequency resulting from an incident electromagnetic plane wave with power density = 10 W/m^2. Volume = 0.07 m^3 with a = 0.875 m and conductivity σ = 0.6 S/m. Left values for b/c = 2, and right values for 10 MHz as a function of aspect ratio b/c. (From Durney et al., ref. 50.)

of man as a function of frequency and aspect ratio, b/c, for 6 standard polarizations. The results again show the order of magnitude difference in average SAR as a result of exposure to the different EKH and HKE polarizations. The results have been verified experimentally in measurements of the average SAR of monkeys exposed to high-frequency fields by Allen, et al. (77). An increase in average SAR for EKH and KEH polarization with increasing b/c is attributed to the increase in the cross-sectional area normal to the incident magnetic field causing an increase in the strength of the magnetically-induced E field and, also with the greater elongation of the ellipsoid the coupling to the incident E field parallel to the long axes of the ellipsoid is increased. The long wavelength approximations are valid for $a/\lambda < 0.1$ corresponding to exposure frequencies <30 MHz for man-sized models and <400 MHz for rat-sized models. For higher frequency exposure, the extended boundary condition method (EBCM), a numerical technique, has been successfully used for determining SAR for prolate spheroidal models exposed to plane waves by Barber (60). His analysis predicts a resonance condition corresponding to a high SAR for the prolate spheroidal model of an average man exposed to a plane wave 70 MHz. The SAR distributions have been calculated for more exact models of man numerically by Chen and Guru (61) and Hagmann, et al. (62). Numerical techniques have been limited to man-sized models up to approximately several hundred MHz. Rowlandson and Barber (78) performed the SAR calculation at high frequencies through a geometric optic technique which is valid at frequencies at wavelengths small compared to the dimensions of the irradiated body, above approximately 6 GHz. Massoudi et al. (79) have been able to make SAR calculations using a cylindrical model of man at frequencies covering the range of invalidity of the EBCM and the geometric optics methods. Through these three methods, the SAR characteristics for $\underline{E}$ and $\underline{H}$ polarization can be calculated at the entire frequency range of 10 MHz through 100 GHz. Values for the mid-frequency range of exposure with $\underline{K}$ polarization were obtained from experimental data by Gandhi (54). Figure 11 shows the combined results for all three polarizations. Based on these calculated and experimental results, Allen et al. (55), and Gandhi, et al., (80) have summarized the results for human exposure to plane waves as follows:

1. The average SAR is greatest for $\underline{E}$ polarization at a resonant frequency corresponding to the height of the subject = 0.4 wavelengths.
2. The average SAR varies at a rate $>f^2$ near but below resonance.
3. The average SAR varies as f^2 when the largest dimension of the model is <0.2 wavelengths.
4. For $\underline{E}$ polarization the average SAR varies at a rate $<f^2$ at frequencies above resonance.

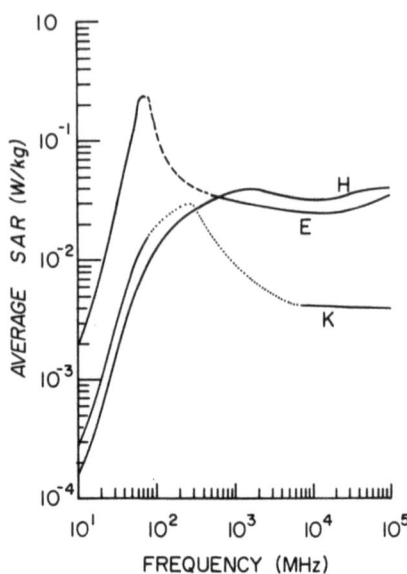

Figure 11. Calculated average specific absorption rate as a function of frequency for an average man. Solid lines are calculated; dotted and dashed lines are estimated. Incident power density is 1 mW/cm^2 (from Durney, et al., ref. 56).

Based on the analyses described above, a considerable amount of data has been calculated on specific absorption rate in man and animals as a result of plane wave exposure. Data covering a broad frequency range from 10 MHz to 100 GHz have been compiled in a Radiofrequency Radiation Dosimetry Handbook by Durney, et al. (56).

VI. FIELD COUPLING FROM FINITE SOURCES

A. Aperture Sources

If other than a plane wave source is used to expose biological tissues, the SAR patterns are also very dependent on source size and distribution. Many applications of microwave power in medicine and studies on the biological effects of microwave power require an understanding of the SAR patterns due to tissues exposed to aperture and waveguide sources. Guy (32) has analyzed the case where a bilayered fat and muscle tissue layer is exposed to a direct contact aperture source of width a and height b. A fat tissue layer of thickness z_1 and dielectric constant ε_f^* in contact with a semi-infinite muscle tissue layer with a dielectric constant ε_m^* is assumed. The origin of the coordinate system is located at the center and in the plane of the aperture with the x axis parallel to height b and the z axis in the direction of propagation into the tissue. The electric fields $E_{f,m}$ in the fat and muscle tissue may be expressed as Fourier integrals

$$E_{f,m}(x,y,z) = (2\pi)^{-2} \int_{-\infty}^{\infty} \int_{-\infty}^{\infty} T_{f,m}(u,v,z) e^{j(ux+vy)} dudv \quad (17)$$

where $T_{f,m}$ are the Fourier transforms of the electric fields at the fat and muscle boundaries, derived from the boundary conditions at $z = 0$ and $z = z_1$ in terms of the Fourier transform of the aperture

$$T_a(u,v) = \int_{-\infty}^{\infty} \int_{-\infty}^{\infty} \hat{z} \times E_f(x,y,0) e^{-j(ux+vy)} dxdy \qquad (18)$$

The aperture field is denoted as $E_f(x,y,0)$ and $\hat{z}$ is a unit vector along the z axis. The expressions may be evaluated numerically and the absorption patterns plotted by means of a digital computer. As an example we may consider a waveguide aperture source and evaluate it as a diathermy applicator for use at 918 MHz. Fig. 12 illustrates the relative SAR patterns in the x-z plane for a = 12 cm and b = 2,4,12, and 26 cm. The SAR at the fat surface for a plane wave exposure is denoted by the dashed line on the figures. The results show that the relative SAR varies from intense levels concentrated near the surface to moderate levels deep in the tissues as aperture size is increased.

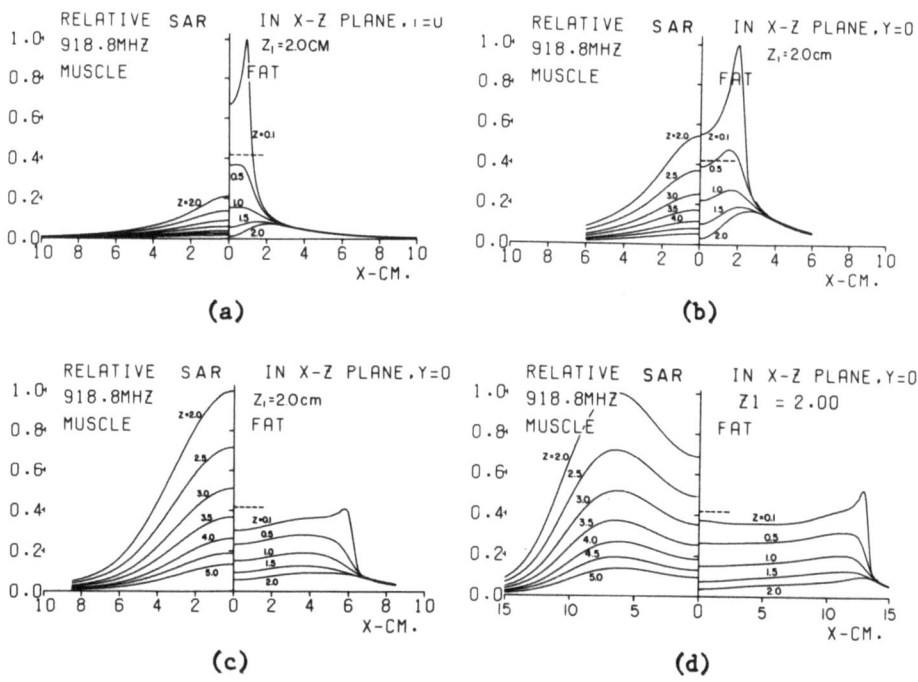

Figure 12. SAR patterns in plane layers of fat and muscle exposed to TE_{10} mode waveguide aperture source with a = 12 cm, f = 918 MHz, and z_1 = 2 cm for various aperture heights. For (a) to (d), the values of b are 2, 4, 12, and 26 cm, respectively. (From Johnson & Guy, ref. 45.)

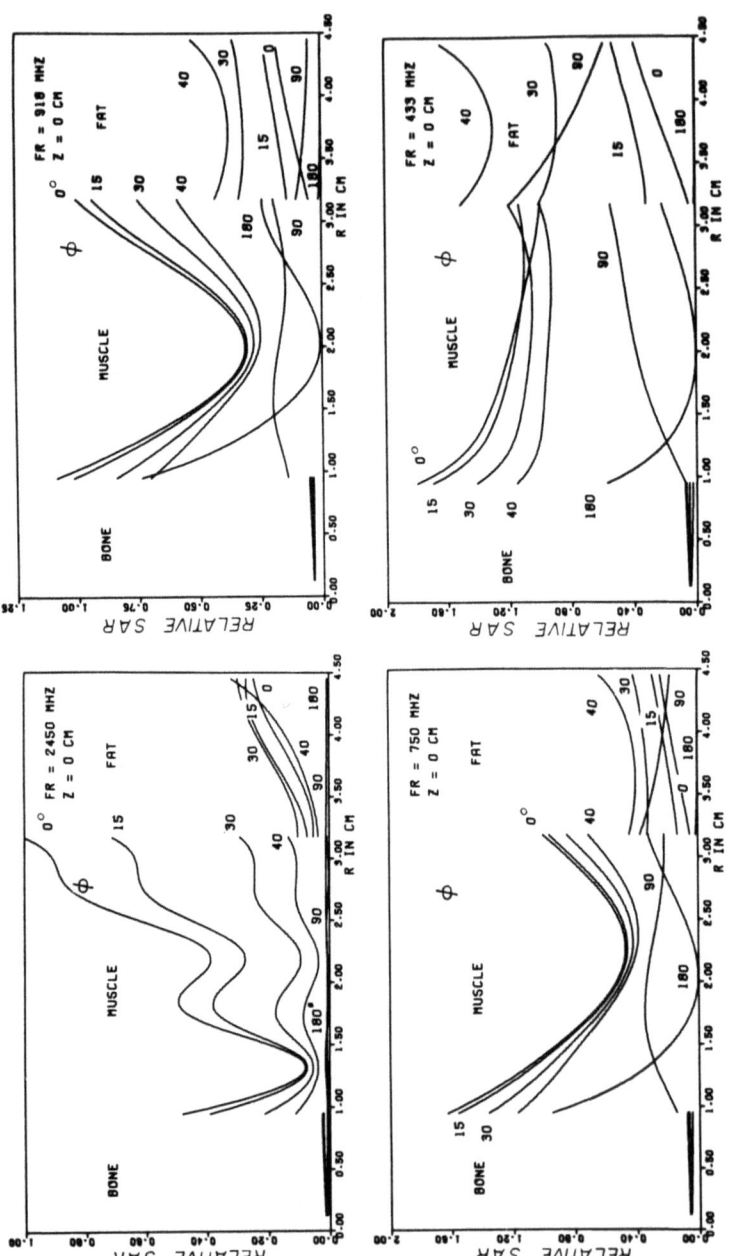

Figure 13. Relative specific absorption rate patterns in a cylindrical model of human arm due to a direct contact cylindrical aperture source (from Johnson and Guy, ref. 45).

The SAR patterns in multilayered cylindrical tissues exposed to an aperture source can also be determined by using a summation of three-dimensional cylindrical waves, expressing the aperture field as a two-dimensional Fourier series and matching the boundary conditions. Ho, et al., (44) have calculated the absorption patterns for a number of different aperture and cylinder sizes. Typical results are shown in Fig. 15 for a human arm-size cylinder exposed to a surface aperture source 12 cm long in the direction of the axis. The patterns are plotted as a function of radial distance from the center of the cylinder for various circumferential angles ϕ from the center of the aperture. The patterns are normalized to the values at $\phi = 0°$ at the muscle interface. The difference between the patterns in the cylindrical tissues and those illustrated for the plane layers indicates the importance of tissue curvature when assessing the effectiveness and safety of devices designed for medical application of microwave energy.

VII. CONCLUSIONS

It has been only within the last decade of the nine decades that man has been concerned with the biological effects and medical applications of electromagnetic fields, that significant strides have been made in quantifying the interaction of the electromagnetic fields and the tissues of living subjects. This is not to say that the techniques previously did not exist for doing this, but engineers and physicists qualified to carry out such work have been involved more with the physical sciences and only recently have turned their attention to problems in the life sciences. Through the theoretical and experimental approaches recently applied to the bioelectromagnetics problem area, we now have a better understanding of the absorption characteristics of energy in the bodies and tissues of man and animals exposed to electromagnetic fields. With crude mathematical and experimental models, we have gained considerable insight and understanding of the SAR patterns in biological subjects exposed under a number of different conditions. Considerable more work, however, needs to be done to improve our understanding of electromagnetic field coupling in the body of man under more common everyday exposure conditions. We must especially characterize the SAR distributions in man that result from partial-body near-zone field exposures encountered in our daily lives, such as with the operation of hand-held radios and transmitting equipment in automobiles, boats, and aircraft, while in an environment of conducting or partially-conducting bodies. We need to develop mathematical and experimental models of more complex geometries to better characterize the true nature of living subjects. We must place more consideration into modeling the more inhomogeneous nature of tissues, as well as anisotropic characteristics of muscle, the circulatory system, and various organs. Models used for assessing the SAR in exposed man, especially at frequencies above 800

MHz, must include considerably more detail to include the effect of clothing, skin, and subcutaneous fat, since the peripheral characteristics of the body play an increasingly important part in controlling the absorption characteristics with increasing frequency above 800 MHz. Present international safety standards have a poor physical basis and are much too conservative over some portions of the spectrum while being outright dangerous over other parts of the spectrum. Improvement of these standards requires considerably more quantitation of the absorption characteristics in laboratory animals and man under a myriad of different exposure conditions.

REFERENCES

1. Licht, S., (ed.), Therapeutic Heat and Cold, Licht, New Haven, Conn., 1965.

2. Williams, N.H., "Production and absorption of electromagnetic waves from 3 cm. to 6 mm. in length," J. Appl. Phys. 8: 655, October, 1937.

3. Southworth, G.C., "New experimental methods applicable to ultra short waves," J. Appl. Phys. 8:660, October, 1937.

4. Hollman, H.E., "Das problem der behandlung biologisher korper in ultrakurz-wellen-strahlungsfel," in Ultrakurz-wellen in Ihren Medizinishe-biologischen Anwendungen. Leipzig, Germany: Thiem, 1938, sec. 4, pp. 232-249.

5. Hollman, H.E., "Zum problem der ultra kurzwellen-behandlung durch anstrahlung," Strahlentherapie, 64:691-702, 1939.

6. Hemingway, A., and K.W. Stenstrom, "Physical characteristics of shortwave diathermy," Handbook of Physical Therapy. Amer. Med. Assoc. Press, Chicago, IL, 1939, pp. 214-229.

7. Danilewsky, B., and A. Worobjew, "Uber die Fernwirkung elektrischer Hochfrequenzstrome auf die Nerven," Arch. f. d. Physiol. 236:443, 1935.

8. Audiat, J., "Action des ondes hertziennes sur l'excitabilite electrique des nerfs: (ondes amorties, entretenues, courtes)," Rev. d'actionol. 8:227, May-June, 1932. Abstract: Compt. rend. Soc. de Biol. 110:876, July 19, 1932.

9. Delherm, L., and H. Fischgold, "Le courant de d'Arsonval diminuent l'excitabilite neuromusculaire," Comp. rend. Acad. d. sc. 199:1688, December 26, 1934.

10. Weissenberg, E., Soc. Francaise d'Electrotherapie et de Radiologie 10:535, December, 1935.

11. Pflomm, E., "Experimentelle und Klinische Untersuchungen uber die Wirkung Ultrakurzer elektrischer Wellen auf die Entzundung," Arch. f. flin. Chir. 166:251, 1931.

12. Hill, L., and H.J. Taylor, "Effect of high-frequency field on some physiologic preparations," Lancet 1:311, February 8, 1936.

13. Mittlemann, E., S.L. Osborne, and J.S. Coulter, "Short wave diathermy power absorption and deep tissue temperature," Arch. Physical Therapy 22:133-139, March, 1941.

14. Osborne, S., and J. Coulter, "Thermal effects of short wave diathermy on bone and muscle," Arch. Phys. Med. pp. 281-284, May, 1938.

15. Johnson, H., "The action of short radio waves on tissues. III. A comparison of the thermal sensitivities of transplantable tumours in vivo and in vitro," Am. J. Cancer 38: 533-550, 1940.

16. Krusen, F.H., J.F. Herrick, U. Leden, and K.G. Wakim, "Microkymatotherapy: preliminary report of experimental studies of the heating effect of microwaves (radar in living tissues," Proc. Staff Meeting Mayo Clin. 22:209-224, 1947.

17. Leden, U.M., J.F. Herrick, K.G. Wakim, and F.H. Krusen, Preliminary studies on the heating and circulating effects of microwaves (radar)," Brit. J. Phys. Med. 10:177-184, 1947.

18. Worden, R.E., J.F. Herrick, K.G. Wakim, and F.H. Krusen, Arch. Phys. Med. 29:751, 1948.

19. Siems, L.L., A.J. Kosman, and S.L. Osborne, Arch. Phys. Med. 29:759, 1948.

20. Richardson, A.W., T.D. Duane, and H.M. Hines, Arch. Phys. Med. 29:765, 1948.

21. Schwan, H.F., and G.M. Piersal, "The absorption of electromagnetic energy in body tissues, Part I," Amer. J. Phys. Med. 33:371-404, 1954.

22. Schwan, H.P., and G.M. Piersal, "The absorption of electromagnetic energy in body tissues, Part II," Amer. J. Phys. Med. 34:425-448, 1955.

23. Schwan, H.P., "Electrical properties of tissues and cells," Advan. Biol. Med. Phys. 5:147-209, 1957.

24. Schwan, H.P., "Survey of micorwave absorption characteristics of body tissues," in Proc. 2nd Tri-Serv. Conf. Biol. Effects of Microwave Energy, 1958, pp. 126-145.

25. Schwan, H.P., "Alternating current spectroscopy of biological substances," Proc. IRE 47:1841-1855, November, 1959.

26. Lehmann, J.F., A.W. Guy, V.C. Johnson, G.D. Brunner, and J.W. Bell, "Comparison of relative heating patterns produced in tissues by exposure to microwave energy at frequencies of 2450 and 900 megacycles," Arch. Phys. Med. 43:69-76, February, 1962.

27. Lehmann, J.F., J.A. McMillan, G.D. Brunner, and A.W. Guy, "A comparative evaluation of temperature distributions produced by microwaves at 2456 and 900 megacycles in geometrically complex specimens," Arch. Phys. Med. 43:502-507, 1962.

28. Lehmann, J.F., et al., "Comparison of deep heating by microwaves at frequencies 2456 and 900 megacycles," Arch. Phys. Med. 46:307-314, April, 1965.

29. Lehmann, J.F., A.W. Guy, C.G. Warren, B.J. DeLateur, and J.B. Stonebridge, "Evaluation of a microwave contact applicator," Arch. Phys. Med. 51:143-147, 1970.

30. Guy, A.W., and J.F. Lehmann, "On the determination of an optimum microwave diathermy frequency for a direct contact applicator," IEEE Trans. Biomed. Eng. BME-13:76-87, April, 1966.

31. DeLateur, B.J., J.F. Lehmann, J.B. Stonebridge, C.G. Warren, and A.W. Guy, "Muscle heating in human subjects with 915 MHz microwave contact applicator," Arch. Phys. Med. 51:147-151, March, 1970.

32. Guy, A.W., "Electromagnetic fields and relative heating patterns due to a rectangular aperture source in direct contact with bilayered biological tissue," IEEE Trans. Microwave Theory Tech. (special issue on Biological Effects of Microwaves) MTT-19:214-233, February, 1971.

33. Guy, A.W., "Analyses of electromagnetic fields induced in biological tissues by thermographic studies on equivalent phantom models," IEEE Trans. Microwave Theory Tech. (special issue on Biological Effects of Microwaves) MTT-19:205-214, February, 1971.

34. Satio, M., et al., "R-F-field-induced forces on microscopic particles,' Dig. Int. Conf. Med. Electronics 21:3, 1961.

35. Satio, M., et al., "The time constants of pearl chain formation," in Biological Effects of Microwave Radiation, vol. 1, Plenum, New York, 1961, p. 85.

36. Furedi, A., and R. Valentine, "Factors involved in the orientation of microscopic particles in suspensions influenced by radio-frequency fields," Biochim. Biophys. Acta. 56:33, 1962.

37. Furedi, A., and I. Ohad, "Effects of high-frequency electric fields on the living cell," Biochim. Biophys. Acta 79:1, 1964.

38. Schwan, H.P., "Biophysics of diathermy," in Therapeutic Heat and Cold, S. Licht, Ed. Licht, New Haven, Conn., 1965, sec. 3, pp. 63-125.

39. Anne, A., "Scattering and absorption of microwaves by dissipative dielectric objects: the biological significance and hazard to mankind," Ph.D. dissertation, Univ. of Pennsylvania, Philadelphia, PA, 106 p., Cont. NONR 551505, ASTIA Doc. 408 997, 1963.

40. Anne, A., M. Satio, O.M. Salati, and H.P. Schwan, "Penetration and thermal dissipation of microwaves in tissues," Univ. of Pennsylvania, Philadelphia, PA, Tech. Rep. RADC-TDR-62-244. Cont. AF 3-(602)-2344, ASTIA Doc. 284 981, 1962.

41. Anne, A., M. Satio, O.M. Salati, and H.P. Schwan, "Relative microwave absorption cross sections of biological significance," in Biological Effects of Microwave Radiation, vol. 1, Plenum, New York, 1960, pp. 153-176.

42. Guy, A.W., J.F. Lehmann, J.A. McDougall, and C.C. Sorensen, "Studies on therapeutic heating by electromagnetic energy," in Thermal Problems in Biotechnology, ASME, New York, 1968, pp. 26-45.

43. Guy, A.W., "Analyses of electromagnetic fields induced in biological tissues by thermographic studies on equivalent phantom models," IEEE Trans. Microwave Theory Tech. (special issue on Biological Effects of Microwaves) MTT-19:205-214, February, 1971.

44. Ho, H.S., A.W. Guy, R.A. Sigelmann, and J.F. Lehmann, "Microwave heating of simulated human limbs by aperture

sources," IEEE Trans. Microwave Theory Tech. MTT-19:224-231, February, 1971.

45. Johnson, C.C., and A.W. Guy, "Nonionizing electromagnetic wave effects in biological materials and systems," Proc. IEEE 60:692-718, June, 1972.

46. Guy, A.W., J.F. Lehmann, and J.B. Stonebridge, "Therapeutic applications of electromagnetic power," Proc. IEEE 62:55-75, January, 1974.

47. Guy, A.W., "Quantitation of induced electromagnetic field patterns and associated biological effects," in <u>Biologic Effects and Health Hazards of Microwave Radiation</u>, Proc. of an International Symp., P. Czerski, ed., Polish Medical Pub., Warsaw, 1974, pp. 203-216.

48. Shapiro, A.R., R.F. Lutomirski, and H.T. Yura, "Induced fields and heating within a cranial structure irradiated by an electromagnetic plane wave, IEEE Trans. Microwave Theory Tech. (special issue on Biological Effects of Microwaves) MTT-19:187-196, February, 1971.

49. Ho, H.S., and A.W. Guy, "Development of dosimetry for RF and microwave radiation. II: Calculations of absorbed dose distributions in two sizes of muscle-equivalent spheres," Health Physics 29:317-324, August, 1975.

50. Durney, C.H., C.C. Johnson, and H. Massoudi, "Long wavelength analysis of plane-wave irradiation of a prolate spheroid model of man," IEEE Trans. Microwave Theory Tech. 23:246, 1975.

51. Johnson, C.C., C.H. Durney, and H. Massoudi, "Long wavelength electromagnetic power absorption in prolate spheroidal models of man and animals," IEE Trans. Microwave Theory Tech. MTT-23:739, 1975.

52. Massoudi, H., C.H. Durney, and C.C. Johnson, "Long-wavelength electromagnetic power absorption in ellipsoidal models of man and animals," IEEE Trans. Microwave Theory Tech. MTT-25:47, 1977.

53. Massoudi, H., C.H. Durney, and C.C. Johnson, "Long-wavelength analysis of plane-wave irradiation of an ellipsoidal model of man," IEEE Trans. Microwave Theory Tech. MTT-25:41, 1977.

54. Gandhi, O.P., "Frequency and orientation effect on whole animal absorption of electromagnetic waves," IEEE Trans.

Biomed. Eng. BME-22:536, 1975.

55. Allen, S.J., C.H. Durney, C.C. Johnson, and H. Massoudi, Comparison of Theoretical and Experimental Absorption of Radiofrequency Power, Report No. SAM-TR-75-52, prepared by the University of Utah for USAF School of Aerospace Medicine, Brooks Air Force Base, Texas, 1975.

56. Durney, C.H., et al., Radiofrequency Radiation Dosimetry Handbook, Second Ed., Report No. SAM-TR-78-22, prepared by the University of Utah for USAF School of Aerospace Medicine, Brooks Air Force Base, Texas, 1978.

57. Johnson, C.C., and T.C. Rozzell, "Liquid crystal fiberoptic RF probes, Part I: Temperature probe for microwave fields," Microwave J. 18(8), 1975.

58. Bowman, R., "A probe for measuring temperature in radiofrequency-heated material," IEEE Trans Microwave Theory Tech. 1976.

59. Cetas, T.C., "A birefringent crystal optical thermometer for measurements of electromagnetically-induced heating," presented at the USNC/International Union of Radio Sciences Annual Meeting, 1975.

60. Barber, P.W., "Electromagnetic power deposition in prolate spheroid models of man and animals at resonance," IEEE Trans. Biomed. Eng. BME-24, 1977.

61. Chen, K.M., and G.S. Guru, "Internal EM field and absorbed power density in human torsos induced by 1-500 MHz EM waves," IEEE Trans. Microwave Theory Tech. MTT-25:746, 1977.

62. Hagmann, M.J., O.P. Gandhi, and C.H. Durney, "Numerical calculations of electromagnetic energy deposition in a realistic model of man," 1977 International Symp. on Biological Effects of Electromagnetic Waves, Oct. 30-Nov. 4, Airlie, VA, p. 55, 1977 (abstract).

63. Emery, A.F., P. Kramar, A.W. Guy, and J.C. Lin, "Microwave induced temperature rise in rabbit eyes in cataract research," J. Heat Trans. 97:123-128, February, 1975.

64. Emery, A.F., R.E. Short, A.W. Guy, K.K. Kraning, and J.C. Lin, "The numerical thermal simulation of the human body when undergoing exercise or nonionizing electromagnetic radiation," J. Heat Trans. 98:284-291, May, 1976.

65. Foster, K.R., and E.E. Finch, "Microwave hearing: evidence for thermoacoustical auditory stimulation by pulsed microwaves, Science 185:256-258, 1974.

66. Guy, A.W., C.K. Chou, J.C. Lin, and D. Christensen, "Microwave induced acoustic effects in mammalian auditory systems and physical materials, Ann. N.Y. Acad. Sci. 247:194-218, 1975.

67. Lin, J.C., Microwave Auditory Effects and Applications, Charles C. Thomas Publisher, Springfield, IL, 1978, 221 p.

68. Chou, C.K., R. Galambos, A.W. Guy, and R.H. Lovely, "Cochlea microphonics generated by microwave pulses," J. Microwave Power 10:361-367, 1975.

69. Vogelhut, P.O., "Interaction of microwave and radio frequency radiation with molecular systems," in Biological Effects and Health Implications of Microwave Radiation, (Proc. of symposium sponsored by Med. College of Virginia and Bureau of Radiological Health., Richmond, VA, Sept. 17-19, 1969), S.F. Cleary, ed., Bureau of Rad. Health Technical Report PB 193 898, June, 1970, pp. 98-100.

70. Grodsky, I.T., "Possible physical substrates for the interaction of electromagnetic fields with biologic membranes," Ann. N.Y. Acad. Sci. 247:117-123, 1975.

71. Fröhlich, H., "Long-range coherence and energy storage in biological systems. Int. J. Quant. Chem. 2:641-649, 1968.

72. Cook, H., "The dielectric behavior of some types of human tissues at microwave frequencies," Brit. J. Appl. Phys. 2:295, 1951.

73. Cook, H., "Dielectric behavior of human blood at microwave frequencies," Nature 168:247, 1951.

74. Cook, H., "A comparison of the dielectric behavior of pure water and human blood at microwave frequencies," Brit. J. App. Phys. 3:249, 1952.

75. Cole, K., and R. Cole, "Dispersion and absorption in dielectrics," J. Chem. Phys. 9:34, 1941.

76. Lin, J.C., A.W. Guy, and C.C. Johnson, "Power deposition in a spherical model of man exposed to 1-20 MHz EM fields," IEEE Trans. Microwave Theory Tech. MTT-21:791-797, December, 1973.

77. Allen, S.J., W.D. Hurt, J.H. Krupp, J.A. Ratliff, C.H. Durney, and C.C. Johnson, "Measurement of Radiofrequency Power Absorption in Monkeys, Monkey Phantoms, and Human Phantoms Exposed to 10-50 MHz Fields, Report No. SAM-TR-76-5 (USAF School of Aerospace Medicine, Brooks Air Force Base, TX), 1976.

78. Rowlandson, G.I., and P.W. Barber, "RF-energy absorption by biological models: calculations based on geometrical optics," in Abstracts of scientific papers, International Symposium on the Biological Effects of Electromagnetic Waves, Airlie, VA, Oct. 30-Nov. 4, 1977, p. 50.

79. Massoudi, H., C.H. Durney, and C.C. Johnson, "Geometrical-optics and exact solutions for internal fields and SAR's in a cylindrical model of man as irradiated by an electromagnetic plane wave, in Abstracts of scientific papers, 1977 International Symp. on the Biological Effects of Electromagnetic Waves, Oct. 30-Nov. 4, Airlie, VA, 1977, p. 49.

80. Gandhi, O.P., E.L. Hunt, and J.A. D'Andrea, "Deposition of electromagnetic energy in animals and in models of man with and without grounding and reflector effects, Radio Science 12(6S):39-47, 1977.

THEORETICAL AND NUMERICAL METHODS OF QUANTITATION OF
ABSORPTION PATTERNS IN MAN AND ANIMAL BODIES

C. Yeh

Electrical Sciences and Engineering Department
University of California, Los Angeles
Los Angeles, California 90024 USA

ABSTRACT. Predicting theoretically the total electromagnetic power absorption and distribution of absorbed power in man and animal bodies is one the most challenging and important problems in electromagnetic biological effects research. This lecture will start with a survey of available analytical/numerical techniques which may be used to treat problems dealing with the interaction of electromagnetic waves and nonspherical three-dimensional dielectric bodies.[1] Then, the two most promising methods will be discussed in detail together with several illustrations. Finally, a discussion will be given on the future research prospects in this area.

I. INTRODUCTION

One of the most serious and challenging problems in electromagnetic biological effects research is that of theoretically predicting the total electromagnetic power absorption and distribution of absorbed power in man and animal bodies. Because of the complexities of these bodies, greatly simplified models must be introduced to approximate them.[2] Even using these simple models, a great deal of analytical and numerical difficulties is still encountered in determining the distribution of absorbed power.

This lecture will first present a brief survey on available analytical techniques which may be used to treat problems dealing with the interaction of electromagnetic waves and nonspherical three-dimensional dielectric bodies. Comments will be given on the prospect of each method in providing accurate results for models of man and animals. Two most promising methods, the extended boundary condition method and the global local finite element method will be discussed in detail. Then several numerical examples on the distribution of absorbed power in several most sophiticated models of man and animals will be presented. Finally, we shall speculate on the future developments in this area.

II. BACKGROUND

The classical solutions of electromagnetic wave scattering are limited to simple objects of separable boundaries such as spheres, cylinders, etc. Recent developments in scattering analysis for scatterers of arbitrary shapes includes the geometric theory of diffraction for high frequency scattering[3]; the method of moment[4]; the perturbation method[5]; the analytic continuation[6]; the point matching method,[7,8] the extended boundary conditions method[9]; and the finite element[10,11] for resonant region scattering. We shall briefly describe those methods in the following.

(1) Geometric Theory of Diffraction

This method is mainly a high frequency technique which treats the smooth part of the scatterer by geometric optics and obtains the scattered fields from edges and glazing surfaces via solutions of canonical problems such as scattering by wedges, tips, etc. The geometric theory of diffraction can sometimes give pretty reasonable solutions even at resonant frequencies, especially when it is combined with the method of moment.[4] But, still it is a technique most conveniently applied to convex and perfectly conducting targets. When the targets are concave or transparent, tracing the multiple scattering rays makes the method overly cumbersome to use. So far, no one has yet successfully applied GTD to scattering by dielectric bodies.

(2) The Method of Moment

The method of moment has been extensively utilized by electromagnetic engineers since it was popularized by Harrington[4]. We shall list some of the highlights of the method of moment as follows:

(a) It is by far the most efficient method to analyze thin and long metal scatterers.

(b) It reduces to a surface integral equation if the target is a perfectly conducting body.

(c) It reduces to two coupled surface integral equation if the target is a piece of homogeneous dielectric body.

(d) It is a volume integral equation if the target is inhomogeneous.

However, when the scatterer is a dielectric volume greater than $0.1 \lambda^3$, the demand of the method of moment for computer memory or time becomes excessive.

(3) Perturbation Method

The perturbation method is an extension of Mie scattering to scatterers of non-spherical geometry. Oguchi[5] successfully applied the perturbation technique to scattering by oblate spheriodal rain drops. The technique is inherently limited to particle shapes which are small perturbations of a sphere. Even at that, the computations are rather complicated. It would be indeed an impossible task for the perturbation technique to handle particle shape more complicated than an oblate spheroid.

(4) Analytic Continuation

The method of analytic continuation was developed by Wilton and Mittra[6]. This method uses harmonic expansions at different origins for the scattered fields at various points of the scatterer boundary. These expansions are analytically continued in terms of harmonic expansions at one special origin by means of the addition theorem. Each analytic continuation results in a linear equation of a truncated series of scattering coefficients, which are solved numerically when the number of linear equations equals or exceeds the number of coefficients. While the authors of the technique claim that the method should apply to scattering by large scatters in 2 or 3 dimensional problems, the sample problems presented in their paper[6] involve very small scatterers in two dimensions. In fact, the analytic continuation involves lengthy computations of the addition formulas of the harmonics which converges slowly when the obstacle is large. Furthermore, the technique does not give any rule of choosing the optimum locations of expansion points and therefore discourages systematic application to large scale problems.

(5) Point Matching Method

The point matching method is an obvious numerical technique in solving nonseparable problems using spherical harmonics. The method has recently been applied by Oguchi[6], and Morrison and Cross[7] to solve oblate spheroidal rain drops. Conceptually the technique is rather straight-forward but the implementation is by no means simple. In particular it involves inversion of large matrices. The point matching procedures used by Oguchi and Morrison both involving Bessel's functions within the drop and Hankel's functions outside the drop, and continuity conditions are enforced at N points on the surface of the drop. This procedure is only approximate, in that the spherical harmonic expansions using Hankel's functions are only valid outside a circumscribing sphere of the rain drop. The rigorous application of the point matching method requires the division of the solutions in three parts. One which is valid inside the drop by using Bessel's functions. One part which is valid between the drop and the circumscribing sphere by using a combination of Bessel's functions and Hankel's function and another part which is valid outside the circumscribing sphere by using Hankel's functions. The boundary conditions are enforced on two surfaces, namely, the surface of the drop and the surfaces of the circumscribing sphere. The approaches used by Oguchi and Morrison are essentially Rayleigh's hypothesis which has been under various criticism in recent years. We contend that Oguchi's results using the Rayleigh's hypothesis is valid for oblate spheroidal of axial ratio 2:3 or less and for convex bodies. But, for spheroidals of higher oblateness or for concave dielectric bodies, the rigorous method has to be used. Using the rigorous point matching technique and depriving it of one degree of symmetry for non-spheroidal particle shape, the number of points to match will be 4 times greater than those needed by Oguchi and Morrison, and the computer time to solve a single particle will be 64 times longer, not to speak of the added complexity in programming resulting from the improved rigor and improved geometries. Therefore, the straightforward point matching technique is not a potentially promising approach to solve the atmosphere particle problem.

(6) Extended Boundary Condition Method

The extended boundary condition method was first suggested by Waterman.[12] It is essentially an integral equation technique. But, instead of point matching or weighting integration on the surface of the obstacle, it integrates over a sphere inside the obstacle. When the fields are expressed in terms of spherical harmonics this technique offers great advantage in reducing the size of the matrix. This method has been applied by Barber and Yeh[9] to solve scattering by arbitrarily shaped dielectric bodies. However, this method is limited to homogeneous or layered dielectric scatterers.

(7) Finite Elements Method

In recent years the field of finite elements has been expanding at an unprecedented rate, especially in structures and mechanics.[13] Some of the earlier "unsolvable" problems have been successfully attacked by the finite elements method. The use of this technique in electromagnetics is still in its infancy, however. That the problems dealing with arbitrarily-shaped inhomogeneous optical fiber or integrated optical waveguides can be solved by the finite elements technique has only been demonstrated by Yeh, Dong and Oliver[10] in 1975. Mei[11] also advocated the use of the finite-element method to solve electromagnetic problems and recently solved the biconical antenna with inhomogeneous dielectric loading problems using this method. As shown by Yeh et al,[10] in their treatment of the guided wave problem, this finite elements technique is very powerful in that it can solve problems dealing with bodies of complex shapes as well as those of general inhomogeniety.

(8) Conclusion

To summarize the above discussion on methods of solving scattering problems, we list the merits and limitations of each method as follows:

  (1) Geometric Theory of Diffraction
      Advantages – Fast, clear physical concept
      Disadvantages – Difficult to apply to concave
                     and dielectric targets

  (2) Method of Moment
      Advantages – Easy to implement, adaptable to
                  arbitrary geometry
      Disadvantages – Limited to targets of small
                     volume

  (3) Perturbation Method
      Advantages – Can be used without a high speed
                  digital computer
      Disadvantages – Limited to bodies which are
                     small perturbations of a
                     sphere

(4) Analytic Continuation
   Advantages – Not clear
   Disadvantages – Cannot be systematically
                    applied

(5) Point Matching Method
   Advantages – Conceptually simple
   Disdvantages – Inaccurate, demands excessive
                    computer time and memory

(6) Extended Boundary Condition Method
   Advantages – Adaptable to targets of large
                 volume and arbitrary shape
   Disadvantages – Difficult to apply to inhomo-
                    geneous dielectric scatterer

(7) Finite Element Method
   Advantages – Adaptable to targets of large
                 volume and to targets of inhomo-
                 geneous material
   Disadvantages – Requires extensive analytical
                    and computational skill

III. TECHNICAL APPROACH

Based on the assessment given in the previous section, it appears that the extended boundary condition method or the finite element method has the greatest potential in solving scattering problems dealing with irregularly shaped bodies, while the finite element method may also be used to solve scattering problems dealing with inhomogeneous bodies. In the following, we shall present the basic formulation for the extended boundary condition method and the Global-Local Finite Element (GLFE) Method as applied to the problem of scattering by inhomogeneous dielectric bodies which may model men or animals.

(1) Extended Boundary Condition Method

The overall goal is to determine the scattered field and the penetrated field when an arbitrary dielectric body is illuminated by a plane electromagnetic wave.

The dielectric body, assumed homogeneous and isotropic, is characterized by the constitutive parameters μ_o, ε where μ_o is the free space permeability and ε is the permittivity of the dielectric material (ε may be complex to account for the lossy case). The surrounding medium is considered to be free space with parameters μ_o, ε_o. For the given incident field, the scattered field must be determined for all locations P on a sphere surrounding the scattering object.

The theoretical development is as follows:[9]

First, the Equivalence Principle is applied, breaking the scattering problem into two separate problems, an exterior part and an interior part. One equation for the unknown field quantities is derived from the external problem where it is found that the scattered field due to the surface currents must completely cancel the incident field throughout the interior volume.

The internal problem is then considered and the fields within the dielectric region are expanded in regular vector spherical wave functions with coefficients to be determined. Superposition is applied and the boundary conditions at the surface lead to a linear system of integral equations for the coefficients of the unknown surface fields in terms of the incident field.

The scattered far field is then determined by evaluating the internal field at the surface and substituting into the original expression which gives the exterior scattered field in terms of the surface fields. The differential scattering cross section is proportional to the square of the far field amplitude.

To avoid writing many lengthy mathematical expression detailed derivation will not be given in this paper. We shall simply state the final results:

$$(K + \sqrt{\varepsilon_r} J) c_\mu + (L + \sqrt{\varepsilon} I) d_\mu = -ja_\nu$$
$$\nu = 1, 2, \ldots, N \quad (1)$$
$$(I + \sqrt{\varepsilon_r} L) c_\mu + (J + \sqrt{\varepsilon_r} K) d_\mu = -jb_\nu$$

where

$$I = \frac{k^2}{\pi} \int_S \bar{n} \cdot \bar{M}_\nu^3 (k\bar{r}') \times \bar{M}_\mu^1 (k'\bar{r}') dS$$

$$J = \frac{k^2}{\pi} \int_S \bar{n} \cdot \bar{M}_\nu^3 (k\bar{r}') \times \bar{N}_\mu^1 (k'\bar{r}') dS$$

$$K = \frac{k^2}{\pi} \int_S \bar{n} \cdot \bar{N}_\nu^3 (k\bar{r}') \times \bar{M}_\mu^1 (k'\bar{r}') dS$$

$$L = \frac{k^2}{\pi} \int_S \bar{n} \cdot \bar{N}_\nu^3 (k\bar{r}') \times \bar{N}_\mu^1 (k'\bar{r}') dS$$

and $\varepsilon_r = \frac{\varepsilon}{\varepsilon_0}$. $\bar{n}$ is a unit vector normal to the boundary surface S and r' is the radial vector from the origin to the boundary surface. a_ν and b_ν are expansion coefficients for the incident wave defined as follows:

$$\bar{E}^i(\bar{r}) = \sum_{\nu=1}^{\infty} D_\nu [a_\nu \bar{M}_\nu^1(k\bar{r}) + b_\nu \bar{N}_\nu^1(k\bar{r})] \tag{2}$$

where ν is a combined index incorporating σ, m and n. D_ν is a normalization constant

$$D_\nu = \varepsilon_m \frac{(2n+1)(n-m)!}{4n(n+1)(n+m)!}, \quad \varepsilon_m = \left. \begin{matrix} 1 \\ 2 \end{matrix} \right\}, \quad m \begin{matrix} = 0 \\ > 0 \end{matrix}$$

The spherical wave harmonics $\bar{M}$ and $\bar{N}$ have been defined by Stratton. These functions, solutions of the vector wave equations are given by,

$$\bar{M}_{\sigma mn}^{1,3}(\bar{r}) = \nabla \times \bar{r} \begin{matrix} \cos m\phi \\ \sin m\phi \end{matrix} P_n^m(\cos\theta) z_n^{1,3}(kr) \tag{3}$$

$$\bar{N}_{\sigma mn}^{1,3}(\bar{r}) = \frac{1}{k} \nabla \times \bar{M}_{\sigma mn}^{1,3}(\bar{r})$$

where σ = even or odd, $P_n^m(\cos\theta)$ = associated Legendre function, and $z_n^{1,3}(kr)$ = an appropriate spherical Bessel function. For solutions of the wave equation which must be finite at $r = 0$, $z_n^1(kr) = j_n(kr)$ and the resulting vector spherical functions $\bar{M}$ and $\bar{N}$ are known as solutions of the first kind. The other vector spherical functions we will be using are obtained by using the spherical Hankel functions, i.e., $z_n^3(kr) = h_n^{(1)}(kr) = j_n(kr) + jn_n(kr)$. This solution, representing outgoing waves, is called the solution of the third kind.

c_μ and d_μ are the expansion coefficients for the field inside the dielectric which are defined as follows:

$$\bar{E}(k'\bar{r}) = \sum_{\mu=1}^{N} [c_\mu \bar{M}_\mu^1(k'\bar{r}) + d_\mu \bar{N}_\mu^1(k'\bar{r})] \tag{4}$$

where μ incorporates the indices σ, m, n and c_μ and d_μ are unknown coefficients. $k' = \sqrt{\epsilon_r} k$ and $k = \omega\sqrt{\mu_0\epsilon_0}$. The H field internal to S is given by

$$\bar{H}(k'\bar{r}) = \frac{1}{j\omega\mu_0} (\nabla \times \bar{E}(k'\bar{r}))$$

$$= \frac{1}{j\omega\mu_0} \sum_{\mu=1}^{N} [c_\mu(\nabla \times \bar{M}_\mu^1(k'\bar{r}) + d_\mu(\nabla \times \bar{N}_\mu^1(k'\bar{r}))] \quad (5)$$

$$= -j\frac{\epsilon}{\mu_0} \sum_{\mu=1}^{N} [c_\mu \bar{N}_\mu^1(k'\bar{r}) + d_\mu \bar{M}_\mu^1(k'\bar{r})] .$$

The coefficients of the scattered fields which are defined as follows:

$$\bar{E}^s(k\bar{r}) = \sum_{\nu=1}^{N} [p_\nu \bar{M}_\nu^3(k\bar{r}) + q_\nu \bar{N}_\nu^3(k\bar{r})] \quad (6)$$

where r is outside a circumscribed sphere, are

$$p_\nu = -jD_\nu \sum_{\mu=1}^{N} \{[K' + \sqrt{\epsilon_r}J'] c_\mu + [L' + \sqrt{\epsilon_r}I'] d_\mu\}$$
$$q_\nu = -jD_\nu \sum_{\mu=1}^{N} [I' + \sqrt{\epsilon_r}L'] c_\mu + [J' + \sqrt{\epsilon_r}K'] d_\mu\} \quad (7)$$

where

$$I' = \frac{k^2}{\pi} \int_S \bar{n}\cdot\bar{M}_\nu^1(k\bar{r}') \times \bar{M}_\mu^1(k'\bar{r}') dS$$

$$J' = \frac{k^2}{\pi} \int_S \bar{n}\cdot\bar{M}^1(k\bar{r};) \times \bar{N}_\mu^1(k'\bar{r}') dS$$

$$K' = \frac{k^2}{\pi} \int_S \bar{n}\cdot\bar{N}^1(k\bar{r}') \times \bar{M}_\mu^1(k'\bar{r}') dS$$

$$L' = \frac{k^2}{\pi} \int_S \bar{n}\cdot\bar{N}^1(k\bar{r}') \times \bar{N}^1(k'\bar{r}') dS.$$

The basic equations which must be solved are given in the previous section. Here we shall briefly outline the steps that we propose to do to convert the theoretical equations to a form suitable for numerical solution on the digital computer.

The exact theoretical problem requires the solution of an infinite set of equations. (The spherical expansions for the various field are all infinite term series.) The equations resulting from the External Problem are such a set. The first concession to practical considerations was made in the derivation of the Internal Problem, where the infinite term spherical expansion for the internal field was trunacatet to N terms. This finite internal field expansion was then substituted into the first 2N equations of the infinite set of Eqs. (1) and the final set of truncated equations was obtained. The scattering results will be obtained numerically by solving the complete set of equations for successive values of N until the final results (the differential scattering cross section) converge to a specified accuracy. This will insure that enough of the expansion terms have been kept to guarantee the correct final result.

The indices μ and ν each run from 1 to N are related to the internal and external field expansions, respectively. As indicated previously, they incorporate the indices σ, m and n where σ is even or odd and m and n are indices of Bessel and Legendre functions. To distinguish between expansions over ν and μ, the indices which they represent will be unprimed and primed, respectively, i.e.,

ν is over σ, m, n

μ is over σ', m', n'

where the expansion scheme is as follows,

σmn = e01, o01, e11, o11, e02, o02, e12, o12, e22, o22, ...

where e = even and o = odd.

There are essentially four main steps in the solution for the differential scattering cross section.

(a) Solution of the system of Eqs. (1) for the coefficients of the internal field c_μ and d_μ in terms of the incident field coefficients as a_μ and b_μ and S is the surface of the scattering object. Note that the solution of the Eqs. (1) will require a matrix inversion procedure.

(b) The internal field can now be computed using c_μ and d_μ.

(2) Global Local Finite Element Method

The term Global-Local Finite Element Method refers to numerical analysis technique in which both the contemporary finite element and classical Rayleigh-Ritz approximations are employed. This hybrid Ritz method not only preserves the finite element modeling capability but adds the advantage of using a priori information regarding the anticipated behavior to represent the total response in a given problem. As a result, substantially fewer degrees of freedom are required in comparison to a problem using finite elements only. This method was first suggested by Mote[14], who demonstrated its feasibility with beam and plate vibration examples. We intend to extend this technique to the EM scattering problems.

The scattering problem is divided into two regions: the exterior region and the interior region as shown in Figure 1. without any sacrifice of rigor we shall assume that the boundary surface between the exterior and interior regions is a sphere for three-dimensional problems and a circular cylinder for two-dimensional problems. It will further be assumed that the boundary surface is so chosen that the medium in the exterior region is homogeneous and that the inhomogenous irregularly-shaped absorbing object is contained within the interior region. Solutions of wave equations in the exterior homogeneous region are well-known. Therefore, knowing the interior fields at the boundary surface will provide the necessary information to solve the exterior fields by using the boundary conditions at the boundary surface. The finite-elements technique is used to find the interior field.

The interior fields must satisfy Maxwell's wave equation which are

$$\nabla \times \underline{E} = -j\omega\mu_0 \underline{H} \tag{8}$$

$$\nabla \times \underline{H} = j\omega\varepsilon(\underline{r})\underline{E} \tag{9}$$

where $\underline{E}$, $\underline{H}$ are the electromagnetic fields in the interior region and $\varepsilon(\underline{r})$ describes the complex non-uniform dielectric variation. It is noted that the irregular surface of the object can be taken into account by assuming that it is shaped as shown in Figure 1 and that it has an inhomogeneous dielectric medium. To make use of the finite-elements method, Maxwell's wave equations must be represented in a functional as follows

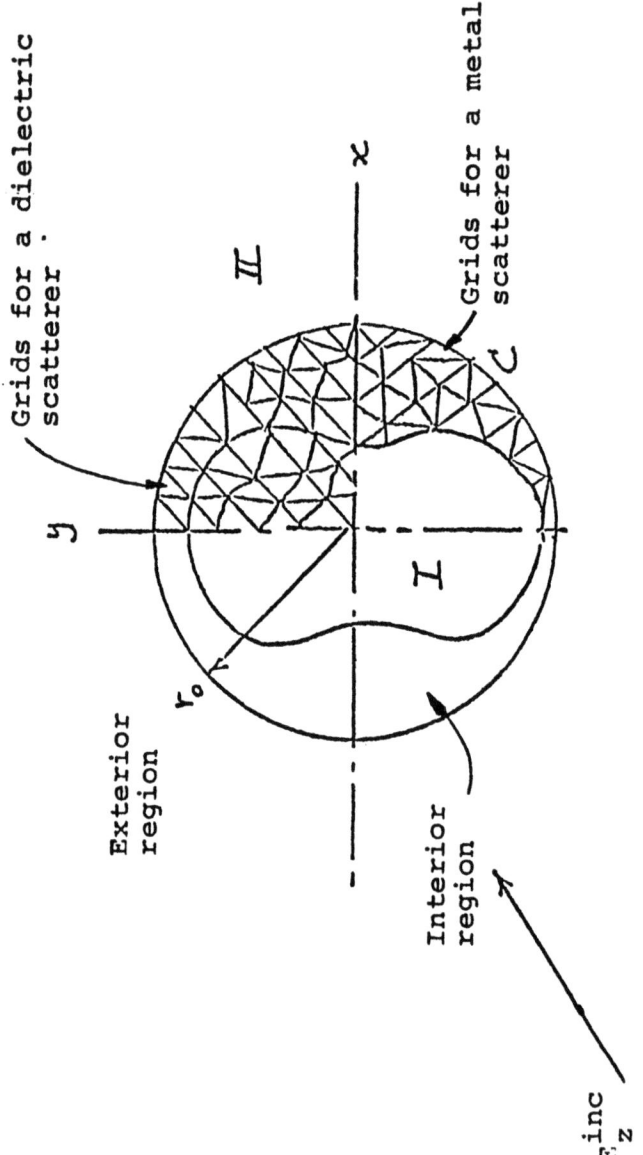

Figure 1. Discretization of the scatterer for the GLFE method.

$$I = \int_S f(\underline{r}, \underline{E}, \underline{H}) \, dS. \tag{10}$$

For example, the functional for an axisymmetric inhomogeneous medium is

$$I = \sum_{p=1}^{P} \int_S \left[\frac{1}{2f_m} \left\{ \cdot \frac{\epsilon_p}{\epsilon_o} R \left[\left(\frac{\partial}{\partial R}(RE_\phi^{(p)}) \right)^2 \right. \right. \right.$$

$$\left. + \left(\frac{\partial}{\partial z}(RE_\phi^{(p)}) \right)^2 \right]$$

$$+ R \left[\left(\frac{\partial}{\partial R}(Z_o RH_\phi^{(p)}) \right)^2 + \left(\frac{\partial}{\partial z}(Z_o RH_\phi^{(p)}) \right)^2 \right]$$

$$+ 2m \left[\frac{\partial}{\partial R}\left(RE_\phi^{(p)}\right) \frac{\partial}{\partial z}\left(Z_o RH_\phi^{(p)}\right) \right.$$

$$\left. - \frac{\partial}{\partial z}\left(RE_\phi^{(p)}\right) \frac{\partial}{\partial R}\left(Z_o RH_\phi^{(p)}\right) \right]$$

$$\left. \left. - \frac{1}{2R} \left[\frac{\epsilon_p}{\epsilon_o} \cdot \left(RE_\phi^{(p)}\right)^2 \pm \left(Z_o RH_\phi^{(p)}\right)^2 \right] \right\} \right] dRdz \tag{11}$$

between them are added via a multiple scattering technique. The multiple scattering process is most conveniently accomplished if the scattered fields of each axially symmetric component of the scatterer can be expressed in spherical harmonics. It is then, the well known addition theorem can be used to compute the interactions. It should be noted that in the previous discussion, the GLFE method has been shown to yield the scattered fields in terms of spherical harmonics, thus the GLFE method is most conveniently applied to take advantage of the situation of symmetry in particle scattering. In other words, the CLFE methods can be used to synthesis a class of scatterers by a composition of a number of axially symmetric ones.

IV. APPLICATIONS - EXAMPLES

It would seem rather primitive indeed to model the complex human or animal body by a homogeneous prolate spheroid or even a layered prolate spheroid. The fact of the matter is that even for such smooth monotonous shapes very sophisticated technique must be used to obtain the distribution of absorbed microwave power in these models. Furthermore, this is but a necessary initial step that one must take in order to reach the goal of being able to predict the absorption characteristics of an actual complex biological body. Of special concern is the possible presence of "hot-spots" within a human body when illuminated by an incident microwave energy.

Several illustrative numerical examples for low as well as resonant frequency absorpion characteristics of prolate spheroidal models or block models of man and animal will be presented at the lecture.

V. CONCLUSIONS AND FUTURE RESEARCH

It would seem rather apparent from the above discussion that there exists an urgent need to provide a means of predicting the absorption characteristics of man. We have selected two most promising approaches to achieve our aim: The EBCM and the GLFEM. One is also aware that research in this area is still in its infancy. It is hope that this lecture will encourage future research interests.

Finally, it should be noted that successful development of the analytical/numerical technique will enhance greatly our ability to deal with problems in related areas, such as, light scattering by bacteria or virus, particle identification via light scattering, and microwave scattering by raindrops, to name only a few.

REFERENCES

1. Johnson, C.C., Radio Science 12, 349 (1977).

2. Barber, P. IEEE Trans. on Biomedical Engineering BME-25, 155 (1978).

3. Keller, J.B., "A Geometric Theory of Diffraction," Proceedings of Symposia in Applied Math., Vol. VIII, American Math. Society, pp. 27-52, 1958.

4. Harrington, R.F., Field Computation by Moment Method, New York, MacMillan, 1968.

5. Oguchi, T., Attenuation of Electromagnetic Wave due to Rain with Distorted Raindrops, J. Radio Res. Lab. (Jap.), 7(33), 467-485, 1960.

6. Wilton, D.R. and R. Mittra, "A New Numerical Approach to the Calculation of Electromagnetic Scattering Properties of Two Dimensional Bodies of Arbitrary Cross Section," IEEE Trans. Antennas and Propagation, Vol. AP-20, pp. 310-317, May 1972.

7. Morrison, J.A. and M.J. Cross, "Scattering of a Plane Electromagnetic Wave by Axisymmetric raindrops," B.S.T.J., Vol. 53, No. 6, pp. 995-1019, 1974.

8. Oguchi, T. "Attenuation and Phase Rotation of Radio Waves due to Rain: Calculations at 19.3 and 34.8 GHz," Radio Science, Vol. 8, No. 1, pp. 31-38, 1973.

9. Barber, P., and C. Yeh, "Scattering of Electromagnetic Waves by Arbitrarily Shaped Dielectric Bodies," Applied Optics, Vol. 14, No. 12, pp. 2864-2872, December 1975.

10. Yeh, C., S.B. Dong and W. Oliver, J., Appl. Phys. 46, 2125 (1975).

11. Mei, K.K., IEEE Trans. on Antennas and Propagation, Vol. AP-22, No. 6, pp. 760-766, Nov. 1974.

12. Waterman, P.C., "Matrix Formulation of Electromagnetic Scattering" Proceedings of IEEE, Special Issue on Radar Reflectivity, Vol. 53, No. 8 pp. 805-811.

13. Whiteman, J.R., Ed., "The Mathematics of Finite elements and Application," Academic Press, New York (1973).

14. Mote, C.D. Int. J. for Numerical Methods in Eng. $\underline{3}$, 565 (1971).

15. van de Hulst, H.C., <u>Light Scattering by Small Particles</u>, Wiley, New York, 1957, Chapter 17.

MEDICAL DIAGNOSIS AND IMAGING USING ELECTROMAGNETIC TECHNIQUES

M. F. Iskander and C. H. Durney

Department of Electrical Engineering
University of Utah
Salt Lake City, Utah 84112

Abstract

A brief review of the potential electromagnetic (EM) methods for medical diagnosis and imaging is presented. This includes electrical impedance methods, microwave methods for measuring lung water content, and electromagnetic imaging. The principles underlying the operation of each method are described along with comments about their adequacy for medical diagnosis. It is shown that while the microwave method was introduced to overcome some of the limitations of the electric impedance method, its potential to estimate the absolute value of the average lung water content still needs further research. Electromagnetic imaging, on the other hand, is expected to overcome some of the difficulties involved in the microwave method. The important theoretical and experimental results that identify the advantages and the limitations of each method are presented. Suggestions for future development and for possible extensions are discussed. Furthermore, other techniques such as microwave radiometry, the electromagnetic flowmeter, and use of microwave doppler radar to monitor arterial wall movement, are briefly mentioned.

I. Introduction

The clinical value of a diagnostic technique is evaluated on the basis of its simplicity, safety, convenience to the patient, and sensitivity for monitoring the early stages of the physiological changes. While it usually starts by a simple study illustrating the feasibility of a new technique, it is often followed by a long engineering struggle to improve the method's sensitivity and safety, and to reduce its complexity considerably. The EM

techniques are certainly not different. For example, the use of
the electrical impedance method for medical diagnostics goes back
to 1926, when Lambert and Gremels first used measurements of electrical resistance across the lung to monitor the development of
pulmonary edema [1]. The method was found to be most sensitive
in the earliest stages of edema formation, which is a decided
advantage in measuring pulmonary edema. Attemps to utilize this
procedure for *in vivo* measurements, however, have achieved marginal success because of the considerable reduction in sensitivity. In parallel with the struggle to reduce these sensitivity
limitations, some efforts were focused on developing microwave
techniques whereby radiation fields rather than conduction currents were used.

The first published suggestion that microwave techniques
might be used in the management of lung disease was made by
Süsskind in 1973 [2]. The idea has been extensively examined and
theoretically and experimentally evaluated by the group at the
University of Utah [3-7]. The initial experience with the microwave transmission method has shown promising results, particularly
in measuring small variations in lung water content. Attempts to
estimate the absolute value of lung water content have also been
reported. The numerical procedure employed for such calculations
takes into account the dimensions and the relative position of the
different organs in the section of the thorax through which most
of the microwave signal is transmitted. Difficulties, however,
were encountered in making an accurate and unique estimation of
the lung water content because the water distribution in the lung
is unknown and is generally affected by both the disease status
and gravity. Attempts to overcome these difficulties have led to
the consideration of electromagnetic imaging. It is shown in Section 4 that a nonlinear reconstruction procedure is required for
EM imaging. The development of such imaging algorithms, although
difficult, are certainly justified, since the use of the well
developed X-ray or ultrasonic imaging is inadequate for some
applications, such as the management of lung disease.

It should be noted that other EM methods have been used in
a variety of ways to monitor parameters of medical relevance.
These include microwave measurements of respiration [8], microwave interrogation of biological targets [9], and measurements of
heart dynamics using microwaves [10]. In this review, we focus
attention on the basic techniques of electromagnetic diagnosis,
their advantages and limitations, rather than restricting the
discussion to historical development. Because of space limitations, no attempt has been made to be all inclusive, but only the
more important and significant techniques are included.

2. Electrical Impedance Methods

The basic idea behind the use of low-frequency electromagnetic methods (20-100 kHz) for measuring changes in the fluid accumulation in the lungs is that most of the current flow through the tissue at these frequencies is due to movement of ions in extracellular water. Because air constitutes the major component in the lung's volume (70 to 80 percent), the resistance of the lung is expected to be high and to be sensitive to changes in blood and extracellular fluid volumes [11]. Attempts to monitor the fluid accumulation in the lung by measuring the electrical impedance (EI) directly across the lungs of cats and dogs were first reported by Lambert and Gremels [1]. The method was found to be most sensitive in the earliest stages, which is a decided advantage in measuring pulmonary edema.

The development of EI plethysmorgraphy [11] in the late 1950's and early 1960's was soon applied to lungs in the thorax. These efforts, however, have been frustrated by the short-circuiting effect of the more conductive surrounding tissues, such as the mediastinum and the chest wall. Because of this shunting effect, the method grossly underestimates true lung impedance, with changes of the order of only a few percent attributed to changes in lung water [12, 13].

The use of guarded electrodes [14, 15] and the focusing electrode bridge have greatly improved the sensitivity of the *in vivo* measurements to changes in lung tissue impedance. Graham first suggested surrounding the detecting electrodes with a guard ring, driven at the same potential as the active electrode [14]. With the guard-ring electrode placed over the lateral chest wall, most of the chest wall current should flow through the guard ring, while the current flowing through the central electrode should be largely directed through the underlying lung. The principle of operation of the guarded electrode system is illustrated in Fig. 1.

Severinghaus, *et al*. [15], further developed the guard-ring principle by adding "focusing" electrodes to obtain a better measurement of lung impedance independently of chest wall and mediastinal structures. This procedure basically minimized the lateral (chest wall) current flow from the active electrode, and also avoided problems arising from skin contact resistance in the guard ring. The sensitivity of this focusing electrode bridge was evaluated by conducting several experiments on dogs and comparing the results with those obtained using the double indicator-dilution technique. The results obtained indicated that the impedance method had about 20 percent the sensitivity of the indicator-dilution method with a reproducibility of ±5 percent. Although the good reproducibility of the experimental measurements is an encouraging advance in the development of the EI method, its poor

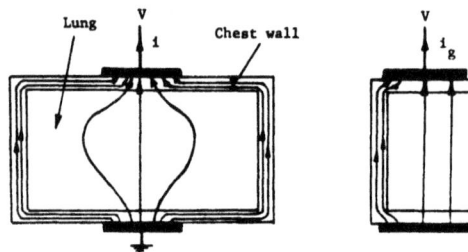

Fig. 1. Impedance measurements across the thorax. (a) With small unguarded electrodes, the major part of the measured current i passes through the low-impedance chest wall section. (b) With the guard electrode, most of the chest wall currents flow through the guard ring while the current flowing through the central electrode is largely directed through the underlying lung.

sensitivity may frustrate widespread clinical use. The possibility of improving the method's sensitivity still exists, but as of now it is not a quantitatively sensitive method.

The use of the EI method as a means of measuring other physiological activities has also been investigated for many years. For example, the technique has been used as a noninvasive measure of stroke volume, myocardial contractility, peripheral vascular disease, and tidal volume [16].

In summary, many potential applications of the EI method are being explored in a parallel effort with the engineering struggle to improve the sensitivity of the methods.

3. Microwave Methods of Lung Water Measurement

At microwave frequencies, changes in the dielectric properties of tissue are closely related to the amount of water present. The microwave method of lung water measurements basically utilizes these changes in dielectric properties of lung tissue to detect changes in its water content. The method is based on a continuous monitoring of the reflection and/or the transmission coefficient to indicate changes in the permittivity of the lung tissue. This method has the advantage of using highly penetrating microwave signals rather than ultrasonic signals that are both highly attenuated and dispersed in the lung. It should be noted that the microwave radiation method, although based on the complex permittivity of the lung, is fundamentally different from the EI method. As indicated in Table 1, the development of the microwave method

Table 1. Comparison between microwave and
electrical impedance methods.

Microwave Method	Electrical Impedance Method
• Utilizes changes in the dielectric properties of tissue.	• Utilizes changes in the dielectric properities of tissue.
• The transmission consists of electromagnetic waves which are attenuated as they travel through the body. 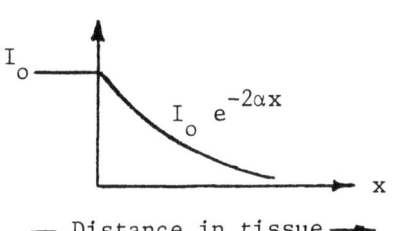	• The transmitted signal is primarily in the form of conduction current. 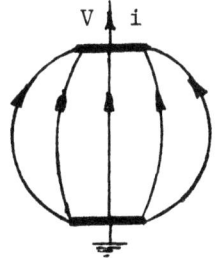
• No short circuiting effects because transmission is dominated by radiation fields.	• Sensitivity is highly reduced because of the short circuiting effect of highly conducting tissue.
• Critically depends on the development of a compact applicator that provides maximum coupling to tissue and minimum leakage radiation.	• Does not require any special applicator design.

critically depends on the development of a compact applicator that provides maximum coupling to tissue and minimum radiation leakage. The minimum external leakage is a critical requirement, since any leakage radiation reaching the microwave receiver could obscure the information carried by the highly attenuated signal traveling through the thorax.

The sensitivity of the microwave method was examined theoretically with a three-layer, planar model representing the front chest wall, lung, and back chest wall [3]. The results showed a superior sensitivity of the transmission coefficient to changes in lung water content compared to that of the reflection coefficient. Typically, a fifty-fold increase in sensitivity was achieved at 915 MHz by utilizing the transmission coefficient (particularly the phase) instead of any other parameter in the scattering matrix [4]. It was also shown that the best compromise between resolution and attenuation was found to be 915 MHz [5].

The development of a suitable microwave applicator is an important factor in determining the feasibility of using the microwave method in clinical applications. Attempts to use radiation-type applicators such as dielectric loaded waveguides [17] have not been successful because of the excessive leakage radiation. It should be noted that the use of transmission rather than reflection measurements complicates further the design problems by requiring consistent relative positions of the applicators during the course of the test. Iskander and Durney recently developed a new and more efficient method of coupling the microwave energy into the thorax [18]. The method utilizes a surface transmission line (coplanar waveguide) applicator that has proven to couple the energy efficiently and with minimum leakage radiation [6], and can also be used as a receiver.

The feasibility of using the microwave transmission method to measure fluid accumulation in the lungs was evaluated by conducting several experiments on dogs, using a printed-circuit version of the applicator. Experimental results from a typical dog experiment are shown in Fig. 2. From these results it is clear that changes in the phase of the microwave transmitted signal agree very well with changes in the pulmonary edema as indicated by the changes in the mean pulmonary arterial pressure. The EM coupler recently developed by the group at the University of Utah has played an important role in the success of this and other similar experiments, since excessive leakage radiation from other applicators produced unsuccessful attempts by others to make similar measurements.

The microwave method was initially used only to detect changes in the lung water content, but recently attempts have been made to calculate the absolute value of the water content of the lung [19]. The method simply involves obtaining a transverse cross-section image of a human thorax, using the computerized axial tomographic X-ray scanner with the microwave applicators in position. This information is used to construct a model of the cross section of the thorax, which is then used in solving for the internal electric fields numerically by the method of moments.

To illustrate the numerical procedure, consider the cross-sectional view of a human thorax shown in Fig. 3. The EM problem is to find the received signal at R due to a transmitter located at T. The incident field at T is assumed to be due to a two-dimensional slit, with the electric field being polarized along

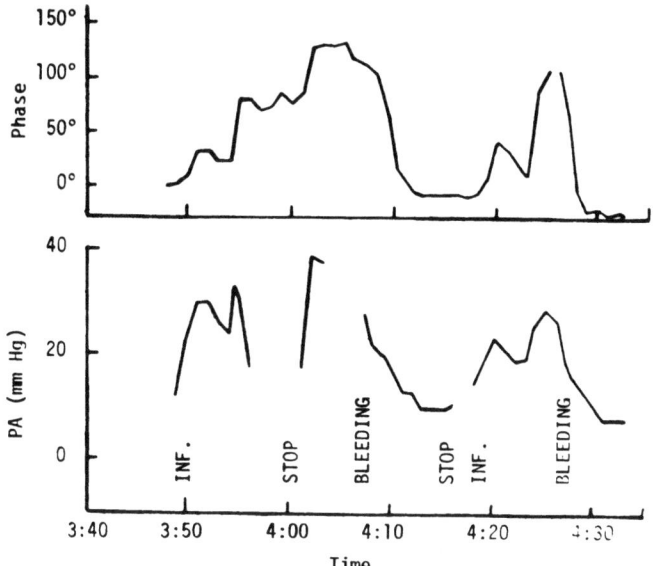

Fig. 2. Experimental results of a dog experiment illustrating the correlation between the phase of the microwave transmitted signal and the PA pressure, with the transfusion of blood and the subsequent bleeding of the dog. (After D. J. Shoff, et al. [4].)

the z-direction (out of the paper). Hence for a TE incident wave,

$$\bar{E}^i = \bar{z}\, E^i(x, y) \tag{1}$$

where $\bar{z}$ is a unit vector in the z direction. A procedure to calculate the incident field in (1) for a given electric field distribution across the transmitting aperture will be described later in this section.

For a TE excitation, the total electric field, on the other hand, is related to the electric polarization current $\bar{J}$ induced inside the dielectric body by:

$$\bar{E} = \frac{1}{j\omega\varepsilon_o (\varepsilon_r - 1)}\, \bar{J} \tag{2}$$

where ε_o is the dielectric constant of free space, ω is the angular frequency, and ε_r is the relative dielectric constant, which is a function of position. The scattered field is defined as the difference between the total and the incident fields, i.e., $\bar{E}^s = \bar{E} - \bar{E}^i$ and is given by [20]:

distribution $E_a(0, y')$ by [21]:

$$\bar{E}^i(x, y) = -\frac{1}{2}\frac{\partial}{\partial x}\int E_a(0, y')H_0^{(1)}\left[k\left\{x^2 + (y-y')^2\right\}^{1/2}\right]dy' \quad (6)$$

where the integration is over the intersection of the aperture with the plane x = 0. In the numerical calculations, we assume a cosine field distribution of unit amplitude across the slit; i.e.,

$$E_a(0, y') = \cos(\pi y'/2a) \quad (7)$$

where 2a is the width of the slit. Furthermore, the one-dimensional integration in (6) is calculated numerically after interchanging the differential and integration operations and using Simpson's rule (3-point Newton Cates' formula and using equally spaced sample points with the interval between the successive points always being less than 0.05λ).

The numerical solution of (5) then involves dividing the cross section into N small cells and assuming the field to be constant in each. By employing the point-matching technique (a delta function is used as a test function [22]), (5) deduces to:

$$\bar{E}_m^i = \bar{E}_m + \frac{jk_0^2}{4}\sum_{n=1}^{N}(\varepsilon_n - 1)\bar{E}_n\iint_{cell} H_0^{(1)}(k_0\rho_m)\,dx_0\,dy_0 \quad (8)$$

where ε_n and $\bar{E}_n$ are, respectively, the complex permittivity and the electric field in the cell n. $\bar{E}_m^i$ and $\bar{E}_m$ are the incident and total electric fields in the cell m [20]. Equation (8) is then a simple system of linear equations with the values of E_n being the unknown coefficients to be determined.

Since the numerical calculations include the size and shape of the internal organs, the complex permittivity of the lung that would produce a measured microwave transmission coefficient can be calculated. In order to calculate realistic data, however, it would be necessary to examine the lung composition, the fractional volume and complex permittivity of each component, and the factors that might alter its normal composition.

The lung is basically composed of three compartments: the lung tissue or parenchyma volume, the air volume, and the blood [3]. A reasonable estimate of the fractional volumes of each of these components is given in Table 2 for the normal, edematous, and emphysematous states of an adult male subject [23]. From Table 2, it is clear that, in each instance, the blood volume remains constant at 400 ml, which is an approximate estimate for the residual value in an adult male. A formula to calculate the

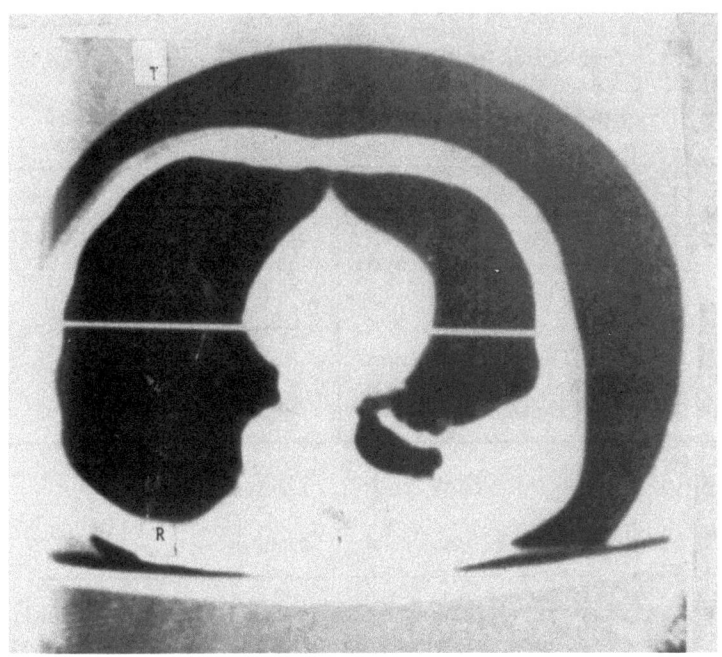

Fig. 3. A cross-section view across the thorax obtained using the computerized axial tomographic scanner.

$$\bar{E}^s(x, y) = -\frac{jk_o^2}{4} \iint (\varepsilon_r - 1)\bar{E}(x_o, y_o) H_0^{(1)}(k_o\rho) \, dx_o \, dy_o \quad (3)$$

where $k_o = 2\pi/\lambda_o$, λ_o is the free space wavelength, $H_0^{(1)}(k_o\rho)$ is the Hankel function of zero order, and ρ is given by:

$$\rho = \sqrt{(x - x_o)^2 + (y - y_o)^2} \quad (4)$$

The integration in (3) is over the cross section of the two-dimensional model. On substituting (3) for the scattered field, we obtain an expression for the total field at any point inside the dielectric model of the cross section in terms of the incident field. Hence,

$$\bar{E}^i = \bar{E} + \frac{jk_o^2}{4} \iint (\varepsilon_r - 1)\bar{E}(x_o, y_o) H_0^{(1)}(k_o\rho) \, dx_o \, dy_o \quad (5)$$

The incident field in (5) is obtained by calculating the fields diffracted by a two-dimensional slit located at T. By employing Green's function of half space, it can be shown that the incident field at point $\bar{E}^i(x, y)$ is given in terms of the aperture field

Table 2. The lung components and their volumes and fractional volumes in normal edematous, and emphysematous lungs of adult male [3].

	Blood Volume		Tissue Volume		Air Volume	
	In ml	Percent F_b	In ml	Percent F_p	In ml	Percent F_a
Normal	400	15.8	500	19.8	1630	64.4
Edema	400	15.8	1000	39.5	1130	44.7
Emphysema	400	10.2	250	6.4	3260	83.4

$$\varepsilon_{lung} = F_b \cdot \varepsilon_{blood} + F_p \cdot \varepsilon_{tissue} + F_a \varepsilon_o$$

$$\sigma_{lung} = F_b \cdot \sigma_{blood} + F_p \cdot \sigma_{tissue}$$

complex permittivity of the lung by employing the fractional volumes of each component is given also in Table 2. This formula is approximate but necessary due to the lack of *in vivo* permittivity measurements for normal, edematous, and emphysematous lungs. Since, in all lung conditions given in Table 2, the blood volume is assumed to be constant, the calculated values of the compelx permittivity can then be simply related to the amount of lung water content by adjusting the relative values of F_p and F_a and keeping the value of F_b constant.

The numerical procedure was evaluated by carrying out the detailed calculations for the human cross section of Fig. 3. The geometry was divided into 221 cells, with the maximum cell size varied according to the local dielectric constant. The digitized cross section is shown in Fig. 4 where it is clear that cells of smaller sizes are generally required for regions of higher dielectric constants, e.g., the chest wall and the heart. In all cases, the size of the cell was maintained between $0.1 \lambda/\sqrt{\varepsilon_r}$ and $0.2 \lambda/\sqrt{\varepsilon_r}$, where ε_o is the free space wavelength and ε_r is the relative permittivity of the cell [24]. This restriction on the cell size is generally necessary for the convergence of the solution and to obtain accurate results. It should be noted that the cell size shown in Fig. 4 is for the calculations carried out at 750 MHz. Absolute values of the calculated field inside each cell is also shown in Fig. 5. This numerical procedure, however, is expensive, simply because of the excessive computer time required. The problem is particularly severe at the higher frequencies where the number of cells increases prohibitively. A procedure to develop an efficient numerical technique by manipulating the method of calculating the matrix elements and by improving the procedure for the matrix inversion has not yet been developed.

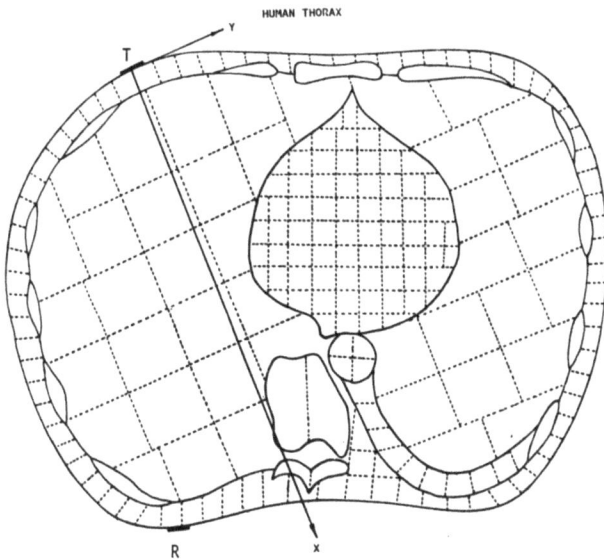

Fig. 4. A digitized cross section illustrating the relative cell sizes in different types of tissue.

This difficulty is a typical characteristic of microwave imaging whereby the interaction between the mathematical cells at larger separation distances cannot be neglected. The small effect of such interaction is a well recognized advantage of X-ray and ultrasound imaging techniques. In contrast to microwave methods, the X-ray and ultrasound energies involve beam paths which are almost straight lines, with the resulting matrix being almost diagonal, hence much easier to handle [25]. The difficulties involved in microwave imaging, however, are certainly justifiable, particularly when applied to lung problems.

4. Electromagnetic Imaging

An additional difficulty involved in estimating the average water content in the lung is the problem of distribution, as affected by both disease states and gravity [26]. It has been shown that water gradients of varying degrees do exist either in the axial direction (apex to bottom) in standing patients or from front to back in prone patients. Therefore, estimating the average lung water content that can satisfy an experimentally predicted signal may be quite complex and not unique. In such cases, the iterative procedure will not only involve changes in the average amount of lung water, but will also require investigating several possible gradients and other problems causing nonuniform distribution of the water in lungs. A simple procedure to obtain a unique solution is to employ several receivers for every

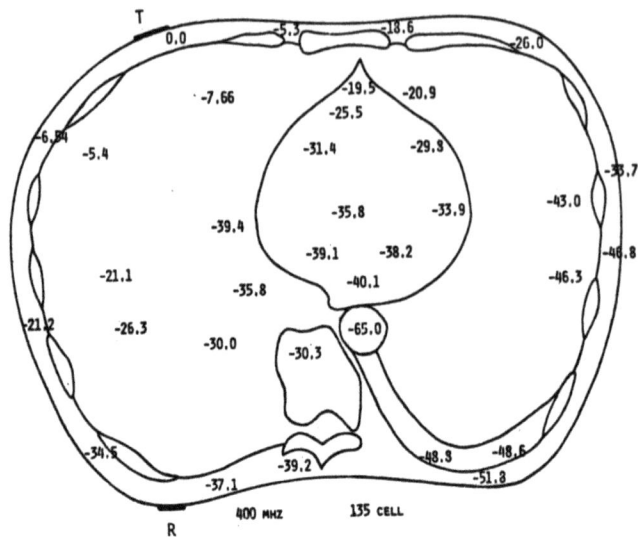

Fig. 5. Magnitude of field distribution across the thorax cross section of a human subject at 400 MHz using 135 cells.

location of the transmitter, which leads to the consideration of electromagnetic imaging. In contrast to X-ray imaging, several problems are involved in developing algorithms for EM imaging. These difficulties occur basically because the path of the eletric current or the microwaves are object-dependent and not always nearly straight lines, as is the case for X-ray beams. Therefore, the coordinates of these paths are not known in advance and can be determined only after identifying the spatial distribtuion of conductivity or complex permittivity of the body.

To illustrate the procedure, consider a two-dimensional object and the problem of reconstructing the density spatial distribution in the given cross section by measuring a finite number of projections [25]. If X-ray imaging is to be used in this case, the mathematical problem is simply to solve a system of integral equations using line integrals over the object's density along straight paths connecting the transmitter and the receiver. By dividing the cross section into small cells and assuming constant density in each cell, the line integrals can be replaced by finite sums and the imaging problem is reduced to finding a solution of a linear system of equations in the form

$$[c][d] = [P] \qquad (9)$$

where [c] is a coefficient matrix, [d] is a vector representing densities at the mathematical cells, and [P] consists of the values of the line integrals measured for discrete rays. The [d] matrix contains the unknown density values, while [P] is known from the measurements, and [c] can be determined from the geometry of the rays passing through or near the grid points of the object [25]. It should be noted that the linearity of the X-ray imaging problem results basically from the fact that the coefficients matrix [c] is not a function of the density d. An excellent review of the commonly used image reconstruction algorithms can be found elsewhere [25].

If now, instead of X-rays, dc currents, low-frequency electric fields, or microwaves are applied to the two-dimensional object, the currents or waves will pass along straight lines only if the object is homogeneous and the electrodes are of idealized geometry as shown in Fig. 1(b). Otherwise, the object will interact with the fields resulting in more severe bending of the current or wave paths. The resulting system of equations will be of the form,

$$[c(d)][d] = [P] \tag{10}$$

In (10), the coefficient matrix [c(d)] depends on the solution d, and hence the problem is nonlinear. Obviously the available algorithms cannot be directly applied in this case because the paths along which values of the line integrals are measured are unknown, and hence a nonlinear reconstruction procedure is required. Tasto and Schomberg [27] proposed a simple extension of the well-known algebraic reconstruction technique (ART) [25] to take into account nonlinearities. For dc or low-frequency conductivity (σ) imaging, the proposed extension involves determining the electric potential V interior to the object by solving Laplace's equation $\nabla \cdot (\sigma \nabla V) = 0$ numerically. In general, wherever there is change in σ, there will be change accumulation in the body. In this case, the conductivity equation will be

$$\sigma \nabla^2 V + \nabla \sigma \cdot \nabla V = -\frac{\partial \rho}{\partial t} \tag{11}$$

where ρ is the free charge density [28]. The solution procedure involves assuming an initial conductivity profile and the potential computation, subject to the given set of boundary conditions, is then performed using the finite difference method [29]. The current entering or leaving each electrode will then follow the gradient of the potential field. Once these stream lines are known, a line integral similar to the line integrals known from X-ray reconstruction can be determined and the reconstruction procedure continued in a manner similar to the X-ray case. It should be noted that the calculation of the electric potential must be performed after each iteration to modify the current stream lines according to the new conductivity profile produced in the previous iteration.

Several experiments with the nonlinear reconstruction technique were carried out using simulated data [27, 29]. The images obtained are generally satisfactory, although there is some blurring and local overshoot [29]. Experiments with measured rather than simulated data revealed that the resulting images of complex objects are not acceptable [27]. A common limitation of the conductivity imaging procedures is the preferred current path through the high-conductivity tissue. For example, it was found very difficult to image a low-conductivity region surrounded by a high-conductivity one because the current genrally takes the path of least resistance around the low-conductivity region [29]. This problem is particularly important in imaging the human thorax because of the low-resistance current path through the highly conducting chest wall. To improve the sensitivity of the measurement to high resistance tissue (e.g., the lungs), Swanson proposed an impedance camera that utilizes an array of guarded electrodes similar to those used in the impedance measurements as described in the second section [30]. Recently, an isoadmittance contour map was obtained by applying the guarded-electrode array to the subject's back [31]. No attempts have been made to use reconstruction algorithms, however, and the map obtained, although promising, clearly indicated that a great deal of work must be done before the impedance camera can be of clinical use. Furthermore, more work should be done toward the development of a complex permittivity camera based on microwave measurements. This camera will have the advantage of providing both the real and imaginary parts of the complex permittivity, which is a step towards the unique identification of the images. On the basis of the microwave method of lung water measurement (see Section 3), it is clear that much improved sensitivity, particularly to high resistance tissue such as the lungs, can be achieved by utilizing the transmission of microwave signals instead of low-frequency currents.

5. Other Electromagnetic Diagnostic Techniques

Among several other diagnostic methods that are based on electromagnetic techniques are the electromagnetic flowmeter and use of microwave radiometry in detecting breast cancer tumors.

The Electromagnetic Flowmeter. The general principle of an electromagnetic flowmeter was first introduced by Kolin [32], and since then has reached a high standard of performance and reliability, and can be easily and safely administered during surgery. The principle of operation is based on the familiar Faraday law of electromagnetic induction. If blood is flowing in a tube or blood vessel oriented at right angles to the magnetic field, then there will be a voltage generated across a diameter perpendicular to the magnetic field and to the direction of flow.

This induced voltage may be picked up from electrodes in contact with the outer surface of the vessel wall without opening the artery. Studies performed on flow models have demonstrated a linear relation between the blood flow and the electrical signal as well as excellent reproducibility [33]. Efforts are presently being focused on utilizing the basic idea of the electromagnetic flowmeter to monitor the blood flow in humans noninvasively [34].

Microwave Radiometry and Its
Potential Application in Medicine

Radiometry originates from the fact that all bodies above absolute zero temperature emit energy in the form of electromagnetic radiation. The frequency at which the maximum radiation occurs depends on the body temperature and, for temperatures of practical interest (T < 1000°K), the maximum radiation occurs in the infrared region. This is why this emitted part of the spectrum has been detected and used for more than twenty years to obtain thermographic images.

Because the radiative properties of objects vary according to their nature, shape, etc., the concept of gray bodies was introduced. The energy radiated by an infinite gray body is related to the energy radiated from a black body by the emissivity factor ε. This factor is, in general, a function of temperature, frequency, orientation, composition, and surface conditions of the body. Therefore, by measuring changes in the emitted energy, variations in the dielectric properties, relative dimensions, and temperature can be detected. While microwave radiometry has been applied to a variety of problems, including radio astronomy, communications, and atmospheric sensing for many years, it has been only recently applied to medical diagnosis. Calculations based on the gray-body theory show that biological systems emit microwave radiation detectable by presently available microwave radiometric techniques. The major advantage of using the emitted radiation at microwave frequencies is simply the greater penetration depth compared to that obtained when using infrared radiation. This permits measurements of microwave radiation emitted from structures within the body. Microwave radiometry, on the other hand, has coarser spatial resolution than infrared thermography. The trade-off between the desired depth of penetration and the necessary resolution is the major consideration in determining the operating frequency range in a radiometer experiment. The feasibility of using the microwave radiometry in detecting breast cancer tumors has been demonstrated by conducting several experiments on phantoms [35]. Results from clinical trials, however, were marginal, but promising [36]. It should be emphasized that more effort is required to improve the response patterns of the antennas. Instead of the commonly used open-ended waveguides with a half-power beam width of approximately 50°, much more directive antennas are required. For example, a pencil-beam antenna would considerably improve the resolution and would therefore increase the usefulness of radiometry in diagnostics. Efforts to develop such a technique for clinical use are certainly commendable because microwave radiometry is not only noninvasive, but is also passive, and hence completely safe.

Conclusions

In this paper, recent advances in the electromagnetic diagnostic techniques are described. From the preceding discussion, one might assume that these techniques represent a large and rapidly growing area of application. Unfortunately, this has been far from the case so far. Apart from the electromagnetic flowmeter, other potential techniques are still in research stages, with their clinical applications being restricted to specialized centers. For example, the sensitivity and potential use of the electrical impedance method, although it was first used in 1926 to measure changes in lung water content, is still being evaluated by researchers.

Among many factors contributing to slow development in this area of research is the complexity and lack of understanding of the nature of the interaction between electromagnetic waves and the human body. For example, attempts to estimate the sensitivity and the accuracy of a given procedure have frequently been frustrated by the complexity of the body structure, limitations on computer storage, and expenses when numerical calculations were performed. This simply forced workers in this area either to base their conclusions on simple and idealized models, with questionable adequacy and accuracy, or to conduct pilot experiments and ignore the ever-necessary basic understanding. The situation is further complicated by the variability of the human body from person to person, not only in structure, but also in reaction to a given experimental procedure.

The interdisciplinary nature of the research in this case is certainly one of the major contributing factors to the slow progress in this area in the past. Additional difficulties in research are sometimes complicated by difficulties encountered in communication among members of interdisciplinary research teams. It is, however, our hope that the encouraging initial experiences with potential electromagnetic diagnostic techniques and the promising results obtained will attract talented engineers, scientists, and physicians to cooperate in developing this important area.

References

[1] R. K. Lambert and H. Gremels, "On the factors concerned in the production of pulmonary edema", *Journal of Physiology*, Vol. 61, London, 1926, pp. 98-112.

[2] C. Süsskind, "Possible use of microwaves in the management of lung disease", *Proceedings of the IEEE (Letters)*, Vol. 61, 1973, p. 673.

[3] P. C. Pedersen, C. C. Johnson, C. H. Durney, and D. G. Bragg, "Microwave reflection and transmission measurements for pulmonary diagnosis and monitoring", *IEEE Transactions on Biomedical Engineering*, Vol. BME-25, 1978, pp. 40-48.

[4] D. J. Shoff, M. F. Iskander, C. H. Durney, and D. G. Bragg, "Noninvasive microwave methods for measuring tissue volume in normal dogs after whole blood infusion", *Medical Research Engineering*, 1979, in press.

[5] M. F. Iskander, C. H. Durney, D. J. Shoff, and D. G. Bragg, "Pulmonary diagnostics using noninvasive microwave methods", presented at the International Symposium on Biological Effects of Electromagnetic Waves, Airlie, Virginia, October 30-November 4, 1977. Also in *Radio Science*, December 1979.

[6] C. H. Durney, M. F. Iskander, and D. G. Bragg, "Noninvasive microwave methods for measuring changes in lung water content", *Proceedings of the IEEE Electro/78*, Session 30, Boston, Massachusetts, May 23-25, 1978.

[7] M. F. Iskander and C. H. Durney, "Electromagnetic techniques for medical diagnosis: a review", *Proceedings of the IEEE*, January 1980, in press.

[8] J. C. Lin, "Noninvasive microwave measurement of respiration", *Proceedings of the IEEE (Letters)*, Vol. 63, 1975, p. 1530.

[9] J. H. Jacobi, L. E. Larson, and C. T. Hast, "Water-immersed microwave antennas and their application to microwave interogation of biological targets", *IEEE Transactions on Microwave Theory and Techniques*, Vol. MTT-27, 1979, pp. 70-78.

[10] I. Yamaura, "Mapping of microwave power transmitted through the human thorax", *Proceedings of the IEEE (Letters)*, Vol. 67, 1979, pp. 1170-1171.

[11] N. C. Staub, "Pulmonary edema", *Physiological Review*, Vol. 54, 1974, pp. 678-811.

[12] M. Pomerantz, F. Delgado, and B. Eiseman, "Clinical evaluation of transthoracic electrical impedance as a guide to intrathoracic fluid volume", *Ann. Surgery*, Vol. 171, 1970, p. 686.

[13] J. V. Van de Water, I. T. Miller, E. N. C. Milne, E. L. Hanson, G. F. Sheldon, and K. S. Kagey, "Impedance plethysmography: a noninvasive means of monitoring the thoracic surgery patient", *Journal of Thoracic and Cardiovascular Surgery*, Vol. 60, 1970, pp. 641-647.

[14] M. Graham, "Guard ring use in physiological measurements", *IEEE Transactions on Biomedical Electronics*, Vol. 12, 1965, pp. 197-198.

[15] J. W. Severinghaus, C. Catron, and W. Noble, "A focusing electrode bridge for unilateral lung resistance", *Journal of Applied Physiology*, Vol. 32, 1972, pp. 526-530.

[16] J. C. Denniston and L. E. Baker, "Measurement of pleural effusion by electrical impedance", *Journal of Applied Physiology*, Vol. 38, 1975, pp. 851-857.

[17] I. Yamaura, "Measurements of 1.8-2.7 GHz microwave attenuation in human torso", *IEEE Transactions on Microwave Theory and Techniques*, Vol. MTT-25, 1977, pp. 707-710.

[18] M. F. Iskander and C. H. Durney, "An electromagnetic energy coupler for medical applications", *Proceedings of the IEEE*, 1979, in press.

[19] M. F. Iskander, C. H. Durney, D. G. Bragg, and B. H. Ovard, "A microwave method of estimating absolute value of average lung water content", presented at the Open Symposium on the Biological Effects of Electromagnetic Waves, Helskinki, Finland, August 1-8, 1978. Also submitted to *Radio Science*.

[20] J. H. Richmond, "Scattering by a dielectric cylinder of arbitrary cross-section shape", *IEEE Transactions on Antennas and Propagation*, Vol. AP-13, 1965, pp. 334-341.

[21] K. Houlberg, "Diffraction by a narrow slit in the interface between two diffraction media", *Canadian Journal of Physics*, Vol. 45, 1967, pp. 51-81.

[22] R. F. Harrington, *Field Computation by Moment Method*, Macmillan Company, New York, 1968.

[23] P. C. Pederson, *Diagnostic Application of Microwave Radiation*,

Ph.D. dissertation, Bioengineering Department, University of Utah, Salt Lake City, Utah 1976.

[24] H. S. Ho, "Dose rate distribution in triple-layered dielectric cylinder with irregular cross section irradiated by plane wave sources", *Journal of Microwave Power*, Vol. 10, 1975, pp. 421-432.

[25] R. A. Brooks and G. D. Chiro, "Principles of computer assisted tomography (CAT) in radiographic and radioisotopic imaging", *Phys. Med. Biol.*, Vol. 21, 1976, pp. 689-632.

[26] J. B. West, *"Stresses" in Regional Differences in the Lung*, edited by J. B. West, Academic Press, New York, 1977.

[27] M. Tasto and H. Schomberg, "Object reconstruction from projections and some nonlinear extensions", presented at NATO Advanced Study Insitute on Pattern Recognition and Signal Processing, June 25-July 4, 1978, Paris, France. Also to appear in *NATO Advanced Study Institute* series.

[28] L. R. Price, "Electrical impedance computed tomography (ICT): a new CT imaging technique", *IEEE Transactions on Nuclear Science*, Vol. NS-26, 1979, pp. 2736-2739.

[29] R. J. Lytle and K. A. Dines, "An impedance camera: a system for determining the spatial variation of electrical conductivity", Lawrence Livermore Laboratory, Report UCRL-52413, 1978.

[30] D. K. Swanson, *Measurement Errors and Origin of Electrical Impedance Changes in the Limb*, Ph.D. dissertation, University of Wisconsin, Madison, Wisconsin, 1976.

[31] R. P. Henderson and J. G. Webster, "An impedance camera for specific measurements of the thorax", *IEEE Transactions on Biomedical Engineering*, Vol. BME-25, 1978, pp. 250-254.

[32] A. Kolin, "Electromagnetic flowmeter: principle of method and its application to blood flow measurements", *Proc. Soc. Experi. Biol. and Med.*, Vol. 35, 1936, pp. 53-56.

[33] C. Cappelen, Jr. and K. V. Hall, "Electromagnetic blood flowmetry in clinical surgery", *Acta. Chir. Scand.*, Suppl. 368, 1967, pp. 3-27.

[34] H. Boccalon, B. Candelon, J. J. Tillie, A. Graulle, H. G. Doll, P. Puel, and A. Enjalbert, "New noninvasive device for pulsatile blood flow measurement", *Digest of the 11th*

International Conference on Medical Biological Engineering, Ottawa, Canada, August 1978, pp. 428-429.

[35] R. A. Porter and H. H. Miller, "Microwave radiometric detection and location of breast cancer", *Proceedings of the IEEE Electro/78*, Session 30, Boston, Massachusetts, May 23-25, 1978.

[36] A. H. Barrett, P. C. Myers, and N. L. Sadowsky, "Detection of breast cancer by microwave radiometry", *Radio Science*, Vol. 12, No. 6S, 1977, pp. 167S-171S.

Panel Discussion

on

EXPERIMENTAL METHODS FOR THE QUANTITATION
OF ABSORPTION PATTERNS IN BIOLOGICAL TISSUES

Chairman: Mr F. Harlen — National Radiological Protection Board, Harwell, England.

Panel:
- Professor A.W. Guy — University of Washington, Seattle, USA.
- Dr E. Burdette — Georgia Institute of Technology, Atlanta, Georgia, USA.
- Dr A D Yaghjian — National Bureau of Standards, Boulder, Colo. USA.
- Professor M.F. Iskander — University of Utah, USA.
- Professor E.H. Grant — Queen Elizabeth College, London, England.
- Dr C.W. Smith — University of Salford, England.
- Dr G. Crosta — Via Dupre, 5, 21013 Gallarate Italy

Contributors:
- Dr. Sinnott — Defence Research Centre, Salisbury, S. Australia
- Prof. R. Mittra — Electrical Engineering Department, University of Illinois, Urbana, Illinois, USA.
- Prof. J. Van Bladel — Laboratorium voor Elektromagnetisme en Acustica, Ghent, Belgium.
- Dr. T.K. Sarkar — Rochester Institute of Technology, Rochester, USA.

Mr F Harlen - This morning we will have two sets of presentations. The first one is on measurements and possibly on implications of some theory as far as measurements are concerned. The speakers will be Professor A.W. Guy, Dr E.C. Burdette, Dr A.D. Yaghjian and Professor M.F. Iskander. The following one will be presented by Professor E.H. Grant, Dr C.W. Smith and Dr G. Crosta around the subject of hypothermia and cancer, and on the implications of their work on cancer.

Professor A.W. Guy - Well, I want to briefly address the topic of the first subject of this session. Until now, most of the discussions were related to the theoretical approaches for quantifying the EM fields and the deposited energy in tissues, but there are many situations in which one cannot apply the theory to biological systems since they are extremely complex not only in shape but also in composition and various other aspects. I have described in my previous lecture various techniques for conducting experimental measurements of energy absorption or specific absorption rate (SAR) of the energy in tissues, showing that the oldest method probably involves the use of internal temperature sensors. The use of such sensors allows the determination of the temperature rise associated with a given time of exposure of tissues to electromagnetic fields. The energy absorption and electric fields in the tissues resulting from the exposure can then be calculated based on the thermal and electrical properties of the tissues. E. Mittleman used this method in his early experiments with diathermy. Internal field sensors were also developed for actually measuring the electric fields in the tissue.

Another method that has gained popularity is whole body calorimetry, which is based on the heat transfer from the body of a subject after it is exposed to a quantified level of microwave or RF radiation. Thermography is also gaining recognition as a valuable tool in the determination of energy absorption patterns and verification of theoretical methods for obtaining the SAR. Another quantitative method is the direct measurement of absorption in an animal by exposing it in a waveguide while monitoring the input and transmitted powers or in a cavity while monitoring the input power. I will now discuss these methods in detail, beginning with the internal sensor method.

Probably the first approach used for quantifying energy absorption was the implantation of thermistors or thermocouples in hypodermic needles inserted in the tissue of exposed animals. Unfortunately, many of the earlier researchers did not recognize the field perturbations that implanted sensors would produce, and often they were simply measuring the temperature rise due to these perturbations. Others recognized this and tried to orient the hypodermic needles containing the implanted thermocouples so that they were perpendicular to the electric fields. Though it is easy to do this in a structure of perfect geometric shape, biological structures are very complex, with irregular boundaries, so that it is nearly impossible to align thermocouples and their leads so that they are perpendicular to the field everywhere in the tissues. Therefore, many errors have resulted from measurements of this type. As they became aware of these problems, researchers managed to make the sensors somewhat transparent by using very fine high-resistance leads to connect them to the recording instruments. This eliminated many of the problems, but not all. If the leads and sensors do not have the same resistivity or dielectric properties as the tissues in which they are imbedded, they still can produce perturbations on a micro scale, if not on a global scale. When leads of high conductivity are inserted in a tissue, whether or not they seem to be transparent in terms of not disturbing the fields around the wire, some of the cells near the tip of the wire or sensor will be subjected to very high field strengths. I am sure that some of the cellular effects resulting from exposure of cell cultures or other _in vitro_ preparations reported in the literature were due to field perturbations resulting from implanted temperature sensors. One way to avoid perturbing the fields in the tissues during measurement of temperature is to deposit the sensor in the tissue only before and after the exposure field is applied. We have successfully used this technique, which I will now describe.

One first selects a glass micropipette with its tip sealed. The pipette is implanted in the tissue with the tip located in the region where one wishes to measure the SAR. Then a thermocouple is placed into the pipette against the tip, where it is allowed to reach equilibrium temperature. This temperature is measured and the thermocouple is quickly withdrawn before the microwave power is applied. After the power is turned off, the thermocouple is reinserted and another temperature measurement is made. Since the rate of change of temperature with time is linear for a short exposure, one can convert the ΔT resulting from the temperatures measured immediately before and after exposure into the SAR, usually expressed in units of W/kg. This method was used to determine the SAR distribution in the eye of a rabbit in connection with studies of microwave-induced cataracts. The results shown in Fig. 1 are based on measurements on five different rabbits, when the eye was illuminated with 2450 MHz radiation from a diathermy applicator (an exposure method used by many researchers for microwave cataract studies). The results show that the SAR is relatively low at the surface of the cornea, but increases very rapidly with increasing depth into the lens, reaching a peak level posterior to the lens. If the resulting

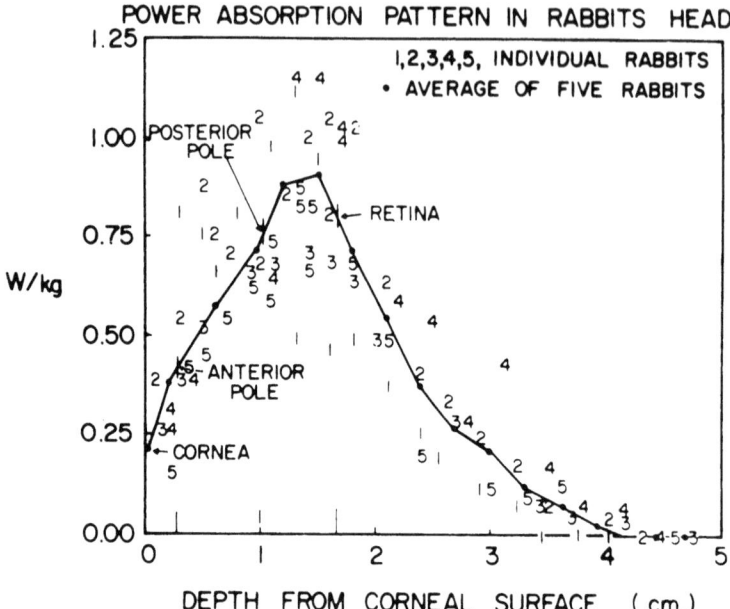

Fig. 1 Measurement of temperature in rabbit eye to determine specific absorption rate (SAR).

data are given to a mechanical engineer well-equipped with the thermal equations and computer methods for determining temperature distributions in materials as a function of energy deposition and various cooling mechanisms, he can determine the temperature distribution pattern in the eye. The typical computer model takes into account the cooling of the tears, the blood flow in the iris and orbit, and other heat transfer mechanisms. An example of the predicted maximum steady-state temperature distribution with a maximum value of 42.8°C is shown in Fig.2. The accuracy of the computer-generated temperature distribution was verified by actual measurements with the thermocouples and micropipette combination described previously. Indeed, cataracts were actually produced in the eyes of exposed rabbits in the region of the hotspots predicted by the computer analysis where temperatures were above 42.8°C. Since these temperatures are somewhat below 47°C, the known temperature for tissue destruction, it appears that the lens of the eye is relatively sensitive to thermal insult.

Another important quantitative approach for determining SAR is the calorimetric method, which gained recognition some years

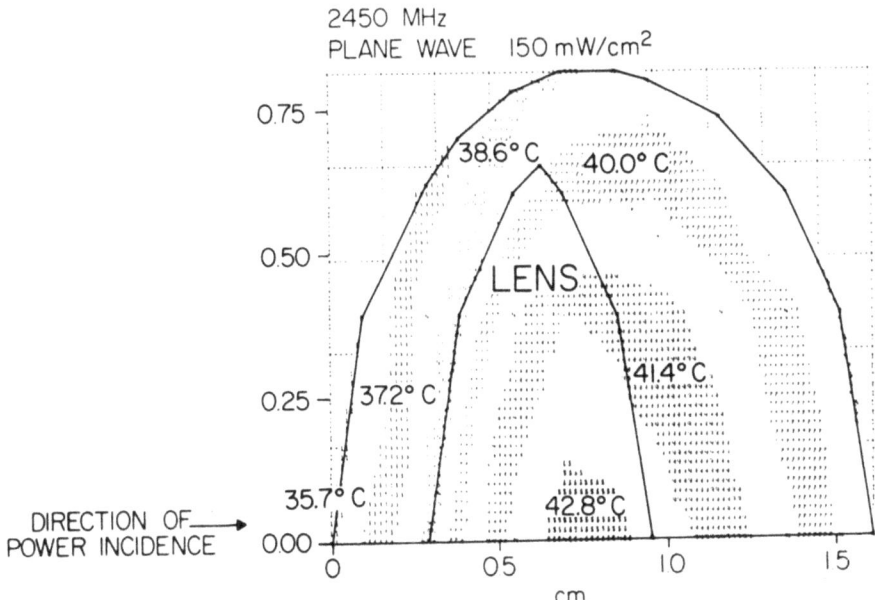

Fig. 2   Computer printout of the steady state temperature distribution for eye irradiated with 150 mW/cm^2. Note 42.8° hot spot at posterior pole of lens where cataracts were noted to form.

back - I think about 10 years ago. I will describe the twin-well calorimeter as an example of a useful instrument for measuring the total power absorbed in the body of a small animal exposed to electromagnetic fields. This device was developed by Dr Richard Phillips and his colleagues at Battelle Pacific Northwest Research Laboratories in Richland, Washington. The inner part of the device consists of two cylindrical wells or shells of copper or aluminum, each surrounded by an array of thermocouples connected in series, with the thermocouples on one cylinder connected in reverse polarity with those of the other cylinder. The two adjacent inner shells are surrounded by another metallic oval shell with the walls separated slightly from those of the inner shells. The temperature of the outside shell is kept constant by circulating fluid or electrical heating elements. Thermal insulation is used to fill the space between the outside walls of the inner cylinders and the inner walls of the outside cylinder. The output of the unit (thermocouples) is connected to a sensitive voltmeter. At equilibrium, the output voltage is 0. When thermal energy is added to the interior of one of the inner wells, however, its temperature will rise and there will be a net output voltage first increasing, then decreasing to 0 over the period of time for the added thermal energy to pass from the inner well to the constant-temperature outside shell. The integral of the output voltage as a function of time is proportional to the added energy. The proportionality constant may be obtained by adding or subtracting a known amount of energy from one of the wells - for example, that contained in an added quantity of hot water or that required to melt a given amount of ice. In order to measure the energy absorption in an animal exposed to electromagnetic radiation, freshly killed animals are brought to temperature equilibrium, and one animal is exposed to the electromagnetic fields. The animals are then placed in the two wells of the calorimeter and the output voltage is integrated over the period of time for temperature equilibrium to be reached. Based on the mass of the animal, the average SAR in the body due to exposure to electromagnetic fields may then be calculated. See Fig. 3. The curves on the graph show the voltages from the thermocouple array for each cylinder in the twin-well calorimeter. Since the thermocouple arrays are connected in reversed polarity, the voltage difference is measured from the twin-well. Thus, the area under the voltage-time curve from the assembly is proportionate to the differential heat loss between the two rat carcasses which is equal to the absorbed microwave power in the exposed animal.

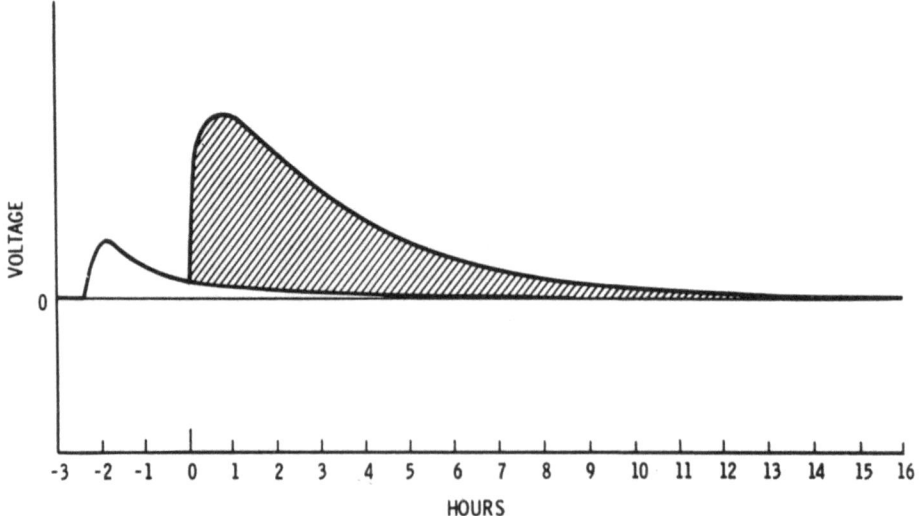

Fig. 3. Output voltage vs. time for twin-well calorimeter.

Another technique used for determining SAR distribution - thermography - was discussed in my previous lecture. To give you some idea of the type of resolution that may be achieved in a thermographic analysis, a 1/11 scale model of a man was exposed to plane wave radiation near the resonant frequency of the body. By using narrow-angle lenses on the thermograph, one can obtain fairly good detail of the SAR patterns in the regions of importance corresponding to the neck, the arms, the legs, and portions of the torso. This information can then be used with numerical computer methods for solving heat flow equations to predict the temperature rise in various locations of the body. It was from these results that Frank Harlen obtained his information concerning the temperature rise in the arms and legs of exposed humans. In addition to whole-body thermography, thermographic studies can be performed on full-scale models representing anatomical portions of the human body. For example, cylindrical models may be constructed of concentric cylinders of synthetic tissues representing bone, muscle tissue, and subcutaneous fat for quantifying the effectiveness of various diathermy applicators. Fig. 4 is an illustration of thermograms taken after exposure of a perfect cylindrical model of the human thigh,

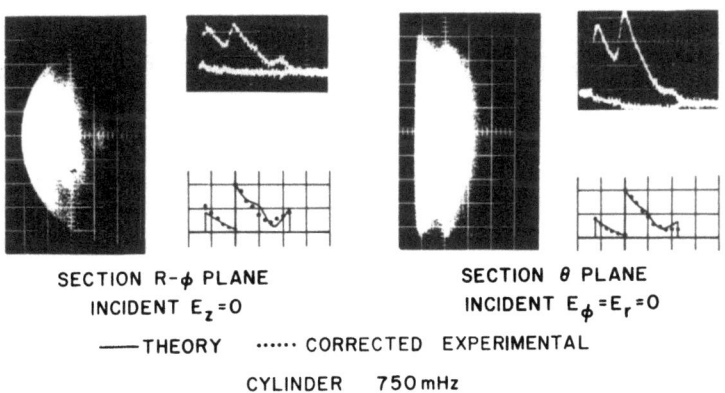

SECTION R-φ PLANE
INCIDENT $E_z = 0$

SECTION θ PLANE
INCIDENT $E_\phi = E_r = 0$

——THEORY ······ CORRECTED EXPERIMENTAL
CYLINDER 750 mHz

Fig. 4.

consisting of concentric cylinders of synthetic bone, muscle, and subcutaneous fat. At the left of the figure is an intensity scan (brightness proportional to temperature and SAR) illustrating the absorbed energy patterns when the cylindrical model was exposed on the left side to a radiation field with the electric field perpendicular to the axis of the cylinder. At the upper right of the intensity scan is a double line scan indicating the temperature or SAR along a horizontal line corresponding to the diameter of the cylinder. The vertical distance between the scans taken before and after exposure is proportional to the temperature and SAR. Below the double scan is a comparison between the values of SAR obtained from the thermograph double scan (denoted by dots) and the SAR as calculated by theory (denoted by a solid line). The thermograms on the right half of the figure correspond to the case where the cylinder is exposed with the electric field parallel to the axis. Very good agreement was obtained between the SAR values obtained thermographically and those calculated theoretically. The SAR variations resulting from standing waves due to reflections from the bone-muscle and the fat-muscle interfaces are clearly seen. This method provides a practical way to analyze the efficiency of various diathermy sources in coupling energy to tissues of complex shape. The model may be used over and over again for various applicators and exposure frequencies, after allowing it to cool between exposures.

The versatility of the thermographic technique is demonstrated by its use with actual animals as well as with synthetic tissues. When used with an animal, the animal is first sacrificed and frozen in the position desired for exposure. The polyfoam (styrofoam) material is then put around the animal and, after the foam is cured, the frozen animal is bisected. A plastic film is then placed over each half of the animal at the surface of bisection. The animal then can be brought back to room temperature, exposed to an electromagnetic source, and thermography applied in the same manner as used for the models. Thermography has proven especially valuable in graphically illustrating the field perturbations resulting from implanted temperature sensors and EEG electrodes, and hotspots in the tissues of animals exposed to electromagnetic fields. This has brought to light many artifactual conditions that existed during the course of previous experiments.

Fig. 5 illustrates the results of a thermographic study of the SAR patterns in a rat exposed to 915 MHz aperture antenna fields. The results illustrate problems that were not apparent to earlier researchers. Note the extreme hotspots and also the areas of low SAR in the body of the animal. Especially intense is the hotspot in the tail at its junction with the body of the rat. Yet in an adjacent region, where a researcher might place a temperature sensor to measure the rectal temperature of the animal, the SAR is considerably lower. This is due in part to the longitudinal

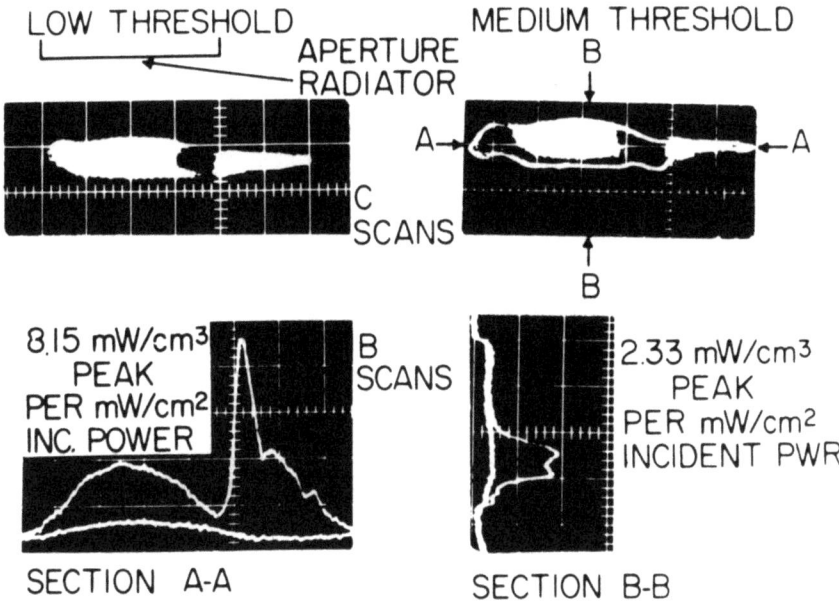

Fig. 5. Thermographic study of rat model exposed to 915 MHz aperture irradiator. Thermograms illustrate regions of high absorption, especially near the base of the tail.

currents generated in the body of the animal being sharply constricted to the smaller cross-sectional region of the tail, producing an SAR value of 8 W/kg (compared to a theoretical value of .5 W/kg as the SAR averaged over the entire body of the animal) for an exposure power density of 1 mW/cm^2.

The versatility of the technique is further demonstrated in the study of induced temperature and SAR patterns at the surface of the body of an exposed pigeon. In exposing the sacrificed bird oriented in a flying position to 2450 MHz plane-wave radiation, one can quantitate the biological impact of a solar power satellite on birds flying through the beam of the microwave transmitting antenna. Such thermograms illustrate the vulnerability of the bird by indicating high SAR in the wings and the neck. This is expected, since the tissues in the wings are quite thin and the neck is long and small in diameter, allowing much stronger coupling than average of the electric fields to these tissues.

The major problem area especially suited for thermographic study is the quantitation of the many different near-zone field exposure situations. Virtually all of the world's safety standards at this time, and most of the new safety standards being proposed, are based on whole-body exposure to plane waves. On the other hand, most of the actual exposures are partial-body in nature and in many cases result from near-zone fields. Personnel occupationally exposed while operating RF sealers may be exposed to high electromagnetic fields localized to the hands and arms. Persons operating hand-held handy-talkies may be exposed to high electromagnetic fields localized in the region of the head and hands. Other highly concentrated exposures may result when tissues come in contact with metal joints in microwave ovens and cavities. With such complex fields incident to complex tissue geometries, it is not difficult to see the many formidable problems one might encounter in seeking theoretical solutions. The thermographic technique, however, lends itself well to quantifying the energy coupling to tissues from such sources.

Fig. 6 illustrates the use of thermography in determining the SAR patterns in the head of a child, based on a full-sized model exposed to a leaky 915 MHz microwave oven. In order to make the

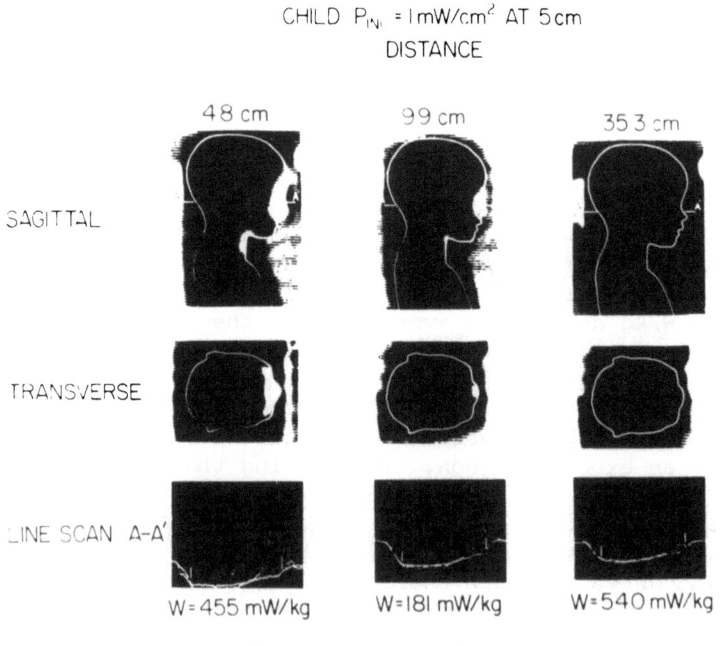

CHILD $P_{IN} = 1 mW/cm^2$ AT 5 cm DISTANCE

Fig. 6. Thermograms of exposed child model corresponding to different distances from the oven, with results normalized for oven leakage of 1 mW/cm^2 measured 5 cm from the oven.

measurements, the microwave oven was modified so that the
leakage fields were allowed to increase to very high levels
sufficient to produce the heating in the model. The illustration
is for the case when the child's face is in a position 4.8 cm from
the oven door, and the measured values of SAR are normalized to
correspond to power density of 1 mW/cm^2 5 cm from the oven door
(the position where most performance standards apply). Based on
these conditions, a maximum SAR of 455 mW/kg is seen to occur in
the area of the nose and the eyes of the phantom model of the
child. At the maximum allowed leakage level of 5 mW/cm^2, these
values would increase to 2 W/kg. The SAR decreases rapidly with
distance between the model and the oven. Normally, plane-wave
exposure at this frequency would produce hotspots in the head,
but due to the rapidly diverging fields of the leaky microwave
oven door, such hotspots do not occur in this case.

I have tried, in my brief discussion, to illustrate some of the
techniques that we have used, and that many other researchers
are now using, in order to quantify average SAR and SAR
distributions in exposed laboratory animals and man. There are
certainly many other techniques that can be used to quantify
this coupling, but there are many limitations depending on the
complexity of the exposure conditions. I think that, in the
future, we will have to be very concerned about quantifying a
large number of near-zone field and partial-body exposure conditions, in order to establish the SAR patterns especially
corresponding to operation of RF sealers, mobile transmitters,
and handy-talkies, in order to develop realistic safety standards
for these RF sources. Thank you.

<u>Mr F. Harlen</u> - Thank you very much Professor Guy! Could we have
Dr Burdette now please?

<u>Dr Burdette</u> - I will discuss briefly two areas involving both
theory and applications. The first of these is an area of
dielectric property measurements. The second area deals with
electromagnetic dosimetry and application of the volume integral
equation.

What I would like to do is to present a dielectric property
measurement technique that has proven to be quite useful for
determining the dielectric characteristics of lossy materials.
In the development of such a technique there were three major
criteria: first, an ability to perform <u>in-situ</u> dielectric
measurements on living tissue (in fact, the technique is very
useful for any liquid dielectric material, lossy or low-loss);
second, a capability of operating over a wide frequency range
either in frequency - or in time-domain; third, an ability to
rapidly collect and process dielectric data. The particular

theorem involved is an antenna modelling theorem. Work utilizing this theorem has been reported by Professor Deschamps (IRE Trans. Antennas & Prop., Sept. 1962, pp. 648-50). The theorem states that for a short open antenna in a lossy dielectric medium, the impedance of that antenna at frequency ω is equal to the impedance of the same antenna in free space at a frequency nω where n is the complex index of refraction. In this case, we consider non-magnetic media ($\mu = \mu_0$):

$$\frac{Z(\omega, \epsilon^*)}{\eta} = \frac{Z(n\omega, \epsilon_0)}{\eta_0} \tag{1}$$

$\eta = \sqrt{\mu_0/\epsilon^*}$ = intrinsic impedance for dielectric

$\eta_0 = \sqrt{\mu_0/\epsilon_0}$ = intrinsic impedance for free space

$\epsilon^* = \epsilon' - j\epsilon''$ = complex permittivity for dielectric

ϵ_0 = permittivity for free space

$n = \sqrt{\epsilon^*/\epsilon_0}$ = complex index of refraction

Using the simple impedance formulation for a short monopole antenna (λ/10 or less in length) in the material and substituting into Equation (1), the result is

$$Z(\omega, \epsilon^*) = A\omega^2 \sqrt{K'(1-j\tan\delta)} + \frac{1}{jC\,K'(1-j\tan\delta)} \tag{2}$$

In Equation (2) the measured impedance in the lossy dielectric medium is related to the dielectric constant K' and loss-tangent tan δ of the unknown dielectric material, where the constants A and C are determined by physical dimensions of the antenna. Equation (2) can be placed in the form Z = R + jX which reduces to two real equations:

$$R = \frac{\sin 2\delta}{2K'\omega C} + A \sqrt{K'} \omega^2 \sqrt{\frac{\sec\delta + 1}{2}}$$

(3)

$$X = \frac{\cos^2\delta}{K'\omega C} + A \sqrt{K'} \omega^2 \sqrt{\frac{\sec\delta - 1}{2}}$$

The low frequency solution will involve only the first term of each equation and by knowing that tan δ = R/X one can solve for δ and, therefore, the dielectric constant and loss tangent. At higher frequencies where the probe impedance is not totally reactive, the second term must be included.

The short monopole probe is basically a monopole antenna of length ℓ, as shown in Figure 1. To insure the validity of the analytical expression for the impedance of a short linear antenna, its length cannot be longer than $\lambda/10$ in the medium under study. This limits severely the upper useful frequency range of this technique. Therefore, we have examined the case of infinitesimal probe in an open-circuit coaxial transmission line. Inside the line the TEM mode exists and beyond the end of the transmission line only a small field exists due to the capacitance of the termination. A very simple impedance expression results for the free space case. In a dielectric medium, this impedance is changed by the medium. The measured impedance change is related to the dielectric properties of the medium through the second term in Equation (2). For the open-circuit probe, the first term of Equation (2) is zero.

A number of probe configurations have been investigated (Burdette, Cain, and Seals, "In-Vivo Probe Measurement Techniques...", IEEE Trans. MTT, 1980). The lengths of the extended center conductor of the probes have ranged from infinitesimal lengths of less than 0.02 inch in length to lengths of 0.4 inch, and probe outside diameters have ranged from the size of a 18-gauge hypodermic needle (approximately 0.042 inch) to 0.141 inch.

The in-vivo measurement probes are each fabricated from a section of open-ended semi-rigid 0.085-inch diameter coaxial cable with a slightly extended center conductor. An SMA connector is attached to the probe by first removing the center conductor and teflon dielectric material. The connector is then soldered to the outer conductor followed by reassembly of the probe using the center conductor as the center pin of the connector, thus avoiding additional soldering. In this manner, it is possible to attach the SMA connector without heating the teflon dielectric. While disassembled, the center and outer conductors of the probe are

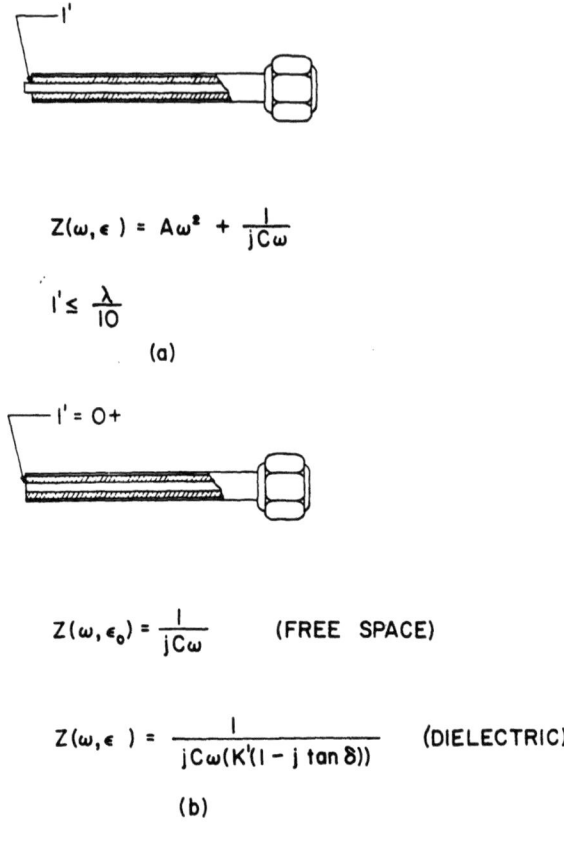

Fig. 1 Probes

(a) Short Monopole Probe   (Upper useful frequency range ≈ 3GHz)

(b) "Infinitesimal" Probe   (Upper useful frequency range > 10GHz)

first flashed with nickel plating and then gold plated. Plating the probe with an inert metal, such as gold, greatly reduces chemical reactions between the probe and the electrolyte in the tissue. This process virtually eliminates oxidation of the probe's metallic surfaces and helps to minimize electrode polarization effects at lower frequencies (0.01 - 0.1 GHz).

More recently, we have been working with smaller probes, similar to hypodermic needles. It should be noted that the lower frequency limitation on this technique is primarily the ability to resolve small changes in the measured complex reflection coefficient, while the upper frequency limitation tends to be either connector problems or radiation from the probe. The useful frequency range of the present dielectric probe is from 0.01 to 10 GHz.

The impedance measurement instrumentation employed to measure the terminal impedance of the probe is schematically illustrated in Figure 2. The key components of the measurement system are the

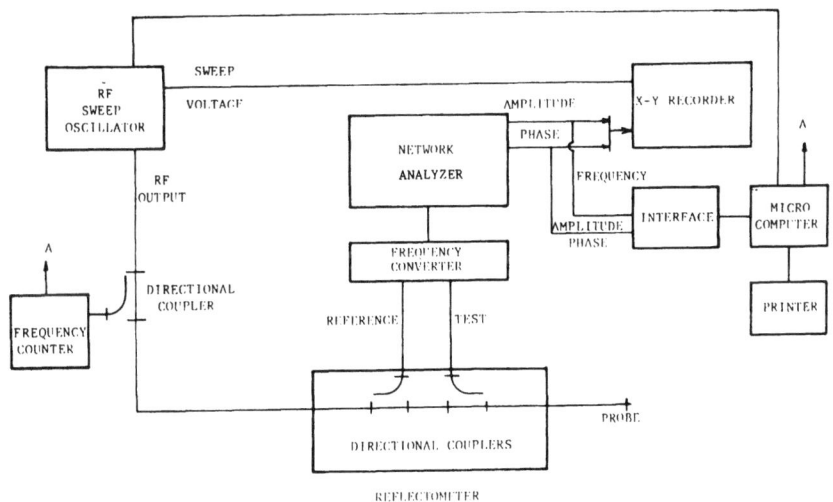

Fig. 2

Block diagram of <u>in-vivo</u> dielectric property measurement system consisting of probe, network analyzer, and associated instrumentation.

probe and the network analyzer, which is a Hewlett-Packard Model 8410B. The relative amplitude and phase difference between the reference and reflected signal channels is measured by the network analyzer, which functions as an amplitude and phase comparator. The network analyzer yields the terminal impedance of the probe in terms of the magnitude and phase angle of the reflection coefficient, and these data are used as input data to a computer algorithm which corrects systemic measurement errors and computes the dielectric property information. A semi-automated data acquisition/data processing system whose key components are an analog/digital converter and micro-processor was also designed and implemented to increase the rate at which in-vivo dielectric data could be acquired and processed.

When the network analyzer system is used for performing microwave measurements, there exist certain inherent measurement errors which can be separated into two categories: instrument errors and test set/connection errors. Instrument errors are measurement variations due to noise, imperfect conversions in such equipment as the frequency converter, cross-talk, inaccurate logarithmic conversion, non-linearity in displays, and overall drift of the system. Test set/connection errors are due to the directional couplers in the reflectometer, imperfect cables, and the use of connector adapters. The instrument errors exhibited by the HP 8410B network analyzer are very small.

The primary source of measurement uncertainty is due to test set/connector errors at UHF and microwave frequencies. These uncertainties are quantified as directivity, source match, and frequency tracking errors.

An analytical model for correcting test set/connection errors was implemented based on the model used by Hewlett-Packard for correcting reflectivity measurements on their semi-automatic network analyzer system.

In any measurement technique, the first thing one must do is validate the technique. This was done in this case by measuring standard dielectric materials. Results using the probe measurement method on water and ethylene glycol are presented in Figures 3 and 4 for relative dielectric constant and relative loss factor. The accuracy of this dielectric measurement technique is approximately ± 5% when compared to reference date (A.R. Von Hippel, DIELECTRIC MATERIALS AND APPLICATIONS, M.I.T. Press, 1954, pp. 36-40, 301-425; F. Buckley and A.A. Maryott, "Tables of Dielectric Dispersion Data for Pure Liquids and Dilute Solutions," National Bureau of Standards Circular 589, November 1958) and within ± 3% when compared to dielectric values predicted by the Debye theory.

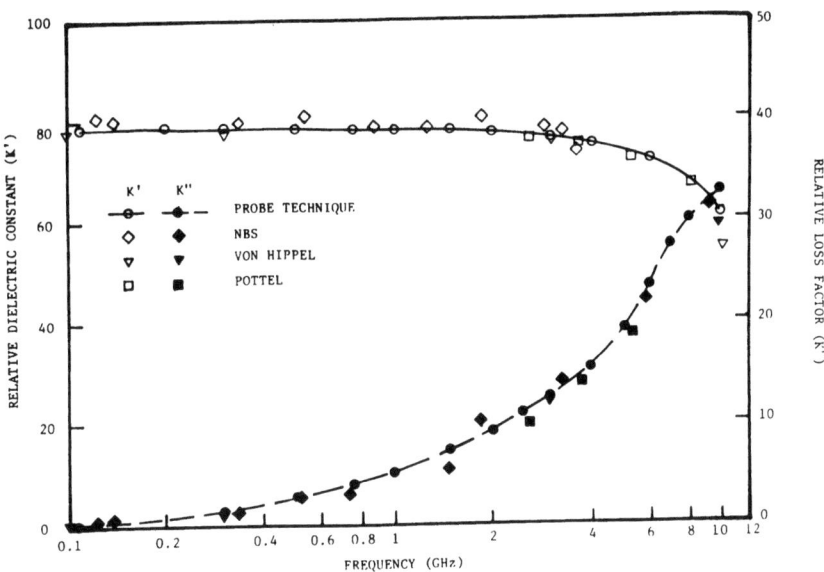

Fig. 3 Deionized water. Maximum SEM for probe measurements
K' = ± 2.7; K" = ± 2.5.

Experimentally determined relative dielectric constant and relative loss factor at 23°C compared to reference data.

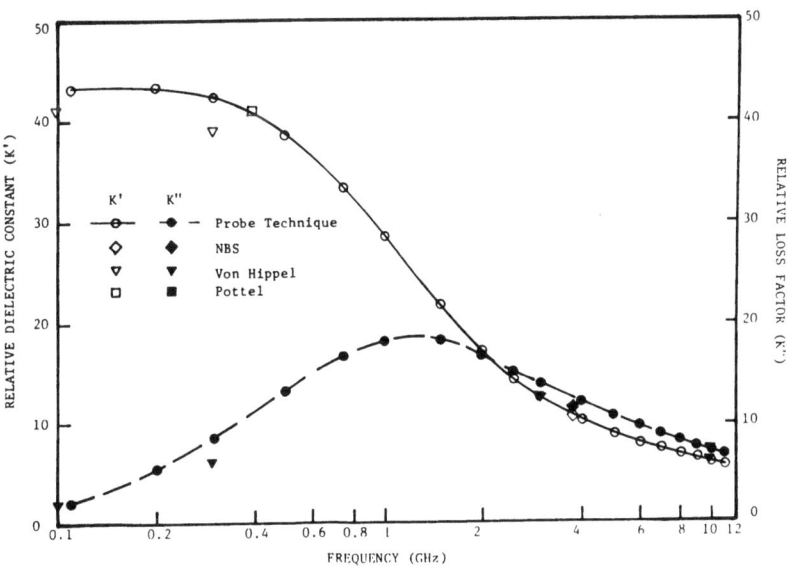

Fig. 4 Ethylene glycol. Maximum SEM for probe measurements:
K' = ± 1.15; K" = ± 1.25.

This technique has proven very useful in measuring the dielectric characteristics of tissue for electromagnetic applications in medical work, such as hyperthermia, organ preservation work, radiometric and/or dielectric imaging, etc. The upper curve in Figure 5 presents _in-vivo_ dielectric data for rat muscle; the other curve is _in-vivo_ canine tissue data, and the unconnected circles are human autopsy material measured by Schwan. The large difference between the _in-vivo_ values is not accountable by the small temperature differential alone, but is attributed to increased perfusion of rat muscle with blood as compared to the canine tissue.

I wish to describe one other area where dielectric property information is essential to accurate analytical predictions of dosimetry and to effective medical application of EM energy. That area is EM-induced hyperthermia as a method for treating cancer. At present, tumor dielectric properties in seven different tumor tissues in mice have been studied with the intention of not only taking advantage of thermal regulatory differences in tumor and in normal tissue but also taking advantage of possible dielectric differences causing differential absorption of EM energy between normal and malignant tissues. If certain frequencies could be determined to be more useful for differential hyperthermia than others, this would provide essential information for the selection of an appropriate frequency or frequencies for a particular tumor type which produces maximum differential heating of the tumor with respect to normal tissues. _In-vivo_ dielectric properties of human and animal tumors and normal tissues have been measured over a frequency range of 0.01 - 4.0 GHz. The results indicate that substantial differences exist between the dielectric properties of the measured tumors (C3HBA, Mendecki, and 16C mammary adenocarcinoma, B16 melanoma, Lewis lung carcinoma, glioblastoma and ependymoblastoma) and the normal tissues which would typically surround the tumors. Differences between normal tissues and tumors of up to 300% are exhibited for some tumor types. These differences were also seen in measurements of human tumors. _In-vivo_ measurements of breast carcinoma and normal skin and breast tissue indicate that over the 0.5 to 4.0 GHz frequency range maximum differential power absorption (30% higher in tumor) occurs between 1.0 and 2.0 GHz. These dielectric measurement results for human breast carcinomas are similar to those obtained for mammary adenocarcinomas in experimental animals. Significant differences (35%) in the dielectric characteristics of melanoma and normal skin were also measured.

This information along with considerations of tumor location, size, and geometry is necessary in the design of effective clinical, electromagnetically-induced, differential hyperthermia procedures for the treatment of malignant tumors. In addition,

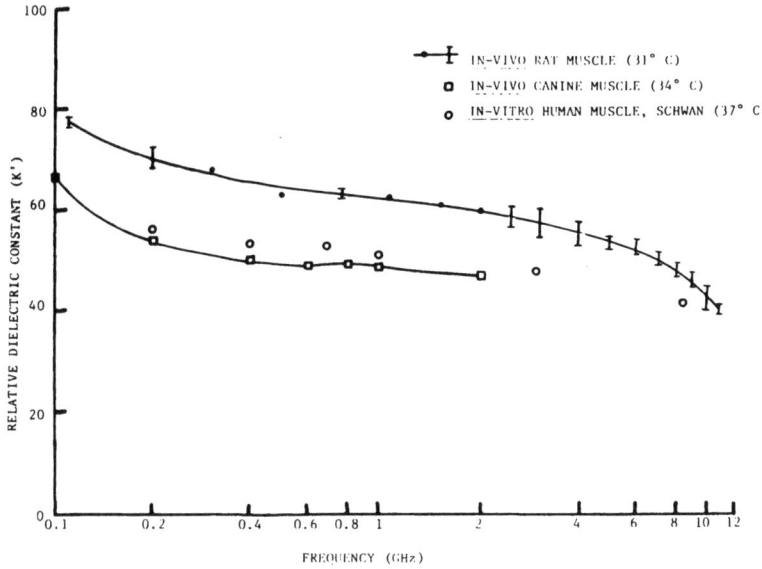

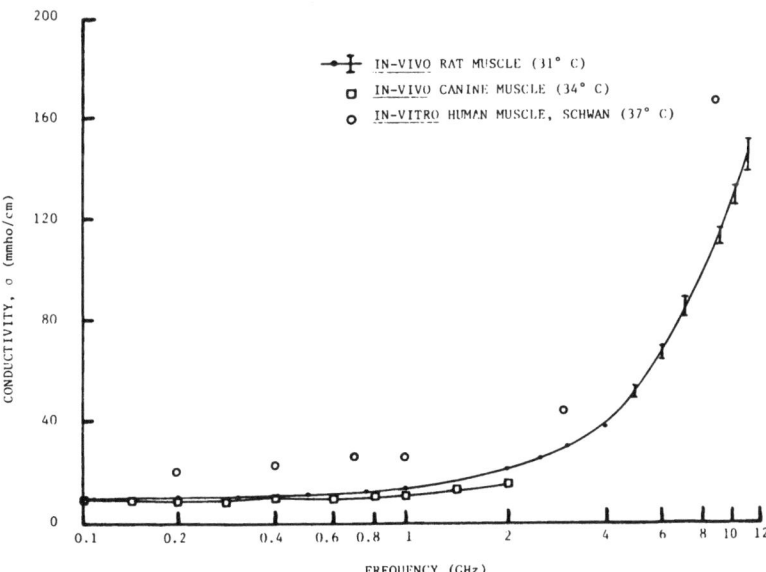

Fig. 5 Experimentally determined relative dielectric constant and conductivity of <u>in-vivo</u> rat muscle and canine muscle compared to reference data.

tissue dielectric properties could be useful in investigations of normal and pathological tissue processes such as malignancy, fibrosis, and edema. Once changes in tissue dielectric properties are correlated to a specific pathological process, then this information can be used as a diagnostic tool.

Obviously, this dielectric information is necessary in the computation of the electromagnetic field distribution inside the tissue. Work on the problem of electromagnetic dosimetry has been done by a number of investigators in the field. The following presentation is only a brief summary of one approach being used at Georgia Tech. For dielectric bodies, we used a volume-cell moment method approach for solving the integral equation. This technique has been used at Georgia Tech for the following applications:

>System Design for EM Thawing of Blood;
>System Design for EM Thawing of Solid Organs;
>EM Hyperthermia in Cancer Treatment (Treatment Planning);
>Enzyme Inactivation (Neurochemistry);
>EM Hazards and Dosimetry.

Obviously, for solid organs only a very small surface area exists relative to total organ volume and conventional conduction heating techniques (water bath) are useless. Attempts to use conventional methods for thawing organs such as kidneys have failed. Only EM thawing has ever resulted in a viable kidney following freezing, storage, and thawing; and the possibility of organ banks exists with further development of EM thawing. Other applications, such as EM hyperthermia, are being investigated and in each case, the success of the medical application of EM energy is dependent upon sound theoretical and experimental development. The possibility of treatment planning for EM hyperthermia depends upon the development of suitable analytical models for computing EM power absorption and field distribution. The approach we examined permits modelling of an arbitrary three-dimensional body with a location-dependent permittivity. The electric field equation is obtained by a formulation using the free space dyadic

$$\underline{E}(\underline{r}) = \int_{V-V_\delta} \underline{J}(\underline{r}') \cdot \underline{\underline{G}}(\underline{r}, \underline{r}') dV - \int_{V_\delta} \underline{J} \cdot \underline{\underline{G}}\, dV \quad (4)$$

Green's function of the electric type which is singular at the source point (Figure 6). There has been some discussion on how

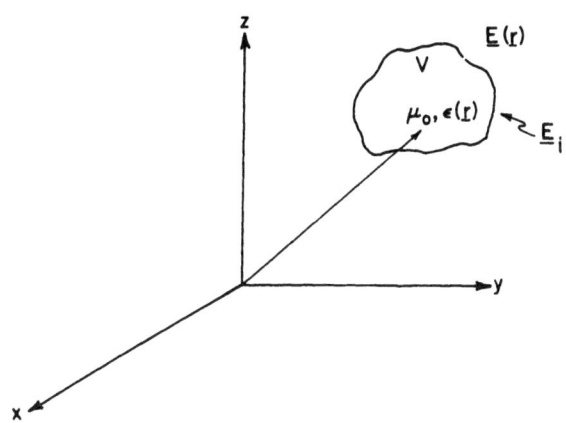

Replace dielectric by equivalent current source:

$$\underline{J}e(\underline{r}) = j\omega(\epsilon(\underline{r})-\epsilon_o)\underline{E}(\underline{r})$$

Electric field:

$$\underline{E}(\underline{r}) = \int_V \underline{\underline{G}}(\underline{r},\underline{r}')\, j\omega(\epsilon(\underline{r})-\epsilon_o)\underline{E}(\underline{r}')\, d\underline{r}' + \underline{E}_i(\underline{r})$$

where $\underline{\underline{G}}$ is free-space electric dyadic Green's function

$$\frac{\underline{J}e(\underline{r})}{j\omega(\epsilon(\underline{r})-\epsilon_o)} - \int_V \underline{\underline{G}}(\underline{r},\underline{r}')\cdot \underline{J}e(\underline{r}')\, d\underline{r}' = \underline{E}_i(\underline{r})$$

Singular when $\underline{r} \to \underline{r}'$

Fig. 6.

Volume Integral Equation Formulation

to handle this singularity and the approach that we have taken is a principal volume (PV) integration. Two different dielectric bodies are shown in Figure 7 with a cell around the source region including the singularity at the source point which is either spherical or cubical in shape.

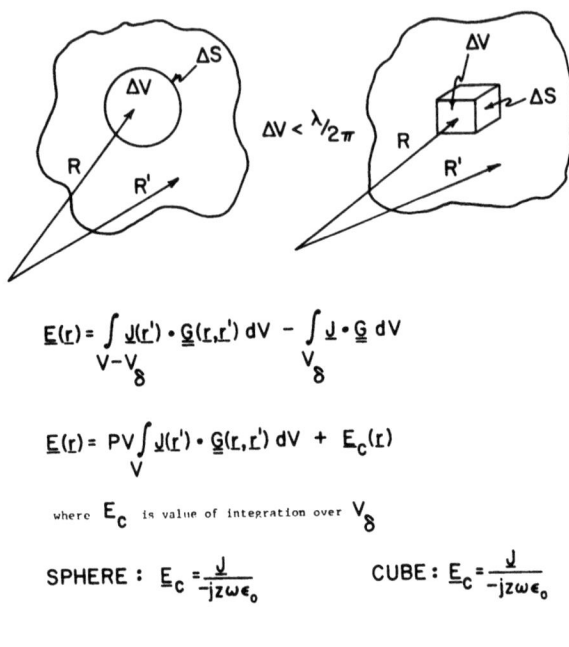

Fig. 7

Principal Value Integration

The integral equation of a PV integral plus the self-cell term is given by

$$\underline{E}(\underline{r}) = PV \int_V \underline{J}(\underline{r}') \cdot \underline{\underline{G}}(\underline{r}, \underline{r}') dV + \underline{E}_c(\underline{r}) \quad (5)$$

where E_c is the value of integration over the body containing the singularity (self-cell). Professor van Bladel looked at this singularity problem some time ago and reported some interesting results for the sphere and also the cube. Dr. Chen at Michigan State has used this result in some of his man models for computing EM dosimetry. (Dr. Yaghjian of the National Bureau of Standards is going to briefly discuss some work he has done related to this problem in a few minutes.) We have been using this formulation at Georgia Tech, where arbitrarily-shaped dielectric bodies are modelled using cubes. However, we were able to use the formulation from the sphere case for removing the source region singularity because, in fact, the results for both are identical (Figure 7). The result for the cylinder formulation in the source region is actually different. The final result is given by Equation (5) which can be solved using moment method techniques involving simple pulse expansion and point matching. The solutions obtained using this approach for EM dosimetry has provided useful results in our applications. The maximum size dielectric body for which we have been able to experimentally verify the analytical solution is approximately $\lambda/2$ free space wavelengths, corresponding to 3λ in the dielectric material (ka $\simeq$ 12). The solution converged rapidly and good agreement with experimental results was obtained. Integration over larger volumes would require that the self-cell be of the order $\lambda/2\pi$.

At this point, I would ask Dr Yaghjian to discuss the work he has done using PV integration and the problems associated with evaluating the integral in the singular region. Following his presentation, I would like to show you a result of an application of the volume integral equation in a biological model.

<u>Dr A.D. Yaghjian</u> - The equation for which a solution is desired is shown in Figure 8. There is a simple method for dealing with

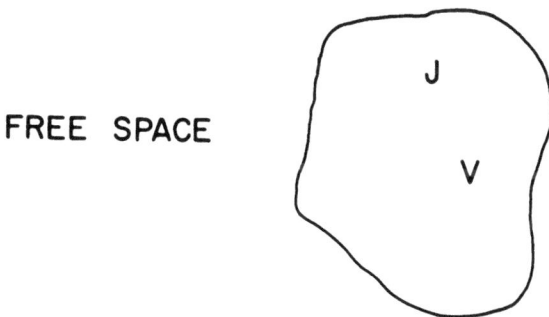

Fig. 8

$$E(\underline{r}) = j\omega\mu_0 \int_V \underline{\underline{G}}_a \cdot \underline{J} \, dV'$$

$$\underline{\underline{G}}_a = \frac{1}{4\pi}\left[\underline{\underline{I}} + \frac{\nabla\nabla}{k^2}\right]\psi$$

the singularity which exists in the source region. Often, people using this equation have problems with the singularity. They will start evaluating it near the singularity and find that depending upon how one approaches the singularity (depending on how that singularity is enclosed) a different value for the integral is obtained. The case illustrated in Figure 8 uses the free space Green's function.

If the solution is obtained rigorously without interchanging limits or interchanging integrals and derivatives, indeed one obtains: (See also Figure 9)

$$\underline{E}(\underline{r}) = i\omega\mu_o \lim_{\delta \to 0} \int_{V-V_\delta} \underline{\underline{G}}_a(\underline{r}, \underline{r}') \cdot \underline{J}(\underline{r}') \, dV' + \frac{\underline{\underline{L}}_\delta \cdot \underline{J}}{i\omega\varepsilon_o} \tag{6}$$

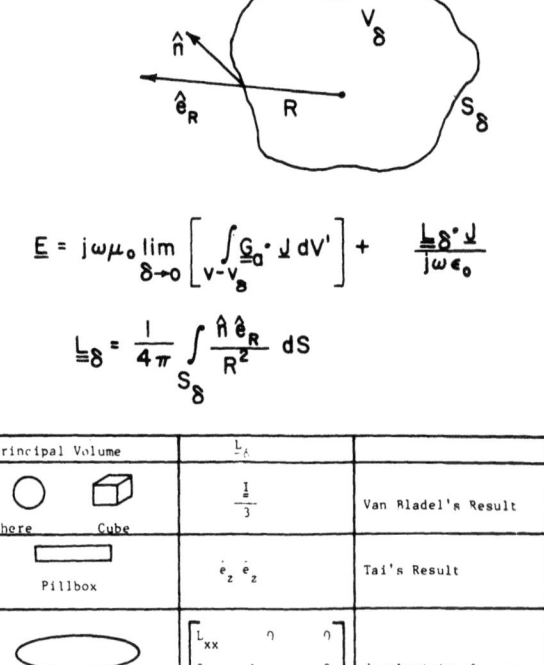

Fig. 9 Electric Dyadic Formulation

where the second term in Equation (6) encloses the singularity with the principal volume (PV) and depending on the geometry of that PV which encloses the singularity, an extra dyadic $\overline{\overline{L}}_\delta$ is obtained which depends only on the shape of that PV about the singularity. The answer obtained for both terms in Equation (6) is dependent upon the PV selected, but the two terms always add to give the same electric field. Evaluating the dyadic $\overline{\overline{L}}_\delta$ with the sphere you get Professor van Bladel's result: $\overline{\overline{I}}/3$. If you use it for a cube you get again $\overline{\overline{I}}/3$. If you use a small pill-box around the singularity you get an easy term which is a result for a couple of problems. In general, this extra dyadic term $\overline{\overline{L}}$ is associated with depolarization factors. If the PV around the singularity is reduced to an ellipsoid, use of the principal axes of that ellipsoid results in three diagonal terms to this $\overline{\overline{L}}_\delta$ dyadic, and these are just equal to the depolarization factors for the ellipsoid. So, $\overline{\overline{L}}_\delta$ is actually a generalized depolarization tensor or dyadic which actually tells you what you remove. Recall the old electrostatics problem of removing a little piece from a dielectric. The perturbation in the electric field created by that removal depends very distinctly on the shape of the small volume that was removed. The dyadic $\overline{\overline{L}}_\delta$ can describe exactly the perturbation on an electric field if you specify the shape of the small volume you remove. The dyadic $\overline{\overline{L}}_\delta$ also describes any internal field perturbation due to the presence of a small metallic probe or implant. This is done by simply including the equivalent source current $\overline{\overline{J}}_{inc}$, or equivalent current and polarization $\overline{\overline{J}}_e$, as illustrated in Figure 10. The second term in $\overline{\overline{J}}_e$ goes to zero outside the body. Place that in the equation in Figure 10 and an integral equation of the second kind is obtained. It applied to bodies of arbitrary shapes, it allows inhomogenous and isotripic lossy media, near field solutions, anything you want. The solution is obtainable by point matching of an integral equation of the second kind which makes the solution easier than some other approaches. If the vector potential approach is used, then more involved basis and/or expansion functions have to be used. The solution is unique; it does satisfy Maxwell's equations inside and outside the dielectric, and it also satisfies the tangential boundary conditions accross the boundary. Of course, this approach has its limitations. The number of cells required, at least in a

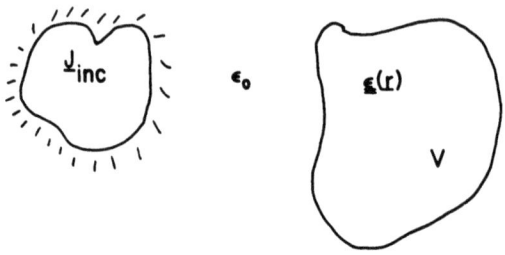

$$\underline{J}_e = \underline{J}_{inc} - j\omega(\underline{\epsilon} - \epsilon_o)\underline{E}$$

$$\underline{E} = \underline{E}_{inc} + k^2 \int_{V-V_\delta} \underline{G}_e \cdot (\tfrac{\underline{\epsilon}}{\epsilon_o} - 1)\underline{E}\, dV'$$

$$-\underline{\underline{L}}_\delta \cdot (\tfrac{\underline{\epsilon}}{\epsilon_o} - 1)\underline{E}$$

Pros: (1) Applies to bodies of arbitrary shape.
(2) Allows inhomogeneous, anisotropic, lossy media.
(3) Arbitrary incident field.
(4) Solution obtainable by point matching integral equation of the 2nd kind.
(5) Solution is unique.

Cons: Limited size; no. of cells $\sim (ka)^3$

Fig. 10   Volume Integral Equation

point matching scheme, is proportional to the volume, or more specifically, to $(ka)^3$. However, if you have a problem suitable for solution using the volume integral equation, don't let the singularity problem prevent your use of this approach.

Dr Burdette - Respective of other techniques such as those presented by Professor Yeh, the Volume Integral Equation is a useful technique having practical application, and at Georgia Tech, we use it for analytical modelling of dielectric bodies, such as tumors and organs. Results from these analyses are then used to optimize hardware designs. Computed results for absorbed power distribution in three dielectric cylinder models are illustrated in Figure 11 for the case where the electric field is parallel to the cylindrical axis. Note the good convergence obtained.

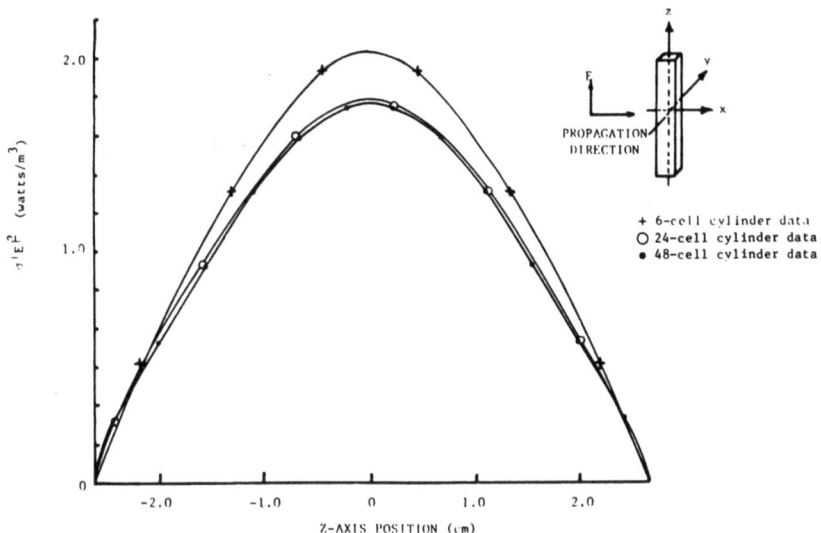

Fig. 11

In Figures 12 and 13, two cases are illustrated where computed results are compared to measurements. The predicted temperature rise is computed from

$$\Delta T_m = \frac{J_m E_m^2}{k \rho_m C_m}, \qquad (7)$$

where C_m and ρ_m are the specific heat and density, respectively, of the material, ΔT_m is the temperature rise, k is a proportionality factor for converting units, and the E-field is the computed solution to the volume integral equation. The use of Equation (7) has been previously described by Dr Guy (<u>IEEE Trans. MTT</u>, Vol. MTT-19(2):205-214, 1971). Results of predicted and measured temperature rise in the presence of an applied EM field are in good agreement for both the simple cylindrical model (Figure 12) and the rat-shaped model (Figure 13).

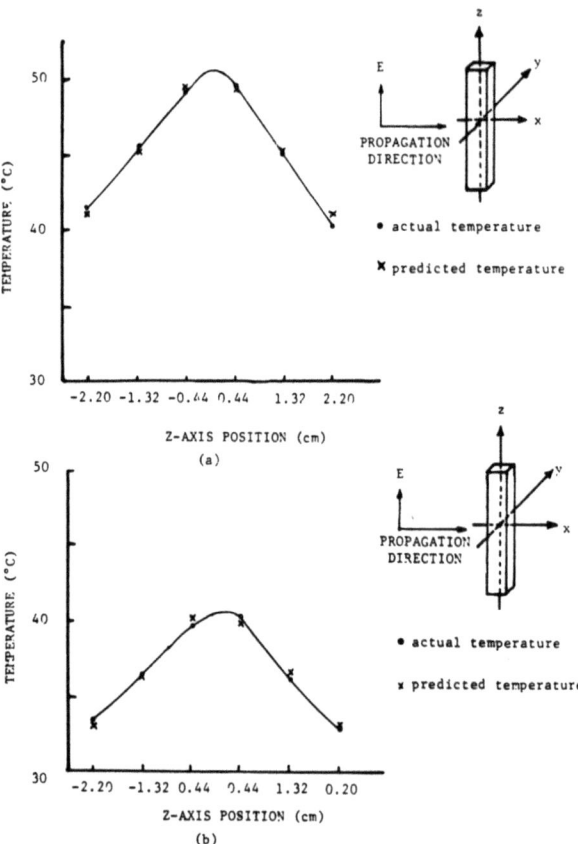

Fig. 12

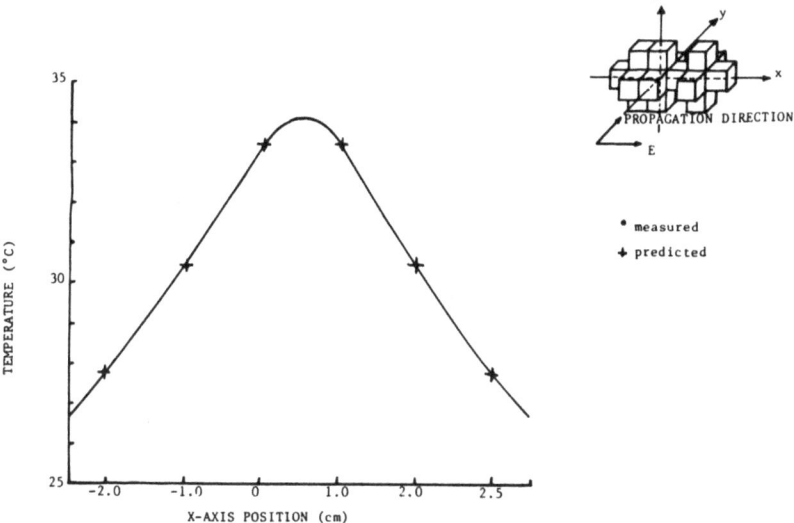

Fig. 13 Comparison of measured and computed temperatures in rat-shaped dielectric model as a function of position along axis (E-field parallel to axis). Origin of axis is located as shown in three-dimensional view of model.

In summary, the volume-cell integral equation method for computing field distributions in dielectric properties, has provided useful solutions to EM dosimetry problems in thawing of solid organs, EM hyperthermia treatment of cancer, and EM radiation bioeffects studies.

Mr Harlen - Let us now turn to Professor Iskander and hear about his work at Utah.

Professor M.F. Iskander - In this presentation I would like to summarize our recent results of the near-field irradiation of biological models. With the availability of significant data about the specific absorption rate (SAR) for plane wave irradiation, attention has been focused on near-field exposure conditions. This is because many industrial and medical applications utilize radio-frequency sources operating at lower frequencies, and therefore the exposures are occurring in the near field, and their effects cannot be adequately explained in terms of plane wave results. Furthermore, it is believed that some biological effects occurring at low power levels (<500 μwatts) are due to the concentration of thermalized energy -- hot spots -- which may conspire to produce thermal stimulation.

At the University of Utah, we are studying the problem of the near-field exposure of biological models both theoretically and experimentally. The following is the summary of both.

Firstly, the theoretical efforts involve the analysis of the near-field irradiation of prolate spheroidal models of humans and animals by simple sources such as a short current element (electric dipole) or a small current loop. The prolate spheroidal model is used since, for plane wave irradiation case, it was found to give an estimate of the average SAR which agrees very well with those obtained from more realistic but complicated models [1]. It should be noted that realistic models are important for analysis dealing with SAR distribution, with emphasis on the location and the level of hot spots and partial body resonances. The near-field radiation from the used simple sources can be calculated exactly and hence used to identify and establish the suitable field parameters necessary to quantify the hazardous level of electromagnetic radiation. The absorption characteristics due to the radiation from these simple sources can then be used as building blocks to describe the absorption due to the radiation from more complicated and realistic sources.

The method of solution involves an integral equation formulation of the problem in terms of the transverse dyadic Green's function and expanding the fields irradiated by the simple sources in terms of the vector spherical harmonics. The extended boundary condition method (EBCM) is employed to solve the integral equation [2]. Based on our initial experience with the EBCM, it

---

[1] C.H. Durney, C.C. Johnson, P.W. Barber, H. Massoudi, M.F. Iskander, J.L. Lords, D.K. Ryser, S.J. Allen, J.C. Mitchell, "Radiofrequency Radiation Dosimetry Handbook: Second Edition", Departments of Electrical Engineering and Bioengineering, University of Utah, Salt Lake City, Utah 84112, 1978.

[2] M.F. Iskander, P.W. Barber, C.H. Durney, and H. Massoudi, "Irradiation of Prolate Spheroidal Models of Humans in the Near-Field of a Short Electric Dipole", *IEE Transactions on Microwave Theory and Techniques*, in press 1980.

was anticipated that the method would be suitable for providing accurate results for frequencies up to resonance.

Let us consider the numerical results obtained when a prolate spheroidal model of man is exposed to near field of a short current element (electric dipole) [2]. Considerable simplification in the analysis and the computer programming were achieved by restricting the dipole location to specific positions of interest (see Fig. 1). This includes locating the dipole at the $\phi = 0$ plane and specifying values of θ at $180°$ and $90°$. These specific values of θ correspond to the K-polarization and the E-polarization, respectively, for the plane wave incident cases [1].

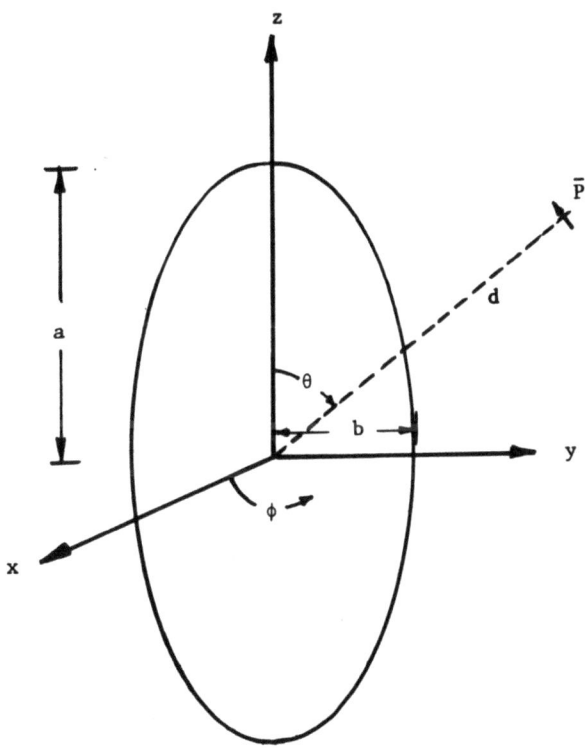

Fig. 1. A dipole $\bar{P}$ located at distance $\bar{r}'$ (d, θ, ϕ) from a prolate spheroid model of man.

Consider first the case where the dipole is placed on the $\phi = 0$ plane and at $\theta = 180°$ (K-polarization case). The power distribution in a spheroidal model of man is shown in Fig.2 for three different separation distances. The semimajor axis of this spheroidal model is a = 0.875 meter and the ratio between the semimajor to the semiminor axes is a/b = 6.34. The complex permittivity is taken equal to 2/3 that of the muscle tissue, i.e., $\varepsilon' = 78.5$ and $\varepsilon'' = 270$ at 27 MHz. From Fig.2 it is clear that while the distribution at large separation distance from

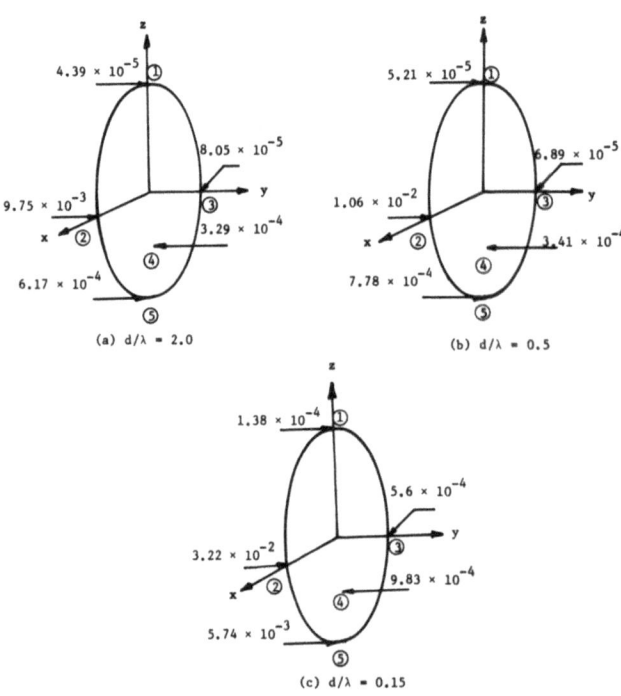

Fig. 2. SAR distribution at 27 MHz as a function of dipole location. $P = P_x$ and $\theta = \pi$. (a) $d/\lambda = 2$. (b) $d/\lambda = 0.5$. (c) $d/\lambda = 0.15$. The indicated locations (r, θ, ϕ) are: ① (a, 0, 0), ② (b, $\pi/2$, 0), ③ (b, $\pi/2$, $\pi/2$), ④ (a/2, π, 0), and ⑤ (a, π, 0).

the dipole d is similar to that of the plane wave case, significant variation in the power distribution is observed for smaller d. In particular for $d/\lambda = 0.15$, the relative power values at both ends of the spheroid were about 40 compared with the ratio of 15 in the plane wave irradiation case. This result simply suggests the possible enhancement of the near-field absorption in regions of small radius of curvature.

Next, we consider a dipole located at the $\phi = 0$ plane and $\Theta = 90°$ (E-polarization case). For this case it is more interesting to examine the average SAR values obtained by integrating the power distribution over the whole volume of the spheroid. From Fig.3 it is clear that the average SAR values oscillate around the

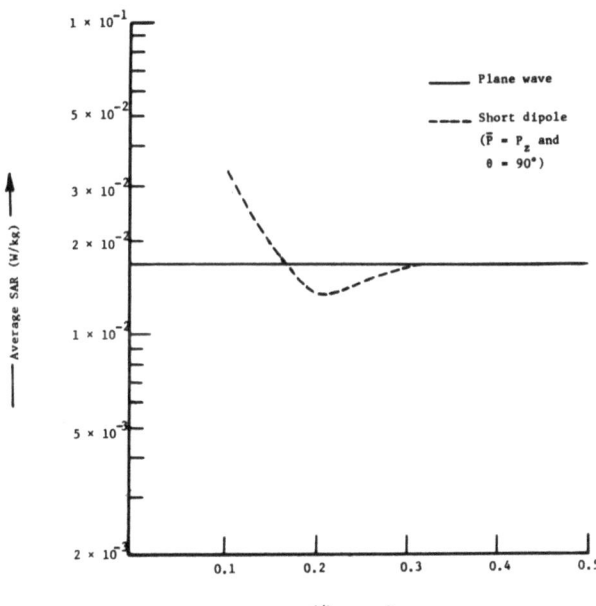

Fig. 3. Average SAR (W/kg) values at 27 MHz of a prolate spheroidal model of man as a function of the dipole location. $\bar{P} = P_z$ and $\Theta = \pi/2$.

constant value obtained from the plane wave irradiation. It
should be noted that this oscillation of the average SAR values
for the E-polarization case and the possible enhancement of the
near-field absorption in regions of small radius of curvature are
general characteristics and are expected to be valid for other
spheroidal models of different dimensions at frequencies below
the resonance frequency. The magnitudes of these changes, however, will obviously depend on the typical dimensions of the
model and the exposure frequency. In an attempt to verify these
results, we employed the experimental procedure described as
follows.

A schematic diagram illustrating the experimental setup is shown
in Fig. 4. A half prolate spheroidal model was placed in direct
contact with an infinite ground plane and was irradiated by a
short electric monopole placed at a distance d from the center of
the spheroidal phantom. From image theory, it can be easily shown
that this arrangement is equivalent to the irradiation of a full

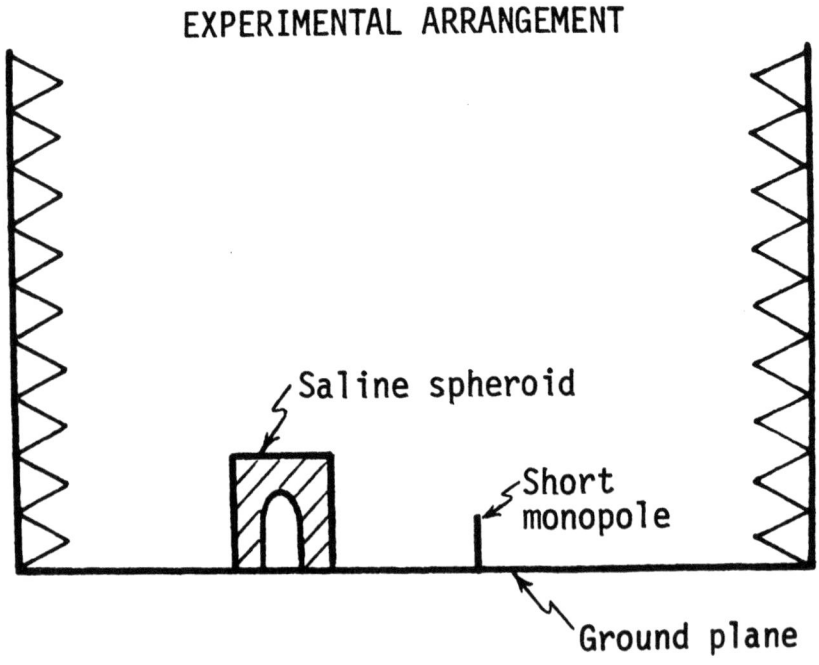

Fig. 4. Schematic diagram illustrating the
experimental arrangement.

spheroid by a short electric dipole. The radiation fields at the
location of the phantom were measured using electric- and
magnetic-field probes. As indicated in the previous section, we
used in the theoretical analysis a short current element (electrical dipole) as a source because its near fields are known
exactly, which considerably simplifies the theoretical analysis.
It is known, however, that the input impedance of an electrically
short monopole is basically a large capacitive reactance. Therefore, it is very difficult to use such an antenna as a radiator
in the experimental measurements because of the difficulties
involved in its impedance matching. A thicker monopole of length
$\ell = 0.1\ \lambda$ and radius $r_o/\lambda = 0.016$ was found to be more efficient
and easier to match and hence was used. The effect of such a
change in dimensions on the nature of the near fields as well as
the measured SARs will be discussed, together with the experimental results.

Design and construction of phantoms - Because of the limited frequency band of the available 1 kW power amplifier (400-800 MHz)
and the limited size of the available anechoic chamber, it was
necessary to use scaled models in order to simulate the exposure
conditions for objects of larger sizes at lower frequencies. The
prolate spheroidal phantoms were constructed of polyurethane foam.
In our experimental arrangement, a half spheroid was cemented on
an aluminum plate to provide good contact with the ground plane.
To relate the measured SAR values in scaled phantoms to those of
full-scale models, a scaling procedure similar to that employed
by Guy, et al. [3], was used. If s is the scale factor, h is
the actual height of the model, f is the exposure frequency of
interest, and ϵ and σ are the dielectric constant and the conductivity of tissue, respectively, then the corresponding parameters in the scaled phantom are $h' = h/s$, $f' = sf$, $\epsilon' = \epsilon$ and
$\sigma' = s\sigma$. Under these conditions, the SAR in the actual model W
is related to the measured SAR in the phantom (W') by $W = W'/s$.

E- and H-field probes - The electric field probe is basically an
electrically short dipole with a low-inductance diode placed across
the dipole. In order to minimize the probe perturbation of the
EM fields and also to avoid any induced current on metallic cables,

---

[3] A.W. Guy, M.D. Webb, and C.C. Sorensen, "Determination
of Power Absorption in Man Exposed to High Frequency
Electromagnetic Fields by Thermographic Measurements
of Scale Models", *IEEE Transactions on Biomedical
Engineering*, Vol. BME-23, 1976, pp. 361-371.

nonmettalic, high-resistance leads of about 20 kΩ/cm were used. The Moebius loop was used as a magnetic field probe. This probe was found to overcome the problem of responding to the electric field during the magnetic field measurements. A loop of 3.8 cm diameter was constructed and the average sensitivity was found to be about 10 mV for each 0.1 mW/cm^2 at 400 MHz.

The near-field exposure experimental results of the average SARs in spheroidal models of man and monkey models are shown in Fig. 5,

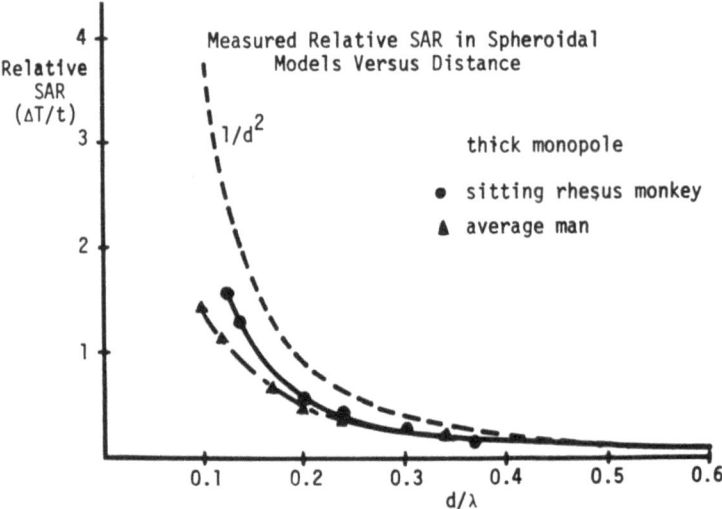

Fig. 5. Measured relative SAR in spheroidal models versus distance.

as a function of the distance from the monopole d/λ. These results were obtained at 400 MHz. However, since different scale factors were used for the man and monkey phantoms, these results are at scaled frequencies equal to the resonant frequencies for both models. It should be noted that the SAR values for the different models were normalized to the same value at d/λ = 0.6 so that the near-field absorption characteristics could be easily compared to the 1/d^2 behaviour. From Fig. 5 it is clear that although the relative SAR increases as the distance from the monopole decreases, the rate of increase of the SAR values is certainly slower that 1/d^2. The 1/d^2 behaviour is of interest because the far-field absorption characteristics vary as 1/d^2, since the far fields of an electrically short dipole vary as 1/d, where d is the distance of the observation point from the center of the source. Therefore, the results given in Fig. 5 simply indicate that the near-field absorption of the spheroidal models ($d/\lambda < 0.4$) is slower than the far-field absorption criteria. In other words, although the near-field radiation increases considerably near the source, these fields do not seem to be absorbed at a rate similar to that of the far fields. To help understand the differences between the absorption characteristics in the near and far fields, we mapped the radiation fields of the source as a function of distance. The results are shown in Fig.6, where it is

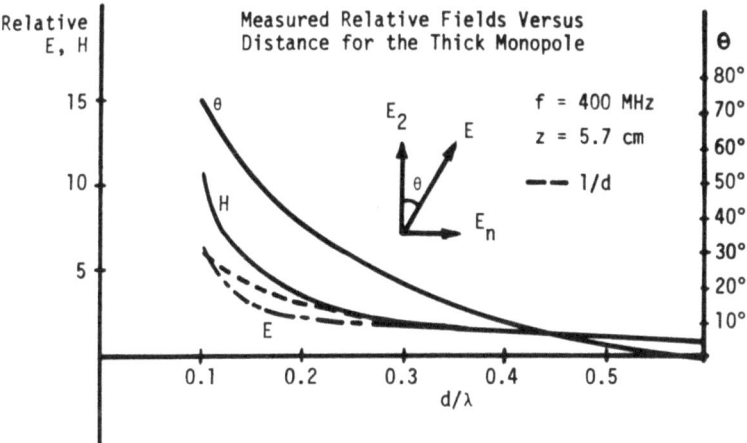

Fig. 6. Measured relative fields versus distance for a monopole of length $\ell = 0.1\ \lambda$ and radius $r_o/\lambda = 0.016$.

clear that in the far field ($d/\lambda > 0.6$), the vector electric field is dominated by the E_z component ($\theta = 0°$), but in the near field ($d/\lambda < 0.2$) the vector electric field is dominated by the E_y component ($\theta > 45°$) which is normal to the spheroid major axis. With this information about the incident fields, the characteristics of the SAR can be explained in terms of the principles shown in Fig. 7, which simply qualitatively relate the internal electric fields in an absorber to the incident EM fields.

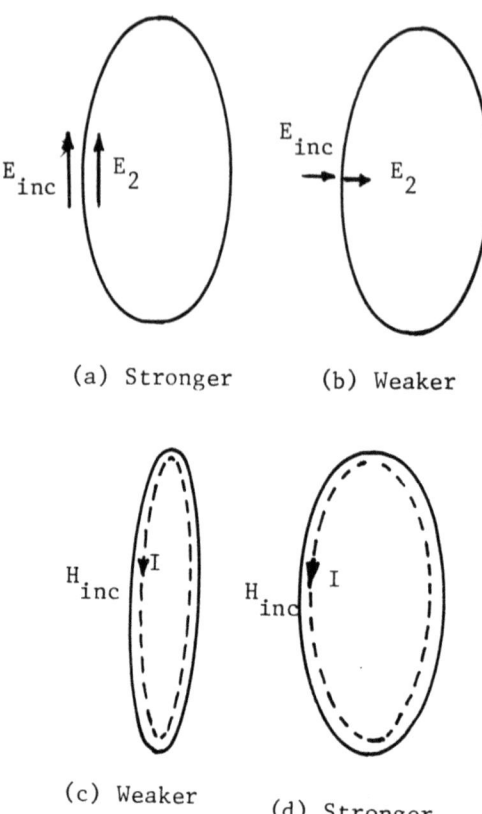

Fig.7. Qualitative explanation of the dependence of the induced internal E fields on the incident fields, E_{inc} and H_{inc}. Conditions (a) and (b) are related to the boundary conditions on electric field. The parallel components are continuous at the boundary ($E_{inc} = E_2$), and the normal components are discontinuous by the ratio of permittivities $E_2 = \varepsilon_1/\varepsilon_2 \, E_{inc}$. The internal electric field E_2 is stronger when E_{inc} is "mostly parallel" to the boundary as in (a) and weaker when E_{inc} is "mostly perpendicular" as in (b). The internal electric fields induced by H_{inc} are related to current I (called eddy currents at low frequencies) that circulate around H_{inc}. Weaker internal E fields are induced when the cross-sectional area intercepted by H_{inc} is smaller as in (c) and stronger when the cross-sectional area is larger as in (d).

From Fig. 7 it is clear that the internal fields induced inside the spheroidal model are stronger when the incident field is mostly parallel to the major axis of the spheroid, and weaker when the incident field is most perpendicular to the major axis. Therefore, the reduction in the average SAR values below the $1/d^2$ curve is mainly because of the variation in the field configuration from mainly parallel to the major axis of the spheroid to mainly perpendicular to it. These results emphasize the importance of identifying the direction as well as the magnitude of the incident radiation if a meaningful evaluation of the possible hazards due to near-field exposure is to be made.

Furthermore, from the results of Fig. 6, it is clear that the rate of change of the SARs as a function of the monopole location for the monkey model is faster than those for the man model. This also can be explained in terms of the field mapping in Fig. 7. Since the increase in θ reflects the fact that the dominant component of the vector electric field of the source has changed its direction from the highly absorbed E_z to the highly scattered E_y, the effect of the increase in θ will be stronger in models of larger a/b ratio than in models of smaller a/b, as illustrated in Fig. 8. The reason for the different electric field coupling in the geometries shown in Fig. 8 is related to the portion of the

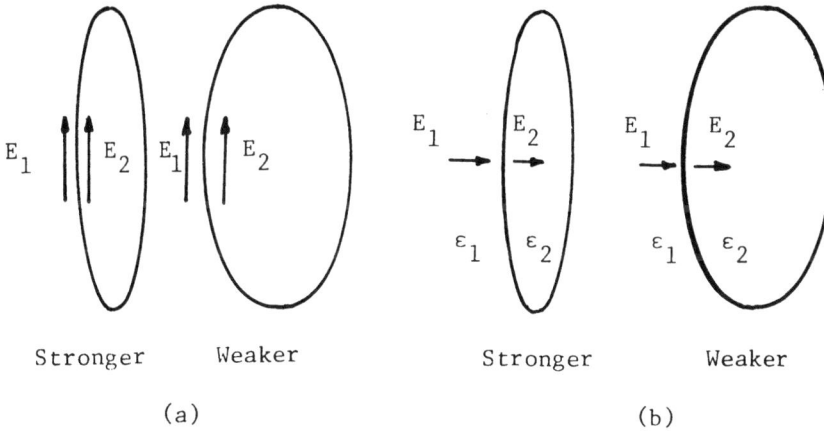

Fig.8. Principles explaining the electric field coupling into a spheroidal model. (a) The internal E fields are stronger when more of the boundary is parallel to the incident E field (models of larger a/b). (b) The effect of the change in the direction of the E field from parallel to the major axis to a different orientation generally results in stronger effects on models of larger a/b. This is because the reduced internal field $E_2 = \varepsilon_1/\varepsilon_2 \, E_1$, ($\varepsilon_2 > \varepsilon_1$) describes the coupling over most of the boundary for models of larger a/b.

surface over which the electric field is parallel (strong coupling) or perpendicular (weak coupling) to the boundary. Obviously in spherical models the change in the electric field orientation has no effect on the total absorption.

Conclusions - The following observations concerning the absorption of the near-field radiation by biological models are specially important:

1. In spite of the complex characteristics of the near-field radiation, including arbitrary angle between $\bar{E}$ and $\bar{H}$, and a wave impedance that is different from 377 ohms, the near-field absorption characteristics can still be explained on the same basis as the far-field SARs [1].

2. It is possible to enhance the near-field absorption in regions of small radius of curvature.

3. The average SAR values obtained in the near-field exposure might oscillate around the constant value obtained from the plane wave radiation. This simply indicates possible increase or reduction in the amount of absorption, depending on the magnitude and direction of the incident radiation fields.

Mr F. Harlen - Any questions, contributions, observations?

Dr D. Sinnott - I am not convinced in this last presentation with the arguments about the field. It seems to me that if you come closer to the dipole, if you integrate $E \wedge H$ over the sphere there may be and there is a radial lead field but there is still the vertical E-field which is Θ-directed no matter how fast you come. I would think that perhaps why there is less power absorbed by this spheroid is that the spheroid is not really oriented in the direction around a circle.

Professor M.F. Iskander - Well I should clarify once and very quickly what I meant by less power absorbed! I meant less than the $1/d^2$, but the rise in temperature is continuously larger than when you are in a far field. When we place the model closer to the dipole for sure it will absorb more power and the rise in temperature is more than what we obtain in the far field. However, the criteria is not formally the $1/d^2$. One would expect more absorption because of the reactive field components. But it turns out that the increase is not as fast as expected. What I am really saying is that, yes, we always have an E_Θ component no matter how close we come to the source. But it is not the dominant component of the electric field vector any more. The dominant component becomes the component perpendicular to the major axis of the spheroid.

Professor R. Mittra - I think that what you are saying is that the power in the neighbourhood of the antenna, the power radiated will be the same whether you go close or far if you integrate around the volume. I am just trying to extend what you argue.

Professor Iskander - That is right.

Professor R. Mittra - It could be that because of the orientation of the dipole, that locally you may have more or less concentrations in the power. It depends on which way you are looking at it but that probably is not the major effect here but you were saying that you had expected the power absorption to rise as you came in closer as function of distance say as a square law power. Another question is why should it increase even though it would appear from the point of view of the field distribution that the field in this near field is rising but the total power if you integrate it around is not.

Professor Iskander - Yes, but you are not integrating over. When we talk about the power density and you know power density is varying as $1/d^2$. Basically, because if you integrate over the area it has a d^2. The power density is the power per unit area that the spheroid is exposed to, and it varies as $1/d^2$. This is why we have a normalisation factor. We assume that the total energy flux coming out of the curved boundary equals to the one incident by the plane wave integrated on a surface area. This takes into account the $1/d^2$ variation in the power density as a function of the separation distances. When we come closer it is not as much as $1/d^2$ because the power radiated by that dipole is a finite quantity and if you follow $1/d^2$ that tends to infinity.

Professor J. van Bladel - But if you come much closer you are going to change the input impedance of your dipole so it will start obviously absorbing from the source.

Professor Iskander - That is right.

Professor van Bladel - You should monitor your power input to the dipole to be sure it remains constant.

Professor Iskander - We did that in the experimental procedure and for distances as close as 0.1λ with a scaled model. We observed about 10% power reflected back. So we increased the input power by 10% in order to maintain it constant. However, it was very difficult to make any experimental measurements for separation distances closer than 0.1λ. Because there was a serious amount of power reflected back and it is also very difficult to keep the generator stable. It shuts off because of the considerable amount of power reflected.

Professor van Bladel - What you call a monopole is that a $\lambda/4$ monopole?

Professor Iskander - No.

Professor van Bladel - You didn't want to have it too short because of the capacity of effect, so it's in-between somewhere?

Professor Iskander - Yes, well we needed to make it very very short to be close to the theoretical assumption of short current element. But very short dipole doesn't radiate because it is just basically a capacitance and it is very difficult to match it while it doesn't radiate! The monopole we used in our experiments is 0.1λ.

Professor van Bladel - You are in the near field because the limit for the far field being $2d^2/\lambda$, might very well be in the far field for the dipole but what occurs then? You have a radiation pattern?

Professor Iskander - We are exposing at 400 MHz, so at 400 MHz if you apply this type of criteria, I think you will find that you are way in the near field. Also the field mapping indicates that there is no way the electromagnetic fields will go from parallel to the Z-axis to perpendicular except in the near field of some sort.

Mr F. Harlen - I have got a couple of slides here which might help at least in showing that the effect is also quite real. That is a diathermy applicator, 2450 MHz, and you can see what we are doing there is heating up some liquid crystal sheet behind an absorbent rubber. We are definitely in the near field and if you do a field plot using an isotropic probe you get quite a decent looking pattern of it, but if you get the actual heating pattern in the liquid crystal sheet in the near field you see that it is horrid. That is a fair description of the far field pattern where you have got the things normalised and radiating, where you are just using dipoles normal to the direction of propagation. If you take a heating pattern in the far field it is more or less the same as the near field so that what we have got as instrumental measurements to begin with isotropic. We have got the inclusion of this field normal to the applicator. It is not interacting at least with the 2450 applicator. It doesn't interact at all well with the body. Thank you. Any other questions?

Dr T. Sarkar - In this presentation primarily you address the problem of biological effects due to thermal heating. I was just curious, are there any non-thermal effects and if so how do you recognise them? How would you go about finding them?

Mr F. Harlen - If you have an effect which appears to be dependent on the amplitude then it is going to be very difficult. That is, an effect which comes in at a given amplitude or field strength and then disappears at the high field strength is going to be very difficult to explain on a firmer basis, and there have been several cases I think where people have claimed an amplitude effect. Two of these are still good as far as I know in that no one has contradicted them. Two of them; we had people coming along and saying: 'well we have tried to do this and we can't do it, we can only find this happens if there is a temperature dependence'. If you found something which has a marked difference between a CW and pulsed radiation, then there is a reasonable chance that that was field-related rather than temperature-related because, say you have got a mark space ratio of 1,000 to 1, you are going to have much higher electric or magnetic field strengths in the tissue for the same average power impulse or CW. But there is very scanty evidence that I find convincing for the difference between CW and pulse radiation. Have you any thoughts on that Professor Guy?

Prof. A.W. Guy - I might comment on the one effect some years ago that was thought to have been non-thermal; in fact it was alleged to have been the interaction of the nervous system with a pulsed microwave. This was the so-called microwave hearing effect. With pulses of certain amplitude you hear clicks in the head, while the same average power of CW has no effect. It turned out, in fact people worked it out theoretically and confirmed it experimentally, that it was just a thermal expansion phenomenon. It was introducing a delta function mechanical disturbance in the head which caused the head to resonate at some resonant frequency. This we heard as a click or a chirp. A temperature rise of only 10^{-4} °C is sufficient to set this off. Now there are other reports on non-thermal effects, but here there is considerable controversy. One is the so called blood-brain barrier effect. Some investigators have shown a more enhanced effect is obtained with pulsed than with CW microwaves. With relatively low levels of exposure not usually causing a temperature rise some experimenters observed a breakdown of the blood-brain barrier. This barrier normally preventing the passage of small molecules from the blood to the interstitial fluid was found to become more permeable when the brain was exposed to microwave radiation.

Mr. F. Harlen - One of the groups who observed this, Oscar and Hawkins, have been unable to repeat their own work. There are at least two groups; one at Brooks Air Force Base, Texas, and a group in Canada have both been trying to repeat that. They can get nothing unless there is an observable temperature rise. The effects are related to temperature in their experiments.

Prof. A.W. Guy - Then there is the effect that Dr. Ross Adey of California has observed; with CW energy at relatively low levels he obtained negative effects, but if he amplitude modulated the signal near 16Hz he was able to get pronounced effects such as an increase in the calcium efflux in the brain. I think this was replicated by Carl Blackman at EPA.

Mr. F. Harlen - But he could only replicate it when he visited Ross Adey so that he could copy his techniques. It is not quite an independent validation, but that would be very difficult to explain thermally.

Prof. A.W. Guy - I think our comments so far relate to research from the West. In the Soviet Union and some East European countries, they are observing effects at much lower levels; but they are not scaling from animal to man as we have talked about here. There are probably uncertainties in absorbed energy ranging over several orders of magnitude.

Prof. E.H. Grant - I would like to say a few words, in the way of amplification, about the differences between the thermal and the non-thermal effects and the reason why one would not expect to observe non-thermal effects in the absence of thermal effects. If you consider the normal average temperature of the body, it is such as to give you a value of kT of about 2.6×10^{-2} eV. This means that if you are going to observe any field effects or any movement of particles, molecules, cells which is an electric field effect rather than a thermal effect, then you are not going to get any response until the electrical energy of the particle is of that order of magnitude. In practice this means that in order to get any direct orientation in a molecule such as haemoglobin, which has a diameter of 100 Å, a field of the order of 10^6 V/m would be required. We have been talking about exposure standards of 10 mw/cm^2 which corresponds with an electric field <u>in air</u> of the order of around 2V/cm or 200V/m. Therefore if one evaluated, using the Langevin function, the percentage of the molecules which would be orientated with a field of this value, it would come to something of the order of 10^{-2}%. With bigger particles the effect occurs for smaller fields but even for a red cell, which has a diameter of around 2 μm, a field of more than 1000 V/m is needed. So it can therefore be concluded, at least as far as mechanical forces are concerned, that if any mechanical force effect can be observed then the thermal effects will be observed beforehand.

Another question worthy of consideration is whether nerve excitation can occur due to direct interaction with the electric field. Following the approach of Schwan a model can be set up of a cell with a membrane where total radius is R. The voltage developed across it is a function of the following variables: the radius R, the field in air E_o, the frequency of the radiation, the resistivities of the inner and the outer parts of the cell, ρ_A and ρB, and the capacitance C which is usually of the order of 1μF/cm^2. If all the relevant values are substituted for a typical nerve membrane it can be shown that for an <u>external</u> field E_o of a few V/cm, the voltage developed across the membrane at microwave frequencies is less than 1 μV Since the voltage required to stimulate a nerve is typically around 1 mV (depending on the nerve) it is clear that no stimulation takes place at microwave frequencies for such values of E_o.

I would now like to pass on to the second part of the session which concerns the medical and biological applications of microwaves. One of these is the use of hyperthermia for treating cancer. As Mr Harlen and Prof. Guy have both said you need to raise a cell to something of the order of 43-44°C before it is destroyed but it has been known for quite some time that, in many cases, cancer cells are more vulnerable to destruction by heat than normal cells. This therefore offers the possibility of treating cancer by hyperthermic methods, a fact which has been known for some time. For example, a hundred years ago, doctors were pioneering the treatment of cancer by deliberately inducing fever in cancer patients in the hope that the elevation in temperature would cause the regression of tumours. Even at that time quite dramatic cases of tumour regression were observed.

The problem with this and other more recent methods is that the heating of a large part of the body, if not the whole part, is involved. What is required is a technique where only the tumour itself is heated, and this is where microwave methods offer hope, in that heat may be localised within the tumour without too much increase of temperature in the surrounding healthy tissue. Recent developments in both interstitial applicators and non-contacting applicators indicate that selective heating can be obtained. Other areas of research concern investigating whether the dielectric properties of malignant tissue are different from normal tissue. If so then it may be possible to identify a frequency band over which cancer tissue heats more rapidly than normal tissue.

<u>Mr F. Harlen</u> - Thank you very much. Could we have Dr Smith please.

<u>Dr C.W. Smith</u> - Before we start this discussion, I will give you the reference to the chapter I did on the Dielectric Properties of Biological Materials for the 1978 Digest of Dielectrics (Vol. 42, Ch. 14 (Washington D.C.: N.R.C.)) which is coming out at the fall meeting (1979) of the Conference on Electrical Insulation and Dielectric Phenomena. The subject of Biological Dielectrics seems to be taking off. There were about as many references in 1978 to this subject as there were to thin films.

Now I think the most useful thing I can do for you is to point you in some of the directions where you can find information on these topics, but I must warn you that there is nothing here which hasn't been criticized in some respect. The first reference is the main Russian one which shows what seems to be millimeter wave non-thermal-effects taking place at radiation levels between the allowed American and Russian radiation levels. (Sov. Phys. - USPEKHI 16 568-579 (1974)). Then there is a collection of papers (Phys. Lett. 60A 267 (1977), 63A 407 1977, 62A 463 (1977)) which shows effects associated with cell division. The trick to enhance the effects observed is to get cells to divide synchronously. The laser Raman spectrum then shows resonances in the sub-millimetre region (round about 100 cm^{-1}). One paper (Int. J. Quantum Chem: Quantum Biol. Symp. 4, 277-284 (1977)) shows both the Raman lines and some microwave resonances. Although the microwave techniques used have been criticized, the results show that in both cases the resonance lines in cancerous material have split into doublets. Then this particular paper (Phyc. Lett. 57A, 102-104 (1976)) shows microwave radiation effects on the rate of yeast cell growth. The frequency was about 40 GHz and a periodicity of about 20 MHz was observed in the results. This work has been criticized on the grounds of the stability of the oscillator and the possibility that intensity window effects could be involved. In the optical region, there is a paper (Phys. Lett. 57A, 102-104 (1976)) which refers to laser induced changes of the enzyme activity of chymotrypsin.

I have been cooperating with Professor Fröhlich for the last five years trying to devise and carry out experiments against which he can check the predictions from his theoretical analysis (Proc. Nat. Acad. Sci (USA) 72, 4211-4215 (1975)). There is a more recent paper which is less mathematical and quite easily readable and from which you can get some of the implications of the possibility of large dipole oscillations in the high GHz to THz region with long range dipole interactions capable of switching from the inverse six power law of Van der Waals forces to an inverse cube power law according to frequency and charge concentrations, and which may well be pumped chemically by metabolic processes (IEE Trans. MTT - 26, 613-617 (Aug. 1978)).

Why do I want to play with enzyme systems? There are about 3,000 different enzymes in an average cell, each one controls a specific biochemical reaction. Cells will normally be in a stable negative feedback situation and electromagnetic stress like the thermal ones we have been talking about, but in the non-equilibrium situation you might expect to get other things happening. Such an instance might be the moment of cell division, in the cancer cell situation or in an isolated enzyme reaction where you deliberately try to avoid any feed-back arrangements. Mostly, I have been using one particular enzyme, lysozyme, although I have also done some measurements on trypsin and ribonuclease. The reason is that lysozyme is stable, it is well characterised

structurally to a resolution of about 2Å. This has been done by
Prof. D.C. Phillips group now at Oxford.

We started off making dielectric measurements (IEE Cont. Publ.
No.129 p.44-48 (1975)) and then found that dielectric properties
were affected by magnetic fields. From there we went on to
measure the diamagnetic susceptibility. There shouldn't be any
appreciable magnetic effects of this kind in such materials.
Diamagnetism can be explained by the equivalent of a short
circuited current loop somewhere in the system. The effects
were followed both by looking at the effect of fields on the
enzyme reaction and by measurements on thin films of the enzyme.
The best results (collective Phenomena $\underline{3}$ 25-33 (1978) were
obtained for a point-plane contact in a waveguide measured at
9 GHz. We used the enzyme as a dielectric lazer between the
electrodes. Any membrane system is more likely to be non-linear
than linear and in that case we found voltage steps of about a
1/25 µV. These were drastically modified by loosely coupling
20 MHz to the system which was the frequency we calculated using
the 500 MHz/µV relation from the concept of Josephson Voltages
used to discuss our results. The work is not uncriticized and
these are a list of the critics: (J. Phys. C: Solid State Phys.
$\underline{9}$ L251 (1976), Solid State Comm. $\underline{19}$, 357 (1976), Phys. Lett. $\underline{60A}$,
491 (1977)). The paper from Physics Letters shows that the magnetic
effects disappear if you have an oxygen free system. Oxygen is a
particularly interesting molecule because ordinarily it is
paramagnetic but the O^{2-}ion is diamagnetic, so that in the
presence of the strong magnetic fields there could be large forces
on the oxygen molecule. Other workers have failed to reproduce
our magnetic susceptibility measurements using uniform fields. I
suspect that there may be a difference between the results of
susceptibility measurements in uniform and non-uniform fields.
The light scattering experiments showed that there was no micelle
formation which, if the effects are really there, must mean that
there is some long range force involvment.

Experimental difficulties in enzyme work include: (1) variation
between different sources of supply, what you would think is
nominally the same material is by no means always so; (2) deposits
on glass, it is well known that proteins stick on glass surfaces
and a paper to be published shortly (Collective Phenomena (early
1980)) shows that magnetic fields effects can be followed over
months using UV spectrometry. These effects can be interpreted
in terms of varying amounts of enzyme sticking to the glass
surfaces; (3) the way you humidify the material matters, some of
the curves of the dielectric properties are just like those for
moist ferro-electric materials; (4) temperature also matters,
the enzyme system remembers what has happened to the temperature
in the past; (5) pH matters, the dielectric and magnetic
susceptibility curves mirror the enzyme activity curves as

function of pH except that you get an additional peak at the
isoelectric point; (6) magnetic fields matter but here I am still
wrestling with the problem of repeatability, one of the things
which does seem to effect the reaction rate of the enzyme system
is how long you leave it in the geomagnetic field, so that in a
few minutes screened material can change back to something like
the original value of activity, the change is exponential and it
can take several hours to return to the final equilibrium value
after you have exposed it to weak magnetic fields; (7) regarding
electrodes for thin films, we agree with Hansma (Phys. Rep. 30C
164 (1977)) that lead seems to be a good electrode for these
materials, he doesn't know why either; (8) electromagnetic field
effects also occur (collective phenomena 2 215-218 (1977)) and
here too I am wrestling with repeatability problems.

Mr F. Harlen - Thank you very much Dr Smith. We now have the
final presentation.

Dr G. Crosta - I shall give a very short account on what control
theory has to do with tumour therapy.

Let us consider the situation shown in figure 19 where we should
like to irradiate with a therapeutically effective electromagnetic
field the tumour inside the body and at the same time avoid
irradiation to healthy tissues.

Let the field sources be located at the boundary Γ ; let $p_d(\vec{x})$
the desired three-dimensional power distribution inside Ω , a
two-dimensional representation of which along a body section is
shown by the line profile of figure 20.

Several questions arise:

1) is the function $p_d(\vec{x})$ physically realizable at all?

ii) if so, can it be achieved by the sources on Γ ?

iii) what about the dynamics of the phenomenon ? i.e. shall we rely
on time-harmonic regime or operate on arbitrary time dependence?

iv) which is the role of cost in implementing a truthful or an
approximate replica of $p_d(\vec{x})$?

Question 1) is an existence problem, as we want to write a field
equation the solution of which, $z_d(\vec{x},t)$ is related to $p_d(\vec{x})$ by

$$p_d(\vec{x}) = \frac{1}{T} \int_0^T dt \left| z_d(\vec{x}, t) \right|^2$$

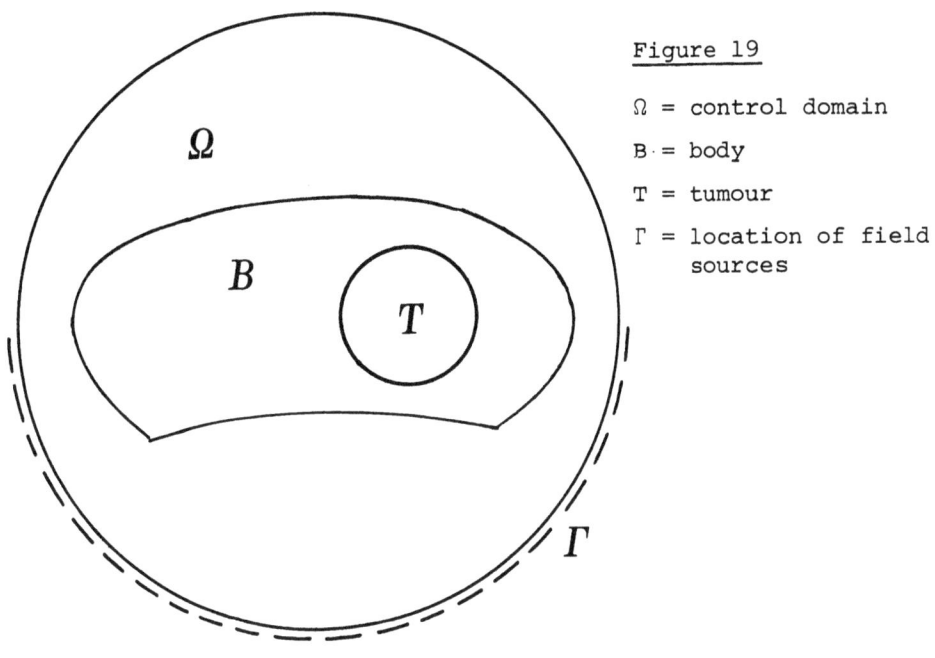

Figure 19

Ω = control domain
B = body
T = tumour
Γ = location of field sources

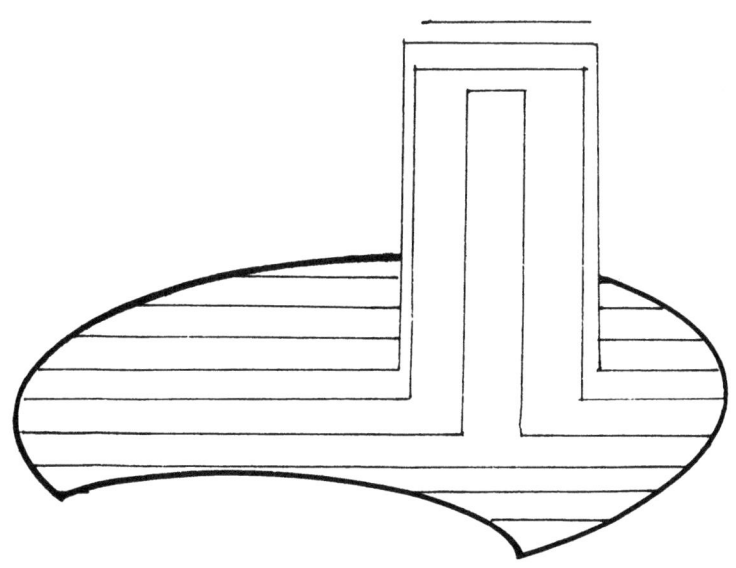

Figure 20 line profile of $p_d(\vec{x})$ along a section

Questions ii) and iii) are a controllability problem, namely a boundary control problem in time T.

Question iv) is an optimal control problem subject to both physical and economical constraints.

The foregoing is an example where system theoretical concepts may help in formulating an electromagnetic interaction problem.

In the following some features of distributed parameter system (DPS) theory will be sketched.

Definition: a DPS is a set

$$(U, \Psi, Y; B_\Omega; B_\Gamma; A_\Omega; A_\Gamma; C)$$

where U is the input set, satisfying (see below) : $U = U_\Omega \times U_\Gamma$

Ψ is the state space

Y is the output set

B_Ω, B_Γ are input maps relating respectively to the distributed and boundary inputs or controls to the state equation to be defined presently

A_Ω, A_Γ are differential and/or spatially dependent operators

C is the output map.

An example of a second order system is the following, where the set of equations relates the above mentioned functional spaces through the above defined operators.

S.E. $\quad \dfrac{\partial^2 \psi(\cdot)}{\partial t^2} = A_\Omega(\vec{x}) \, \psi(\vec{x},t) + B_\Omega u(\vec{x},t); \; t \geq 0; \; \vec{x} \in \Omega; \; \psi \in \Psi; \; u \in U_\Omega$

B.C. $\quad A_\Gamma(\vec{x}) \psi(\cdot) = B_\Gamma(\vec{x}) \, W(\vec{x},t); \; \vec{x} \in \Gamma = \partial\Omega; \; w \in U_\Gamma$

I.C. $\quad \psi(\vec{x}, t=0) = \psi_0(\vec{x}); \; \dfrac{\partial \psi}{\partial t}(\vec{x}, t=0) = \psi_1(\vec{x}); \; \vec{x} \in \Omega \cup \Gamma$

O.T. $\quad z = C\psi; \quad z \in Y$

We have a differential equation for ψ containing partial derivatives because ψ depends both on spatial position $\vec{x}$ and on time t. It is called the state equation (S.E.) because it says how the system state vector ψ inside Ω (see figure 21) evolves in time as a consequence of internal dynamics and external distributed control $u \varepsilon U_\Omega$.

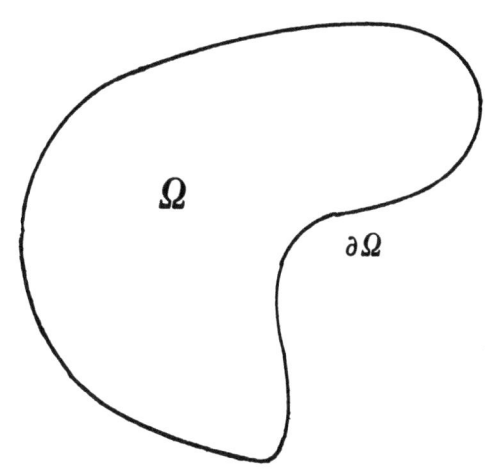

Fig. 21.
domain geometry

$\Omega \subset \mathbb{R}^n$; $n \geq 2$

We also need a boundary condition (B.C.) containing some other source term w acting on ∂Ω, and initials conditions (I.C.).

To complete the picture the output transform (O.T.) C is defined, which enables us to get z when we measure ψ.

Let us then identify a simple electromagnetic problem [1] in the formalism of DPS theory: the higher dimensional scalar wave equation in a bounded domain Ω (see figure 22).

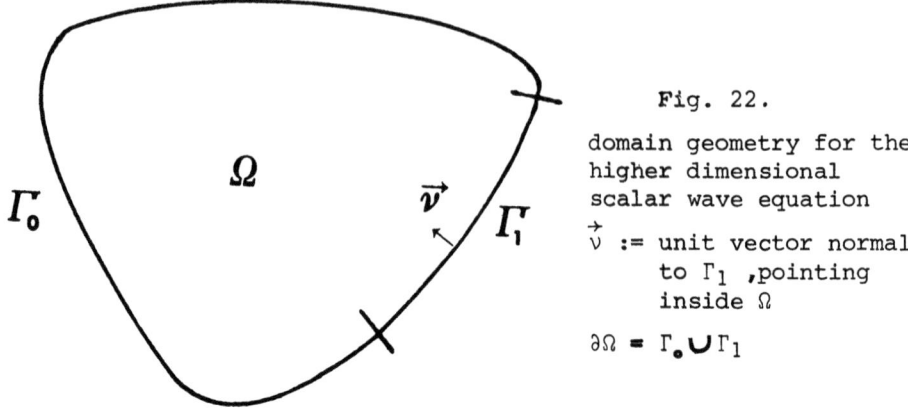

Fig. 22.

domain geometry for the higher dimensional scalar wave equation

$\vec{v}$:= unit vector normal to Γ_1, pointing inside Ω

$\partial\Omega = \Gamma_o \cup \Gamma_1$

S.E. $\quad \dfrac{\partial^2 \psi}{\partial t^2}(\cdot) = \nabla^2 \psi(\vec{x},t); \quad B_\Omega = 0; \quad \vec{x} \in \Omega \subset CIR^n; \quad n \geq 2; \quad t \geq 0$

B.C.1. $\quad \psi(\vec{x},t)\big|_{\Gamma_o} = 0; \quad B_{\Gamma_o} = 0; \quad \vec{x} \in \Gamma_o$

B.C.2. $\quad \dfrac{\partial \psi}{\partial \nu}\bigg|_{\Gamma_1} = f; \quad \vec{x} \in \Gamma_1; \quad f \in U_\Gamma$

I.C. $\quad \psi\big|_t = 0 = \dfrac{\partial \psi}{\partial t}\bigg|_{t=0} = 0 \quad \forall\, x$

O.T. $\quad z = \psi(\vec{x},t); \quad C = 1$

The S.E. describes the free field. The B.C.s are mixed: Dirichlet type on Γ_o and Neumann with source term f on Γ_1.

We suppose the system to be initially at rest and that we can measure the field amplitude inside Ω at every time.

I do not mean thereby to model the process of tumour therapy by such an oversimplified equation but just give the guidelines for assessing existence and uniqueness, controllability, optimal control.

<u>Existence and uniqueness of the solution $\psi(\vec{x}, t)$</u>:

They usually require suitable hypotheses on operators, B.Cs, I.Cs, contour geometry. For the scalar wave equation these conditions hold, i.e. the problem is well-posed.

Controllability at time T:

We know the solution ψ belongs to Ψ which is a Hilbert space; a priori we don't know if, as we change the control f, $\psi(\vec{x}, t=T; f)$ spans a linear subspace dense in Ψ [2].

When this is the case, we say the DPS to be "controllable at time T". For the scalar wave equation controllability can be easily proved for suitable contour geometry [3].

Then:
$$\overline{\{\psi(\vec{x}, T; f) \mid f \in U_\Gamma\}} = \Psi$$

Observability:

It holds if C is injective or one-to-one map from the state to the output space.

$$C: \Psi \longrightarrow Y \qquad \text{is one-to-one}$$
$$\psi \longmapsto C\psi = z$$

This is self evident when C equals the identity operator.

Some remarks are needed at this point.

We may relate the output to the input without making reference to the state: this is the "empirical" approach. The inner dynamical structure of the system would thus be ignored. To take it into account we adopt the "axiomatic" approach which leads us to define controllability and observability.

These concepts arise because we can act on ψ only through B_Ω, B_Γ and look at ψ only through C.

Optimal control:

Let $z_d(\vec{x})$ be the desired output at time T: then we construct the following functional [4]:

$$J(\underline{u}: z=z_d) := P(\psi(\underline{u})) + Z(\underline{u})$$

where $\underline{u}$ is the input vector including both distributed and boundary controls. Our task is to find an input $\hat{\underline{u}}$ which minimises J:

$$J(\hat{\underline{u}}) = \text{Inf } J(\underline{u}) \ \forall \ \underline{u} \in U_{ad} \subset U,$$

where U_{ad} is the admissible input set.

For the scalar wave equation J reads:

$$J(f) = \int_\Omega d\vec{x} \left| \psi(\vec{x},T;f) - z_d(\vec{x}) \right|^2 + \int_0^T dt \int_{\Gamma_1} d\vec{x}\, N(\vec{x}) \left| f \right|^2$$

i.e. we have a physical term accounting for the accuracy with which we want $\psi(\vec{x},T)$ to approach $z_d(\vec{x})$ by a quadratic criterion. In addition we have an economical functional $E(\underline{u})$ in the general case. In our example $E(f)$ increases with the energy expenditure at the boundary Γ_1; moreover the values of $f(\vec{x};\vec{x}\epsilon\Gamma_1)$ are weighted by $N(\vec{x})>1$, $\forall\, \vec{x}\epsilon\Gamma_1$; if $N(\vec{x})$ is "large" at some location, we may be discouraged to apply a control there.

Now to <u>applications</u> : this theory is helpful is solving time-optimal control for wave phenomena, where the field amplitude is relevant.

In order to relate our scheme more strictly to the tumour therapy problem, we just sketch what the functional $J(.)$ might be:

$$J(f) = \frac{1}{T} \int_0^T dt \int_\Omega d\vec{x}\, W(\vec{x}) \left[\left| \psi(\vec{x},t;f) \right|^2 - \left| z_d(\vec{x}) \right|^2 \right]^2 +$$

$$\frac{1}{T} \int_0^T dt \int_{\Gamma_1} d\vec{x}\, N(\vec{x}) \left| f \right|^2 ,$$

where: $W(\vec{x}) \geq 0$; $\vec{x}\epsilon\Omega$; $N(\vec{x}) \geq 1$, $\vec{x}\epsilon\Gamma_1$.

Let us operate in the quasi-stationary regime; there is a simple relationship between $p_d(\vec{x})$ and $z_d(\vec{x},t)$:

$$p_d(\vec{x}) = \frac{1}{T} \int_0^T dt \left| z_d(\vec{x},t) \right|^2$$

Then the desired intensity distribution is $\left| z_d \right|^2$, whereas $\left| \psi \right|^2$ is the obtained intensity distribution.

Their difference squared is the local error signal power. It is weighted by $W(\vec{x})$. $W(\vec{x})$ "high" means high accuracy required.

The time integration averages power over the interval $[0,T]$ both of the physical and economical term.

We have to find $\hat{f} \in U$ which minimises J (f). This is the task we are working at. Thank you for your attention.

Mr. F. Harlen Thank you very much, Dr. Crosta. Are there any questions?

Dr. Burdette In your "Cost-effectiveness" analysis terms, what is the determination of $N(\vec{x})$? On what does it rely primarily? What are you using as the basis of "cost"?

Dr. Crosta For "cost-effectiveness" I mean an optimisation problem where: - for "cost" I mean the role of the economical term:

$$E(f) = \frac{1}{T} \int_0^T dt \int_{\Gamma_1} dx \; N(\vec{x}) \left| f \right|^2$$

where N(.) is a known function, independent of f, which is assigned a priori and may take into account all experimental difficulties met in implementing this particular control f on Γ_1. $N(\vec{x})$ depends on the geometry of our arrangement. - for "effectiveness" I mean the accuracy with which we approximate the desired output with the one we obtain by applying the boundary control f.

References

[1] G. Chen: "Energy Decay Estimates and Boundary Value Controllability for the Wave Equation in a Bounded Domain" J. Math. Pures et Appliquées, to appear 1979.

[2] H.O. Fattorini "Boundary Control Systems" SIAM J. Contr. 6 (68) 349-385

[3] D.L. Russel "Controllability and Stabilizability. Theory for Linear Partial Differential Equations: Recent Progress and Open Questions" SIAM Rev. 20 (78) 639-739.

[4] J.L. Lions "Optimal Control of Systems Governed by Partial Differential Equations" Springer: Berlin 1971.

LIST OF LECTURERS AND DELEGATES

AAS, J.A., Dr., The University of Trondheim, Norway.
ADAM, D. Dr., Laboratoire Central de Telecommunications, Villacoublay, France.
ADATIA, N.A. Dr., ERA Ltd., Leatherhead, England.
ARMOUR, T.W., A.W.R.E., Aldermaston, England.
BACH, H. Prof., Technical University, Lyngby, Denmark.
BARBER, P.W., University of Utah, Salt Lake City, USA.
BEAULIEU, R., Laboratoire Central de Telecommunications, Villacoublay, France.
BENNETT, C.L., Dr., Sperry Rand Research Centre, Sudbury, Massachusetts, USA.
BESNAULT, B., L.S.S. - E.S.E., Gif-Sur-Yvette, France.
BLADEL, J. van, Prof. Laboratorium voor Elektromagnetisme en Acustica, Ghent, Belgium.
BRAIN, D.J., ESTEC, Noordwijk, The Netherlands.
BREWSTER, D.C. Dr., GEC-Marconi Electronics, Great Baddow, England.
BURDETTE, E. Dr., Georgia Institute of Technology, Atlanta, Georgia, USA.
BUTLER, C. Prof., University of Mississippi, USA.
CALVO, M.L., University Complutense, Madrid, Spain.
CANATAN, F.Y. Prof. Dr., Middle East Technical University, Ankara, Turkey.
CHEVALIER, M., Universita Complutense, Madrid, Spain.
CHIGNELL, R.J. Dr., ERA, Leatherhead, England.
CHRISTIANSEN, P.L. Prof., Technical University of Denmark, Lyngby, Denmark.
CORNBLEET, S. Dr., University of Surrey, Guildford, England.
CROSTA, G. Dr., Via Dupre, 5, 21013 Gallarate, Italy.
DAVENPORT, E.M. Dr., British Aerospace, Filton, England.
DESCHAMPS, G.A. Prof., University of Illinois, USA.
DJORDJEVIC, A.R. Dr., University of Belgrade, Yugoslavia.
DOBRICH, M., Universitat des Saarlandes, Saarbrucken, West Germany.
EGIDO, P., C. Universitaria, Madrid-3, Spain.
EXCELL, P.S., Bradford University, England.
FELSEN, L.B. Prof., Polytechnic Institute of New York, USA.
FER, A. Dr., Middle East Technical University, Ankara, Turkey.
FINNIE, J.S., Directorate of Radio Technology, London, England.
FISCHER, M., Universitat des Saarlandes, Saarbrucken, West Germany.
FOULDS, K.W.H. Dr., University of Surrey, Guildford, England.
FRANDSEN, A., Technical University of Denmark, Lyngby, Denmark.
FRENKEL, A. Dr., P.O.B. 2250, Haifa, Israel.
GARTHWAITE, A.W.R.E. Aldermaston, England.
GRANT, E.H. Prof., Queen Elizabeth College, London, England.
GRIBBLE, J.J., University of Surrey, Guildford, England.

GRIFFITHS, D., Ferranti Ltd., Edinburgh, Scotland.
GUIRAUD, J.L. Dr., O.N.E.R.A./C.E.R.T., Toulouse,, France.
GUY, A.W. Prof., University of Washington, Seattle, USA.
HALD, J., Ermelundsvej 67, 2820 Gentofte, Denmark.
HARBOTTLE, P.K., GEC-Marconi Electronics, Great Baddow, England.
HARLEN, F., National Radiological Protection Board, Harwell, England.
HARRINGTON, R.F. Prof., Syracuse University, New York, USA.
HENDERSON, A. Dr., R.M.C.S., Shrivenham, England.
HERMAN, G.C., University of Technology, Delft, The Netherlands.
HUGHES, K.A., Home Office (DRT), London, England.
IDEMEN, M. Prof. Dr., Technical University of Istanbul, Turkey.
ISKANDER, M.F. Prof., University of Utah, USA.
ISRAEL, M., Raanan 5, Haifa, Israel.
JAGGARD, D.L. Prof., University of Utah, USA.
JECKO, B., U.E.R. des Sciences, Limoges, France.
JONES, W.L., RMCS, Shrivenham, England.
KORNIEWICZ, H.R. Dr., Zakład Zagrożeń Fizycznych, Warszawa, Poland.
KOUYOUMJIAN, R.G. Prof., Ohio State University, Columbus, USA.
KUBINA, S.J. Prof., Concordia University, Montreal, Canada.
LIER, E., Technical University of Norway, Trondheim, Norway.
LIL, E. Van., Catholic University of Leuven, Belgium.
MARTIN, P., GEC-Marconi Electronics, Great Baddow, England.
MILLER, E.K., Lawrence Livermore Laboratory, CA 94550, USA.
MITTRA, R. Prof. University of Illinois, Urbana, USA.
MOLINET, F. Dr., Laboratoire Central de Telecommunications, Velizy, France.
MONTROSSET, I. Prof., Politecnico, Torino, Italy.
MOODY, D.A., A.W.R.E., Aldermaston, England.
MORRIS, R.G., British Aerospace, Filton, England.
MURESAN, L.V., Miss, 3212 Kent Avenue, Montreal, Quebec H3S 1N1, Canada.
NEBAT, J. Dr., P.O.B. 2250, Haifa, Israel.
NIVER, E. Dr., Polytechnic Institute of New York, USA.
ORTA, R.G., RAI, Centro Ricerche, Torino, Italy.
ORTON, R.S. Dr., GEC-Marconi Electronics, Great Baddow, England.
OWEN, J.I.R., Royal Aircraft Establishment, Farnborough, England.
PACELLO, E.A., GEC-Marconi Electronics, Great Baddow, England.
PATHAK, P.H., The Ohio State University, Columbus, USA.
PEARSON, L.W., University of Kentucky, Lexington, USA.
POLITCH, J. Dr., Technion City, Haifa, Israel.
PONTOPPIDAN, K. Dr., TICRA ApS, Copenhagen, Denmark.
PUES, H.F., M.I.L., Heverlee, Belgium.
RAHMAT-SAMII, Y. Dr., Jet Propulsion Laboratory, Pasadena, USA.
RAMSDALE, P.A. Dr., Royal Military College of Science, Shrivenham, England.
RAO, C.R.A., University of Technology, Delft, The Netherlands.
ROBERTS, R., c/o 14 Coleridge Ave., St. Helens, Mersyside, England.

ROBIN-JOUAN, Y. Laboratoire Central de Telecommunications, Velizy, France.
ROUMIGUIRES, J.L. Dr., Laboratoire Central de Telecommunications, Velizy, France.
ROUSSELOT, J.L., Laboratoire Central de Telecommunications, Velizy, France.
SADLER, M. Miss, GEC-Marconi Electronics, Great Baddow, England.
SANDER, K.F. Prof., Bristol University, England.
SARKAR,T.K.,Rochester Institute of Technology, New York, USA.
SENSIPER, S., P.O. Box 3102, Culver City, CA 90230, USA.
SERBEST, H., Technical University of Istanbul, Istanbul, Turkey.
SINNOTT, D. Dr., Defence Research Centre, Salisbury, S. Australia.
SKWIRZYNSKI, J.K., GEC-Marconi Electronics, Great Baddow, England.
SLIVNIK, T., Fakulteta za Elektrotehniko, Ljubljana, Yugoslavia.
SMITH, C.W., University of Salford, England.
STORM, B. Prof., University of Trondheim, Norway.
STROM, S., Institute of Theoretical Physics, Goteborg, Sweden.
SUMMERS, J.E., Signals & Radar Establishment, Malvern, England.
TEKEOGLU, I., Cirpan Mah, Altiparmak Sok, No. 38/6, Bursa, Turkey
TIJHUIS, A.J., University of Technology, Delft, The Netherlands.
TOSUN, H.Y. Prof. Dr., Middle East Technical University, Ankara, Turkey.
UZGOREN, G., Technical University of Istanbul, Istanbul, Turkey.
WEBB, J.P., Cambridge University, England.
WHITMAN, G.M., Prof. Dr., New Jersey Institute of Technology, Newark, USA.
YAGHJIAN, A.D., National Bureau of Standards, Boulder, Colo. USA.
YEH, C. Prof. University of California, Los Angeles, USA.

MIX
Papier aus verantwortungsvollen Quellen
Paper from responsible sources
FSC® C105338

If you have any concerns about our products,
you can contact us on
ProductSafety@springernature.com

In case Publisher is established outside the EU,
the EU authorized representative is:
**Springer Nature Customer Service Center GmbH
Europaplatz 3, 69115 Heidelberg, Germany**

Printed by Libri Plureos GmbH
in Hamburg, Germany